中国高等植物

Higher Plants of China in Colour

《中国高等植物彩色图鉴》编委会　主编

Edited by
Editorial Committee of
Higher Plants of China in Colour

科学出版社
北京

中国高等植物
彩色图鉴

Higher Plants of China in Colour

《中国高等植物彩色图鉴》编委会　主编
Edited by Editorial Committee of Higher Plants of China in Colour

第 4 卷　被子植物　罂粟科 – 毒鼠子科

Volume Ⅳ　Angiosperms　Papaveraceae—Dichapetalaceae

卷编辑　于胜祥
Edited by Shengxiang YU

内 容 简 介

本套图鉴精选中国境内野生高等植物和重要栽培植物1万余种，配以图片近2万张，每一物种以中英文形式简要介绍植物的中文名、拉丁学名、形态特征、花果期、生境和分布。图鉴共分为9卷，收载苔藓植物100科、蕨类植物40科、裸子植物11科、被子植物232科，共计383科，且除苔藓植物之外，已收全所有科。本套图鉴是继《中国高等植物图鉴》、《中国植物志》、Flora of China之后，又一部大型植物分类学巨著。本卷为第4卷。

本书适合植物学领域的科研人员、管理人员及爱好植物学的普通大众阅读和收藏。

This set of pictorial books contains nearly 20 thousand photographs, presenting the cream of wild higher plants and important cultivated plants in China, the species of which number more than 10 thousand. Each of the species is concisely introduced in both Chinese and English from such aspects as Chinese name, Latin name, morphological features, flowering and fruiting season, habitat and distribution. Divided into nine volumes, this work includes 100 bryophyte families, 40 pteridophyte families, 11 gymnosperm families and 232 angiosperm families, 383 families altogether; the inclusion of all the said families is complete except for the bryophytes. The set of pictorial books is another monumental work on plant taxonomy, after *Iconographia Cormophytorum Sinicorum*, *Flora Reipublicae Popularis Sinicae*, and *Flora of China*. This is volume Ⅳ of the series.

This work is intended for scientific researchers and administrators in the field of botany and also for botany enthusiasts. As well as for reading, the work can be a classic collection.

图书在版编目（CIP）数据

中国高等植物彩色图鉴＝Higher Plants of China in Colour. 第4卷，被子植物. 罂粟科—毒鼠子科：汉英 / 《中国高等植物彩色图鉴》编委会主编；于胜祥分册主编. —北京：科学出版社，2016.1

ISBN 978-7-03-047064-5

Ⅰ. ①中… Ⅱ. ①中… ②于… Ⅲ. ①高等植物-中国-图集 ②罂粟科-中国-图集 ③毒鼠子科-中国-图集 Ⅳ. ①Q949.4-64

中国版本图书馆CIP数据核字（2016）第013534号

责任编辑：王 静 付 聪 马 俊 / 责任校对：郑金红
责任印制：肖 兴 / 书籍设计：北京美光设计制版有限公司

科学出版社 出版
北京东黄城根北街16号
邮政编码：100717
http://www.sciencep.com

北京汇瑞嘉合文化发展有限公司 印刷

科学出版社发行 各地新华书店经销

*

2016年1月第 一 版 开本：787×1092 1/16
2016年1月第一次印刷 印张：34 1/2
字数：1 579 000

定价：510.00元

（如有印装质量问题，我社负责调换）

#《中国高等植物彩色图鉴》编委会

编委会主任

王文采

编委会副主任

吴声华　李振宇

编委（按姓氏汉语拼音排序）

陈　彬　陈又生　成　晓　费　勇　金效华　李锡文　李振宇　林秦文　刘　冰
刘　博　刘怡涛　卢　刚　彭镜毅　覃海宁　陶国达　王文采　吴声华　夏念和
于胜祥　张　力　张树仁　张宪春　左　勤

总策划

吴声华

协同策划

费　勇　陶国达　成　晓　覃海宁　李振宇　王　静　刘怡涛　臧　穆

作者

费　勇　刘怡涛　刘　冰　陶国达　成　晓　陈　彬　陈又生　傅立国　何国生
金效华　李振宇　林秦文　刘　博　王文采　吴声华　徐松芝　徐晔春　于胜祥
喻勋林　张　力　张代贵　张树仁　张宪春　朱鑫鑫　左　勤

编委会秘书

钟小红　徐松芝　孙久琼

责任编辑

王　静　付　聪　马　俊

学术支持单位

中国科学院植物研究所
中国科学院昆明植物研究所
中国科学院华南植物园
深圳市中国科学院仙湖植物园

Editorial Committee of Higher Plants of China in Colour

Authors

Yong FEI　Yitao LIU　Bing LIU　Guoda TAO　Xiao CHENG
Bin CHEN　Yousheng CHEN　Liguo FU　Guosheng HE　Xiaohua JIN
Zhenyu LI　Qinwen LIN　Bo LIU　Wentsai WANG　Shenghua WU
Songzhi XU　Yechun XU　Shengxiang YU　Xunlin YU　Li ZHANG
Daigui ZHANG　Shuren ZHANG　Xianchun ZHANG　Xinxin ZHU　Qin ZUO

Collaborators

Institute of Botany, Chinese Academy of Sciences
Kunming Institute of Botany, Chinese Academy of Sciences
South China Botanical Garden, Chinese Academy of Sciences
Fairylake Botanical Garden, Shenzhen & Chinese Academy of Sciences

丛书图片主要拍摄者

(按姓氏汉语拼音排序)

阿不都拉·阿巴斯　白鹭　白重炎　毕延超　邴艳红　曹同　车晋滇　陈彬　陈高
陈丽　陈庆　陈鑫　陈炳华　陈世品　陈贤兴　陈又生　陈志雄　成晓　程文达
迟敏杰　戴攀峰　邓涛　邓云飞　丁炳扬　丁学欣　董仕勇　杜诚　杜巍　杜玉芬
段长虹　段士民　方振东　方振兴　费勇　冯君茹　傅连中　甘啟良　高贤明　高信芬
高云东　葛斌杰　耿玉英　古训铭　顾余兴　管开云　郭世伟　韩国营　郝加琛　郝云庆
何海　何理　何春梅　何国生　和兆荣　侯元同　胡光万　胡国雄　华国军　黄健
黄江华　黄圣卓　黄向旭　黄俞淞　惠肇祥　季定乾　贾渝　姜林　蒋宏　蒋蕾
蒋日红　金伟涛　金孝锋　金效华　康世昌　赖阳均　郎楷永　黎斌　黎兴江　李东
李恒　李凯　李敏　李攀　李不言　李策宏　李东辉　李家湘　李建民　李剑武
李良千　李文军　李先源　李晓东　李小杰　李新华　李新伟　李学东　李泽贤　李振宇
李志奇　李智选　李中阳　梁同军　廖明林　廖云标　林敏　林祁　林维　林广旋
林建勇　林俊杰　林茂祥　林秦文　林哲丽　刘冰　刘博　刘静　刘军　刘坤
刘夙　刘翔　刘鑫　刘演　刘莹　刘大伟　刘光裕　刘海桑　刘红梅　刘伦辉
刘全儒　刘晟源　刘怡涛　刘正宇　刘宗才　柳永红　卢刚　卢元　罗柳青　骆适
吕碧凤　吕志学　马林　马炜梁　马文章　马欣堂　莫水松　牟善杰　沐先运　慕泽泾
南程慧　倪静波　倪素碧　农东新　潘勃　潘建斌　彭博　彭镜毅　彭日成　乔明明
秦卫华　秦祥堃　覃海宁　邱志敬　仁琛　任飞　任丽华　任明波　任昭杰　尚策
邵剑文　沈阳肇　施忠辉　石硕　寿海洋　税玉民　宋纬文　宋柱秋　买买提明·苏来曼
苏丽飞　苏享修　孙航　孙苗　孙观灵　孙明洲　孙卫邦　孙小美　谭运洪　陶国达
田乾福　田新民　童毅　童毅华　汪远　王辰　王东　王泓　王晖　王健
王进　王强　王颖　王耘　王喆　王长荣　王钧杰　王慷林　王清隆　王秋美
王文卿　王雅琼　王亚玲　王英伟　王玉兵　王正元　王祝年　韦宏金　韦毅刚　韦玉梅
卫然　魏来　温韩东　温九良　翁茂伦　吴丰　吴磊　吴双　吴棣飞　吴凤琴
吴光弟　吴国晞　吴林芳　吴声华　吴望辉　吴问舫　吴永红　吴增源　伍凯　武全安
武素功　武玉东　夏念和　肖翠　肖亮　肖艳　肖红菊　谢磊　辛夷　辛晓伟
辛益群　辛宇明　熊源新　徐徭　徐锦泉　徐克学　徐连升　徐申健　徐文斌　徐晔春
徐永福　许为斌　寻路路　严新富　严岳鸿　阳文静　杨浩　杨永　杨成梓　杨建昆
杨金财　杨科明　杨青山　杨世雄　杨奕绯　杨增宏　姚永飚　叶德平　叶建飞　叶喜阳
叶幸儿　易思荣　殷建涛　尹志坚　于胜祥　郁文彬　喻勋林　袁彩霞　曾孝濂　曾云保
张力　张良　张强　张伟　张莹　张勇　张彩飞　张重岭　张代贵　张凤秋
张海华　张宏伟　张金龙　张金政　张守君　张淑梅　张树仁　张维柱　张宪春　张霄林
张志翔　赵宏　赵伟　赵大昌　郑宝江　郑希龙　郑小明　钟智明　周繇　周重建
周海成　周家宝　周兰平　周喜乐　周小林　周浙昆　朱弘　朱大海　朱仁斌　朱淑霞
朱维明　朱鑫鑫　David E. Boufford　Dmitry Sokoloff　Jan Thomas Johansson
Jozef Lemmens　Kirill Tkachenko　Pavel Novák　Ralf Knapp　Richard Ree
Susan Kelley

香港植物标本室(免费提供)

Major Photographers of the Series

(in the order of Chinese pinyin)

Abdulla ABASI	Lu BAI	Chongyan BAI	Yanchao BI	Yanhong BING	Tong CAO
Jindian CHE	Bin CHEN	Gao CHEN	Li CHEN	Qing CHEN	Xin CHEN
Binghua CHEN	Shipin CHEN	Xianxing CHEN	Yousheng CHEN	Chihhsiung CHEN	Xiao CHENG
Wenda CHENG	Minjie CHI	Panfeng DAI	Tao DENG	Yunfei DENG	Bingyang DING
Xuexin DING	Shiyong DONG	Cheng DU	Wei DU	Yufen DU	Changhong DUAN
Shimin DUAN	Zhendong FANG	Zhenxing FANG	Yong FEI	Junru FENG	Lianzhong FU
Qiliang GAN	Xianming GAO	Xinfen GAO	Yundong GAO	Binjie GE	Yuying GENG
Xunming GU	Yuxing GU	Kaiyun GUAN	Shiwei GUO	Guoying HAN	Jiachen HAO
Yunqing HAO	Hai HE	Li HE	Chunmei HE	Guosheng HE	Zhaorong HE
Yuantong HOU	Guangwan HU	Guoxiong HU	Guojun HUA	Jian HUANG	Jianghua HUANG
Shengzhuo HUANG	Xiangxu HUANG	Yusong HUANG	Zhaoxiang HUI	Dingqian JI	Yu JIA
Lin JIANG	Hong JIANG	Lei JIANG	Rihong JIANG	Weitao JIN	Xiaofeng JIN
Xiaohua JIN	Shihchang KANG	Yangjun LAI	Kaiyong LANG	Bin LI	Xingjiang LI
Dong LI	Heng LI	Kai LI	Min LI	Pan LI	Buyan LI
Cehong LI	Donghui LI	Jiaxiang LI	Jianmin LI	Jianwu LI	Liangqian LI
Wenjun LI	Xianyuan LI	Xiaodong LI	Xiaojie LI	Xinhua LI	Xinwei LI
Xuedong LI	Zexian LI	Zhenyu LI	Zhiqi LI	Zhixuan LI	Zhongyang LI
Tongjun LIANG	Minglin LIAO	Yunbiao LIAO	Min LIN	Qi LIN	Wei LIN
Guangxuan LIN	Jianyong LIN	Junjie LIN	Maoxiang LIN	Qinwen LIN	Zheli LIN
Bing LIU	Bo LIU	Jing LIU	Jun LIU	Kun LIU	Su LIU
Xiang LIU	Xin LIU	Yan LIU	Ying LIU	Dawei LIU	Guangyu LIU
Haisang LIU	Hongmei LIU	Lunhui LIU	Quanru LIU	Shengyuan LIU	Yitao LIU
Zhengyu LIU	Zongcai LIU	Yonghong LIU	Gang LU	Yuan LU	Liuqing LUO
Shi LUO	Pifong LU	Zhixue LÜ	Lin MA	Weiliang MA	Wenzhang MA
Xintang MA	Shuisong MO	Shannjye MOORE	Xianyun MU	Zejing MU	Chenghui NAN
Jingbo NI	Subi NI	Dongxin NONG	Bo PAN	Jianbin PAN	Bo PENG
Ching-I PENG	Richeng PENG	Mingming QIAO	Weihua QIN	Xiangkun QIN	Haining QIN
Zhijing QIU	Chen REN	Fei REN	Lihua REN	Mingbo REN	Zhaojie REN
Ce SHANG	Jianwen SHAO	Yangzhao SHEN	Zhonghui SHI	Shuo SHI	Haiyang SHOU

Yumin SHUI	Weiwen SONG	Zhuqiu SONG	Mamtimin SULAYMAN	Lifei SU	Xiangxiu SU
Hang SUN	Miao SUN	Guanling SUN	Mingzhou SUN	Weibang SUN	Xiaomei SUN
Yunhong TAN	Guoda TAO	Qianfu TIAN	Xinmin TIAN	Yi TONG	Yihua TONG
Yuan WANG	Chen WANG	Dong WANG	Hong WANG	Hui WANG	Jian WANG
Jin WANG	Qiang WANG	Ying WANG	Yun WANG	Zhe WANG	Changrong WANG
Junjie WANG	Kanglin WANG	Qinglong WANG	Chiumei WANG	Wenqing WANG	Yaqiong WANG
Yaling WANG	Yingwei WANG	Yubing WANG	Zhengyuan WANG	Zhunian WANG	Hongjin WEI
Yigang WEI	Yumei WEI	Ran WEI	Lai WEI	Handong WEN	Jiuliang WEN
Maolun WENG	Feng WU	Lei WU	Shuang WU	Difei WU	Fengqin WU
Guangdi WU	Guoxi WU	Linfang WU	Shenghua WU	Wanghui WU	Wenfang WU
Yonghong WU	Zengyuan WU	Kai WU	Quan'an WU	Sugong WU	Yudong WU
Nianhe XIA	Cui XIAO	Liang XIAO	Yan XIAO	Hongju XIAO	Lei XIE
Yi XIN	Xiaowei XIN	Yiqun XIN	Yuming XIN	Yuanxin XIONG	Yao XU
Jinquan XU	Kexue XU	Liansheng XU	Shenjian XU	Wenbin XU	Yechun XU
Yongfu XU	Weibin XU	Lulu XUN	Hsinfu YEN	Yuehong YAN	Wenjing YANG
Hao YANG	Yong YANG	Chengzi YANG	Jiankun YANG	Jincai YANG	Keming YANG
Qingshan YANG	Shixiong YANG	Yifei YANG	Zenghong YANG	Yongbiao YAO	Deping YE
Jianfei YE	Xiyang YE	Xing'er YE	Sirong YI	Jiantao YIN	Zhijian YIN
Shengxiang YU	Wenbin YU	Xunlin YU	Caixia YUAN	Xiaolian ZENG	Yunbao ZENG
Li ZHANG	Liang ZHANG	Qiang ZHANG	Wei ZHANG	Ying ZHANG	Yong ZHANG
Caifei ZHANG	Chongling ZHANG	Daigui ZHANG	Fengqiu ZHANG	Haihua ZHANG	Hongwei ZHANG
Jinlong ZHANG	Jinzheng ZHANG	Shoujun ZHANG	Shumei ZHANG	Shuren ZHANG	Weizhu ZHANG
Xianchun ZHANG	Xiaolin ZHANG	Zhixiang ZHANG	Hong ZHAO	Wei ZHAO	Dachang ZHAO
Baojiang ZHENG	Xilong ZHENG	Xiaoming ZHENG	Zhiming ZHONG	You ZHOU	Chongjian ZHOU
Haicheng ZHOU	Jiabao ZHOU	Lanping ZHOU	Xile ZHOU	Xiaolin ZHOU	Zhekun ZHOU
Hong ZHU	Dahai ZHU	Renbin ZHU	Shuxia ZHU	Weiming ZHU	Xinxin ZHU
David E. Boufford	Dmitry Sokoloff	Jan Thomas Johansson		Jozef Lemmens	Kirill Tkachenko
Pavel Novák	Ralf Knapp	Richard Ree	Susan Kelley		

Hong Kong Herbarium (free of charge)

丛书文字主要编写者

(按姓氏汉语拼音排序)

陈世龙　陈又生　陈之端　成　晓　崔逸群　邓云飞　杜　宁　段士民　樊　杰　方瑞征
费　勇　傅立国　高　凡　高　乞　谷粹芝　郭　慧　韩　宇　郝加琛　何　理　侯学良
侯元同　胡光万　黄向旭　黄俞淞　蒋　宏　金孝锋　金效华　赖阳均　雷立公　黎　斌
李　恒　李　嵘　李秉滔　李宏哲　李剑武　李梦华　李文军　李锡文　李晓贤　李新华
李新伟　李章海　李振宇　李中阳　廖文波　林秦文　刘　冰　刘　博　刘　演　刘大伟
刘海桑　刘全儒　刘衍男　卢金梅　马欣堂　潘　勃　覃海宁　邱志敬　萨　仁　尚　策
税玉民　孙　苗　孙久琼　万　涛　王　东　王　晖　王　健　王　强　王德艺　王文采
王英伟　卫　然　魏　来　吴　磊　吴鹏程　吴声华　向巧萍　向小果　谢　磊　徐松芝
徐晓婷　薛大伟　闫瑞亚　严岳鸿　杨世雄　叶建飞　游旨价　于胜祥　袁　慊　张　力
张　梅　张　强　张重岭　张钢民　张红瑞　张树仁　张宪春　张志翔　赵　宏　赵存峰
周兰平　左　勤

Major Textwriters of the Series

(in the order of Chinese pinyin)

Shilong CHEN	Yousheng CHEN	Zhiduan CHEN	Xiao CHENG	Yiqun CUI
Yunfei DENG	Ning DU	Shimin DUAN	Jie FAN	Ruizheng FANG
Yong FEI	Liguo FU	Fan GAO	Qi GAO	Cuizhi GU
Hui GUO	Yu HAN	Jiachen HAO	Li HE	Xueliang HOU
Yuantong HOU	Guangwan HU	Xiangxu HUANG	Yusong HUANG	Hong JIANG
Xiaofeng JIN	Xiaohua JIN	Yangjun LAI	Ligong LEI	Bin LI
Heng LI	Rong LI	Bingtao LI	Hongzhe LI	Jianwu LI
Menghua LI	Wenjun LI	Xiwen LI	Xiaoxian LI	Xinhua LI
Xinwei LI	Zhanghai LI	Zhenyu LI	Zhongyang LI	Wenbo LIAO
Qinwen LIN	Bing LIU	Bo LIU	Yan LIU	Dawei LIU
Haisang LIU	Quanru LIU	Yannan LIU	Jinmei LU	Xintang MA
Bo PAN	Haining QIN	Zhijing QIU	Ren SA	Ce SHANG
Yumin SHUI	Miao SUN	Jiuqiong SUN	Tao WAN	Dong WANG
Hui WANG	Jian WANG	Qiang WANG	Deyi WANG	Wentsai WANG
Yingwei WANG	Ran WEI	Lai WEI	Lei WU	Pengcheng WU
Shenghua WU	Qiaoping XIANG	Xiaoguo XIANG	Lei XIE	Songzhi XU
Xiaoting XU	Dawei Xue	Ruiya YAN	Yuehong YAN	Shixiong YANG
Jianfei YE	Zhijia YOU	Shengxiang YU	Qian YUAN	Li ZHANG
Mei ZHANG	Qiang ZHANG	Chongling ZHANG	Gangmin ZHANG	Hongrui ZHANG
Shuren ZHANG	Xianchun ZHANG	Zhixiang ZHANG	Hong ZHAO	Cunfeng ZHAO
Lanping ZHOU	Qin ZUO			

丛书前言

中国是世界上植物最丰富的国家之一，已知有三万五千多种野生和重要栽培的高等植物，其中特有种达一万五千多种，形成复杂而独具特色的植物区系。中国的先人们创造了古老而辉煌的农业文明，选育出水稻、大豆、茶、枣、桃、柿等重要作物，其中水稻的栽培历史可追溯到约七千年前新石器时期的河姆渡文化，如今稻米已成为世界上近一半人口的粮食。丰富的植物资源和灿烂的历史文化，使中国成为“花园之母”和世界农作物七大起源中心之一。

中国植物学家为了系统地展示中国植物的多样性，历经艰辛，相继编研了《中国高等植物图鉴》和《中国植物志》，并与外国专家合作出版Flora of China等大型志书，这些著作在国内外应用广泛、影响巨大，客观地展现了不同时期的植物分类学研究和植物资源调查的成果，成为植物分类学领域最重要的大型经典著作。但是，它们都有一个共同的缺憾，即仅有黑白线条图，难以充分表达植物各器官的质地和颜色等自然状态下的外貌特征，其效果难以满足部分读者鉴赏植物的需要。

大多数发达国家都有自己的植物彩色图鉴，这些图鉴不仅展示了本国的生物多样性，还兼备工具书功能和富有感染力的艺术效果，具有很高的应用和收藏价值。迄今为止，国内出版的植物彩色图书多为地区性的，或局限于某一类植物的，如观赏植物、栽培作物和药用植物。作为世界生物多样性大国，中国应当拥有一套全面体现本国野生植物多样性的的大型鉴赏类彩色图册。

将灿烂的瞬间变为永恒是广大植物爱好者和摄影爱好者的追求。为了填补上述空白，台湾吴声华研究员策划并启动了这项工作。在海峡两岸学者的共同努力下，本书的规模在不断扩大，从最初的云南植物写真集扩展到全国性大型彩色植物图鉴。中国科学院植物研究所王文采院士出任丛书编委会主任，吴声华研究员和中国科学院植物研究所李振宇研究员任副主任。编委会遴选国内从事植物分类学研究的专家担任各卷卷编辑，邀请中国大陆、台湾和香港近200位植物学家承担各科的编写和审稿工作，卷编辑在专家审稿的基础上，再次对本卷内容进行核查。近400位摄影作者提供了大量精美的植物彩色照片。丛书还采用了著名的动植物科学画大师曾孝濂先生绘制的20余幅优雅而灵动的彩色图片。

本丛书划分为九卷，共收录中国高等植物1万余种，种类以野生植物为主，同时收载重要的栽培植物，精选图片近2万张。本丛书中科的系统排列如下：苔藓植物主要参考《中国苔藓志》中的系统；蕨类植物按张宪春2015年在《石松类和蕨类名词及名称》提出的系统；裸子植物和被子植物的系统排列按第尔斯(L. Diels, 1936)于A. Engler's Syllabus der Pflanzenfamilien中采用的系统。仅第三卷将毛茛科分为星叶草科、毛茛科和芍药科，将木兰科分为木兰科、八角科、五味子科和水青树科。全书收载中国高等植物383科，其中苔藓植物100科，占全国苔藓科总数的大多数；其余是蕨类植物40科，裸子植物11科，被子植物232科，分别代表了国产三大门类所有的科。本丛书收载的植物中有一些是Flora of China出版后发表的国产新种，如香港鹅耳枥(*Carpinus insularis*)、球柱楼梯草(*Elatostema globosostigmatum*)和西藏小囊兰(*Micropera tibetica*)，以及中国分布新记录，如轮叶三棱栎(*Trigonobalanus verticillata*)和格力兜兰(*Paphiopedilum gratrixianum*)。

为了方便更多的读者阅读，本丛书的文字采用中英文，简要介绍各种植物的中文名、拉丁学名、形态特征、花果期、生境和分布。

本书在编写过程中，承中国科学院植物研究所中国植物图像库和中国自然标本馆提供了许多方便和帮助，在此向他们表示衷心的感谢。

感谢国家出版基金和科学出版社对本丛书出版的大力支持。

由于编著者的业务水平有限、错漏之处，欢迎批评指正。

《中国高等植物彩色图鉴》编委会

2015年10月31日

Preface to the Series

As one of the countries with the richest diversity of plant species in the world, China has more than 35 000 known species of wild and important cultivated higher plants, among which there are over 15 000 endemic species, forming a complex and unique flora. The ancestors of the Chinese people created an ancient and splendid agricultural civilization. They selected and cultivated significant crops like rice, soya bean, tea, jujube, peach, and persimmon. Among these crops, the cultivation history of rice can be traced back to the Hemudu culture of the Neolithic Period around 7000 years ago. Nowadays, rice has become the staple food for nearly half of the world's population. With abundant plant resources and a long history and great culture, China is renowned as 'the mother of gardens' and is one of the seven important centers of origin for crops in the world.

In order to present the diversity of China's plants systematically, botanists from China have made pain-taking efforts to compile a series of large-volume floras including *Iconographia Cormophytorum Sinicorum*, *Flora Reipublicae Popularis Sinicae* (Chinese version) by themselves, and *Flora of China* (English version) with the collaboration of international specialists. These books are well known both in China and abroad and have been used extensively for studying Chinese plants and plants from adjacent areas, these works present the results of plant taxonomic study and study of plant resources in China at different periods, and constitute some of the most important large classic volumes in the field of plant taxonomy. However, in all these works the plants are only partly illustrated by black and white line-drawings, unable to present the texture and colour of flowers and leaves fully in their natural state, and they hardly reveal the spectacular beauty and fascination of the wealth of plant species.

Most developed countries have colour pictorial books of their plants, which form greatly desirable works, because they are not only a presentation of the plant diversity of the countries, but are also an attractive record of the beauty of the nature. So far, most of the Chinese colour pictorial books of plants are regional treatments, or concentrate on particular groups, such as ornamental plants, cultivated crops and medicinal plants or certain taxonomic groups. As a country with a high level of biodiversity, China merits a large-scale colour pictorial book with high appreciation value featuring the wild plants that occur within its territory.

It is the goal of every lover of plants and plant photography to capture the essence of plant beauty and make it permanent. In order to fill the above-mentioned gap, Professor Shenghua WU from Taiwan, planned and launched the present project. With the joint effort of specialists from all over China, the scale of the book has expanded from the initial pictorial book of plants of Yunnan to a many-volume colour pictorial book of plants of the whole country. Academician Wentsai WANG of the Institute of Botany, Chinese Academy of Sciences, took up the post of the chairman of the editorial committee, and the positions of vice chairmen of the editorial committee were assumed by Prof. Shenghua WU, Taiwan, and Prof. Zhenyu LI of the Institute of Botany, Chinese Academy of Sciences. The editorial committee then selected experienced plant taxonomists as volume editors for each volume, and invited nearly 200 botanists from mainland China, Taiwan and Hong Kong to undertake the compilation and reviewing work for each plant family by volumes. Nearly 400 photographers provided numerous beautiful full colour plant photos. The well known zoological and botanical artist, Xiaolian ZENG, kindly allowed the use of more than twenty of his elegant and vivid plant portraits in this series.

This whole work is divided into nine volumes, depicting more than 10 thousand species of higher plants from China, dealing mainly with wild plants, but also including some important cultivated plants, and has involved the careful selection of nearly 20 thousand photographs. The system arrangement for plant families are as follows: bryophytes are mainly arranged according to the system used in *Flora Bryophytorum Sinicorum*; pteridophytes are arranged according to the system proposed by Professor Xianchun ZHANG in *A Glossary of Terms and Names of Lycopods and Ferns* (2015); gymnosperms and angiosperms are arranged according to the system used in *A. Engler's Syllabus der Pflanzenfamilien* (L. Diels, 1936), with the difference that in Volume III, Ranunculaceae is divided into Circaesteraceae, Ranunculaceae, and Paeoniaceae, and Magnoliaceae is divided into Magnoliaceae, Illiciaceae, Schisandraceae, and Tetracentraceae. The higher plants of China included in this work comprise 383 families; with 100 families of bryophytes, which represent the majority of the bryophyte families in China; the others are 40 pteridophyte families, 11 gymnosperm families and 232 angiosperm families, which represent all the families distributed in China respectively. The work includes some new additions of species published since *Flora of China*, such as *Carpinus insularis*, *Elatostema globosostigmatum*, *Micropera tibetica*, and new distribution records for China, such as *Trigonobalanus verticillata* and *Paphiopedilum gratrixianum*.

To facilitate and attract readers from both China and abroad, the text of this book series is bilingual in Chinese and English, providing the Chinese name, Latin name, morphological features, flowering and fruiting season, habitat and distribution.

In the process of compiling this work, Plant Photo Bank of China (PPBC) and Chinese Field Herbarium, both of which are under the Institute of Botany of Chinese Academy of Sciences, provided great help with the selection of photographs, for which we express our gratitude.

We also thank National Publication Foundation and Science Press, Beijing, for their great support for the publication of this book series.

It will be appreciated if mistakes and omissions are brought to our attention.

Editorial Committee of *Higher Plants of China in Colour*
31 October, 2015

关于本图鉴

1995年夏天，我参加由中国科学院昆明植物研究所臧穆教授带领的云南野外工作，同行的还有国际真菌学会理事长德籍的Franz Oberwinkler教授与法国的学者。臧教授爽朗好客，外国人都喜欢他的热情。那年去丽江，再去南部的西双版纳。西双版纳热带植物园的陶国达先生带领我们的野外工作，他是当地植物鉴定首席专家，知道好的树林在何处。一天，在傣族传统农家的木架房子吃中饭。臧教授建议陶先生既然喜欢摄影，何不出一本版纳植物图鉴，问我能不能帮忙在台湾找出版。我答应回去问问。

先问自然科学博物馆的李家维馆长，他对植物研究及保育充满热诚，对这项工作有兴趣。但未久他感觉这项工作所需时间过久，博物馆经费也不足以出版。我又问其他出版公司，没有得到响应。我想应该先有成果再问出版吧，就请陶先生持续植物拍摄。臧教授和夫人黎兴江教授推荐了费勇帮忙这项工作。1997年夏天，我在昆明机场与臧教授和费勇会合，一同飞去版纳。费勇年纪与我相当，长得瘦黑，话不太多。陶先生带领我们野外工作。回程时费勇说他想找几个同事一起负责滇西北的植物拍摄工作，与陶先生滇南的工作结合成为云南植物图鉴。回台后看陶先生给我的幻灯片，感觉质量不是太好，问他才知道所用的相机是正牌，镜头却是小厂牌。我汇钱请他购买一套相机，以利拍摄质量。

1998年，我到昆明植物所，和陶国达、费勇及孙航，讨论植物图鉴工作。费勇对此工作充满兴趣，人缘也好，决定由他征集昆明植物所人员拍摄的植物照片，并且中、英文字也由他撰写。翌年臧穆夫妇介绍昆明植物所的著名画家曾孝濂先生。曾先生长期考察云南山野的植物与动物，画作结合了科学性与艺术美感，是中国写实花鸟画得最好的。

2000年，费勇在日本富山县中央植物园半年，其间拍摄植物园栽培的中国植物。那年秋天费勇带我去大理点苍山和楚雄紫溪山。一天，我们在大理古城一间白族旧庭院吃风味晚餐，他兴致好，畅所欲言。费勇起初给我的印象是有些木讷，几次往来后就把我当熟人。几次的讨论，感觉他满心想做好这件事，并不在意条件。大理巷弄中有摊贩卖当地特产乳扇，他说闺女爱吃，买了两大张带走。

2001年年初一个早上，臧教授发来邮件，通知我费勇前一日在丽江不幸去世。一个年轻健康生命的突然离去，令人难以承受。出席完上午的会议后即打电话到昆明。黎教授说费勇到丽江出差，半夜室友听到声响，见他口吐白沫急送医院。地方医院初以为是癫痫，到清晨就不治了。

几个月后我有事联络曾孝濂先生，他告诉我费勇太太想与我联系。费勇太太姓向，我们称小向。她电话中希望植物图鉴工作能够继续，而且费勇的几个同事愿意帮忙。当年夏天在昆明的一个晚上，小向同昆明植物所的成晓、孙航、周浙昆一起和我见面，商讨后续的工作。成晓说他与费勇是同学，同时毕业，同时上班，他一定会帮忙。他确实尽力后续工作的联系与推动。2002年在昆明，几个朋友见面，小向带初中的女儿同来。女儿乖巧懂事，我说长得像费勇，她眼眶微微红了。成晓研究蕨类，他的岳父武素功先生及岳母方瑞征女士也是昆明植物所学者，两位在图片提供及文稿修订均提供协助。昆明植物所李锡文教授对植物分类的造诣比较全面，负责图片和文稿审查。昆明植物所还有多位专家对本书工作做出贡献，不在此逐一罗列。

2001年，曾孝濂先生介绍昆明植物所的画家刘怡涛先生。刘先生在版纳热带植物园待过，建立独特的版纳风光绘画风格，也喜好摄影，带过我几次野外工作。他建议我把植物图鉴工作扩大到全中国。艺术家天生具有美感，曾、刘两位画家拍摄的植物图，构图与取景皆有独到之处。2003年我到河北与吉林进行野外工作，2004年到新疆与吉林时决定把植物图鉴范围扩大到全中国。我和小向说明书的分量和质量要到位，才能彰显费勇的努力精神。费勇原本即有中国植物图鉴的梦想，干脆一次到位。

2004年，中国科学院植物研究所覃海宁博士来台，我们是1994年在英国邱园认识的。中国植物图像库在海宁领导下建立得有声有色。海宁总是满脸笑容，热诚谦虚，听我说植物图鉴的事，立即寻思找人帮忙。他人面广，介绍不少人，拍摄较好的有福建的何国生、四川的吴光弟、广西的刘演、广东的李泽贤。刘演的图片色彩饱满令人赞叹。我去爱丁堡皇家植物园时知道David Chamberlain博士是杜鹃花科专家，他同意审查杜鹃花科及小檗科图片。彭镜毅介绍哈佛大学David E. Boufford博士，他的图片是从中国西南的横断山脉植物调查工作所拍摄。David又推荐Susan Kelley及Richard Ree提供植物图片。

中科院植物所吴鹏程教授是苔藓专家，1990年我在芬兰赫尔辛基大学即将取得博士学位时他在赫大待了几个月。吴教授介绍几位中科院植物所的专家帮忙图片审查及文字撰写。台湾真菌学前辈吕理燊博士介绍昆明市农业局副局长惠肇祥先生提供杂草图片，惠先生又介绍北京的车晋滇先生提供华北的杂草图片。台湾赖明洲教授介绍上海自然博物馆的秦祥堃先生提供华东植物图片，又介绍中国科学院沈阳应用生态研究所的赵大昌先生提供长白山植物图片。我2004年到乌鲁木齐开会，组织会议的新疆大学阿不都拉教授拍了不少新疆植物图片，也提供给我。

大学同学康世昌是植物及计算机高手，拍摄的植物图片也提供给我。他早预想到网络世界的影响力，不推荐大部头实体书的出版构想。多年前他写个网址要我去看，那是我不知道的“Google”，可以查询信息。网络上图片的数量越来越多，趋势是如此。我在芬兰的指导教授Tuomo Niemelä出版过大型真菌的小书，亲自编排，图片与文字搭配得美感十足，我每翻阅总是心情愉悦。我向Tuomo请教对这套植物实体书的意见。他说网络的数据有时会消失，且许多没经过审查。我想这套书终要完成，无法顾及趋势与新世代人类的想法。

2007年年初，我在网络发现中科院植物所的中国植物图像库有影像部分。负责的是李敏，我问他图片提供者，他推荐几位拍摄较好的。多数是中科院植物所的年轻人，有刘冰、林秦文、于胜祥、李敏、高贤明、郦艳红，还有陕西的王秐。我当时已收集中国植物5000种的图片，工作超过10年理应收尾。然不加入这批有许多北方植物的图片实在不舍。刘冰是植物分类奇葩，这么年轻就拍到数量惊人的植物图片。刘冰和刘怡涛是给这项工作提供图片最多的两位。刘博帮忙不少文字撰写及图片审查，工作积极。当年年底，图片收集到6500种以上，接着准备文字、图片审查等出书的各项工作。

2010年在台湾“中央研究院”召开一项研讨会，覃海宁和李敏也来了，他们的报告显示中国植物图像库已收到数十万张图片。我如果再搜寻一次图片，能收到更好及更多的图片，但面对许多人殷盼这套书问世，时间的延长，压力更大。终究，我相信费勇会支持这最后一批图片的征求。湖南喻勋林及张代贵两位教授寄来许多华中植物图片，浙江张宏伟先生及安徽施忠辉先生也送来图片。吉林通化的周繇教授寄来他辛苦拍摄的长白山植物图片。近三年送来较多图片的还有朱鑫鑫、陈又生、徐晔春、陈彬、陈世品、何海、周喜乐，以及蕨类的张宪春和兰科的金效华。好友张力负责苔藓部分。

“中央研究院”彭镜毅博士提供了许多秋海棠科图片，也修订这科的文稿。牟善杰是台湾的蕨类学家，提供一些蕨类图片给本书，也审查过蕨类图片及文字。我在台湾大学念博士时，善杰是大学生，见他圆圆的笑脸，成天在标本馆研究。2010年11月，44岁的他突然中风走了，令人感慨！吕碧凤小姐是台湾优秀的业余蕨类专家，提供一些好的蕨类图片。还要感谢提供及审查图片的几位同事：王秋美、陈志雄、胡维新、黄俊霖、严新富和邱少婷。

早期收到的是幻灯片及少数印好的照片，2005年以后送来的是数码影像。数码图片干净，缺点是饱和度、清晰度和锐利度表现稍差，绿色部分有时偏黄。图效调整可改善这些问题。幻灯片的影像则会受到底片、冲洗、保存、扫描等质量的影响，好质量的并不多。图片须裁切出重点部位，再调整影像效果，这些工作大多是我处理。商请到一批人分别撰写文字。虽然有范本给撰写人参考，但各人的写法与仔细程度难免不一，有疏漏或小错误的情形普遍存在。起初我自己参考文献逐一查核，修订了约两千种的文字，但工作量太大，无法继续亲为。文字工作贡献较多的有费勇、刘博、萨仁、谷粹芝、成晓、杜宁、李锡文、徐晓婷和方瑞征等。

吴鹏程教授与科学出版社生物分社社长王静女士提及这项工作，王静有兴趣了解出版的可能，我们2010年在北京见面。自己过于深入这项工作，甚至如排版形式、字形等都亲自研究。像是自己养大的小孩，不放心交给他人处理，而且书稿已经在台湾找设计公司开始排版了。王静有毅力，持续两年逐渐消减我的疑虑。二十年来两岸的社会经济形势改变，使得这套书在大陆出版成为自然。德高望重的王文采院士及植物分类权威李振宇教授鼎力相助、组织动员，国家出版基金给予资助，促使整体工作能顺利完成。

吴声华

2015年10月20日

About the Pictorial Series

In the summer of 1995, I took part in the Yunnan fieldwork led by Prof. Mu ZANG from Kunming Institute of Botany, Chinese Academy of Sciences. Joining us were German professor Franz Oberwinkler, director general of International Mycological Association, and some French scholars. Prof. ZANG was candid, cordialand hospitable, which impressed everyone, especially the foreign guests. We first went to Lijiang and then Sipsongpanna in the south. During this fieldwork, we were guided by Mr. Guoda TAO from Xishuangbanna Tropical Botanical Garden, Chinese Academy of Sciences. He was the chief expert of plant identification in the area, knowing which areas of the woods were worth this field inspection of ours. One day, when we were having lunch together in a traditional wooden house of an ethnic Dai family, Prof. ZANG proposed to Mr. TAO: "Since you are so fond of photography, why not compile a pictorial book of Banna's plants?" Prof. ZANG then turned to me, asking whether I could give help in getting the book published in Taiwan, and I promised to give it a try after returning to Taiwan.

I first contacted Dr. Chiawei LI, director of Museum of Natural Science, who was passionate about plant research and conservation and interested in the project. But before long, his passion faded due to his sense that the project was likely to take too long a time, and the Museum did not have sufficient fund to support the publishing. I then inquired of other publishing companies, but none of them gave a positive response. These setbacks sent me thinking that perhaps we should make some tangible achievements first before our work could be accepted for publication. So I asked Mr. TAO to proceed with shooting plants. Prof. ZANG and his wife Prof. Xingjiang LI recommended Yong FEI to provide assistance to the work. In the summer of 1997, I met Prof. ZANG and Yong FEI at Kunming Airport, and we flew to Sipsongpanna together for the fieldwork led by Mr. TAO. Of the same age as mine, Yong FEI was a thin and swarthy man, not very talkative. On our way back, Yong FEI said he was considering asking several of his colleagues to join him in shooting plants of the northwest of Yunnan so that the pictures taken in the two areas (the SouthYunnan and the Northwest Yunnan) could be combined to make a single pictorial book that could be called "Plants of Yunnan". After returning to Taiwan, I browsed the slides given by Mr. TAO, feeling that their quality was not ideal. Having asked Mr. TAO about this, I learned that it had been caused by his camera whose main body was of good brand and quality but whose lens was made by a mediocre producer. I remitted money to him for purchasing a new camera set, hoping that the quality of photos could be ensured by a high-quality camera.

In 1998, I visited Kunming Institute of Botany to discuss with Guoda TAO, Yong FEI and Hang SUN about the work of pictorial book for plants. Yong FEI was full of enthusiasm on the work, and had good relations with people, so we decided to commission him to collect plant photos taken by staff from the Kunming Institute, and to compose text both in English and Chinese. The next year, Prof. Mu ZANG and Prof. LI introduced to me Mr. Xiaolian ZENG, who had been engaging in the investigation of plants and animals in the wilds of Yunnan Province for a long time and was also a famous painter from the Kunming Institute of Botany. His paintings are the best realistic bird-and-flower works in China, blending scientificity with artistic beauty.

In 2000, Yong FEI spent half a year in Botanic Gardens of Toyama (Japan), taking photos of Chinese plants grown in the Gardens. In autumn of the same year, with Yong FEI as my guide, we went to Diancang Mountain in Dali and Zixi Mountain in Chuxiong. One day, when we were having local delicacies for super in an old courtyard of Bai nationality, Yong FEI got into high spirit and chatted with me without restraint. My first impression of Yong FEI was that he was a bit unapproachable, but after several rounds of conversations, he regarded me as his close friend. After several discussions with him, I felt that he very much concentrated on doing the work well, paying no attention to remuneration. In a lane of Dali, we found a vendor selling milk fan cake, a kind of local specialty, and he bought two big pieces, saying that they were for his daughter who liked such food.

One early morning in the early 2001, Prof. ZANG sent me an email, saying sadly that Yong FEI passed away in Lijiang the day before. It was really unbearable to hear of the sudden passing of such a young life. As soon as the meeting in that morning ended, I called to Kunming. The call was answered by Prof. LI who said that Yong FEI had been on a working trip at the time. At midnight, his roommates were awakened by some noises and found him foaming at the mouth. He was rushed to a local hospital and initially diagnosed as only having a fit of epilepsy, but no amount of treatment took effect on him; he passed away just as dawn came.

Several months later, when contacting Mr. Xiaolian ZENG, I was told that Yong FEI's wife was looking for me. The family name of Yong FEI's wife was XIANG, so we called her Little XIANG, a traditional way of Chinese people addressing their acquaintances who were younger than themselves. Little XIANG expressed her wish over phone that the project of the pictorial book should go on as usual and she also said that several of Yong FEI's colleagues were willing to help. One summer evening of the same year in Kunming, Little XIANG, together with Xiao CHENG, Hang SUN and Zhekun ZHOU all from the Kunming Institute of Botany, had a meeting with me to discuss about the remaining work of the project. Xiao CHENG said he and Yong FEI were classmates, graduating and first getting employed at the same time, so he would definitely offer his help. And in fact he did try his best to facilitate the progress of the work through networking. In 2002, we had a gathering in Kunming. Little

XIANG brought her daughter there, who was then a junior-secondary-school student. The girl was both clever and well-behaved, and when I said to her that she looked like her father, her eyes moistened slightly. Xiao CHENG was a fern researcher. His father-in-law Mr. Sugong WU and mother-in-law Mrs. Ruizheng FANG were also scholars of the Kunming Institute of Botany, both of whom offered their assistance in providing photos and editing texts. Prof. Xiwen LI also from the Kunming Institute of Botany, who had comprehensive attainments in plant taxonomy, was responsible for examining photos and texts. There were many other experts from the Kunming Institute of Botany who made contributions to this book, but due to space constraint, their names are not listed here one by one.

In 2001, Mr. Xiaolian ZENG introduced to me Mr. Yitao LIU, another painter from Kunming Institute of Botany. Mr. LIU used to stay in Xishuangbanna Tropical Botanical Garden, where he developed his distinctive painting style with which to depict typical Sipsongpanna's landscape. He was also a lover of photography, and used to be my fieldwork guide for several times. Mr. LIU suggested that I expand the pictorial plant book project to cover the whole territory of China. Due to the innate aesthetic sense of artist, the plant photos taken by the two painters - Mr. ZENG and Mr. LIU - had unique characteristics both in picture composing and view finding. My fieldwork in Hebei and Jilin in 2003 and then my travelling in Xinjiang and Jilin in 2004 prompted my final decision to expand the pictorial plant book project to the whole country. I explained to Little XIANG that only when the book was comprehensive enough and of high quality, could Yong FEI's hardworking spirit and aspiration in this regard be fully manifested. And only in this way could Yong FEI's cherished dream of compiling a pictorial book on plants of China be realized without unnecessary pre-steps.

2004 saw Dr. Haining QIN's visit to Taiwan. Dr. QIN was from Institute of Botany, Chinese Academy of Sciences, and we got to know each other at British Kew Gardens in 1994. Under the leadership of Haining, the construction of Plant Photo Bank of China was making marvelous progress. Haining was a cordial and modest man, with his face always shining with smile. Upon knowing that I was conducting the project of pictorial plant book, he offered to give help. Taking advantage of his wide network, he brought in many talents, among whom Guosheng HE from Fujian, Guangdi WU from Sichuan, Yan LIU from Guangxi, and Zexian LI from Guangdong were all good at photography. In terms of color, Yan LIU's photos were particularly good, which was admirable. In addition, Dr. David Chamberlain, an expert in Ericacea, whom I got to know when I visited Royal Botanic Garden Edinburgh, agreed to review the photos of Ericaceae and Berberidaceae. Besides, Dr. Ching-I PENG introduced Dr. David E. Boufford from Harvard University who provided photos taken when he was investigating the plants of the Hengduan Mountains in the southwest of China. And David also recommended Susan Kelley and Richard Ree who both offered their plant photos.

Prof. Pengcheng WU from Institute of Botany, Chinese Academy of Sciences. was an expert in bryophytes. In 1990, he stayed in University of Helsinki, Finland for a few months when I was about to obtain my doctorate awarded by the University. Prof. WU introduced several experts from Institute of Botany to help review photos and write text. Dr. Liisin LEU, a Taiwan veteran in mycology, recommended Mr. Zhaoxiang HUI, deputy director of Kunming Municipal Bureau of Agriculture, to provide photos of weeds. And Mr. HUI invited Mr. Jindian CHE from Beijing to provide photos of weeds in Northern China. Prof. Mingjou LAI from Taiwan involved Mr. Xiangkun QIN from Shanghai Natural History Museum in contributing photos of plants in Eastern China, and then recommended Mr. Dachang ZHAO from Shenyang Institute of Applied Ecology, Chinese Academy of Sciences to offer photos of plants in Changbai Mountains. In 2004, I went to Urumqi to attend a conference whose organizer, Prof. Abdulla from Xinjiang University, gave me many photos of Xinjiang plants taken by himself.

My college classmate Shihchang KANG, an expert in plants and computer, also sent me plant photos taken by himself. Having long foreseen the power of internet, he did not quite agree with the idea of publishing a bulky physical book. Years ago, he wrote down a website address and asked me to visit it. The website, which I had never heard of before, was 'Google', a 'search engine' enabling us to search for information easily. And it turned out that this became a strong upward trend, with more and more photos being uploaded onto internet for people to view or download. However, Prof. Tuomo Niemelä, my Finnish adviser, had a different view on this phenomenon. He had published a handbook about large fungi, whose formatting was done by himself. The photos and text were arranged so well that a full sense of beauty permeated the entire book, and this always made me in a good mood each time I read it. When being consulted about the idea of publishing a physical plant book like this one, he encouraged me to continue doing so, saying that sometimes online data and materials would vanish for no reason and many online materials could not be said to be authentic because they had not undergone necessary review and approval. With this encouragement, I decided to carry out this project through to the end, paying no attention to the trends and fashionable ideas of new generations.

At the beginning of 2007, I found on internet that the Plant Photo Bank of China owned by Institute of Botany contained image data being managed by Min LI. So I asked him for sources of these photos.

Min LI recommended several persons whose photos in the Database were regarded as excellent. Most of these photo-takers were young people from the Institute of Botany. They were Bing LIU, Qinwen LIN, Shengxiang YU, Min LI, Xianming GAO, Yanhong BING. Besides, Yun WANG from Shaanxi was also added to the list of recommendation. By that time, I had already collected photos of 5000 species of plants in China through over 10 years of my hardwork which could have very well wound up. However, it would have been regrettable if I had not added so many fine photos of plants in Northern China to this important book. Bing LIU was a wonder in plant taxonomy - so young as he was, he had taken astonishingly large number of plant photos. It was Bing LIU and Yitao LIU who provided the largest number of photos for this work. Bo LIU, who was very active in work, helped a lot in writing text and reviewing photos. By the end of the same year, we had collected photos of more than 6500 species, paving the way for doing other publication-related preparatory work such as text writing and photo reviewing.

In 2010, "Academia Sinica" held a seminar in Taiwan, at which Haining QIN and Min LI delivered their reports which revealed that the Plant Photo Bank of China had collected hundreds of thousands of photos. In this circumstance, one more round of photo searching and collecting would certainly make more and better photos available for this upcoming book. Only, it would take more time. With so many people looking forward to the publication of the book, the longer time we took in publishing, the heavier pressure we would face. But in final analysis, I believed that Yong FEI, if he were still alive, would support this last round of photo searching and collecting. Prof. Xunlin YU and Prof. Daigui ZHANG from Hunan sent me many photos of plants of Central China. Mr. Hongwei ZHANG from Zhejiang and Mr. Zhonghui SHI from Anhui also sent photos to me. Prof. You ZHOU from Tonghua of Jilin contributed the photos of plants of Changbai Mountain that he took with great efforts. In the recent three years, a lot of photos were also provided by Xinxin ZHU, Yousheng CHEN, Yechun XU, Bin CHEN, Shipin CHEN, Hai HE, and Xile ZHOU. Xianchun ZHANG offered many photos of ferns. Xiaohua JIN submitted many photos of Orchidaceae plants. My good friend Li ZHANG was responsible for bryophytes.

Dr. Ching-I PENG from "Academia Sinica" provided a lot of pictures of Begoniaceae and edited the draft for this family. Shannjye MOORE, an expert of ferns from Taiwan, contributed some pictures of ferns to this book, and reviewed the pictures and text for the fern part. When I studied for doctorate in Taiwan University, Shannjye was still an undergraduate of the University. With a lovely round face often with smile, he was always seen studying in herbarium. In November 2010, however, he suddenly died of a stroke at the age of 44, making us very sad and regretful. Miss Pifong LU, an excellent amateur expert of ferns from Taiwan, contributed some good pictures of ferns. I also would like to express my thanks to the following colleagues who provided and reviewed pictures for me: Chiumei WANG, Chihhsiung CHEN, Weihsin HU, Chunlin HUANG, Hsinfu YEN and Shauting CHIU.

What we received in earlier stages were slides and a small number of prints,and after 2005, all contributions were in the form of digital image. Digital photos are clean, but their saturation, definition and sharpness are not very ideal, with green parts tending to turn slightly yellowish. Fortunately, these problems could be solved through photo-effect modification. As to the slides, high-quality ones were not many, as the quality of such images hinged on such factors as: quality of the film, developing process, storage condition and scanning, etc. The photos first needed some trimming so as to highlight their essential parts and then required modification to the image effect; most of the work was done by myself. In the meantime, we engaged a group of people to do text writing. Although templates were provided to text writers, inconsistency still appeared in some places due to different writing styles of different writers. There were also not a few oversights or slips caused by some writers who were not conscientious enough. At first, I myself did the correction and revision one by one against reference literature, finishing the work on about two thousand species, but as the amount of this kind of work was so big that I could not continue to do it all by myself. Here, I would like to list those who made greater contributions to the text. They are: Yong FEI, Bo LIU, Ren SA, Cuizhi GU, Xiao CHENG, Ning DU, Xiwen LI, Xiaoting XU and Ruizheng FANG, etc.

Prof. Pengcheng WU mentioned this work to Ms. Jing WANG, director of Biological Division of Science Press, who was interested in exploring the possibility of publishing the work, so we met each other in Beijing in 2010. Before this, I had devoted myself to the work so deeply that even small details like typesetting and font were studied and arranged by myself. Therefore, the work was like a child brought up by myself, so I would feel uneasy if I put it in the care of someone else, and moreover, we had already commissioned a design company in Taiwan to start typesetting the draft. However, my concern and worry were gradually dispelled by Jing WANG's sincerity and her perseverance in persuasion and explanation over two successive years. And the changes in social and economic situations across the Straits also made it natural for the book to be published in the Mainland. Also worthy of mentioning are: the generous support, organization and mobilization given or conducted by both Wentsai WANG, a renowned academician and Prof. Zhenyu LI, an expert on plant taxonomy, as well as the funding by National Publication Foundation. All this facilitated the smooth completion of the entire work.

Shenghua WU

20 October, 2015

第4卷编审者分工

罂粟科	崔逸群	于胜祥	张代贵
白花菜科	徐松芝	于胜祥	彭　华
十字花科	李锡文	于胜祥	崔逸群
木犀草科	于胜祥		
辣木科	刘　冰		
钟萼木科	刘　博	于胜祥	
猪笼草科	于胜祥		
茅膏菜科	王　晖		
景天科	崔逸群	于胜祥	
虎耳草科	谷粹芝	张红瑞	叶建飞
海桐花科	于胜祥	张志耘	
金缕梅科	赖阳均	张志耘	
杜仲科	于胜祥	彭　华	
悬铃木科	林秦文		
蔷薇科	谷粹芝	徐松芝	于胜祥
牛栓藤科	刘　博	于胜祥	
豆科	潘　勃	徐松芝	黄星凡
酢浆草科	向小果	李振宇	
牻牛儿苗科	刘全儒	郭　慧	
金莲花科	于胜祥		
亚麻科	刘　冰		
古柯科	于胜祥		
蒺藜科	段士民	彭　华	
芸香科	刘　博	魏　来	何东辑
苦木科	刘　博	于胜祥	彭　华
橄榄科－楝科	刘　博	于胜祥	
金虎尾科	于胜祥		
远志科	樊　杰	陈书坤	
毒鼠子科	李振宇		

Authors and Reviewers of Volume Ⅳ

Papaveraceae	Yiqun CUI	Shengxiang YU	Daigui ZHANG
Capparidaceae	Songzhi XU	Shengxiang YU	Hua PENG
Cruciferae	Xiwen LI	Shengxiang YU	Yiqun CUI
Resedaceae	Shengxiang YU		
Moringaceae	Bing LIU		
Bretschneideraceae	Bo LIU	Shengxiang YU	
Nepenthaceae	Shengxiang YU		
Droseraceae	Hui WANG		
Crassulaceae	Yiqun CUI	Shengxiang YU	
Saxifragaceae	Cuizhi GU	Hongrui ZHANG	Jianfei YE
Pittosporaceae	Shengxiang YU	Zhiyun ZHANG	
Hamamelidaceae	Yangjun LAI	Zhiyun ZHANG	
Eucommiaceae	Shengxiang YU	Hua PENG	
Platanaceae	Qinwen LIN		
Rosaceae	Cuizhi GU	Songzhi XU	Shengxiang YU
Connaraceae	Bo LIU	Shengxiang YU	
Fabaceae	Bo PAN	Songzhi XU	Xingfan HUANG
Oxalidaceae	Xiaoguo XIANG	Zhenyu LI	
Geraniaceae	Quanru LIU	Hui GUO	
Tropaeolaceae	Shengxiang YU		
Linaceae	Bing LIU		
Erythroxylaceae	Shengxiang YU		
Zygophyllaceae	Shimin DUAN	Hua PENG	
Rutaceae	Bo LIU	Lai WEI	Dongji HE
Simarubaceae	Bo LIU	Shengxiang YU	Hua PENG
Burseraceae—Meliaceae	Bo LIU	Shengxiang YU	
Malpighiaceae	Shengxiang YU		
Polygalaceae	Jie FAN	Shukun CHEN	
Dichapetalaceae	Zhenyu LI		

目录 | Contents

第 4 卷
Volume Ⅳ

被子植物
罂粟科—毒鼠子科
Angiosperms
Papaveraceae—Dichapetalaceae

罂粟科
Papaveraceae

蓟罂粟
Argemone mexicana L.

一年生或偶为多年生短命草本，植株被刺。基生叶密；叶覆白粉，脉上具蓝绿色斑块，下面灰绿色，阔倒披针形或倒卵形至椭圆形。花黄色；萼片先端距状。蒴果长圆形至阔椭圆体形。花果期3-10月。生海拔850-1200米的田间或河边。产云南、广东、台湾和福建。原产中美洲和南美洲；栽培于全世界大部分地区。

Herbs annual or occasionally short-lived perennial, with spines. Basal leaves dense; leaves glaucous with blue-green markings on veins, paler abaxially, broadly oblanceolate or obovate to elliptic. Flowers yellow; sepal apex spurred. Capsules oblong to broadly ellipsoid. Fl. and fr. Mar-Oct. Fields or by rivers at 850-1200 m. Distributed in Yunnan, Guangdong, Taiwan and Fujian. Native to Central and South America; cultivated worldwide.

蓟罂粟 *Argemone mexicana*

椭果绿绒蒿 *Meconopsis chelidonifolia*

椭果绿绒蒿
Meconopsis chelidonifolia Bur. et Franch.

多年生草本，高50-150厘米。茎直立，具分枝。基生叶和下部茎生叶卵状长圆形或宽卵形，长7-8厘米，宽6.5-7厘米，羽状分裂，裂片3-5；上部茎生叶宽卵形羽状3全裂或3深裂。聚伞状圆锥花序；花瓣黄色。蒴果椭圆形，长1-1.5厘米，无毛，自顶端向下微裂。花期5-8月。生海拔1400-2700米的林下阴处或溪边路旁。产四川西部至北部和云南北部。

Perennial herbs, 50-150 cm tall. Stems erect, branched. Basal and lower cauline leaves: ovate-oblong or broadly ovate, 7-8 × 6.5-7 cm, runcinatus, lobes 3-5; upper cauline leaves broadly ovate, 3-pinnatisect or 3-pinnatipartite. Cymose panicles; petals yellow. Capsule elliptic, 1-1.5 cm long, glabrous, valvate for a short distance from apex. Fl. May-Aug. Shade of forest understories, creek sides or roadsides at 1400-2700 m. Distributed in W to N Sichuan, and N Yunnan.

锥花绿绒蒿
Meconopsis paniculata (D. Don) Prain

一年生草本。基生叶常绿密集，莲座状丛生；叶形多变，常近基部羽状全裂，近顶端羽状浅裂。花序下部圆锥状，上部总状；花大而美丽，蓝色、紫色、红色或黄色；萼片2；花瓣4(5-10)；雄蕊多数；子房球形，被金色须状绒毛。花果期6-8月。生海拔3000-4400米草坡、路边或灌丛。产云南西北部至西部和四川西部。缅甸北部亦有。

Herbs, monocarpic. Basal leaves in a dense evergreen rosette; leaves variously shaped, usually near base pinnatisect, near apex pinnatifid. Inflorescences paniculate below, racemose above; flowers large and pretty, blue, purple, red or yellow; sepals 2; petals 4(5-10); stamens many; ovary globose, golden barbellate-tomentose. Fl. and fr. Jun-Aug. Grassy slopes, roadsides or shrublands at 3000-4400 m. Distributed in NW to W Yunnan, and W Sichuan. Also in N Myanmar.

锥花绿绒蒿 *Meconopsis paniculata*

尼泊尔绿绒蒿

Meconopsis wilsonii Grey-Wils

一次性结果草本，高70-150厘米。茎直立。叶密集莲座状；基生叶披针形至披针状椭圆形，羽状全裂至近全缘。总状圆锥花序；花瓣紫色至酒红色；花丝与花瓣同色；柱头紫色；花柱宿存。蒴果卵形至椭圆形，长14-20毫米，通常5瓣裂。花期6-9月。生海拔2700-4000米的森林和灌丛的边缘、多石地带、草地或悬崖。产四川和云南。缅甸北部亦有。

Monocarpic herbs, 70-150 cm tall. Stem erect. Leaves densely rosette; basal leaves lanceolate to lanceolate-elliptic, pinnatisect to subentire. Panicles racemose; petals purple to wine-red; filaments same color as petals; stigmas purple; style persistent. Capsule ovoid to ellipsoid, 14-20 mm long, generally 5-valved. Fl. Jun-Sep. Forest and scrub margins, stony places, grassy places, cliffs at 2700-4000 m. Distributed in Sichuan and Yunnan. Also in N Myanmar.

贡山绿绒蒿

Meconopsis smithiana (Hand.-Mazz.) Tayl. ex Hand.-Mazz.

多年生草本。茎生叶疏离，最下部叶具柄；叶羽状3小叶。约4花成总状花序；萼片被毛；花瓣4，黄色；子房近球形；柱头5裂。蒴果密被褐色长硬毛。花果期6-8月。生海拔3100-3400米的潮湿林中、林缘或山坡湿草地。产云南西北部。缅甸东北部亦有。

Herbs perennial. Cauline leaves distant, lowermost petiolate; leaves pinnately 3-foliolate. Flowers ca. 4 in racemes; sepals pilose; petals 4, yellow; ovary globose; stigmas 5-lobed. Capsules densely brown villous. Fl. and fr. Jun-Aug. Wet forests, forest edges or wet grasslands on slopes at 3100-3400 m. Distributed in NW Yunnan. Also in NE Myanmar.

尼泊尔绿绒蒿 *Meconopsis wilsonii*

贡山绿绒蒿 *Meconopsis smithiana*

全缘叶绿绒蒿 *Meconopsis integrifolia*

总状绿绒蒿

Meconopsis racemosa Maxim.

一年生草本。叶片全缘或波状，被黄褐色或淡黄色平展或紧贴的刺毛。单生总状花序，具花多至14朵；花瓣5-8，蓝色或蓝紫色。蒴果卵球形或狭卵球形，密被平展刚毛，4-6瓣裂。花期5-8月，果期7-11月。生海拔3000-4600(-4900)米的草坡、石坡或林下。产云南、四川、西藏、甘肃和青海。

Herbs, monocarpic. Leaves margin entire or undulate, both surfaces with fulvous or yellowish, spreading spines. Inflorescences a simple raceme with up to 14 flowers; petals 5-8, blue or bluish purple. Capsules ovoid or narrowly ovoid, with dense, spreading bristles, 4-6-valvate. Fl. May-Aug. Fr. Jul-Nov. Grassy slopes, stony slopes or forests at 3000-4600(-4900) m. Distributed in Yunnan, Sichuan, Xizang, Gansu and Qinghai.

全缘叶绿绒蒿

Meconopsis integrifolia (Maxim.) Franch.

一年生草本。基生叶莲座状，早落，其间混生鳞片状叶；叶全缘。花常3-5；花瓣6-8，黄色或稀白色。蒴果阔椭圆体形或长圆形至椭圆体形，被金色或褐色柔毛，4-7瓣裂。花期5-8月，果期7-11月。生海拔2700-5100米的灌丛或草地。产云南、四川、西藏、甘肃和青海。缅甸东北部亦有。

Herbs, monocarpic. Basal leaves in a deciduous rosette, among often mixed scalelike leaves; leaves margin entire. Flowers usually 3-5; petals 6-8, yellow or rarely white. Capsules broadly ellipsoid or oblong to ellipsoid, golden or brown hirsute, 4-7-valvate. Fl. May-Aug. Fr. Jul-Nov. Grassy slopes or under forests at 2700-5100 m. Distributed in Yunnan, Sichuan, Xizang, Gansu and Qinghai. Also in NE Myanmar.

藿香叶绿绒蒿

Meconopsis betonicifolia Franch.

多年生草本，偶尔二年生。茎直立，高30-150厘米。基生叶卵状披针形或卵形，边缘宽缺刻状齿裂。花3-6朵，生于最上部茎生叶腋内；花瓣蓝色或紫色，具明显的纵条纹；花丝白色；柱头淡绿色。蒴果长圆状椭圆形，长2-4.5厘米，无毛，自顶端微裂。花期6-8月。生海拔3000-4000米的林下或草坡。产云南西北部和西藏东南部。缅甸北部亦有。

Perennial herbs, occasionally biennial. Stems erect, 30-150 cm tall. Basal leaves ovate-lanceolate or ovate, margin broadly incised-toothed. 3-6 flowers at upper leaf axil; petals, blue or purple, with conspicuously longitudinal stripes; filaments white; stigmas virescent. Capsule oblong-elliptic, 2-4.5 cm long, glabrous, slightly valvate from apex. Fl. Jun-Aug. Forest understories, grassy slopes at 3000-4000 m. Distributed in NW Yunnan and SE Xizang. Also in N Myanmar.

藿香叶绿绒蒿 *Meconopsis betonicifolia*

总状绿绒蒿 *Meconopsis racemosa*

丽江绿绒蒿
Meconopsis forrestii Prain

一年生草本。主根圆锥形或萝卜状。叶通常全部基生，两面疏被紧贴、亮褐色的长硬毛。花序单花葶，直立，花葶被亮褐色、稍反曲的长硬毛；花3-7；花柱无。蒴果近狭圆柱形；无毛或疏被长硬毛，2-4瓣裂。花果期5-7月。生海拔(3100-)3400-4300米的草坡、岩石区或林缘。产云南西北部和四川西南部。

Herbs, monocarpic. Taproots conical or radishlike. Leaves all basal, both surfaces sparsely appressed bright brown hirsute. Inflorescences simple scapose, erect, scape with straw-colored, deflexed bristles; flowers 3-7; styles absent. Capsules narrowly subcylindrical, glabrous or sparsely hirsute, 2-4-valvate. Fl. and fr. May-Jul. Grassy slopes, rocky places or woodland edges at (3100-)3400-4300 m. Distributed in NW Yunnan and SW Sichuan.

长叶绿绒蒿
Meconopsis lancifolia (Franch.) Franch. ex Prain

一年生草本。主根萝卜状。叶多数或全部基生，倒披针形、匙形、倒卵形或椭圆披针形至狭倒披针形，全缘，两面无毛或被黄褐色卷曲硬毛。花瓣4-8，蓝色或紫色；花柱明显。蒴果无毛或疏被开展茶色刚毛，3-6瓣裂。花果期6-8月。生海拔3300-4800米的林下、多石区或高山草地。产云南、四川、西藏和甘肃。缅甸东北部亦有。

Herbs, monocarpic. Taproots radishlike. Leaves mostly or all basal, oblanceolate, spatulate, obovate or elliptic-lanceolate to narrowly oblanceolate, margin entire, both surfaces glabrous or brownish crispate-hirsute. Petals 4-8, blue or purple; styles conspicuous. Capsules glabrous or sparsely spreading fulvous setose, 3-6-valvate. Fl. and fr. Jun-Aug. Forests, rocky places or alpine meadows at 3300-4800 m. Distributed in Yunnan, Sichuan, Xizang and Gansu. Also in NE Myanmar.

丽江绿绒蒿 *Meconopsis forrestii*

长叶绿绒蒿 *Meconopsis lancifolia*

长叶绿绒蒿 *Meconopsis lancifolia*

红花绿绒蒿

Meconopsis punicea Maxim.

多年生草本。叶全部基生，莲座状，全缘，两面密被淡黄色或棕褐色、具多短分枝的刚毛。花葶1-6；花单生基出花葶上；花瓣4(-6)，红色。蒴果椭圆状长圆形，无毛或密被淡黄色须状刚毛，4-6瓣裂。花期6-9月。生海拔2800-4300米的山坡草地。产四川、西藏、甘肃和青海。

Herbs perennial. Leaves all basal, rosette, margin entire, both surfaces with dense, shortly branched, yellowish or brown, barbellate setae. Scapes 1-6; flowers solitary on basal scapes; petals 4(-6), red. Capsules ellipsoidal-oblong, glabrous or densely yellowish barbellate-setose, 4-6-valvate. Fl. Jun-Sep. Grassy slopes at 2800-4300 m. Distributed in Sichuan, Xizang, Gansu and Qinghai.

单叶绿绒蒿

Meconopsis simplicifolia (D. Don) Walp.

一次性结果草本，高20-50厘米。叶全部基生，莲座状，长达16厘米，边缘全缘或具不规则的锯齿，两面被长柔毛。花单生于基生花葶上；花瓣5-8，紫色至天蓝色；花丝与花瓣同色。蒴果狭长圆形至长圆状圆形，长4.2-6.5厘米，被刚毛，自顶端开裂。花期6-9月。生海拔3300-4500米的山坡灌丛草地或石缝中。产西藏南部。不丹、印度北部和尼泊尔中部亦有。

Monocarpic herbs, 20-50 cm tall. Leaves all basal, forming a rosette, 16 cm long, margin entire or irregularly serrate, both surfaces villous. Flowers solitary on scape; petals 5-8, purple to sky blue; filaments of same color as petals. Capsule narrowly oblong to oblong-rounded, 4.2-6.5 cm long, setose, valvate from apex. Fl. Jun-Sep. Grasslands on slopes, among shrubs or rock crevices at 3300-4500 m. Distributed in S Xizang. Also in Bhutan, N India and C Nepal.

五脉绿绒蒿

Meconopsis quintuplinervia Regel

多年生草本。叶全基生，莲座状，早落；叶3-5脉。花单生花葶，每个叶丛至多3朵，下垂；花瓣4-6，淡蓝色或紫色；柱头3-6裂；子房近球形。蒴果密被紧贴的刚毛，3-6瓣裂。花果期6-9月。生海拔2300-4600米的阴坡灌丛中或高山草地。产四川、西藏、湖北、陕西、甘肃和青海。

Herbs perennial. Leaves all basal, forming a rosette, deciduous, 3-5-veined. Flowers solitary on scape, up to 3 per leaf rosette, pendent; petals 4-6, pale lilac-blue or purple; stigmas 3-6-lobed; ovary globose. Capsules densely appressed barbellate-setose, 3-6-valvate. Fl. and fr. Jun-Sep. Thickets of shady slopes or alpine grasslands at 2300-4600 m. Distributed in Sichuan, Xizang, Hubei, Shaanxi, Gansu and Qinghai.

五脉绿绒蒿 *Meconopsis quintuplinervia*

红花绿绒蒿 *Meconopsis punicea*

单叶绿绒蒿 *Meconopsis simplicifolia*

多刺绿绒蒿 *Meconopsis horridula*

多刺绿绒蒿
Meconopsis horridula Hook. f. et Thoms.

一年生草本。主根肥厚而延长，圆柱状。叶全部基生，两面被茶色或淡黄色平展的刺。花单生于花葶，每植物具多达29花；花瓣5-10，靛蓝、浅至深蓝、淡紫或紫堇蓝。蒴果倒卵球形或椭圆状长圆形，具平展刺，刺基部增粗，常3-5瓣裂。花果期6-9月。生海拔3600-5400米的草坡。产四川、西藏、甘肃和青海。印度东北部、尼泊尔、不丹和缅甸亦有。

Herbs monocarpic. Taproots plump, prolonged, terete. Leaves all basal, both surfaces with fulvous or yellowish compressed spines. Flowers solitary on scape, to 29 per plant; petals 5-10, indigo, pale to deep blue, purplish or violet-blue. Capsules obovoid or elliptic-oblong, with spreading spines, spine bases thickened, usually 3-5-valvate. Fl. and fr. Jun-Sep. Grassy slopes at 3600-5400 m. Distributed in Sichuan, Xizang, Gansu and Qinghai. Also in NE India, Nepal, Bhutan and Myanmar.

白花绿绒蒿
Meconopsis argemonantha Prain

草本。茎很短。基生叶背面具白霜，狭长圆形至椭圆形长圆形，中部和上部叶片较大，4.5-7 × 1.6-2厘米，羽状浅裂。花单生于叶腋，半下垂；花瓣4-8，白色，先端圆形，具凹槽；花丝白色；子房密被紧贴的皮刺。蒴果圆柱状纺锤形，15-25 × 6-7毫米，有稀疏向上的硬毛。花期7-8月。生海拔3700-4600米的潮湿苔岸、崖壁或开阔的林中河岸。产西藏东南部。

Herbs. Stem very short. Basal leaves glaucous abaxially, narrowly oblong to elliptic-oblong, middle and upper leaves generally larger, 4.5-7 × 1.6-2 cm, pinnately lobed. Flowers solitary at leaf axils, half-nodding; petals 4-8, white, apex rounded, fluted; filaments white; ovary densely appressed setose. Capsule cylindric-fusiform, 15-25 × 6-7 mm, with sparse ascending stiff hairs. Fl. Jul-Aug. Moist mossy banks, cliff ledges and open leafy banks at 3700-4600 m. Distributed in SE Xizang.

白花绿绒蒿 *Meconopsis argemonantha*

罂粟
Papaver somniferum L.

一年生草本。叶互生，卵形或长圆形，两面无毛，表面具白粉且呈光滑蜡质。花单生，深杯状；萼片2；花瓣4；雄蕊多数。蒴果直径4-5厘米，球形或

罂粟 *Papaver somniferum*

长圆状椭圆体形，无毛。花果期3-8月。原产南欧。中国在药用研究单位有栽培。南亚一些地区亦有栽培。

Herbs annual. Leaves alternate, ovate or oblong, both surfaces glabrous, glaucous and rather waxy. Flowers solitary, deeply cup-shaped; sepals 2; petals 4; stamens many. Capsules 4-5 cm diam, globose or oblong-ellipsoid, glabrous. Fl. and fr. Mar-Aug. Native to S Europe. In China only cultivated for medicinal research. Also cultivated in some regions of S Asia.

野罂粟

Papaver nudicaule L.

多年生草本。叶基生，两面具白粉，密被或疏被刚毛。花单生，下垂，常被褐色刚毛；萼片2；花瓣4，黄色、黄橙色。蒴果狭倒卵球形、倒卵球形或倒卵球状长圆形，密被紧贴浅白色或红褐色刚毛，具4-8宽棱。花果期5-9月。生海拔(200-)1000-2500(-3500)米的林中、林缘、沟谷、河岩砾石地或山坡草地。产中国西南、华北、华西和东北；也栽培在其他地区。中亚和东北亚亦有。

Herbs perennial. Leaves basal, pruinose on both surfaces, densely or sparsely bristly. Flowers solitary, pendulous, often brown-bristly; sepals 2; petals 4, yellow, yellow-orange. Capsules narrowly obovoid, obovoid or obovoid-oblong, densely appressed whitish- or red-brown setose, slightly broadly 4-8-costate. Fl. and fr. May-Sep. Forests, forest edges, valleys, river gravel or grasslands on slopes at (200-)1000-2500 (-3500) m. Distributed in SW, N, W and NE China; also cultivated in other regions of China. Also in C and NE Asia.

野罂粟 *Papaver nudicaule*

长白山罂粟

Papaver radicatum Rottb. var. **pseudo-radicatum** (Kitag.) Kitag.

多年生草本，小且成簇。叶全基生，两面被紧贴糙毛，羽状分裂或二回羽状分裂。花单生；萼片2；花瓣4，黄色；雄蕊多数；子房长圆形，密被紧贴的糙毛；柱头约6，辐射状。蒴果倒卵球形，密被紧贴或斜展糙毛。花果期6-8月。生海拔1600米以上的砾石地、高山苔原、砂地或岩石坡。产吉林。朝鲜半岛亦有。

Herbs perennial, small and tufted. Leaves all basal, both surfaces appressed setose, pinnatifid or bipinnatifid. Flowers solitary; sepals 2; petals 4, yellow; stamens many; ovary oblong, densely appressed setose; stigmas ca. 6, actinomorphic. Capsules obovoid, appressed or inclined-spreading setose. Fl. and fr. Jun-Aug. Gravelly slopes alpine tundra, sands or rocky slopes above 1600 m. Distributed in Jilin. Also in Korean Peninsula.

长白山罂粟 *Papaver radicatum* var. *pseudo-radicatum*

宽果秃疮花
Dicranostigma platycarpum

宽果秃疮花

Dicranostigma platycarpum C. Y. Wu et H. Chuang

草本，高1-2米。茎直立，无毛。基生叶大头羽状分裂，叶柄基部扩大成鞘；茎生叶抱茎，边缘有粗齿。花1-3朵；萼片舟状宽卵形，先端延长成距；花瓣黄色，倒卵形，长2-3厘米。蒴果圆柱形，长6-8毫米，粗5-8毫米，无毛，自顶端2瓣裂。种子卵珠形。花果期7-9月。生海拔3300-4000米的高山草地或沟边岩石隙。产云南西北部和西藏南部。

Herbs, 1-2 m tall. Stems erect, glabrous. Basal leaves several pinnatifid, petiolar base with inflated sheath; cauline leaves amplexicaul, margin with coarse teeth. Flowers 1-3; sepals cymbiform, broadly ovate, extending into spur; petals yellow, obovate, 2-3 cm long. Capsule cylindrical, 6-8 × 5-8 mm, glabrous, 2-valvate from apex nearly to base. Seeds ovoid. Fl. and fr. Jul-Sep. Alpine meadows or rock crevices at ditch sides at 3300-4000 m. Distributed in NW Yunnan and S Xizang.

秃疮花

Dicranostigma leptopodum (Maxim.) Fedde

两年生或短期多年生草本。基生叶丛生；叶狭倒披针形，羽状浅裂；裂片4-6对，裂片再羽状浅裂或羽状瓣裂。萼片2；花瓣黄色；雄蕊多数；柱头2裂。蒴果条形，绿色。花期3-7月，果期6-9月。生海拔400-2900(-3700)米的草坡、路边、田埂、墙头或屋顶。产中国西南、华北、华西和西北。

Herbs, biennial to short-lived perennial. Basal leaves in a rosette; leaves narrowly oblanceolate, pinnatipartite; lobes 4-6 pairs, pinnatipartite or pinnatilobate again. Sepals 2; petals yellow; stamens many; stigmas

秃疮花 *Dicranostigma leptopodum*

2-divided. Capsules linear, green. Fl. Mar-Jul. Fr. Jun-Sep. Grassy slopes, roadsides, field ridges, top of wall or housetops at 400-2900(-3700) m. Distributed in SW, N, W and NW China.

海罂粟

Glaucium fimbrilligerum Boiss

一年生草本，高30-60厘米。茎直立，不分枝。基生叶狭倒披针形，长7-10厘米，大头羽状深裂；茎生叶宽长圆形，长1-3厘米，羽状分裂。花瓣橙黄色，长2-2.5厘米，基部具斑点或无。蒴果线状圆柱形，长12-18厘米，疏被皮刺，成熟时自先端开裂，果梗极长。花果期6-10月。生荒漠或干旱山坡。产新疆。中亚和伊朗等地亦有。

Annual herbs, 30-60 cm tall. Stems erect, unbranched. Basal leaves narrowly oblanceolate, 7-10 cm long, pinnatipartite; cauline leaves broadly oblong, 1-3 cm long, pinnatifid. Petals orange-yellow, 2-2.5 cm long, blotched at base or not. Capsule linear-terete, 12-18 cm long, sparsely aculeate, dehiscing from apex to base when mature, carpopodium very long. Fl. and fr. Jun-Oct. Deserts or dry slopes. Distributed in Xinjiang. Also in C Asia and Iran.

金罂粟

Stylophorum lasiocarpum (Oliv.) Fedde

草本，高30-100厘米，具橙红色液汁。茎直立，通常不分枝，无毛。基生叶狭倒卵形，大头羽状深裂，长13-25厘米，顶生裂片宽卵形；茎生叶近对生或近轮生。伞形花序；花瓣黄色，倒卵状圆形，长约2厘米。蒴果狭圆柱形，长5-8厘米，被短柔毛。种子具鸡冠状种阜。花期4-8月。生海拔600-1800米的林下或沟边。产湖北西部、陕西南部和四川东部。

Herbs, 30-100 cm tall, orange-red lactiferous. Stems erect, usually unbranched, glabrous. Basal leaves narrowly obovate, with a large pinnatipartite terminal lobe, 13-25 cm long, apical lobes broadly ovate; cauline leaves almost opposite or whorled. Flowers in an umbellike cluster; petals yellow, obovate-orbicular, ca. 2 cm long. Capsule narrowly terete, 5-8 cm long, pubescent. Seeds cristately caruncúlate. Fl. Apr-Aug. Forest understories or ditch sides at 600-1800 m. Distributed in W Hubei, S Shaanxi and E Sichuan.

海罂粟 *Glaucium fimbrilligerum*

金罂粟 *Stylophorum lasiocarpum*

荷青花 *Hylomecon japonica*

荷青花
Hylomecon japonica (Thunb.) Prantl et Kündig

多年生草本。基生叶少数，具长柄，羽状全裂；裂片2或3对。花1-2(-3)朵成伞房花序，顶生，有时腋生；花梗纤细，直立，长3.5-7厘米；柱头2裂。蒴果2瓣裂。花期4-7月，果期5-8月。生海拔300-2400米的林下、林缘或沟边。产中国西南、华北、华西、华东和东北。俄罗斯(东西伯利亚)、朝鲜半岛和日本亦有。

Herbs perennial. Basal leaves few, long petiolate, pinnatisect; lobes 2 or 3 pairs. Flowers 1-2(-3) in corymb, terminal, sometimes axillary; pedicels tenuous, erect, 3.5-7 cm; stigmas 2-lobed. Capsules 2-valvate. Fl. Apr-Jul. Fr. May-Aug. Under forests, forest edges or by streams at 300-2400 m. Distributed in SW, N, W, E and NE China. Also in Russia (E Siberia), Korean Peninsula and Japan.

白屈菜
Chelidonium majus L.

多年生草本。主根圆锥状，粗壮，侧根多。基生叶少量，早凋落；叶下具白粉，倒卵状长圆形或阔倒卵形，羽状全裂；裂片2-4对。伞形花序多花。蒴果狭圆柱状。花果期4-9月。生海拔500-2200米的山坡、林缘、草地、路边或石缝中。产中国西南、华中、华北、华西、华东、西北和东北。俄罗斯、朝鲜半岛、日本和欧洲亦有。

Herbs perennial. Taproots conical, stout, lateral roots many. Basal leaves few, caducous; leaves glaucous abaxially, obovate-oblong or broadly obovate, pinnatisect; lobes 2-4 pairs. Umbels multiflorous. Capsules narrowly terete. Fl. and fr. Apr-Sep. Slopes, forest edges, grasslands, roadsides or crevices at 500-2200 m. Distributed in SW, C, N, W, E, NW and NE China. Also in Russia, Korean Peninsula, Japan and Europe.

血水草
Eomecon chionantha Hance

多年生无毛草本。根橘黄色。根茎匍匐。叶心形，全部基生。花葶蓝灰色且略带淡紫色，具3-5花；萼片合生成佛焰苞状；花瓣白色。蒴果狭椭圆体形。花期3-6月，果期6-10月。生海拔1400-1800米的林下、路边、潮湿地或水边。产中国西南、东南、华中和华东。

Herbs perennial, glabrous. Roots orange. Rootstock stoloniferous. Leaves cordate, all basal. Scapes blue-gray and slightly mauve, 3-5-flowered; sepals united into a spathe; petals white. Capsules narrowly ellipsoid. Fl. Mar-Jun. Fr. Jun-Oct. Under forests, roadsides, wet places or by waters at 1400-1800 m. Distributed in SW, SE, C and E China.

博落回
Macleaya cordata (Willd.) R. Br.

直立草本，基部木质化，具乳黄色浆汁。叶下面具白粉，上

白屈菜 *Chelidonium majus*

血水草 *Eomecon chionantha*

博落回 *Macleaya cordata*

面灰绿色，阔卵形或近圆形，常7或9浅裂至深裂。大型圆锥花序，多花；无花瓣；萼片黄白色；柱头2裂。蒴果狭倒卵形或倒披针形。花期6-11月。生海拔100-800米丘陵或林中，灌丛或草地。产中国长江以南，南岭以北的大部分地区。日本亦有。

Herbs, erect, basally lignified, yellow lactiferous. Leaves glaucous abaxially, gray-green adaxially, broadly ovate or suborbicular, usually 7- or 9-parted or lobed. Panicles large, many flowered; petals absent; sepals yellowish-white; stigmas 2-lobed. Capsules narrowly obovate or oblanceolate. Fl. Jun-Nov. Hills, forests, thickets or grasslands at 100-800 m. Distributed throughout the provinces to the south of Yangtze River, and to the north of Nanling Mountain. Also in Japan.

小果博落回

Macleaya microcarpa (Maxim.) Fedde

直立草本，基部木质化，具乳黄色浆汁。叶下面具白粉，上面灰绿色，阔卵形或近圆形，常7或9深裂或浅裂。大型圆锥花序，多花；花蕾圆柱状。蒴果近球形；直径约5毫米。花果期6-10月。生海拔400-1600米的山坡、路边、草地或灌丛中。产中国西南、华北、华西和华东。

Herbs erect, basally lignified, yellow lactiferous. Leaves glaucous abaxially, green adaxially, broadly ovate or suborbicular, usually 7- or 9-parted or -lobed. Panicles large, multiflorous; flower buds terete. Capsules subglobose, ca. 5 mm diam. Fl and fr. Jun-Oct. Slopes, roadsides, grasslands or thickets at 400-1600 m. Distributed in SW, N, W and E China.

小果博落回 *Macleaya microcarpa*

角茴香 *Hypecoum erectum*

角茴香

Hypecoum erectum L.

一年生草本。叶极多数，倒披针形，二或三回羽裂，裂片再深裂。二歧聚伞花序多花；花瓣黄色，有时具深色斑点或条纹。蒴果直立，狭圆柱状，2瓣裂。花果期4-10月。生海拔400-1200(-4500)米的山坡、草地、河边或砾石质沙地。产华中、华北、华西和东北。俄罗斯(西伯利亚)和蒙古亦有。

Herbs annual. Leaves very numerous, oblanceolate, 2 or 3 pinnate with deeply divided segments. Dichotomous cymes many flowered; petals yellow, sometimes with darker spots or streaks. Capsules erect, narrowly terete, dehiscent with 2 valves. Fl. and fr. Apr-Oct. Slopes, grasslands, riversides or gravel sands at 400-1200(-4500) m. Distributed in C, N, W and NE China. Also in Russia (Siberia) and Mongolia.

荷包牡丹

Dicentra spectabilis (L.) Lem.

多年生无毛草本。茎近直立，具分枝。多叶，极厚，肉质；小裂片全缘或具2-3裂片。总状花序顶生及腋生上部叶腋，几水平，长且疏松，具7-15花；外花瓣紫红色至粉红色。蒴果长圆形，具2-8粒种子。花期4-6月。生海拔800-2800米的湿润草地或山坡。产黑龙江、吉林和辽宁，栽培于中国大部分地区。俄罗斯和朝鲜半岛亦有。

Herbs perennial, glabrous. Stems erect, branched. Leafy, rather thick, juicy; leaflets entire or 2-3-lobed. Inflorescences terminal and axillary from upper leaves, racemelike, almost horizontal,

荷包牡丹 *Dicentra spectabilis*

大花荷包牡丹 *Dicentra macrantha*

紫金龙 *Dactylicapnos scandens*

long, lax, 7-15-flowered; outer petals purple-red to pink. Capsules oblong, 2-8-seeded. Fl. Apr-Jun. Wet grasslands or slopes at 800-2800 m. Distributed in Heilongjiang, Jilin and Liaoning, also cultivated in most parts of China. Also in Russia and Korean Peninsula.

大花荷包牡丹

Dicentra macrantha Oliv.

多年生直立草本，无毛。茎生叶阔三角形，二至常三出复叶至近三出复叶。总状花序聚伞状，下垂；萼片2，鳞片状；花长4-5厘米；雄蕊6，合成2束。蒴果条形至椭圆体形，2瓣裂。花果期4-7月。生海拔1500-2700米湿润林下或次生受干扰植被。产云南、四川、贵州和湖北。缅甸北部亦有。

Herbs perennial, erect, glabrous. Cauline leaves broadly triangular, bi- to usually triternate to subtripinnate. Racemes cymose, pendulous; sepals 2, scalelike; flowers 4-5 cm long; stamens 6, connate in 2 fascicles. Capsules linear to ellipsoid, 2-valvate. Fl. and fr. Apr-Jul. Under wet forests or secondary and disturbed vegetation at 1500-2700 m. Distributed in Yunnan, Sichuan, Guizhou and Hubei. Also in N Myanmar.

紫金龙

Dactylicapnos scandens (D. Don) Hutch.

多年生草质藤本。叶片轮廓三角形或卵形，三回三出复叶。花瓣黄色至白色，先端粉红色或淡紫红色。蒴果成熟后紫色、红色、白色或浅黄色，卵球形至长圆状狭卵形。花期7-11月，果期8-12月。生海拔1600-2500米的林下、山坡、石隙或水沟边。产云南、西藏和广西。南亚亦有。

Climbers perennial, herbaceous. Leaves triangular or ovate, triternate. Petals yellow to white, apically pink or pale purplish red. Capsules purple, red, whitish or pale yellow when mature, ovoid to lanceolate. Fl. Jul-Nov. Fr. Aug-Dec. Forests, slopes, rock crevices or by streams at 1600-2500 m. Distributed in Yunnan, Xizang and Guangxi. Also in S Asia.

宽果紫金龙

Dactylicapnos roylei (Hook.f. et Thoms.) Hutch.

藤本。茎具翅状棱，多分枝。叶片二回三出复叶。总状花序伞房状；苞片两侧带紫色；花瓣淡黄色，外花瓣基部囊状，顶端向两侧叉开。蒴果线状长圆形，长4-5厘米。自交亲和。花期7-10月，果期8-12月。生海拔2000-3000米的林下、山坡灌丛、蕨类丛中或路边等。产四川、西藏和云南。不丹、印度和尼泊尔亦有。

Vines. Stems winged-ridged, much branched. Leaves twice ternately divided. Raceme corymbose; bracts usually purplish on both sides; petals yellow, outer petals base broadly saccate, apex slightly divergent. Capsule linear-oblong, 4-5 cm long. Self-compatible. Fl. Jul-Oct. Fr. Aug-Dec. Forest understories, scrub on slopes, among ferns or roadsides at 2000-3000 m. Distributed in Sichuan, Xizang and Yunnan. Also in Bhutan, India and Nepal.

宽果紫金龙 *Dactylicapnos roylei*

扭果紫金龙 *Dactylicapnos torulosa*

扭果紫金龙

Dactylicapnos torulosa (Hook. f. et Thoms.) Hutch.

藤本。茎分枝，具锐棱，脆嫩易折断。叶轮廓长圆状卵形，羽状分裂，具3-5个互生的羽片。伞状花序，具2-7朵下垂花；花瓣淡黄色。蒴果念珠状。种子具光泽，外种皮具细网纹。花期6-10月，果期8月至翌年1月。生海拔1200-2500米的疏林、灌丛或山谷。产云南、四川、西藏和贵州。印度、不丹、孟加拉国和缅甸亦有。

Climbers. Stems branched, sharply angular, very weak and squashy. Leaves oblong-ovate in outline, pinnate, with 3-5 alternate pinnae. Inflorescences corymbose, 2-7-flowered, nutant; petals yellowish. Capsules moniliform. Seeds shiny, testa reticulate. Fl. Jun-Oct. Fr. Aug to next Jan. Open forests, thickets or valleys at 1200-2500 m. Distributed in Yunnan, Sichuan, Xizang and Guizhou. Also in India, Bhutan, Bangladesh and Myanmar.

鸡血七

Corydalis temulifolia Franch. subsp. **aegopodioides** (H. Lévl. et Van.) C. Y. Wu

多年生草本。苞片卵形或倒卵形，边缘上部具浅圆齿，下部全缘。总状花序疏松，具8-15花；花蓝紫色；距与冠檐近等长，细；柱头宽。蒴果劲直，条状圆柱形，近念珠状。花果期3-6月。生海拔1300-2700米的林下、路边或水边。产云南、四川、贵州和广西。越南北部亦有。

Herbs perennial. Bracts ovate or obovate, upper margin crenulate, lower part entire. Racemes lax, 8-15-flowered; flowers blue-purple; spurs ca. as long as limbs, thin; stigma broader. Capsules straight, linear, almost moniliform. Fl. and fr. Mar-Jun. Forests, roadsides or by waters at 1300-2700 m. Distributed in Yunnan, Sichuan, Guizhou and Guangxi. Also in N Vietnam.

金钩如意草

Corydalis taliensis Franch.

多年生草本，无毛。叶片轮廓近圆形或楔状菱形，二至三回三出全裂。总状花序具7-12花，密生，果期显著伸长；花瓣紫色、蓝紫色、红色或粉红色。蒴果条形。花果期3-11月。生海拔(1500-)1900-2300米的林下、灌丛或草丛中。产云南。

Herbs perennial, glabrous. Leaves suborbicular or cuneate-rhombic, 2-3-ternatisect. Racemes 7-12-flowered, dense, much elongating in fruit; petals purple,

鸡血七 *Corydalis temulifolia* subsp. *aegopodioides*

金钩如意草 *Corydalis taliensis*

大叶紫堇 *Corydalis temulifolia*

blue-purple, red or pink. Capsules linear. Fl. and fr. Mar-Nov. Forests, thickets or grassy places at (1500-)1900-2300 m. Distributed in Yunnan.

大叶紫堇

Corydalis temulifolia Franch.

多年生草本，高20-90厘米。基生叶基部膨大，三角形，二回三出羽状全裂；茎生叶与基生叶同形。总状花序；花瓣淡紫红色，上花瓣边缘开展，背部突起矮且短或无，下花瓣背部具突起，爪狭楔形；柱头具10个乳突。蒴果线形，近念珠状。花果期3-6月。生海拔1300-2700米的常绿阔叶林或混交林下、灌丛中或溪边。产重庆、甘肃、湖北和陕西。

Perennial herbs, 20-90 cm tall. Basal leaves base very broad and thick, deltoid, biternate; cauline leaves like radical leaves. Racemes; petals pale amaranth, upper petal margin spreading, usually with short and narrow abaxial crest or not, lower petal with abaxially crested; stigma with 10 papillae. Capsule linear, submoniliform. Fl. and fr. Mar-Jun. Evergreen broad-leaved and mixed forests, among shrubs or brook sides at 1300-2700 m. Distributed in Chongqing, Gansu, Hubei and Shaanxi.

穆坪紫堇

Corydalis flexuosa Franch.

无毛草本，高20-50厘米。茎通常不分枝。基生叶具叶鞘，三角形、卵形至近圆形，长3.5-8厘米，二至三回三出分裂；茎生叶互生，近圆形或宽卵形，二至三回三出全裂。总状花序；花瓣天蓝色或蓝紫色，上花瓣背部无突起。蒴果线形，长15-22毫米。花果期4-6月。生海拔1300-2700米的山坡水边或岩石边。产四川西部。

Glabrous herbs, 20-50 cm tall. Stems usually unbranched. Radical leaves sheathed, deltoid or ovate to suborbicular, 3.5-8 cm long, bi- to triternate; cauline leaves suborbicular to broadly ovate, biternatisect to triternatisect. Raceme; petals pale blue to indigo, upper petal without dorsal crest. Capsule linear, 15-22 mm long. Fl. and fr. Apr-Jun. Riversides, or wet rocks at 1300-2700 m. Distributed in W Sichuan.

穆坪紫堇 *Corydalis flexuosa*

师宗紫堇 *Corydalis duclouxii*

细果紫堇 *Corydalis leptocarpa*

师宗紫堇

Corydalis duclouxii H. Lévl. et Van.

多年生草本。须根多数。茎1-4，直立至斜升。丛生叶，叶三角形，下面具白粉，无毛或具细小乳突，二或三回三出复叶(至近二回羽状)；羽片具柄。总状花序生茎或分枝顶端；花瓣紫色。蒴果圆柱状。花果期3-8月。生海拔1500-3000米的林中、灌丛中、山谷、路旁或岩石下。产云南、贵州和重庆。

Herbs perennial. Fibrous roots many. Stems 1-4, erect to ascending. Rosette leaves, deltoid, abaxially pruinose, glabrous or very finely papillose, 2(-3) ternate (to sub-bipinnate); pinnae petiolulate. Racemes terminal on stems and branches; petals purple. Capsules terete. Fl. and fr. Mar-Aug. Forests, shrubs, valleys, roadsides or under rocks at 1500-3000 m. Distributed in Yunnan, Guizhou and Chongqing.

细果紫堇

Corydalis leptocarpa Hook. f. et Thoms.

多年生草本，无毛。叶片轮廓宽三角形或三角形，二至三回三出分裂。总状花序具2-7(-9)花；花瓣紫色或白而先端紫红色。蒴果具10-20粒种子。花果期4-12月。生海拔1200-2600米间常绿阔叶林、山谷草地或路边石隙中。产云南。印度(阿萨姆邦)、尼泊尔、不丹、缅甸北部和泰国北部亦有。

Herbs perennial, glabrous. Leaves broadly triangular or triangular in outline, 2-3-ternate-lobate. Racemes 2-7(-9)-flowered; petals purple or white with purple-red apices. Capsules 10-20-seeded. Fl. and fr. Apr-Dec. Evergreen broad-leaved forests, grassy places in valleys or rock crevices by roads at 1200-2600 m. Distributed in Yunnan. Also in India (Assam), Nepal, Bhutan, N Myanmar and N Thailand.

南黄紫堇

Corydalis davidii Franch.

多年生草本。茎少数至数个，具狭翅状脊，稀分枝，共具3-6叶。叶片轮廓宽三角形，三回三出全裂；茎生叶数枚。总状花序密生5-15花，生一侧；花瓣黄色。蒴果圆柱形。花果期4-10月。生海拔2000-3500米林下、林缘或草坡。产云南、四川和重庆。缅甸东部亦有。

Herbs perennial. Stems few to several, narrowly winged-ridged, sparingly branched, with 3-6 leaves throughout. Leaves broadly triangular in outline, triternate; cauline leaves several. Racemes dense, 5-15-flowered, secund; petals yellow. Capsules cylindric. Fl. and fr. Apr-Oct. Forests, forest edges or grassy slopes at 2000-3500 m. Distributed in Yunnan, Sichuan and Chongqing. Also in E Myanmar.

飞燕黄堇

Corydalis delphinioides Fedde

多年生草本。基生叶较少；叶片轮廓近三角形，三回羽状分裂。花小，排成圆锥花序，顶生和最上部叶腋生，由多数总状花序组成；花瓣黄色。蒴果长圆状梨形。花果期5-9月。生海拔3000-4000米林下、灌丛、林缘或草地。产云南西北部和四川西南部。

南黄紫堇 *Corydalis davidii*

飞燕黄堇 *Corydalis delphinioides*

Herbs perennial. Radical leaves few; leaves subtriangular, 3-pinnate. Flowers small in panicles, constituted by many racemes terminal and axillary from uppermost leaves; petals yellow. Capsules oblong-pyriform. Fl. and fr. May-Sep. Forests, shrubs, forest edges or grasslands at 3000-4000 m. Distributed in NW Yunnan and SW Sichuan.

钩距黄堇

Corydalis hamata Franch.

多年生草本。根状茎粗短。基生叶数枚，长圆形，二回羽状复叶，具长柄；羽片3-5对；基生叶茎生叶具短柄或无柄。总状花序近穗状，具20-40花，密集。蒴果长圆形。花果期7-9月。生海拔3300-5200米的山坡草地或溪边。产云南西北部、四川西部和西藏东部。

Herbs perennial. Rhizomes short, stout. Rosette leaves several, oblong, bipinnate, long petiolate; pinnae 3-5 pairs; cauline leaves short petiolate or sessile. Racemes subspicate, 20-40-flowered, very dense. Capsules oblong. Fl. and fr. Jul-Sep. Grassy slopes or streamsides at 3300-5200 m. Distributed in NW Yunnan, W Sichuan and E Xizang.

钩距黄堇 *Corydalis hamata*

糙果紫堇

Corydalis trachycarpa Maxim.

直立草本，高7-20厘米。茎分枝。基生叶宽卵形，二至三回羽状分裂；茎生叶1-4枚。总状花序多花密集；花瓣淡紫色，上花瓣先端钝，背部具突起，距弯曲，下花瓣稍呈囊状。蒴果倒卵形，具多数淡黄色的小瘤密集排列成6条纵棱。花果期4-9月。生海拔3500-5200米的高山流石滩。产甘肃、青海、四川和西藏。

Erect herbs , 7-20 cm tall. Stems branched. Radical leaves broadly ovate, bi- to tripinnate; cauline leaves 1-4. Racemes very dense; petals purplish, upper petal apex obtuse, with dorsal crest, spur downcurved, lower petal shallowly saccate. Capsule obovoid, with 6 raised lines of dense papillae. Fl. and fr. Apr-Sep. Alpine scree at 3500-5200 m. Distributed in Gansu, Qinghai, Sichuan and Xizang.

糙果紫堇 *Corydalis trachycarpa*

黑顶黄堇

Corydalis nigro-apiculata C. Y. Wu

多年生草本，高10-30厘米。茎分枝。基生叶宽卵形，长2-7厘米，三回羽状分裂。总状花序排列密集；花瓣白色或乳白色；上花瓣长距末端显著下弯，圆筒形；柱头具8个乳突。蒴果狭披针形至长椭圆形，长1.2-1.5厘米。花果期7-9月。生海拔3500-4600米的山坡、林下或高山草甸。产青海、四川西北部和西藏东部。

Perennial herbs, 10-30 cm tall. Stems branched. Radical leaves broadly ovate, 2-7 cm long, tripinnate. Racemes very dense; corolla white to creamy white; upper petal spur distinctly downcurved at apex, cylindric; stigma with 8 papillae. Capsule narrowly lanceolate to oblong, 1.2-1.5 cm long. Fl. and fr. Jul-Sep. Mountain slopes, forests or alpine meadows at 3500-4600 m. Distributed in Qinghai, NW Sichuan and E Xizang.

粗糙黄堇

Corydalis scaberula Maxim.

多年生草本，高7-12厘米。茎1-4条。基生叶卵形，三回羽状分裂。总状花序头状；花瓣淡黄色，上花瓣背部具绿色的突起；距显著弓形弯曲，圆柱形，渐狭，先端钝，下花瓣背部具突起。蒴果长圆形，长7-8毫米。花果期6-9月。生海拔3500-5600米的高山流石滩。产青海、四川西部和西北部、西藏东北部。

Perennial herbs, 7-12 cm tall. Stems 1-4. Radical leaves ovate, tripinnate. Raceme capitate; corolla pale yellow, upper petal with green abaxial crest; spur strongly arcuately downcurved, cylindric to slightly tapering to obtuse apex, lower petal with abaxial crest. Capsule obovoid, 7-8 mm long. Fl. and fr. Jun-Sep. Alpine scree at 3500-5600 m. Distributed in Qinghai, W and NW Sichuan, and NE Xizang.

粗糙黄堇 *Corydalis scaberula*

陕西紫堇

Corydalis shensiana Lidén

多年生草本，17-35厘米。茎1-4条，直立，不分枝。基生叶近圆形，掌状全裂；茎生叶掌状全裂。总状花序；花瓣蓝色，上花瓣背部突起极矮，下花瓣背部无突起，内花瓣背部有显著突起；柱头具8个乳突。蒴果长圆状线形，长1-1.3厘米。花果期6-8月。生海拔1300-3300米的林下、灌丛下或山顶。产河南西部、陕西和山西。

Perennial herbs, 17-35 cm tall. Stems 1-4, straight, unbranched. Radical leaves orbicular, palmatisect; cauline leaves palmatisect. Raceme; petals blue, upper petal with low dorsal crest, lower petal without dorsal crest, inner petals with conspicuous dorsal crest; stigma with 8 papillae. Capsule oblong-linear, 1-1.3 cm long. Fl. and fr. Jun-Aug. Forests, among shrubs or mountain summits at 1300-3300 m. Distributed in W Henan, Shaanxi and Shanxi.

黑顶黄堇 *Corydalis nigro-apiculata*

陕西紫堇 *Corydalis shensiana*

金雀花黄堇

Corydalis cytisiflora (Fedde) Liden

多年生草本，高15-40厘米。茎不分枝。基生叶近圆形，直径1.5-3厘米，3全裂，再次2-3深裂；茎生叶1-2枚，掌状全裂。总状花序；萼片小；花瓣黄色，下花瓣背部有或无突起，内花瓣7-11毫米，背部突起宽阔，先端明显高于花瓣顶部。蒴果长椭圆形，长达15毫米。花果期5-8月。生海拔2600-4500米的山坡灌丛下或草地草甸中。产甘肃南部和四川西部。

Perennial herbs, 15-40 cm tall. Stems unbranched. Radical leaves orbicular, 1.5-3 cm diam, ternate to biternate; cauline leaves 1-2, palmatisect. Raceme; sepals mi-

金雀花黄堇 *Corydalis cytisiflora*

曲花紫堇 *Corydalis curviflora*

大海黄堇 *Corydalis feddeana*

nute; petals yellow, lower petal dorsal crest present or absent, inner petals 7-11 mm, dorsal crest broad, clearly overtopping apex. Capsule oblong, 15 mm long. Fl. and fr. May-Aug. Among shrubs, grasslands, meadows at 2600-4500 m. Distributed in S Gansu and W Sichuan.

曲花紫堇

Corydalis curviflora Maxim.

无毛草本，高7-50厘米。茎不分枝。基生叶圆形或肾形，3全裂，全裂片2-3深裂；茎生叶掌状全裂。总状花序；花瓣淡蓝色；上花瓣距向上弯曲，圆筒形，粗壮；下花瓣宽倒卵形，背部突起较矮。蒴果线状长椭圆形至椭圆形。花果期5-8月。生海拔2400-4600米的山坡灌丛下或草地草甸中。产甘肃、宁夏、青海和四川。

Glabrous herbs, 7-50 cm tall. Stems unbranched. Radical leave orbicular to reniform, sub-biternate to biternatisect; cauline leaves palmatisect. Raceme; petals light blue; upper petal spur upwardly reflexed, cylindric, thick; lower petal broadly obovate, abaxial crest short. Capsule oblong to elliptic. Fl. and fr. May-Aug. Among shrubs, grasslands, meadows at 2400-4600 m. Distributed in Gansu, Ningxia, Qinghai and Sichuan.

浪穹紫堇

Corydalis pachycentra Franch.

多年生粗壮草本。茎不分枝。茎生叶着生茎中部，无柄，掌状5-11深裂。总状花序具4-12花；花淡蓝色、淡紫色或紫红色。蒴果倒卵球形。花果期5-9月。生海拔(2700-)3500-4200(-5200)米的山坡、草地、高山草甸或林下。产云南、四川和西藏。

Herbs perennial, stout. Stems unbranched. Cauline leaves inserted on the middle of stem, sessile, palmately 5-11-parted. Racemes 4-12-flowered; flowers pale blue, pale purple or purple-red. Capsules obovoid. Fl. and fr. May-Sep. Slopes, grasslands, alpine meadows or forests at (2700-)3500-4200(-5200) m. Distributed in Yunnan, Sichuan and Xizang.

大海黄堇

Corydalis feddeana H. Lévl.

多年生草本。叶小，二回三出复叶，小裂片条形至条状披针形。总状花序顶生，多花；花瓣黄色；上花瓣长1.8-3.2厘米，瓣片卵形；下花瓣长约1厘米。蒴果倒卵球形至窄倒卵形。花果期6-8月。生海拔3200-4100米的山顶杜鹃灌丛下或草坡。产云南东北部和四川南部。

Herbs perennial. Leaves small, biternate, segments linear to linear-lanceolate. Racemes terminal, many flowered; petals yellow; upper petals 1.8-3.2 cm, limbs ovate; lower petals ca. 1 cm. Capsules obovoid to narrowly ovoid. Fl. and fr. Jun-Aug. Thickets of *Rhododendron* at top of mountains or grassy slopes at 3200-4100 m. Distributed in NE Yunnan and S Sichuan.

浪穹紫堇 *Corydalis pachycentra*

扇苞黄堇 *Corydalis rheinbabeniana*

扇苞黄堇

Corydalis rheinbabeniana Fedde

多年生草本，高15-45厘米。茎1-3条，不分枝。茎生叶2-5枚，互生，宽卵形至近圆形，一回奇数羽状全裂。总状花序；花瓣柠檬黄色，外花瓣突起先端突出于花瓣之外；柱头具6乳突。蒴果狭倒卵形，长8-10毫米。花果期6-9月。生海拔3100-4100米的灌丛下或草坡。产甘肃西南部、青海东部和东南部、四川西北部。

Perennial herbs, 15-45 cm tall. Stems 1-3, unbranched. Cauline leaves 2-5, alternate, broadly ovate to subrounded, pinnate. Raceme; petals lemon yellow, outer petal crest overtopped apex; stigma with 6 papillae. Capsule narrowly obovoid, 8-10 mm long. Fl. and fr. Jun-Sep. Among shrubs or grassy slopes at 3100-4100 m. Distributed in SW Gansu, E and SE Qinghai, and NW Sichuan.

秦岭紫堇

Corydalis trisecta Franch.

多年生草本，高10-28厘米。茎不分枝或稀分枝。基生叶近于5全裂；茎生叶2枚，互生，明显具柄，奇数羽状分裂。总状花序；花瓣黄色，上花瓣先端急尖，背部高突起，距稍向下弯曲，下花瓣先端急尖，背部具短突起。蒴果长椭圆形，长达1-1.6厘米。花果期7-8月。生海拔1400-3800米的山顶、山坡草丛或岩石缝。产重庆、河南、湖北、陕西和四川。

Perennial herbs, 10-28 cm tall. Stems simple or rarely branched. Radical leaves subquinate; cauline leaves 2, alternate, conspicuously stalked, imparipinnate. Raceme; petals yellow, upper petal apex acute, with abaxial crest, spur slightly downcurved, lower petal apex acute, abaxial crest short. Capsule oblong, 1-1.6 cm long. Fl. and fr. Jul-Aug. Mountain summits, grassy slopes or stony crevices at 1400-3800 m. Distributed in Chongqing, Henan, Hubei, Shaanxi and Sichuan.

条裂黄堇

Corydalis linarioides Maxim.

多年生草本，高10-50厘米。茎通常不分枝。基生叶长2-4厘米，宽2-5厘米，羽状分裂；茎生叶羽状全裂。总状花序；花瓣黄色，上花瓣背部突起自花瓣片先端延伸至距；柱头具2乳突。蒴果长圆形，长1-1.4厘米。花果期6-9月。生海拔2100-4500米的林下、林缘、灌丛下或草坡中。产甘肃、宁夏、青海、陕西、山西、四川和西藏。

Perennial herbs, 10-50 cm tall. Stems usually unbranched. Radical leaves 2-4 × 2-5 cm, pinnate; cauline leaves pinnate. Raceme; petals yellows; upper petal abaxial crest from apex extended to spur; stigma with 2 papillae. Capsule oblong, 1-1.4 cm. Fl. and fr. Jun-Sep. Forests, forest margins, shrubs or meadows at 2100-4500 m. Distributed in Gansu, Ningxia, Qinghai, Shaanxi, Shanxi, Sichuan and Xizang.

暗绿紫堇

Corydalis melanochlora Maxim.

多年生草本，具灰粉。鳞茎较大，鳞片椭圆形；茎2-5条，常具2个近对生叶。基生叶2-5枚，三回羽状全裂，具灰粉，有时具小深色斑点，卵状长圆

秦岭紫堇 *Corydalis trisecta*

条裂黄堇 *Corydalis linarioides*

暗绿紫堇 *Corydalis melanochlora*

西藏宽花紫堇 *Corydalis latiflora* subsp. *gerdae*

形，肉质。总状花序伞房状，具4-10(-15)花；花瓣蓝色。蒴果阔长圆形至倒卵球形。花果期6-9月。生海拔4000-5000米的草地或石坡。产云南、四川、西藏、甘肃和青海。

Herbs perennial, gray-glaucous. Bulbs large, scales elliptic; stems 2-5, usually with 2 subopposite leaves. Radical leaves 2-5, tripinnatisect, gray-glaucous, sometimes with small dark spots, ovate-oblong, fleshy. Racemes corymbose, 4-10(-15)-flowered; flowers blue. Capsules broadly oblong to obovoid. Fl. and fr. Jun-Sep. Grasslands or rocky slopes at 4000-5000 m. Distributed in Yunnan, Sichuan, Xizang, Gansu and Qinghai.

西藏宽花紫堇

Corydalis latiflora Hook. f. et Thoms. subsp. **gerdae** (Fedde) Liden

多年生草本，高5-15厘米，疏松的垫状植物。茎上部具2枚对生叶。基生叶二回羽状全裂；茎生叶与基生叶同形。花序近伞形；花灰蓝色或淡紫色，具浓烈香味；上花瓣突起伸达距末端；柱头具4乳突。蒴果倒卵形，约长10毫米。花期7-9月。生海拔4300-5500米的高山流石滩。产西藏南部。不丹、印度(锡金)和尼泊尔亦有。

Perennial herbs, 5-15 cm tall, lax cushion plant. Stems above with 2 opposite leaves. Radical leaves bipinnate; cauline leaves like radical leaves. Inflorescences subumbellate; corolla pale grayish blue to lavender, with strong pleasant scent; upper petal crest gradually attenuate to spur end; stigma with 4 papillae. Capsule obovoid, ca. 10 mm long. Fl. Jul-Sep. Alpine scree at 4300-5500 m. Distributed in S Xizang. Also in Bhutan, India (Sikkim) and Nepal.

粗梗黄堇

Corydalis pachypoda (Franch.) Hand.-Mazz.

多年生草本，高5-15厘米。茎不分枝，无叶或基部具1叶。基生叶具长鞘，一回羽状全裂。总状花序；花冠橙黄色，平展；外花瓣急尖，具突起，宽展；柱头近扁四方形，顶端2裂，具2乳突。蒴果倒卵形，长8-15毫米。花期5-7月。果期6-8月。生海拔2300-4700米的石缝或沙石地。产云南西北部。

Perennial herbs, 5-15 cm tall. Stems unbranched, without leaves or base with 1 leaf. Radical leaves long vaginate, pinnate. Raceme; corolla orange-yellow, patent; outer petals acute, broadly crested, with spreading margin; stigma square, apex 2-lobed with 2 papillae. Capsule obovoid, 8-15 mm long. Fl. May-Jul. Fr. Jun-Aug. Stone crevices or limestone scree at 2300-4700 m. Distributed in NW Yunnan.

新疆黄堇

Corydalis gortschakovii Schrenk

多年生草本，高10-40厘米。茎不分枝，具1-3叶。基生叶二回羽状全裂；茎生叶与基生叶同形。总状花序；花冠橙黄色；外花瓣具高而伸出瓣片顶端的突起；柱头扁四方形，具2突起。蒴果狭倒卵形，长10-15毫米。花果期6-9月。生海拔2100-3600米的云杉林缘或多石阴湿地。产新疆北部。中亚亦有。

Perennial herbs, 10-40 cm tall. Stems unbranched, with 1-3 leaves. Radical leaves bipinnate; cauline leaves like radical leaves. Racemes; corolla orange-yellow; outer petals with very high crest extended beyond petal apex; stigma square, with 2 papillae. Capsule narrowly obovoid, 10-15 mm long. Fl. and fr. Jun-Sep. *Picea* forest margins or stony shaded moist areas at 2100-3600 m. Distributed in N Xinjiang. Also in C Asia.

粗梗黄堇 *Corydalis pachypoda*

新疆黄堇 *Corydalis gortschakovii*

拟锥花黄堇
Corydalis hookeri Prain

多年生草本，高8-50厘米。茎分枝。基生叶具鞘，二回羽状全裂。总状花序；花污黄色；外花瓣渐尖，具宽或极狭的突起；距与瓣片等长；柱头扁四方形，具4突起。蒴果卵圆形或长圆形，长6-8毫米。花期5-9月，果期6-10月。生海拔 3700-5000米的高山草原或流石滩。产西藏西部和西南部。不丹、印度北部和尼泊尔亦有。

Perennial herbs, 8-50 cm tall. Stems branched. Radical leaves vaginate, bipinnate. Racemes; flowers dirty yellow; outer petals acuminate, broadly to very narrowly crested; stigma flat square, with 4 papillae. Capsule ovoid or oblong, 6-8 mm long. Fl. May-Sep. Fr. Jun-Oct. Alpine grasslands or stony screes at 3700-5000 m. Distributed in W and SW Xizang. Also in Bhutan, N India and Nepal.

半荷包紫堇
Corydalis hemidicentra Hand-Mazz.

多年生草本。茎具分枝。叶簇生基部，三出，具长叶柄；小叶肉质，圆形至椭圆形，厚，全缘，三条脉。花蓝白色、蓝色至蓝紫色。蒴果俯垂于直立的花梗上，倒卵球形。花果期7-9月。生海拔3500-5300米的高山流石滩上。产云南西北部和西藏东南部。

Herbs perennial. Stems branched. Leaves crowded at base, trifoliolate, long petiolate; leaflets fleshy, orbicular or elliptic, thick, margin entire, trinerved. Flowers blue-white, blue or blue-purple. Capsules nutant on erect pedicel, obovoid. Fl. and fr. Jul-Sep. Alpine screes at 3500-5300 m. Distributed in NW Yunnan and SE Xizang.

半荷包紫堇 *Corydalis hemidicentra*

囊距紫堇
Corydalis benecincta W. W. Smith

多年生草本。叶三出，叶柄长；小叶两面具白粉，常有不规则斑驳条纹，倒卵形。花序无明显梗，疏松伞房状，具5-15花，常自基部分枝；花粉红色至淡紫色；距粗大，囊状，约与花瓣等长。蒴果椭圆体形。花果期7-9月。生海拔4000-6000米的高山流石滩。产云南西北部、四川西南部和西藏东南部。

Herbs perennial. Leaves trifoliolate, long petiolate; leaflets glaucous on both surfaces, often mottled with irregular streaks, obovate. Inflorescences without a distinct stalk, loosely corym-

拟锥花黄堇 *Corydalis hookeri*

囊距紫堇 *Corydalis benecincta*

刻叶紫堇 *Corydalis incisa*

bose, 5-15-flowered, often branched at base; flowers pink to pale purple; spurs thick, saccate, the length nearly equal to petals. Capsules ellipsoid. Fl. and fr. Jul-Sep. Alpine screes at 4000-6000 m. Distributed in NW Yunnan, SW Sichuan and SE Xizang.

刻叶紫堇
Corydalis incisa (Thunb.) Pers.

一年或二年生草本。茎多叶，自基部分枝。叶片二至三回三出复叶。总状花序具6-17花，初密集，后疏离；花紫红色至紫色；上花瓣长2-2.5厘米；距圆筒形。蒴果条形至长圆形。花果期4-9月。生海拔1800米以下的林缘、路边或疏林下。产中国西南、东南、华中、华北、华西和华东。朝鲜半岛和琉球群岛亦有。

Herbs annual or biennial. Stems leafy and branched from base. Leaves bi- to triternate. Racemes 6-17-flowered, at first dense, then lax; flowers purplish red to purple; upper petals 2-2.5 cm; spurs cylindric. Capsules linear to oblong. Fl. and fr. Apr-Sep. Forest edges, roadsides or sparse forests below 1800 m. Distributed in SW, SE, C, N, W and E China. Also in Korean Peninsula and Ryukyu Islands.

紫堇
Corydalis edulis Maxim.

一年生草本，高20-50厘米。基生叶具长柄，一至二回羽状全裂。总状花序具3-10花；花粉红色至紫红色；上花瓣长1.5-2厘米。蒴果条形，下垂。花果期4-7月。生海拔400-1200米的丘陵、沟边或多石地。产中国西南、东南、华中、华北、华西和华东。日本亦有。

Herbs annual, 20-50 cm tall. Basal leaves long petiolate, pinnate to bipinnate. Racemes 3-10-flowered; flowers pink to amaranth; upper petals 1.5-2 cm. Capsules linear, drooping. Fl. and fr. Apr-Jul. Hills, by streams or rocky places at 400-1200 m. Distributed in SW, SE, C, N, W and E China. Also in Japan.

地丁草
Corydalis bungeana Turcz.

草本。茎自基部铺散分枝。基生叶多数，二至三回羽状复叶。花粉红色至淡紫色；苞片叶状；上花瓣长1.1-1.4厘米；距长4-5毫米；内花瓣顶端深紫色；柱头小，圆肾形。生海拔1500米以下的多石坡或河边。产华北、华西、华东和东北。俄罗斯(远东地区)、蒙古东南部和朝鲜半岛北部亦有。

Herbs. Stems diffusely spreading and branched from base. Basal leaves many, 2-3-pinnate. Flowers pink to lavender; bracts leaflike; upper petals 1.1-1.4 cm; spurs 4-5 mm; inner petals darker lavender at tip; stigmas small, orbicular-reniform. Rocky slopes or river banks below 1500 m. Distributed in N, W, E and NE China. Also in Russia (Far East), SE Mongolia and N Korean Peninsula.

紫堇 *Corydalis edulis*

地丁草 *Corydalis bungeana*

石生黄堇 *Corydalis saxicola*

石生黄堇
Corydalis saxicola Bunting

多年生草本，淡绿色。叶羽状全裂。总状花序多花；苞片大，明显长于花梗；花黄色或金黄色；柱头二叉状分裂，各枝顶端具2裂的乳突。蒴果反折，条形。花果期5-7月。生海拔600-1400(-3900)米石灰岩缝隙中。产中国西南。

Herbs perennial, pale green. Leaves pinnate-sect. Racemes many flowered; bracts large, longer than pedicels; flowers yellow or golden yellow; stigma arms forward-arcuate, each with 2 indistinct papillae at apex. Capsules reflexed, linear. Fl. and fr. May-Jul. Rock crevices of limestones at 600-1400(-3900) m. Distributed in SW China.

蛇果黄堇
Corydalis ophiocarpa Hook. f. et Thoms.

草本。茎分枝。基生叶边缘具膜质翅，二回羽状全裂；茎生叶与基生叶同形。总状花序；花淡黄色至苍白色；内花瓣顶端暗紫红色，具伸出顶端的突起；柱头具4乳突。蒴果线形，长2-3厘米，蛇形弯曲。花果期5-8月。生海拔1100-2700米的沟谷林缘。除中国东北与新疆外，几乎遍布全国。不丹、印度(锡金)和日本亦有。

Herbs. Stems branched. Basal leaves margin winged, bipinnate; cauline leaves like basal leaves. Racemes; flowers pale yellow to whitish; inner petals tipped with dark purple-red, with crest extending beyond apex; stigma with 4 papillae. Capsule linear, 2-3 cm long. Fl. and fr. May-Aug. River valleys and forest margins at 1100-2700 m. Distributed all over China, except NE China and Xinjiang. Also in Bhutan, India (Sikkim) and Japan.

小花黄堇 *Corydalis racemosa*

小花黄堇
Corydalis racemosa (Thunb.) Pers.

丛生草本。茎多叶，多分枝，细弱，铺散。茎生叶具柄；叶下面灰色，三角形，二回羽状复叶。总状花序密具多花；花黄色至淡黄色。蒴果劲直，条形。花果期2-9月。生海拔400-2000米的林缘、溪边或灌丛中。产中国西南、华南、东南、华中、华西和华东。日本亦有。

Herbs clustered. Stems leafy,

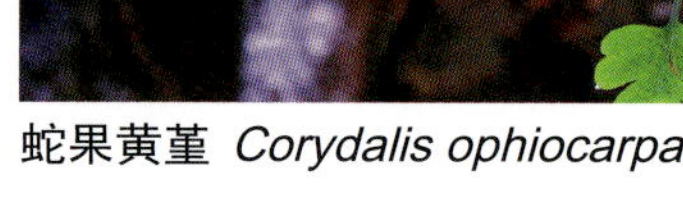

蛇果黄堇 *Corydalis ophiocarpa*

黄堇 *Corydalis pallida*

珠果黄堇 *Corydalis speciosa*

much branched, weak, diffuse. Cauline leaves petiolate; leaves gray abaxially, deltoid, bipinnate. Racemes densely many flowered; flowers yellow to pale yellow. Capsules straight, linear. Fl. and fr. Feb-Sep. Forest edges, streamsides or thickets at 400-2000 m. Distributed in SW, S, SE, C, W and E China. Also in Japan.

黄堇

Corydalis pallida (Thunb.) Pers.

多年生草本。基生叶多数，丛生，二回羽状全裂，长圆形，羽片2-4对。总状花序顶生或腋生，具10-25花；花黄色至淡黄色。蒴果条形。种子龙骨状，具小刺。花期4-6月，果期5-7月。生林缘、河岸或多石地。产中国东北、华北、华东和华中。俄罗斯(远东地区)、朝鲜半岛和日本亦有。

Herbs perennial. Basal leaves many, rosette, 2-pinnatisect, oblong; pinnae 2-4 pairs. Racemes terminal or axillary, 10-25-flowered; flowers yellow to yellowish. Capsules linear. Seeds keeled, with smaller spines. Fl. Apr-Jun. Fr. May-Jul. Forest edges, river banks or rocky places. Distributed in NE, N, E and C China. Also in Russia (Far East), Korean Peninsula and Japan.

珠果黄堇

Corydalis speciosa Maxim.

两年生草本(有时为短期多年生)，具白粉。基生叶长圆形，二回羽状分裂(至近三回羽状)；羽片4或5对。总状花序，多花；花金黄色。蒴果条形，长约3厘米，念珠状。花果期4-9月。生海拔500米以下的林下、路边或水边多石地。产华中、东南、华北和东北。俄罗斯(远东地区)、蒙古、朝鲜半岛和日本亦有。

Herbs, biennial (sometimes short-lived perennial), glaucous. Basal leaves oblong, bipinnate (to almost tripinnate); pinnae 4 or 5 pairs. Racemes many flowered; flowers golden yellow. Capsules linear, ca. 3 cm long, moniliform. Fl. and fr. Apr-Sep. Forests, roadsides or rocky places by streams below 500 m. Distributed in C, SE, N and NE China. Also in Russia (Far East), Mongolia, Korean Peninsula and Japan.

北越紫堇

Corydalis balansae Prain

一年生草本。枝条花葶状，常对叶生。叶二回羽状全裂，二回羽片卵圆形，3-5深裂。总状花序具3-15花；上花瓣长15-17毫米，距短囊状，约占花瓣全长的1/4；柱头横向伸出2臂，各枝顶端具3乳突。蒴果线形，约长3厘米，具1列种子。花果期5-8月。生海拔200-700米的山谷、潮湿地方或山脚灌丛。产华南、华东和华中。日本、越南和老挝亦有。

Annual herbs. Branches scapiform, usually opposite with leaves. Leaves bipinnatisect, pinnules ovate-rounded, 3-5-parted. Raceme 3-15-flowered; upper petal 15-17 mm, spur saccate, ca. 1/4 of petal; stigma horizontally extended 2-armed, each arm with 3 terminal papillae. Capsule linear, ca. 3 cm long, with a row of seeds. Fl. and fr. May-Aug. Valleys, wet areas, thickets on foothills at 200-700 m. Distributed in S, E and C China. Also in Japan, Vietnam and Laos.

北越紫堇 *Corydalis balansae*

夏天无 *Corydalis decumbens*

夏天无

Corydalis decumbens (Thunb.) Pers.

多年生草本。茎高10-25厘米，细长，不分枝。叶为二回羽状复叶。总状花序疏具3-10花；花近白色、粉红色或淡蓝色。蒴果条形，多少扭曲，长13-18毫米。种子6-14粒。花果期2-5月。生海拔100-300米的山坡或路边。产华中、东南和华东。日本南部亦有。

Herbs perennial. Stems 10-25 cm tall, slender, simple. Leaves 2-pinnate. Racemes lax, 3-10-flowered; flowers whitish, pink or pale blue. Capsules linear, slightly tortuous, 13-18 mm long. Seeds 6-14. Fl. and fr. Feb-May. Slopes or roadsides at 100-300 m. Distributed in C, SE and E China. Also in S Japan.

小药八旦子

Corydalis caudata (Lam.) Pers.

多年生草本。茎于节间膝曲，细弱。叶为二至三回三出复叶，背面灰色，上面绿色。总状花序疏具3-8花；萼片小，早落；花蓝色或紫蓝色。蒴果卵球形至椭圆体形。花果期4-5月。生海拔100-1200米的山坡或林缘。产华中、华北、华西和华东。

Herbs perennial. Stems geniculate at nodes, slender, weak. Leaves bi- to triternate, glaucous abaxially, green adaxially. Racemes lax, 3-8-flowered; sepals small, caducous; flowers blue or indigo. Capsules ovoid to ellipsoid. Fl. and fr. Apr-May. Slopes or forest edges at 100-1200 m. Distributed in C, N, W and E China.

小药八旦子 *Corydalis caudata*

堇叶延胡索

Corydalis fumariifolia Maxim.

多年生草本，高8-28厘米。块茎圆球形。茎直立或上升。叶二至三回三出，全缘至深裂。总状花序；花淡蓝色或蓝紫色；内花瓣近白色；上花瓣瓣片多少上弯；柱头具4-6乳突。蒴果线形，常呈红棕色，长15-30毫米。花果期3-5月。生海拔600米左右的林缘或灌木丛中。产中国东北。俄罗斯(远东地区)亦有。

堇叶延胡索 *Corydalis fumariifolia*

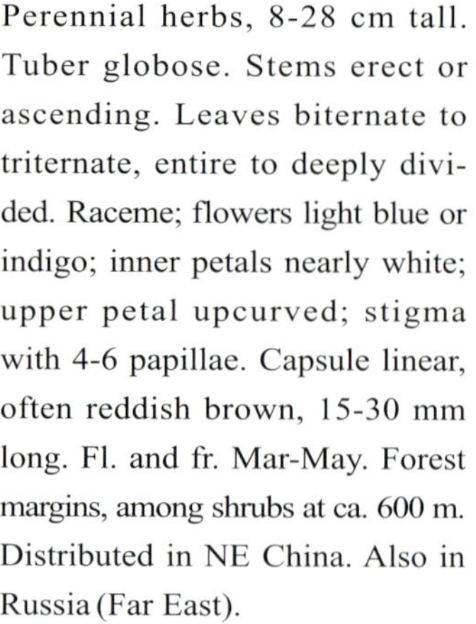

Perennial herbs, 8-28 cm tall. Tuber globose. Stems erect or ascending. Leaves biternate to triternate, entire to deeply divided. Raceme; flowers light blue or indigo; inner petals nearly white; upper petal upcurved; stigma with 4-6 papillae. Capsule linear, often reddish brown, 15-30 mm long. Fl. and fr. Mar-May. Forest margins, among shrubs at ca. 600 m. Distributed in NE China. Also in Russia (Far East).

全叶延胡索

Corydalis repens Mandl et Muehld.

多年生草本，高8-20厘米。块茎球形。茎自先出叶分枝。叶二回三出。总状花序；花淡蓝

全叶延胡索 *Corydalis repens*

齿瓣延胡索 *Corydalis turtschaninovii*

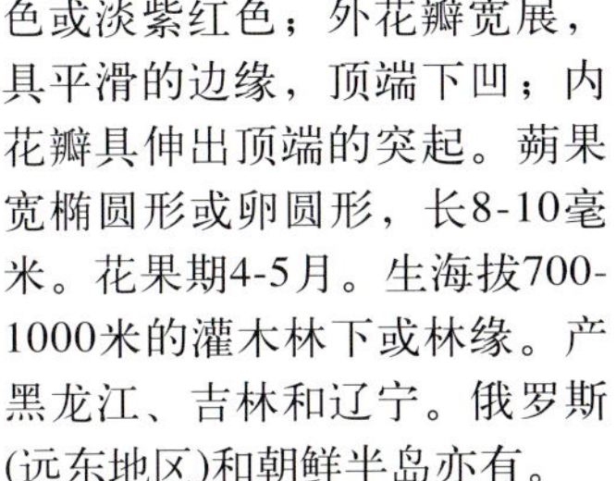

色或淡紫红色；外花瓣宽展，具平滑的边缘，顶端下凹；内花瓣具伸出顶端的突起。蒴果宽椭圆形或卵圆形，长8-10毫米。花果期4-5月。生海拔700-1000米的灌木林下或林缘。产黑龙江、吉林和辽宁。俄罗斯(远东地区)和朝鲜半岛亦有。

Perennial herbs, 8-20 cm tall. Tuber globose. Stems often with several branches from prophyll. Leaves biternate. Raceme; flowers pale blue or pale amaranth; outer petals broad, flat, with smooth margin, apex emarginate; inner petals with dorsal crests prolonged beyond petal apex. Capsule broadly elliptic or ovoid, 8-10 mm long. Fl. and fr. Apr-May. Shrubs, understories or forest margins at 700-1000 m. Distributed in Heilongjiang, Jilin and Liaoning. Also in Russia (Far East) and Korean Peninsula.

齿瓣延胡索

Corydalis turtschaninovii Besser

多年生草本。块茎黄色，圆球形。茎生叶2枚，一至三回三出。总状花序花密生；花蓝色、白色或品蓝，有时具一淡紫色斑；外面花瓣顶端凹缺，具一明显的小尖。蒴果条形。花果期4-6月。生开旷林地、林缘或溪边。产河北、内蒙古、黑龙江、吉林和辽宁。俄罗斯(西伯利亚)、朝鲜半岛和日本亦有。

Herbs perennial. Tubers yellow, rounded. Cauline leaves 2, once to three ternate. Racemes densely flowered; flowers blue, white or royal blue, sometimes with a purplish tint; outer petals apically emarginate with a distinct mucro in notch. Capsules linear. Fl. and fr. Apr-Jun. Open forests, forest edges or by streams. Distributed in Hebei, Neimenggu, Heilongjiang, Jilin and Liaoning. Also in Russia (Siberia), Korean Peninsula and Japan.

延胡索

Corydalis yanhusuo W. T. Wang ex Z. Y. Su et C. Y. Wu

多年生草本，高10-30厘米。块茎圆球形。茎直立，常分枝，基部以上具1鳞片。叶二回三出或近三回三出，小叶三深裂。总状花序；花紫红色；萼片小，早落；外花瓣顶端微凹，具短尖，距上弯，下花瓣具短爪，内花瓣爪长于瓣片。蒴果线形，长2-2.8厘米。花果期4-6月。生丘陵草地。产安徽、河南、湖北、江苏和浙江。

Perennial herbs, 10-30 cm tall. Tuber globose. Stems erect, usually branched, with 1 scale leaves. Leaves biternate or nearly triternate, leaflets deeply trifid. Raceme; flowers amaranth; sepals small, caducous; outer petals emarginate with a mucro, spur of upper petal upcurved, lower petal with short claw, inner petals claw longer than petal lobes. Capsule linear, 2-2.8 cm long. Fl. and fr. Apr-Jun. Grasslands on highlands. Distributed in Anhui, Henan, Hubei, Jiangsu and Zhejiang.

延胡索 *Corydalis yanhusuo*

白花菜科 Capparidaceae

树头菜
Crateva unilocularis Buch.-Ham.

乔木。小叶椭圆形，薄革质，先端渐尖至急尖。花序总状或伞房状，着生在小枝顶端，具13-25(-35)花，基部具少量叶；花瓣白色或黄色。果球形，顶端粗糙，有近圆形的小灰黄色斑点。花期3-7月，果期7-8月。生海拔1500米以下的湿润地区。产云南、广西、广东、海南和福建。印度、尼泊尔、缅甸、老挝、越南和柬埔寨亦有。

Trees. Leaflets elliptic, thinly leathery, apex acuminate to abruptly acuminate. Inflorescences racemes or corymbs at top of branchlets, 13-25(-35)-flowered, with a few leaves on basal part; petals white or yellow. Fruits globose, apically scabrous, with nearly circular small ash-yellow flecks. Fl. Mar-Jul. Fr. Jul-Aug. Damp sites below 1500 m. Distributed in Yunnan, Guangxi, Guangdong, Hainan and Fujian. Also in India, Nepal, Myanmar, Laos, Vietnam and Cambodia.

沙梨木
Crateva magna (Lour.) DC.

乔木或灌木。掌状3小叶；小叶纸质至薄革质，先端渐尖至长渐尖。总状聚伞花序，具(3-)20-30(-40)花，由数叶所托；花瓣白色。浆果长圆状椭圆体形至长圆状卵形，干后淡黄灰色。花期3-4月，果期8-9月。生海拔1000米以下的溪边、湖畔或平地。产云南、西藏、广西、广东和海南。南亚和东南亚亦有。

Trees or shrubs. Leaves palmately 3-foliolate; leaflets papery to thinly leathery, apex acuminate to long acuminate. Inflorescences corymbose racemes, (3-)20-30(-40)-flowered, subtended by several leaves; petals white. Berries oblong-ellipsoid to oblong-ovate, yellowish gray when dry. Fl. Mar-Apr. Fr. Aug-Sep. By streams, lakes or plains below 1000 m. Distributed in Yunnan, Xizang, Guangxi, Guangdong and Hainan. Also in S and SE Asia.

沙梨木 *Crateva magna*

树头菜 *Crateva unilocularis*

野香橼花
Capparis bodinieri H. Lévl.

常绿灌木或小乔木。小枝有刺，常被淡褐色或灰色毛。叶卵形至披针形，基部圆形至楔形，不下延至叶柄。花序排成纵列，腋上生，具(1或)2-6(或7)花；花瓣白色；雄蕊(18-)20-37。果球形，成熟时黑色。花期3-4月，果期8-10月。生海拔700-1700(-2300)米的林中或石灰山。产云南、四川、贵州和广西。印度北部、不丹和缅甸北部亦有。

野香橼花 *Capparis bodinieri*

Shrubs or small trees, evergreen. Branchlets spiny, with pale brown or gray hairs. Leaves ovate to lanceolate, base rounded to cuneate but not decurrent on petiole. Inflorescences superaxillary rows, (1 or)2-6(or 7)-flowered; petals white; stamens (18-)20-37. Fruits globose, black when ripe. Fl. Mar-Apr. Fr. Aug-Oct. Forests or limestone hills at 700-1700(-2300) m. Distributed in Yunnan, Sichuan, Guizhou and Guangxi. Also in N India, Bhutan and N Myanmar.

雷公橘

Capparis membranifolia Kurz

藤本、灌木或稀小乔木。茎具多托叶刺。叶狭椭圆状披针形；幼叶密被锈色短绒毛；侧脉5-7对。花序排成纵列，腋上生，具2-5花；花瓣白色；雄蕊20-28(-35)。果黑色至紫黑色，球形，直径8-15毫米。花期1-4月，果期5-8月。生海拔1800米以下的石山灌丛中、山谷疏林、林缘、山坡、路旁或溪边。产中国西南和华南。南亚亦有。

Vines, shrubs or rarely small trees. Stems with many stipular spines. Leaves narrowly elliptic-lanceolate; young leaves densely rust-colored shortly tomentose; lateral veins 5-7 per side. Inflorescences superaxillary rows, 2-5-flowered; petals white; stamens 20-28(-35). Fruits black to purplish black, globose, 8-15 mm diam. Fl. Jan-Apr. Fr. May-Aug. Thickets on rocky mountains, sparse forests in valleys, forest edges, slopes, roadsides or by streams below 1800 m. Distributed in SW and S China. Also in S Asia.

独行千里

Capparis acutifolia Sweet

藤本或灌木，无毛或小枝、叶柄和花梗有时被污黄色短绒毛，后立即变无毛。叶纸质或近革质，侧脉8-10对。花序排成纵列，具(1-)2-4花，腋上生；花瓣白色；雄蕊(19或)20-30。果近球形，成熟后鲜红色。花期4-5月，果期全年有记载。生海拔300-1100米的旷野、山坡、路边、石山上或林中。产广东、台湾、福建、浙江、湖南和江西。印度、不丹、泰国和越南亦有。

Vines or shrubs, glabrous or twigs, petiole and pedicels sometimes dirty yellow shortly tomentose but soon glabrescent. Leaves papery or subleathery, lateral veins 8-10 each side. Inflorescences superaxillary rows, (1-)2-4-flowered; petals white; stamens (19 or)20-30. Fruits subglobose, bright red when ripe. Fl. Apr-May. Fr. almost all year. Open places, slopes, roadsides, rocky mountains or forests at 300-1100 m. Distributed in Guangdong, Taiwan, Fujian, Zhejiang, Hunan and Jiangxi. Also in India, Bhutan, Thailand and Vietnam.

雷公橘 *Capparis membranifolia*

独行千里 *Capparis acutifolia*

小绿刺 *Capparis urophylla*

小绿刺
Capparis urophylla F. Chun

乔木或灌木。托叶刺生茎上，基部膨大；叶卵形至椭圆形，宽1-2.5厘米。花序排成纵列，具1-3花，腋上生；花瓣白色；雄蕊12-20。果实黄色至橘红色，球形，近光滑。花期3-6月，果期8-12月。生海拔300-1900米的山坡、溪边、沟谷、疏林或灌丛。产云南南部、广西、湖南南部和东南部。老挝亦有。

Trees or shrubs. Stipular spines on stems, base inflated; leaves ovate to elliptic, 1-2.5 cm wide. Inflorescences superaxillary rows, 1-3-flowered; petals white; stamens 12-20. Fruits yellow to orangish red, globose, nearly smooth. Fl. Mar-Jun. Fr. Aug-Dec. Slopes, by streams, valleys, open forests or thickets at 300-1900 m. Distributed in S Yunnan, Guangxi, S and SE Hunan. Also in Laos.

无柄山柑
Capparis subsessilis B. S. Sun

灌木。小枝基部无钻形鳞片。叶无柄或近无柄，椭圆形至微倒卵状椭圆形，长9-12.5厘米，约为宽的2.5倍，顶端渐尖至长渐尖，基部心形，两面无毛，中脉与侧脉在表面均凹入，背面均凸起。花序腋上生。果单生，直径8-9毫米；花梗和雌蕊柄果时均不木质化增粗，纤细，无毛。果期8-10月。生海拔500-1000米的山谷林中。产广西。越南亦有。

Shrubs. Twigs base without subulate scales. Leaves sessile or subsessile, elliptic to slightly obovate-elliptic, 9-12.5 cm long, ca. 2.5 times as long as wide, apex acuminate to long acuminate, base cordate, both surfaces glabrous, midvein and secondary veins adaxially impressed and abaxially raised. Inflorescences superaxillary. Fruit solitary, 8-9 mm diam; fruiting pedicel and fruiting gynophore not lignified and thickened, slender, glabrous. Fr. Aug-Oct. Valley forests at 500-1000 m. Distributed in Guangxi. Also in Vietnam.

山柑
Capparis spinosa Linnaeus

匍匐灌木。托叶刺先端下弯；叶卵形、倒卵形、宽椭圆形或近圆形，长1.3-3厘米，为宽的1-1.7倍，先端刺状，基部圆形；叶柄长1-4毫米。花单生于上部叶腋；花瓣两型。果具6-8条细纵棱；花梗和雌蕊柄果时成直角。花期6-7月，果期8-9月。生海拔1100米以下的平原、沙地和向阳空地。产新疆和西藏。北非、南亚、东南亚、澳大利亚和欧洲南部亦有。

Prostrate shrubs. Stipular spines apex recurved; leaves ovate, obovate, broadly elliptic or suborbicular, 1.3-3 cm long, 1-1.7 times as long as wide, apex spine-tipped, base rounded; petiole 1-4 mm long. Flowers solitary in upper axils; petals dimorphic. Fruit with 6-8 lengthwise thin ridges; fruiting pedicel and gynophore

无柄山柑 *Capparis subsessilis*

山柑 *Capparis spinosa*

荚蒾叶山柑 *Capparis viburnifolia*

毛果山柑 *Capparis trichocarpa*

forming a right angle with each other. Fl. Jun-Jul. Fr. Aug-Sep. Plains, desert flats, open and sunny areas below 1100 m. Distributed in Xinjiang and Xizang. Also in N Africa, S and SE Asia, Australia and S Europe.

荚蒾叶山柑

Capparis viburnifolia Gagnep.

灌木或木质藤本。小枝密被锈色长柔毛。托叶刺粗壮，常外弯；叶椭圆形、长圆形或有时倒卵形，革质。花序顶生，伞房状，具3-10花；花瓣白色至紫色；雄蕊约75；胎座4；胚珠多数。花期2-3月。生海拔1000-1300米的干旱山坡或湿润林中。产云南南部。泰国和越南亦有。

Shrubs or woody vines. Twigs densely rust-colored villous. Stipular spines thick, often recurved; leaves elliptic, oblong or sometimes obovate, leathery. Inflorescences terminal, corymbs, 3-10-flowered; petals white to purple; stamens ca. 75; placentae 4; ovules many. Fl. Feb-Mar. Dry slopes or moist forests at 1000-1300 m. Distributed in S Yunnan. Also in Thailand and Vietnam.

毛果山柑

Capparis trichocarpa B. S. Sun

藤本。小枝浅灰色。托叶刺至少自中部至基部密被淡红色绒毛；叶椭圆形至有时倒卵形，革质。伞房状或短总状果序；果实近球形至椭圆体形，密被绒毛及锈色绒毛，顶端具一极短喙。果期5月以后。生海拔1200-1600米的灌丛。产云南南部。

Vines. Twigs grayish. Stipular spines densely reddish tomentose at least from middle to base; leaves elliptic to sometimes obovate, leathery. Infructescences corymbs or short racemes; fruits sub globose to ellipsoid, densely tomentose with rusty trichomes, apex with an extremely short beak. Fr. after May. Thickets at 1200-1600 m. Distributed in S Yunnan.

勐海山柑

Capparis fohaiensis B. S. Sun

木质藤本。小枝干后红褐色，无毛。叶厚革质，两面无毛，全缘，侧脉5-8对。花序为短总状或伞房状，腋生或/种在枝端再组成圆锥花序；花瓣膜质；雄蕊约110。果常近黑红色，椭圆体形，直径5-7.5厘米。花期6月，果期10-11月。生海拔400-1000米的河边次生森林中。产云南南部。

Vines, woody. Twigs reddish brown when dry, glabrous. Leaves thickly leathery, both surfaces glabrous, margin entire, lateral veins 5-8 pairs. Inflorescences axillary short racemes or corymbs or axillary and at twig apex thus forming a panicle; petals membranous; stamens ca. 110. Fruits often almost dark red, ellipsoid, 5-7.5 cm diam. Fl. Jun. Fr. Oct-Nov. Secondary forests by rivers at 400-1000 m. Distributed in S Yunnan.

勐海山柑 *Capparis fohaiensis*

马槟榔 *Capparis masaikai*

苦子马槟榔

Capparis yunnanensis Craib et W. W. Smith

灌木或藤本。枝干后呈灰紫褐色，密被短茶色柔毛，后渐无毛。叶薄革质。亚伞形花序在花枝中上部腋生及在顶部再组成圆锥花序，具3-7花；花瓣白色，膜质；雄蕊(60-)85-95。果实干后黄褐色，椭圆体形至近球形。花期3-4月，果期10-12月。生海拔1200-2300米的疏林。产云南和广东。缅甸、泰国和越南亦有。

Shrubs or vines. Branches pale purplish brown when dry, densely short tan pubescent, later glabrescent. Leaves thinly leathery. Inflorescences axillary subumbels or on flowering twig central and apical part together forming a terminal panicle, 3-7-flowered; petals white, membranous; stamens (60-)85-95. Fruits tan when dry, ellipsoid to nearly globose. Fl. Mar-Apr. Fr. Oct-Dec. Open forests at 1200-2300 m. Distributed in Yunnan and Guangdong. Also in Myanmar, Thailand and Vietnam.

马槟榔

Capparis masaikai Lévl .

灌木或攀援植物。幼枝密被锈色短绒毛。叶近革质，干后常呈暗红褐色，背面幼时密被锈色短绒毛。亚伞形花序腋生或在枝端再组成圆锥花序；萼片长8-12毫米，外轮革质；雄蕊45-50。果紫红褐色，具4-8条纵行鸡冠状肋棱，顶端具喙；花梗及雌蕊柄果时木质化增粗。花期5-6月，果期11-12月。生海拔1600米的沟谷或山坡密林中。产广西、贵州和云南。

Shrubs or climbing plants. Young branches densely rusty pubescent. Leaves subleathery, usually dark red-brown when dry, abaxially densely rusty pubescent when young. Subumbels axillary or formed panicles at the end of branches; sepals 8-12 mm long, outer sepals leathery; stamens 45-50. Fruits red-brown, with 4-8 longitudinally cristate ribs, beaked at apex; pedicels and gynophore lignified and thickened in fruit. Fl. May-Jun. Fr. Nov-Dec. Valleys or dense forests on slopes at 1600 m. Distributed in Guangxi, Guizhou and Yunnan.

文山山柑

Capparis fengii B. S. Sun

攀援灌木。小枝具黄色半球形突起。叶长圆披针形。伞房状花序腋生或顶生；花初时白色后转红色；萼片花后宿存；雄蕊约35；子房球形，无毛。果近球形，直径约5厘米，表面密被细疣状突起。花期4-5月，果期10月。生海拔400-1300米的沟谷、湿润灌丛或林中。产云南东南部。

Shrubs, scandent. Twigs with yellow hemispheric raised areas. Leaves orbicular-lanceolate. Inflorescences axillary and terminal, corymbs; petals at first white, then becoming red; sepals persistent after flowering; stamens ca. 35; ovary globose, glabrous. Fruits globose, ca. 5 cm diam, with dense thin verrucose raised areas. Fl. Apr-May. Fr. Oct. Valleys, wet thickets or forests at 400-1300 m. Distributed in SE Yunnan.

广州山柑

Capparis cantoniensis Lour.

攀援灌木。小枝被灰黄色柔毛。托叶刺先端常深黑色；叶长圆形、长圆状披针形或有时卵形。圆锥花序顶生，由数至多个亚伞形花序组成，每亚伞形花序有花数至11朵；花芳香或无香；花瓣白色；雄蕊20-45；胎座2。果实球形至椭圆体形。花果期几乎全年。生海拔800(-1100)米的山沟或疏林中。产云南、贵州、广西、广东、海南和福建。南亚和东南亚亦有。

Shrubs, scandent. Twigs pale yellow pubescent. Stipular spines apex often dark black; leaves oblong, oblong-lanceolate or sometimes ovate. Inflorescences terminal and subumbellate or axillary and terminal thus forming a panicle, to 11-flowered;

苦子马槟榔 *Capparis yunnanensis*

文山山柑 *Capparis fengii*

广州山柑 *Capparis cantoniensis*

flowers fragrant or not; petals white; stamens 20-45; placentae 2. Fruits spheroid to ellipsoid. Fl. and fr. almost all year. Wet and shaded water sides, hillsides, shrub thickets, open forests at 800(-1100) m. Distributed in Yunnan, Guizhou, Guangxi, Guangdong, Hainan and Fujian. Also in S and SE Asia.

斑果藤
Stixis suaveolens (Roxb.) Pierre

木质大藤本。小枝被短柔毛，立即变无毛。叶椭圆形、长圆形或长圆状披针形，革质。总状花序腋生，有时分枝或成圆锥花序，初时直立，后则下垂；花瓣浅黄色。果实成熟时橘黄色，椭圆体形，表面有淡黄色疣状斑点。花期4-5月，果期8-10月。生海拔1500米以下的林中或灌丛。产云南、西藏、广西、广东和海南。印度、缅甸、老挝、泰国、越南和柬埔寨亦有。

Vines large, woody. Twigs shortly pubescent, soon glabrescent. Leaves elliptic oblong or oblong-lanceolate, leathery. Inflorescences axillary, racemes or sometimes branched or forming panicles, at first erect then drooping; sepals pale yellow. Fruits orange when mature, ellipsoid, surface with thin yellow verrucose flecks. Fl. Apr-May. Fr. Aug-Oct. Forests or thickets below 1500 m. Distributed in Yunnan, Xizang, Guangxi, Guangdong and Hainan. Also in India, Myanmar, Laos, Thailand, Vietnam and Cambodia.

节蒴木
Borthwickia trifoliata W. W. Smith

常绿灌木或小乔木。小枝密被白色短柔毛，后渐无毛，鲜时及干后长期均有芳香气味。小叶膜质。花蕾象牙色，锥状；花萼连生成帽状；花瓣5-8，白色。蒴果念珠状，基部渐狭，干后黑褐色。花期4-6月，果期8-9月。生海拔300-1400米的林中或山谷。产云南南部。缅甸亦有。

Shrubs or small trees, evergreen. Twigs with dense short white pubescence, later glabrescent, fragrant when fresh and after drying. Leaflets membranous. Flower buds ivory-colored, awl-shaped; sepals connate, cap-shaped; petals 5-8, white. Capsules moniliform, base attenuate, drying blackish brown. Fl. Apr-Jun. Fr. Aug-Sep. Forests or valleys at 300-1400 m. Distributed in S Yunnan. Also in Myanmar.

黄花草
Arivela viscosa (L.) Raf.

一年生草本，全株密被黏质腺毛和淡黄色的柔毛。小叶3或5；小叶卵形至倒披针状椭圆形。花序具3-6花；花瓣亮黄色，基部稍紫色，花期前上面为半圆形，但是花期后为辐射状排列；雄蕊多数。果直立，稍镰刀状。7月果熟。多生海拔300米的干燥条件下的草坡、旱荒地、路边或田间。产中国西南、华南、东南、华中和华东。全球热带和亚热带亦有。

Herbs, annual, whole plant densely glandular hirsute, viscous and yellowish-pubescent. Leaflets 3 or 5; leaflets ovate to oblanceolate-elliptic. Inflorescences 3-6-flowered; petals bright yellow, basally sometimes purple, arranged in an adaxial semicircle before anthesis but radially arranged after anthesis; stamens numerous. Fruits erect, slightly falcate. Fr. ripe in Jul. Grassy slopes, dry wastelands, roadsides or fields at 300 m. Distributed in SW, S, SE, C and E China. Also in tropical and subtropical regions of the world.

斑果藤 *Stixis suaveolens*

节蒴木 *Borthwickia trifoliata*

黄花草 *Arivela viscosa*

十字花科 Cruciferae

花椰菜
Brassica oleracea L. var. **botrytis** L.

二年生或多年生草本。茎基部伸长，圆柱形。茎基部和下部的叶子绿色，少到多数。花序白色，紧缩，常球形，具肉质花序梗、小花梗和花；花黄色。果圆柱状。花期4月，果期5月。栽培于中国各地。亦广泛栽培于世界各地。

Herbs biennial or perennial. Stem base elongated, cylindric. Basal and lower cauline leaves green, few to several. Inflorescences white, compact, often globose, with fleshy peduncle, pedicels and flowers; flowers yellow. Fruits terete. Fl. Apr. Fr. May. Cultivated throughout China. Widely cultivated elsewhere.

甘蓝
Brassica oleracea L. var. **capitata** L.

二年生或多年生草本。茎基部和下部叶绿色，多数，层层包裹形成一个紧缩的、球形或长圆形、闭合的、顶端圆形或平坦的球体。总状花序顶生或腋生。果圆柱状。花期4月，果期5月。栽培于中国各地。

花椰菜 *Brassica oleracea* var. *botrytis*

甘蓝 *Brassica oleracea* var. *capitata*

青菜 *Brassica rapa* var. *chinensis*

Herbs biennial or perennial. Basal and lower cauline leaves green, numerous, strongly overlapping into a compact, globose or oblong, closed, apically rounded or flattened head. Racemes terminal or axillary. Fruits terete. Fl. Apr. Fr. May. Cultivated throughout China.

青菜
Brassica rapa L. var. **chinensis** (L.) Kitam.

一年生稀二年生草本。主根圆柱状。基部叶常多于20，明显丛生，不形成紧密头状；叶柄肉质或增厚，横切面半圆柱形或长圆形，无翅；小叶全缘或浅波状。花瓣明黄色。果实线形。花期4-5月，果期5-6月。栽培于中国各地。

Herbs annual or rarely biennial. Taproots cylindric. Basal leaves usually more than 20, strongly rosulate, not forming compact heads; petioles fleshy or thickened, semiterete or transversely oblong in cross section, wingless; leaflets margin entire or repand. Petals bright yellow. Fruits linear. Fl. Apr-May. Fr. May-Jun. Cultivated throughout China.

白菜
Brassica rapa L. var. **glabra** Regel

一年生或二年生草本。主根不为肉质，圆柱形。基生叶通常约20，明显丛生，形成长圆形或倒卵形紧密头状；叶柄极扁平，有具细齿或圆齿的翅；叶具圆齿。花瓣明黄色。果实线形。花期5-6月，果期6-7月。长久栽培于中国和世界其他各地。

Herbs annual or biennial. Taproots not fleshy, cylindric. Basal leaves usually ca. 20, strongly rosulate, forming oblong or subovoid compact heads; petioles strongly flattened, with incised or dentate wings; leaves dentate. Petals bright yellow. Fruits linear. Fl. May-Jun. Fr. Jun-Jul. Long cultivated throughout China and worldwide.

芸薹 (油菜)
Brassica rapa L. var. **oleifera** DC.

一年生或二年生草本。主根圆柱形。基生叶少，不丛生；叶柄纤细，无翅；叶轮廓卵形、长圆形或披针形，边缘全缘、锯齿或波状，有时羽状半裂或全裂。花瓣明黄色。果实线

白菜 *Brassica rapa* var. *glabra*

形。花期3-4月，果期5-6月。栽培于中国各地和世界其他各地。

Herbs annual or biennial. Taproots cylindric. Basal leaves several, not rosulate; petioles slender, not winged; leaves ovate, oblong or lanceolate in outline, margin entire, dentate or sinuate, sometimes pinnatifid or pinnatisect. Petals bright yellow. Fruits linear. Fl. Mar-Apr. Fr. May-Jun. Cultivated throughout China and worldwide.

芥菜

Brassica juncea (L.) Czern.

一年生草本。主根纤细，圆柱形，直径约1.5厘米。基生叶的叶柄细；叶长4-30厘米。边缘变化较大，提琴状羽状深裂或全裂。花瓣黄色，先端圆或凹缺。果线形。花期3-6月，果期4-7月。生田野、荒地或路边。广泛栽培于中国和世界其他各地。

Herbs annual. Taproots slender, cylindric, ca. 1.5 cm diam. Basal leaf petioles slender; leaves 4-30 cm long, margin variable, lyrate-pinnatifid or pinnatisect. Petals yellow, apex rounded or emarginate. Fruits linear. Fl. Mar-Jun. Fr. Apr-Jul. Fields, waste places or roadsides. Cultivated throughout China and worldwide.

芥菜 *Brassica juncea*

芸薹（油菜）*Brassica rapa* var. *oleifera*

欧洲油菜 *Brassica napus*

欧洲油菜

Brassica napus L.

一年生草本。基生叶和下部茎叶羽状浅裂或大头羽裂，有时不分裂，叶柄长达15厘米；中部及上部茎生叶无柄，披针形、卵形或长圆形，长达8厘米，基部抱茎，耳状，边缘全缘或浅波状。总状花序伞房状；花直径10-15毫米；花瓣浅黄色，爪长5-9毫米。果瓣具1中脉。花期3-6月，果期4-7月。中国各地均有栽培。世界各地广泛栽培并归化。

Annual herbs. Basal and lowermost cauline leaves pinnately lobed or lyrate, sometimes undivided, petiole to 15 cm long; middle and upper cauline leaves sessile, lanceolate, ovate or oblong, to 8 cm long, base amplexicaul, auriculate, margin entire or repand. Racemes corymbiform; flowers 10-15 mm diam; petals pale yellow, claw 5-9 mm long. Valves with a prominent midvein. Fl. Mar-Jun. Fr. Apr-Jul. Cultivated throughout China. Also widely cultivated and naturalized elsewhere.

白芥

Sinapis alba L.

一年生草本，具外折硬单毛，稀变无毛。基生叶和下部茎叶大头羽裂、羽状半裂或羽状全裂，叶柄长1-3(-6)厘米；上部茎叶具短柄，卵形或长圆状卵形，长2-4.5厘米。花淡黄色，直径约1厘米。果披针形，果瓣有3-5(-7)脉。花果期5-9月。生路边、田间或牧场。辽宁、山西、山东、安徽、新疆、四川、甘肃、河北和青海有栽培。原产欧洲。

Annual herbs, retrorsely hispid, rarely glabrescent. Basal and lower cauline leaves lyrate, pinnatifid or pinnatisect, petiole 1-3(-6) cm long; upper cauline leaves shortly petiolate, ovate or oblong-ovate, 2-4.5 cm long. Flowers yellowish, ca. 1 cm diam. Fruit lanceolate, valvular segment 3-5(-7)-veined. Fl. and fr. May-Sep. Roadsides, fields or pastures. Cultivated in Liaoning, Shanxi, Shandong, Anhui, Xinjiang, Sichuan, Gansu, Hebei and Qinghai. Native to Europe.

芝麻菜

Eruca vesicaria (L.) Cav. subsp. **sativa** (Mill.) Thell.

一年生草本。基生叶不丛生，叶片大头羽状分裂或二回羽状全裂。花瓣黄色，后变白色，具深褐色或紫色脉纹。果实条形、长圆形或椭圆体形。花期5-7月，果期6-8月。生海拔3800米以下的荒地、田间、路边或山坡。产中国西南、华

白芥 *Sinapis alba*

芝麻菜 *Eruca vesicaria* subsp. *sativa*

萝卜 *Raphanus sativus*

南、华北、华西、西北和东北。西南亚、中亚、非洲西北部和欧洲亦有；也逸生到各处。

Herbs annual. Basal leaves not rosulate, leaf blade lyrate pinnatifid or bipinnatisect. Petals yellow turning white, with dark brown or purplish veins. Fruits linear, oblong or ellipsoid. Fl. May-Jul. Fr. Jun-Aug. Waste areas, fields, roadsides or slopes below 3800 m. Distributed in SW, S, N, W, NW and NE China. Also in SW and C Asia, NW Africa and Europe; naturalized elsewhere.

萝卜

Raphanus sativus L.

一年生或二年生草本。根肉质。叶轮廓长圆形、倒卵形、倒披针形或匙形，大头羽状分裂或羽状全裂，有时不分裂，边缘具圆齿。总状花序顶生或侧生；花瓣紫色、粉红色，有时白色，常具深色脉纹。果实纺锤形或披针形，有时卵球形或圆柱形。花果期因耕种时间而异。生田间、路边或荒地。栽培于中国各地。原产地中海地区，栽培遍及全球。

Herbs annual or biennial. Roots fleshy. Leaves oblong, obovate, oblanceolate or spatulate in outline, lyrate or pinnatisect, sometimes undivided, margin dentate. Racemes terminal or lateral; petals purple, pink, sometimes white, often with darker veins. Fruits fusiform or lanceolate, sometimes ovoid or cylindric. Fl. and fr. depending on cultivation time. Fields, roadsides, or waste areas throughout China. Native to Mediterranean, cultivated worldwide.

诸葛菜

Orychophragmus violaceus (L.) Schulz

一年生或二年生草本。基生叶不丛生；叶心形、肾形、阔卵形或近圆形，边缘具粗锯齿有牙齿。总状花序顶生；萼片线形；花瓣深紫色。果窄线形。花期3-6月，果期5-7月。生海拔1500米以下的路边、田园、林下、田间、灌丛、山脚或山坡。产中国西南、华中、华北、华西和华东。朝鲜半岛亦有；归化于日本。

Herbs annual or biennial. Basal leaves not rosulate; leaves cordate, reniform, broadly ovate or suborbicular, margin coarsely crenate with teeth. Racemes terminal; sepals linear; petals deeply purple. Fruits narrowly linear. Fl. Mar-Jun. Fr. May-Jul. Roadsides, gardens, forests, fields, thickets, hillsides or slopes below 1500 m. Distributed in SW, C, N, W and E China. Also in Korean Peninsula; naturalized in Japan.

诸葛菜 *Orychophragmus violaceus*

碱独行菜
Lepidium cartilagineum (J. Mayer) Thell.

多年生草本，被乳突或曲星状毛状柔毛。茎直立，分枝。基生叶丛生，肉质，宿存；叶卵形、长圆形或椭圆形。总状花序在果期长达7-10厘米；花瓣白色。短角果卵状椭圆体形。花期8月。生海拔400-1000米的盐土上。产内蒙古和新疆。西南亚、中亚、东北亚、欧洲南部和中部亦有。

Herbs perennial, puberulent with papillate or curved trichomes. Stems erect, branched. Basal leaves rosulate, fleshy, persistent; leaves ovate, oblong or elliptic. Racemes extending to 7-10 cm in fruiting; petals white. Silicles ovoid-ellipsoid. Fl. Aug. Saline steppe at 400-1000 m. Distributed in Neimenggu and Xinjiang. Also in SW, C and NE Asia, and S and C Europe.

宽叶独行菜
Lepidium latifolium L.

多年生草本。茎上部多分枝。叶革质，基生叶及茎下部叶椭圆状卵形或长圆形，边缘通常有锯齿；上部茎生叶椭圆状卵形、椭圆形或披针形。总状花序圆锥状；花瓣白色。短角果长圆形或卵状椭圆形。花期5-9月，果期6-10月。生海拔1800-4300米的田边、路旁、山坡及盐化草甸。产内蒙古和西藏。非洲北部、西南亚和欧洲南部亦有。

Perennial herbs. Stems much branched above. Leaves leathery, basal and lower cauline leaves elliptic-ovate or oblong, margin usually serrate; upper cauline leaves elliptic-ovate, elliptic or lanceolate. Raceme paniculate; petals white. Silicle oblong or ovate-elliptic. Fl. May-Sep. Fr. Jun-Oct. Fields, roadsides, slopes or saline meadows at 1800-4300 m. Distributed in Neimenggu and Xizang. Also in N Africa, SW Asia and S Europe.

头花独行菜
Lepidium capitatum Hook.f. et Thomson

头花独行菜 *Lepidium capitatum*

一年生或二年生草本。茎匍匐，分枝，披散，具腺毛。基生叶及下部茎生叶椭圆形、匙形或披针形，通常无毛，羽状半裂；上部茎生叶与最基部的相似，向上逐渐变小。总状花序头状；萼片长圆形；花瓣白色，狭倒卵形。短角果卵形，无毛，顶端具翅。花果期6-9月。生海拔2700-5000米的山坡。产青海、四川、云南和西藏。不丹、印度、克什米尔地区和尼泊尔亦有。

Annual or biennial herbs. Stems prostrate, branched, diffuse, glandular. Basal and lower cauline leaves elliptic, spatulate or lanceolate, usually glabrous, pinnatifid; upper cauline leaves similar to lowermost leaves, progressively smaller upward. Racemes capitate; sepals oblong; petals white, narrowly obovate. Silicle ovate, glabrous, wing apical. Fl. and fr. Jun-Sep. Mountain slopes at 2700-5000 m. Distributed in Qinghai, Sichuan, Yunnan and Xizang. Also in Bhutan, India, Kashmir and Nepal.

楔叶独行菜
Lepidium cuneiforme C. Y. Wu

二年生草本。茎直立，不分枝或分枝。茎上部叶倒卵形或倒

碱独行菜 *Lepidium cartilagineum*

宽叶独行菜 *Lepidium latifolium*

楔叶独行菜 *Lepidium cuneiforme*

披针形，基部近耳状或楔形，上半部边缘具锯齿。花瓣白色，倒卵形或近矩圆形。果阔椭圆体形。花果期3-8月。生海拔600-2700米的山坡、路边或河滩。产中国西南和华西。

Herbs biennial. Stems erect, simple or branched. Upper cauline leaves obovate or oblanceolate, base subauriculate or cuneate, margin serrate along distal half. Petals white, obovate or suboblong. Fruits broadly ellipsoid. Fl. and fr. Mar-Aug. Slopes, roadsides or river beaches at 600-2700 m. Distributed in SW and W China.

独行菜

Lepidium apetalum Willd.

一年生或两年生草本，具棍棒状或头状柔毛。叶长圆形、披针形或倒披针形，羽状深裂，波状或具锯齿。花瓣缺；雄蕊2。果宽椭圆体形，上部有短翅。花期4-8月，果期5-9月。生海拔400-4800米路边、山坡、荒地、河边或田间。产中国除华南以外大部分地区。哈萨克斯坦、蒙古和朝鲜半岛亦有。

Herbs annual or biennial, puberulent with clavate or capitate trichomes. Leaves oblong, lanceolate or oblanceolate, pinnatifid, sinuate or dentate. Petals absent; stamens 2. Fruits broadly ellipsoid, apex narrowly winged. Fl. Apr-Aug. Fr. May-Sep. Roadsides, slopes, waste places, by streams or fields at 400-4800 m. Distributed in most parts of China, except S China. Also in Kazakhstan, Mongolia and Korean Peninsula.

独行菜 *Lepidium apetalum*

臭荠

Coronopus didymus (L.) Smith

一年生稀二年生草本。基生叶不丛生，一回或二回羽状深裂。总状花序顶生或侧生且与叶对生；萼片卵形；花瓣白色，椭圆形至线形；雄蕊2，稀4。果实成对，压扁状，顶端和基部具凹缺。花期3-5月。生海拔1000米以下的路边、荒地或田间。产中国西南、华南、东南、华东和西北。原产南美洲；逸生各处。

Herbs annual or rarely biennial. Basal leaves not rosulate, pinnatisect or bipinnatisect. Racemes terminal or lateral and leaf opposed; sepals ovate; petals white, elliptic to linear; stamens 2, rarely 4. Fruits didymous, compressed, emarginate at apex and base. Fl. Mar-May. Roadsides, waste places or fields below 1000 m. Distributed in SW, S, SE, E and NW China. Native to South America; naturalized elsewhere.

臭荠 *Coronopus didymus*

群心菜 *Cardaria draba*

群心菜
Cardaria draba (L.) Desv.

多年生草本。基生叶有柄，倒卵状匙形，边缘有波状齿，开花时枯萎；茎生叶无柄，基部戟形，抱茎或具耳。萼片长圆形，无毛；花瓣白色，倒卵形，有短爪。果心形、卵形或近球形，基部心形，果瓣有明显网脉，无毛。花期5-6月，果期7-8月。生海拔1600米的山坡、路边、田间、河滩及牧场。产辽宁、山东和新疆。欧洲、亚洲和北美洲亦有。

Perennial herbs. Basal leaves petiolate, obovate-spatulate, margin sinuate-dentate, withered by anthesis; cauline leaves sessile, base sagittate, amplexicaul or auriculate. Sepals oblong, glabrous; petals white, obovate, shortly clawed. Fruit cordate, ovoid or subglobose, base cordate, valves prominently reticulate, glabrous. Fl. May-Jun. Fr. Jul-Aug. Mountain slopes, roadsides, fields, river banks and pastures at 1600 m. Distributed in Liaoning, Shandong and Xinjiang. Also in Europe, Asia and North America.

毛果群心菜
Cardaria pubescens (C. A. Mey.) Jarm.

多年生草本，通常密被柔毛。茎直立，上部分枝。基部和下部的茎生叶倒披针形或倒卵形，上部茎生叶长圆形或披针形，基部箭头形。花瓣白色，倒卵形。果球形至近球形，先端和基部较圆或基部稍心形。花果期5-7月。生海拔400-1600米的沟边、农田和牧场。产甘肃、内蒙古、宁夏、青海、陕西和新疆。中亚、蒙古、巴基斯坦和俄罗斯亦有。

Perennial herbs, usually densely pubescent. Stems erect, branched above. Basal and lower cauline leaves oblanceolate or obovate, upper cauline leaves oblong or lanceolate, base sagittate. Petals white, obovate. Fruit globose to subglobose, apex and base rounded or slightly cordate basally. Fl. and fr. May-Jul. Along ditches, fields and pastures at 400-1600 m. Distributed in Gansu, Neimenggu, Ningxia, Qinghai, Shaanxi and Xinjiang. Also in C Asia, Mongolia, Pakistan and Russia.

毛果群心菜 *Cardaria pubescens*

菘蓝 (板蓝根)
Isatis tinctoria L.

二年生草本，无毛。茎上部分枝，常圆锥状分枝。基生叶丛生；叶长圆形或倒披针形，基

菘蓝（板蓝根）*Isatis tinctoria*

沙芥 *Pugionium cornutum*

部箭形或耳状，全缘。花瓣黄色。短角果近长圆形。花期4-6月，果期5-7月。生海拔600-2800米的田地、牧场、路边或荒地。产中国西南、东南、华北、华西和华东。西南亚、中亚、东北亚和欧洲亦有；世界各地归化。

Herbs biennial, glabrous. Stems branched above, often paniculately branched. Basal leaves rosulate; leaves oblong or oblanceolate, base sagittate or auriculate, margin entire. Petals yellow. Silicles nearly oblong. Fl. Apr-Jun. Fr. May-Jul. Fields, pastures, roadsides or waste places at 600-2800 m. Distributed in SW, SE, N, W and E China. Also in SW, C and NE Asia, and Europe; naturalized worldwide.

沙芥

Pugionium cornutum (L.) Gaertn.

一年生草本。基生叶稍肉质，羽状分裂，长8-25厘米，叶柄长2-6厘米；茎上部叶倒披形或线形，长3-6厘米，基部渐狭，全缘。花瓣白色，条形至条状披针形，长1.2-1.5厘米，爪长5-7毫米。果横长圆形或卵形，先端渐尖，翅剑形，长(2-)3-5厘米，具3条纵脉，有8-10个刺。花期6-8月，果期7-9月。生海拔1000-1100米的荒漠沙丘。产内蒙古、陕西和宁夏。

Annual herbs. Basal leaves slightly fleshy, pinnatisect, 8-25 cm long, petiole 2-6 cm long; uppermost cauline leaves oblanceolate or linear, 3-6 cm long, base attenuate, margin entire. Petals white, linear to linear-lanceolate, 1.2-1.5 cm long, claw 5-7 mm long. Fruit transversely oblong or ovoid, apex acuminate, wings ensiform, (2-)3-5 cm long, longitudinally 3-veined, spines 8-10. Fl. Jun-Aug. Fr. Jul-Sep. Desert dunes at 1000-1100 m. Distributed in Neimenggu, Shaanxi and Ningxia.

高河菜

Megacarpaea delavayi Franch.

多年生草本。茎直立，上部分枝。叶轮廓长圆状倒披针形，羽状复叶，具疏或密柔毛。花瓣淡紫或深紫色。短角果裂瓣歪倒卵形。花期5-8月，果期7-9月。生海拔3300-4800米的山坡、草地、岩石裂隙或湖边。产甘肃和青海。

Herbs perennial. Stems erect, branched above. Leaves oblong-oblanceolate in outline, appearing pinnately compound, sparsely to densely pubescent. Petals lavender or deep purple. Silicle lobes oblique-obovate. Fl. May-Aug. Fr. Jul-Sep. Slopes, grasslands, rock crevices or by the lakes at 3300-4800 m. Distributed in Gansu and Qinghai.

高河菜 *Megacarpaea delavayi*

菥蓂（遏蓝菜）*Thlaspi arvense*

荠 *Capsella bursa-pastoris*

菥蓂（遏蓝菜）

Thlaspi arvense L.

一年生草本，全株无毛。叶倒披针形、匙形或倒卵形。总状花序顶生；花瓣白色，匙形，基部渐狭成爪状；子房具6-16个胚珠。果倒卵形或近圆形，顶端具深凹缺。花果期3-10月。生海拔100-5000米的路边、草坡、田中或荒地。产中国大部分地区。南亚、中亚、东北亚和非洲亦有；澳大利亚和美洲引进。

Herbs annual, glabrous throughout. Leaves oblanceolate, spatulate or obovate. Racemes terminal; petals white, spatulate, narrowed to a clawlike base; ovules 6-16 per ovary. Fruits obovate or almost orbicular, apex deeply emarginate. Fl. and fr. Mar-Oct. Roadsides, grassy slopes, fields or waste places at 100-5000 m. Distributed in most parts of China. Also in S, C and NE Asia, and Africa; introduced from Australia and America.

荠

Capsella bursa-pastoris (L.) Medic.

草本。茎直立，单生或分枝。基生叶丛生；叶长圆形或倒披针形，大头羽状分裂、羽状全裂、羽状深裂或倒向羽裂。花白色，稀粉红色。果倒三角形或近心形。花果期4-7月。生田间、路旁、山坡或荒地。产中国各地。西南亚和欧洲广布。

Herbs. Stems erect, simple or branched. Basal leaves rosulate; leaves oblong or oblanceolate, margin lyrate, pinnatisect, pinnatifid or runcinate. Petals white, rarely pinkish. Fruits obtriangular or subcordate. Fl. and fr. Apr-Jul. Fields, roadsides, mountain slopes or wastelands. Distributed throughout China. Also widely in SW Asia and Europe.

藏荠

Hedinia tibetica (Thoms.) Ostenf.

草本。茎匍匐或斜升，基部密被单毛。叶一或二回羽状深裂。花瓣白色，长于萼片；雄蕊6，稍四强；蜜腺4，侧生。果开裂，短角果，稀长角果。花期6-8月，果期7-9月。生海

藏荠 *Hedinia tibetica*

拔3900-5200米的高山草甸、开旷草原、山麓碎石或多沙山坡。产四川、西藏、甘肃、青海和新疆。印度、尼泊尔、不丹和塔吉克斯坦亦有。

Herbs. Stems procumbent or ascending, densely hirsute basally with primarily simple trichomes. Leaves 1- or 2-pinnatisect. Petals white, longer than sepals; stamens 6, slightly tetradynamous; nectar glands 4, lateral. Fruits dehiscent, silicles, rarely siliques. Fl. Jun-Aug. Fr. Jul- Sep. Alpine meadows, steppes, screes or sandy slopes at 3900-5200 m. Distributed in Sichuan, Xizang, Gansu, Qinghai and Xinjiang. Also in India, Nepal, Bhutan and Tajikistan.

小叶半脊荠

Hemilophia rockii O. E. Schulz

多年生草本。茎自根茎中发生。基生叶披针形至椭圆状披针形，密被腺毛；茎生叶倒披针形或狭椭圆形，稀倒卵形，被悉数腺毛，全缘。总状花序；萼片长圆形或卵形，绿色，早落；花瓣淡黄色至乳白色，早落，倒心形；花丝白色。角果瓣膜纸质。花期6-7月，果期7-8月。生海拔3900-4900米的松散的石灰岩砾石、流石滩。产四川和云南。

Perennial herbs. Stems originate from rhizomes. Basal leaves lanceolate to elliptic-lanceolate, densely covered with trichomes; cauline leaves oblanceolate or narrowly elliptic, rarely ovate, sparsely covered with trichomes, margin entire. Racemes; sepals oblong or ovate, green, caduceus; petals yellowish to creamy white, caducous, obcordate; filaments white. Fruit valves papery. Fl. Jun-Jul. Fr. Jul-Aug. Loose limestone gravel, scree at 3900-4900 m. Distributed in Sichuan and Yunnan.

蛇头荠

Dipoma iberideum Franch.

多年生草本。茎不分枝，有毛。基生叶倒卵形或倒披针形，稀全缘；茎生叶无柄或基部渐狭为叶柄状；全缘。萼片粉色或绿色，长椭圆形，边缘膜质；花瓣白色，宽倒卵形；花丝白色，花药紫色。短角果一侧败育，裂瓣纸质，无毛。种子红棕色。花期4-7月，果期7-9月。生海拔3000-4600米的高山流石滩间。产四川和云南。

Perennial herbs. Stems simple, pubescent. Basal leaves obovate or oblanceolate, rarely entire; cauline leaves sessile or base attenuate to a petiolelike; margin entire. Sepals pink or green, oblong, margin membranous; petals white, broadly obovate; filaments white, anthers purple. Silicle with 1 side aborting, valves papery, glabrous. Seeds red-brown. Fl. Apr-Jul. Fr. Jul-Sep. Alpine gravel at 3000-4600 m. Distributed in Sichuan and Yunnan.

蛇头荠 *Dipoma iberideum*

盐泽双脊荠

Dilophia salsa Thomson

无毛草本。茎直立。基生叶莲座状，叶线形，顶端圆钝。伞房花序；萼片直立或上升，宽卵形；花瓣白色或粉色，干后淡紫色，匙状线形，顶端略凹；花丝白色。种子褐色或黑色，广椭圆形。花期6-8月，果期7-9月。生海拔2200-5500米的盐沼泽地。产甘肃、青海、新疆和西藏。中亚、尼泊尔和巴基斯坦亦有。

Glabrous herbs. Stems erect. Basal leaves attenuate, linear, apex obtuse. Corymbs; sepals erect or ascending, broadly ovate; petals white or pink, drying purplish, spatulate-linear, apex emarginate; filaments white. Seeds brown or black, broadly elliptic. Fl. Jun-Aug. Fr. Jul-Sep. Salty marshes at 2200-5500 m. Distributed in Gansu, Qinghai, Xinjiang and Xizang. Also in C Asia, Nepal and Pakistan.

小叶半脊荠 *Hemilophia rockii*

盐泽双脊荠 *Dilophia salsa*

山芥 *Barbarea orthoceras*

山芥

Barbarea orthoceras Lédeb.

二年生草本，高25-60厘米，全株无毛。基生叶及茎下部叶长2-5.5厘米，宽1-3厘米，宽椭圆形或近圆形；茎上部叶较小，宽披针形或长卵形。总状花序顶生；萼片椭圆状披针形；花瓣黄色，长倒卵形，长3-4.5毫米，宽0.7-1.2毫米，基部具爪。长角果线状四棱形，长2-3.5厘米。种子椭圆形，长约1.5毫米，宽约0.5毫米，深褐色。花果期5-8月。生海拔450-2100米的草甸、河岸、溪谷、河滩湿草地及山地潮湿处。产黑龙江、吉林、辽宁、内蒙古和新疆北部。蒙古、俄罗斯、朝鲜半岛和日本亦有。

Herbs biennial, 25-60 cm tall, stems glabrous throughout. Basal and lowermost cauline leaves 2-5.5 × 1-3 cm, wide ellipse or suborbicular; cauline leaves small, wide lanceolate or long ovoid. Racemes terminal; sepals elliptic-lanceolate; petals yellow, obovate, 3-4.5 × 0.7-1.2 mm, attenuate to base. Silique linear four prismatic, 2-3.5 cm long. Seeds oblong, ca. 1.5 × 0.5 mm, brown. Fl. and fr. May-Aug. Open grasslands, scree, temperate mixed forests, river and streamsides, moist grassy slopes at 450-2100 m. Distributed in Heilongjiang, Jilin, Liaoning, Neimenggu and N Xinjiang. Also in Mongolia, Russia, Korean Peninsula and Japan.

疣果匙荠

Bunias orientalis L.

二年生草本，高25-150厘米，具毛。茎直生，上部分枝。基生叶羽状全裂，长8-12厘米；上部茎生叶近无柄，披针形，具齿或近全缘。萼片淡黄色，长椭圆形，伸展，无毛；花瓣黄色，倒卵形，爪细长；花丝淡黄色，四强雄蕊。短角果卵形，长6-8毫米，宽3-4毫米。花果期5-7月。生田野。产中国东北。中亚和欧洲亦有。

Biennial herbs, 25-150 cm tall, pilose. Stems erect, branched above. Basal leaves pinnatifid, 8-12 cm long; upper cauline leaves subsessile, lanceolate, dentate or entire. Sepals yellowish, oblong, spreading, glabrous; petals yellow, obovate, claw slender; filaments yellowish, tetradynamous. Silicle ovoid, 6-8 × 3-4 mm. Fl. and fr. May-Jul. Fields. Distributed in NE China. Also in C Asia and Europe.

北方庭荠

Alyssum lenense Adams

多年生草本，灰绿色，被7-16辐星状毛。茎叶条状披针形或条状长圆形，长0.4-2(-3)厘米，向上渐变小，先端急尖，基部渐狭，无柄。花序伞房状；花瓣黄色，宽匙形；每室2枚胚珠；花柱宿存。果瓣无脉，无毛或疏被星毛。花期5-6月，果期7-8月。生草坡、林中或沙地。产黑龙江、内蒙古、河北、甘肃和新疆。哈萨克斯坦、蒙古和俄罗斯亦有。

Perennial herbs, canescent, 7-16-rayed stellate trichomes. Cauline leaves linear-lanceolate or linear-oblong, 0.4-2(-3) cm long, gradually smaller upward, apex acute, base attenuate, subsessile.

疣果匙荠 *Bunias orientalis*

北方庭荠 *Alyssum lenense*

Racemes corymbose; petals yellow, broadly spatulate; ovules 2 per locule; style persistent. Valves not veined, glabrous or sparsely stellate. Fl. May-Jun. Fr. Jul-Aug. Grassy slopes, forests or sandy places. Distributed in Heilongjiang, Neimenggu, Hebei, Gansu and Xinjiang. Also in Kazakhstan, Mongolia and Russia.

香雪球

Lobularia maritima (L.) Desv.

多年生草本或半灌木，全珠被银灰色毛。叶条形、披针形或倒披针形，基部渐狭，边缘全缘。萼片绿色或紫色；花瓣白色或深紫色，倒卵形或近圆形，基部突然变窄成长达1毫米的爪；花丝白色或紫色，长1.2-2毫米；花柱宿存。果瓣被毛，中脉清晰。花果期全年。生海拔2000米以下的石地、荒地或庭院。中国多省均有栽培。原产地中海西部。

Perennial herbs or subshrubs, silvery pubescent. Leaves linear, lanceolate or oblanceolate, base attenuate, margin entire. Sepals green or purple; petals white or deep purple, obovate or suborbicular, abruptly narrowed to claw at base, claw to 1 mm; filaments white or purple, 1.2-2 mm; style persistent. Valves pubescent, with a distinct midvein. Fl. and fr. throughout the year. Stony areas, waste grounds or yards below 2000 m. Cultivated in most parts of China. Native to W Mediterranean.

团扇荠

Berteroa incana (L.) DC.

一年生或二年生草本，密被贴伏星状毛并混生单毛。基生叶具柄，倒披针形，开花时枯萎；中间和上部茎叶无柄，边缘全缘。花瓣白色，狭倒心形，裂片长圆形，先端钝；花丝白色。果长(4-)5-8.5(-10)毫米；果瓣星状。花期5-8月，果期6-9月。生海拔700-1900米的山坡、农田及河边等。产甘肃、辽宁、内蒙古和新疆。欧洲、俄罗斯和中亚亦有。

Annual or biennial herbs, densely pubescent with appressed stellate trichomes mixed with some simple ones. Basal leaves petiolate, oblanceolate, withered by flowering time; middle and upper cauline leaves sessile, margin entire. Petals white, narrowly obcordate, lobes oblong, apex obtuse; filaments white. Fruit (4-)5-8.5(-10) mm long; valves stellate. Fl. May-Aug. Fr. Jun-Sep. Mountain slopes, fields and river banks at 700-1900 m. Distributed in Gansu, Liaoning, Neimenggu and Xinjiang. Also in Europe, Russia and C Asia.

香雪球 *Lobularia maritima*

团扇荠 *Berteroa incana*

辣根 *Armoracia rusticana*

辣根

Armoracia rusticana (Lam.) P. Gaertner et Schreb.

多年生草本。基生叶边缘具粗齿，稀羽状半裂，叶柄长达60厘米，基部膨大；下部和中间茎叶具短柄，羽状半裂或羽状全裂；上部茎叶无柄或有短柄，条形或条状披针形，基部楔形或渐狭，边缘具锯齿或圆齿，稀全缘。花瓣倒卵形或倒披针形，爪长达1.5毫米。花期5-7月。产黑龙江、吉林、辽宁、河北和江苏。原产欧洲，广泛栽培和归化。

Perennial herbs. Basal leaves coarsely crenate, rarely pinnatifid, petiole to 60 cm long, broadly expanded at base; lower and middle cauline leaves shortly petiolate, pinnatifid or pinnatisect; upper cauline leaves sessile or shortly petiolate, linear to linear-lanceolate, base cuneate or attenuate, margin serrate or crenate, rarely entire. Petals obovate or oblanceolate, claw to 1.5 mm long. Fl. May-Jul. Distributed in Heilongjiang, Jilin, Liaoning, Hebei and Jiangsu. Native to Europe, cultivated and naturalized elsewhere.

刚毛葶苈

Draba setosa Royle

多年生丛生草本，植株无毛。茎基多分枝，上部叶莲座状丛生；茎直立，不分枝。基生叶莲座状，宿存，线形或线状长椭圆形，全缘；无茎生叶。总状花序；萼片有单毛；花瓣黄色，狭倒卵形，顶端微凹，脉较密。角果椭圆状卵形，具宽隔膜。花期6-7月，果期7-8月。生海拔3200-4600米的山坡或流石滩。产西藏。克什米尔地区亦有。

Perennial cespitose herbs, glabrous throughout. Caudex often many branched, ultimate branches terminated in rosettes; stems erect, simple. Basal leaves rosulate, persistent, linear or linear-oblong, margin entire; cauline leaves absent. Racemes; sepals pubescent; petals yellow, narrowly obovate, apex emarginated, with dense veins. Fruit elliptic-ovate, latiseptate. Fl. Jun-Jul. Fr. Jul-Aug. Mountain slopes or scree at 3200-4600 m. Distributed in Xizang. Also in Kashmir.

喜山葶苈

Draba oreades Schrenk

多年生草本，丛生，具花葶。基生叶丛生，倒披针形。总状花序具(2-)4-15(-25)花，无苞片，近伞形，果期不伸长；花瓣黄色，无爪。果卵球形。花果期6-8月。生海拔2300-5500米的高山草甸、草坡、冰川或石隙中。产中国西南、华西和西北。南亚、中亚和东北亚亦有。

Herbs perennial, cespitose, scapose. Basal leaves rosulate, oblanceolate. Racemes (2-)4-15(-25)-flowered, ebracteate, subumbellate and not elongated in fruit; petals yellow, claw absent. Fruits ovoid-globose. Fl. and fr. Jun-Aug. Alpine meadows, grassy slopes, glacier margins or rock crevices at 2300-5500 m. Distributed in SW, W and NW China. Also in S, C and NE Asia.

衰老葶苈

Draba senilis O. E. Schulz

多年生草本，丛生，具根出条和花葶。茎具多数细弱分枝顶生成莲座状；基生叶丛生。总状花序具3-7(-12)花，无苞片，果期稍伸长；花瓣黄色。果卵

刚毛葶苈 *Draba setosa*

喜山葶苈 *Draba oreades*

衰老葶苈 *Draba senilis*

球形。花期5-7月，果期7-9月。生海拔4000-4900米的高山草甸或岩隙。产云南、四川、西藏和青海。

Perennial herbs, cespitose, surculose, scapose. Caudex with many, slender branches terminated in rosettes; basal leaves rosulate. Racemes 3-7(-12)-flowered, ebracteate, elongated slightly in fruit; petals yellow. Fruits ovate. Fl. May-Jul. Fr. Jul-Sep. Alpine meadows or rock crevices at 4000-4900 m. Distributed in Yunnan, Sichuan, Xizang and Qinghai.

总苞葶苈

Draba involucrata (W. W. Sm.) W. W. Sm.

多年生丛生草本。根茎上部密生莲座状叶。茎直立，不分枝，密被绒毛。基生叶莲座状，宿存，近圆形或倒卵形，背面着生绒毛、叉状毛，近于星状的分枝毛；无茎生叶。总状花序；花瓣黄色，倒卵形，顶端微凹。果实近圆形或椭圆形，无毛。花期6-8月，果期8-9月。生海拔3300-5100米的悬岩上或山坡沟谷。产青海、四川、西藏东部和云南。

Perennial cespitose herbs. Ultimate branches of caudex with dense rosettes. Stems erect, simple, densely tomentose. Basal leaves rosulate, persistent, suborbicular or obovate, abaxially tomentose with subsessile, forked and rayed stellate trichomes with unbranched ray; cauline leaves absent. Racemes; petals yellow, obovate, apex emarginate. Fruit suborbicular to elliptic, glabrous. Fl. Jun-Aug. Fr. Aug-Sep. Crevices or montane ravines at 3300-5100 m. Distributed in Qinghai, Sichuan, E Xizang and Yunnan.

愉悦葶苈

Draba jucunda W. W. Smith

多年生草本，密集丛生，有花葶。基生叶丛生，无柄；无茎生叶。总状花序2-7(-10)花，无苞片；花瓣黄色，倒卵形。果实长圆形至椭圆体形或近圆形，侧向开裂。花期6-8月，果期8-9月。生海拔3400-4600米的砾石沙地或灌丛中。产云南和西藏。

Herbs perennial, densely cespitose, scapose. Basal leaves rosulate, sessile; cauline leaves absent. Racemes 2-7(-10)-flowered, ebracteate; petals yellow, obovate. Fruits oblong to ellipsoid or suborbicular, latiseptate. Fl. Jun-Aug. Fr. Aug-Sep. Sandy areas or thickets at 3400-4600 m. Distributed in Yunnan and Xizang.

总苞葶苈 *Draba involucrata*

愉悦葶苈 *Draba jucunda*

高茎葶苈

Draba elata Hook. f. et Thoms.

高茎葶苈 *Draba elata*

多年生草本，高20-45(-60)厘米。基生叶丛生；叶狭倒披针形至匙形；茎生叶3-6(-9)，无柄。总状花序具10-25花，无苞片，果期剧烈伸长；花瓣黄色。果卵形至卵圆状披针形。花果期6-8月。生海拔3400-4900米的山坡、路边或潮湿的草地。产西藏。印度北部亦有。

Herbs perennial, 20-45(-60) cm tall. Basal leaves rosulate; leaves narrowly oblanceolate to spatulate; cauline leaves 3-6(-9), sessile. Racemes 10-25-flowered, ebracteate, elongated considerably in fruit; petals yellow. Fruits ovoid to ovoid-lanceolate. Fl. and fr. Jun-Aug. Mountain slopes, roadsides or wet grasslands at 3400-4900 m. Distributed in Xizang. Also in N India.

抱茎葶苈

Draba amplexicaulis Franch.

抱茎葶苈 *Draba amplexicaulis*

多年生草本，丛生，不变灰白色。茎基不延伸。茎生叶(6-)10-25(-30)；叶卵形，狭长圆形或披针形，耳状或半抱茎，顶端尖。总状花序具30-80(-100)花；花瓣黄色。果实椭圆体形、长圆形或长圆状条形。花期6-8月，果期7-9月。生海拔2500-4700米的山坡草地、碎石地、石质山坡或林缘。产云南、四川和西藏。

Perennial herbs, cespitose, not canescent. Caudex not extended. Cauline leaves (6-)10-25(-30); leaves ovate, narrowly oblong or lanceolate, auriculate or amplexicaul, apex acute. Racemes 30-80 (-100)-flowered; petals yellow. Fruits ellipsoid, oblong or oblong-linear. Fl. Jun-Aug. Fr. Jul-Sep. Grassy areas, rocky cliffs, stony slopes, or forest edge at 2500-4700 m. Distributed in Yunnan, Sichuan and Xizang.

山菜葶苈

Draba surculosa Franch.

多年生草本。根状茎延长。叶长圆形、披针形或卵形，先端钝，基部钝，常耳状或抱茎。总状花序具20-60花；花瓣黄色。果实长圆形、椭圆体形或椭圆状条形。花期6-8月，果期7-9月。生海拔2600-4600米的草坡、岩隙或高山草甸。产云南、四川和西藏。

山菜葶苈 *Draba surculosa*

云南葶苈 *Draba yunnanensis*

棉毛葶苈 *Draba winterbottomii*

Herbs perennial. Rhizomes elongate. Leaves oblong, lanceolate or ovate, apex obtuse, base obtuse, often auriculate or amplexicaul. Racemes 20-60-flowered; petals yellow. Fruits oblong, ellipsoid or elliptic-linear. Fl. Jun-Aug. Fr. Jul-Sep. Grassy slopes, pastures, rock crevices, alpine meadows, thickets, ravines or scree at 2600-4600 m. Distributed in Yunnan, Sichuan and Xizang.

云南葶苈

Draba yunnanensis Franch.

多年生草本。茎基不分枝。基生叶丛生，宿存；茎生叶(4-)6-12(-18)，狭长圆形。总状花序具(12-)20-60(-80)花；萼片长2.5-3毫米；花瓣黄色；短角果无毛。花期5-7月，果期6-9月。生海拔2300-5500米的山坡、石灰岩碎石滩、草地或岩隙。产云南、四川和西藏。

Herbs perennial. Caudex not branched. Basal leaves rosulate, persistent; cauline leaves (4-)6-12(-18), narrowly oblong. Racemes (12-)20-60(-80)-flowered; sepals 2.5-3 mm long; petals yellow; silicles glabrous. Fl. May-Jul. Fr. Jun-Sep. Slopes, limestone scree, grasslands or rock crevices at 2300-5500 m. Distributed in Yunnan, Sichuan and Xizang.

棉毛葶苈

Draba winterbottomii (Hook. f. et Thomson) Pohle

多年生丛生草本。茎基多分枝，上部叶莲座状丛生。茎直立，不分枝，被绒毛。基生叶莲座状，成柱形，叶倒卵形或长椭圆形，密被绒毛，全缘；无茎生叶。总状花序；花瓣白色，匙状。短角果狭椭圆形，披针形或长圆形，具假隔膜，假隔膜无毛。花期6-8月，果期7-9月。生海拔4000-5900米的高山草甸。产青海和西藏。克什米尔地区亦有。

Perennial cespitose herbs. Caudex many branched, ultimate branches terminated in rosettes, cylindrical. Stems erect, simple, tomentose. Basal leaves rosulate, obovate or oblong, densely tomentose, margin entire; cauline leaves absent. Racemes; petals white, spatulate. Fruit narrowly elliptic, lanceolate or oblong, latiseptate, valves glabrous. Fl. Jun-Aug. Fr. Jul-Sep. Alpine meadow at 4000-5900 m. Distributed in Qinghai and Xizang. Also in Kashmir.

葶苈

Draba nemorosa L.

一年生草本。基生叶丛生，长圆状倒卵形或倒披针形；茎生叶(2-)3-12(-15)，阔卵形至长圆形。总状花序具15-90花，无苞片；萼片卵形；花瓣黄色。果矩圆状或椭圆体形。花果期3-6月。生海拔4800米以下的草丛、潮湿山谷、林缘、溪边、路边或山坡。产中国除华南以外大部分地区。中亚和东北亚亦有。

Herbs annual. Basal leaves clustered, oblong-obovate or oblanceolate; cauline leaves (2-)3-12(-15), broadly ovate to oblong, racemes 15-90-flowered, ebracteate; sepals ovate; petals yellow. Fruits oblong or ellipsoid. Fl. and fr. Mar-Jun. Grasslands, wet valleys, forest margins, streamsides, roadsides or slopes below 4800 m. Distributed in most parts of China, except S China. Also in C and NE Asia.

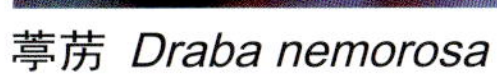
葶苈 *Draba nemorosa*

唐古碎米荠 *Cardamine tangutorum*

唐古碎米荠

Cardamine tangutorum O. E. Schulz.

多年生草本。根状茎细长，不匍匐。茎单一，直立。侧生小叶3-5(或6)对，不下延至基部。总状花序具10-15花；侧生萼片基部囊状；花瓣紫色，匙形。果线形。花期5-7月，果期6-8月。生海拔1300-4400米的山沟、草甸、河岸或林下。产中国西南、华北和华西。

Herbs perennial. Rhizomes creeping, not stoloniferous. Stems simple, erect. Lateral leaflets 3-5 (or 6) pairs, not decurrent at base. Racemes 10-15-flowered; base of lateral pair of sepals saccate; petals purple, spatulate. Fruits linear. Fl. May-Jul. Fr. Jun-Aug. Mountain ditches, meadows, river banks or forests at 1300-4400 m. Distributed in SW, N and W China.

水田碎米荠

Cardamine lyrata Bunge

多年生草本，全株无毛。生于匍匐茎上的叶为单叶，近圆形、心形或肾形，叶柄长3-12毫米；茎生叶无柄，具小叶2-9对。花瓣白色，倒卵形，无爪，顶端圆形或微凹。果瓣平滑，无毛。种子边缘具宽1毫米的翅。花期4-6月，果期5-7月。生海拔0-1000米的湿润地和溪边。产中国东北、华东、广西、贵州、河北、河南、湖南和四川。朝鲜半岛、日本和俄罗斯亦有。

Perennial herbs, glabrous throughout. Leaves on stolons simple, suborbicular, cordate or reniform, petiole 3-12 mm long; cauline leaves sessile, with 2-9-paired leaflets. Petals white, obovate, not clawed, apex rounded or emarginate. Valves smooth, glabrous. Seeds winged all around, wing to 1 mm wide. Fl. Apr-Jun. Fr. May-Jul. Moist places and streamsides at 0-1000 m. Distributed in NE and E China, Guangxi, Guizhou, Hebei, Henan, Hunan and Sichuan. Also in Korean Peninsula, Japan and Russia.

裸茎碎米荠

Cardamine scaposa Franch.

多年生草本，具花葶，全株无毛。根状茎纤细。叶肾形或近圆形；无茎生叶。总状花序顶生，具2-10花；花瓣白色，阔倒卵形。果实线形。花期4-6月，果期6-7月。生海拔1400-2500米的山坡或林下潮湿处。产四川、河北、山西、内蒙古和陕西。

Herbs perennial, scapose, glabrous throughout. Rhizomes slender. Leaves reniform or suborbicular; cauline leaves absent. Racemes terminal, 2-10-flowered; petals white, broadly obovate. Fruits linear. Fl. Apr-Jun. Fr. Jun-Jul. Slopes or moist areas under forests at 1400-2500 m. Distributed in Sichuan, Hebei, Shanxi, Neimenggu and Shaanxi.

水田碎米荠 *Cardamine lyrata*

裸茎碎米荠 *Cardamine scaposa*

大叶碎米荠 *Cardamine macrophylla*

白花碎米荠 *Cardamine leucantha*

大叶碎米荠

Cardamine macrophylla Willd.

多年生草本。根状茎细长，不具鳞，无匍匐茎。侧生小叶(1或)2-6对；茎生叶3-12(-18)，长(1-)2-15(-25)厘米。总状花序具10-30花；花瓣紫红或淡紫色，倒卵形或匙形。果实线形。花期(3-)4-10月，果期5-10月。生海拔500-4200米的溪边疏林下、苔原、岩隙、灌丛、沟谷或草坡上。产中国西南、华中、华北、华西、华东、西北和东北。南亚、中亚和东北亚亦有。

Herbs perennial. Rhizomes slender, not scaly, not stoloniferous. Lateral leaflets (1 or)2-6 pairs; cauline leaves 3-12(-18), (1-)2-15(-25) cm long. Racemes 10-30-flowered; petals purple or lilac, obovate or spatulate. Fruits linear. Fl. (Mar-)Apr-Oct. Fr. May-Oct. Streamsides, open forests, tundra, rock crevices, thickets, valleys or grassy slopes at 500-4200 m. Distributed in SW, C, N, W, E, NW and NE China. Also in S, C and NE Asia.

白花碎米荠

Cardamine leucantha (Tausch) O. E. Schulz

多年生草本。根状茎匍匐，纤细。侧生小叶2或3对。总状花序具12-24花；萼片矩圆形，边缘膜质；花瓣白色，匙形至长圆状倒披针形。果线形。花期4-7月，果期5-8月。生海拔100-2000米的路边、林中、溪边或路边。产中国西南、华北、华西、华东和东北。俄罗斯(远东地区)、蒙古、朝鲜半岛和日本亦有。

Herbs perennial. Rhizomes creeping, slender. Lateral leaflets 2 or 3 pairs. Racemes 12-24-flowered; sepals oblong, margin membranous; petals white, spatulate to oblong-oblanceolate. Fruits linear. Fl. Apr-Jul. Fr. May-Aug. Roadsides, forests, along streams or roadsides at 100-2000 m. Distributed in SW, N, W, E and NE China. Also in Russia (Far East), Mongolia, Korean Peninsula and Japan.

碎米荠

Cardamine hirsuta L.

一年生草本。茎直立、斜升或匍匐，自基部1至数个伸出，上部分枝或不分枝，不弯曲。基生叶丛生，大头羽状全裂。萼片矩圆形；花瓣白色，匙形。果线形。种子具狭边。花期2-5月，果期4-7月。生海拔3000米以下的山坡、路边、田间、荒地或草地。产中国各地。世界各地广布。

Herbs annual. Stems erect, ascending or decumbent, 1 to several from base, simple or branched above, not flexuous. Basal leaves rosulate, lyrate-pinnatisect. Sepals oblong; petals white, spatulate. Fruits linear. Seeds narrowly margined. Fl. Feb-May. Fr. Apr-Jul. Mountain slopes, roadsides, fields, wastelands or grassy areas below 3000 m. Distributed throughout China. Also widely in the world.

碎米荠 *Cardamine hirsuta*

弯曲碎米荠 *Cardamine flexuosa*

弯曲碎米荠

Cardamine flexuosa With.

一年生或二年生草本。茎直立、斜升或匍匐，1至数个从基部伸出，弯曲或直立。基生叶不丛生，常于花期枯萎，大头羽状分裂；茎生叶3-15，顶生裂片3-5裂。花瓣白色，匙形；雄蕊6。果线形。花期2-5月，果期4-6月。生海拔3600米以下的田间、路旁、草地、溪边、潮湿灌丛或干燥处。产中国各地。全世界普遍分布。

Herbs annual or biennial. Stems erect, ascending or decumbent, 1 to several from base, flexuous or straight. Basal leaves not rosulate, often withered by anthesis, lyrated; cauline leaves 3-15, terminal lobe 3-5-lobed. Petals white, spatulate; stamens 6. Fruits linear. Fl. Feb-May. Fr. Apr-Jun. Fields, roadsides, grasslands, by streams, thickets in damp places or dry sites below 3600 m. Distributed throughout China. Also widely in the world.

宽翅碎米芥

Cardamine franchetiana Diels

多年生草本。根茎基部丛生白色小鳞茎。茎单生，直立。侧生小叶(2或)3-6对。总状花序顶生，有3-8花；花瓣紫红或白色。果实线形。花期6-7月，果期8-9月。生海拔2300-4800米的山坡、草甸、湿润牧场、沟谷或石缝。产云南、四川、西藏和青海。

Herbs perennial. Rhizomes with clustered white small bulbs at base. Stems simple, erect. Lateral leaflets (2 or)3-6 pairs. Racemes terminal, 3-8-flowered; petals purplish red or white. Fruits linear. Fl. Jun-Jul. Fr. Aug-Sep. Slopes, meadows, moist pastures, valleys or rock cracks at 2300-4800 m. Distributed in Yunnan, Sichuan, Xizang and Qinghai.

单花荠

Pegaeophyton scapiflorum (Hook. f. et Thoms.) Marq. et Shaw

草本，具细弱或粗壮、顶端分枝，稀不分枝的茎。花瓣白色、粉红色或蓝色，宽倒卵形，长5-8毫米，宽3-7毫米。果实具宽隔膜。花期5-9月，果期7-10月。生海拔3500-5400(-5600)米的高山湿地、草地、岩隙、湖边沼泽地、溪边沙地或石坡。产云南、四川、西藏、甘肃、青海和新疆。印度、尼泊尔、不丹和缅甸亦有。

Herbs with slender or stout, apically branched or unbranched caudex. Petals white, pink or blue, broadly obovate, 5-8 × 3-7 mm broad. Fruits latiseptate. Fl. May-Sep. Fr. Jul-Oct. Alpine wet places, grasslands, rock crevices, boggy ground by lakes, sandy stream edges or stony slopes at 3500-5400(-5600) m. Distributed in Yunnan, Sichuan, Xizang, Gansu, Qinghai and Xinjiang. Also in India, Nepal, Bhutan and Myanmar.

宽翅碎米芥 *Cardamine franchetiana*

单花荠 *Pegaeophyton scapiflorum*

堇叶芥

Neomartinella violifolia (Levl.) Pilger

一年生矮小草本。叶全部基生，心形、肾形或近圆形，长(0.8-)1.5-4(-5)厘米，基部心形，边缘具波状圆齿或近全缘；叶柄长(1.5-)3-10(-14)厘米。萼片卵形，长1.5-2毫米；花瓣白色，倒心形；花丝白色；无宿存花柱，稀长达0.2毫米。果瓣具不明显中脉。花期2-4月，果期3-5月。生海拔800-1600米的多石地区。产湖南、湖北、贵州、四川和云南。

Annual and small herbs. Leaves all basal, cordate, reniform or suborbicular, (0.8-)1.5-4(-5) cm long, base cordate, margin crenate-repand or rarely subentire, petiole (1.5-)3-10(-14) cm long. Sepals ovate, 1.5-2 mm; petals white, obcordate; filaments white; persistent style absent or rarely to 0.2 mm long. Valves with an obscure midvein. Fl. Feb-Apr. Fr. Mar-May. Rocky areas at 800-1600 m. Distributed in Hunan, Hubei, Guizhou, Sichuan and Yunnan.

堇叶芥 *Neomartinella violifolia*

垂果南芥

Arabis pendula L.

二年生草本。茎直立，单生，常上部圆锥状分枝。叶披针形、长圆形或椭圆形。总状花序无苞片；花瓣白色，稀粉色。果实开展或向一侧反折。花期6-8月，果期7-9月。生海拔4300米以下的岩石山坡、草甸、荒地、灌丛、草丛、林缘、河岸或荒漠。产中国西南、华北、华西、西北和东北。哈萨克斯坦、俄罗斯、蒙古、朝鲜半岛、日本和欧洲亦有。

Herbs biennial. Stems erect, simple, often paniculate branched above. Leaves lanceolate, oblong or elliptic. Racemes ebracteate; petals white or rarely pink. Fruits spreading or unilaterally deflexed. Fl. Jun-Aug. Fr. Jul-Sep. Rocky slopes, meadows, waste places, thickets, grasses, forest edges, river banks or deserts below 4300 m. Distributed in SW, N, W, NW and NE China. Also in Kazakhstan, Russia, Mongolia, Korean Peninsula, Japan and Europe.

垂果南芥 *Arabis pendula*

抱茎南芥 *Arabis amplexicaulis*

抱茎南芥

Arabis amplexicaulis Edgew.

二年生草本，被刚毛或粗毛。茎直立。基生叶莲座状，倒卵形或长椭圆形；茎生叶卵形或长椭圆形，叶缘具锯齿或全缘。总状花序；萼片长椭圆形，无毛；花瓣白色，长椭圆形或倒披针形，末端钝圆。长角果直立，分叉。花期4-6月，果期5-7月。生海拔1800-3200米的林缘或林阴处。产西藏。巴基斯坦、阿富汗、伊朗、喜马拉雅和印度亦有。

Biennial herbs, hispid or hirsute. Stems erect. Basal leaves rosulate, obovate or oblong; cauline leaves ovate or oblong, margin dentate or entire. Racemes; sepals oblong, glabrous; petals white, oblong or narrowly oblanceolate, apex obtuse. Fruit erect, divaricate. Fl. Apr-Jun. Fr. May-Jul. Forest margins or shady places at 1800-3200 m. Distributed in Xizang. Also in Pakistan, Afghanistan, Iran, Himalaya and India.

鼠耳芥

Arabidopsis thaliana (L.) Heynh.

一年生草本。基生叶倒卵形、匙形、卵形或椭圆形，长0.8-3.5(-4.5)厘米，上面被单毛和1叉毛，具短柄；茎生叶无柄。花瓣白色，匙形，基部渐狭为短爪；花丝白色，长1.5-2毫米。果瓣具明显中脉。花期4-6月。生海拔0-2000米的平地、山坡、河边或路边。产华东、华南、西北和华西。东北亚、中亚、俄罗斯、印度、伊朗、欧洲、非洲和北美洲亦有。

Annual herbs. Basal leaves obovate, spatulate, ovate or elliptic, 0.8-3.5(-4.5) cm long, adaxially with predominantly simple and stalked 1-forked trichomes, shortly petiolate; cauline leaves sessile. Petals white, spatulate, base attenuate to a short claw; filaments white, 1.5-2 mm long. Valves with a distinct midvein. Fl. Apr-Jun. Plains, mountain slopes, river banks or roadsides at 0-2000 m. Distributed in E, S, NW and W China. Also in NE and C Asia, Russia, India, Iran, Europe, Africa and North America.

须弥扇叶芥

Desideria himalayensis (Cambess.) Al-Shehbaz

多年生草本。茎不分枝。基生叶宿存，宽倒卵形或匙形，先端锐尖；茎生叶与基生叶相似，常全缘。总状花序；萼片早落，具柔毛，边缘膜质；花瓣紫色或淡紫色，具淡黄中心，宽匙形；花丝白色。角果披针形至披针状线形。花期6-8月，果期7-10月。生海拔4300-5700米的高山苔原、开阔的丘陵或流石滩。产青海和西藏。印度、克什米尔地区和尼泊尔亦有。

Perennial herbs. Stems simple. Basal leaves persistent, broadly obovate or spatulate, apex acute; cauline leaves similar to basal, often entire. Racemes; sepals caducous, pilose, margin membranous; petals purple or lilac with yellowish center, broadly spatulate; filaments white. Fruit lanceolate to lanceolate-linear, Fl. Jun-Aug. Fr. Jul-Oct. Alpine tundra, open hills or sandstone scree at 4300-5700 m. Distributed in Qinghai and Xizang. Also in India, Kashmir and Nepal.

鼠耳芥 *Arabidopsis thaliana*

须弥扇叶芥 *Desideria himalayensis*

旗杆芥 *Turritis glabra*

旗杆芥
Turritis glabra L.

二年生草本，稀多年生。基生叶匙形、倒披针形或长圆形，长4-15厘米，被短柔毛，稀无毛，具柄；茎生叶无柄，基部戟形或具耳，边缘具齿或全缘。花瓣淡黄色或乳白色。果近圆柱形或呈四棱形。花期4-7月，果期5-8月。生海拔100-3500米的山坡、林缘，河谷、草甸或路旁等。产辽宁、山东、浙江、江苏和新疆。欧洲、亚洲、北美洲和大洋洲亦有。

Biennial herbs, rarely perennial. Basal leaves spatulate, oblanceolate or oblong, 4-15 cm long, pubescent or rarely glabrous, petiolate; cauline leaves sessile, base sagittate or auriculate, margin dentate or entire. Petals pale yellow or creamy white. Fruit subterete or quadrangular. Fl. Apr-Jul. Fr. May-Aug. Mountain slopes, forest margins, valleys, meadows or roadsides at 100-3500 m. Distributed in Liaoning, Shandong, Zhejiang, Jiangsu and Xinjiang. Also in Europe, Asia, North America and Oceania.

焯菜
Rorippa indica (L.) Hiern

一年生草本，无毛。茎常基部和顶部分枝。基生叶大头羽状深裂或不裂；叶披针形或长圆形，全缘。总状花序顶生或侧生；花瓣黄色。果线形，上部常弯曲。花果期全年。生海拔3200米以下的路边、田野、菜园或河边。产中国大部分地区。南亚、东南亚和东北亚亦有；归化于美洲。

Herbs annual, glabrous. Stems often branched basally and apically. Basal leaves lyrate or undivided; leaves lanceolate or oblong, margin entire. Racemes terminal or lateral; petals yellow. Fruits linear, often curved upward. Fl. and fr. all year. Roadsides, fields, gardens or by rivers below 3200 m. Distributed in most parts of China. Also in S, SE and NE Asia; naturalized in America.

焯菜 *Rorippa indica*

无瓣蔊菜 *Rorippa dubia*

广州蔊菜 *Rorippa cantoniensis*

无瓣蔊菜

Rorippa dubia (Pers.) Hara

一年生草本，无毛。叶大头羽状浅裂或不分裂，倒卵形、长圆形或披针形。总状花序无苞片；花瓣大多缺少，若存在则常短于萼片。种子每室1行。花果期全年。生海拔3700米的山坡、路边、田间、草地或水边。产中国大部分地区。南亚、东南亚和东北亚亦有；归化于南美洲和北美洲。

Herbs annual, glabrous. Leaves lyrate-pinnatipartite or undivided, obovate, oblong or lanceolate. Racemes ebracteate; petals mostly absent, often shorter than sepals if present. Seeds uniseriate. Fl. and fr. throughout the year. Slopes, roadsides, fields, grassy places or by waters at 3700 m. Distributed in most parts of China. Also in S, SE and NE Asia; naturalized in South and North America.

广州蔊菜

Rorippa cantoniensis (Lour.) Ohwi

一年生草本。茎直立或匍匐，基部和上部分枝。基生叶具柄，丛生，迅速枯萎，大头羽状分裂、羽状全裂或二回羽状全裂。总状花序顶生；每花具苞片；花瓣黄色，倒卵形或匙形。果实阔或狭长圆形。花期3-4月，果期4-5月。生海拔1800米以下的田间、路边、山谷、河边或潮湿处。产中国大部分地区。越南、俄罗斯、朝鲜半岛和日本亦有。

Herbs annual. Stems erect or decumbent, branched basally and above. Basal leaves petiolate, rosulate, soon withered, lyrate, pinnatisect or bipinnatisect. Racemes terminal; flowers bracteate; petals yellow, obovate or spatulate. Fruits broadly or narrowly oblong. Fl. Mar-Apr. Fr. Apr-May. Fields, roadsides, valleys, by rivers or damp places below 1800 m. Distributed in most parts of China. Also in Vietnam, Russia, Korean Peninsula and Japan.

风花菜

Rorippa globosa (Turcz. ex Fisch. et Mey.) Hayek

一年生草本或短期多年生草本。基生叶丛生，不久即枯

风花菜 *Rorippa globosa*

萎，大头羽状分裂或者近倒向羽裂，边缘具不规则牙齿或细齿。总状花序无苞片；花瓣黄色。果球形或近球形。花果期9-11月。生海拔2500米以下的潮湿或干燥地、河岸、路边或草地。产中国大部分地区。越南、俄罗斯、蒙古、朝鲜半岛和日本亦有。

Herbs annual or short-lived perennial. Basal leaves rosulate, soon withered, lyrate or subruncinate, margin irregularly dentate or serrate. Racemes ebracteate; petals yellow. Fruits globose or subglobose. Fl. and fr. Apr-Nov. Moist or dry areas, river banks, roadsides or grasslands below 2500 m. Distributed in most parts of China. Also in Vietnam, Russia, Mongolia, Korean Peninsula and Japan.

沼生蔊菜

Rorippa palustris (L.) Besser

一年生草本。基生叶丛生，不久即枯萎，大头羽状分裂。总状花序顶生或侧生；花瓣黄色或浅黄色，匙形。果实长圆形、椭圆体形或长圆状卵球形，常稍弯曲。花果期3-10月。生海拔4000米以下的牧场、草地、草甸、路边、河边或灌丛。产中国除华南以外大部分地区。南亚、中亚、东北亚、欧洲和北美洲亦有；也引种到澳大利亚、南美洲和其他地方。

Herbs annual. Basal leaves rosulate, withered early, lyrate. Racemes terminal or lateral; petals yellow or pale yellow, spatulate. Fruits oblong, ellipsoid or oblong-ovoid, often slightly curved. Fl. and fr. Mar-Oct. Pastures, grasslands, meadows, roadsides, by rivers or thickets below 4000 m. Distributed in most parts of China, except S China. Also in S, C and NE Asia, Europe and North America; introduced to Australia, South America and elsewhere.

豆瓣菜 *Nasturtium officinale*

沼生蔊菜 *Rorippa palustris*

豆瓣菜

Nasturtium officinale R. Br.

多年生水生草本，高10-200厘米，全体光滑无毛。茎匍匐，多分枝，节上生不定根。奇数羽状复叶。总状花序；萼片长椭圆形，侧裂片略呈囊状；花瓣白色或粉色，倒卵形，先端圆；花丝白色。长角果圆柱形。花期4-9月，果期5-9月。生海拔850-3700米的溪流中、水沟边或沼泽地中。广布中国。世界广布。

Perennial aquatic herbs, 10-200 cm tall, glabrous throughout. Stems decumbent, much branched, rooting at proximal nodes. Leaves odd-pinnate. Racemes; sepals oblong, lateral lobes slightly saccate; petals white or pink, obovate, apex rounded; filaments white. Fruit cylindric. Fl. Apr-Sep. Fr. May-Sep. Streams, ditches or swamps at 850-3700 m. All over the China. Widely naturalized elsewhere.

花旗杆 *Dontostemon dentatus*

花旗杆

Dontostemon dentatus (Bunge) Ledeb.

一年生草本，无腺体。茎直立，常单生，上部分枝。叶披针形至披针状线形，边缘具小齿至粗齿。萼片矩圆形；花瓣淡蓝色或淡紫色，倒卵形。果实无毛，直伸、直立或稍斜展。花果期6-9月。生海拔200-1900米的岩石山坡、沙地或路边。产中国西南、华北、华东、华西和东北。俄罗斯、朝鲜半岛和日本亦有。

Herbs annual, eglandular. Stems erect, often simple, branched above. Leaves lanceolate to lanceolate-linear, margin minutely to coarsely dentate. Sepals oblong; petals lilac or purplish, obovate. Fruits glabrous, straight, erect or slightly ascending. Fl. and fr. Jun-Sep. Rocky slopes, sandy areas or roadsides at 200-1900 m. Distributed in SW, N, E, W and NE China. Also in Russia, Korean Peninsula and Japan.

羽裂花旗杆

Dontostemon pinnatifidus (Willd.) Al-Shehbaz et H. Ohba

一年生或二年生草本。茎直立，常单生，上部分枝。叶披针形、椭圆形或长圆形，边缘具锯齿、细齿或羽状深裂。花瓣白色，阔倒卵形，(5-)6-8 × (2.5-)3-4(-5)毫米，先端略凹，具爪，1-3毫米。花果期6-8月。生海拔1100-4600米的草原、沙丘、山腰、岩石坡或路边。产中国西南、华北、西北和东北。印度、尼泊尔、俄罗斯和蒙古亦有。

Herbs annual or biennial. Stems erect, often simple, branched above. Leaves lanceolate, elliptic or oblong, margin dentate, serrate or pinnatifid. Petals white, broadly obovate, (5-)6-8 × (2.5-)3-4(-5) mm, apex emarginated, claws 1-3 mm. Fl. and fr. Jun-Aug. Grassy plains, sand dunes, hillsides, rocky slopes or roadsides at 1100-4600 m. Distributed in SW, N, NW and NE China. Also in India, Nepal, Russia and Mongolia.

丛菔

Solms-laubachia pulcherrima Muschl.

草本。叶片披针形、倒披针形或线形，无毛或疏被杂有皱曲柔毛，具缘毛；无茎生叶。花单生；花瓣粉色或浅蓝色至天蓝色，倒卵形至阔倒卵形。果实披针形。花期5-7月，果期6-8月。生海拔3300-5200米的山坡、巨砾、溪边或岩石裂隙。产云南、四川和西藏。

Herbs. Leaves lanceolate, oblanceolate or linear, glabrous or sparsely curved-pubescent, ciliate; cauline leaves absent. Flowers solitary; petals pink or light to turquoise blue, obovate to broadly obovate. Fruits lanceolate. Fl. May-Jul. Fr. Jun-Aug. Slopes, boulders, by streams or rock crevices at 3300-5200 m. Distributed in Yunnan, Sichuan and Xizang.

弯角四齿芥

Tetracme recurvata Bunge

一年生草本，全株被绒毛。基生叶和最下部茎叶长圆状条形或条状披针形，基部渐狭，边缘羽状半裂、羽状深裂或具波状齿，叶柄长0.3-2厘米；上部茎叶较小，近无柄，边缘近全缘。花瓣白色，匙形，基部渐狭为爪；花丝白色。果上半部弯曲或反折；果瓣被绒毛。花果期4-6月。生海拔200-600米

羽裂花旗杆 *Dontostemon pinnatifidus*

丛菔 *Solms-laubachia pulcherrima*

的沙地及冲积平原。产新疆。中亚和西南亚亦有。

Annual herbs, tomentose throughout. Basal and lowermost cauline leaves oblong-linear or linear-lanceolate, base attenuate, margin pinnatifid, pinnatipartite or sinuate-dentate, petioles 0.3-2 cm long; upper cauline leaves similar to basal, smaller, subsessile, sometimes entire. Petals white, spatulate, attenuate to clawlike base; filaments white. Fruit distal half arcuate or recurved; valves tomentose. Fl. and fr. Apr-Jun. Sandy deserts and plains at 200-600 m. Distributed in Xinjiang. Also in C and SW Asia.

紫罗兰

Matthiola incana (L.) R. Br.

二年生或多年生草本，高达60厘米，全株密被灰白色具柄的柔毛。叶长圆形至倒披针形或匙形，连叶柄长6-14厘米，基部渐狭成柄。萼片直立，长椭圆形，边缘膜质；花瓣紫红、淡红色或白色，近卵形，顶端浅2裂或微凹，下部具长爪。长角果圆柱形，长7-8厘米，果瓣中脉明显，顶端浅裂。花期4-5月。中国多省常有引种。原产欧洲南部。

Biennial or perennial herbs, to 60 cm tall, throughout grey and stalked pubescent. Leaves oblong to oblanceolate or spatulate, 6-14 cm long including petiole, base tapering into petiole. Sepals erect, long-elliptic, margin membranous; petals purple-red, reddish or white, subovate, apex 2-lobed or retuse, with long claw at base. Siliqua terete, 7-8 cm long, valves with a distinct midvein, apex lobed. Fl. Apr-May. Cultivated in many provinces of China. Native to the S Europe.

弯角四齿芥 *Tetracme recurvata*

紫罗兰 *Matthiola incana*

离子芥 *Chorispora tenella*

涩荠 *Malcolmia africana*

离子芥
Chorispora tenella (Pall.) DC.

一年生草本。茎直立，上部分枝。基生叶不丛生，常于花期枯萎；叶倒披针形或长圆形，具腺点，边缘具波状齿至锯齿。花排成总状；萼片线形，带紫色；花瓣紫色，倒披针形。果实线状圆柱形。花期4-6月，果期5-8月。生海拔100-2200米的牧场、路边、田间或荒地。产华北、华西、华东和西北。南亚、中亚、东北亚、非洲北部和欧洲亦有。

Annual herbs. Stems erect, branched above. Basal leaves not rosulate, often withered by anthesis; leaves oblanceolate or oblong, glandular, margin sinuate-dentate to dentate. Flowers in racemes; sepals linear, purplish; petals purple, oblanceolate. Fruits linear-cylindric. Fl. Apr-Jun. Fr. May-Aug. Pastures, roadsides, fields or waste places at 100-2200 m. Distributed in N, W, E and NW China. Also in S, C and NE Asia, N Africa and Europe.

涩荠
Malcolmia africana (L.) R. Br.

二年生草本，密生单毛或叉状硬毛。叶长圆形、倒披针形或近椭圆形，边缘有波状齿或全缘。花瓣紫色或粉红色。长角果圆柱形或近圆柱形，近4棱，常密生短或长分叉毛或二者间生，或具刚毛；柱头圆锥状；果梗长1-2毫米。花果期6-8月。生路边荒地或田间。产中国西北、河北、山西、河南、安徽、江苏和四川。亚洲、欧洲和非洲亦有。

Biennial herbs, densely simple hairy or forked bristle. Leaves oblong, oblanceolate or subelliptic, margin undulate-toothed or entire. Petals purple or pink. Siliqua terete or subterete, nearly 4-angled, usually densely short or long forked hairy or mixed, or bristle; stigma conic; fruiting pedicel 1-2 mm long. Fl. and fr. Jun-Aug. Roadsides, waste places or fields. Distributed in NW China, Hebei, Shanxi, Henan, Anhui, Jiangsu and Sichuan. Also in Asia, Europe and Africa.

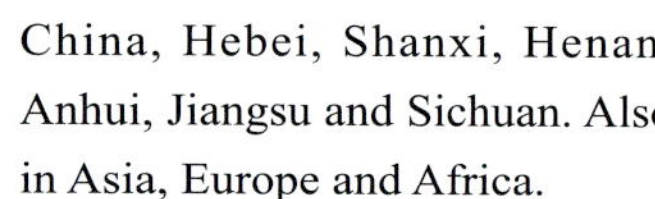

北香花芥
Hesperis sibirica L.

多年生或二年生草本，常密被腺毛、长硬毛和长达3毫米的单毛。中间和上部茎叶狭或宽披针形，无柄或近无柄，基部楔形，边缘有小牙齿或近全缘。花瓣深紫色、淡紫色或白色，倒卵形，爪长7-10毫米。果瓣疏或密具腺毛。种子之间缢缩。花果期5-9月。生海拔900-2900米的山坡、灌丛、平原和河边。产河北、辽宁和新疆。俄罗斯、蒙古和中亚亦有。

北香花芥 *Hesperis sibirica*

小花离子芥 *Chorispora macropoda*

砂生离子芥 *Chorispora sabulosa*

Perennial or biennial herbs, often densely glandular, hirsute and simple trichomes to 3 mm. Middle and upper cauline leaves narrowly to broadly lanceolate, sessile or subsessile, base cuneate, margin denticulate or subentire. Petals deep purple, lavender or white, obovate, claw 7-10 mm long. Valves sparsely or densely glandular. Constricted between seeds. Fl. and fr. May-Sep. Mountains slopes, shrubby areas, plains or near rivers at 900-2900 m. Distributed in Hebei, Liaoning and Xinjiang. Also in Russia, Mongolia and C Asia.

小花离子芥

Chorispora macropoda Trautv.

多年生草本，高4-28厘米，遍布腺毛。基生叶莲座状，广倒披针形至线状披针形，具腺体；无茎生叶。总状花序；萼片淡黄色，卵形；花瓣黄色，宽倒卵形，顶端微凹。长角果线状圆柱形，念珠状，具腺毛。种子棕色，长椭圆形。花果期5-7月。生海拔2200-4500米的沙砾地。产新疆。巴基斯坦、阿富汗和伊朗亦有。

Perennial herbs, 4-28 cm tall, glandular hairy throughout. Basal leaves rosulate, broadly oblanceolate to linear-lanceolate, glandular; cauline leaves absent. Racemes; sepals yellowish, ovate; petals yellow, broadly obovate, apex emarginate. Fruit linear-cylindric, torulose, glandular. Seeds brown, oblong. Fl. and fr. May-Jul. Gravelly areas at 2200-4500 m. Distributed in Xinjiang. Also in Pakistan, Afghanistan and Iran.

砂生离子芥

Chorispora sabulosa Cambess.

多年生草本，高3-15厘米，具腺体。基生叶莲座状，倒披针形或长椭圆形，无毛或具腺毛，羽状深裂，深波状齿或全缘；无茎生叶。总状花序；萼片略带紫色，卵形；花瓣紫色，宽倒卵形，先端钝。角果线状圆筒形，念珠状。花果期6-7月。生海拔2900-4800米的山坡。产西藏。印度、克什米尔地区、哈萨克斯坦、巴基斯坦、塔吉克斯坦和乌兹别克斯坦亦有。

Perennial herbs, 3-15 cm tall, glandular. Basal leaves rosulate, oblanceolate or oblong, glabrous or glandular, margin pinnatipartite, sinuate-dentate or entire; cauline leaves absent. Racemes; sepals purplish, ovate; petals purple, broadly obovate, apex obtuse. Fruit linear-cylindric, torulose. Fl. and fr. Jun-Jul. Mountain slopes at 2900-4800 m. Distributed in Xizang. Also in India, Kashmir, Kazakhstan, Pakistan, Tajikistan and Uzbekistan.

无茎光籽芥

Leiospora exscapa (C. A. Mey.) Dvorák

草本，高5-10厘米，丛生。茎粗，少分枝，密被残存叶柄。花每簇1-4；花瓣倒卵形，顶端圆；雄蕊6，4强；蜜腺2。果为开裂的长角果，线形至线状披针形。花果期7-9月。生山坡。产新疆北部(阿尔泰山脉)。哈萨克斯坦、俄罗斯和蒙古亦有。

Herbs, 5-10 cm tall, cespitose. Stems thick, few-branched, densely covered with petiolar remains of previous year. Flowers 1-4 per rosette; petals obovate, apex rounded; stamens 6, tetradynamous; nectar glands 2. Fruits dehiscent siliques, linear to linear-lanceolate. Fl. and fr. Jul-Sep. Mountain slopes. Distributed in N Xinjiang (Altay Mountains). Also in Kazakhstan, Russia and Mongolia.

无茎光籽芥 *Leiospora exscapa*

毛萼香芥 *Clausia trichosepala*

毛萼香芥
Clausia trichosepala (Turcz.) Dvorák

二年生草本。茎直立，多单一。叶椭圆形，边缘具粗锯齿。总状花序顶生；花萼远轴部分密被柔毛；花瓣紫色，倒卵形。长角果线形。花果期5-8月。生海拔1100-1700米的山坡。产河北、内蒙古、山西、山东和吉林。蒙古和朝鲜半岛亦有。

Herbs biennial. Stems erect, usually simple. Leaves elliptic, margin coarsely serrate. Racemes terminal; sepals densely hirsute distally; petals purple, obovate. Siliques linear. Fl. and fr. May-Aug. Mountain slopes at 1100-1700 m. Distributed in Hebei, Neimenggu, Shanxi, Shandong and Jilin. Also in Mongolia and Korean Peninsula.

棒果芥
Sterigmostemum caspicum (Lam.) Rupr.

多年生草本，被绒毛。基生叶条形、披针形或倒卵形，长(3-)4-8(-13)厘米，宽5-10毫米，基部渐狭，叶柄长0.5-2厘米；最上部茎生叶近无柄，较小。花瓣黄色，基部渐狭成爪；花柱长1-3毫米，增粗。果条形，直或稍弯；果瓣增厚，被绒毛。花期4-6月，果期5-7月。生海拔500-1200米的山坡、沙漠或干旱地区。产新疆。俄罗斯和哈萨克斯坦亦有。

Perennial herbs, tomentose. Basal leaves linear, lanceolate or obovate, (3-)4-8(-13) cm × 5-10 mm, base attenuate, petiole 0.5-2 cm long; uppermost cauline leaves subsessile, smaller. Petals yellow, base attenuate into a claw; style 1-3 mm long, thickened. Fruit linear, straight or curved; valves thickened, tomentose. Fl. Apr-Jun. Fr. May-Jul. Mountain slopes, deserts or arid areas at 500-1200 m. Distributed in Xinjiang. Also in Russia and Kazakhstan.

棒果芥 *Sterigmostemum caspicum*

糖芥
Erysimum amurense Kitag.

多年生草本。茎直立，单生或基部分枝。叶狭线形至线状披针形。总状花序伞房状，多花密集；花瓣橙黄色。果线形，近直立或平展；柱头锤状，突起，2裂。花期5-8月，果期6-10月。生海拔100-2800米的平原、山谷、干燥砂或多石山坡、路边、灌丛、丘陵或干枯河边。产河北、山西、内蒙古、陕西、江苏和辽宁。俄罗斯和朝鲜半岛亦有。

Herbs perennial. Stems erect, simple or branched basally. Leaves narrowly linear to linear-lanceolate. Racemes corymbose, densely flowered; petals orange-yellow. Fruits linear, sub-

糖芥 *Erysimum amurense*

erect or flattened; stigmas capitate, prominently 2-lobed. Fl. May-Aug. Fr. Jun-Oct. Plains, valleys, dry-sandy or stony slopes, roadsides, thickets, hillsides or dry river banks at 100-2800 m. Distributed in Hebei, Shanxi, Neimenggu, Shaanxi, Jiangsu and Liaoning. Also in Russia and Korean Peninsula.

小花糖芥

Erysimum cheiranthoides L.

一年生草本，高7-150厘米。茎直立，通常上部分枝。基生叶莲座状；茎生叶披针形或线形，边缘近全缘或具齿。总状花序伞房状；花瓣黄色，狭匙形，顶端圆形。长角果线形，念珠状。花期5-8月，果期6-9月。生海拔800-3000米的干河床或潮湿地区。产黑龙江、吉林、内蒙古和新疆。蒙古、朝鲜半岛、欧洲、非洲和北美洲亦有。

Annual herbs, 7-150 cm tall. Stems erect, often branched above. Basal leaves rosulate; cauline leaves lanceolate or linear, margin subentire or denticulate. Racemes corymbose; petals yellow, narrowly spatulate, apex rounded. Fruit linear, moniliform. Fl. May-Aug. Fr. Jun-Sep. Dry beds or moist areas at 800-3000 m. Distributed in Heilongjiang, Jilin, Neimenggu and Xinjiang. Also in Mongolia, Korean Peninsula, Europe, Africa and North America.

小花糖芥 *Erysimum cheiranthoides*

白马芥

Baimashania pulvinata Al-Shehbaz

多年生草本。基生叶莲座状，卵形或长圆形，长2-4毫米，密被长柔毛，先端钝，基部近渐狭，边缘全缘；叶柄长2-5毫米，基部膨大至宽0.5-1毫米，具缘毛，宿存。花单生；花瓣粉色，匙形，爪长1.5-2毫米；花柱长0.4-1毫米。果条形，长4-8毫米；果瓣具纵条纹，无明显中脉。花期6-7月，果期7-8月。生海拔4200-4600米的湿润多石草地或石灰岩石缝。产云南。

Perennial herbs. Basal leaves rosulate, ovate or oblong, 2-4 mm long, densely pilose, apex obtuse, base subattenuate, margin entire; petiole 2-5 mm long, expanded base 0.5-1 mm wide, ciliate, persistent. Flowers solitary; petals pink, spatulate, claw 1.5-2 mm long; style 0.4-1 mm long. Fruit linear, 4-8 mm long; valves longitudinally striate, without a distinct midvein. Fl. Jun-Jul. Fr. Jul-Aug. Moist gravelly meadows or limestone rock crevices at 4200-4600 m. Distributed in Yunnan.

白马芥 *Baimashania pulvinata*

葱芥

Alliaria petiolata M. Bieb.

二年生草本。基生叶莲座状，结果时枯萎，边缘具圆齿或牙齿，叶柄长3-10(-16)厘米；茎叶具较短柄，卵形、心形或三角形，边缘具尖或钝齿。花瓣白色，倒披针形，基部渐狭为爪状。长角果四棱柱状或近圆柱形；果瓣无毛。花期4-6月，果期5-7月。生路边、田间、木林或河边。产新疆和西藏。原产东南亚和欧洲；中亚、南亚和俄罗斯亦有。

Biennial herbs. Basal leaves rosulate, withered by fruiting, margin crenate or dentate, petiole 3-10(-16) cm long; cauline leaves with much shorter petioles, ovate, cordate or deltoid, margin acutely to obtusely toothed. Petals white, oblanceolate, attenuate to clawlike base. Siliqua quadrangular or subterete; valves glabrous. Fl. Apr-Jun. Fr. May-Jul. Roadsides, fields, woodlands or river banks. Distributed in Xinjiang and Xizang. Native to SE Asia and Europe; also in C and S Asia, and Russia.

葱芥 *Alliaria petiolata*

外折糖芥 *Erysimum deflexum*

外折糖芥

Erysimum deflexum Hook. f. et Thom.

多年生草本，高2-15厘米。茎匍匐或直立。基生叶莲座状，宿存，线状披针形或长圆形，全缘或具齿，先端急尖；茎生叶极少或无。总状花序伞房状；花瓣黄色，匙形或倒卵形；花丝黄色。长角果线形，近圆柱状，念珠状。花期5-7月，果期7-8月。生海拔3700-5200米的高山碎石堆上。产新疆和西藏。印度亦有。

Perennial herbs, 2-15 cm tall. Stems decumbent or erect. Basal leaves rosulate, persistent, linear-lanceolate or oblong, margin entire or denticulate, apex acute; cauline leaves few or absent. Racemes corymbose; petals yellow, spatulate or obovate; filaments yellow. Fruit linear, subterete, torulose. Fl. May-Jul. Fr. Jul-Aug. Gravelly areas at 3700-5200 m. Distributed in Xinjiang and Xizang. Also in India.

红紫糖芥

Erysimum roseum (Maxim.) Polatsch.

多年生草本。茎少分枝；茎自基部单个伸出，多叶。基部叶丛生；叶长圆形、长圆状倒卵形、倒披针状线形或线形。总状花序伞房状，果期伸长，仅最下部花具苞片；花瓣粉红色或紫色。长角果线形，具4棱。花期5-7月，果期7-9月。生海拔3200-4900米的高山草甸或石灰岩砾石堆上。产云南、四川、西藏、甘肃和青海。

Herbs perennial. Caudex few branched; stems single from base, leafy. Basal leaves rosulate; leaves oblong, oblong-obovate, oblanceolate-linear or linear. Racemes corymbose, elongated in fruit, only lowermost flowers bracteates; petals pink or purple. Siliques linear, 4-angled. Fl. May-Jul. Fr. Jul-Sep. Alpine meadows or limestone screes at 3200-4900 m. Distributed in Yunnan, Sichuan, Xizang, Gansu and Qinghai.

沟子荠

Taphrospermum altaicum C. A. Mey.

草本，高4-25厘米，无毛。茎多分枝，直立。叶近圆形或近心形。总状花序密集多花，果期伸长；萼片长圆形，边缘膜质，宿存或早落；花瓣白色，倒卵形。短角果狭圆锥形，念珠状，基部具假隔膜。花期6-8月，果期7-9月。生海拔2000-4000米的山坡草甸或路旁。产甘肃、青海、新疆和西藏。中亚与俄罗斯亦有。

Herbs, 4-25 cm tall, glabrous. Stems much branched, erect. Leaves suborbicular or subcordate. Racemes densely flowered, elongated in fruit; sepals oblong, margin membranous, persistent or caducous; petals white, obovate. Fruit narrowly conical, torulose, angustiseptate at least basally. Fl. Jun-Aug. Fr. Jul-Sep. Mountain meadows or roadsides at 2000-4000 m. Distributed in Gansu, Qinghai, Xinjiang and Xizang. Also in C Asia and Russia.

泉沟子荠

Taphrospermum fontanum (Maxim.) Al-Shehbaz et G. Yang

草本，疏生柔毛。茎单生，分枝。叶卵形或长椭圆形，基部钝或楔形，全缘或具波状缘。总状花序密集多花；萼片长圆形，宿存，边缘膜质；花瓣白色或淡紫色，倒卵形或匙形，先端微缺；花丝白色或淡紫色。角果倒心形，具假隔膜。生海拔3200-5300米的矮灌木、受干扰的高山草甸或退化的高山牧场。产甘肃、青海、四川、新疆和西藏。

Herbs, sparsely to pubescent. Stems solitary, branched. Leaves ovate or oblong, base obtuse or

红紫糖芥 *Erysimum roseum*

沟子荠 *Taphrospermum altaicum*

泉沟子荠 *Taphrospermum fontanum*

cuneate, margin entire or repand. Racemes densely flowered; sepals oblong, persistent, margin membranous; petals white or lavender, obovate or satulate, apex slightly emarginated; filaments white or lavender. Fruit obcordate, angustiseptate. Dwarf bushes, disturbed alpine meadows or degraded alpine pastures at 3200-5300 m. Distributed in Gansu, Qinghai, Sichuan, Xinjiang and Xizang.

南山萮菜

Eutrema yunnanense Franch.

草本，高达80厘米。茎直立，单生，常部分自基部伸出。基生叶丛生；叶心形或肾形；果序疏松总状，无苞片；花瓣白色，长圆状匙形。角果长圆筒状。花期3-5月，果期4-6月。生海拔400-3500米的林下、山腰、草地或溪边。产中国西南、华中、西北和华东。

Herbs, to 80 cm tall. Stems erect, simple, often a few from base. Basal leaves rosulate; leaves cordate or reniform; infructescences lax raceme, ebracteate; petals white, oblong-spatulate. Siliques long-cylindrical. Fl. Mar-May. Fr. Apr-Jun. Forests, hillsides, grasslands or streamsides at 400-3500 m. Distributed in SW, C, NW and E China.

垂果大蒜芥

Sisymbrium heteromallum C. A. Meyer

一年生草本。茎直立，近基部疏或密被短柔毛，上部常无毛。基生叶莲座状，叶柄长1-3(-5)厘米；叶大头羽裂或倒向羽状深裂，顶端裂片披针形，边缘具齿。花瓣淡黄色，长3-5毫米。果瓣无毛，略隆起。花期5-8月，果期6-9月。生海拔900-4500米的石坡、路边、林中或草地等。产中国大部分省区。东北亚、印度、巴基斯坦和哈萨克斯坦亦有。

南山萮菜 *Eutrema yunnanense*

Annual herbs. Stems erect, sparsely to densely pubescent at least near base, usually glabrous above. Basal leaves rosulate, petiole 1-3(-5) cm long; blades lyrate- or runcinate-pinnatipartite, terminal lobe lanceolate, margin dentate. Petals pale yellow, 3-5 mm long. Valves glabrous, torulose. Fl. May-Aug. Fr. Jun-Sep. Rocky slopes, roadsides, forests or grassy areas at 900-4500 m. Distributed in most provinces of China. Also in NE Asia, India, Pakistan and Kazakhstan.

垂果大蒜芥 *Sisymbrium heteromallum*

尖果寒原荠 *Aphragmus oxycarpus*

尖果寒原荠

Aphragmus oxycarpus (J. D. Hooker et Thoms.) Jafri

多年生草本。基生叶无毛，边缘全缘；叶柄宿存，基部增粗至3毫米宽；茎叶和苞片与基生叶相似，但较窄，无柄或具短柄，向上依次变小。萼片常淡紫色；花瓣深紫色或白色，宽倒卵形或匙形，爪长1-2.5毫米。短角果披针形或椭圆形；果瓣脉不明显。花期5-8月，果期7-9月。生海拔3300-5600米的山顶草丛等。产华西部分省区。南亚和中亚亦有。

Perennial herbs. Basal leaves glabrous, margin entire; petioles persistent, base broadly expanded and to 3 mm wide; cauline leaves and bracts similar to basal leaves but narrower and sessile or shortly petiolate, reduced in size upward. Sepals often purplish; petals deep purple or white, broadly obovate or spatulate, claw 1-2.5 mm long. Silicula lanceolate or elliptic; valves obscurely veined. Fl. May-Aug. Fr. Jul-Sep. Alpine pastures at 3300-5600 m. Distributed in some provinces of W China. Also in S and C Asia.

密序山萮菜

Eutrema heterophyllum (W. W. Sm.) H. Hara

草本，高2-25厘米，全体无毛。茎直立，不分枝。基生叶莲座状，卵形或近圆形，全缘；茎生叶披针形、卵形或线状披针形。萼片倒卵形，宿存，长约3毫米；花瓣白色，匙状；花丝白色。角果线形或椭圆形；果瓣具明显中脉。花期6-7月，果期7-8月。生海拔2500-5400米的山坡草丛中。产中国西北、四川、云南和西藏。不丹、尼泊尔和中亚亦有。

Herbs, 2-25 cm tall, glabrous throughout. Stems erect, simple. Basal leaves rosulate, ovate or suborbicular, margin entire; cauline leaves lanceolate, ovate or linear-lanceolate. Sepals obovate, persistent, ca. 3 mm long; petals white, spatulate; filaments white. Fruit linear or oblong; valve with a prominent midvein. Fl. Jun-Jul. Fr. Jul-Aug. Alpine meadows at 2500-5400 m. Distributed in NW China, Sichuan, Yunnan and Xizang. Also in Bhutan, Nepal and C Asia.

密序山萮菜 *Eutrema heterophyllum*

蚓果芥

Neotorularia humilis (C. A. Meyer) Hedge et J. Léonard

多年生草本。茎基部分枝。基生叶丛生；叶阔匙形或狭长圆形。花序呈紧密伞房状；花瓣白色。长角果筒状，念珠状。花期4-6月。生海拔1000-4200米的林下、河滩或草地。产华北、华西和西北。中亚、俄罗斯(西伯利亚)、蒙古、朝鲜半岛和北美洲亦有。

Herbs perennial. Caudex basally branched. Basal leaves rosulate; leaves broadly spatulate or narrowly oblong. Inflorescences densely umbellate; petals white. Siliques tubular, moniliform. Fl. Apr-Jun. Forests, river beaches or grasslands at 1000-4200 m. Distributed in N, W and NW China.

蚓果芥 *Neotorularia humilis*

盐芥 *Thellungiella salsuginea*

Also in C Asia, Russia (Siberia), Mongolia, Korean Peninsula and North America.

盐芥

Thellungiella salsuginea (Pallas) O. E. Schulz

一年生草本。基生叶倒卵形、匙形或长圆形，边缘通常全缘，叶柄长5-10毫米；茎生叶无柄，基部深戟形抱茎，边缘全缘或波状。花瓣白色，倒卵形，长2-3毫米。长角果长0.7-1.6(-2)厘米，无柄；果瓣脉不明显。花果期4-7月。生盐渍化土壤、水沟旁和干草原。产内蒙古、新疆、吉林、河北、河南、江苏和山东。中亚、蒙古、俄罗斯和北美洲亦有。

Annual herbs. Basal leaves obovate, spatulate or oblong, margin usually entire, petiole 5-10 mm long; cauline leaves sessile, base deeply sagittate-amplexicaul, margin entire or repand. Petals white, obovate, 2-3 mm long. Siliqua 0.7-1.6(-2) cm long, sessile; valves obscurely veined. Fl. and fr. Apr-Jul. Saline flats, river banks or steppe. Distributed in Neimenggu, Xinjiang, Jilin, Hebei, Henan, Jiangsu and Shandong. Also in C Asia, Mongolia, Russia and North America.

小果亚麻荠

Camelina microcarpa Andrz

一年生草本。茎下部密被长硬毛。茎生叶披针形、狭长圆形或条状披针形，边缘全缘，稀具细牙齿。花瓣淡黄色，长3-4毫米。短角果倒梨形至狭倒梨形，先端急尖；果瓣中脉明显，侧脉不显著。花期4-6，果期5-8月。生海拔700-1600米的农田、林缘、路边或山坡。产中国东北、内蒙古、甘肃、河南、山东和新疆。俄罗斯、蒙古、中亚、东南亚和欧洲亦有。

Annual herbs. Stems densely hirsute basally. Cauline leaves lanceolate, narrowly oblong or linear-lanceolate, margin entire or rarely remotely denticulate. Petals pale yellow, 3-4 mm long. Silicula obpyriform to narrowly so, acute at apex; valves with a distinct midvein and less prominent lateral veins. Fl. Apr-Jun. Fr. May-Aug. Farms, forest margins, roadsides or mountain slopes at 700-1600 m. Distributed in NE China, Neimenggu, Gansu, Henan, Shandong and Xinjiang. Also in Russia, Mongolia, C and SE Asia, and Europe.

小果亚麻荠 *Camelina microcarpa*

播娘蒿
Descurainia sophia (L.) Webb ex Prantl

一年生草本，无腺体。茎直立，基部单生，上部常分枝。基生叶和最下部茎生叶二至三回羽状全裂，轮廓卵形或长圆形。萼片淡黄色，长圆状线形；花瓣黄色。果实狭线形。花果期4-6月。生海拔4200米以下的路边、荒野、田间或牧场。产中国各地，除广西、广东、海南和台湾外均有。亚洲、非洲北部和欧洲亦有；引种各地。

Herbs annual, eglandular. Stems erect, simple basally, often branched above. Basal and lowermost cauline leaves 2- or 3-pinnatisect, ovate or oblong in outline. Sepals yellowish, oblong-linear; petals yellow. Fruits narrowly linear. Fl. and fr. Apr-Jun. Roadsides, waste places, fields or pastures below 4200 m. Distributed throughout China except Guangxi, Guangdong, Hainan and Taiwan. Also in Asia, N Africa and Europe; introduced to elsewhere.

芹叶荠
Smelowskia calycina (Stephan) C. A. Meyer

多年生草本。基生叶一至二回羽状全裂，长1.5-8厘米，叶柄长(0.5-)1-5(-7)厘米；中间和上部茎生叶无柄或近无柄，较基生叶较小，较少分裂。花瓣白色或淡黄色，近圆形或倒卵形，先端圆，下部渐狭成长1-2.5毫米的爪。果瓣中脉明显。花期6-8月，果期7-9月。生海拔2500-4900米的石坡、石缝或高山草甸。产新疆。蒙古、俄罗斯、南亚、中亚和北美洲亦有。

Perennial herbs. Basal leaves 1- or 2-pinnatisect, 1.5-8 cm long, petioles (0.5-)1-5(-7) cm long; middle and upper cauline leaves sessile or subsessile, smaller and less divided than basal ones. Petals white or pale yellow, suborbicular or obovate, rounded at apex, narrowed to claw 1-2.5 mm long. Valves with a prominent midvein. Fl. Jun-Aug. Fr. Jul-Sep. Rocky slopes, rocky crevices or alpine meadows at 2500-4900 m. Distributed in Xinjiang. Also in Mongolia, Russia, S and C Asia, and North America.

播娘蒿 *Descurainia sophia*

芹叶荠 *Smelowskia calycina*

木犀草科 Resedaceae

黄木犀草
Reseda lutea L.

一年生或多年生草本。叶纸质，无柄或具短柄，3-5深裂或羽状分裂。花黄色或黄绿色，排列成顶生的总状花序；萼片通常6；花瓣通常6，有圆形的瓣爪；雄蕊12-20，子房1室，有3个合生的心皮。蒴果直立，圆筒形，有时卵形或近球形。种子肾形。花果期6-8月。产辽宁。欧洲、西亚和非洲北部亦有。

Herbs annual or perennial. Leaves 3-5-parted to pinnatifid, papery Flowers in terminal racemes, yellow to yellowish green; sepals 6, linear, unequal, shorter than pedicel; petals 6; stamens 12-20; carpels 3. Capsule erect, cylindric, sometimes ovoid to subglobose. Seeds reniform. Fl. and fr. Jun-Aug. Distributed in Liaoning. Also in Europe, W Asia and N Africa.

黄木犀草 *Reseda lutea*

辣木科 Moringaceae

辣木
Moringa oleifera Lam.

乔木。叶互生，三回羽状，长25-60厘米，小叶卵形、椭圆形或长圆形，长1-2厘米。圆锥花序腋生，长10-30厘米；花两性，两侧对称，白色，芳香。蒴果细长，长20-50厘米，直径1-3厘米，下垂。花果期全年。中国南部各省引种栽培，在海南逸为野生。原产印度。

Trees. Leaves alternative, 3-pinnate, 25-60 cm long, leaflets ovate, elliptic or oblong, 1-2 cm long. Inflorescences axillary panicles, 10-30 cm long; flowers bisexual, zygomorphic, white, aromatic. Capsule slender, 20-50 cm long, 1-3 cm diam. Fl. and fr. year round. Introduced to S China and naturalized in Hainan. Native to India.

辣木 *Moringa cleifera*

钟萼木科 Bretschneideraceae

伯乐树（钟萼木）
Bretschneidera sinensis Hemsl.

乔木，高10-20米。叶长25-75厘米；小叶7-15片。总状花序顶生，长20-36厘米，具褐色柔毛；花瓣白色至淡红色，后颜色逐渐变深，具红色条纹。蒴果椭圆状球形至近球形至卵球形或倒卵形。花期3-9月，果期8月至翌年4月。生海拔300-1700米的密林中。产中国西南、华南、东南和华中。泰国北部和越南北部亦有。

Trees, 10-20 m tall. Leaves 25-75 cm long; leaflets 7-15. Racemes terminal, 20-36 cm long, brown-villose; petals white to light red, becoming darker with age, red striate. Capsules ellipsoid-globose to subglobose to ovoid or obovoid. Fl. Mar-Sep. Fr. Aug to next Apr. Dense forests at 300-1700 m. Distributed in SW, S, SE and C China. Also in N Thailand and N Vietnam.

伯乐树（钟萼木） *Bretschneidera sinensis*

猪笼草科 Nepenthaceae

猪笼草
Nepenthes mirabilis (Lour.) Druce

直立或攀援草本。瓶状体狭卵形至近圆柱状，大小不一。总状花序；花被片4，红色至紫色，腹面被近圆形腺体；雄花雄蕊柱状；子房椭圆形，密被淡黄色柔毛或星状毛。蒴果栗色。花期3-12月，果期8-12月(翌年3月)。生海拔400米以下的沼泽地、山地、路边、灌丛、草地上或林下。产广东和海南。老挝、泰国、越南、柬埔寨至大洋洲亦有。

Herbs erect or climbing. Pitchers narrowly ovoid to subcylindric, variable in size. Racemes; tepals 4, red to purple, abaxially with suborbicular glands; male flowers stamens terete; ovary elliptic, densely pubescent with light yellow villi or stellate hairs. Capsules brown. Fl. Mar-Dec. Fr. Aug-Dec (next Mar). Swamps, mountains, roadsides, thickets, grassy places or forests below 400 m. Distributed in Guangdong and Hainan. Also in Laos, Thailand, Vietnam, Cambodia to Oceania.

茅膏菜科 Droseraceae

锦地罗
Drosera burmannii Vahl

一年生或二年生草本。茎短缩。叶基生，密集；托叶膜质；叶黄绿色或淡红色至紫色，楔形或倒卵形，下面无毛或被短腺毛，上面及边缘被淡红色头状黏腺毛。螺状聚伞花序1-3，腋生，花葶状，具花2-19朵；苞片戟形，小；花瓣通常白色，有时粉红色至紫红色，倒卵形。蒴果(5-)6瓣裂。花果期几全年。生海拔1500米以下的潮湿荫蔽处、田埂边及潮湿的土壤上。产中国东南、华南和西南。东亚、东南亚和澳大利亚亦有。

Annual or biannual herbs. Stem short. Leaves basal, dense; stipule scarious; leaf blades yellowish green or light red to reddish violet, cuneate or obovate, abaxially glabrous or with glandular hairs, adaxially and margin with red, capitated, sticky, glandular hairs. Cincinnus 1-3, axillary, scapiform, 2-19-flowered; bracts hastate, small; petals usually white, sometimes pink or reddish violet, obovate. Capsules (5-)6-valved. Fl. and fr. all year. Shaded wet places, ridges between rice fields below 1500 m. Distributed in SE, S and SW China. Also in E and SE Asia, and Australia.

茅膏菜
Drosera peltata Smith ex Willd.

多年生草本。茎直立或攀援，具分枝。叶无托叶；基生叶有或无，有时退化呈钻状；茎生叶互生，盾状，叶新月形至半圆形，下面无毛，上面及边缘被头状黏腺毛。螺状聚伞花序顶生，具花3-22朵；苞片楔形、倒披针形或近钻形；萼片无毛或被腺毛；花瓣通常白色，稀粉红色至红色。蒴果近球形，(2-)3(-5)瓣裂。花果期6-9月。生海拔3700米以下的疏林、灌丛下、山坡、草地、草甸上及溪边。产华东、华中、华南、西南和华西。东亚、东南亚和澳大利亚亦有。

Perennial herbs. Stems erect or climbing, branched distally. Stipule absent; basal leave present or absent, often reduced to subulate shape; cauline leaves alternate, peltate, lunate to semicircle, abaxially glabrous, adaxial-

猪笼草 *Nepenthes mirabilis*

锦地罗 *Drosera burmannii*

茅膏菜 *Drosera peltata*

ly and margin with capitated, sticky, glandular hairs. Cincinnus terminal, 3-22-flowered; bracts cuneate, oblanceolate or subulate; sepals glabrous or glandular; petals usually white, rare pink to red. Capsules subglobose, (2-)3(-5)-valved. Fl. and fr. Jun-Sep. Open forests, thickets, slopes, grasslands, meadows or streamsides below 3700 m. Distributed in E, C, S, SW and W China. Also in E, SE Asia and Australia.

圆叶茅膏菜

Drosera rotundifolia L.

多年生草本。茎短缩。叶基生，密集；托叶膜质；叶黄绿色、绿色至红色，圆形、近圆形或略呈肾形，下面无毛，上面及边缘被红色头状黏腺毛。螺状聚伞花序1-2，腋生，花葶状，具花3-30朵；苞片钻形至条形，小；花瓣白色或淡粉色，匙形。蒴果3-4瓣裂。花期夏季至秋季，果期秋季至冬季。生海拔1500米以下的潮湿林下或草甸、沼泽的潮湿向阳处。产华南、西北和东北。亚洲、欧洲中部和北部、北美洲亦有。

Perennial herbs. Stem short. Leaves basal, dense; stipule scarious; leaf blades yellowish-green, green to red, suborbicular or weakly reniform, abaxially glabrous, adaxially and margin with red, long capitated, sticky, glandular hairs. Cincinnus 1-2, axillary, scapiform, 3-30-flowered; bracts subulate to linear, small; petals white or tinged with pink, spatulate. Capsules 3-4-valved. Fl. summer to autumn. Fr. autumn to winter. Wet forests, wet meadows, bogs or wet sunny open places below 1500 m. Distributed in S, NW and NE China. Also in Asia, C and N Europe, and North America.

圆叶茅膏菜 *Drosera rotundifolia*

匙叶茅膏菜

Drosera spatulata Labill.

多年生草本。茎短缩。叶基生，密集；托叶膜质；叶黄绿色，倒卵形或匙形，下面无毛或疏被腺毛，上面及边缘被红色头状黏腺毛。螺状聚伞花序1-6，腋生，花葶状，具花10-20朵；苞片条形或倒披针形；花瓣紫红色或红色，倒卵形。蒴果3(-4)瓣裂。花果期3-9月。生海拔1000米以下的潮湿阳坡、草地及沼泽上。产中国东南、华南和西南。东亚、东南亚、南亚、澳大利亚和新西兰亦有。

Perennial herbs. Stem short. Leaves basal, dense; stipule scarious; leaf blades yellowish-green, obovate or spatulate, abaxially sparsely glandular haired or glabrous, adaxially and margin with red, capitated, sticky, glandular hairs. Cincinnus 1-6, axillary, scapiform, 10-20-flowered; bracts linear or oblanceolate. Capsules 3(-4)-valved. Fl. and fr. Mar-Sep. Wet sunny slopes, meadows, bogs below 1000 m. SE, S and SW China. Also in E, SE and S Asia, Australia and New Zealand.

匙叶茅膏菜 *Drosera spatulata*

景天科
Crassulaceae

落地生根
Bryophyllum pinnatum (L. f.) Oken

草本，高40-150厘米。羽状复叶具3-5小叶。花序顶生，圆锥状，具多花；花下垂；花萼浅裂，裂片卵形；花丝着生在花冠管基部。蓇葖果包被于花萼和花冠筒内。花期1-3月。栽培且归化于云南、广西、广东、台湾和福建。原产非洲；归化各地。

Herbs, 40-150 cm tall. Leaves pinnately compound with 3-5 leaflets. Inflorescences terminal, paniculate, many flowered; flowers pendulous; calyx lobed with ovate lobes; filaments inserted basally on corolla tubes. Follicles included in calyx and corolla tube. Fl. Jan-Mar. Cultivated and naturalized in Yunnan, Guangxi, Guangdong, Taiwan and Fujian; native to Africa; naturalized elsewhere.

匙叶伽蓝菜
Kalanchoe integra (Medik.) Kuntze

多年生草本。叶匙状长圆形，边缘不规则浅裂，稀近全缘，几无柄，抱茎。花序总状，长约10厘米，果期伸长；花冠黄色；裂片先端渐尖。花期5-8月。生海拔300米左右的疏林下。产云南、西藏、广东、台湾和福建。南亚和东南亚亦有。

匙叶伽蓝菜 *Kalanchoe integra*

Herbs perennial. Leaves spatulate-oblong, margin irregularly lobed, rarely subentire, subsessile, amplexicaul. Inflorescences cymose, ca. 10 cm long, elongated in fruit; corolla yellow; lobes acuminate at apex. Fl. May-Aug. Open forests at ca. 300 m. Distributed in Yunnan, Xizang, Guangdong, Taiwan and Fujian. Also in S and SE Asia.

瓦松
Orostachys fimbriata (Turcz.) A. Berger

二年生肉质草本。丛生叶条形，短。花序总状或圆锥形；花瓣红色或白色，披针状椭圆形；花药紫色。蓇葖果矩圆

落地生根 *Bryophyllum pinnatum*

瓦松 *Orostachys fimbriata*

形，顶端具细喙。种子多数，卵球形。花期8-9月，果期9-10月。生海拔1600米以下(甘肃和青海达3500米)的岩石、房顶或树干上。产中国东南、华北、华西、华东、西北和东北。俄罗斯、蒙古和朝鲜半岛亦有。

Herbs biennial, fleshy. Rosette leaves linear, short. Inflorescences racemose or conical; petals red or white, lanceolate-elliptic; anthers purple. Follicles oblong, apical beak slender. Seeds numerous, ovoid. Fl. Aug-Sep. Fr. Sep-Oct. Rocks on slopes, house roofs or mossy tree trunks below 1600 m (to 3500 m in Gansu and Qinghai). Distributed in SE, N, W, E, NW and NE China. Also in Russia, Mongolia and Korean Peninsula.

黄花瓦松

Orostachys spinosa (L.) Sweet

两年生肉质草本。基生叶丛生，密集，长圆形；附属物白色，近圆形，软骨质，顶端有长2-4毫米的刺；茎生叶顶端刺软骨质。花序顶生，穗状或总状；萼片有红斑；花瓣黄绿色；雄蕊微长于花瓣。蓇葖果直立，椭球状披针形。花期7-8月，果期9月。生海拔600-2900米的干山坡石隙。产中国西南、西北和东北。俄罗斯、蒙古和朝鲜半岛亦有。

Herbs biennial, fleshy. Rosette leaves crowded, oblong; appendage white, suborbicular, cartilaginous, apical spine 2-4 mm long; apical prickles of stem leaves cartilaginous. Inflorescences terminal, spicate or racemose; sepals red-spotted; petals yellowish green; stamens slightly longer than petals.

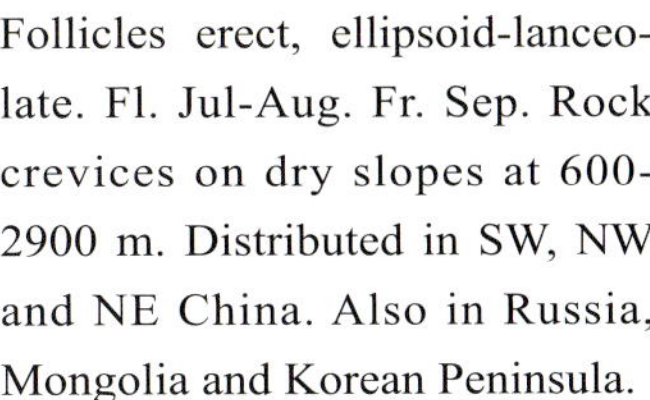
Follicles erect, ellipsoid-lanceolate. Fl. Jul-Aug. Fr. Sep. Rock crevices on dry slopes at 600-2900 m. Distributed in SW, NW and NE China. Also in Russia, Mongolia and Korean Peninsula.

圆叶八宝

Hylotelephium ewersii (Ledeb.) H. Ohba

多年生草本。根纤细。根状茎分枝，木质。茎多数，自基部分枝，淡紫褐色。叶对生，常具褐色斑点，阔卵形至近圆形。聚伞花序伞形。蓇葖果直立。花期7-8月。生海拔1800-2500米的林中或石缝中。产内蒙古、新疆和西藏。印度、巴基斯坦、阿富汗、俄罗斯、吉尔吉斯斯坦、塔吉克斯坦和蒙古亦有。

Herbs perennial. Roots slender. Rhizomes branched, woody. Stems many, branched near base, purplish brown. Leaves opposite, usually with brown spots, broadly ovate to suborbicular. Cymes umbellike. Follicles erect. Fl. Jul-Aug. Forests or rock crevices at 1800-2500 m. Distributed in Neimenggu, Xinjiang and Xizang. Also in India, Pakistan, Afghanistan, Russia, Kyrgyzstan, Tajikistan and Mongolia.

黄花瓦松 *Orostachys spinosa*

圆叶八宝 *Hylotelephium ewersii*

华北八宝 *Hylotelephium tatarinowii*

华北八宝

Hylotelephium tatarinowii
(Maxim.) H. Ohba

多年生草本。根块状，常有小形胡萝卜状的根。叶互生，条状披针形、披针形、狭倒披针形或长圆形，近具柄。聚伞花序平顶；花瓣浅红色；花药紫色；心皮直立，卵球状披针形。花期7-8月，果期9月。生海拔1000-3000米的石缝中。产河北、内蒙古和陕西。蒙古亦有。

Herbs perennial. Roots tuberous, usually with carrot-shaped rootlets. Leaves alternate, linear-lanceolate, lanceolate, narrowly oblanceolate or oblong, pseudopetiolate. Cymes flat topped; petals reddish; anthers purple; carpels erect, ovoid-lanceolate. Fl. Jul-Aug. Fr. Sep. Rock crevices at 1000-3000 m. Distributed in Hebei, Neimenggu and Shaanxi. Also in Mongolia.

八宝

Hylotelephium erythrostictum
(Miq.) H. Ohba

多年生草本。茎不分枝，直立。叶对生，无柄，长圆形至卵状长圆形，边缘疏具锯齿。聚伞花序平顶，密具花；花白色或粉红色；心皮近离生，直立，椭球形。花期8-10月。生海拔400-1800米的草地或沟谷边。产中国西南、华北、华东和东北。俄罗斯、朝鲜半岛和日本亦有。

Herbs perennial. Stems simple, erect. Leaves opposite, sessile, oblong to ovate-oblong, margin sparsely serrate. Cymes flat topped, densely flowered; petals white or pink; carpels nearly free, erect, ellipsoid. Fl. Aug-Oct. Grasslands or sides of ravines at 400-1800 m. Distributed in SW, N, E and NE China. Also in Russia, Korean Peninsula and Japan.

长药八宝

Hylotelephium spectabile
(Boreau) H. Ohba

多年生草本。叶对生或3枚轮生；叶狭椭圆状长圆形、长圆状卵形、卵形或阔卵形。聚伞花序密集；花瓣紫红色；雄蕊长于花冠；心皮狭椭球形；花药紫色。蓇葖果直立。花期8-9月，果期9-10月。生林缘或岩石山坡。产华北、华东和东北。朝鲜半岛亦有。

Herbs perennial. Leaves opposite or 3-verticillate, narrowly elliptic-oblong, oblong-ovate, ovate or broadly so. Cymes dense; petals purplish red; stamens longer

八宝 *Hylotelephium erythrostictum*

长药八宝 *Hylotelephium spectabile*

than corolla; carpels narrowly ellipsoid; anthers violet. Follicles erect. Fl. Aug-Sep. Fr. Sep-Oct. Forest edges or rocky slopes. Distributed in N, E and NE China. Also in Korean Peninsula.

德钦石莲

Sinocrassula techinensis (S. H. Fu) S. H. Fu

多年生草本，植株无毛或少有被极短毛。花茎单一，直立，高7-10厘米。茎生叶互生，线形或披针形，长1.5-2.5厘米，宽2-5毫米。伞房状花序；萼片披针形；花瓣5，红色，长圆形，先端钝至急尖；心皮直立，卵状披针形。蓇葖有多数种子。种子卵形，有条纹，有狭翅。花期9月。生海拔2700米的山坡石上。产云南西北部。

Perennial herbs, glabrous, rarely very shortly hairy. Flowering stem solitary, erect, 7-10 cm tall. Cauline leaves alternate, linear to lanceolate, 1.5-2.5cm × 2-5 mm. Inflorescences corymbiform; sepals lanceolate; petals 5, red, oblong, apex obtuse to acute; carpels erect, ovate-lanceolate. Follicles many seeded. Seeds ovoid, striate, narrowly winged. Fl. Sep. Rocks on slopes at 2700 m. Distributed in NW Yunnan.

石莲

Sinocrassula indica (Decne.) Berger

二年生草本，无毛。基生叶匙状长圆形；茎生叶宽倒披针形、近倒卵形或卵状圆形。花序圆锥状，常伞房状；花瓣5，红色；心皮卵球形。蓇葖果顶端具反折的喙。花期7-10月。生海拔500-4000米的山谷多石处、山坡或河边。产中国西南、华中和华西。印度、尼泊尔、不丹和巴基斯坦亦有。

Herbs biennial, glabrous. Basal leaves spatulate-oblong; stem leaves broadly oblanceolate, subobovate or ovate-orbicular. Inflorescences paniculate, often corymbiform; petals 5, red; carpels ovoid. Follicles apically with a recurved beak. Fl. Jul-Oct. Rocky places in valleys, slopes or by rivers at 500-4000 m. Distributed in SW, C and W China. Also in India, Nepal, Bhutan and Pakistan.

德钦石莲 *Sinocrassula techinensis*

石莲 *Sinocrassula indica*

山飘风 *Sedum majus*

山飘风

Sedum majus (Hemsl.) Migo

草本。茎单一，高约10厘米。叶2对，交互对生，但表现为4叶轮生，具假柄或近无柄。伞房状聚伞花序；花为不等的5基数；雄蕊10；鳞片长圆形；心皮直立。蓇葖果具少数种子。花期7-10月。生海拔1000-4300米的山坡林下石上。产云南、四川、西藏、湖北和陕西。

Herbs. Stems simple, ca. 10 cm tall. Leaves in 2 pairs, decussate, but appearing 4-verticillate, pseudopetiolate or subsessile. Cymes corymbiform; flowers unequally 5-merous; stamens 10; nectar scales oblong; carpels erect. Follicles few seeded. Fl. Jul-Oct. Rocks on forested slopes at 1000-4300 m. Distributed in Yunnan, Sichuan, Xizang, Hubei and Shaanxi.

火焰草

Sedum stellariifolium Franch.

一年生或二年生草本。茎单一或簇生。叶互生，三角形至阔三角状卵形。聚伞花序疏散；花瓣5，黄色；雄蕊10，短于花瓣；心皮近直立，长圆形。蓇葖果基部合生，顶部散开。花期6-8月，果期8-9月。生海拔400-3400米的山谷或山坡的土地或岩缝。产中国西南、东南、华中、华北、华西和东北。

Herbs annual or biennial. Stems solitary or tufted. Leaves alternate, triangular to broadly triangular-ovate. Cymes lax; petals 5, yellow; stamens 10, shorter than petals; carpels suberect, oblong. Follicles basally connate, apically divergent. Fl. Jun-Aug. Fr. Aug-Sep. Soils or rock crevices in valleys or on slopes at 400-3400 m. Distributed in SW, SE, C, N, W and NE China.

细叶景天

Sedum elatinoides Franch.

一年生草本，无毛。茎单生或丛生，高5-30厘米。3-6叶轮生，狭倒披针形，先端急尖，全缘。花序圆锥状或伞房状；5基数；萼片狭三角形至卵状披针形，先端近急尖；花瓣白色，披针状卵形，先端急尖；雄蕊10，较花瓣短；心皮近直立，下部合生。花期5-7月，果期8-9月。生海拔400-3400米的山坡石上。产云南、四川、湖北、陕西和甘肃。缅甸亦有。

Annual herbs, glabrous. Stems solitary or tufted, 5-30 cm tall. Leaves 3-6-verticillate, narrowly oblanceolate, apex acute, margin entire. Inflorescences paniculate or corymbiform; flowers unequally 5-merous; sepals narrowly triangular to ovate-lanceolate, apex subacute; petals white, lanceolate-ovate, apex acute; stamens 10, shorter than petals; carpels

火焰草 *Sedum stellariifolium*

细叶景天 *Sedum elatinoides*

山景天 *Sedum oreades*

宽萼景天 *Sedum platysepalum*

suberect, base connate. Fl. May-Jul. Fr. Aug-Sep. Rocks on slopes at 400-3400 m. Distributed in Yunnan, Sichuan, Hubei, Shaanxi and Gansu. Also in Myanmar.

山景天

Sedum oreades (Decne.) Hamet

一年生草本。叶互生，披针形至阔长圆形。聚伞花序伞房状，具1至数花；花瓣黄色，狭倒卵形或倒披针形，长6-13毫米；心皮直立，披针形；胚珠多数。花期7-8月，果期9-10月。生海拔3000-4500米的林下、灌丛、草地或岩隙。产云南和西藏。印度、巴基斯坦和马来西亚亦有。

Herbs annual. Leaves alternate, lanceolate to broadly oblong. Cymes corymbiform, 1- to several flowered; petals yellow, narrowly obovate or oblanceolate, 6-13 mm long; carpels erect, lanceolate; ovules many. Fl. Jul-Aug. Fr. Sep-Oct. Forests, thickets, grasslands or rock crevices at 3000-4500 m. Distributed in Yunnan and Xizang. Also in India, Pakistan and Malaysia.

宽萼景天

Sedum platysepalum Franch.

一年生或二年生草本。花茎自基部多分枝，高6-10厘米，近密生叶。叶线形至线状披针形，有宽距，先端急尖。花序密伞房状；花为不等的5基数；萼片长圆形，有距，先端渐尖；花瓣黄色，线状披针形，先端有短突尖头，基部稍狭而合生；雄蕊10。花期8-9月，果期10月。生海拔3200-4000米的林缘、高山顶部岩石上。产四川西南部和云南西北部。

Annual or biennial herbs. Flowering stems many branched from base, 6-10 cm tall, subleafy. Leaves linear to linear-lanceolate, base broadly spurred, apex acute. Inflorescences corymbiform, dense; flowers unequally 5-merous; sepals oblong, base spurred, apex acuminate; petals yellow, linear-lanceolate, apex shortly mucronate, base slightly narrowed and connate; stamens 10. Fl. Aug-Sep. Fr. Oct. Forest margins, rocks on alpine summits at 3200-4000 m. Distributed in SW Sichuan and NW Yunnan.

钝瓣景天

Sedum obtusipetalum Franch.

二年生草本，高5-10(-20)厘米。叶阔披针形至披针状长圆形。聚伞花序伞房状，密集；花为不等的5基数；花瓣黄色，长圆形，先端有近突尖头和乳头状突起；雄蕊10。蓇葖果具多粒种子。花期8-10月，果期9-11月。生海拔1900-3700米的疏林下岩隙或灌丛。产云南。尼泊尔亦有。

Herbs biennial, 5-10(-20) cm tall. Leaves broadly lanceolate to lanceolate-oblong. Cymes corymbiform, dense; flowers unequally 5-merous; petals yellow, oblong, apex subemarginate and mammillate; stamens 10. Follicles many seeded. Fl. Aug-Oct. Fr. Sep-Nov. Rock crevices in open forests or thickets at 1900-3700 m. Distributed in Yunnan. Also in Nepal.

钝瓣景天 *Sedum obtusipetalum*

大花景天 *Sedum magniflorum*

大花景天

Sedum magniflorum K. T. Fu

一年生草本，丛生，无毛。叶狭长圆形，侧脉每边3-5。伞序状聚伞花序，疏松，3-5花；苞片叶状；花瓣淡黄色；雄蕊10，2轮；心皮近直立，长圆形；胚珠10-12。花期8月，果期9月。生海拔约3800米的冷杉林下岩石上、杂草中或海滩石地。产云南西北部。

Herbs annual, tufted, glabrous. Leaves narrowly oblong, lateral veins 3-5 on each side. Cymes corymbiform, lax, 3-5-flowered; bracts foliaceous; petals yellowish; stamens 10, 2-whorled; carpels suberect, oblong; ovules 10-12. Fl. Aug. Fr. Sep. Rocks in *Abies* forests, among weeds or pebble beaches at ca. 3800 m. Distributed in NW Yunnan.

镘瓣景天

Sedum trullipetalum Hook. f. et Thomson

多年生草本，无毛。不育茎密丛生；花茎高2.8-8厘米。叶半长圆形至狭三角形，先端渐尖。花序伞房状；花几无花梗，为不等的5基数；萼片半长圆形或长卵形，先端渐尖；花瓣黄色，镘状，长6-10毫米，先端有小突尖头；雄蕊10。花期8-10月，果期10-11月。生海拔2700-4400米的山顶草地上。产四川、云南和西藏。尼泊尔和印度北部亦有。

Perennial herbs, glabrous. Sterile stems densely tufted; flowering stems 2.8-8 cm tall. Leaves suboblong to narrowly triangular, apex acuminate. Cymes corymbiform; flowers subsessile, unequally 5-merous; sepals suboblong or long-ovate, apex acuminate; petals yellow, trullate, 6-10 mm long, apex mucronate; stamens 10. Fl. Aug-Oct. Fr. Oct-Nov. Grassy meadows on alpine summits at 2700-4400 m. Distributed in Sichuan, Yunnan and Xizang. Also in Nepal and N India.

镘瓣景天 *Sedum trullipetalum*

藓茎景天

Sedum dugueyi Raym.-Hamet

多年生草本。不育茎直立，藓状；花茎通常单生。叶三角状披针形，中脉常红褐色。伞房状聚伞花序，具少花；花瓣黄色；心皮近长圆形。蓇葖果具10-12粒种子。花期9-10月，果期10-11月。生海拔2000-3600米的山谷多石地或山坡冷杉林下。产云南西北部和四川西北部。

Herbs perennial. Sterile stems erect, mosslike; flowering stems

藓茎景天 *Sedum dugueyi*

大炮山景天 *Sedum erici-magnusii*

usually solitary. Leaves triangular-lanceolate, midvein often reddish brown. Cymes corymbiform, few flowered; petals yellow; carpels suboblong. Follicles 10-12-seeded. Fl. Sep-Oct. Fr. Oct-Nov. Rocky places in valleys or *Abies* forests at 2000-3600 m. Distributed in NW Yunnan and NW Sichuan.

大炮山景天
Sedum erici-magnusii Fröd.

一年生草本。茎从基部分枝。叶长圆形。聚伞花序顶生或腋生，具1或2花；花为不等的4-5基数或不规则；花瓣离生，淡黄色；雄蕊8-10；心皮3或4，稀5，卵圆形。蓇葖果具种子4-8粒。花期8月，果期8-9月。生海拔3800-4900米的草场、山谷、沙滩或岩缝中。产甘肃。

Herbs annual. Stems branched from the base. Leaves oblong. Cymes terminal or axillary, 1- or 2-flowered; flowers unequally 4-5-merous or irregular; petals free, yellowish; stamens 8-10; carpels 3 or 4, rarely 5, ovoid. Follicles 4-8-seeded. Fl. Aug. Fr. Aug-Sep. Pastures, valleys, sandy beaches or rock crevices at 3800-4900 m. Distributed in Gansu.

轮叶景天
Sedum chauveaudii Raym.-Hamet

多年生草本，无毛。叶3数轮生，上面具锈色斑点，匙形。伞房状聚伞花序，疏松，有多数花；苞片叶形；花具短梗；花瓣5，黄色；雄蕊10。蓇葖果长圆形，具多粒种子。花期8-11月，果期10-12月。生海拔1700-3800米的冷杉林岩隙或山坡。产云南、四川和贵州。尼泊尔亦有。

Herbs perennial, glabrous. Leaves 3-whorled, adaxially brown spotted, spatulate. Cymes corymbose, lax, many flowered; bracts leaflike; flowers shortly pedicellate; petals 5, yellow; stamens 10. Follicles oblong, many seeded. Fl. Aug-Nov. Fr. Oct-Dec. Rock crevices of *Abies* forests or slopes at 1700-3800 m. Distributed in Yunnan, Sichuan and Guizhou. Also in Nepal.

费菜
Phedimus aizoon (L.) 't Hart

多年生草本。叶互生；叶缘具不整齐细齿。花序平展分枝，多花；苞片叶状；花瓣黄色；鳞片近四方形；雄蕊10；心皮卵状长圆形。蓇葖果星状。花期6-7月，果期8-9月。生海拔1000-3100米的灌丛中、山谷河边、田边或石隙。产中国西南、华中、华北、华西、华东和东北。

轮叶景天 *Sedum chauveaudii*

Herbs perennial. Leaves alternate; leaves margin irregularly serrate. Inflorescences horizontally branched, many-flowers; bracts leaflike; petals yellow; nectar scales subquadrangular; stamens 10; carpels ovoid-oblong. Follicles stellate. Fl. Jun-Jul. Fr. Aug-Sep. Scrub, ravine edges in valleys, field banks or rock crevices at 1000-3100 m. Distributed in SW, C, N, W, E and NE China.

费菜 *Phedimus aizoon*

吉林费菜 *Phedimus middendorffianus*

藓状景天 *Sedum polytrichoides*

吉林费菜

Phedimus middendorffianus (Maxim.) 't Hart

多年生草本。叶互生；叶条状匙形，边缘先端具细齿。花序常具多歧分枝，多花；花瓣黄色，披针形至条状披针形；雄蕊10；心皮幼时直立，后微弯曲。蓇葖果近横星形。花期6-8月，果期8-9月。生海拔300-1000米的山地林下多石地。产吉林和辽宁。俄罗斯、朝鲜半岛和日本亦有。

Herbs perennial. Leaves alternate; leaves linear-spatulate, margin apically serrate. Inflorescences often with divergent branches, many flowered; petals yellow, lanceolate to linear-lanceolate; stamens 10; carpels initially erect, later slightly curved. Follicles stellately subhorizontal. Fl. Jun-Aug. Fr. Aug-Sep. Among rocks in forests in mountain areas at 300-1000 m. Distributed in Jilin and Liaoning. Also in Russia, Korean Peninsula and Japan.

多茎景天

Sedum multicaule Wall. ex Lindl.

多年生草本。茎下部分枝。叶互生，覆瓦状排列；叶条形，基部有短距。聚伞花序具数个蝎尾状分枝；花无梗或具极短梗；花瓣黄色；心皮幼时直立，果期横展。花期7-8月，果期8-9月。生海拔1300-3500米的多石山坡、石隙或石壁上。产云南、四川、西藏、陕西和甘肃。印度亦有。

Herbs perennial. Stems basally branched. Leaves alternate, imbricate; linear, base shortly spurred. Cymes with several scorpioid branches; flowers sessile or very shortly pedicellate; petals yellow; carpels initially erect, horizontally spreading in fruit. Fl. Jul-Aug. Fr. Aug-Sep. Rocky slopes, rock crevices or rocky walls at 1300-3500 m. Distributed in Yunnan, Sichuan, Xizang, Shaanxi and Gansu. Also in India.

藓状景天

Sedum polytrichoides Hemsl.

多年生草本。有多数不育枝；茎斜升，丛生，细弱。叶互生；叶条形至条状披针形，基部具矩。聚伞花序具2-4分枝，多花；花具短梗；花瓣5，黄色。蓇葖果卵状矩圆形。花期7-8月，果期8-9月。生海拔约

多茎景天 *Sedum multicaule*

1000米的岩石山坡。产华西、华东、华中和东北。朝鲜半岛和日本亦有。

Perennial herbs. Sterile shoots numerous; stems ascending, tufted, slender. Leaves alternate, linear to linear-lanceolate, base spurred. Cymes 2-4-branched, many flowered; flowers shortly pedicellate; petals 5, yellow. Follicles ovoid-oblong. Fl. Jul-Aug. Fr. Aug-Sep. Rocky slopes at ca. 1000 m. Distributed in W, E, C and NE China. Also in Korean Peninsula and Japan.

佛甲草 *Sedum lineare*

珠芽景天 *Sedum bulbiferum*

佛甲草

Sedum lineare Thunb.

多年生草本。叶3(-4)轮生，条形，无柄。聚伞花序2或3分枝；分枝常再2分枝；花瓣黄色，披针形；鳞片5，宽楔形至近四方形。蓇葖果叉开，先端具短喙。花期4-5月，果期6-7月。生低山、草坡石上或平原。产中国西南、东南、华中、华西和华东。日本亦有。

Perennial herbs. Leaves 3(-4)-verticillate, linear, sessile. Cymes 2- or 3-branched; branches 2-forked; petals yellow, lanceolate; nectar scales 5, broadly cuneate to subquadrangular. Follicles divergent, apex shortly beaked. Fl. Apr-May. Fr. Jun-Jul. Low mountains, rocks on grassy slopes or plains. Distributed in SW, SE, C, W and E China. Also in Japan.

垂盆草

Sedum sarmentosum Bunge

多年生草本。茎匍匐，节处生根；3叶轮生。叶倒披针形至长圆形。三至五回的伞房状聚伞花序；花瓣黄色；心皮分叉，长圆形。花期5-7月，果期8月。生海拔1600米以下的阴地或多石山坡。产中国西南、东南、华中、华北、华西、华东和东北。

Herbs perennial. Stems creeping and rooting at nodes; leaves 3-verticillate. Leaves oblanceolate to oblong. Cymes 3-5-branched, corymbiform; petals yellow; carpels divergent, oblong. Fl. May-Jul. Fr. Aug. Shady places or rocky slopes below 1600 m. Distributed in SW, SE, C, N, W, E and NE China.

垂盆草 *Sedum sarmentosum*

珠芽景天

Sedum bulbiferum Makino

多年生草本。根须状。叶腋珠芽白色；叶匙状倒披针形。聚伞花序具3分枝，多花；常再二歧分枝；花瓣黄色，披针形；雄蕊10；鳞片倒卵形，长约0.6毫米。花期4-7月。生海拔1000米以下的山地树林下阴处或平原。产中国西南、华南、东南和华东。

Herbs perennial. Roots fibrous. Bulbils in leaf axils white; leaves spatulate-oblanceolate. Cymes 3-branched, many flowered; branches 2-forked; petals yellow, lanceolate; stamens 10; nectar scales obovate, ca. 0.6 mm long. Fl. Apr-Jul. Shady places under trees on mountains or plains below 1000 m. Distributed in SW, S, SE and E China.

凹叶景天 *Sedum emarginatum*

凹叶景天
Sedum emarginatum Migo

多年生草本。叶对生，匙状倒卵形至阔倒卵形，先端圆且具凹缺。聚伞花序通常3分枝；花瓣黄色，条状披针形至披针形；鳞片长圆形；心皮长圆形。蓇葖果叉开。花期5-6月，果期7月。生海拔600-1800米的阴湿山坡。产中国西南、华中、华西和华东。

Herbs perennial. Leaves opposite; leaves spatulate-obovate to broadly obovate, apex rounded and emarginate. Cymes usually 3-branched; petals yellow, linear-lanceolate to lanceolate; nectar scales oblong; carpels oblong. Follicles divergent. Fl. May-Jun. Fr. Jul. Shady and moist slopes at 600-1800 m. Distributed in SW, C, W and E China.

岷江景天
Ohbaea balfourii (Raym.-Hamet) V. V. Byalt et I. V. Sokolova

多年生草本，无毛。基生叶丛生，近长圆形至长圆形；茎生叶长圆形至披针形。花序伞房状蝎尾状，花序梗弓形；花瓣几离生，铜黄色，长圆形。蓇葖果具多粒种子。花期9月，果期10月。生海拔2700-4000米的灌丛、向阳草地、石坡或山谷石隙中。产云南和四川。

Herbs perennial, glabrous. Rosette leaves suboblong to oblong; stem leaves oblong to lanceolate. Inflorescences cymose-scorpioid, peduncles arcuate; petals free, brassy yellow, oblong. Follicles many seeded. Fl. Sep. Fr. Oct. Thickets, sunny grasslands, rocky slopes or rock crevices in valleys at 2700-4000 m. Distributed in Yunnan and Sichuan.

背药红景天
Rhodiola hobsonii (Prain ex Raym.-Hamet) S. H. Fu

多年生草本。茎基直立；花茎长5.5-13.5厘米。基生叶有长柄，倒披针形至长圆形，长达18毫米；茎生叶互生，狭倒披针形至长圆形，无柄，全缘。花序螺状聚伞状；5基数；萼片披针形；花瓣红色，卵形，边缘上部稍呈流苏状；雄蕊10。蓇葖直立，基部合生。花期7月。生海拔2600-4100米的林下、灌丛中和岩缝中。产西藏。不丹亦有。

Perennial herbs. Caudex erect; flowering stems 5.5-13.5 cm long. Basal leaves long petiolate, oblanceolate to oblong, to 18 mm long; cauline leaves alternate, narrowly oblanceolate to oblong, sessile, margin entire. Inflorescences helicoid cymose; flowers unequally 5-merous; sepals lanceolate; petals red, ovate, margin slightly fimbriate toward apex; stamens 10. Follicles erect, base connate. Fl. Jul. Forests, thickets or rock crevices at 2600-4100 m. Distributed in Xizang. Also in Bhutan.

矮生红景天
Rhodiola humilis (Hook. f. et Thomson) S. H. Fu

多年生草本。茎基不分枝，直立；花茎少数，长2.5厘米。基

岷江景天 *Ohbaea balfourii*

背药红景天 *Rhodiola hobsonii*

矮生红景天 *Rhodiola humilis*

报春红景天 *Rhodiola primuloides*

生叶宿存，线状倒披针形至线状菱形；茎生叶互生，线状椭圆形，全缘。花单生；5基数；萼片卵状长圆形，先端钝或急尖；花瓣卵形，向上渐狭；雄蕊10；与萼对生的与花瓣等长或稍短于花瓣；心皮直立，花柱短。花果期9月。生海拔3900-4500米的高山草甸。产西藏。印度北部亦有。

Perennial herbs. Caudex simple, erect; flowering stems few, to 2.5 cm long. Basal leaves persistent, linear-oblanceolate to linear-rhomboid; cauline leaves alternate, linear-elliptic, margin entire. Flowers solitary; flowers unequally 5-merous; sepals ovate-oblong, apex obtuse or acute; petals ovate, apex attenuate; stamens 10; antesepalous ones equaling or slightly shorter than petals; carpels erect, styles short. Fl. and fr. Sep. Alpine meadows at 3900-4500 m. Distributed in Xizang. Also in N India.

报春红景天

Rhodiola primuloides (Franch.) S. H. Fu

多年生草本。茎分枝，粗厚，叶密生；花茎退化，花仅1-2。茎生叶叶状；基生叶倒披针形、倒卵形或阔卵形。萼片狭披针形至卵状披针形；花瓣白色，顶端啮蚀状。花期10月。生海拔2500-4400米的多石山坡或山谷。产云南、西藏和青海。

Herbs perennial. Caudex branched, thick, densely leafy; Flowering stems much reduced, flowers 1-2. Caudex leaves leaf-like; rosette leaves oblanceolate, obovate or broadly ovate. Sepals narrowly lanceolate to ovate-lanceolate; petals white, apically erose. Fl. Oct. Rocky slopes or in valleys at 2500-4400 m. Distributed in Yunnan, Xizang and Qinghai.

四轮红景天

Rhodiola prainii (Raym.-Hamet) H. Ohba

多年生草本。花茎单生，直立，老的枝茎凋落，高达8厘米。茎生叶4枚，轮状着生，假柄长1-3厘米，基部急狭至长渐狭。伞房状花序；花两性，5基数；萼片狭三角状卵形；花瓣白色、粉色或红色，边缘啮蚀状；雄蕊10，较花瓣为短。花期7月，果期9月。生海拔2200-3600米的处山脚石缝中或河谷阔叶林石上。产西藏。尼泊尔和印度北部亦有。

Perennial herbs. Flowering stem simple, erect, old shoots and stems deciduous, to 8 cm tall. Cauline leaves 4, verticillate, pseudopetiole 1-3 cm long, base abruptly narrowed to long attenuate. Inflorescences corymbiform; flowers bisexual, unequally 5-merous; sepals narrowly triangular-ovate; petals white, pink or red, margin erose; stamens 10, shorter than petals. Fl. Jul. Fr. Sep. Rock crevices at bases of mountains or rocks in broad-leaved forests in valleys at 2200-3600 m. Distributed in Xizang. Also in Nepal and N India.

四轮红景天 *Rhodiola prainii*

托花红景天 *Rhodiola stapfii*

小丛红景天 *Rhodiola dumulosa*

托花红景天

Rhodiola stapfii (Raym.-Hamet) S. H. Fu

多年生草本。花茎不分枝，直立，高1.4-3.5厘米。茎中部有5-6叶轮生，卵形或卵状长圆形，基部急狭，全缘，先端稍钝。伞形聚伞花序；花单性，5基数；无苞片；萼片三角形，基部合生，先端钝；花瓣红色，倒卵形至长圆形，边缘稍啮蚀状；心皮基部合生。种子褐色，长圆形。花期8-9月。生海拔2900-5000米的山坡草地上。产西藏南部。不丹亦有。

Perennial herbs. Flowering stems simple, erect, 1.4-3.5 cm tall. Middle stem leaves 5-6-verticillate, ovate to ovate-oblong, base abruptly narrowed, margin entire, apex subobtuse. Inflorescences umbelliform cymes; flowers unisexual, unequally 5-merous; bracts absent; sepals triangular, base connate, apex obtuse; petals red, obovate to oblong, margin erose; carpels base connate. Seeds brown, oblong. Fl. Aug-Sep. Grassland slopes at 2900-5000 m. Distributed in S Xizang. Also in Bhutan.

小丛红景天

Rhodiola dumulosa (Franch.) S. H. Hu

多年生草本。茎分枝。茎生叶互生，无柄，条形至阔条形。聚伞花序具花4-7朵；花瓣5，直立，白色或红色；心皮直立，卵状长圆形。种子具狭翅。花期6-7月，果期8月。生海拔1600-4100米的山坡岩石上。产中国西南、华中、华北、华西、西北和东北。不丹和缅甸亦有。

Herbs perennial. Caudex branched; stem leaves alternate, sessile, linear to broadly so. Cymes with 4-7 flowers; petals 5, erect, white or red; carpels erect, ovoid-oblong. Seeds narrowly winged. Fl. Jun-Jul. Fr. Aug. Rocks on slopes at 1600-4100 m. Distributed in SW, C, N, W, NW and NE China. Also in Bhutan and Myanmar.

粗糙红景天

Rhodiola coccinea (Royle) Boriss. subsp. **scabrida** (Franch.) H. Ohba

多年生草本。茎生叶条形或披针状条形，先端渐尖至长渐尖。萼片披针形至三角状长圆形，先端锐尖至渐尖；花瓣红或黄色，披针形至阔长圆形；雄蕊比花瓣短。蓇葖果成熟时红色，卵球形至长圆状卵球形。花期6-7月，果期7-9月。生海拔2200-5300米的高山地区或岩石裂隙。产云南、四川和西藏。印度北部亦有。

Herbs perennial. Stem leaves linear-lanceolate or lanceolate, apex acuminate to long acuminate. Sepals lanceolate to triangular-oblong, apex acute to acuminate; petals red or yellow, lanceolate to broadly oblong; stamens shorter than petals. Follicles red when mature, ovoid to oblong-ovoid. Fl. Jun-Jul. Fr. Jul-Sep. Alpine regions or rock crevices at 2200-5300 m. Distributed in Yunnan, Sichuan and Xizang. Also in N India.

西藏红景天

Rhodiola tibetica (Hook. f. et Thomson) S. H. Fu

多年生草本。花茎长30厘米。茎生叶线形至狭卵形，基部宽三角形，全缘或有牙齿，先端长芒状渐尖。伞房状花序花紧密；花单性；萼片近长圆形；

粗糙红景天 *Rhodiola coccinea* subsp. *scabrida*

西藏红景天 *Rhodiola tibetica*

花瓣紫色至红色，椭圆状披针形；雄蕊10，长与花瓣略等或稍长；心皮直立，先端稍外弯。花期7-8月，果期9月。生海拔4100-5400米的山沟碎石坡或山沟边。产西藏。阿富汗、印度和巴基斯坦亦有。

Perennial herbs. Flowering stems to 30 cm long. Cauline leaves linear to narrowly ovate, base broadly triangular, margin entire or dentate, apex long acuminate and awn-shaped. Inflorescences corymbiform, compact; flowers unisexual; sepals suboblong; petals purple to red, elliptic-lanceolate; stamens 10, subequaling or somewhat longer than petals; carpels erect, apex recurved. Fl. Jul-Aug. Fr. Sep. Stony slopes on sides of ravines at 4100-5400 m. Distributed in Xizang. Also in Afghanistan, India and Pakistan.

长鞭红景天

Rhodiola fastigiata (Hook. f. et Thoms.) S. H. Fu

多年生草本。基生叶鳞状，三角形；茎生叶互生，条状长圆形、条状披针形、椭圆形或倒披针形。花序伞房状，密集；花单性；花瓣红色。蓇葖果先端稍反折。花期6-8月，果期9月。生海拔3500-5400米的多石山坡。产四川、西藏、青海和云南。印度北部亦有。

Herbs perennial. Rosette leaves scalelike, triangular; caudex leaves alternate, linear-oblong, linear-lanceolate, elliptic or oblanceolate. Inflorescences corymbiform, dense; flowers unisexual, petals red. Follicles apex recurved. Fl. Jun-Aug. Fr. Sep. Rocky slopes at 3500-5400 m. Distributed in Sichuan, Xizang, Qinghai and Yunnan. Also in N India.

美花红景天

Rhodiola calliantha (H. Ohba) H. Ohba

多年生草本，雌雄异株。茎基叶片淡棕色，鳞片状。茎生叶互生，具柄，菱状卵形或狭菱状卵形，长4-5.5厘米，宽1-2.2厘米，边缘上部有圆齿。花序顶生，复生聚伞花序；花单性；5基数；萼片三角形至卵状三角形；花瓣淡粉红色至紫色，狭倒卵形；雄蕊10；心皮直立，披针形。花期6月。生海拔3600米的阴坡岩石上。产西藏。尼泊尔亦有。

Perennial herbs, dioecious. Caudex leaves brownish, scalelike. Cauline leaves alternate, stalked, rhombic-ovate or narrowly so, 4-5.5 × 1-2.2 cm, margin apically crenate. Inflorescences terminal, compound corymbs; flowers unisexual; 5-merous; sepals triangular to ovate-triangular; petals pinkish to purple, narrowly obovate; stamens 10; carpels erect, lanceolate. Fl. Jun. Rocks on shady slopes at 3600 m. Distributed in Xizang. Also in Nepal.

长鞭红景天 *Rhodiola fastigiata*

美花红景天 *Rhodiola calliantha*

大花红景天

Rhodiola crenulata (Hook. f. et Thoms.) H. Ohba

多年生草本。茎少分枝，短。茎生叶鳞片状，倒披针形；叶椭圆形至近圆形。花序伞房状，多花；花瓣5，红色至淡紫红色，倒披针形。蓇葖果直立，干后红色。花期6-8月，果期8-9月。生海拔2800-5600米的草坡、岩隙、高山灌丛或碎石滩。产云南、四川、西藏和青海。印度北部亦有。

Herbs perennial. Caudex few branched, short. Caudex leaves scalelike, oblanceolate; leaves elliptic to subcircular. Inflorescences corymbiform, many flowered; petals 5, red to purplish red, oblanceolate. Follicles erect, red when dry. Fl. Jun-Aug. Fr. Aug-Sep. Grassy slopes, rock crevices, alpine thickets or screes at 2800-5600 m. Distributed in Yunnan, Sichuan, Xizang and Qinghai. Also in N India.

异齿红景天 *Rhodiola heterodonta*

大花红景天 *Rhodiola crenulata*

异齿红景天

Rhodiola heterodonta (Hook. f. et Thomson) Boriss.

多年生草本，雌雄异株。茎基分枝；花茎长30-40厘米。叶互生，边缘有粗锯齿。花序紧密伞房状，不具苞片；萼片线形；花单性，4基数；花瓣黄绿色，线形，长达7毫米；雄蕊8，带红色，长超出花瓣。蓇葖果直立。花期5-6月，果期7月。生海拔2800-4700米的山坡沟边冰积石中。产西藏和新疆。伊朗、阿富汗、巴基斯坦和蒙古亦有。

Perennial herbs, dioecious. Caudex branched; flowering stems 30-40 cm long. Leaves alternate, margin coarsely serrate. Inflorescences corymbiform, compact, ebracteate; sepals linear; flowers unisexual, 4-merous; petals greenish yellow, linear, 7 mm long; stamens 8, reddish, much longer than petals. Follicles erect. Fl. May-Jun. Fr. Jul. Slopes, sides of ravines or glacial rocks at 2800-4700 m. Distributed in Xizang and Xinjiang. Also in Iran, Afghanistan, Pakistan and Mongolia.

狭叶红景天

Rhodiola kirilowii (Regel) Maxim.

多年生草本。茎生叶对生或近轮生，无柄，条形至条状披针形。花单性，有时两性；花瓣绿色，淡黄绿色或红色；心皮直立。蓇葖果披针形，顶端喙反折。花期5-9月。生海拔2000-5600米的草坡或林缘。产中国西南、华北、华西和西北。

Herbs perennial. Stem leaves alternate or subverticillate, sessile, linear to linear-lanceolate. Flowers unisexual, sometimes bisexual; petals green, greenish yellow or red; carpels erect. Follicles lanceolate, apical beaks recurved. Fl. May-Sep. Grassy slopes or forest edges at 2000-5600 m. Distributed in SW, N, W and NW China.

狭叶红景天 *Rhodiola kirilowii*

紫绿红景天

Rhodiola purpureoviridis (Praeg.) S. H. Fu

多年生草本。茎生叶多数，互生，无柄，狭长圆披针形，下面有腺毛。花瓣带绿色或带红色，条状倒披针形至狭倒卵形；心皮直立，披针形。蓇葖果顶端喙反折。花期6-8月。生海拔2500-4100米的林中、林缘、草地或多石地。产云南、四川和西藏。

Herbs perennial. Stem leaves numerous, alternate, sessile, narrowly oblong-lanceolate-elliptic, abaxially glandular hairy. Petals greenish or reddish, linear-oblanceolate to narrowly obovate; carpels erect, lanceolate. Follicles apical beak recurved. Fl. Jun-Aug. Forests, forest edges, grasslands or stony places at 2500-4100 m. Distributed in Yunnan, Sichuan and Xizang.

粗茎红景天

Rhodiola wallichiana (Hook.) S. H. Fu

多年生草本。花茎少数。茎生叶披针形至线状披针形，两端渐狭，两侧上部各有1-3个疏锯齿。花序伞房状，顶生；花两性；萼片线形，先端钝；花瓣淡红色或淡绿色或黄白色，倒卵状椭圆形，先端钝；雄蕊10；心皮直立。蓇葖果披针形。花期8-9月，果期10月。生海拔2500-3800米的山坡石上或林下。产西藏和云南。不丹、缅甸、尼泊尔和印度北部亦有。

Perennial herbs. Flowering stems few. Cauline leaves lanceolate to linear-oblanceolate, attenuate on both ends, margin apically 1-3-serrate. Inflorescences corymbiform, terminal; flowers bisexual; sepals linear, apex obtuse; petals reddish, greenish or yellowish white, obovate-elliptic, apex obtuse; stamens 10; carpels erect. Follicles lanceolate. Fl. Aug-Sep. Fr. Oct. Rocks on slopes or forests at 2500-3800 m. Distributed in Xizang and Yunnan. Also in Bhutan, Myanmar, Nepal and N India.

紫绿红景天 *Rhodiola purpureoviridis*

粗茎红景天 *Rhodiola wallichiana*

圣地红景天 *Rhodiola sacra*

圣地红景天
Rhodiola sacra (Prain ex Raym.-Hamet) S. H. Fu

多年生草本。茎基短；花茎少数，不分枝，直立，高8-16厘米。茎生叶互生，倒卵形或倒卵状长圆形，边缘有4-5个浅裂。伞房状花序花少数；两性；5基数；萼片狭披针状三角形；花瓣白色，狭长圆形，全缘或略啮蚀状；雄蕊10，花丝淡黄色，花药紫色。蓇葖果直立。花期8-9月，果期9月。生海拔2700-5000米的山坡石缝中。产青海和西藏。尼泊尔亦有。

Perennial herbs. Caudex short; flowering stems few, simple, erect, 8-16 cm tall. Cauline leaves alternate, obovate or obovate-oblong, margin with 4-5 dentate lobes on each side. Inflorescences corymbiform, few flowered; flowers bisexual, unequally 5-merous; sepals narrowly lanceolate-triangular; petals white, narrowly oblong, margin entire or somewhat erose; stamens 10; filaments yellowish, anthers purple. Follicles erect. Fl. Aug-Sep. Fr. Sep. Rock crevices on slopes at 2700-5000 m. Distributed in Qinghai and Xizang. Also in Nepal.

六叶红景天
Rhodiola sexifolia S. H. Fu

多年生草本。花茎斜上，高18厘米。6叶轮生，椭圆状长圆形，羽状浅裂；裂片全缘或有疏牙齿，先端钝，基部楔形，无柄，被微乳头状突起。伞房状花序；花两性，5基数；萼片线形；花瓣披针形，先端急尖；雄蕊10，较花瓣稍短。蓇葖果直立，分离几至基部。果期9月。生海拔3500-4100米的云杉林石上或山坡上。产西藏。

Perennial herbs. Flowering stems ascending, 18 cm tall. Leaves 6-verticillate, elliptic-oblong, margin pinnately lobed; lobes margin entire or remotely dentate, apex obtuse, base cuneate, sessile, finely to minutely mammillate. Inflorescences corymbiform; flowers bisexual, unequally 5-merous; sepals linear; petals lanceolate, apex acute; stamens 10, slightly shorter than petals. Follicles erect, free almost to base. Fr. Sep. Rocks in *Picea* forests or slopes at 3500-4100 m. Distributed in Xizang.

柴胡红景天
Rhodiola bupleuroides (Wall. ex Hook. f. et Thoms.) S. H. Fu

多年生草本。叶长圆状卵形、卵形至近圆形，边缘不反卷，无柄。花序顶生，伞房状，具7-100花；苞片叶状；花单性；花瓣5，暗紫红色。种子10-16。花期6-8月，果期8-9月。生海拔2400-5700米的林下、灌丛或岩石裂隙。产云南、四川和西藏。

六叶红景天 *Rhodiola sexifolia*

柴胡红景天 *Rhodiola bupleuroides*

羽裂红景天 *Rhodiola pinnatifida*

Herbs perennial. Leaves oblong-ovate, ovate to subcircular, margin never revoluted, sessile. Inflorescences terminal, corymbiform, 7-100-flowered; bracts leaflike; flowers unisexual; petals 5, dark purplish red. Seeds 10-16. Fl. Jun-Aug. Fr. Aug-Sep. Forests, thickets or rock crevices at 2400-5700 m. Distributed in Yunnan, Sichuan and Xizang.

羽裂红景天
Rhodiola pinnatifida Boriss.

多年生草本。花茎少，直立。基生叶鳞状；茎生叶互生，近对生或者几乎3叶轮生，条状披针形至披针形。花单性，雄株不等4基数；花瓣黄色；雄蕊8。蓇葖果长圆披针形，长达8毫米，顶端有喙。花期6-8月，果期7-8月。生海拔约2200米处。产新疆北部。蒙古和俄罗斯亦有。

Herbs perennial. Flowering stems few, erect. Rosette leaves scale-like. Stem leaves alternate, subopposite or nearly 3-verticillate, linear-lanceolate to lanceolate. Flowers unisexual, unequally 4-merous in male plant; petals yellow; stamens 8. Follicles oblong-lanceolate, to 8 mm long, with apex beak. Fl. Jun-Aug. Fr. Jul-Aug. At ca. 2200 m. Distributed in N Xinjiang. Also in Mongolia and Russia.

云南红景天
Rhodiola yunnanensis (Franch.) S. H. Fu

多年生草本。花茎单一或少数几条。茎生叶3叶轮生，稀对生。花序聚伞圆锥状，分枝轮状；花多数，单性；雄花浅黄绿色至黄色；雌花紫色；心皮喙反折。蓇葖果星状。花期5-7月，果期6-8月。生海拔1000-4000米的山坡林中。产中国西南、华北和华西。印度和缅甸亦有。

Herbs perennial. Flowering stems solitary or few. Stem leaves 3-verticillate, rarely opposite. Inflorescences cymose-paniculate, branches verticillate; flowers numerous, unisexual; petals of male flowers greenish yellow to yellow; those of female flowers purple; carpel beaks reflexed. Follicles stellate. Fl. May-Jul. Fr. Jun-Aug. Forests on slopes at 1000-4000 m. Distributed in SW, N and W China. Also in India and Myanmar.

菊叶红景天
Rhodiola chrysanthemifolia (H. Lévl.) S. H. Fu

多年生草本。叶聚生茎的顶端；叶长圆形、卵状长圆形或卵形，边缘羽状分裂。伞房花序，紧密；花两性；苞片圆匙形；花瓣长圆状卵形。蓇葖果直立，披针形。花期8月，果期9-10月。生海拔3200-4200米的高山岩石裂隙。产云南和四川。

Herbs perennial. Leaves aggregate toward stem apex, oblong, ovate-oblong or ovate, margin pinnately lobed. Inflorescences corymbiform, compact; flowers bisexual; bracts orbicular-spatulate; petals oblong-ovate. Follicles erect, lanceolate. Fl. Aug. Fr. Sep-Oct. Alpine rock crevices at 3200-4200 m. Distributed in Yunnan and Sichuan.

云南红景天 *Rhodiola yunnanensis*

菊叶红景天 *Rhodiola chrysanthemifolia*

虎耳草科
Saxifragaceae

扯根菜
Penthorum chinense Pursh

多年生草本。叶互生，披针形至狭披针形，先端渐尖，边缘具细重锯齿，无毛。聚伞花序具多花，被褐色腺毛；花黄白色；萼片5；无花瓣；雄蕊10；心皮5(-6)，下部合生；胚珠多数；花柱较粗。蒴果红紫色。花果期7-8月。生海拔90-2200米的林下、灌丛草甸及水边。除新疆、西藏和青海，几遍全国。东南亚、俄罗斯、日本、蒙古和朝鲜半岛亦有。

大叶子 *Astilboides tabularis*

Perennial herbs. Leaves alternate, lanceolate to narrowly so, apex acuminate, margin doubly serrulate, glabrous. Cymes many flowered, brown glandular hairy; flowers yellow-white; sepals 5; petals absent; stamens 10; carpels 5(-6), connate near base; ovules numberous; style thick. Capsule red-purple. Fl. and fr. Jul-Aug. Forests, scrub meadows and wet places along rivers at 90-2200 m. Distributed almost throughout China, except Xinjiang, Xizang and Qinhai. Also in SE Asia, Russia, Japan, Mongolia and Korean Peninsula.

扯根菜 *Penthorum chinense*

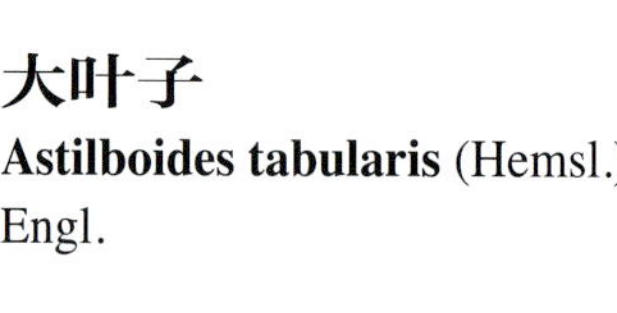

大叶子
Astilboides tabularis (Hemsl.) Engl.

多年生草本，高1-1.5米。叶柄具刺状腺毛；基生叶1，盾状着生，掌状浅裂。圆锥花序长15-20厘米，多花；花瓣4或5，白色或淡紫色；雄蕊(6-)8；子房半下位，2-4室。蒴果2-4裂。种子有翅。花期8-9月。生山坡林中或山谷。产吉林和辽宁。朝鲜半岛亦有。

Herbs perennial, 1-1.5 m tall. Petioles spinose-glandular-hairy; leaves peltate, palmatilobed. Panicles 15-20 cm long, many flowered; petals 4 or 5, white or lilac; stamens (6-)8; ovary semi-inferior, 2-4-loculed. Capsules 2-4-valved. Seeds winged. Fl. Aug-Sep. Forests on slopes or valleys. Distributed in Jilin and Liaoning. Also in Korean Peninsula.

七叶鬼灯檠
Rodgersia aesculifolia Batalin

多年生草本。茎具棱，近无毛。掌状复叶；小叶5-7，倒卵形至倒披针形。多歧聚伞花序圆锥状，长约26厘米；萼片4或5，外面和边缘被疏毛和腺毛；子房半下位；花柱2。蒴果卵球形，有喙。花果期5-10月。生海拔1100-3800米的林下、灌丛、草甸或石缝。产华中和西南。

Herbs perennial. Stems angular, subglabrous. Leaves palmately compound; leaflets 5-7, obovate to oblanceolate. Pleiochasium

七叶鬼灯檠 *Rodgersia aesculifolia*

paniculate, ca. 26 cm long; sepals 4 or 5, abaxially and marginally pilose and shortly glandular-hairy; ovary semi-inferior; styles 2. Capsules ovoid, rostrate. Fl. and fr. May-Oct. Forests, scrub areas, meadows or rock clefts at 1100-3800 m. Distributed in C and SW China.

西南鬼灯檠

Rodgersia sambucifolia Hemsl.

多年生草本。茎无毛。羽状复叶；小叶3-9(-10)，倒卵形或长圆形至披针形，腹面被糙伏毛。多歧聚伞花序圆锥状，长13-38厘米；无花瓣；心皮2，近轴处合生；子房半下位；花柱2。花果期5-10月。生海拔1800-3700米的林下、灌丛、草甸或岩石裂隙。产云南、四川和贵州。

Herbs perennial. Stems glabrous. Leaves pinnately compound; leaflets 3-9(-10), obovate or oblong to lanceolate, abaxially strigous. Pleiochasium paniculate, 13-38 cm long; petals absent; carpels 2, proximally connate; ovary semi-inferior; styles 2. Fl. and fr. May-Oct. Forests, thickets, meadows or rock crevices at 1800-3700 m. Distributed in Yunnan, Sichuan and Guizhou.

羽叶鬼灯檠

Rodgersia pinnata Franch.

多年生草本。茎无毛。叶近羽状复叶；基部和近轴叶常具6-9小叶；小叶椭圆形或长圆形至狭倒卵形，边缘具重锯齿。多歧聚伞花序圆锥状，长12-31厘米，多花；无花瓣；心皮2，基部贴生；子房半下位；花柱2。蒴果紫色。花果期6-8月。生海拔2000-3800米的林下、灌丛、高山草甸或石缝。产云南、四川和贵州。

Herbs perennial. Stems glabrous Leaves subpinnately compound; basal and proximal cauline leaves usually with 6-9 leaflets; leaflets elliptic or oblong to narrowly obovate, margin doubly serrate. Pleiochasium paniculate, 12-31 cm long, many flowered; petals absent; carpels 2, connate at base; ovary semi-inferior styles 2. Capsules purple. Fl. and fr. Jun-Aug. Forests, scrub areas, alpine meadows or rock clefts at 2000-3800 m. Distributed in Yunnan, Sichuan and Guizhou.

羽叶鬼灯檠 *Rodgersia pinnata*

西南鬼灯檠 *Rodgersia sambucifolia*

落新妇

Astilbe chinensis (Maxim.) Franch. et Savat.

多年生草本。基生叶为二至三回三出羽状复叶。圆锥花序密具花；近轴分枝长4-11.5厘米，密具褐色长卷毛；花瓣5，淡紫色至紫色；心皮2，基部贴生。花果期6-9月。生海拔400-3600米的溪边、林下、路边、林缘或草地。产中国西南、华南、华中、华北、华西、华东和东北。俄罗斯、朝鲜半岛和日本亦有。

Herbs perennial. Basal leaves 2-3 ternately compound. Panicles densely flowered; proximal branches 4-11.5 cm long, densely brown long crisped hairy; petals 5, lilac to purple, carpels 2, base connate. Fl. and fr. Jun-Sep. By streams, forests, forest edges, riversides or grasslands at 400-3600 m. Distributed in SW, S, C, N, W, E and NE China. Also in Russia, Korean Peninsula and Japan.

落新妇 *Astilbe chinensis*

大落新妇

Astilbe grandis Stapf ex E. H. Wils

草本。叶二或三回三出羽状复叶。圆锥花序顶生，长16-40厘米；分枝具腺毛；花瓣5，白色或紫色，线形，具单脉，先端锐尖；心皮2，基部连合；子房半下位。花果期6-9月。生海拔400-2000米的林中、灌丛或潮湿峡谷。产中国西南、华南、华北、华东和东北。朝鲜半岛亦有。

大落新妇 *Astilbe grandis*

Herbs. Leaves 2 or 3 ternately to pinnately compound. Panicles terminal, 16-40 cm long; branches glandular hairy; petals 5 white or purple, linear, 1-veined, apex acute; carpels 2, base connate; ovary semi-inferior. Fl. and fr. Jun-Sep. Forests, thickets or damp ravines at 400-2000 m. Distributed in SW, S, E and NE China. Also in Korean Peninsula.

槭叶草

Mukdenia rossii (Oliv.) Koidz.

多年生草本。叶均基生，长10-14.3厘米，掌状5-7(-9)浅裂至深裂，裂片近卵形，先端急尖，边缘有锯齿，两面均无毛；叶柄长7-15.5厘米，无毛。多歧聚伞花序具多花；花梗与萼筒均被黄褐色腺毛；萼片长3-5毫米，单脉；花瓣披针形，长约2.5毫米，单脉；心皮2，下部合生，子房半下位。花果期5-7月。生山谷石隙。产吉林和辽宁。朝鲜半岛亦有。

Perennial herbs. Leaves all basal, 10-14.3 cm long, palmately 5-7(9) lobed to parted, lobes subovate, apex acute, margin serrate, both surfaces glabrous; petiole 7-15.5 cm long, glabrous. Pleiochasium many flowered; pedicels and hypanthium yellow-brown glandular hairy; sepals 3-5 mm long, 1-veined; petals lanceolate, ca. 2.5 mm long, 1-veined; carpels 2, lower part connate, ovary semi-inferior. Fl. and fr. May-Jul. Rocky slopes, ravines. Distributed in Jilin and Liaoning. Also in Korean Peninsula.

槭叶草 *Mukdenia rossii*

独根草 *Oresitrophe rupifraga*

独根草

Oresitrophe rupifraga Bunge

多年生草本。叶基生，2-3枚；叶背和边缘被腺毛，基部心形，边缘具不规则锯齿。花序长5-16厘米，密被腺毛，多花；萼片卵形至狭卵形，不等大，无毛，具多脉，先端锐尖至短渐尖。花期5-9月。生海拔600-2100米的深谷、峭壁或岩石缝中。产河北、山西和辽宁。

Perennial herbs. Leaves 2 or 3; leaves abaxially and marginally glandular hairy, base cordate, margin irregularly dentate. Inflorescences 5-16 cm long, densely glandular hairy, many flowered; sepals ovate to narrowly so, unequal, glabrous, many veined, apex acute to shortly acuminate. Fl. May-Sep. Ravines, cliffs or rock crevices at 600-2100 m. Distributed in Hebei, Shanxi and Liaoning.

厚叶岩白菜

Bergenia crassifolia (L.) Fritsch

多年生草本。叶基生，革质。花序聚伞状；花梗和总花梗被腺毛；托杯外面被腺毛；萼片直立，外面被腺；花瓣紫色，基部具爪；花柱2。蒴果。花期5-9月。生海拔1100-1800米的林下或阴山坡石隙。产新疆。俄罗斯、蒙古北部和朝鲜半岛北部亦有。

Perennial herbs. Leaves all basal, leathery. Inflorescences cymose; pedicels and peduncles glandular; hypanthium outside glandular; sepals erect, outside glandular; petals purple, base narrowed into a claw; styles 2. Capsules. Fl. May-Sep. Forests or rock crevices on shaded slopes at 1100-1800 m. Distributed in Xinjiang. Also in Russia, N Mongolia and N Korean Peninsula.

厚叶岩白菜 *Bergenia crassifolia*

岩白菜 *Bergenia purpurascens*

岩白菜

Bergenia purpurascens (Hook. f. et Thomson) Engl.

多年生草本。叶基生，革质，两面无毛且具腺点。花序聚伞状；分枝和花梗均密被具长柄的腺毛；花萼狭卵形，先端钝；花瓣紫色，阔卵形，脉多数，基部渐狭成爪。花果期5-10月。生海拔2700-4800米的林下、灌丛、高山草甸或岩石上。产云南、四川和西藏。印度、尼泊尔、不丹和缅甸亦有。

Perennial herbs. Leaves all basal, leathery, both surfaces glabrous and glandular pitted. Inflorescences cymose; branches and pedicels densely long glandular hairy; sepals narrowly ovate, apex obtuse; petals purple, broadly ovate, veins many, base narrowed into a claw. Fl. and fr. May-Oct. Forests, thickets, alpine meadows or rocks at 2700-4800 m. Distributed in Yunnan, Sichuan and Xizang. Also in India, Nepal, Bhutan and Myanmar.

峨眉岩白菜

Bergenia emeiensis C. Y. Wu

多年生草本。叶全部基生；叶柄鞘基部边缘有硬纤毛，有时顶端带腺；叶狭倒卵形，革质，无毛。花序聚伞状；分枝和花梗疏具近无柄腺点；花萼开展，近卵形；花瓣白色或淡红色，窄倒卵形，基部渐狭成爪。蒴果。花期5-9月。生海拔1600-4200米林中、阴山坡或石隙。产四川。

Perennial herbs. Leaves all basal; sheath of petiole base rigidly ciliate at margin, cilia sometimes glandular tipped; leaves narrowly obovate, leathery, glabrous. Inflorescences cymose; branches and pedicels sparsely subsessile glandular; sepals spreading, subovate; petals white or reddish, narrowly obovate, base gradually narrowed into a claw.Capsules. Fl. May-Sep. Forests, shaded slopes or rock crevices at 1600-4200 m. Distributed in Sichuan.

短柄岩白菜

Bergenia stracheyi (Hook. f. et Thoms.) Engl.

多年生草本。叶倒卵形，边缘有锯齿或重锯齿，基部楔形或稍圆。花序分枝，花梗和托杯外面均被腺毛；萼片先端无齿，边缘有锯齿状硬睫毛，背面被腺毛；花瓣红色，近匙形，基部渐狭成爪。花果期6-10月。生海拔3900-4500米的林下或岩坡石隙。产西藏。印度、尼泊尔、巴基斯坦、阿富

峨眉岩白菜 *Bergenia emeiensis*

短柄岩白菜 *Bergenia stracheyi*

涧边草 *Peltoboykinia tellimoides*

汗、塔吉克斯坦、俄罗斯和克什米尔地区亦有。

Perennial herbs. Leaves obovate, margin serrate or doubly so, base cuneate or rounded. Inflorescences branches, pedicels and hypanthium abaxially long glandular hairy; sepals toothless at apex, margin denticulate-ciliate, abaxially glandular hairy; petals red, subspatulate, base gradually narrowed into claw. Fl. and fr. Jun-Oct. Forests, rock crevices at 3900-4500 m. Distributed in Xizang. Also in India, Nepal, Pakistan, Afghanistan, Tajikistan, Russia and Kashmir.

涧边草
Peltoboykinia tellimoides (Maxim.) Hara

多年生草本。基生叶圆状心形，直径15-25厘米，掌状7-9浅裂，边缘有不规则粗锯齿，叶柄长20-35厘米；茎生叶与基生叶同形，较小。聚伞花序顶生；苞片线形，长1-3毫米；萼片背面疏生腺毛；花瓣淡黄色，近顶部具疏齿，疏生腺毛；子房半下位，2室，花柱2。蒴果疏生腺毛。花果期7-10月。生海拔1100-1900米的沟谷林下阴湿处。产福建。日本亦有。

Perennial herbs. Basal leaves rounded-cordate, 15-25 cm in diam, palmately 7-9-lobed, margin irregularly coarsely serrate, petiole 20-35 cm long; cauline leaves same shape with basal leaves, much smaller. Cymes terminal; bracts linear, 1-3 mm long; sepals sparsely glandular hairy abaxially; petals pale yellow, margin sparsely dentate distally, sparsely glandular hairy; ovary semi-inferior, 2-locular, style 2. Capsules sparsely glandular hairy. Fl. and fr. Jul-Oct. Shaded places in ravines and forest understories at 1100-1900 m. Distributed in Fujian. Also in Japan.

双喙虎耳草
Saxifraga davidii Franch.

多年生草本。茎和花梗被白色腺柔毛。叶均基生，倒卵形，先端钝，基部楔形，边缘具钝齿，两面和边缘均被腺柔毛，具羽状达缘脉序。多歧聚伞花序总状或圆锥状，具7-30花；萼片反曲，无毛，3脉；花瓣白色，基部具1黄色斑点，3-4脉；花丝棒状；两心皮近离生，子房近上位。花期4-5月。生海拔1500-2400米的山沟石隙。产四川西部。缅甸亦有。

Perennial herbs. Stems and pedicels white glandular villous. Leaves all basal, obovate, apex obtuse, base cuneate, margin crenate, both surfaces and margin glandular pubescent, pinnately craspedodroma. Pleiochasium racemose or paniculate, 7-30-flowered; sepals reflexed, glabrous, veins 3; petals white, with a yellow spot at base, 3-4-veined; filaments clavate; 2 carpels nearly free, ovary subsuperior. Fl. Apr-May. Rock crevices in ravines at 1500-2400 m. Distributed in W Sichuan. Also in Myanmar.

双喙虎耳草 *Saxifraga davidii*

棒蕊虎耳草

Saxifraga clavistaminea Engl. et Irmsch.

多年生草本。茎被长腺毛。叶边缘具重锯齿和腺睫毛。花序具2或3花；萼片反折，卵形；花瓣白色，下面具2个黄色和3个紫色斑点；花丝棒状。蓇葖果瓶形。花果期5-7月。生海拔2300-3600米的林下或山谷石隙。产云南和四川。

Perennial herbs. Stems glandular villose. Leaves margin double serrate and glandular ciliate. Inflorescences 2- or 3-flowered; sepals reflexed, ovate; petals white, proximally with 2 yellow and 3 purple spots; filaments clavate. Follicles bottle-shaped. Fl. and fr. May-Jul. Forests or rock crevices in valleys at 2300-3600 m. Distributed in Yunnan and Sichuan.

邓波虎耳草

Saxifraga dungbooi Engler et Irmscher

多年生草本。茎和花梗被淡褐色腺柔毛。叶卵状长圆形至长圆状椭圆形，长1-1.5厘米，先端近急尖，基部楔形，边缘全缘，两面无毛或疏被毛。花序圆锥状，具3-7花；分枝长1.5-3厘米；萼片开展，无毛，3脉于先端汇合成1疣点；花瓣白色，3-5脉；花丝钻形；子房半下位，具1个不明显环状蜜腺；柱头近无柄。花果期8月。产西藏。印度北部亦有。

邓波虎耳草 *Saxifraga dungbooi*

Perennial herbs. Stems and pedicels pale brown glandular villous. Leaves ovate-oblong to oblong-elliptic, 1-1.5 cm long, apex subacute, base cuneate, margin entire, both surfaces glabrous or sparsely hairy. Inflorescence paniculate, 3-7-flowered; branches 1.5-3 cm long; sepals spreading, glabrous, veins 3, confluent into a verruca at apex; petals white, 3-5-veined; filaments subulate; ovary semi-inferior, with an indistinct annular nectary; stigmas subsessile. Fl. and fr. Aug. Distributed in Xizang. Also in N India.

棒蕊虎耳草 *Saxifraga clavistaminea*

多叶虎耳草

Saxifraga pallida Wall. ex Ser.

多年生草本。叶均基生，叶狭卵形、卵形至阔卵形，腹面被柔毛，背面常无毛。聚伞花序圆锥状，具4-11花；花序分枝和花梗均被柔毛；萼片背面无毛，3-7脉；花瓣白色，卵形，3-7脉。花果期7-10月。生海拔3000-5000米的针叶林下、高山灌丛、高山草甸和高山碎石隙。产甘肃、四川、云南和西藏。印度、不丹、尼泊尔和克什米尔地区亦有。

Perennial herbs. Leaves basal, leaf narrow-ovate, ovate to broad ovate, adaxially piliferous, abaxially glabrous. Inflorescence paniculate, 4-11-flowered; branches and pedicels pubescent; sepals abaxially glabrous, veins 3-7; petals white, ovate, 3-7-veined. Fl. and fr. Jul-Oct. In aciculiailvae, alpine scrub, alpine meadows and alpine rock crevices at 3000-5000 m. Distributed in Gansu, Sichuan, Yunnan and Xizang. Also in India, Bhutan, Nepal and Kashmir.

多叶虎耳草 *Saxifraga pallida*

镜叶虎耳草 *Saxifraga fortunei* var. *koraiensis*

红毛虎耳草

Saxifraga rufescens Balf. f.

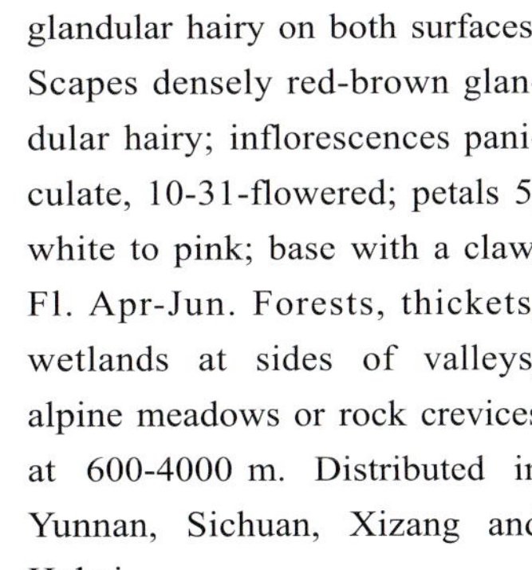

多年生草本。叶肾形或圆肾形至心形，边缘9-11浅裂，基生叶两面被腺毛。花葶密具红褐色腺毛；圆锥花序，具10-31花；花瓣5，白色至粉红色；基部具爪。花期4-6月。生海拔600-4000米的林下、灌丛、沟谷侧边湿地、高山草甸或岩石裂隙。产云南、四川、西藏和湖北。

Perennial herbs. Leaves reniform or orbicular-reniform to cordate, margin 9-11-lobed, basal leaves glandular hairy on both surfaces. Scapes densely red-brown glandular hairy; inflorescences paniculate, 10-31-flowered; petals 5, white to pink; base with a claw. Fl. Apr-Jun. Forests, thickets, wetlands at sides of valleys, alpine meadows or rock crevices at 600-4000 m. Distributed in Yunnan, Sichuan, Xizang and Hubei.

红毛虎耳草 *Saxifraga rufescens*

镜叶虎耳草

Saxifraga fortunei Hook. f. var. **koraiensis** Nakai

多年生草本。叶柄被长腺柔毛；叶肾形至近心形，基部心形，边缘具7-11浅裂。圆锥状花序，约具35花；萼片两面和边缘有腺毛，单脉；花瓣5，白色至浅红色，全缘无毛；心皮2。花期6-7月。生海拔2700-2900米的林中、石隙或小河边。产吉林和辽宁。朝鲜半岛亦有。

Perennial herbs. Petioles glandular-villose; leaves reniform to subcordate, base cordate, margin 7-11-lobed. Inflorescences paniculate, ca. 35-flowered; sepals glandular on both surfaces and at margin, 1-veined; petals 5, white to reddish, entire and glabrous; carpels 2. Fl. Jun-Jul. Forests, rock crevices or brooksides at 2700-2900 m. Distributed in Jilin and Liaoning. Also in Korean Peninsula.

虎耳草 *Saxifraga stolonifera*

虎耳草
Saxifraga stolonifera Curtis

多年生草本。茎被长腺毛；鞭匐枝细长，被卷曲腺毛和鳞片状叶。叶具点，近心形或肾形至圆形，具(5-)7-11浅裂，具腺毛；花两侧对称；花瓣5，白色；心皮2；下部合生。花果期4-11月。生海拔400-4500米的林中、灌丛、草甸或阴处石隙。产中国西南、华南、华中、华北和华东。朝鲜半岛和日本亦有。

Perennial herbs. Stems glandular-villose and stoloniferous, stolons filiform, crisped-glandular-villose and with scaly leaves. Leaves spotted, subcordate or reniform to orbicular, (5-)7-11-lobed, glandular hairy; flowers zygomorphic; petals 5, white; carpels 2; below connate. Fl. and fr. Apr-Nov. Forests, scrubby areas, meadows or shaded rock crevices at 400-4500 m. Distributed in SW, S, C, N and E China. Also in Korean Peninsula and Japan.

球茎虎耳草
Saxifraga sibirica L.

多年生草本，具鳞茎。茎密被腺柔毛。叶肾形，边缘7-9浅裂。聚伞花序伞房状，具2-13花，稀花单生；花萼直立，具3-5脉；花瓣白色，具3-8脉；心皮在下面贴生。花果期5-11月。生海拔800-5100米的林中、灌丛、高山草甸或岩峭壁。产中国西南、华中、华北、华西和东北。印度、尼泊尔、克什米尔地区、俄罗斯和蒙古亦有。

Perennial herbs, with bulbs. Stems densely glandular pilose. Leaves reniform, margin 7-9-lobed. Cymes corymbose, 2-13-flowered, flowers rarely solitary; sepals erect, veins 3-5; petals white, 3-8 veined; carpels connate proximally. Fl. and fr. May-Nov. Forests, scrubby areas, alpine meadows or rock clefts at 800-5100 m. Distributed in SW, C, N, W and NE China. Also in India, Nepal, Kashmir, Russia and Mongolia.

珠芽虎耳草
Saxifraga granulifera Harry Smith

多年生草本。基生叶肾形至近圆形，腹面无毛或具稀疏腺毛，边缘具腺毛；茎生叶肾形或阔卵形，具腺毛，基部肾形

球茎虎耳草 *Saxifraga sibirica*

珠芽虎耳草 *Saxifraga granulifera*

或楔形至截形。萼片直立，背面和边缘具腺毛，3-5脉；花瓣白色或黄色，狭倒卵形至楔形，3-8脉。花果期6-9月。生海拔3100-4600米的高山草甸、崖壁及岩石上。产四川、西藏和云南。不丹、印度和尼泊尔亦有。

Perennial herbs. Basal leaves reniform to suborbicular, adaxially glabrous or sparsely glandular pilose, margin glandular pilose; cauline leaves reniform or broadly ovate, glandular pilose, base reniform or cuneate to truncate. Sepals erect, abaxially and marginally glandular pubescent, veins 3-5; petals white or yellowish, narrowly obovate to cuneate, 3-8-veined. Fl. and fr. Jun-Sep. Alpine grasslands, cliff ledges and mossy rocks at 3100-4600 m. Distributed in Sichuan, Xizang and Yunnan. Also in Bhutan, India and Nepal.

长白虎耳草

Saxifraga laciniata Nakai et Takeda

多年生草本。茎被疏腺毛。叶常匙形，上面具腺毛，上部边缘具5-8个粗锯齿。花序伞序状，有花5-7朵；花瓣白色，基部具2黄色斑点，具3-5脉；心皮2；子房卵球形。花果期7-8月。生海拔2300-2600米的草甸或石隙。产吉林。萨哈林岛(库页岛)、朝鲜半岛和日本亦有。

Perennial herbs. Stems glandular-pilose.Leaves usually spatulate, adaxially glandular piliferous, coarsely 5-8-serrate distally. Inflorescences corymbose, with 5-7 flowers; petals white, proximally with 2 yellow spots, 3-5-veined; carpels 2; ovary ovoid. Fl. and fr. Jul-Aug. Meadows or rock crevices at 2300-2600 m. Distributed in Jilin. Also in Sakhalin, Korean Peninsula and Japan.

黑蕊虎耳草

Saxifraga melanocentra Franch.

多年生草本。叶均基生，菱状卵形或卵形至长圆形，边缘有钝齿和腺睫毛。花序伞房状，具2-17花；花瓣白色，基部具2个黄色腺点，具3-9(-14)脉；花药黑色。蒴果卵球形。花果期7-9月。生海拔3000-5300米的高山灌丛、草甸、石隙、河边或沼泽。产云南、四川、西藏、陕西、甘肃和青海。印度北部、尼泊尔、不丹和克什米尔地区亦有。

Perennial herbs. Leaves basal, rhombic-ovate or ovate to oblong, crenate-serrate and glandular-ciliate at margin. Inflorescences corymbose, 2-17-flowered; petals white, proximally with 2 yellow spots, 3-9(-14)-veined; anthers black. Capsules ovoid. Fl. and fr. Jul-Sep. Alpine thickets, meadows, rock crevices, by streams or bogs at 3000-5300 m. Distributed in Yunnan, Sichuan, Xizang, Shaanxi, Gansu and Qinghai. Also in N India, Nepal, Bhutan and Kashmir.

长白虎耳草 *Saxifraga laciniata*

黑蕊虎耳草 *Saxifraga melanocentra*

条叶虎耳草 *Saxifraga linearifolia*

条叶虎耳草

Saxifraga linearifolia Engl. et Irmsch.

多年生草本，植株高3-5厘米。基生叶密集呈莲座状，长圆形至椭圆形，腹面和边缘具腺毛，背面无毛；茎生叶长圆形至倒披针形，两面和边缘均具腺毛。单花生于茎顶；萼片在花期开展，边缘啮蚀状，背面和边缘被腺毛，4脉；花瓣黄色，3-5脉，无痂体。花果期7-9月。生海拔3900-4200米的高山草甸或石隙。产四川(康定)和云南(丽江)。

Perennial herbs, 3-5 cm tall. Basal leaves aggregated into a rosette, oblong to elliptic, adaxially and marginally glandular hairy, abaxially glabrous; cauline leaves oblong to oblanceolate, both surfaces and margin glandular hairy. Flower solitary; sepals spreading, glabrous or glandular hairy abaxially and margin, veins 4; petals yellow, 3-5-veined, not callose. Fl. and fr. Jul-Sep. Alpine meadows or rock crevices at 3900-4200 m. Distributed in Sichuan (Kangding), Yunnan (Lijiang).

紫花虎耳草 *Saxifraga bergenioides*

紫花虎耳草

Saxifraga bergenioides Marquand

多年生草本，密丛。茎不分枝，密被褐色卷曲柔毛。基生叶和茎生叶两面和边缘均具褐色卷曲柔毛。聚伞花序具2-4花或单花；萼片背面被褐色卷曲柔毛，5脉；花瓣紫红色，倒披针形至狭倒披针形，先端微凹，边缘基部具卷曲柔毛，5脉，无痂体。花期7-9月。生海拔4200-5000米的灌丛、高山草甸和高山碎石隙。产西藏。不丹亦有。

Perennial herbs, densely cespitose. Stem simple, densely brown crisped villous. Basal and cauline leaves both surfaces and margin brown crisped villous. Cymes 2-4-flowered or solitary; sepals abaxially with marginally brown crisped brown villous, veins 5; petals purple-red, oblanceolate to narrowly so apex retuse, base crisped villous, 5-veined, not callose. Fl. Jul-Sep. Scrub, alpine meadows and boulder screes, rock crevices at 4200-5000 m. Distributed in Xizang. Also in Bhutan.

小芒虎耳草

Saxifraga aristulata Hook. f. et Thoms.

多年生草本。基生叶狭卵形、长圆形或线形，先端具芒状毛(有时带腺头)，基部具褐色长柔毛。花梗被腺毛；萼片无毛或具褐色卷曲柔毛；花瓣橙黄色或黄色，3-5脉，无毛，或有时具褐色卷曲柔毛，具2痂体。花果期7-10月。生海拔3000-5000米的林下、林缘、高山灌丛草甸、高山草甸和岩壁石隙。产四川、云南和西藏。不丹、印度和尼泊尔亦有。

Perennial herbs. Basal leaves narrowly ovate, oblong or linear, apex acute, often aristate, sometimes glandular, base brown villose. Pedicel dark brown glandular hairy; sepals glabrous, rarely glandular at margin; petals orange-yellow or yellow, 2-callose near base, 3-5-veined, glabrous, sometimes brown crispate villous. Fl. and fr. Jul-Oct. Forests, forest

小芒虎耳草 *Saxifraga aristulata*

山地虎耳草 *Saxifraga sinomontana*

金丝桃虎耳草 *Saxifraga hypericoides*

margins, alpine meadows, rocky hillsides, stony ground and rock crevices at 3000-5000 m. Distributed in Sichuan, Yunnan and Xizang. Also in Bhutan, India and Nepal.

山地虎耳草

Saxifraga sinomontana J. T. Pan et Gornall

多年生草本。茎疏被褐色卷曲柔毛。基生叶椭圆形至线状长圆形，基部扩大，边缘具褐色卷曲长柔毛；茎生叶披针形至线形。萼片在花期直立，5-8脉；花瓣黄色，5-15脉，具2痂体。花果期5-10月。生海拔2700-5300米的灌丛、高山草甸、高山沼泽化草甸和高山碎石隙。产陕西、甘肃、青海、四川、新疆、云南和西藏。不丹至克什米尔地区亦有。

Perennial herbs. Stem sparsely brown crisped villous. Basal leaves elliptic to linear-oblong, margin brown crisped villous; cauline leaves lanceolate to linear. Sepals erect, veins 5-8; petals yellow, 5-15-veined, 2-callose near base. Fl. and fr. May-Oct. Scrub, alpine meadows, marshy meadows, rock crevices, calcareous rocks at 2700-5300 m. Distributed in Shaanxi, Gansu, Qinghai, Sichuan, Xinjiang, Yunnan and Xizang. Also in Bhutan to Kashmir.

金丝桃虎耳草

Saxifraga hypericoides Franch.

多年生草本，密丛生。基生叶密集，具长柄，叶长圆形至线形，两面和边缘均具褐色柔毛；茎生叶倒披针形至狭长圆形，先端无芒状毛。萼片先端钝圆且有时啮蚀状，背面具腺毛，3脉；花瓣黄色，狭卵形至长圆形，3-4(-5)脉，具2痂体。花果期7-9月。生海拔2700-5300米的林下、林缘、高山灌丛草甸、高山草甸和石隙。产四川和云南。

Perennial herbs, densely cespitose. Basal leaves dcnsc, long petiolate, Leaves oblong to linear, both sides and margin brown pubescent; cauline leaves oblanceolate to narrowly oblong, apex acute or mucronate. Sepals apex obtuse, sometimes erose, abaxially glandular hairy, 3-veined; petals yellow, narrowly ovate to oblong, 3-4(5)-veined, 2-callose. Fl. and fr. Jul-Sep. Forests, forest margins, alpine scrub meadows, alpine meadows and rock crevices at 2700-5300 m. Distributed in Sichuan and Yunnan.

叶枝虎耳草

Saxifraga yezhiensis C. Y. Wu

多年生草本，高6.5-11厘米。基生叶通常早枯；茎生叶近无柄，长圆形，两面和边缘均具腺毛，基部边缘疏生长腺毛。花梗被卷曲长腺毛并杂有直腺毛；萼片在花期直立状开展，边缘具腺睫毛，8-11脉；花瓣黄色，长圆形至狭卵形，先端急尖，5脉，具2痂体。花期7-8月。生海拔3600米左右的山坡。产云南(维西县)。

Perennial herbs, 6.5-11 cm tall. Basal leaves caducous; cauline leaves subsessile, oblong, both surfaces and margin glandular hairy, basal margin sparsely glandular villous. Pedicels brown crisped glandular villous and shorter straight glandular hairy; sepals erect-spreading, glandular hairy marginally, veins 8-11; petals yellow, oblong to narrowly ovate, apex acute, 5-veined, 2-callose near base. Fl. Jul-Aug. Slopes at ca. 3600 m. Distributed in Yunnan (Weixi County).

叶枝虎耳草 *Saxifraga yezhiensis*

优越虎耳草
Saxifraga egregia Engl.

多年生草本。基生叶基部心形，边缘和下面被褐色长柔毛；叶柄被卷曲腺毛。多歧聚伞花序伞房状，具3-9花；萼片反折，脉3-6；花瓣黄色，具(2-)4-6(-10)胼质齿，具3-6(或7)脉。花果期7-10月。生海拔2000-4600米的林下、灌丛、林下湿润处、高山草甸或石隙。产云南西南部、四川西部、西藏东部、甘肃南部和青海东部及南部。

Perennial herbs. Basal leaves cordate at base, marginally and abaxially brown villose; petioles crisply glandular-villose. Pleiochasium corymbose, 3-9-flowered; sepals reflexed, veins 3-6; petals yellow, (2-)4-6(-10)-callose, 3-6(or 7)-veined. Fl. and fr. Jul-Oct. Forests, thickets, wet places in forest understories, alpine meadows or rock crevices at 2000-4600 m. Distributed in SW Yunnan, W Sichuan, E Xizang, S Gansu, and E and S Qinghai.

无斑虎耳草
Saxifraga omphalodifolia Hand. -Mazz.

多年生草本。茎通常扭曲，密被带黑褐色腺头的卷曲柔毛。基生叶于花期枯凋；茎生叶心形至卵状心形，两面和边缘均具粗腺毛。聚伞花序圆锥状，具9-11花；萼片在花期反曲，背面被腺毛；花瓣黄色，狭卵形至近长圆形，3-5脉，具不明显6痂体。花期7-8月。生海拔3800-4200米的松林下和高山草甸。产四川、西藏和云南。

Perennial herbs. Stem zigzagged, densely dark brown crisped villous, glandular distally. Basal leaves caducous by anthesis; cauline leaves cordate to ovate-cordate, both surface and margin glandular hispid. Cymes paniculate, 9-11-flowered; sepals reflexed, abaxially glandular hairy; petals yellow, narrowly ovate to suboblong, 3-5-veined, obscurely 6-callose. Fl. and fr. Jul-Aug. *Pinus* forests and alpine meadows at 3800-4200 m. Distributed in Sichuan, Xizang and Yunnan.

异叶虎耳草 *Saxifraga diversifolia*

异叶虎耳草
Saxifraga diversifolia Wall. ex Ser.

多年生草本。基生叶与茎生叶二型；茎生叶8-12；叶柄被褐色卷曲长柔毛。花5-17朵呈伞序状聚伞花序；花萼反折，脉3(-5)；花瓣黄色，常无小痂体，具(3-)5-7(-9)脉；心皮2。花果期8-10月。生海拔2700-4300米的林中、林缘、高山草甸或石隙。产云南北部、四川西部和西藏东南部。印度北部、尼泊尔、不丹、克什米尔地区和缅甸亦有。

Perennial herbs. Basal leaves and cauline leaves dimorphic; cauline leaves 8-12; petioles rusty-crisped-villose. Flowers 5-17 in corymbose-cyme; sepals reflexed, veins 3(-5); petals yellow, usually not callose, (3-)5-7(-9)-veined; carpels 2. Fl. and fr. Aug-Oct. Forests, forest edges, alpine meadows or rock crevices at 2700-4300 m. Distributed in N Yunnan, W Sichuan and SE Xizang. Also in N India, Nepal, Bhutan, Kashmir and Myanmar.

梅花草叶虎耳草
Saxifraga parnassiifolia D. Don

多年生草本。基生叶心状卵形；茎生叶卵形至心形，无柄且半抱茎；叶两面和边缘均具腺毛。聚伞花序具6-11花；花梗被短腺毛；萼片腹面上部和背面及边缘均具腺毛，5-7脉；花瓣黄色，倒卵形至阔卵形，5-7脉，具4-6痂脉。花期7-9月。生海拔2700-4000米的山坡。产西藏(聂拉木县、察隅县)。印度、不丹和尼泊尔亦有。

Perennial herbs. Basal leaves cordate-ovate; cauline leaves ovate to cordate, sessile, semiamplexicaul; glandular villous on both surfaces and margin. Pleiochasium 6-11-flowered; pedicels shortly glandu-

优越虎耳草 *Saxifraga egregia*

无斑虎耳草 *Saxifraga omphalodifolia*

梅花草叶虎耳草 *Saxifraga parnassiifolia*

小泡虎耳草 *Saxifraga bulleyana*

lar hairy; sepals brown glandular hairy abaxially, adaxially distally, and at margin, veins 5-7; petals yellow, obovate to broadly ovate, 5-7-veined, 4-6-callose. Fl. Jul-Sep. *Abies* forest margins, slopes at 2700-4000 m. Distributed in Xizang (Nyalam County, Zayü County). Also in India, Bhutan and Nepal.

小泡虎耳草

Saxifraga bulleyana Engl. et Irmsch.

多年生草本。茎生叶无柄，卵状椭圆形，两面具白色腺毛，基部圆形。聚伞花序具(1或)2-5花；花瓣黄色，卵形、椭圆形、阔椭圆形至阔卵形，具2-4小痂体，3-6条脉。花期7-8月。生海拔3000-4600米的高山草甸或岩石裂隙。产云南西北部。

Perennial herbs. Cauline leaves sessile, ovate-elliptic, both surfaces white glandular hairy, base rounded. Cymes (1 or)2-5-flowered; petals yellow, ovate, elliptic, broadly elliptic to broadly ovate, 2-4-callose, 3-6-veined. Fl. Jul-Aug. Alpine meadows or rock crevices at 3000-4600 m. Distributed in NW Yunnan.

短柄虎耳草

Saxifraga brachypoda D. Don

多年生草本。基生叶缺；茎生叶沿茎密生，光亮，披针形至线状披针形，革质，坚硬，两面无毛，顶端有硬芒。花单生或2或3花形成聚伞状；花瓣黄色，具5-8脉；花药紫黑色。花果期7-9月。生海拔3000-5000米的林下、高山草甸或石隙中。产四川、云南和西藏。印度北部、尼泊尔、不丹和缅甸亦有。

Perennial herbs. Basal leaves absent; cauline leaves crowded along stems, shiny, lanceolate to linear-lanceolate, leathery, rigid, both surfaces glabrous, rigidly aristate at apex. Flowers solitary or cymes 2- or 3-flowered; petals yellow, 5-8-veined; anthers atropurpureus. Fl. and fr. Jul-Sep. Forests, alpine meadows or rock crevices at 3000-5000 m. Distributed in Sichuan, Yunnan and Xizang. Also in N India, Nepal, Bhutan and Myanmar.

短柄虎耳草 *Saxifraga brachypoda*

腺瓣虎耳草 *Saxifraga wardii*

腺瓣虎耳草

Saxifraga wardii W. W. Smith

多年生草本。茎不分枝。叶在茎上密生，无柄，卵形或长圆形至线状长圆形，两面无毛，边缘具腺状骨质糙缘毛。花单生；花瓣黄色，宽卵形至圆形，边缘具腺睫毛。花期7-9月。生海拔1200-4800米的高山草甸、灌丛或石隙中。产云南西北部和西藏东南部。

Perennial herbs. Stems simple. Leaves crowded along stems, sessile, ovate or oblong to linear-oblong, both surfaces glabrous, margin cartilaginous eglandular setose-ciliate. Flowers solitary; petals yellow, broadly ovate to orbicular, margin glandular-ciliate. Fl. Jul-Sep. Alpine meadows, thickets or rock crevices at 1200-4800 m. Distributed in NW Yunnan and SE Xizang.

流苏虎耳草

Saxifraga wallichiana Sternb.

多年生草本。茎基部和叶腋部均具芽。基生叶不存在；茎生叶密集，卵形、狭卵形至披针形，基部心形。聚伞花序具2-4花或单花生于茎顶；花瓣黄色，卵形、倒卵形至椭圆形，3-9脉，具2痂体。花果期7-11月。生海拔2000-5000米的林下、林缘、灌丛、高山草甸及石隙。产四川、云南和西藏。缅甸、不丹、尼泊尔和印度亦有。

Perennial herbs. Leaf buds present in axils of proximal, median or (especially) distal cauline leaves. Basal leaves absent; cauline leaves crowded along stem, ovate, narrowly ovate to lanceolate, base cordate. Cymes 2-4-flowered or flower solitary; petals yellow, ovate, obovate to elliptic, 3-9-veined, 2-callose near base. Fl. and fr. Jul-Nov. Forests, forest margins, scrub, alpine meadows and rock crevices at 2000-5000 m. Distributed in Sichuan, Yunnan and Xizang. Also in Myanmar, Bhutan, Nepal and India.

流苏虎耳草 *Saxifraga wallichiana*

猬状虎耳草 *Saxifraga erinacea*

猬状虎耳草

Saxifraga erinacea H. Smith

多年生草本。茎无毛。无基生叶；茎生叶密集呈莲座状，长圆形至线状长圆形，先端急尖且具芒状短尖头，边缘具软骨质硬睫毛。花单生茎顶；萼片花期直立，4-5脉于先端汇合；花瓣黄色，近倒披针形，基部狭缩成爪，5-6脉，无痂体。花期7-8月。生海拔4000-4600米的高山草甸和高山碎石隙。产西藏(琼结县、错那县)。不丹亦有。

Perennial herbs. Stem glabrous. No basal leaves; cauline leaves crowded along stem, oblong to linear-oblong, apex acute, aristate-mucronate, margin cartilaginous eglandular setose-ciliate. Flower solitary; sepals erect, veins 4-5, confluent at apex; petals yellow, oblong to oblanceolate, base cordate with a claw, 5-6-veined, not callose. Fl. Jul-Aug. Alpine meadows and rock clefts at 4000-4600 m. Distributed in Xizang (Qonggyai County, Cona County). Also in Bhutan.

马耳山虎耳草

Saxifraga balfourii Engl. et Irmsch.

多年生草本。茎生叶椭圆形，两面密被无色糙伏毛，边缘具硬腺睫毛。花梗被褐色腺毛；萼片在花期由直立变开展，腹面和边缘无毛，背面被褐色腺毛，4-7脉；花瓣黄色，椭圆形，基部狭缩成爪，3-7脉，具4-6(-8)痂体，有时痂体不明显。花果期7-9月。生海拔2300-4600米的杂木林下、高山草甸和岩壁石隙。产云南(丽江、鹤庆)。

Perennial herbs. Cauline leaves elliptic, both surfaces densely strigose, margin entire, eglandular or glandular hairy. Pedicels glandular hairy; sepals erect, then spreading, margin and adaxially glabrous, abaxially glandular hairy, veins 4-7; petals yellow, elliptic 3-7-veined, 4-6(-8)-callose, sometimes obscurely so. Fl. and fr. Jul-Sep. Mixed forests, alpine meadows and rock crevices at 2300-4600 m. Distributed in Yunnan (Lijiang, Heqing).

马耳山虎耳草 *Saxifraga balfourii*

线茎虎耳草 *Saxifraga filicaulis*

芽生虎耳草
Saxifraga gemmipara Franch.

多年生草本。茎多分枝。中部茎生叶常密集成莲座状，边缘有腺缘毛；上部茎生叶鳞状，小于中部的1/2。聚伞花序常伞房状，疏松分枝，具2-12花；花瓣白色，具黄色或紫色斑，具2(-4)小痂体，3-7脉。花果期6-11月。生海拔1700-4900米的林下、林缘、灌丛、草甸或石隙。产云南和四川。泰国北部亦有。

Perennial herbs. Stems many-branched. Median cauline leaves often aggregated into rosette, margin glandular-ciliate; upper cauline leaves often scalelike, <1/2 size of median ones. Cymes usually corymbose, laxly branched, 2-12-flowered; petals white, spotted yellow or purple, 2(-4)-callose, 3-7-veined. Fl. and fr. Jun-Nov. Forests, forest edges, thickets, meadows or rock crevices at 1700-4900 m. Distributed in Yunnan and Sichuan. Also in N Thailand.

线茎虎耳草
Saxifraga filicaulis Wall. ex Ser.

多年生草本。茎多分枝。中上部茎生叶线形至剑形；茎生叶顶端具芒状尖头。花单生或2至3花组成聚伞状；花梗被腺毛；花瓣黄色，具2-4小痂体，3-7脉。花果期6-10月。生海拔2200-4800米的林下、灌丛、高山草甸、崖面或石隙中。产云南、四川、西藏和陕西。印度、尼泊尔、不丹和克什米尔地区亦有。

Perennial herbs. Stems many branched. Median and distal cauline leaves linear to ensiform, cauline leaves with aristate mucros at apex. Flowers solitary or cymes 2- or 3-flowered; pedicels glandular pubescent; petals yellow, 2-4-callose, 3-7-veined. Fl. and fr. Jun-Oct. Forests, thickets, alpine meadows, cliff faces or rock crevices at 2200-4800 m. Distributed in Yunnan, Sichuan, Xizang and Shaanxi. Also in India, Nepal, Bhutan and Kashmir.

芽生虎耳草 *Saxifraga gemmipara*

伏毛虎耳草 *Saxifraga strigosa*

伏毛虎耳草

Saxifraga strigosa Wall. ex Ser.

多年生草本。叶两面被糙伏毛，在茎中部密集呈莲座状，卵形、倒卵形或椭圆形至长圆形。花单生或3-10朵呈聚伞状；花瓣白色，具红棕色斑，具2-4个小痂体，3-7脉。花果期7-10月。生海拔1800-4200米的林中、林边、灌丛、草甸或石隙。产四川、西藏和云南。印度北部、尼泊尔、不丹和缅甸亦有。

Perennial herbs. Leaves strigose on both surfaces, median cauline leaves aggregated in a rosette, ovate, obovate or elliptic to oblong. Flowers solitary or 3-10 in a cyme; petals white, reddish brown spotted, 2-4-callose, 3-7-veined. Fl. and fr. Jul-Oct. Forests, forest edges, crubby areas, meadows or rock crevices at 1800-4200 m. Distributed in Sichuan, Xizang and Yunnan. Also in N India, Nepal, Bhutan and Myanmar.

疏叶虎耳草

Saxifraga substrigosa J. T. Pan

多年生草本。基生叶不存在；茎生叶卵形、倒卵形至长圆形，边缘具5-6疏锯齿，两面和边缘均具糙伏毛。聚伞花序具3-9花；萼片在花期直立，腹面和边缘无毛，背面被腺毛，5-8脉；花瓣黄色，倒卵形，3-8脉，具4-6(-9)痂体。花果期7-9月。生海拔2700-4200米的栎、桦、云杉混交林和云杉林下或林缘石上。产云南和西藏。尼泊尔亦有。

Perennial herbs. Basal leaves absent; cauline leaves ovate, obovate to oblong, margin sparsely 5-6-serrate, both surfaces strigose. Cymes 3-9-flowered or flower solitary; sepals erect spreading, adaxially and marginally glabrous, abaxially glandular hairy, veins 5-8; petals yellow, obovate, 3-8-veined, 4-6(-9)-callose. Fl. and fr. Jul-Sep. Mixed forests, *Picea* forests, forest margins, alpine meadows, rock crevices at 2700-4200 m. Distributed in Yunnan and Xizang. Also in Nepal.

齿叶虎耳草

Saxifraga hispidula D. Don

多年生草本。茎生叶近椭圆形至卵形，边缘先端常具3-5齿牙，两面和边缘均具糙伏毛。单花生于茎顶或枝端，或聚伞花序；萼片在花期直立或稍开展，腹面上部、背面和边缘具腺毛，3-8脉；花瓣黄色，倒卵形、椭圆形至阔椭圆形，3-10脉，具2-16痂体。花期7-9月。生海拔2300-5600米的林下、林缘、灌丛、高山草甸及石隙。产中国西南。东南亚和印度亦有。

Perennial herbs. Cauline leaves subelliptic to ovate, both surfaces strigose, margin with 3-5 acute lobes toward apex. Flower usually solitary or cymes 2-4-flowered; sepals erect or somewhat spreading, glandular or eglandular hairy abaxially, adaxially distally and

疏叶虎耳草 *Saxifraga substrigosa*

齿叶虎耳草 *Saxifraga hispidula*

marginally, veins 3-8; petals yellow, elliptic to broadly so or obovate, 3-10-veined, 2-16-callose. Fl. and fr. Jul-Sep. Rocks and rock crevices in forests, forest margins, scrub, alpine meadows, and on cliffs at 2300-5600 m. Distributed in SW China. Also in SE Asia and India.

小伞虎耳草

Saxifraga umbellulata Hook. f. et Thoms.

多年生草本，高5.5-10厘米。基生叶密集，呈莲座状，匙形，无毛；茎生叶长圆形至近匙形，两面和边缘均具褐色腺毛，或腹面无毛。萼片直立，腹面无毛，背面和边缘多少具褐色腺毛，3脉；花瓣黄色，提琴状长圆形至提琴形，3-5脉，具2痂体。花期6-9月。生海拔3000-4700米的沼泽地和岩壁石隙。产西藏。尼泊尔和印度亦有。

Perennial herbs, 5.5-10 cm tall. Basal leaves aggregated into a rosette, leaf blade spatulate, glabrous; cauline leaves oblong to subspatulate, both surfaces and margin brown glandular hairy or adaxially glabrous. Sepals usually erect, abaxially and marginally brown glandular hairy, veins 3; petals yellow, pandurate to pandurate-oblong, 3-5-veined, 2-callose. Fl. Jun-Sep. In forests, scrub, marshlands, alpine rock crevices, sunny cliffs at 3000-4700 m. Distributed in Xizang. Also in Nepal and India.

近加拉虎耳草

Saxifraga llonakhensis W. W. Smith

多年生草本。茎分枝，具丛生叶；开花茎具深褐色腺毛。茎生叶长圆形至线形，肉质。花单生，稀聚伞花序具花2-3朵；花瓣黄色，微具橙黄色斑，具4-6小痂体，3脉。花果期7-9月。生海拔3700-4600米的林中或岩石缝中。产云南和西藏。印度北部、尼泊尔和缅甸亦有。

Perennial herbs. Stems branched, with leaf rosettes; flowering stems dark brown glandular hairy. Cauline leaves oblong to linear, carnose. Flowers solitary, rarely cymes 2-3-flowered; petals yellow, obscurely orange-yellow spotted, 4-6-callose, 3-veined. Fl. and fr. Jul-Sep. Forests or rock crevices at 3700-4600 m. Distributed in Yunnan and Xizang. Also in N India, Nepal and Myanmar.

近加拉虎耳草 *Saxifraga llonakhensis*

加拉虎耳草

Saxifraga gyalana Marquand et Airy-Shaw

多年生草本。小主轴分枝，具莲座叶丛。基生叶两面和边缘均具刚毛；茎生叶腹面上部和边缘具褐色腺毛，背面无毛或疏生腺毛。花梗被黑褐色腺毛；萼片在花期开展至反曲，腹面凹陷无毛，边缘具腺睫毛，3脉；花瓣黄色，披针形至长圆形，基部近心形，具6-8痂体。花果期6-9月。生海拔2300-4100米的林下和石隙。产西藏。

Perennial herbs. Shoots branched, with leaf rosettes. Rosette leaves setose abaxially distally, adaxially, and marginally; cauline leaves adaxially distally and marginally brown glandular hairy, abaxially glabrous or sparsely glandular hairy. Pedicels dark brown glandular hairy; sepals spreading to reflexed, adaxially concave glabrous, margin glandular ciliate, veins 3; petals yellow, lanceolate to oblong, base subcordate, 6-8-callose. Fl. and fr. Jun-Sep. Forests and rock crevices at 2300-4100 m. Distributed in Xizang.

小伞虎耳草 *Saxifraga umbellulata*

加拉虎耳草 *Saxifraga gyalana*

芽虎耳草 *Saxifraga gemmigera*

光缘虎耳草 *Saxifraga nanella*

芽虎耳草

Saxifraga gemmigera Engl.

多年生草本。小主轴分枝，具莲座叶丛。莲座叶肉质肥厚，倒卵形至近匙形，边缘疏生软骨质刚毛状睫毛；茎生叶革质，卵形至狭长圆形，边缘具腺睫毛。单花生于茎顶；萼片肉质，在花期反曲，腹面和边缘无毛，背面具腺毛，3-5脉；花瓣黄色，椭圆形，3-4脉，具2痂体。花果期7-9月。生海拔3100-4700米的高山草甸。产甘肃、青海、陕西和四川。

Perennial herbs. Stem branched, with leaf rosettes. Rosette leaves carnose, obovate to subspatulate, margin sparsely cartilaginous setose-ciliate; cauline leaves leathery, to oblong-linear, margin glandular ciliate or setose-ciliate. Flower solitary; sepals carnose, reflexed, adaxially and marginally glabrous, abaxially glandular hairy, veins 3-5; petals yellow, elliptic to subovate or narrowly ovate, 3-4-veined, 2-callose. Fl. and fr. Jul-Sep. Alpine meadows at 3100-4700 m. Distributed in Gansu, Qinghai, Shaanxi and Sichuan.

光缘虎耳草

Saxifraga nanella Engl. et Irmsch.

多年生草本。叶密集呈莲座状，近匙形至近长圆形，先端钝圆且无毛，两面无毛，边缘具腺睫毛。萼片在花期开展至反曲，背面被腺毛，3-5脉；花瓣黄色，中下部具橙色斑点，椭圆形至卵形，5脉，具2痂体或无痂体。花期7-8月。生海拔3000-5800米的高山草甸、高山灌丛草甸和高山碎石隙。产新疆、青海、云南和西藏。尼泊尔亦有。

Perennial herbs. Leaves aggregated into a rosette. Leaves subspatulate to suboblong or subovate, apex obtuse, glabrous, both surfaces glabrous, margin glandular ciliate. Sepals spreading to reflexed, abaxially glandular hairy, veins 3-5; petals yellow, proximally orange spotted, elliptic to ovate, veins 5, 2-callose or not callose. Fl. Jul-Aug. Alpine meadows, alpine scrub meadows and rock crevices at 3000-5800 m. Distributed in Xinjiang, Qinghai, Yunnan and Xizang. Also in Nepal.

爪瓣虎耳草

Saxifraga unguiculata Engl.

多年生丛生草本。莲座状叶匙形至狭倒卵形，近肉质。花枝具叶；花单生或聚伞花序伞房状，具花2-8；花瓣黄色，具橙色斑，近基部具1或2个小痂体，3-7脉。花期7-8月。生海拔1800-5600米的林中、灌丛、高山草甸或岩峭壁。产中国西南、华北、华西和西北。

Perennial herbs, caespitose. Rosette leaves spatulate to narrowly obovate, subcarnose. Flowering stems leafy; flowers soli-

爪瓣虎耳草 *Saxifraga unguiculata*

拟黄花虎耳草 *Saxifraga chrysanthoides*

tary or cymes corymbose, 2-8-flowered; petals yellow, orange spotted, 1- or 2-callose near base, 3-7-veined. Fl. and fr. Jul-Aug. Forests, crubby areas, alpine meadows or rock crevices at 1800-5600 m. Distributed in SW, N, W and NW China.

拟黄花虎耳草

Saxifraga chrysanthoides Engl. et Irmsch.

多年生草本。基生叶莲座状，叶近倒披针形，边缘上半部具刚毛状睫毛，两面无毛，3脉；茎生叶线形至狭倒披针形。花单生；花瓣黄色，不具小痂体，具5脉。花期7-8月。生海拔2700-5300米的石隙或石滩。产云南西北部。

Perennial herbs. Basal leaves rosette, blades suboblanceolate, margin setose-ciliate distally, both surfaces glabrous, 3-veined; cauline leaves linear to narrowly oblanceolate. Flowers solitary; petals yellow, not callose, 5-veined. Fl. Jul-Aug. Rock crevices or screes at 2700-5300 m. Distributed in NW Yunnan.

红虎耳草

Saxifraga sanguinea Franch.

多年生草本。茎紫红色，密被紫色腺毛。基生叶密集，呈莲座状，肉质，匙形至近匙形；茎生叶近倒披针形至线状倒披针形，两面和边缘具紫褐色腺毛。花序分枝较细弱；花瓣腹面黄白色且中下部具紫红色斑点，背面红色或全部红色，披针形，基部圆形，3脉。花期7-8月。生海拔3300-4500米的山坡草甸和石灰岩缝隙。产青海、四川、云南和西藏。

Perennial herbs. Stem purple, densely purple glandular hairy. Basal leaves aggregated into a rosette, carnose, spatulate to sub-spatulate; cauline leaves, sub-oblanceolate to linear-lanceolate, adaxially and marginally purple-brown glandular hairy, Cymes branches slender; petals adaxially pale yellow, proximally purple spotted, abaxially red or red on both surfaces, lanceolate to narrowly elliptic, 3-veined. Fl. Jul-Aug. Rocky hillside meadows and limestone crevices at 3300-4500 m. Distributed in Qinghai, Sichuan, Yunnan and Xizang.

西南虎耳草

Saxifraga signata Engl. et Irmsch.

多年生草本。茎被黑褐色腺毛。基生叶莲座状，无毛，边有刚毛状睫毛。花序伞房状，具4-24花，具深褐色腺毛；花瓣黄色，上面基部具紫斑，具2个小痂体，3-7脉。花果期7-9月。生海拔2800-4600米的山地草甸或石隙。产云南、四川、西藏和青海。

Perennial herbs. Stems blackish glandular-hairy. Basal leaves rosette, glabrous, margin setose-ciliate. Inflorescences corymbose, 4-24-flowered, dark brown glandular hairy; petals yellow, adaxially purple spotted proximally, 2-callose, 3-7-veined. Fl. and fr. Jul-Sep. Alpine meadows or rock crevices at 2800-4600 m. Distributed in Yunnan, Sichuan, Xizang and Qinghai.

红虎耳草 *Saxifraga sanguinea*

西南虎耳草 *Saxifraga signata*

灯架虎耳草 *Saxifraga candelabrum*

灯架虎耳草

Saxifraga candelabrum Franch.

多年生草本。基生叶聚生成莲座状；叶匙形至倒卵形，边缘上部具3-7个钝齿。圆锥花序，具19-29花；花瓣浅黄色，中下部具紫色斑点，基部心形，具2个小痂体，3-5脉。花果期7-9月。生海拔2000-4200米的林下、林缘、高山草甸或岩石裂隙。产云南西北部和四川西部。

Perennial herbs. Basal leaves aggregated into a rosette; leaves spatulate to obovate, margin 3-7-dentate distally. Inflorescences paniculate, 19-29-flowered; petals light yellow, purple-spotted below middle, base cordate, 2-callose, 3-5-veined. Fl. and fr. Jul-Sep. Forests, forest edges, alpine meadows or rock crevices at 2000-4200 m. Distributed in NW Yunnan and W Sichuan.

喜马拉雅虎耳草

Saxifraga brunonis Wall.

多年生草本。基生叶密集，呈莲座状，长圆状剑形；茎生叶较疏；叶先端具软骨质芒，两面无毛，边缘具软骨质刚毛状睫毛。萼片在花期开展；花瓣黄色，椭圆形、长圆形至披针形，3-5脉，具2痂体。花果期6-10月。生海拔2800-4000米的林下、高山草甸和岩坡石隙。产四川、云南和西藏。不丹、印度、尼泊尔、克什米尔地区和缅甸亦有。

Perennial herbs. Basal leaves aggregated into a rosette, oblong-ensiform; cauline leaves remote; leaf blade apex cartilaginous aristate, both surfaces glabrous, margin cartilaginous setose-ciliate. Sepals spreading; petals yellow, elliptic, oblong to lanceolate, 3-5-veined, obscurely 2-callose. Fl. and fr. Jun-Oct. Mossy rocks in woods, hillsides, alpine gullies at 2800-4000 m. Distributed in Sichuan, Yunnan and Xizang. Also in Bhutan, India, Nepal, Kashmir and Myanmar.

太白虎耳草

Saxifraga josephii Engl.

多年生草本。莲座叶革质而较硬，长圆状剑形至近剑形，先端具软骨质芒，两面无毛，边缘具软骨质刚毛状腺毛；茎生叶较疏，长圆状剑形。花梗纤细，疏生褐色短腺毛；萼片在花期开展至反曲，无毛，3-5脉；花瓣黄色，狭卵形至卵形，3脉，具2-3痂体，有时痂体不明显。花果期

喜马拉雅虎耳草 *Saxifraga brunonis*

太白虎耳草 *Saxifraga josephii*

7-9月。生海拔1300-2100米的阴湿石隙。产陕西和河南。

Perennial herbs. Rosette leaves leathery, rigid, oblong-ensiform to subensate, apex cartilaginous aristate, both surfaces glabrous, margin cartilaginous eglandular setose-ciliate; cauline leaves remote, oblong-ensiform. Pedicels sparsely brown glandular pubescent; sepals spreading to reflexed, glabrous, veins 3-5; petals yellow, ovate to narrowly so, 3-veined, 2-3-callose, sometimes obscurely so. Fl. and fr. Jul-Sep. Shaded damp rock crevices at 1300-2100 m. Distributed in Shaanxi and Henan.

大花虎耳草

Saxifraga stenophylla Royle

多年生草本。基生叶密集，呈莲座状，革质，狭椭圆形至近匙形；茎生叶较疏，革质，长圆形。聚伞花序具2-3花；花梗密被腺毛；萼片花期直立，具5-9脉；花瓣黄色，倒卵形至倒阔卵形，具8-11脉，无爪，无痂体。花期7-8月。生海拔3700-5000米的高山草甸和高山灌丛草甸。产四川、云南和西藏。南亚、塔吉克斯坦和克什米尔地区亦有。

Perennial herbs. Basal leaves aggregated into a rosette, leathery, narrowly elliptic to subspatulate; cauline leaves remote, leathery, oblong. Cymes 2-3-flowered; pedicels densely glandular hairy; sepals erect, veins 5-9; petals yellow, obovate to broadly so, 8-11-veined, clawless, not callose. Fl. Jul-Aug. Scrub, alpine meadows, among rocks at 3700-5000 m. Distributed in Sichuan, Yunnan and Xizang. Also in S Asia, Tajikistan and Kashmir.

垂头虎耳草

Saxifraga nigroglandulifera Balakr.

多年生草本。茎不分枝，上部具黑褐色腺毛。叶披针形至长圆形，两面近无毛，边缘具褐色腺状长柔毛。聚伞花序总状，具2-14花；花下垂并偏向一方；花梗密被褐色腺毛；花瓣黄色，不具小痂体，具3-5脉。花果期7-10月。生海拔2700-5000(-5400)米的林下、林缘、灌丛或高山草甸。产云南、四川和西藏。印度北部、尼泊尔和不丹亦有。

Perennial herbs. Stems simple, distally dark-brown-glandular-hairy. Leaves lanceolate to oblong, both surfaces subglabrous, margin brown glandular villous. Cymes racemiform, 2-14-flowered; flowers usually nodding and secund; pedicels densely brown-glandular-hairy; petals yellow, not callose, 3-5-veined. Fl. and fr. Jul-Oct. Forests, forest edges, thickets or alpine meadows at 2700-5000(-5400) m. Distributed in Yunnan, Sichuan and Xizang. Also in N India, Nepal and Bhutan.

唐古特虎耳草

Saxifraga tangutica Engl.

多年生草本。基生叶两面无毛，边缘具褐色卷曲长柔毛。聚伞花序具(2-)8-24花；萼片在花期由直立变开展至反曲，有时背面下部具卷曲柔毛；花瓣黄色，或腹面黄色而背面紫色，3-5(-7)脉，具2痂体。花果期6-10月。生海拔2900-5600米的林下、灌丛、高山草甸和高山碎石隙。产甘肃、青海、四川和西藏。不丹至克什米尔地区亦有。

Perennial herbs. Basal leaves both surfaces glabrous, margin brown crisped villous. Inflorescence (2-)8-24-flowered; sepals erect, then spreading to reflexed, abaxially sometimes brown crisped villous proximally; petals yellow on both surfaces or purple abaxially and yellow adaxially, 3-5(-7)-veined, 2-callose. Fl. and fr. Jun-Oct. Forests, scrub, alpine meadows and rock crevices at 2900-5600 m. Distributed in Gansu, Qinghai, Sichuan and Xizang. Also in Bhutan to Kashmir.

大花虎耳草 *Saxifraga stenophylla*

垂头虎耳草 *Saxifraga nigroglandulifera*

唐古特虎耳草 *Saxifraga tangutica*

狭瓣虎耳草 *Saxifraga pseudohirculus*

小短尖虎耳草 *Saxifraga mucronulata*

狭瓣虎耳草
Saxifraga pseudohirculus Engl.

多年生草本。基生叶先端无芒，两面和边缘具腺毛；茎生叶近长圆形至倒披针形。聚伞花序具2-12花，或单花生于茎顶；萼片腹面疏生腺毛或无毛，背面和边缘密生腺毛，3-5(-7)脉；花瓣黄色，披针形、狭长圆形至剑形，3-5(-7)脉。花果期7-9月。生海拔3100-5600米的林下、灌丛、高山草甸和高山碎石隙。产陕西、甘肃、青海、四川和西藏。

Perennial herbs. Basal leaves apex subobtuse, glandular hairy both surface and margin; cauline leaves suboblong to oblanceolate. Cymes 2-12-flowered or flower solitary; sepals adaxially sparsely glandular hairy or glabrous, abaxially and marginally dark brown glandular hairy, veins 3-5(-7); petals yellow, lanceolate, narrowly oblong to ensiform, 3-5(-7)-veined. Fl. and fr. Jul-Sep. Forests, scrubs, alpine meadows and rock crevices at 3100-5600 m. Distributed in Shaanxi, Gansu, Qinghai, Sichuan and Xizang.

小短尖虎耳草
Saxifraga mucronulata Royle

多年生草本。茎密被腺柔毛。基生叶密集呈莲座状，匙形至线状匙形，两面无毛，边缘具刚毛状腺睫毛，肉质；茎生叶线形，先端具短尖头，背面和边缘具腺毛。萼片在花期直立，3脉；花瓣黄色，倒卵形，5-6脉，无痂体。花期7-8月。生海拔2800-5400米的岩坡石隙。产四川、西藏和云南。尼泊尔至克什米尔地区亦有。

Perennial herbs. Stem densely glandular pubescent. Basal leaves aggregated into a rosette, spatulate to linear-spatulate, both surfaces glabrous, margin glandular setose-ciliate, carnose; cauline leaves linear, apex mucronate, abaxially and marginally glandular hairy. Sepals erect, veins 3; petals yellow, obovate, 5-6-veined, not callose. Fl. Jul-Aug. Rocky alpine meadows, cliff ledges, boulders at 2800-5400 m. Distributed in Sichuan, Xizang and Yunnan. Also in Nepal to Kashmir.

黄水枝
Tiarella polyphylla D. Don

多年生草本。茎单生，密被腺毛。叶多基生；叶心形，掌状3-5浅裂。总状花序密具腺毛；花白色，小；无花瓣；心皮不等大，基部贴生；雄蕊10；子房近上位。花果期4-11月。生海拔1000-3800米的林下、灌丛或水边。产中国西南、华南、华中和华西。印度、尼泊尔、不丹、缅甸、老挝、越南、柬埔寨和日本亦有。

Perennial herbs. Stems simple, densely glandular hairy. Leaves mostly basal; leaves cordate, palmately 3-5-lobed. Racemes densely glandular hairy; flowers white, small; petals absent; carpels unequal, connate proximally; stamens 10; ovary subsuperior. Fl. and fr. Apr-Nov. Forests, thickets or by waters at 1000-3800 m. Distributed in SW, S, C and W China. Also in India, Nepal, Bhutan, Myanmar, Laos, Vietnam, Cambodia and Japan.

黄水枝 *Tiarella polyphylla*

台湾唢呐草 *Mitella formosana*

台湾唢呐草

Mitella formosana (Hayata) Masamune

多年生草本。叶全部基生，卵状心形，长3.5-7厘米，不明显5-7裂，先端急尖，两面被腺伏毛，边缘具不规则小齿，基部心形；叶柄长7-13厘米，密被褐色卷曲长腺毛。总状花序具多花，被腺状长柔毛；萼片阔卵形，背面被褐色腺毛，单脉；花瓣羽状5深裂，裂片线形，具腺点；雄蕊5；子房半下位。花期4-8月。生海拔2900-3000米的沟谷林下。产台湾。

Perennial herbs. Leaves all basal, ovate-cordate, 3.5-7 cm long, obscurely 5-7-lobed, apex acute, both surfaces glandular strigose, margin irregularly dentate, base cordate; petiole 7-13 cm long, densely brown crisped glandular villous. Raceme many flowered, glandular villous; sepals broadly ovate, abaxially brown glandular hairy, 1-veined; petals margin pinnately 5-parted, segments linear, glandular dotted; stamens 5; ovary semi-inferior. Fl. Apr-Aug. Forests along ravines at 2900-3000 m. Distributed in Taiwan.

峨屏草

Tanakaea radicans Franchet et Savatier

多年生草本。叶、叶柄和花序均被褐色腺毛，有的带腺头。叶均基生，椭圆形、卵形至阔卵形，先端急尖，基部圆形或近心形，边缘具粗锯齿；叶柄长1.1-6.5厘米。花序总状；萼片狭卵形至披针形，单脉；无花瓣；雄蕊8-10；心皮2，下部合生，子房近上位，花柱2。蒴果2瓣裂。花果期4-10月。生海拔900-1100米的阴湿石隙。产四川。日本亦有。

Perennial herbs. Leaves, petioles and inflorescences brown glandular hairy, sometimes hairs glandular tipped. Leaves all basal, elliptic, ovate to broadly ovate, apex acute, base rounded or subcordate, margin coarsely serrate; petiole 1.1-6.5 cm long. Inflorescence racemose; sepals narrowly ovate to lanceolate, 1-veined; petals absent; stamens 8-10; carpels 2, connate near base, ovary subsuperior, styles 2. Capsule 2-valvate. Fl. and fr. Apr-Oct. Wet shaded rocks at 900-1100 m. Distributed in Sichuan. Also in Japan.

峨屏草 *Tanakaea radicans*

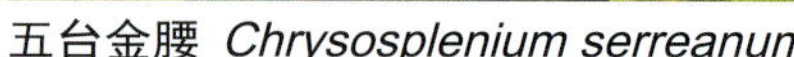
五台金腰 *Chrysosplenium serreanum*

锈毛金腰 *Chrysosplenium davidianum*

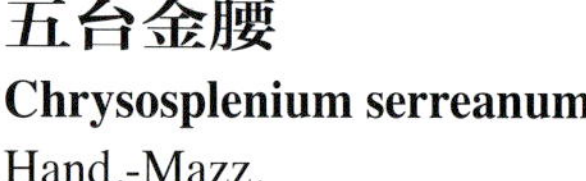

五台金腰
Chrysosplenium serreanum Hand.-Mazz.

多年生草本，无单宁质斑纹。鞭匐枝常在地下，具鳞片状叶。基生叶腹面疏生柔毛，肾形至圆状肾形；茎生叶常1枚，具柔毛。花黄色，花梗无毛或疏生褐色柔毛；萼片近圆形至阔卵形，无毛；无花盘。种子光滑无毛，有光泽。花果期5-7月。生海拔1700-2800米的林区湿地或溪畔。产黑龙江、内蒙古、河北和山西。俄罗斯、蒙古、朝鲜半岛和日本亦有。

Perennial herbs, not brown spotted. Stolons usually subterranean, scaly. Basal leaves pilose adaxially, long petiolate, reniform to orbicular-reniform; cauline leaf usually 1, pubescent. Flowers yellow, pedicel glabrous or pilose; sepals erect, broadly ovate to orbicular, glabrous; disc absent. Seeds glabrous, smooth. Fl. and fr. May-Jul. Forests, riversides at 1700-2800 m. Distributed in Heilongjiang, Neimenggu, Hebei and Shanxi. Also in Russia, Mongolia, Korean Peninsula and Japan.

肾叶金腰
Chrysosplenium griffithii Hook. f. et Thomson

丛生草本。茎单生，无毛或具褐色短柔毛。茎生叶肾形，两面无毛，边缘具11-15浅裂。聚伞花序上部多花；花萼开展，边缘常全缘；花黄色；雄蕊8。蒴果先端近截形；心皮横生，近等大。花果期5-9月。生海拔2500-4800米的林下、林缘、高山岩隙或高山草甸。产中国西南。印度北部、尼泊尔、不丹和缅甸北部亦有。

Cespitose herbs. Stems simple, glabrous or brown pilose. Cauline leaves reniform, both surfaces glabrous, margin 11-15-lobed. Cymes remotely many flowered; sepals spreading, margin usually entire; flowers yellow; stamens 8. Capsules apex subtruncate; carpels horizontal, subequal. Fl. and fr. May-Sep. Forests, forest edges, alpine rock clefts or alpine meadows at 2500-4800 m. Distributed in SW China. Also in N India, Nepal, Bhutan and N Myanmar.

肾叶金腰 *Chrysosplenium griffithii*

锈毛金腰
Chrysosplenium davidianum Decne. ex Maxim.

丛生草本。茎具褐色长曲柔毛。茎生叶互生，阔卵形至阔近椭圆形。聚伞花序多花；花黄色；萼片通常圆形；花瓣缺；雄蕊8；子房半下位；无花盘；心皮横生，近等大。蒴果。花果期4-8月。生海拔1500-4100米的湿地或林下岩石裂隙。产云南、四川和贵州。

Cespitose herbs. Stems brown crisped villous. Cauline leaves alternate, broadly ovate to broadly subelliptic. Cymes many flowered; flowers yellow; sepals usually orbicular; petals absent; stamens 8; ovary semi-inferior; disc absent; carpels horizontal, subequal. Fruits a capsule. Fl. and fr. Apr-Aug. Wet grasslands or forests rock crevices at 1500-4100 m. Distributed in Yunnan, Sichuan and Guizhou.

大叶金腰
Chrysosplenium macrophyllum Oliv.

多年生草本。基生叶倒卵形，先

大叶金腰 *Chrysosplenium macrophyllum*

端钝圆，全缘或具微波状小圆齿，基部楔形，腹面疏生褐色柔毛，背面无毛；茎生叶常1枚，狭椭圆形。萼片近卵形至阔卵形，先端微凹；雄蕊明显高出萼片；无花盘。果喙较长。种子密被微乳头突起。花果期4-6月。生海拔1000-2200米的林下或沟旁阴湿处。产华南。

Perennial herbs. Basal leaves obovate, apex obtuse, margin entire or undulate-crenulate, base cuneate, brown pilose adaxially, glabrous abaxially; cauline leaf usually 1, narrowly elliptic. Sepals subovate to broadly ovate, apex retuse; stamens distinctly longer than sepals; disc absent. Capsule rostrums long. Seeds densely papillose. Fl. and fr. Apr-Jun. Forests or shaded and wet places in ravines at 1000-2200 m. Distributed in S China.

天胡荽金腰

Chrysosplenium hydrocotylifolium Lévl. et Vaniot

多年生草本。茎多少具褐色柔毛。基生叶具长柄；叶近圆形或圆状肾形至肾形，具褐色单宁质斑点，两面无毛；叶柄基部具褐色柔毛。花序分枝疏生褐色柔毛；萼片在花期开展，具褐色单宁质斑点；子房近下位；花盘8裂。种子具微乳头突起，有光泽。花果期4-7月。生海拔1300-2400米的石灰岩隙。产广东、广西、贵州、四川和云南。

Perennial herbs. Stems brown pilose. Basal leaves long petiolate; leaf blade orbicular or orbicular-reniform to reniform, brown spotted, both surfaces glabrous; basal petiole and pleiochasium branches brown villous. Sepals spreading, brown spotted; ovary subinferior; disc 8-lobed. Seeds papillose, shiny. Fl. and fr. Apr-Jul. Forests, shaded places on limestone hills, limestone clefts at 1300-2400 m. Distributed in Guangdong, Guangxi, Guizhou, Sichuan and Yunnan.

蔓金腰

Chrysosplenium flagelliferum Fr. Schmidt.

多年生草本。鞭匐枝出自基生叶腋。基生叶肾形至圆状肾形，边缘具12-18钝齿，腹面疏生柔毛；茎生叶近扁圆形，边缘具5钝齿，无毛。花序分枝和花梗无毛；花较疏，花盘常8裂，周围无褐色乳头突起。种子具微乳头突起或柔毛。花果期5-7月。生海拔400-500米的林下阴湿处或溪边。产河北、黑龙江、吉林、辽宁。俄罗斯、蒙古、朝鲜半岛和日本亦有。

Perennial herbs. Stolons arising from basal leaf axils. Basal leaves reniform to orbicular-reniform, margin obtusely 12-18-dentate, pilose adaxially; cauline leaves broadly suborbicular, margin obtusely 5-dentate, glabrous. Inflorescence branches and pedicels glabrous; cyme remotely flowered, disc usually 8-lobed, not surrounded by brown papillae. Seeds sparsely puberulous. Fl. and fr. May-Jul. Shaded and wet places in forest understories and riversides at 400-500 m. Distributed in Hebei, Heilongjiang, Jilin, Liaoning. Also in Russia, Mongolia, Korean Peninsula and Japan.

天胡荽金腰 *Chrysosplenium hydrocotylifolium*

蔓金腰 *Chrysosplenium flagelliferum*

微子金腰 *Chrysosplenium microspermum*

微子金腰

Chrysosplenium microspermum Franch.

多年生草本。叶腋、花序分枝和花梗具褐色乳头突起。茎无毛。基生叶宽卵形至近肾形，边缘具8圆齿，两面无毛，叶柄无毛；茎生叶常3枚，互生，常近阔卵形，边缘具7-8圆齿，两面无毛，叶柄无毛。聚伞花序具5-9花；子房半下位；花盘8裂。种子褐色，具微瘤突。花果期4-9月。生海拔1800-2900米的沟谷湿地。产陕西、湖北和四川。

Perennial herbs. Leaf axils, inflorescences branches and pedicels brown papillose. Stems glabrous. Basal leaves broadly ovate to subreniform, margin 8-crenate, both surfaces glabrous, petioles glabrous; cauline leaves usually 3, alternate, usually broadly subovate, margin 7-8-crenate, both surfaces glabrous, petiole glabrous. Cymes 5-9-flowered; ovary semi-inferior; disc 8-lobed. Seeds brown, slightly tuberculate. Fl. and fr. Apr-Sep. Wet places in ravines at 1800-2900 m. Distributed in Shaanxi, Hubei and Sichuan.

山溪金腰

Chrysosplenium nepalense D. Don

草本。茎无毛。叶卵形至阔卵形，上面有时具褐色乳突，边缘具6-16圆齿。聚伞花序具8-18花；花黄绿色；萼片直立，阔倒卵形；雄蕊8；子房半下位；无花盘；心皮近等大。花果期5-7月。生海拔1500-5900米的林中、草甸或岩壁。产四川、云南和西藏。印度北部、尼泊尔、不丹和缅甸北部亦有。

Herbs. Stems glabrous. Leaves ovate to broadly so, adaxially sometimes brown papillose, margin 6-16-crenate. Cymes 8-18-flowered; flowers yellow-green; sepals erect, broadly ovate; stamens 8; ovary semi-inferior; disc absent; carpels subequal. Fl. and fr. May-Jul. Forests, meadows or rock clefts at 1500-5900 m. Distributed in Sichuan, Yunnan and Xizang. Also in N India, Nepal, Bhutan and N Myanmar.

中华金腰

Chrysosplenium sinicum Maxim.

多年生草本。不育枝发育良好；茎无毛。叶圆形至阔卵形，边缘具12-16钝齿。聚伞花序具4-10花；花黄绿色；雄蕊8；无花盘；心皮二叉状，明显不等大。花果期4-8月。生海拔500-3600米的林中或深谷中阴湿处。产中国西南、东南、华北、华西、华东和东北。俄罗斯、蒙古和朝鲜半岛亦有。

Perennial herbs. Sterile branches well developed; stems glabrous. Leaves orbicular to broadly ovate, margin obtusely 12-16-dentate. Cymes 4-10-flowered; flowers yellow-green; stamens 8; disc absent; carpels divaricate, distinctly unequal. Fl. and fr. Apr-Aug. Forests or shaded and wet

山溪金腰 *Chrysosplenium nepalense*

中华金腰 *Chrysosplenium sinicum*

林金腰 *Chrysosplenium lectus-cochleae*

毛金腰 *Chrysosplenium pilosum*

places in ravines at 500-3600 m. Distributed in SW, SE, N, W, E and NE China. Also in Russia, Mongolia and Korean Peninsula.

林金腰

Chrysosplenium lectus-cochleae Kitag.

多年生草本。茎生叶对生，具褐色斑点，近扇形，边缘具5-9圆齿。聚伞花序；花黄绿色；萼片直立，宽近卵形，顶端钝；雄蕊8；子房半下位，无花盘；花柱2，离生。蒴果，具2明显不等的心皮。花果期5-8月。生海拔400-1800米的林中、阴湿的地方或林缘岩石裂隙。产黑龙江、吉林和辽宁。

Perennial herbs. Cauline leaves opposite, brown spotted, subflabellate, margin 5-9-crenate. Cymes; flowers yellow-green; sepals erect, broadly subovate; stamens 8; ovary semi-inferior; disc absent; styles 2, free. Fruits a capsule, with 2 distinctly unequal carpels. Fl. and fr. May-Aug. Forests, shaded and wet places or rock clefts at forest edges at 400-1800 m. Distributed in Heilongjiang, Jilin and Liaoning.

毛金腰

Chrysosplenium pilosum Maxim.

多年生草本。茎生叶和苞叶背面和边缘具褐色柔毛，腹面无毛，边缘具钝齿；叶扇形，边缘具不明显6个波状齿或不明显钝齿。花萼具褐色腺点；雄蕊8；无花盘；心皮不等大。种子纵沟较浅。花果期4-6月。生海拔1500-3500米的阴湿林下或深谷。产中国西南、华南、东南、华中、华北、华西和东北。俄罗斯和朝鲜半岛亦有。

Perennial herbs. Cauline and bracteal leaves brown pilose abaxially and marginally, glabrous adaxially, margin distinctly obtusely dentate; leaves flabellate, margin obscurely 6-undulate-crenate or distinctly obtusely dentate. Sepals brown spotted; stamens 8; disc absent; carpels unequal. Seeds shallowly sulcate. Fl. and fr. Apr-Jun. Shady and wet places under forests or ravines at 1500-3500 m. Distributed in SW, S, SE, C, N, W and NE China. Also in Russia and Korean Peninsula.

肾萼金腰

Chrysosplenium delavayi Franch.

多年生草本。叶对生，阔卵形或圆形至扇形，边缘具7-12圆齿。花单生或2-5花组成聚伞状；花黄绿色；萼片开展，近阔圆形，具一褐色乳突；花盘8裂；心皮2，不等大。花果期3-6月。生海拔400-2800米的林下、灌丛或山谷石隙。产中国西南、华南、华中和华东。缅甸北部亦有。

Perennial herbs. Leaves opposite, broadly ovate or orbicular to flabellate, margin 7-12-crenate. Flowers solitary or cyme 2-5-flowered; flowers yellow-green; sepals spreading, broadly suborbicular, brown 1-papillate; disk 8-lobed; carpels 2, unequal. Fl. and fr. Mar-Jun. Forests, thickets or rock crevices in valleys at 400-2800 m. Distributed in SW, S, C and E China. Also in N Myanmar.

肾萼金腰 *Chrysosplenium delavayi*

长瓣梅花草

Parnassia longipetala Hand.-Mazz.

多年生草本。基生叶常1枚；茎生叶无柄，肾形。花瓣绿色，披针形或长披针形；退化雄蕊深紫色，短于雄蕊；子房上位，扁球形，具3或4个稍加厚的棱。蒴果扁球形。花期7月，果期8月。生海拔2400-3900米的林缘或高山草甸。产云南和西藏。

Perennial herbs. Basal leaves usually 1; cauline leaves sessile, reniform. Petals green, lanceolate or long-lanceolate; staminodes dark purple, shorter than stamens; ovary superior, depressed globose, with 3 or 4 slightly thickened angles. Capsules depressed globose. Fl. Jul. Fr. Aug. Forest edges or alpine meadows at 2400-3900 m. Distributed in Yunnan and Xizang.

青铜钱

Parnassia tenella Hook. f. et Thoms.

多年生草本。基生叶肾形，基部深心形，边全缘；茎生叶无柄半抱茎，肾形。萼筒陀螺状，萼片半圆形至卵形，边缘膜质，啮蚀状，有紫色小斑点；花瓣绿色，扇形；退化雄蕊锤状，顶端不裂，圆形或盘状。花期8月，果期9月。生海拔2800-3400米的杂木林下或林边。产云南西北部、四川西部和西藏。尼泊尔和印度亦有。

Perennial herbs. Basal leaves reniform, base deeply cordate, apex rounded; cauline leaves sessile, semiamplexicaul, reniform. Hypanthium turbinate; sepals semiorbicular, ovate, or triangular-ovate, margin erose, sparsely purple punctate; petals green, flabellate; staminodes terete, apex of lamina entire, occasionally minutely crenate. Fl. Aug. Fr. Sep. Mixed forests or forest margins at 2800-3400 m. Distributed in NW Yunnan, W Sichuan and Xizang. Also in Nepal and India.

白花梅花草

Parnassia scaposa Mattf.

多年生草本。基生叶椭圆形或倒卵形，叶脉常5条并行；茎上无叶。花单生于茎顶；萼片卵状披针形；花瓣白色，倒卵形，顶端通常圆钝，基部楔形，有短爪；退化雄蕊扁平，3个裂片宽短，为全长的1/5，中间裂片高出两侧裂片。花期8月。生海拔3700-4500米的河谷、高山草甸或灌丛中。产青海、四川和西藏。

Perennial herbs. Basal leaves elliptic or obovate, 5-veined; cauline leafless. Flower solitary on apex of stem; sepals ovate-lanceolate; petals white, obovate, apex roun-

长瓣梅花草 *Parnassia longipetala*

青铜钱 *Parnassia tenella*

白花梅花草 *Parnassia scaposa*

中国梅花草 *Parnassia chinensis*

凹瓣梅花草 *Parnassia mysorensis*

ded or emarginated, base cuneate into a claw; staminodes flat, divided for 1/5 its length into 3 lobes, central lobe longer than lateral ones. Fl. Aug. Scrub, alpine meadows at 3700-4500 m. Distributed in Qinghai, Sichuan and Xizang.

中国梅花草

Parnassia chinensis Franch.

小草本。茎在近中部有一片叶；叶背灰绿色，具紫色腺点，常肾形或卵状肾形。花单生于茎顶，花瓣白色；退化雄蕊阔匙形，扁平；子房扁球形，先端急缩成花柱；柱头3浅裂。花期7-8月。生海拔3600-4200米的灌丛、高山草甸或山坡。产云南、四川和西藏。印度北部、尼泊尔、不丹和缅甸亦有。

Small herbs. Stems with 1 leaf near middle; leaves abaxially gray-green, purple punctate, usually reniform or ovate-reniform. Flowers solitary at stem apices, petals white; staminodes broadly spatulate, flat; ovary depressed globose, apex abruptly contracted into style; stigma 3-lobed. Fl. Jul-Aug. Thickets, alpine meadows or slopes at 3600-4200 m. Distributed in Yunnan, Sichuan and Xizang. Also in N India, Nepal, Bhutan and Myanmar.

凹瓣梅花草

Parnassia mysorensis Heyne ex Wight et Arn.

多年生草本。茎生叶1；叶卵状心形、阔卵形或卵状长圆形。花瓣白色，先端凹或2裂；退化雄蕊扁平，宽匙形，3浅裂，约长度的1/3，裂片近等长；子房上位。蒴果卵球形，3瓣裂。花期7-8月，果期9月。生海拔2500-3600米的混生林、灌丛、草甸、山坡草地或山坡开阔地。产云南、四川、西藏和贵州。印度北部亦有。

Perennial herbs. Cauline leaves one; leaves ovate-cordate, broadly ovate or ovate-oblong. Petals white, apex emarginate or 2-cleft; staminodes flat, broadly spatulate, 3-lobed for ca. 1/3 its length, lobes subequal; ovary superior. Capsules ovoid, 3-valved. Fl. Jul-Aug. Fr. Sep. Mixed forests, scrub areas, meadows, grassy places or open slopes at 2500-3600 m. Distributed in Yunnan, Sichuan, Xizang and Guizhou. Also in N India.

近凹瓣梅花草

Parnassia submysorensis J. T. Pan

多年生草本。基生叶卵状长圆形或宽卵形，基部深心形，7-9脉；近中部或稍偏上具1茎生叶，无柄半抱茎，基部有铁锈色附属物。花单生；萼筒管陀螺状；花瓣白色，倒卵形，边缘呈稀疏啮蚀状和极短流苏状毛；退化雄蕊3(-6)浅裂，裂片顶端平，中间裂片窄，两侧宽。花期7月。生海拔3400-3600米的林下阴湿草坡灌丛中。产云南。

Perennial herbs. Basal leaves ovate-oblong or broadly ovate, base cordate or deeply so, arcuate 7-9-veined; cauline leaf sessile, semi-amplexicaul, often with rusty brown appendages at base. Flower solitary; sepals turbinate, densely punctate; petals white, obovate, margin erose-dentate and fimbriate; staminodes flat, 3(-6)-lobed, central lobe narrower than lateral ones. Fl. Jul. Shrubs in forests, shaded slopes at 3400-3600 m. Distributed in Yunnan.

近凹瓣梅花草 *Parnassia submysorensis*

细叉梅花草 *Parnassia oreophila*

细叉梅花草
Parnassia oreophila Hance

多年生草本。基生叶具柄，茎生叶无柄，半抱茎；叶卵状长圆形或三角状卵形。花单生于茎顶；花白色，具紫褐色脉纹；退化雄蕊扁平，3裂至其(1/2-)2/3；子房半下位。蒴果长卵球形。花期7-8月，果期9月。生海拔1600-3000米的高山草甸、林缘、湿润山坡或路边。产中国西南、西北和华北。

Perennial herbs. Basal leaves with petioles, cauline leaves sessile, semiamplexicaul; leaves ovate-oblong or triangular-ovate. Flowers solitary at stem apex; petals white, purple-brown veined; staminodes flat, 3-parted for (1/2-) 2/3 its length; ovary semi-inferior. Capsules long ovoid. Fl. Jul-Aug. Fr. Sep. Alpine meadows, forest edges, moist slopes or roadsides at 1600-3000 m. Distributed in SW, NW and N China.

绿花梅花草
Parnassia viridiflora Batalin

多年生草本。茎生叶无柄，有或无紫色腺点，半抱茎。花单生顶端，绿色；萼筒管陀螺状；雄蕊5；退化雄蕊5，扁平，具柄，顶端3裂，裂片顶端截形；子房半下位。蒴果。花期8月，果期9月。生海拔3600-4100米的灌丛或高山草甸。产云南、四川、陕西和青海。

Perennial herbs. Cauline leaves sessile, purple punctate or not, semiamplexicaul. Flowers solitary, terminal, green; hypanthium turbinate; stamens 5; staminodes 5, flat, with stalk, apex 3-lobed, lobes truncate at apex; ovary semi-inferior. Fruits a capsule. Fl. Aug. Fr. Sep. Scrubby areas or alpine meadows at 3600-4100 m. Distributed in Yunnan, Sichuan, Shaanxi and Qinghai.

指裂梅花草
Parnassia cooperi W. E. Evans

多年生草本。基生叶肾形或卵状心形；茎生叶无柄，半抱茎。萼片质地粗糙，内面有紫色小斑点；花瓣淡黄绿色，窄披针形，边缘具疏流苏状毛；退化雄蕊3裂，两侧裂片长，呈弯月形，中间裂片很短小，直立，呈三角形。花期7-8月，果期9月。生海拔2400-2800米的山坡铁杉林下。产西藏。印度和不丹亦有。

Perennial herbs. Basal leaves reniform or ovate-cordate; cauline leaf sessile, semiamplexicaul. Sepals scabrous, adaxially purple punctate; petals yellowish green, narrowly lanceolate, margin sparsely long fimbriate; staminode 3-lobed, lateral lobes lanceolate, central lobe short, triangular. Fl. Jul-Aug. Fr. Sep. *Tsuga* forests on slopes at 2400-2800 m. Distributed in Xizang. Also in India and Bhutan.

短柱梅花草
Parnassia brevistyla (Brieg.) Hand.-Mazz.

多年生草本。基生叶卵状心形或卵形；茎生叶与基生叶同形，基部有铁锈色附属物。花单生茎顶；花瓣白色，具爪；

绿花梅花草 *Parnassia viridiflora*

指裂梅花草 *Parnassia cooperi*

药隔连合并伸长呈匕首状；退化雄蕊3浅裂，中间裂片窄，两侧裂片宽；花柱极短。花期7-8月，果期9月。生海拔2800-4400米的山坡阴湿的林下和林缘、云杉林间空地、山顶草坡下或河滩草地。产四川、西藏、云南、甘肃和陕西。

Perennial herbs. Basal leaf blade ovate-cordate or ovate; cauline leaf similar to basal ones, base often with several rusty brown appendages. Petals white, base attenuate into claw; anthers connective projected at apex into a lanceolate appendage; staminodes 3-lobed, central lobe narrower than lateral ones; styles usually short. Fl. Jul-Aug. Fr. Sep. Moist forests, canopy openings in *Picea* forests, forest margins, grassy slopes, riverside meadows at 2800-4400 m. Distributed in Sichuan, Xizang, Yunnan, Gansu and Shaanxi.

突隔梅花草
Parnassia delavayi Franch.

多年生草本。基生叶3或4(-7)；叶肾形或近圆形；茎生叶常于基部具2或3个早落的锈棕色附属物。花白色；花瓣5；药隔连合伸长，呈披针形附属物；退化雄蕊扁平，3裂。蒴果倒卵球形。花期7-8月，果期9月。生海拔1800-3800米的溪边疏林中、冷杉林、杂木林、砾石地或潮湿草地。产云南、四川、湖北、陕西和甘肃。不丹亦有。

Perennial herbs. Basal leaves 3 or 4(-7); leaves reniform or suborbicular; cauline leaves often with 2 or 3 caducous, rusty brown appendages at base. Flowers white; petals 5; connective apex projected into a lanceolate appendage; staminodes flat, 3-lobed. Capsules obovoid. Fl. Jul-Aug. Fr. Sep. Open forests by streams, *Abies* forests, mixed forests, gravelly slopes or moist grasslands at 1800-3800 m. Distributed in Yunnan, Sichuan, Hubei, Shaanxi and Gansu. Also in Bhutan.

鸡肫草
Parnassia wightiana Wall. ex Wight et Arn.

多年生草本。基生叶2-5；叶近三角状卵形或肾形。花瓣白色，边缘下半部具长流苏状毛，上半部啮蚀状之齿，稀啮蚀状；退化雄蕊扁平，5浅裂达1/2，顶端偶有不明显腺体。蒴果倒卵球形。花期7-8月，果期9月。生海拔600-2000米的山谷疏林中、山谷、草地区或路旁。产中国西南、华南和华中。印度北部、尼泊尔、不丹和泰国北部亦有。

Perennial herbs. Basal leaves 2-5; leaves subtriangular-ovate or reniform. Petals white, margin long fimbriate proximally, erose dentate or rarely erose distally; staminodes flat, 5-lobed for up to 1/2 their lengths, occasionally with inconspicuous glands at apex. Capsules obovoid. Fl. Jul-Aug. Fr. Sep. Open forests in valleys, valleys, grassy areas or roadsides at 600-2000 m. Distributed in SW, S and C China. Also in N India, Nepal, Bhutan and N Thailand.

短柱梅花草 *Parnassia brevistyla*

突隔梅花草 *Parnassia delavayi*

鸡肫草 *Parnassia wightiana*

龙胜梅花草 *Parnassia longshengensis*

龙胜梅花草
Parnassia longshengensis T. C. Ku

多年生草本。基生叶通常6-8，丛生；叶阔圆形；茎生叶基部常具数个锈褐色附属物。花单生于茎顶；花瓣白色，密具紫色腺点；退化雄蕊分裂至其1/2至5分枝，顶部具球形腺点。蒴果三棱形。花期7-8月，果期9-11月。生林下或潮湿处。产广西东北部。

Perennial herbs. Basal leaves often 6-8, forming a rosette; leaves broadly orbicular; cauline leaves base often with several rusty brown appendages. Flowers solitary at stem apex; petals white, densely purple punctate; staminodes divided for up to 1/2 their length into 5 branches with globose glands at apex. Capsules trigonous. Fl. Jul-Aug. Fr. Sep-Nov. Under forests or moist places. Distributed in NE Guangxi.

梅花草
Parnassia palustris L.

多年生草本。基生叶通常3至多数；叶下常具紫色腺点，卵形或长卵形。花单生于茎顶；花瓣白色，常具紫色腺点；退化雄蕊具(7-)9-11(-13)分枝，顶端具球形腺点。蒴果卵球形。花期7-9月，果期10月。生海拔1200-2200米的潮湿草坡、溪边或沟谷阴湿处。产新疆北部。哈萨克斯坦、俄罗斯、蒙古、朝鲜半岛、日本、欧洲和北美洲亦有。

Perennial herbs. Basal leaves 3 to numerous; leaves abaxially often purple punctate, ovate or long ovate. Flowers solitary at stem apex; petals white, often purple punctate; staminodes (7-)9-11 (-13)-branched with globose glands at apex. Capsules ovoid. Fl. Jul-Sep. Fr. Oct. Moist grassy slopes, streamsides or shaded moist places in valleys at 1200-2200 m. Distributed in N Xinjiang. Also in Kazakhstan, Russia, Mongolia, Korean Peninsula, Japan, Europe and North America.

多枝梅花草
Parnassia palustris L. var. **multiseta** Ledeb.

本变种与梅花草的区别在于本变种的退化雄蕊比雄蕊长，具13-21分枝。花期8月。生海拔1200-2200米的山谷、溪边或草地。产华北、东北和西北。哈萨克斯坦、俄罗斯、蒙古、朝鲜半岛、日本、欧洲和北美洲亦有。

梅花草 *Parnassia palustris*

多枝梅花草 *Parnassia palustris* var. *multisetu*

黄山梅 *Kirengeshoma palmata*

This variety differs from the typical variety in its sterile anthers longer than anthers, 13-21-branched; Fl. Aug Valleys, streamsides or grassy fields at 1200-2200 m. Distributed in N, NE and NW China. Also in Kazakhstan, Russia, Mongolia, Korean Peninsula, Japan, Europe and North America.

黄山梅

Kirengeshoma palmata Yatabe

多年生草本。叶圆心形，掌状7-10裂，基部近心形，两面被糙伏毛。聚伞花序常具3花，中部最大，两侧较小，苞片披针形；花黄色；萼筒半球形，被柔毛，裂片三角形；花瓣长圆状倒卵形或近狭倒卵形。蒴果阔椭圆形或近球形。种子周围具膜质斜翅。花期3-4月，果期5-8月。生海拔700-1800米的山谷林中阴湿处。产安徽和浙江。日本和朝鲜半岛亦有。

Perennial herbs. Leaf blade near base of stem orbicular, palmately veined, base subcordate, margin 3-10-lobed, with appressed hairy on both surfaces. Inflorescences 3-flowered; flowers at center of inflorescence largest, lateral flowers with a linear bracteole; petals yellow; calyx tube hemispheric, lobes deltoid; petals, irregularly oblong-obovate or subovate. Capsule broadly ellipsoid or subglobose. Seeds surrounded by an oblique wing. Fl. Mar-Apr. Fr. May-Aug. Moist forests, valleys at 700-1800 m. Distributed in Anhui and Zhejiang. Also in Japan and Korean Peninsula.

光萼溲疏

Deutzia glabrata Kom.

灌木。花枝红褐色，具4-6叶，无毛。叶卵形至卵状披针形，膜质，上面无毛或疏被3-4(-5)辐线星状毛，下面无毛。花序伞房状，具5-20(-30)花，无毛；萼筒杯状，无毛。蒴果球形，无毛。花期6-7月，果期8-9月。生海拔 300-1300米的混交林中、灌丛或山坡。产华中、华北、华西和东北。俄罗斯(西伯利亚)和朝鲜半岛亦有。

Shrubs. Flowering branchlets red-brown, 4-6-leaved, glabrous. Leaves ovate to ovate-lanceolate, membranous, adaxially glabrous or 3-4(-5)-rayed stellate hairy, abaxially glabrous. Inflorescences corymbose, 5-20(-30)-flowered, glabrous; calyx tubes cupular, glabrous. Capsules globose, glabrous. Fl. Jun-Jul. Fr. Aug-Sep. Mixed forests, thickets or mountain slopes at 300-1300 m. Distributed in C, N, W and NE China. Also in Russia (Siberia) and Korean Peninsula.

光萼溲疏 *Deutzia glabrata*

小花溲疏 *Deutzia parviflora*

粉红溲疏 *Deutzia rubens*

小花溲疏

Deutzia parviflora Bunge

灌木。花枝具4-6叶，被星状毛。叶椭圆状卵形或卵状披针形，纸质，下面具6-12辐线星状毛。伞房花序通常多花；萼筒杯状，密具星状毛；花瓣覆瓦状，白色。蒴果半球形。花期4-6月，果期8-10月。生海拔300-1800米的林中、灌丛、山坡或山谷。产华北、华中、华西和东北。俄罗斯和朝鲜半岛亦有。

Shrubs. Flowering branchlets 4-6-leaved, stellate hairy. Leaves elliptic-ovate or ovate-lanceolate, papery, abaxially 6-12-rayed stellate hairy. Inflorescences corymbose, usually many flowered; calyx tubes cupular, densely stellate hairy; petals imbricate, white. Capsules hemispheric. Fl. Apr-Jun. Fr. Aug-Oct. Forests, thickets, slopes or valleys at 300-1800 m. Distributed in N, C, W and NE China. Also in Russia and Korean Peninsula.

粉红溲疏

Deutzia rubens Rehder

灌木。花枝具4叶，被星状毛。叶长圆形或卵状长圆形，膜质，下面具5或6(或7)辐线星状毛。聚伞花序5-10花；花萼裂片紫色；花瓣重瓣，粉红色，倒卵圆形。蒴果半球形。花期4-7月，果期8-10月。生海拔2100-3000米的山坡或灌丛中。产四川、湖北、陕西和甘肃。

Shrubs. Flowering branchlets 4-leaved, stellate hairy. Leaves oblong or ovate-oblong, membranous, abaxially 5- or 6(or 7)-rayed stellate hairy. Cymes 5-10-flowered; calyx lobes purple; petals induplicate, pink, obovate. Capsules hemispheric. Fl. Apr-Jul. Fr. Aug-Oct. Slopes or thickets at 2100-3000 m. Distributed in Sichuan, Hubei, Shaanxi and Gansu.

密序溲疏

Deutzia compacta Craib

灌木。花枝褐色或红褐色，被星状毛。叶纸质，卵状披针形或长圆状披针形，下面疏被6-8辐线星状毛，星状毛大小不等，小者密，大者散生于叶脉上。伞房花序顶生，花序轴被具疣状体的星状毛；花瓣粉红色，阔倒卵形或近圆形。蒴果近球形。花期4-5月，果期6-7月。生海拔2000-4200米的山坡林缘。产云南和西藏。尼泊尔、不丹、印度和缅甸亦有。

密序溲疏 *Deutzia compacta*

西藏溲疏 *Deutzia hookeriana*

纸质，上面散生5或6(-8)辐线星状毛。圆锥花序狭，多花；萼筒有时具紫色斑点，无毛或仅裂片疏具星状毛；花重瓣，白色。蒴果半球形。花期5-6月，果期8-9月。生海拔600-1200米的混交林中、山坡或山谷。产浙江、湖北、河南、安徽和江西。

Shrubs. Flowering branchlets 4-6-leaved, glabrous. Leaves papery, adaxially sparsely 5 or 6 (-8)-rayed stellate hairy. Panicles narrow, many flowered; calyx tubes sometimes purple spotted, glabrous or only lobes sparsely stellate hairy; petals induplicate, white. Capsules hemispheric. Fl. May-Jun. Fr. Aug-Sep. Mixed forests, mountain slopes or valleys at 600-1200 m. Distributed in Zhejiang, Hubei, Henan, Anhui and Jiangxi.

Shrubs. Flowering branchlets brown or reddish brown, stellate hairy. Leaf blade papery, ovate-lanceolate or oblong-lanceolate, abaxially 6-8-rayed stellate hairy, hairs of 2 types (numerous, minute, contiguous ones and less numerous, larger ones with long central rays when along leaf veins). Inflorescences terminal, corymbose, stellate hairy, hairs with papilliform base; petals, pink, broadly obovate or suborbicular, Capsule hemispheric. Fl. Apr-May. Fr. Jun-Jul. Mixed forest margins, mountain slopes at 2000-4200 m. Distributed in Yunnan and Xizang. Also in Nepal, Bhutan, India and Myanmar.

西藏溲疏

Deutzia hookeriana (Schneid.) Airy-Shaw

灌木。花枝红褐色，被具疣状体的星状毛。叶纸质，卵状披针形或长圆状披针形，叶下面密被8-10(-15)辐线星状毛，星状毛大小近相等。伞房花序顶生，被具疣状体的星状毛；花瓣白色，阔倒卵形，外面疏被星状毛，内面无毛，花蕾时覆瓦状排列。花期5-8月，果期7-9月。生海拔2000-3500米的林下、林缘或灌丛中。产西藏和云南。不丹、印度和缅甸亦有。

Shrubs. Flowering branchlets reddish brown, stellate hairy, hairs with papilliform base. Leaf blade papery, ovate- or oblong-lanceolate, abaxially densely 8-10 (-15)-rayed stellate hairy, equal. Inflorescences corymbose, stellate hairy, hairs with papilliform base; petals white, broadly ovate, abaxially sparsely with stellate hairy, imbricate. Fl. May-Aug. Fr. Jul-Sep. Mixed forest margins, thickets at 2000-3500 m. Distributed in Xizang and Yunnan. Also in Bhutan, India and Myanmar.

黄山溲疏

Deutzia glauca Cheng

灌木。花枝具4-6叶，无毛。叶

黄山溲疏 *Deutzia glauca*

异色溲疏 *Deutzia discolor*

异色溲疏

Deutzia discolor Hemsl.

灌木。花枝具2-4(-6)叶，疏具星状毛。叶纸质，椭圆状披针形或长圆状披针形。聚伞花序具花12-20朵；萼筒密具10-12辐线星状毛；花瓣重瓣，白色，椭圆形。蒴果褐色，半球形。花期6-7月，果期8-10月。生海拔1000-2500米的山坡或溪边灌丛中。产四川、湖北、河南、陕西和甘肃。

Shrubs. Flowering branchlets 2-4(-6)-leaved, sparsely stellate hairy. Leaves papery, elliptic-lanceolate or oblong-lanceolate. Cymes with flowers 12-20; calyx tubes densely 10-12-rayed stellate hairy; petals induplicate, white, elliptic. Capsules brown, hemispheric. Fl. Jun-Jul. Fr. Aug-Oct. Slopes or bushes by rivers at 1000-2500 m. Distributed in Sichuan, Hubei, Henan, Shaanxi and Gansu.

大花溲疏

Deutzia grandiflora Bunge

灌木。花枝具2-4叶，被星状毛。叶卵状菱形或椭圆状卵形。聚伞花序1-5花；萼筒密具灰黄色星状毛；花白色；外轮雄蕊长6-7毫米；花丝先端2裂。蒴果半球形。花期4-6月，果期9-11月。生海拔800-1600米的灌丛、山坡或山谷。产华中、华北、华西和华东。

Shrubs. Flowering branchlets 2-4-leaved, stellate hairy. Leaves ovate-rhomboid or elliptic-ovate. Cymes 1-5-flowered; calyx tubes densely gray-yellow stellate hairy; petals white; outer stamens 6-7 mm long; filaments 2-dentate at apex. Capsules hemispheric. Fl. Apr-Jun. Fr. Sep-Nov. Thickets, mountain slopes or valleys at 800-1600 m. Distributed in C, N, W and E China.

灌丛溲疏

Deutzia rehderiana C. K. Schneid.

灌木。花枝褐色，具2-4叶。叶下面密被5-8(-10)辐线星状毛，上面疏具4-6(-7)辐线星状毛。聚伞花序具(1-)3-5(-11)花；萼筒外面密被8-10辐线星状毛；花白色。蒴果半球形。花期5月，果期7-8月。生海拔500-2000米的灌丛或山坡。产云南、四川和贵州。

Shrubs. Flowering branchlets brown, 2-4-leaved. Leaves abaxially densely 5-8(-10)-rayed stellate hairy, adaxially sparsely 4-6(-7)-rayed stellate hairy; Cymes (1-)3-5(-11)-flowered; calyx tubes densely 8-10-rayed stellate hairy; petals white. Capsules

大花溲疏 *Deutzia grandiflora*

灌丛溲疏 *Deutzia rehderiana*

钩齿溲疏 *Deutzia baroniana*

球花溲疏 *Deutzia glomeruliflora*

hemispheric. Fl. May. Fr. Jul-Aug. Thickets or mountain slopes at 500-2000 m. Distributed in Yunnan, Sichuan and Guizhou.

钩齿溲疏
Deutzia baroniana Diels

灌木。花枝浅褐色，具2-4叶，具星状毛。叶卵状菱形或卵状椭圆形，边缘具不规则细齿或具间隔的大小齿。聚伞花序具1-3花；萼筒密具4-6辐线星状毛；花瓣重瓣，白色。蒴果半球形。花期4-5月，果期9-10月。生海拔500-1200米的灌丛或山坡。产华北和华东。

Shrubs. Flowering branchlets brownish, 2-4 leaved, stellate hairy. Leaves ovate-rhomboid or ovate-elliptic, margin irregularly serrulate or with alternating large and small teeth. Cymes 1- 3-flowered; calyx tubes densely 4-6-rayed stellate hairy; petals induplicate, white. Capsules hemispheric. Fl. Apr-May. Fr. Sep-Oct. Thickets or slopes at 500-1200 m. Distributed in N and E China.

球花溲疏
Deutzia glomeruliflora Franch.

灌木。花枝红褐色，具4-6叶，具星状毛。叶纸质，上面疏被具4-5辐线星状毛，下面被具4-7辐线星状毛。聚伞花序聚集，具花3-18朵；萼筒密具星状毛；裂片淡紫色或绿色；花瓣重瓣，白色。蒴果半球形。花期4-6月，果期8-10月。生海拔2000-3600米的灌丛、杂木林或云杉林下。产云南和四川。

Shrubs. Flowering branchlets red-brown, 4-6-leaved, stellate hairy. Leaves papery, adaxially sparsely covered with 4-5-rayed stellate hairs, abaxially with 4-7-rayed stellate hairs. Cymes aggregate, 3-18-flowered; calyx tubes densely stellate hairy; lobes purplish or green; petals induplicate, white. Capsules hemispheric. Fl. Apr-Jun. Fr. Aug-Oct. Thickets, mixed forests or *Picea* forests at 2000-3600 m. Distributed in Yunnan and Sichuan.

紫花溲疏
Deutzia purpurascens (Franch. ex L. Henry) Rehder

灌木。叶阔卵状披针形或卵状披针形，上面疏被具3-5辐线星状毛，下面浅绿色，疏被具4-8(-10)辐线星状毛。伞房状聚伞花序具3-12花；萼筒革质；花瓣粉红色。蒴果半球形。花期4-6月，果期6-10月。生海拔2600-3500米的灌丛或杂木林中。产云南、四川和西藏东南部。印度和缅甸亦有。

Shrubs. Leaves broadly ovate-lanceolate or ovate-lanceolate, adaxially sparsely covered with 3-5-rayed stellate hairs, abaxially greenish and sparsely 4-8(-10)-rayed stellate hairs. Corymbose cymes 3-12-flowered; calyx tubes leathery; petals pink. Capsules hemispheric. Fl. Apr-Jun. Fr. Aug-Oct. Thickets or mixed forests at 2600-3500 m. Distributed in Yunnan, Sichuan and SE Xizang. Also in India and Myanmar.

紫花溲疏 *Deutzia purpurascens*

大萼溲疏 *Deutzia calycosa*

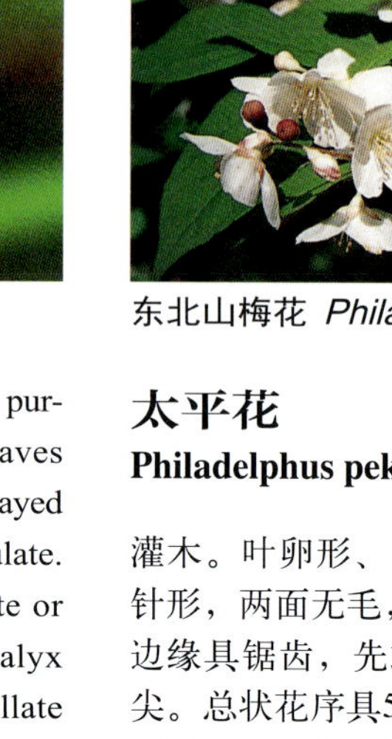

东北山梅花 *Philadelphus schrenkii*

大萼溲疏

Deutzia calycosa Rehder

灌木。花枝紫褐色，具2-4叶。叶下疏具7-10(-12)辐线星状毛，边缘具细齿。伞房状聚伞花序聚集或开展，具9-12花；萼筒密具8-10辐线星状毛；花瓣重瓣，白色或粉红色。蒴果球形。花期3-4月，果期7-8月。生海拔2000-3600米的林中或灌丛。产云南西部和四川西南部。

Shrubs. Flowering branchlets purplish brown, 2-4-leaved. Leaves abaxially sparsely 7-10(-12)-rayed stellate hairy, margin serrulate. Corymbose cymes aggregate or spreading, 9-12-flowered; calyx tubes densely 8-10-rayed stellate hairy; petals induplicate, white or pink. Capsules globose. Fl. Mar-Apr. Fr. Jul-Aug. Forests or thickets at 2000-3600 m. Distributed in W Yunnan and SW Sichuan.

太平花

Philadelphus pekinensis Rupr.

灌木。叶卵形、阔椭圆形或披针形，两面无毛，离基3-5脉，边缘具锯齿，先端渐尖或长渐尖。总状花序具5-9花；花萼干后黄绿色，无毛；花瓣白色。蒴果近球状或倒圆锥状。花期5-7月，果期8-10月。生海拔700-900米的林中、灌丛中或山坡。产湖北、河北、山西、陕西和辽宁。朝鲜半岛亦有。

Shrubs. Leaves ovate, broadly elliptic or lanceolate, both surfaces glabrous, veins 3-5 basifugal, margin serrate, apex acuminate or long acuminate. Racemes 5-9-flowered; calyx yellowish green when dry, glabrous; corolla white. Capsules subglobose or obconical. Fl. May-Jul. Fr. Aug-Oct. Forests, thickets or slopes at 700-900 m. Distributed in Hubei, Hebei, Shanxi, Shaanxi and Liaoning. Also in Korean Peninsula.

东北山梅花

Philadelphus schrenkii Rupr.

灌木。叶卵形或椭圆状卵形，

太平花 *Philadelphus pekinensis*

边缘具细齿，先端渐尖。总状花序具5-7花；花萼干后黄绿色，疏具柔毛；花冠碟形；花瓣白色；花柱分裂达1/2或超过；柱头槌状。蒴果椭圆体形。花期6-7月，果期8-9月。生海拔100-1500米的杂木林中。产河北、陕西、黑龙江、吉林和辽宁。俄罗斯东南部和朝鲜半岛亦有。

Shrubs. Leaves ovate or elliptic ovate, margin serrate, apex acuminate. Racemes 5-7-flowered; calyx yellowish green when dry, sparsely pubescent; corolla discoid; petals white; styles divided for 1/2 its length or more; stigmas mallet-shaped. Capsules ellipsoid. Fl. Jun-Jul. Fr. Aug-Sep. Mixed forests at 100-1500 m. Distributed in Hebei, Shaanxi, Heilongjiang, Jilin and Liaoning. Also in SE Russia and Korean Peninsula.

紫萼山梅花

Philadelphus purpurascens (Koehne) Rehder

灌木。叶卵形或椭圆形，两面无毛，具离基3出脉。总状花序具5-7(-9)花；共轴紫色；花萼紫色或暗紫色；冠筒灰色，具斑点，坛形，疏具柔毛或渐无毛；花瓣白色。蒴果卵球形。花期5-6月，果期7-9月。生海拔2200-3500米的杂木林中、灌丛或山坡。产云南和四川。

Shrubs. Leaves ovate or elliptic, both surfaces glabrous, triplinerved. Racemes 5-7(-9)-flowered; rachis purple; calyx tubes purple or dark purple; tubes glaucous, spotted, urn-shaped, sparsely pubescent to glabrescent; petals white. Capsules ovoid. Fl. May-Jun. Fr. Jul-Sep. Mixed forests, thickets, or slopes at 2200-3500 m. Distributed in Yunnan and Sichuan.

云南山梅花

Philadelphus delavayi L. Henry

灌木。叶长圆形或卵状披针形，下面密具灰色长柔毛，脉3-5，稍离基，边缘具细齿或近全缘，先端渐尖。总状花序具花5-21朵；花萼紫色、黑紫色或褐色；萼筒灰色，无毛；花瓣白色。蒴果倒卵形。花期6-8月，果期9-11月。生海拔700-3800米的林中、灌丛中或山坡。产云南、四川和西藏。缅甸亦有。

云南山梅花 *Philadelphus delavayi*

Shrubs. Leaves oblong- or ovate-lanceolate, abaxially densely gray villous, veins 3-5 slightly basifugal, margin serrulate or subentire, apex acuminate. Racemes 5-21-flowered; sepals purple, black-purple or brown; Calyx gray, glabrous; corolla white. Capsules obovoid. Fl. Jun-Aug. Fr. Sep-Nov. Forests, thickets or slopes at 700-3800 m. Distributed in Yunnan, Sichuan and Xizang. Also in Myanmar.

毛柱山梅花

Philadelphus subcanus Koehne

灌木。叶卵形或阔卵形，纸质，下面具近绵毛状长柔毛或仅沿脉具长柔毛，离基3-5脉，边缘疏具细齿，先端锐尖或渐尖。总状花序具9-11花；萼筒被金黄色或灰黄色柔毛；花冠碟形；花瓣白色。蒴果倒卵球形。花期6-7月，果期8-10月。生海拔500-2300米的林中或灌丛。产云南、四川和湖北。

Shrubs. Leaves ovate or broadly so, papery, abaxially sublanate-villous or villous only along veins, veins 3-5 basifugal, margin sparsely serrulate, apex acute or acuminate. Racemes 9-11-flowered; calyx tubes golden yellow or gray-yellow villous; corolla discoid; petals white. Capsules obovoid. Fl. Jun-Jul. Fr. Aug-Oct. Forests or thickets at 500-2300 m. Distributed in Yunnan, Sichuan and Hubei.

紫萼山梅花 *Philadelphus purpurascens*

毛柱山梅花 *Philadelphus subcanus*

滇南山梅花 *Philadelphus henryi*

绢毛山梅花 *Philadelphus sericanthus*

滇南山梅花

Philadelphus henryi Koehne

灌木。叶卵形或卵状长圆形，叶上面有刚毛，下面沿主脉和侧脉疏被长硬毛，基出3-5脉，边缘具细锯齿，先端急渐尖。花序总状，具5-22花；萼筒白色，具糙毛或刚毛；花冠近碟形；花瓣白色。蒴果倒卵球形。花期6-7月，果期8-10月。生海拔1300-2500米的灌丛或山坡。产云南和贵州。

Shrubs. Leaves ovate or ovate-oblong, adaxially setose, abaxially sparsely hirsute along costa and nerves, veins 3-5 basifugal, margin serrate, apex abruptly acuminate. Inflorescences racemose, 5-22-flowered; calyx tubes white, strigose or bristly; corolla subdiscoid; petals white. Capsules obovoid. Fl. Jun-Jul. Fr. Aug-Oct. Thickets or mountain slopes at 1300-2500 m. Distributed in Yunnan and Guizhou.

绢毛山梅花

Philadelphus sericanthus Koehne

灌木。叶椭圆形、椭圆状披针形，下面仅沿脉具长柔毛，具离基3-5脉，边缘具细齿或9-12糙粗齿，先端渐尖。总状花序或圆锥花序具3-15(-30)花；花萼外面疏被糙状毛；花冠碟形；花瓣白色。蒴果倒卵球形。花期5-6月，果期8-9月。生海拔300-3000米的山坡灌丛中或杂木林下。产中国西南、东南、华中、华北、华西和华东。

Shrubs. Leaves elliptic, elliptic-lanceolate, abaxially villous only along veins, veins 3-5 basifugal, margin serrate or coarsely 9-12-dentate, apex acuminate. Inflorescences racemose or paniculate, 3-15(-30)-flowered; calyx tubes sparsely strigose; corolla discoid; petals white. Capsules obovoid. Fl. May-Jun. Fr. Aug-Sep. Thickets on slopes or mixed forests at 300-3000 m. Distributed in SW, SE, C, N, W and E China.

赤壁木

Decumaria sinensis Oliver

攀援灌木。叶革质，倒卵形至椭圆形，先端钝或急尖，边全缘或上部具疏锯齿或波状；叶柄长1-2厘米。花序梗长1-3厘米；萼筒陀螺状；花白色；雄蕊20-30，子房5-10室。蒴果钟状或陀螺状，先端截形，具宿存花柱和柱头，有10-12条棱。种子有白翅。花期3-5月，果期8-10月。生海拔600-1300米的山坡灌丛或岩石缝中。产陕西、甘肃、湖北、四川和贵州。

Climbing shrubs. Leaves leathery, obovate to elliptic, apex obtuse to acute, margin entire or sparsely serrate or undulate distally from middle; petiole 1-2 cm long. Peduncle 1-3 cm long; calyx tubes turbinate; petals white; stamens 20-30, ovary 5-10-loculed. Capsule campanulate to turbinate, apex truncate, with persistent style and stigma, 10-12-angled. Seeds with white wings. Fl. Mar-May. Fr. Aug-Oct. Thickets on mountain slopes or rock crevices at 600-1300 m. Distributed in Shaanxi, Gansu, Hubei, Sichuan and Guizhou.

冠盖藤

Pileostegia viburnoides Hook. f. et Thoms

常绿攀援状灌木。小枝、叶和花序无毛或极少被星状绒毛或柔毛。叶椭圆状倒披针形或长椭圆形，基部楔形至阔楔形。伞房状圆锥花序顶生；苞片和小苞片线状披针形，花白色；萼筒圆锥状，裂片三角形；花瓣卵形。蒴果圆锥形，5-10肋纹或棱。花期7-8月，果期9-12月。生海拔600-1000米的山谷

赤壁木 *Decumaria sinensis*

林中。产华南。印度、越南和日本亦有。

Shrubs. Branchlets, leaves and inflorescences glabrous or rarely stellate pubescent. Leaf blade elliptic-oblanceolate or long elliptic, base truncate. Inflorescences terminal; bracts linear-lanceolate; calyx tube lobes 4 or 5, deltoid; petals white, ovate. Capsule conical-turbinate, 5-10-ribbed. Fl. Jul-Aug. Fr. Sep-Dec. Forests in valleys at 600-1000 m. Distributed in S China. Also in India, Vietnam and Japan.

冠盖藤 *Pileostegia viburnoides*

常山

Dichroa febrifuga Lour.

灌木。叶两面无毛或仅叶脉被卷曲短柔毛，稀下面被长柔毛。花序为伞房圆锥状；萼裂片4-6；花瓣成熟后反折，蓝色或白色，稍肉质；花瓣蓝色或白色；子房3/4下位；花柱4(-6)；花丝线形。浆果成熟后深蓝色。花期2-4月，果期5-8月。生海拔200-2000米的林中。产中国西南、华南、华中、华西和华东。南亚和东南亚亦有。

Shrubs. Leaves glabrous on both surfaces or crisped-pubescent along veins, rarely villose abaxially. Inflorescences a corymbose panicle; calyx lobes 4-6; petals reflexed at maturity, blue or white, slightly fleshy; ovary 3/4 inferior; styles 4(-6); filaments linear. Berries dark blue when mature. Fl. Feb-Apr. Fr. May-Aug. Forests at 200-2000 m. Distributed in SW, S, C, W and E China. Also in S and SE Asia.

常山 *Dichroa febrifuga*

云南常山

Dichroa yunnanensis S. M. Hwang

灌木。叶纸质，上面疏被卷曲短柔毛和散生长粗毛，下面被卷曲短柔毛，沿脉更密。花瓣5，蓝色或白色；子房半下位；花柱(2-)3(-4)；柱头近圆形；花丝钻形。浆果成熟后深蓝色，球形。花期5-6月，果期7-8月。生海拔2000米的林中。产云南。

Shrubs. Leaves papery, adaxially sparsely crisped-pubescent and hirsute, abaxially crisped-pubescent, especially along veins. Petals 5, blue or white; ovary semi-inferior; styles (2-)3(-4); stigmas subglobose; filaments subulate. Berries dark blue when mature, globose. Fl. May-Jun. Fr. Jul-Aug. Forests at 2000 m. Distributed in Yunnan.

云南常山 *Dichroa yunnanensis*

草绣球 *Cardiandra moellendorffii*

草绣球

Cardiandra moellendorffii (Hance) Migo

亚灌木。叶常分散互生于茎上，边缘有粗牙齿状锯齿。伞房状聚伞花序顶生；不育花萼片2-3，阔卵形至近圆形，基部近截平，膜质；孕性花萼筒杯状，萼齿阔卵形，先端钝；子房近下位，3室，花柱3。种子两端具不透明或半透明的翅，翅比种子颜色深。花果期7-10月。生海拔700-1500米的山谷密林或山坡疏林。产中国东南和华中。日本亦有。

Subshrubs. Leaves sparsely scattered along stem, alternate, margin roughly dentate-serrate. Corymbiform cymes terminal; sterile flowers with sepals 2 or 3, broadly ovate to suborbicular, base subtruncate, membranous; fertile flowers with calyx tube cupular, teeth broadly ovate, apex obtuse; ovary subinferior, 3- locular, style 3. Seeds with opaque or translucent wings, darker than color of seed body. Fl. and fr. Jul- Oct. Dense to sparse forests in valleys or on mountain slopes at 700-1500 m. Distributed in SE and C China. Also in Japan.

蛛网萼

Platycrater arguta Sieb. et. Zucc.

落叶灌木。叶披针形至椭圆形，先端尾状渐尖，基部稍下延，边缘有粗锯齿或小齿，两面被短柔毛或正面近无毛。不育花具萼片3-4，中部以下合生，轮廓三角形或四方形；孕性花萼齿4-5，先端长渐尖；雄蕊极多数，子房下位。蒴果倒圆锥状。花果期7-10月。生海拔400-1800米的山谷疏林或灌丛、溪边和山坡。产安徽、浙江、江西和福建。日本亦有。

Deciduous shrubs. Leaves lanceolate to elliptic, apex caudate-acuminate, base slightly decurrent, margin roughly serrate to serrulate, both surfaces pubescent or adaxially subglabrous. Sterile flowers with sepals 3-4, connate from base to middle and forming a triangle or square; fertile flowers with teeth 4-5, apex long-acuminate; stamens extremely numerous, ovary inferior. Capsule obconical. Fl. and fr. Jul-Oct. Sparse forests or thickets in valleys, stream banks and mountain slopes at 400-1800 m. Distributed in Anhui, Zhejiang, Jiangxi and Fujian. Also in Japan.

钻地风

Schizophragma integrifolium Oliv.

攀援植物。叶椭圆形、狭椭圆形或阔卵形，下面无毛或有时沿脉具柔毛或脉腋具髯毛。花序伞房状；不育花具增大的1(-3)花萼，黄白色；花瓣狭卵形，先端钝。蒴果钟状或陀螺

蛛网萼 *Platycrater arguta*

钻地风 *Schizophragma integrifolium*

状，基部宽楔形。花期6-7月，果期10-11月。生海拔200-2000米的沟谷或林中。产中国西南、华南、东南、华中和华东。

Climbers. Leaves elliptic, narrowly so or broadly ovate, abaxially glabrous or sometimes sparsely pubescent along veins or barbate at vein axils. Inflorescences corymbose; sterile flowers with expanded sepal 1(-3), yellow-white; petals narrowly ovate, apex obtuse. Capsules campanulate or turbinate, base broadly cuneate. Fl. Jun-Jul. Fr. Oct-Nov. Valleys or forests at 200-2000 m. Distributed in SW, S, SE, C and E China.

叉叶蓝

Deinanthe caerulea Stapf

多年生草本。叶常4片聚集于茎顶部，近轮生，阔椭圆形、卵形或倒卵形，先端具尾状尖头，不分裂或2裂，裂片长5-6厘米，边缘具粗的锐尖齿，两面疏被单毛。伞房状聚伞花序顶生；不育花花梗纤细；萼片3-4；孕性花花梗粗壮；萼筒宽陀螺状，萼齿5；子房半下位，花柱合生，顶端5裂。花期6-7月。生海拔700-1600米的山谷潮湿林中。产湖北。

Perennial herbs. Leaves usually 4, crowded apically on stem, subverticillate, broadly elliptic, ovate or obovate, apex with caudate cusp, apex entire or deeply 2-lobed with lobes 5-6 cm long, margin roughly and acutely dentate, both surfaces with sparsely simple hairs. Corymbiform cymes terminal; sterile flowers with slender pedicel; sepals 3-4; fertile flowers with stout pedicel; calyx tube turbinate, lobes 5; ovary semi-inferior, styles connate, apex 5-lobed. Fl. Jun-Jul. Moist forests in valleys at 700-1600 m. Distributed in Hubei.

中国绣球

Hydrangea chinensis Maxim.

灌木。叶披针形、狭椭圆形或倒卵形，两面疏具柔毛或除沿脉外无毛。花序为伞形或近伞房状；不育花具3或4萼片；花瓣黄色。蒴果纺锤状至卵球状球形。花期3-8月，果期5-10月。生海拔300-2000米的林中、山坡或山谷。产华南、东南和华东。日本亦有。

Shrubs. Leaves lanceolate, narrowly elliptic or obovate, both surfaces sparsely pubescent or glabrous except along veins. Inflorescences umbellate or subcorymbose; sterile flowers with sepals 3 or 4; petals yellow. Capsules fusiform to ovoid-globose. Fl. Mar-Aug. Fr. May-Oct. Forests, slopes or valleys at 300-2000 m. Distributed in S, SE and E China. Also in Japan.

叉叶蓝 *Deinanthe caerulea*

中国绣球 *Hydrangea chinensis*

西南绣球 *Hydrangea davidii*

西南绣球
Hydrangea davidii Franch.

灌木。叶长圆形至狭椭圆形，上面除脉外近无毛，脉腋处具簇生柔毛。伞房状聚伞花序；不育花具3或4萼片；花瓣蓝色；子房半下位；花柱3-4。蒴果近球状。种子无翅。花期4-6月，果期9-10月。生海拔1400-2400米的山坡杂木林或山谷。产云南、四川和贵州。

Shrubs. Leaves oblong to narrowly elliptic, abaxially subglabrous except pilose along veins and fasciculate pubescent at vein axils. Inflorescences corymbose cymes; sterile flowers with 3 or 4 sepals; petals blue; ovary semi-inferior; styles 3-4. Capsules subglobose. Seeds wingless. Fl. Apr-Jun. Fr. Sep-Oct. Mixed forests on slopes or valleys at 1400-2400 m. Distributed in Yunnan, Sichuan and Guizhou.

福建绣球
Hydrangea chungii Rehder

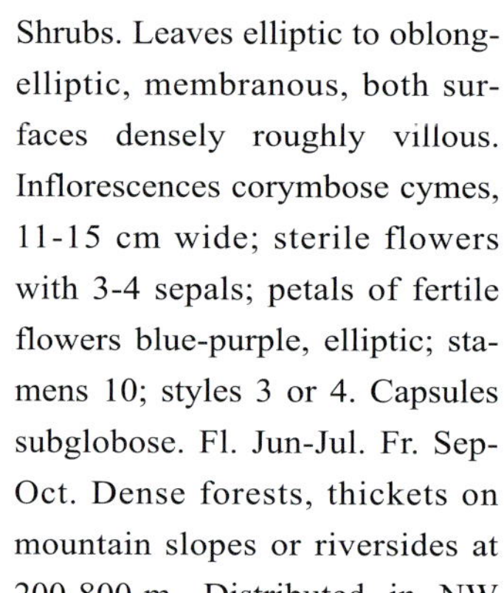

灌木。叶椭圆形至长圆状椭圆形，膜质，两面密具粗糙长柔毛。花序为伞房状聚伞花序，宽11-15厘米；不育花具3-4萼片；能育花的花瓣蓝紫色，椭圆形；雄蕊10；花柱3或4。蒴果近球形。花期6-7月，果期9-10月。生海拔200-800米的密林下、山坡灌丛中或河边。产福建西北部。

Shrubs. Leaves elliptic to oblong-elliptic, membranous, both surfaces densely roughly villous. Inflorescences corymbose cymes, 11-15 cm wide; sterile flowers with 3-4 sepals; petals of fertile flowers blue-purple, elliptic; stamens 10; styles 3 or 4. Capsules subglobose. Fl. Jun-Jul. Fr. Sep-Oct. Dense forests, thickets on mountain slopes or riversides at 200-800 m. Distributed in NW Fujian.

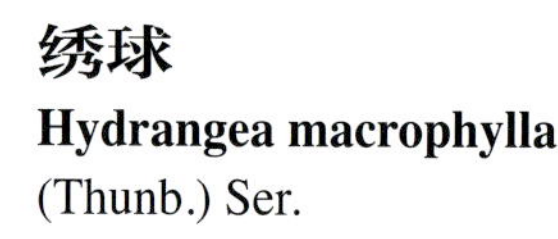

绣球
Hydrangea macrophylla (Thunb.) Ser.

灌木。叶倒卵形或广椭圆形，边缘具疏锯齿，先端骤尖。伞房状聚伞花序近球形，直径8-20厘米，大多为辐射状花；花美丽，白色、粉红色或蓝色，为不孕花。蒴果近卵球形。花期6-8月。生海拔400-2400米的沟谷或疏林中。栽培。原产于日本。

福建绣球 *Hydrangea chungii*

绣球 *Hydrangea macrophylla*

圆锥绣球 *Hydrangea paniculata*

Shrubs. Leaves obovate or broadly elliptic, margins sparsely crenate, apex abrupt acute. Corymbose cymes subglobose, 8-20 cm diam, mostly radiate flowers; flowers beautiful, white, pink or blue, sterile. Capsules subovoid. Fl. Jun-Aug. Valleys or open forests at 400-2400 m. Cultivated. Native to Japan.

圆锥绣球
Hydrangea paniculata Siebold

灌木。叶2片对生或3片轮生，卵形至椭圆形，纸质，下面沿脉具平伏长柔毛。圆锥状聚伞花序，金字塔形，长26厘米；花梗和分枝密具柔毛；不育花具4枚萼片，白色；花瓣白色。蒴果椭圆体形。花期7-8月，果期10-11月。生海拔300-2100米的山谷、疏林或灌丛中。产中国西南、东南、华中和华西。俄罗斯和日本亦有。

Shrubs. Leaves 2-opposite or 3-verticillate, ovate to elliptic, papery, abaxially appressed villous along veins. Paniculate cymes, pyramidal, 26 cm long; peduncles and branches densely pubescent; sterile flowers with sepals 4, white; petals white. Capsules ellipsoid. Fl. Jul-Aug. Fr. Oct-Nov. Valleys, open forests or thickets at 300-2100 m. Distributed in SW, SE, C and W China. Also in Russia and Japan.

东陵绣球
Hydrangea bretschneideri Dippel

灌木。叶纸质，侧脉每边7-8条，下面密具灰白色曲至近直立的长柔毛。花序伞房聚伞花序；不育花萼片4；孕性花萼筒杯状；萼齿三角形；花瓣白色。蒴果卵圆状球形。花期6-7月，果期9-10月。生海拔1200-2800米的山谷沟边林中或山坡。产华北至华西。

Shrubs. Leaves papery, lateral veins 7-8 per side, abaxially densely gray-white crisped to suberect villous, Inflorescences corymbose cymes; sterile flowers with 4 sepals; fertile flowers with cupular calyx tube; teeth triangular; petals white. Capsules ovoid-globose. Fl. Jun-Jul. Fr. Sep-Oct. Forests along streams banks in valleys or mountain slopes at 1200-2800 m. Distributed in N to W China.

东陵绣球 *Hydrangea bretschneideri*

微绒绣球 *Hydrangea heteromalla*

微绒绣球

Hydrangea heteromalla D. Don

灌木。叶下面密被绒毛。伞房状聚伞花序，宽15-20厘米，果期宽达27厘米；不育花萼片4，白色或带黄色，全缘；子房半下位；花柱3或4。蒴果倒卵状球形或近球形。花期6-7月，果期9-10月。生海拔2400-3400米的山坡或近山顶林中或灌丛。产云南、四川和西藏。印度东北部、尼泊尔和不丹亦有。

Shrubs. Leaves abaxially densely gray-white velutinous. Inflorescences corymbose cymes, 15-20 cm wide, to 27 cm wide in fruit; sterile flowers with 4 sepals, white or yellowish, entire; ovary up to 1/2 superior; styles 3 or 4. Capsules ovoid-globose or subglobose. Fl. Jun-Jul. Fr. Sep-Oct. Forests or thickets on mountain slopes or tops at 2400-3400 m. Distributed in Yunnan, Sichuan and Xizang. Also in NE India, Nepal and Bhutan.

马桑绣球

Hydrangea aspera D. Don

灌木。叶上面疏或密被糙伏毛，下面密被灰白色短柔毛或长柔毛。花序为伞房状聚伞花序，宽8-25厘米；不育花萼片4，稀5，淡红色；花瓣紫蓝色或紫红色；花柱2-3；子房下位。蒴果坛状。花期7-9月，果期9-11月。生海拔700-4000米的山谷灌丛、密林或山坡。产中国西南、华中和华西。印度东北部、尼泊尔和越南亦有。

Shrubs. Leaves adaxially sparsely to densely strigose, abaxially densely gray-white pubescent or villose. Inflorescences corymbose cymes, 8-25 cm wide; sepals of sterile flowers 4 or rarely 5, pale red; petals purple-blue or purple-red; styles 2 or 3, ovary inferior. Capsules urn-shaped. Fl. Jul-Sep. Fr. Sep-Nov. Dense forests or thickets in valleys or mountain slopes at 700-4000 m. Distributed in SW, C and W China. Also in NE India, Nepal and Vietnam.

蜡莲绣球

Hydrangea strigosa Rehder

灌木。叶下面密被黄褐色颗粒状腺体和灰白色糙伏毛。伞房状聚伞花序，宽达28厘米；不育花萼片4-5，白色至紫红色；孕性花萼钟状；花瓣紫红色；花柱2。蒴果坛形。花期7-8月，果期10-12月。生海拔500-1800米的山谷林中或灌丛。产四川、贵州、湖北、湖南和陕西。

Shrubs. Leaves abaxially dense fulvous granular glands and gray-white strigose. Inflorescences corymbose cymes, to 28 cm wide; sterile flowers with 4 or 5 sepals, white to purplish red; fertile flowers with campanulate calyx tube; petals purplish red; styles 2. Capsules urn-shaped. Forests or thickets in valleys at 500-1800 m. Fl. Jul-Aug. Fr. Oct-Dec. Distributed in Sichuan, Guizhou, Hubei, Hunan and Shaanxi.

冠盖绣球

Hydrangea anomala D. Don

攀援灌木。叶干后两面黄褐色，两面无毛或下面有时沿脉疏具浅褐色柔毛和脉腋具髯毛，纸质。伞房状聚伞花序，果期宽达30厘米，具柔毛；不育花具4萼片；花瓣连合成冠盖状；子房下位。种子周边有翅。花期5-6月，果期9-10月。生海拔500-2900米的山谷溪边、多石山坡或疏或密林中。产长江以南各省。印度、不丹、尼泊尔和缅甸亦有。

马桑绣球 *Hydrangea aspera*

蜡莲绣球 *Hydrangea strigosa*

Shrubs climbing. Leaves yellow-brown on both surfaces when dry, both surfaces glabrous or abaxially sometimes sparsely brownish pubescent along veins and barbate at vein axils papery. Inflorescences corymbose cymes, to 30 cm wide in fruit, pubescent; sterile flowers with sepals 4; petals apex connected and forming a calyptra; ovary inferior. Seeds surrounded by wings. Fl. May-Jun. Fr. Sep-Oct. By streams in valleys, rocky mountain slopes or open or dense forests at 500-2900 m. Distributed throughout the provinces to the south of Yangtze River. Also in India, Bhutan, Nepal and Myanmar.

滇鼠刺

Itea yunnanensis Franch.

灌木。叶卵形或椭圆形，5-10厘米长，2.5-5厘米宽，边缘具刺状锯齿。总状花序顶生，稍下垂至下垂；花多数，常3朵簇生；花萼浅杯形；花瓣花时直立，顶端稍弯曲；雄蕊常短于花瓣。蒴果圆锥形。花果期5-12月。生海拔1100-3000米的林中、岩石地或河岸。产云南、四川、西藏、贵州和广西。

Shrubs. Leaves ovate or elliptic, 5-10 cm long, 2.5-5 cm broad, margin spiny-serrate. Racemes terminal, nodding to pendulous; flowers numerous, often 3-clustered; calyx shallowly cupular; petals erect at anthesis, slightly curved apically; stamens always shorter than petals. Capsules conical. Fl. and fr. May-Dec. Forests, rocky places or riversides at 1100-3000 m. Distributed in Yunnan, Sichuan, Xizang, Guizhou and Guangxi.

大叶鼠刺 *Itea macrophylla*

冠盖绣球 *Hydrangea anomala*

大叶鼠刺

Itea macrophylla Wall. ex Roxb.

乔木。叶宽卵形或宽椭圆形，10-20厘米长，5-12厘米宽，边缘具腺锯齿。总状花序腋生，常2或3朵簇生；花萼杯形；花瓣花时反卷；雄蕊约花瓣长度一半。蒴果开展或下垂，具棱，狭圆锥形。花果期4-6月。生海拔500-1500米的林中或山坡路边。产云南南部、广西和海南。南亚和东南亚亦有。

Trees. Leaves broadly ovate or broadly elliptic, 10-20 cm long, 5-12 cm broad, margin glandular serrate. Racemes axillary, usually 2- or 3-clustered; calyx cupular; petals reflexed at anthesis; stamens ca. 1/2 as long as petals. Capsules spreading or pendulous, striate, narrowly conical. Fl. and fr. Apr-Jun. Forests or roadsides on mountain slopes at 500-1500 m. Distributed in S Yunnan, Guangxi and Hainan. Also in S and SE Asia.

滇鼠刺 *Itea yunnanensis*

毛脉鼠刺 *Itea indochinensis* var. *pubinervia*

毛脉鼠刺

Itea indochinensis Merr. var. **pubinervia** (Chang) C. Y. Wu

灌木或小乔木。小枝被密柔毛。叶纸质，椭圆形，边缘有细锯齿，上面无毛，下面脉腋内有短柔毛。总状花序簇生于叶腋，少于4个，且较短；花序轴及花梗被长柔毛；萼筒杯状，被长柔毛；花瓣白色，披针形。蒴果被毛，成熟时基部开裂。花期3-5月，果期5-12月。生海拔1000-2100米的疏林中、水边、路边石旁。产广东、广西、贵州和云南。

Shrubs or trees. Young branchlets densely pubescent. Leaf blade papery, elliptic, margin serrulate, abaxially barbellate only at vein axils. Racemes axillary, less than 4-fascicled, short; rachis and pedicel densely villous; calyx cupular, villous; petals white, lanceolate. Capsule puberulous, dehiscing from base when ripe. Fl. Mar-May. Fr. May-Dec. Sparse forests, streamsides, roadsides at 1000-2100 m. Distributed in Guangdong, Guangxi, Guizhou and Yunnan.

长刺茶藨子

Ribes alpestre Wall. ex Decne.

灌木。小枝节处刺3枚，轮生，粗壮，长1-3厘米。叶阔卵形，3-5裂。花腋生，单生或2或3个形成短总状，两性；花萼淡绿色或红褐色；花瓣白色；子房和果实有腺毛。果实紫色，球形至椭圆体形。花期4-6月，果期6-9月。生海拔1000-3900米的阳坡疏林中、林缘、河谷草地或河边。产中国西南和华西。不丹、克什米尔地区和阿富汗亦有。

Shrubs. Branches nodal spines 3, verticillate, stout, 1-3 cm long. Leaves broadly ovate, 3-5-lobed. Flowers axillary, solitary or 2 or 3 in short racemes, bisexual; calyx greenish or reddish brown; petals white; ovary and fruits glandular hairy. Fruits purple, globose to ellipsoid. Fl. Apr-Jun. Fr. Jun-Sep. Open forests on sunny slopes, forest edges, grasslands in ravines or riversides at 1000-3900 m. Distributed in SW and W China. Also in Bhutan, Kashmir and Afghanistan.

刺果茶藨子

Ribes burejense Fr. Schmidt

灌木。节间具3-7粗状刺。叶阔卵形，幼时具柔毛，渐无毛。花腋生，单生或2-3组成短总状花序；花萼棕色至红褐色；花瓣白色至淡粉红色。果球状，具多数小刺。花期5-6月，果期7-8月。生海拔900-2300米山上的林中、林缘、灌丛或溪边。产华北、华西和东北。俄罗斯、蒙古和朝鲜半岛亦有。

长刺茶藨子 *Ribes alpestre*

刺果茶藨子 *Ribes burejense*

长果茶藨子 *Ribes stenocarpum*

香茶藨子 *Ribes odoratum*

Shrubs. Nodal spines 3-7, stout. Leaves broadly ovate, pubescent when young, gradually glabrescent. Flowers axillary, solitary or 2-3 in short racemes; calyx brown to reddish brown; petals white to pinkish. Fruits globose, with numerous minute spines. Fl. May-Jun. Fr. Jul-Aug. Forests, forest edges, thickets or streamsides in mountains at 900-2300 m. Distributed in N, W and NE China. Also in Russia, Mongolia and Korean Peninsula.

长果茶藨子
Ribes stenocarpum Maxim.

灌木。节间刺1-3。叶阔卵形至近圆形，具柔毛，渐无毛。花腋生，单生或2-3组成短总状花序；花萼绿色至绿褐色，无毛；花瓣白色。果红色或绿色带红色，矩圆状。花期5-6月，果期7-8月。生海拔2300-3300米的林中、山坡灌丛或山沟。产四川、陕西、甘肃和青海。

Shrubs. Nodal spines 1-3. Leaves broadly ovate to suborbicular, pubescent, gradually glabrescent. Flowers axillary, solitary or 2-3 in short racemes; calyx green to greenish brown, glabrous; petals white. Fruits red or green tinged red, oblong. Fl. May-Jun. Fr. Jul-Aug. Forests, thickets on mountain slopes or ravines at 2300-3300 m. Distributed in Sichuan, Shaanxi, Gansu and Qinghai.

香茶藨子
Ribes odoratum H. L. Wendl.

灌木。叶倒卵圆形至圆肾形，3浅裂，具柔毛及腺点，后渐无毛。总状花序常下垂，5-10花；花两性，芳香；花萼黄色，无毛；花瓣浅红色。果熟后黑色，球形至宽椭圆体形。花期5月，果期7-8月。栽培于黑龙江、辽宁或东北其他地区。原产北美洲。

Shrubs. Leaves obovate to orbicular-reniform, 3-lobed, pubescent and glandular, later glabrescent. Racemes often nodding, 5-10-flowered; flowers bisexual, fragrant; calyx yellow, glabrous; petals pinkish. Fruits black, globose to broadly ellipsoid. Fl. May. Fr. Jul-Aug. Cultivated in Heilongjiang, Liaoning and probably also elsewhere in NE China. Native to North America.

糖茶藨子
Ribes himalense Royle ex Decne. var. **himalense**

灌木。小枝、叶面、叶柄和花序均无腺毛。叶卵圆形或近圆形，3-5裂。总状花序密，具8-20花；花两性；花萼绿色带紫色或紫红色；花瓣红色或绿色带紫色，近匙形至扇形。果实红色，成熟后变为紫黑色，球形。花期4-6月，果期7-8月。生海拔1200-4100米的林中、灌丛、路边或河岸。产中国西南、华北和华西。印度北部、尼泊尔、不丹和克什米尔地区亦有。

Shrubs. Branchlets, leaf surfaces, petioles and inflorescences without glandular hairs. Leaves ovate or suborbicular, 3-5-lobed. Racemes dense, 8-20-flowered; flowers bisexual; calyx green tinged purple or purplish red; petals red or green tinged purple, subspatulate to flabellate. Fruits red, turning purplish black after maturity, globose. Fl. Apr-Jun. Fr. Jul-Aug. Forests, thickets, roadsides or river banks at 1200-4100 m. Distributed in SW, N and W China. Also in N India, Nepal, Bhutan and Kashmir.

糖茶藨子 *Ribes himalense* var. *himalense*

瘤糖茶藨子 *Ribes himalense* var. *verruculosum*

天山茶藨子 *Ribes meyeri*

瘤糖茶藨子

Ribes himalense Royle ex Decne. var. **verruculosum** (Rehder) L. T. Lu

本变种与糖茶藨子的区别在于本变种叶背脉上和叶柄有显著瘤状突起或混生少数短腺毛。总状花序较小；花萼光滑。花期4-6月，果期7-8月。生海拔1200-4100米的林中、林缘、山谷或山坡。产中国西南、华北和华西。印度北部、尼泊尔、不丹和克什米尔地区亦有。

This variety differs from the typical variety in its abaxial leaf blade veins and petiole distinctly tuberculate or sometimes sparsely shortly stalked glandular. Inflorescences short. calyx glabrous. Fl. Apr-Jun. Fr. Jul-Aug. Forests, forest edges, valleys or slopes at 1200-4100 m. Distributed in SW, N and W China. Also in N India, Nepal, Bhutan and Kashmir.

天山茶藨子

Ribes meyeri Maxim.

灌木。小枝无刺。叶近圆形，无毛或下面具柔毛，裂片(3-)5，边缘具粗糙锯齿。花两性；总状花序下垂，密集，具7-17花；花轴和花梗具柔毛或近无毛；萼筒钟状至短筒状。果紫黑色，球形。花期5-6月，果期7-8月。生海拔1200-3900米的山坡疏林中、河边针叶林下或路边灌丛中。产新疆。俄罗斯和蒙古亦有。

Shrubs. Branchlets unarmed. Leaves suborbicular, glabrous or abaxially pubescent, lobes (3-)5, margin coarsely serrate. Flowers bisexual; racemes nodding, dense, 7-17-flowered; rachis and pedicels pubescent or subglabrous; calyx tubes campanulate to shortly cylindric. Fruits purplish black, globose. Fl. May-Jun. Fr. Jul-Aug. Open forests on mountain slopes, coniferous forests on river banks or thickets along roads at 1200-3900 m. Distributed in Xinjiang. Also in Russia and Mongolia.

东北茶藨子 *Ribes mandshuricum*

东北茶藨子

Ribes mandshuricum (Maxim.) Kom.

灌木。叶两面具平伏柔毛，渐无毛，裂片3(-5)，边缘具不规则糙锐齿或重锯齿。花两性；总状花序长，后下垂，具40-50花；萼筒盆状；萼片反折；花瓣黄绿色。果实球形，红色。花期4-6月，果期7-8月。生海拔300-1900米的山坡针叶或混交林中或灌丛中，或山谷。产华北、华西和东北。俄罗斯(西伯利亚)和朝鲜半岛北部亦有。

Shrubs. Leaves appressed-pubescent on both surfaces, glabrescent, lobes 3(-5), margin irregularly coarsely sharply serrate or doubly serrate. Flowers bisexual; racemes long, then nodding, 40-50-flowered; calyx tubes pelviform; sepals reflexed; petals yellowish green. Fruits globose,

桂叶茶藨子 *Ribes laurifolium*

长白茶藨子 *Ribes komarovii*

red. Fl. Apr-Jun. Fr. Jul-Aug. Coniferous or mixed forests or thickets on mountain slopes or in valleys at 300-1900 m. Distributed in N, W and NE China. Also in Russia (Siberia) and N Korean Peninsula.

桂叶茶藨子

Ribes laurifolium Jancz.

常绿灌木，稀小乔木。叶卵圆形、长卵圆形或椭圆形，革质，无毛，3-5脉。雄花序悬垂，多达12花，雌花序直立；花萼黄绿色；花瓣楔状匙形。果紫色，椭圆体形至矩圆形。花期4-6月，果期8-10月。生海拔2100-3600米的林中、山坡、河岸或岩石上。产云南、四川和贵州。

Evergreen shrubs, rarely small trees. Leaves ovate, ovate-oblong or elliptic, leathery, glabrous, 3-5-veined. Male racemes pendulous, up to 12-flowered, female ones erect; calyx yellowish green; petals cuneate-spatulate. Fruits purple, ellipsoid to oblong. Fl. Apr-Jun. Fr. Aug-Oct. Forests, slopes, riverbanks or rocks at 2100-3600 m. Distributed in Yunnan, Sichuan and Guizhou.

长白茶藨子

Ribes komarovii Rojark.

灌木，雌雄异株。叶阔卵形至近圆形，有时较狭。总状花序直立，多于10花；花轴和花梗具短柄状腺体；花单性；萼筒杯状；花柱2裂。果红色，倒卵球形至球形，无毛。花期5-6月，果期8-9月。生海拔400-2100米的路边林下、灌丛或石坡。产华北、华西和东北。俄罗斯和朝鲜半岛北部亦有。

Shrubs, dioecious. Leaves broadly ovate to suborbicular, sometimes narrower. Racemes erect, more than 10-flowered; rachis and pedicels shortly stalked glandular; flowers unisexual; calyx tubes cupular; styles 2-lobed. Fruits red, obovoid-globose to globose, glabrous. Fl. May-Jun. Fr. Aug-Sep. Forests by roadsides, thickets or stony slopes at 400-2100 m. Distributed in N, W and NE China. Also in Russia and N Korean Peninsula.

冰川茶藨子

Ribes glaciale Wall.

灌木，雌雄异株。叶窄卵圆形，稀为近圆形，3-5裂，无毛或疏具带柄腺体。雄总状花序具10-30花，雌花序具4-10花；花瓣短于萼片；花萼红褐色，无毛。果实红色，酸味，近球形至倒卵球状球形。花期4-6月，果期7-9月。生海拔1900-3000米的山坡、山谷丛林中或岩石上。产中国西南、华北和华西。印度北部、尼泊尔、不丹、克什米尔地区和缅甸北部亦有。

Shrubs, dioecious. Leaves narrowly ovate, rarely suborbicular, 3-5-lobed, glabrous or sparsely stalked glandular. Male racemes

10-30-flowered, female ones 4-10-flowered; petals shorter than sepals; calyx brownish red, glabrous. Fruits red, sour tasting, subglobose to obovoid-globose. Fl. Apr-Jun. Fr. Jul-Sep. Mountain slopes, thickets in valleys or on rocks at 1900-3000 m. Distributed in SW, N and W China. Also in N India, Nepal, Bhutan, Kashmir and N Myanmar.

冰川茶藨子 *Ribes glaciale*

华蔓茶藨子

Ribes fasciculatum Sieb. et Zucc. var. **chinense** Maxim.

灌木，雌雄异株。嫩枝、叶和花梗均密被柔毛。小枝无刺。叶直径达10厘米，边缘掌状3-5裂，具粗钝单锯齿。花单性，组成几无总梗的伞形花序；雄花序具花2-9朵；雌花2-4(6)朵簇生，稀单生；苞片长圆形；花萼黄绿色；萼片花期反折。果无毛。花期4-5月，果期7-9月。生海拔700-1300米的林下、林缘或石质坡地。产华中和华东。日本和朝鲜半岛亦有。

Shrubs, dioecious. Branchlets, leaves and pedicels densely pubescent. Branchlets unarmed. Leaves up to 10 cm diam, margin palmately 3-5-lobed, coarsely obtusely serrate. Flowers unisexual, arranged in nearly sessile umbels, staminate ones 2-9 in umbels; pistillate ones 2-4(-6) in fascicles, rarely solitary; bracts oblong; calyx yellowish green; sepals reflexed at anthesis. Fruits glabrous. Fl. Apr-May. Fr. Jul-Sep. Forests, forest margins, stony slopes at 700-1300 m. Distributed in C and E China. Also in Japan and Korean Peninsula.

华蔓茶藨子 *Ribes fasciculatum* var. *chinense*

海桐花科 Pittosporaceae

光叶海桐

Pittosporum glabratum Lindl.

灌木，高2-3米。叶长圆形或倒披针形，背面光滑，薄革质，侧脉5-8对。花序1-4枝生小枝顶端叶腋，伞形或聚伞状，多花；花瓣离生，倒披针形。蒴果椭圆体形，3瓣裂开。花期3-8月，果期7-12月。生海拔200-2000米的林中、山谷、山坡或河边。产中国西南、华南和华中。越南亦有。

Shrubs, 2-3 m tall. Leaves oblong or oblanceolate, abaxially glabrous, thinly leathery, lateral veins 5-8-paired. Inflorescences 1-4 in leaf axils at branchlet apex, umbellate or corymbose, many flowered; petals free, oblanceolate. Capsules ellipsoid, dehiscing by 3 valves. Fl. Mar-Aug. Fr. Jul-Dec. Forests, valleys, slopes or riversides at 200-2000 m. Distributed in SW, S and C China. Also in Vietnam.

光叶海桐 *Pittosporum glabratum*

异叶海桐

Pittosporum heterophyllum Franch.

灌木。叶簇生于小枝顶端，二年生，倒披针形、狭披针形、带状或条形，薄革质。花序顶生，伞状，具1-5花；花瓣合生，披针形。蒴果近球形，稍压扁，2片裂开。花期3-5月，果期6-11月。生海拔800-3000米的林中、山谷或山坡上。产云南北部、四川和西藏东南部。

Shrubs. Leaves clustered at branchlet apex, biennial, oblanceolate, narrowly lanceolate, beltlike or linear, thinly leathery. Inflorescences terminal, umbel-like, 1-5-flowered; petals connate, lanceolate. Capsules subglobose, slightly compressed, dehiscing by 2 valves. Fl. Mar-May. Fr. Jun-Nov. Forests, valleys or slopes at 800-3000 m. Distributed in N Yunnan, Sichuan and SE Xizang.

海金子

Pittosporum illicioides Makino

灌木。叶倒卵状披针形、倒披针形或狭披针形，薄革质，侧脉6-8对。伞形花序顶生，有花2-10朵；子房柄短。蒴果近球形，多少三角形或有3条纵沟，3片裂开；果片薄木质。花期3-5月，果期6-11月。生海拔100-2200米的林下、灌丛、山谷或溪边。产中国西南、华南、东南、华中和华东。日本亦有。

Shrubs. Leaves obovate-lanceolate, oblanceolate or narrowly lanceolate, thinly leathery, lateral veins 6-8-paired. Inflorescences terminal, umbellate, 2-10-flowered; ovary stipe short. Capsules subglobose, triangular or with 3 longitudinal folds, dehiscing by 3 valves; pericarp thinly woody. Fl. Mar-May. Fr. Jun-Nov. Forests, thickets, valleys or streamsides at 100-2200 m. Distributed in SW, S, SE, C and E China. Also in Japan.

羊脆木 *Pittosporum kerrii*

异叶海桐 *Pittosporum heterophyllum*

海金子 *Pittosporum illicioides*

羊脆木

Pittosporum kerrii Craib

小乔木。嫩枝被锈色柔毛。叶倒披针形至倒卵状披针形或长椭圆形，厚革质，长6-15厘米，宽2-5厘米。圆锥花序长4-6厘米；花瓣离生，黄白色。蒴果扁圆形，2片裂开；果片薄木质。花期3-5月，果期6-10月。生海拔700-2300米的林中或山坡上。产云南南部。缅甸、老挝和泰国亦有。

Trees small. Branchlets ferruginous pubescent. Leaves oblanceolate to obovate-lanceolate or long elliptic, thickly leathery, 6-15 × 2-5 cm. Panicles 4-6 cm long; petals free, yellow-white. Capsules oblate, compressed, dehiscing by 2 valves; pericarp thinly woody. Fl. Mar-May. Fr. Jun-Oct. Forests or slopes at 700-2300 m. Distributed in S Yunnan. Also in Myanmar, Laos and Thailand.

峨眉海桐 *Pittosporum omeiense*

峨眉海桐

Pittosporum omeiense H. T. Chang et S. Z. Yan

灌木。叶2-4丛生在小枝顶端；侧脉约5对。伞房状伞形花序顶生，有花9-17；子房壁极薄；胚珠8或9个；侧膜胎座2或3个。蒴果椭圆体形，果梗长2-6厘米。种子7或8，长3-5毫米。花期3-5月，果期6-11月。生海拔900-1800米的林下、灌丛或山谷沟边。产四川、贵州和湖北西北部。

Shrubs. Leaves 2-4-clustered at apex of branchlets; lateral veins ca. 5-paired. Inflorescences terminal, corymbose-umbellate, 9-17-flowered; ovary wall very thin; ovules 8 or 9; placentas 2 or 3. Capsules ellipsoid; stipes 2-6 cm long. Seeds 7 or 8, 3-5 mm long. Fl. Mar-May. Fr. Jun-Nov. Forests, thickets or streamsides in valleys at 900-1800 m. Distributed in Sichuan, Guizhou and NW Hubei.

海桐

Pittosporum tobira (Thunb.) W. T. Aiton

灌木或小乔木。叶倒卵形或倒卵状披针形，革质。苞片及萼片密被褐色短柔毛；花瓣白色，后变黄色。蒴果球形，具棱，直径约1.2厘米，3片裂开，多少被柔毛。花期3-5月，果期5-10月。栽培供观赏或可能为自然归化，生海拔1800米以下的林中、石灰岩区域、山坡、海边沙地或路边。原产台湾北部；引入中国西南、华南、东南和华东。亦原产日本南部和朝鲜半岛南部。

Shrubs or small trees. Leaves obovate or obovate-lanceolate, leathery. Bracts and sepals densely brown pubescent; petals white at first, becoming yellow later. Capsules globose, angular, ca. 1.2 cm diam, dehiscing by 3 valves, pubescent. Fl. Mar-May. Fr. May-Oct. Cultivated for ornament and possibly naturalized, forests, limestone areas, slopes, sandy seashores or roadsides below 1800 m. Native in N Taiwan; introduced to SW, S, SE and E China. Native to S Japan and S Korean Peninsula.

卵果海桐

Pittosporum lenticellatum Chun ex H. Peng et Y. F. Deng

常绿灌木，高约3米。嫩枝被短柔毛。叶厚革质，倒卵状披针形，长7-10厘米，初两面被柔毛，以后变无毛。伞形花序生于枝顶叶腋；萼片披针形；子房被毛，侧膜胎座2个。蒴果强烈压扁，直径1.3-1.6厘米，2片裂开；果片厚2-3毫米，具突起横格。种子16-24。花期3-5月，果期6-10月。生海拔200-1100米的石灰岩山地常绿林和灌丛。产广西和贵州。

Evergreen shrubs, ca. 3 m tall. Young branchlets pubescent. Leaves thickly leathery, obovate-lanceolate, 7-10 cm long, pubescent on both surfaces at first, later glabrate. Umbels in leaf axils at branchlet apex; sepals lanceolate; ovary pubescent, parietal placentas 2. Capsule strongly compressed, 1.3-1.6 cm diam, dehiscing by 2 valves; pericarp 2-3 mm thick, horizontally convex striate. Seeds 16-24. Fl. Mar-May. Fr. Jun-Oct. Evergreen forests and thickets in limestone mountains at 200-1100 m. Distributed in Guangxi and Guizhou.

扁片海桐

Pittosporum planilobum Chang et Yan

常绿小乔木。嫩枝被柔毛。叶革质，倒披针形，长4-6厘米，下面初时有毛，不久变无毛。伞形花序顶生，具花8-17朵；萼片线状披针形，被毛；花柱长2毫米，无毛，侧膜胎座2个，胚珠12个。蒴果扁圆形，直径10-12毫米，2片裂开；果片木

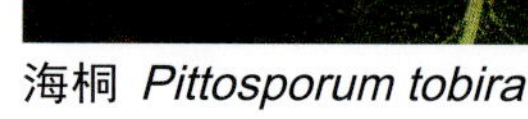

海桐 *Pittosporum tobira*

卵果海桐 *Pittosporum lenticellatum*

质，有横格10条。种子10-12，长3.5毫米。花期2-5月，果期5-10月。生海拔200-1300米的石灰岩林地和山坡。产广西。

Evergreen small trees. Young branchlets pubescent. Leaves leathery, oblanceolate, 4-6 cm long, pubescent abaxially at first, soon glabrate. Umbels terminal, 8-17-flowered; sepals linear-lanceolate, pubescent; style 2 mm long, glabrous, parietal placentas 2, ovules 12. Capsule compressed globose, 10-12 mm diam, dehiscing by 2 valves; pericarp woody, horizontally 10-striate. Seeds 10-12, 3.5 mm long. Fl. Feb-May. Fr. May-Oct. Forests and slopes in limestone areas at 200-1300 m. Distributed in Guangxi.

扁片海桐 *Pittosporum planilobum*

秀丽海桐

Pittosporum pulchrum Gagnep.

常绿灌木，高3米。嫩枝无毛。叶厚革质，倒卵形或倒披针形，长3-6厘米，先端圆，侧脉干后在上面陷下，边缘反卷。伞房花序顶生，被毛；萼片卵形，被毛。蒴果圆球形，直径7-8毫米，2片裂开；果片薄木质，内侧有横格，胎座2，位于果片的基部到中部。种子15。花期1-4月，果期3-10月。生海拔400-500米的树林山坡。产广西。越南亦有。

Evergreen shrubs, 3 m tall. Young branchlets glabrous. Leaves thickly leathery, obovate or oblanceolate, 3-6 cm long, apex rounded, lateral veins concave adaxially after drying, margin reflexed. Umbels terminal, pubescent; sepals ovate, pubescent. Capsule globose, 7-8 mm diam, dehiscing by 2 valves; pericarp thinly woody, horizontally striate adaxially; parietal placentas 2, at base to middle part of valves. Seeds 15. Fl. Jan-Apr. Fr. Mar-Oct. Slopes in forests at 400-500 m. Distributed in Guangxi. Also in Vietnam.

秀丽海桐 *Pittosporum pulchrum*

金缕梅科
Hamamelidaceae

马蹄荷 *Exbucklandia populnea*

长柄双花木
Disanthus cercidifolius Maxim. subsp. **longipes** (H. T. Chang) K. Y. Pan

灌木。叶柄长3.5厘米；叶阔卵圆形，膜质，干时下面灰白色，上面绿色。花序具2个对生的花；萼片5，卵形；花瓣红色，线形，每一个基部具2蜜腺；雄蕊5，明显短于花瓣。蒴果木质，倒卵形。花期10-11月，果期翌年9-10月。生海拔450-1200米的常绿或落叶林中。产浙江、湖南和江西。

Shrubs. Petiole 3.5 cm long; leaves broadly ovate-rounded, membranous, drying gray-white abaxially and green adaxially. Inflorescence with 2 opposite flowers; sepals 5, ovoid; petals red, linear, each with 2 basal nectaries; stamens 5, much shorter than petals. Capsules woody, obovoid. Fl. Oct-Nov. Fr. next Sep-Oct. Mixed evergreen or deciduous forests at 450-1200 m. Distributed in Zhejiang, Hunan and Jiangxi.

马蹄荷
Exbucklandia populnea (R. Br. ex Griff.) R. W. Br.

乔木。托叶椭圆形或倒卵形；叶宽卵圆形，基部心形。头状花序或总状花序，头状花序具8-12花；花瓣常缺无；雄蕊长约5毫米。蒴果椭圆体形。种子具狭翅。花期5-7月，果期8-10月。生海拔1200-1300米的山坡常绿林或混交林中。产中国西南。南亚和东南亚亦有。

Trees. Stipules elliptic or obovate; leaves broad ovate, base cordate. Inflorescences capitate, sometimes arranged in racemes, heads 8-12-flowered; petals usually absent; stamens ca. 5 mm long. Capsules ellipsoid. Seeds narrowly winged. Fl. May-Jul. Fr. Aug-Oct. Evergreen or mixed forests on slopes at 1200-1300 m. Distributed in SW China. Also in S and SE Asia.

长柄双花木 *Disanthus cercidifolius* subsp. *longipes*

大果马蹄荷
Exbucklandia tonkinensis (Lecomte) H. T. Chang

乔木，高达30米。托叶狭长圆形；叶宽卵形或幼时3掌裂。花序头状，具7-9朵花，有时排成总状；无花瓣；雄蕊约13枚，长约8毫米。蒴果卵球形，具疣凸。种子6粒。花期5-7月，果期8-9月。生海拔800-1500米的常绿林中的山坡或山谷。产中国西南、华南、东南、华中和华东。老挝和越南北部亦有。

Trees, to 30 m tall. Stipules narrowly oblong; leaves broadly ovate or sometimes palmately 3-lobed in young leaves. Inflorescences capitate, 7-9-flowered, sometimes arranged in racemes; petals absent; stamens ca. 13, ca. 8 mm long. Capsules ovoid, tuberculate. Seeds 6. Fl. May-Jul. Fr. Aug-Sep. Slopes or valleys in evergreen forests at 800-1500 m. Distributed in SW, S, SE, C and E China. Also in Laos and N Vietnam.

红花荷
Rhodoleia championii Hook. f.

常绿乔木。叶卵形至长卵形，基部不明显三出脉，基部阔楔形，常无毛。头状花序常下垂，总花梗长2-3.8厘米，具数个鳞状苞片；花瓣匙形，宽4-8毫米。蒴果卵球形。花期2-4月，果期5-8月。生海拔约1000米的林中。产贵州、广东和海

南。缅甸、印度尼西亚、马来西亚和越南亦有。

Evergreen trees. Leaves ovate to broadly ovate, obscurely 3-veined at base, base broadly cuneate, usually glabrous. Capitulum usually pendulous, peduncles 2-3.8 cm long, with several scalelike bracts; petals spatulate, 4-8 mm wide. Capsules ovoid. Fl. Feb-Apr. Fr. May-Aug. Forests at ca. 1000 m. Distributed in Guizhou, Guangdong and Hainan. Also in Myanmar, Indonesia, Malaysia and Vietnam.

小花红花荷

Rhodoleia parvipetala Tong

常绿乔木。叶互生，具长柄；叶革质，矩圆形，下面灰白色，无毛；三出脉不很强烈。花序头状，具柄，长2-2.5厘米；总苞片5-7，卵圆形；花瓣匙形，红色；雄蕊6-8。头状果序宽2-2.5厘米；蒴果卵圆形，上部裂成4片。花期3-4月，果期4-9月。生海拔约1000米的山坡常绿林。产广西、贵州和云南。越南亦有。

Evergreen trees. Leaves alternate, long petiolate; leaf blade coriaceous, oblong, abaxially whitish-gray, glabrous; obscurely 3-veined at base. Inflorescences capitate, pedunculate, 2-2.5 cm long; involucral bracts 5-7, ovate-rounded; petals spatulate, red; stamens 6-8. Infructescences 2-2.5 cm diam; capsules ovoid, upper part 4-valved. Fl. Mar-Apr. Fr. Apr-Sep. Slopes in evergreen forests at ca. 1000 m. Distributed in Guangxi, Guizhou and Yunnan. Also in Vietnam.

大果马蹄荷 *Exbucklandia tonkinensis*

红花荷 *Rhodoleia championii*

小花红花荷 *Rhodoleia parvipetala*

显脉红花荷
Rhodoleia henryi Tong

乔木。叶卵状椭圆形，具离基三出脉，侧脉明显，无毛，基部阔楔形。花序具5花；花梗长1-1.5厘米；总苞多数，被锈褐色绒毛；花瓣深红色，匙形。蒴果长1.6-2厘米。花期3-5月，果期8-9月。生海拔2000-2450米的山顶密林中。产云南东南部。

Trees. Leaves ovate-elliptic, triple-nerved, lateral nerves distinct, glabrous, base broadly cuneate. Inflorescences5-flowered; peduncles 1-1.5 cm long; involucral bracts many, rusty tomentose; petals dark red, spatulate. Capsules 1.6-2 cm long. Fl. Mar-May. Fr. Aug-Sep. Dense forests on mountain tops at 2000-2450 m. Distributed in SE Yunnan.

壳菜果
Mytilaria laosensis Lec.

乔木。叶革质，宽卵形，全缘，先端3浅裂，下面黄绿色或浅灰色，无毛，基部心形。穗状花序稠密，顶生或腋生；花瓣白色；子房下位。蒴果黄褐色，外果皮厚。花期3-6月，果期7-9月。生海拔约1000米的山谷常绿阔叶林。产云南、广西和广东。老挝和越南北部亦有。

Trees. Leaves coriaceous, broadly ovate, margin entire, apex 3-lobed, abaxially yellow-green or grayish, glabrous, base cordate. Spikes dense, terminal or axillary; petals white; ovary inferior. Capsules yellow-brown, with thick exocarp. Fl. Mar-Jun. Fr. Jul-Sep. Evergreen broad-leaved forests in valleys at 1000 m. Distributed in Yunnan, Guangxi and Guangdong. Also in Laos and N Vietnam.

山铜材 *Chunia bucklandioides*

山铜材
Chunia bucklandioides H. T. Chang

常绿乔木。小枝绿色，光滑无毛。叶互生，具长柄，叶厚革质，阔卵圆形，掌状脉5条，两面无毛。肉穗状花序生于新枝侧面，纺锤形；花瓣不存在；雄蕊8，花药红色。果序长3-4厘米；蒴果卵圆形，木质，开裂。花期3-6月，果期7-9月。生海拔300-600米的潮湿山谷或热带雨林。产海南。

Evergreen trees. Branchlets green, glabrous. Leaves alternate, long petiolate, leaf blade thickly coriaceous, broadly ovate, veins 5, palmate; both surfaces glabrous. Spikes born at lateral surfaces of twigs, fleshy, fusiform; petals absent; stamens 8, anthers red. Infructescences 3-4 cm long; capsules ovoid-globose, ligneous, dehiscent. Fl. Mar-Jun. Fr. Jul-Sep. Wet valleys or rain forests at 300-600 m. Distributed in Hainan.

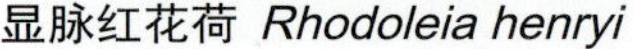

显脉红花荷 *Rhodoleia henryi*

壳菜果 *Mytilaria laosensis*

枫香
Liquidambar formosana Hance

落叶乔木。叶掌状3裂，边缘具腺锯齿，基部心形；托叶线形。雄花序短穗状，数个排成总状，雄蕊多数；雌花序具24-43花，萼齿4-7个，针形。头状果序圆球形，木质；蒴果有宿存花柱。花期3-6月，果期7-9月。生海拔500-800米的开阔地、村庄附近或山区林中。产华南、东南和华中。老挝、越南和朝鲜半岛南部亦有。

Deciduous trees. Leaves palmately 3-lobed, margin glandular-serrate, base cordate; stipules linear. Male inflorescences a short spike, several arranged in a raceme, stamens many; female inflorescences 24-43-flowered, staminode teeth 4-7, needlelike. Infructescences globose, woody; capsules with persistent staminodes. Fl. Mar-Jun. Fr. Jul-Sep. Sunny places, near villages or montane forests at 500-800 m. Distributed in S, SE and C China. Also in Laos, Vietnam and S Korean Peninsula.

枫香 *Liquidambar formosana*

缺萼枫香树

Liquidambar acalycina H. T. Chang

落叶乔木，高约25米。叶宽卵形。雄花序球形，多数排成总状；雌花序在短枝叶腋处单生，具15-26花；萼齿缺，或为鳞片状，有时极短。蒴果无宿存萼齿；可育种子具翅，不可育种子多数，棕色，具不规则棱。花期5-6月，果期7-9月。生海拔600-1000米的山地常绿林中。产中国西南、华南、华中和华东。

缺萼枫香树 *Liquidambar acalycina*

Deciduous trees, ca. 25 m tall. Leaves broadly ovate. Male inflorescences globose, several, arranged in racemes; female inflorescences solitary in leaf axils of short shoots, 15-26-flowered; staminode teeth absent or scalelike and sometimes very short. Capsules without persistent staminode teeth; fertile seeds winged, sterile seeds many, brown, irregularly angular. Fl. May-Jun. Fr. Jul-Sep. Montane evergreen forests at 600-1000 m. Distributed in SW, S, C and E China.

半枫荷

Semiliquidambar cathayensis H. T. Chang

常绿或半常绿乔木，高15-20米。叶异型，厚革质，叶柄粗壮；不分裂的叶椭圆形、卵状椭圆形、长圆形、卵状长圆形；分裂的叶掌状3裂。雄花序约6厘米长；雌花序单生。果序球形；具22-28个蒴果，具宿存萼齿。花期5-6月，果期7-8月。生海拔1000米的林中。产贵州南部、广西北部、广东、福建、海南和江西南部。

Evergreen or semievergreen trees, 15-20 m tall. Leaves dimorphic, firmly leathery, petiole stout; entire leaves elliptic, ovate-elliptic, oblong, ovate-oblong; lobed leaves palmately 3-lobed. Male inflorescences ca. 6 cm long; female inflorescences solitary. Infructescences globose; capsules 22-28, with persistent staminode teeth. Fl. May-Jun. Fr. Jul-Aug. Forests at 1000 m. Distributed in S Guizhou, N Guangxi, Guangdong, Fujian, Hainan and S Jiangxi.

半枫荷 *Semiliquidambar cathayensis*

细柄半枫荷 *Semiliquidambar chingii*

细柄半枫荷

Semiliquidambar chingii (Metc.) H. T. Chang

常绿或半常绿乔木。叶柄细弱，长2-4.5厘米；叶两型，薄革质，掌状3出脉，侧脉3-7对。雌花：萼齿宿存，长1-2毫米。果序球形；蒴果具4-6毫米长的宿存花柱，先端反折。花期3-6月，果期7-9月。生海拔约1000米的林下。产贵州、广东、福建和江西。

Evergreen or semievergreen trees. Petioles slender, 2-4.5 cm long; leaves dimorphic, thinly coriaceous, palmately 3-veined, lateral veins 3-7 on each side. Female flowers: staminode teeth persistent, 1-2 mm long. Infructescence globose; capsules with persistent styles 4-6 mm long, apex recurved. Fl. Mar-Jun. Fr. Jul-Sep. Forests at ca. 1000 m. Distributed in Guizhou, Guangdong, Fujian and Jiangxi.

云南蕈树 *Altingia yunnanensis*

云南蕈树

Altingia yunnanensis Rehd. et E. H. Wils

常绿乔木。叶常长圆形，近革质，边缘具锯齿。雄花序椭圆体形，几个排列成圆锥花序；雌花序通常为总状花序，具16-24花。果序近球形，宽1.5-2.5厘米。花期3-5月，果期5-7月。生海拔约1000米的林中。产云南东南部。

Evergreen trees. Leaves usually oblong, nearly leathery, margin serrate. Male inflorescences ellipsoid, several arranged in panicles; female inflorescences usually in racemes, 16-24-flowered. Infructescences subglobose, 1.5-2.5 cm wide. Fl. Mar-May. Fr. May-Jul. Forests at ca. 1000 m. Distributed in SE Yunnan.

蕈树

Altingia chinensis (Champ.) Oliv.ex Hance

常绿乔木。树皮灰色。叶倒卵状长圆形，革质，边缘具细圆齿，先端锐尖。雄花序多个排列成圆锥花序；雌花序单生或多个排列成圆锥花序，具15-26花。果序近球形，宽1.7-2.8厘米，基部楔形。花期3-6月，果期7-9月。生海拔600-1000米的林中。产中国西南、华中和东南。越南亦有。

Evergreen trees. Bark gray. Leaves obovate-oblong, leathery, margin crenate-serrate, apex acute. Male inflorescences many arranged in panicles; female inflorescences solitary or many in panicles, 15-26-flowered. Infructescences subglobose, 1.7-2.8 cm wide, base truncate. Fl. Mar-Jun. Fr. Jul-Sep. Forests at 600-1000 m. Distributed in SW, C and SE China. Also in Vietnam.

海南蕈树

Altingia obovata Merr. et Chun

常绿乔木。叶互生，具短柄，叶革质，倒卵形，先端圆形或钝，基部窄楔形，边缘有小钝齿。雄花短穗状花序椭圆形，苞片卵形；雄蕊多数，花丝极短，花药倒卵圆形，红色；雌花头状花序通常单生，具长柄，有花16-28；花柱弯曲。头状果序近于圆球形，直径2厘米。花期3-6月，果期7-9月。生海拔800-1400米的山地常绿林。产海南。

蕈树 *Altingia chinensis*

海南蕈树 *Altingia obovata*

细柄蕈树 *Altingia gracilipes*

Evergreen trees. Leaves alternate, shortly petiolate, leaf blade coriaceous, obovate, apex rounded or obtuse, base narrowly cuneate, margin crenate-serrulate. Male inflorescences shortly ellipsoid, bracts ovate; male flowers numerous, filaments very short, anthers obovoid, red; female inflorescences usually solitary, long pedunculate, 16-28-flowered; styles recurved. Infructescences subglobose, ca. 2 cm diam. Fl. Mar-Jun. Fr. Jul-Sep. Montane evergreen forests at 800-1400 m. Distributed in Hainan.

细柄蕈树

Altingia gracilipes Hemsl.

乔木。叶柄细弱，长1-3厘米；叶卵状披针形，4-7 × 1.5-2.5厘米，革质，常全缘。雄花头状花序圆球形，常多个排成圆锥花序；雌花头状花序生于当年枝的叶腋里，单个或数个排成总状式。果序倒圆锥形。种子棕色有棱。花期4-6月，果期7-9月。生海拔400-1000米的常绿林中。产广东、海南、福建和浙江。

Trees. Petioles slender, 1-3 cm long; leaves ovate-lanceolate, 4-7 × 1.5-2.5 cm, leathery, margin usually entire. Male capitulum orbicular, usually many ranged to panicles; female capitulum axillary at current growth, unique or many ranged to racemose. Infructescences obconical. Seeds brown, angular. Fl. Apr-Jun. Fr. Jul-Sep. Evergreen forests at 400-1000 m. Distributed in Guangdong, Hainan, Fujian and Zhejiang.

檵木

Loropetalum chinense (R. Br.) Oliv.

灌木或小乔木。叶卵形，先端锐尖。花白色或淡黄色；萼筒杯状，被星状毛。花期3-4月，果期5-7月。生海拔1000-1200米的山谷灌丛。产中国西南、华南、华中和东南。印度东部及北部和日本亦有。

Shrubs or small trees. Leaves ovate, apex acute. Petals white or pale yellow; calyx tube cupular, stellately pubescent. Fl. Mar-Apr. Fr. May-Jul. Thickets in valleys at 1000-1200 m. Distributed in SW, S, C and SE China. Also in E and N India, and Japan.

檵木 *Loropetalum chinense*

红花檵木 *Loropetalum chinense* var. *rubrum*

红花檵木

Loropetalum chinense (R. Br.) Oliv. var. **rubrum** Yieh

本变种与檵木的区别在于本变种的花瓣常紫红色或红色。生灌木丛中。产广西和湖南。华南广泛栽培。

This variety differs from the typical variety in its petals usually purple-red or red. Thickets. Distributed in Guangxi and Hunan. Widely cultivated in S China.

金缕梅

Hamamelis mollis Oliv.

灌木或小乔木。嫩枝有星状绒毛，老枝秃净。叶阔倒卵状圆形或长圆形，下面密具星状绒毛及灰毛。花序梗长约5毫米；花瓣黄色；退化雄蕊先端平截；子房有绒毛。蒴果卵球形。花期4-5月，果期6-8月。生海拔300-800米的林中或灌丛中。产中国西南和华南。

Shrubs or small tree. Young branches with stellate tomenta, old branches glabrous. Leaves broadly obovate-rounded or oblong, abaxially densely stellately tomentose with gray hairs. Inflorescences peduncles ca. 5 mm long; petals yellow; reduced stamens truncate at apex; ovary tomentose. Capsules ovoid. Fl. Apr-May. Fr. Jun-Aug. Forests or among bushes at 300-800 m. Distributed in SW and S China.

瑞木

Corylopsis multiflora Hance

半常绿灌木或小乔木。叶卵形。花序为总状花序，长2-4厘米；总花梗和总苞无毛或有柔毛，1-5基生叶；花瓣倒披针形或狭倒披针形，宽1-2毫米；子房半下位，无毛。果序5-6厘米；蒴果木质。花期2-4月，果期5-7月。生海拔1000-1500米的林中。产中国西南、华南和华中。

Semievergreen shrubs or small trees. Leaves ovate. Inflorescences racemose, 2-4 cm long; peduncles and general bracts glabrous or pubescent, with 1-5 basal leaves; petals oblanceolate or narrowly so, 1-2 mm wide; ovary semi-inferior, glabrous. Infructescences 5-6 cm long; capsules woody. Fl. Feb-Apr. Fr. May-Jul. Forests at 1000-1500 m. Distributed in SW, S and C China.

蜡瓣花

Corylopsis sinensis Hemsl.

灌木。叶倒卵形或倒卵状圆形，基部卵形或长圆状倒卵形，基部心形或近截形。花序3-4厘米长；退化雄蕊2裂；花瓣匙形。蒴果被星状毛。生海

金缕梅 *Hamamelis mollis*

瑞木 *Corylopsis multiflora*

拔1000-1500米的林中或山地。产中国西南、华南、东南、华中和华东。

Shrubs. Leaves obovate or obovate-rounded, broadly ovate or oblong-obovate, base cordate or subtruncate. Inflorescences 3-4 cm long; staminodes 2-lobed; petals spatulate. Capsules stellately pubescent. Forests or mountains at 1000-1500 m. Distributed in SW, S, SE, C and E China.

红药蜡瓣花
Corylopsis veitchiana Bean

落叶灌木。嫩枝无毛。叶互生，具短柄，叶倒卵形或椭圆形，先端急尖，基部不等侧心形，下面带灰色，第一对侧脉离基部稍远，分枝强烈，边缘有锯齿，齿尖刺毛状。总状花序基部有1-2叶；总苞状鳞片卵圆形；萼筒有星毛；花瓣匙形；雄蕊稍突出花冠外，花药红褐色；子房有星状绒毛。果序长5-6厘米；蒴果近圆卵形，有星毛。花期4-6月，果期6-8月。生海拔约1200米的林中。产安徽、湖北和四川。

Deciduous shrubs. Young branches glabrous. Leaves alternate, shortly petiolate, leaf blade obovate or elliptic, apex acute, base asymmetrical, subcordate, abaxially grayish, the 2 lowermost lateral veins slightly distant from base, with tertiary veins, margin serrate, teeth mucronate. Inflorescences with 1-2 leaves at base; basal bracts of inflorescences ovate; floral cup tomentose; petals spatulate; stamens slightly exserted from corolla, anthers red-brown; ovary stellately tomentose. Infructescences 5-6 cm long; capsules subglobose, stellately pubescent. Fl. Apr-Jun. Fr. Jun-Aug. Forests at ca. 1200 m. Distributed in Anhui, Hubei and Sichuan.

蜡瓣花 *Corylopsis sinensis*

红药蜡瓣花 *Corylopsis veitchiana*

滇蜡瓣花
Corylopsis yunnanensis Diels

灌木。叶坚纸质，背面初时被丝状长柔毛，老时仅脉上被长柔毛；托叶干膜质，外面常呈紫色，无毛，内面密被长丝毛。花序长1.5-2.5厘米；花瓣匙形，长6-7毫米；雄蕊4-5毫米；柱头短于花瓣。花期4-5月，果期6-8月。生海拔2400-2800(-3300)米的沟谷混交林或灌丛中。产云南西部。

滇蜡瓣花 *Corylopsis yunnanensis*

Shrubs. Leaves chartaceous, abaxially silky-villose at first, silky-villose only along nerves when mature; stipules dry-membranaceous, abaxially always purple, glabrous, adaxially densely sericeous-villose. Inflorescences 1.5-2.5 cm long petals spatulate, 6-7 mm long; stamens 4-5 mm long; styles shorter than petals. Fl. Apr-May. Fr. Jun-Aug. Mixed forests or thickets in valleys at 2400-2800(-3300) m. Distributed in W Yunnan.

腺蜡瓣花
Corylopsis glandulifera Hemsl.

落叶灌木。嫩枝秃净无毛。叶互生，具短柄，叶倒卵形或倒卵圆形，先端略尖，基部不等侧微心形，下面有星状柔毛，侧脉在下面突起，边缘上半部有锯齿。总状花序生于具有1-2叶的侧枝顶端；总苞状鳞片近圆形；萼筒无毛；花瓣匙形；子房无毛。果序长4-6厘米；蒴果长6-8毫米，无毛。花期3-5月，果期6-8月。生海拔约1300米的山坡或路边。产安徽、江西和浙江。

Deciduous shrubs. Young branches glabrous. Leaves alternate, shortly petiolate, leaf blade obovate or obovate-rounded, apex subacute, base asymmetrical, subcordate, abaxially stellately pubescent, lateral veins raised on lower surface, margin serrate above middle part. Inflorescences terminal on lateral shoots with only 1-2 leaves; basal bracts of inflorescences subrounded; floral cup glabrous; petals spatulate; ovary glabrous. Infructescences 4-6 cm long; capsules 6-8 mm long, glabrous. Fl. Mar-May. Fr. Jun-Aug. Slopes or roadsides at ca. 1300 m. Distributed in Anhui, Jiangxi and Zhejiang.

腺蜡瓣花 *Corylopsis glandulifera*

四川蜡瓣花
Corylopsis willmottiae Rehd. et Wils.

落叶灌木或小乔木。嫩枝纤细，无毛。叶互生，具短柄，叶倒卵形或广倒卵形，先端急短尖，基部不等侧微心形或圆形，第一对侧脉分枝不强烈，边缘上半部有不显著的小齿突。总状花序生于具有1-3叶的侧枝顶端；总苞状鳞片卵圆形；萼筒无毛；花瓣广倒卵形；子房无毛。果序长4-5厘米；蒴果长7-8毫米，无毛。花期3-6月，果期6-8月。生海拔约

四川蜡瓣花 *Corylopsis willmottiae*

牛鼻栓 *Fortunearia sinensis*

椭圆形，干后褪色，叶背有毛，叶腹面叶脉处被稀疏柔毛。花多为单性；花无柄；花瓣缺或仅存极小的退化部分。蒴果和宿存萼筒被灰黄色长丝毛。花期3-5月，果期6-8月。生海拔800-1500米的林中。产四川、河南、陕西、山西和甘肃。

Shrubs or small trees. Leaves obovate, rarely elliptic, drying discolorous, abaxially pubescent, adaxially sparsely pubescent along veins. Flowers mostly unisexual; flowers sessile; petals lacking or present as tiny rudiments. Capsules and persistent floral cup villous with gray-yellow hairs. Fl. Mar-May. Fr. Jun-Aug. Forests at 800-1500 m. Distributed in Sichuan, Henan, Shaanxi, Shanxi and Gansu.

1200米的林中。产四川西部。

Deciduous shrubs or small trees. Young branches slender, glabrous. Leaves alternate, shortly petiolate, leaf blade obovate or broadly obovate, apex shortly acute, base asymmetrical, subcordate or rotundate, the 2 lowermost lateral veins with obscure tertiary veins, margin obscurely mucronate above middle part. Inflorescences terminal on lateral shoots with only 1-3 leaves; basal bracts of inflorescences ovate; floral cup glabrous; petals broadly obovate; ovary glabrous. Infructescences 4-5 cm long; capsules 7-8 mm long, glabrous. Fl. Mar-Jun. Fr. Jun-Aug. Forests at ca. 1200 m. Distributed in W Sichuan.

牛鼻栓

Fortunearia sinensis Rehd. et Wils.

落叶灌木或小乔木。嫩枝有灰褐色柔毛。叶互生，具短柄，叶膜质，倒卵形或倒卵椭圆形，先端锐尖，基部稍偏斜，边缘有锯齿。总状花序长4-8厘米，被绒毛；花瓣小，狭披针形；花药卵形，红色；花柱2，反卷。蒴果卵圆形，无毛，显具皮孔，2片裂开。花期3-4月，果期5-8月。生海拔800-1000米的山地林中。产华东和华中。

Deciduous shrubs or small trees. Young branches with gray-brown hairs. Leaves alternate, shortly petiolate, leaf blade membranaceous, obovate or obovate-elliptic, apex acute, base slightly oblique, margin serrate. Inflorescences 4-8 cm long, tomentose; petals small, narrowly lanceolate; anthers ovoid, red; styles 2, reflexed. Capsules ovoid-globose, glabrous, distinctly lenticellate, dehiscing loculicidally by two 2-lobed valves. Fl. Mar-Apr. Fr. May-Aug. Montane forests at 800-1000 m. Distributed in E and C China.

山白树

Sinowilsonia henryi Hemsl.

灌木或小乔木。叶倒卵形，稀

山白树 *Sinowilsonia henryi*

蚊母树 *Distylium racemosum*

杨梅蚊母树 *Distylium myricoides*

蚊母树

Distylium racemosum Siebold et Zucc.

灌木或乔木。芽裸露无鳞片状苞片。叶椭圆形或倒卵状椭圆形，下面渐无毛，全缘。花序长1.8-2厘米。蒴果先端尖，上半部两片裂开，每片2浅裂。花期4-6月，果期6-8月。生海拔1000-1300米的林中。产福建、海南、台湾和浙江。朝鲜半岛和日本南部亦有。

Shrubs or trees. Buds exposed without scaly bract. Leaves elliptic or obovate-elliptic, abaxially glabrescent, margin entire. Inflorescences 1.8-2 cm long. Capsules acute at apex, upper half dehiscent to 2 parts, valvules 2-lobed. Fl. Apr-Jun. Fr. Jun-Aug. Forests at 1000-1300 m. Distributed in Fujian, Hainan, Taiwan and Zhejiang. Also in Korean Peninsula and S Japan.

杨梅蚊母树

Distylium myricoides Hemsl.

灌木或小乔木。幼枝具星状鳞片。叶柄具鳞片状毛；叶长圆形或倒披针形，干时褪色，两面无毛，全缘或中上部具牙齿。花序长1-3厘米；子房被星状柔毛。蒴果卵球形或近卵球形。花期4-6月，果期6-8月。生海拔500-800米的常绿林。产中国西南、中南和东南。

Shrubs or small trees. Young branches stellately lepidote. Petioles lepidote hairy; leaves oblong or oblanceolate, drying discolorous, both surfaces glabrous, margin entire or toothed above middle. Inflorescences 1-3 cm long; ovary stellately pubescent. Capsules ovoid or subovoid. Fl. Apr-Jun. Fr. Jun-Aug. Evergreen forests at 500-800 m. Distributed in SW, SC and SE China.

小叶蚊母树 *Distylium buxifolium*

小叶蚊母树

Distylium buxifolium (Hance) Merr.

灌木。幼枝细弱，渐无毛，老枝无毛。叶柄长1-3毫米；叶倒披针形、长圆状披针形或倒卵形，两面无毛，全缘或近顶端有不明显1-2齿。蒴果，花柱宿存。花期4-5月，果期6-8月。生海拔1000-1200米的河边或沟边。产中国西南、华南、东南和华中。

Shrubs. Young branches slender, glabrescent, older growth glabrous. Petiole 1-3 mm long; leaves oblanceolate, oblong-lanceolate or obovate, both surfaces glabrous, entire or with 1-2 obscure subapical teeth. Capsules, styles persistent. Fl. Apr-May. Fr. Jun-Aug. Riversides or streamsides at 1000-1200 m. Distributed in SW, S, SE and C China.

中华蚊母树

Distylium chinense (Franch. ex Hemsl.) Diels

灌木，高达1米。叶柄密被鳞片；叶椭圆形至倒披针形，两面无毛，基部宽楔形，全缘或每侧近顶端具2或3个锯齿。花序长1-1.5厘米。蒴果沿4个瓣膜开裂。花期4-6月，果期6-8月。生海拔1000-1300米的湿地、河边或溪边。产四川和湖北。

Shrubs, to 1 m tall. Petioles densely lepidote; leaves elliptic to oblanceolate, both surfaces glabrous, base broadly cuneate, margin entire or with 2 or 3 teeth on each side near apex. Inflorescences 1-1.5 cm long. Capsules dehiscing by 4 valves. Fl. Apr-Jun. Fr. Jun-Aug. Wet places, riversides or streamsides at 1000-1300 m. Distributed in Sichuan and Hubei.

中华蚊母树 *Distylium chinense*

樟叶假蚊母树

Distyliopsis laurifolia (Hemsl.) P. K. Endress

灌木，高达3米。叶柄密被星状鳞片；叶卵形或狭卵形，背面被星状绒毛，正面光亮，基部楔形或钝，具离基三出脉。花序长1-2厘米，雄花无柄；两性花具柄；萼筒与蒴果等长。蒴果长10-12毫米，密被棕黄色绒毛。生海拔1300-1500米的林中。产云南东南部和贵州西南部。

Shrubs, to 3 m tall. Petioles densely stellately lepidote; leaves ovate or narrowly ovate, abaxially stellately tomentose, adaxially shiny, base cuneate or obtuse, with 3 main veins from base. Inflorescences 1-2 cm long; male flowers sessile; bisexual flowers pedicellate; floral cup as long as capsule. Capsules 10-12 mm long, densely yellow-brown villous. Forests at 1300-1500 m. Distributed in SE Yunnan and SW Guizhou.

钝叶假蚊母树

Distyliopsis tutcheri (Hemsl.) P. K. Endress

灌木或小乔木。嫩枝和叶无柄，被鳞片，渐脱落。叶椭圆形或倒卵形，两面不同色，渐无毛，顶端钝或近圆形。宿存花杯被固着鳞片；宿存柱头极短。蒴果黄褐色长柔毛。花期4-6月，果期6-9月。生海拔800-1000米的山地常绿林中。产广东、海南和福建。

Shrubs or small trees. Young branches and petioles sessile, with peltate scales, glabrescent. Leaves elliptic or obovate, discolorous, both surfaces glabrescent, apex obtuse or subrounded. Persistent floral cup, with sessile scales; persistent styles very short. Capsules yellow-brown villous. Fl. Apr-Jun. Fr. Jun-Sep. Evergreen forests at 800-1000 m. Distributed in Guangdong, Hainan and Fujian.

樟叶假蚊母树 *Distyliopsis laurifolia*

钝叶假蚊母树 *Distyliopsis tutcheri*

银缕梅 *Parrotia subaequalis*

银缕梅

Parrotia subaequalis (H. T. Chang) R. M. Hao et H. T. Wei

落叶小乔木。叶互生，具柄，叶薄革质，倒卵形，下面浅褐色，有星状柔毛；近先端边缘具波状浅齿。头状花序腋生，具柄；花无梗；苞片卵形，大；最下面1或2花雄性，雄蕊3-10；两性花4或5，萼片卵形，雄花蕾时直立，花时弯曲，花丝丝状，长可达2厘米，白色。蒴果近圆形，干后2片裂。花期5-6月，果期6-8月。生海拔600-700米的山地林中。产安徽、江苏和浙江。

Small deciduous trees. Leaves alternate, petiolate, leaf blade thinly coriaceous, obovate, abaxially stellately pubescent with pale brown hairs, margin sparsely sinuate-dentate near apex. Inflorescences axillary, capitate, flowers impedicellate; floral bracts ovate, large; lowermost 1 or 2 flowers male, stamens 3-10; bisexual flowers 4 or 5, sepals ovate, stamens erect in buds, pendent at anthesis, filaments filiform, to 2 cm long, white. Capsules subglobose, dehiscing by 2 valves. Fl. May-Jun. Fr. Jun-Aug. Montane forests at 600-700 m. Distributed in Anhui, Jiangsu and Zhejiang.

水丝梨

Sycopsis sinensis Oliv.

乔木，高14米。叶常绿，狭卵形或披针形，脉弓形，背面被稀疏星状柔毛。花序具7或8花；雄花萼筒短，退化雌蕊具丝毛；花丝细长。蒴果具毛，不规则开裂，具宿存萼筒和花柱。花期4-6月，果期7-9月。生海拔1300-1500米的山地、灌丛或常绿林中。产中国西南、华南、东南、华中、华西和华东。

Trees, 14 m tall. Leaves evergreen, narrowly ovate or lanceolate, venation brochidodromous, abaxially sparsely stellately pubescent. Inflorescences 7- or 8-flowered; male flowers with short floral cup, reduced pistils pubescent; filaments slender. Capsules villous, dehiscing irregularly, with persistent floral cup and persistent styles. Fl. Apr-Jun. Fr. Jul-Sep. Mountain, thickets, evergreen forests at 1300-1500 m. Distributed in SW, S, SE, C, W and E China.

水丝梨 *Sycopsis sinensis*

杜仲科 Eucommiaceae

杜仲
Eucommia ulmoides Oliv.

乔木。叶椭圆形、卵圆形或长圆形，薄革质，起初具褐色柔毛，后仅脉具柔毛，侧脉6-9对。雄花：雄蕊长约1厘米，无毛；雌花：子房长约1厘米，无毛。翅果扁椭圆形，先端2裂。花期3-5月，果期6-11月。生海拔100-2000米的林中、灌丛、低山区或山谷，亦广泛栽培。产中国西南、华中、华北和华东。

Trees. Leaves elliptic, ovate or oblong, thinly coriaceous, brown pubescent at first, later pubescent only along veins, lateral veins 6-9 paired. Male flowers: stamens ca. 1 cm long, glabrous; female flowers: ovary ca. 1 cm long, glabrous. Samaras flattened-ellipsoidal, apex 2-lobed. Fl. Mar-May. Fr. Jun-Nov. Forests, thickets, lower mountains or valleys at 100-2000 m, also widely cultivated. Distributed in SW, C, N and E China.

杜仲 *Eucommia ulmoides*

悬铃木科 Platanaceae

三球悬铃木
Platanus orientalis L.

落叶乔木，高达30米。小枝被黄褐色绒毛。叶互生；叶柄长3-8厘米，基部常包裹腋芽；叶大，宽卵形，5-7深裂，基出脉3-5条，边缘具粗齿。花小，单性，4数。果枝具(2-)3-5枚果序。果序头状，直径2-2.5厘米。瘦果具宿存刺状柱头，柱头长3-4毫米，基生刚毛黄色，二者均超出果序。花期3-5月，果期6-10月。中国各大城市栽培。原产亚洲西部和欧洲东南部。

Deciduous trees, to 30 m tall. Twigs yellow-brown tomentose. Leaves alternate; petioles 3-8 cm long, usually enclosing axillary bud at base; leaf blades large, broadly ovate, deeply 5-7-lobed, principal veins 3-5, arising from base, margin coarsely dentate. Flowers small, monoecious, 4-merous. Fruiting branchlets with (2-)3-5 infructescences. Infructescence capitate, 2-2.5 cm diam. Achenes with persistent style spiniform, 3-4 mm long, basal bristle yellow, both style and hairs exserted from infructescence. Fl. Mar-May. Fr. Jun-Oct. Cultivated in many cities of China. Native to W Asia and SE Europe.

三球悬铃木 *Platanus orientalis*

一球悬铃木
Platanus occidentalis L.

落叶乔木，高达40米。叶柄长4-7厘米；叶宽卵形，3或5浅裂，主脉3条，侧边2条常自基部以上1厘米处发出。花4-6数。果枝具1(-2)枚果序。果序圆球状，直径约3厘米。瘦果先端钝，宿存柱头极短；基生刚毛长约为瘦果一半，不超出果序。花期3-5月，果期6-10月。华中和华北栽培。原产北美洲。

Deciduous trees, to 40 m tall. Petioles 4-7 cm long; leaf blades broadly ovate, 3- or 5-lobed, principal veins 3, lateral 2 arising from midvein ca. 1 cm above base. Flowers 4-merous to 6-merous. Fruiting branchlets with 1(-2) infructescences. Infructescences globose, ca. 3 cm diam. Achenes obtuse at apex, with persistent style very short; basal bristle ca. 1/2 as long as achene, not exserted from infructescences. Fl. Mar-May. Fr. Jun-Oct. Cultivated in C and N China. Native to North America.

一球悬铃木 *Platanus occidentalis*

二球悬铃木
Platanus × acerifolia (Aiton) Willd.

落叶乔木，高达30米。叶柄长3-10厘米；叶宽卵形，(3-)5(-7)浅裂，主脉3或5条，自基部发出或侧边2(-4)条自基部以上发出。花常4数。果枝具(1-)2(-3)枚果序。果序头状，直径约2.5厘米。瘦果具宿存刺状柱头，长2-3毫米；基生刚毛无或极短，不超出果序。花期3-5月，果期6-10月。华中、东北和华南栽培。原栽培于亚洲西南部和欧洲。

Deciduous trees, to 30 m tall. Petiole 3-10 cm long; leaf blades broadly ovate, (3-)5(-7)-lobed, principal veins 3 or 5, arising from base or lateral 2(-4) from midvein above base. Flowers usually 4-merous. Fruiting branchlets with (1-)2(-3) infructescences. Infructescence capitate, ca. 2.5 cm diam. Achenes with persistent style spiniform, 2-3 mm long; basal bristle absent or very short, not exserted from infructescences. Fl. Mar-May. Fr. Jun-Oct. Cultivated in C, NE and S China. Originally cultivated in SW Asia or Europe.

二球悬铃木 *Platanus × acerifolia*

蔷薇科 Rosaceae

蕤核
Prinsepia uniflora Batal.

灌木。叶长圆状披针形至狭披针形，边缘呈浅波状或有不明显锯齿。花序为1-3花簇生；花梗长3-15毫米，无毛；花瓣白色，具紫色脉纹，先端啮蚀状。核果红褐色至黑褐色，球形。花期5月，果期6-9月。生海拔800-2200米的向阳坡或山脚处。产中国西南、华北和西北。

Shrubs. Leaves oblong-lanceolate to narrowly lanceolate, margin repand or inconspicuously serrulate. Inflorescences 1-3-flowered fascicle; pedicels 3-15 mm long, glabrous; petals white with purple veins, apically erose. Drupes reddish brown to blackish brown, globose. Fl. May. Fr. Jun-Sep. Sunny slopes or bases of hills at 800-2200 m. Distributed in SW, N and NW China.

绣线菊
Spiraea salicifolia L.

直立灌木。叶长圆披针形至披针形。圆锥花序长圆形或金字塔形，多花，密集；花序轴和花梗被薄层柔毛，萼筒钟状，微被柔毛；心皮5，散生柔毛。蓇葖果直立，近平行。花期6-8月，果期8-9月。生海拔200-900米的河岸、山坡、潮湿草地、山沟或沟谷草地。产华北和东北。俄罗斯、蒙古、朝鲜半岛、日本、欧洲和北美洲亦有。

Shrubs erect. Leaves oblong-lanceolate to lanceolate. Panicles oblong or pyramidal, densely many flowered; rachises and pedicels thinly pubescent, hypanthium campanulate, puberulous; carpels 5, sparsely pubescent. Follicles erect, almost parallel. Fl. Jun-Aug. Fr. Aug-Sep. River banks, slopes, damp grasslands, gullies or meadows in valleys at 200-900 m. Distributed in N and NE China. Also in Russia, Mongolia, Korean Peninsula, Japan, Europe and North America.

绣线菊 *Spiraea salicifolia*

渐尖粉花绣线菊
Spiraea japonica L. f. var. **acuminata** Franch.

直立灌木。叶长卵形至披针形，基部楔形，叶缘有尖锐重锯齿，下面被短柔毛。伞房花序顶生，直径10-14厘米；花瓣常粉红色，有时白色。蓇葖果半开张，腹缝线无毛或具柔毛。花期6-7月，果期8-9月。生海拔900-4000米的杂木林中空地、山谷或河岸。产中国西南、华南、东南和华中。

Shrubs erect. Leaves long ovate to lanceolate, base cuneate, margin sharply doubly serrate, pubescent abaxially. Inflorescences corymbs, terminal, 10-14 cm diam; petals usually pink, sometimes white. Follicles divergent, glabrous or pilose on adaxial suture. Fl. Jun-Jul. Fr. Aug-Sep. Clearings in mixed forests, mountain valleys or river banks at 900-4000 m. Distributed in SW, S, SE and C China.

蕤核 *Prinsepia uniflora*

渐尖粉花绣线菊 *Spiraea japonica* var. *acuminata*

光叶粉花绣线菊 *Spiraea japonica* var. *fortunei*

光叶粉花绣线菊

Spiraea japonica L. var. **fortunei** (Planchon) Rehder

直立灌木。叶缘有尖锐重锯齿，先端短渐尖。伞房花序顶生，直径4-8厘米，被柔毛；花粉红色；花盘环状，有不规则圆锯齿。蓇葖果半开张，无毛或沿腹缝线被疏毛。花期6-7月，果期8-9月。生海拔700-3000米的山坡或向阳混交林下。产中国西南、东南、华中、华北和华西。

Shrubs erect. Leaves margin sharply doubly serrate, apex shortly acuminate. Corymbs terminal, 4-8 cm diam, pubescent; flowers pink; disk annular, irregularly crenulate. Follicles divergent, glabrous or pilose on adaxial suture. Fl. Jun-Jul. Fr. Aug-Sep. Slopes or open places in mixed forests at 700-3000 m. Distributed in SW, SE, C, N and W China.

紫花绣线菊

Spiraea purpurea Hand.-Mazz.

灌木。叶卵圆形，两面无毛，下面无乳突，边缘有重锯齿，无毛。复伞房花序，密具多花；萼筒宽陀螺状；花瓣深紫红色，圆形，远长于萼片。蓇葖果直立，无毛。花期5-7月，果期8-9月。生海拔2800-3300米的杂木林或山坡灌丛中。产云南、四川和西藏。

Shrubs. Leaves ovate, glabrous on both surfaces, without papillae abaxially, margin doubly serrate, glabrous. Corymbs compound, densely numberous flowered; hypanthium broadly turbinate; petals dark purple-red, orbicular, much longer than sepals. Follicles erect, glabrous. Fl. May-Jul. Fr. Aug-Sep. Mixed forests or thickets on slopes at 2800-3300 m. Distributed in Yunnan, Sichuan and Xizang.

华北绣线菊

Spiraea fritschiana C. K. Schneid.

灌木。冬芽卵形，具数个褐色鳞片。叶卵形、椭圆状卵形或椭圆状长圆形，边缘具不整齐的重锯齿或单锯齿。复伞房花序顶生，多花；花瓣卵形，白色，芽时淡粉色。蓇葖果近直立，开张。花期5-6月，果期8-9月。生海拔100-2400米的山谷林中或岩石坡地。产中国西南、华中、华北和华东。朝鲜半岛亦有。

Shrubs. Buds ovoid, with several brown scales. Leaves ovate, elliptic-ovate or elliptic-oblong, margin

紫花绣线菊 *Spiraea purpurea*

华北绣线菊 *Spiraea fritschiana*

irregularly doubly or singly serrate. Corymbs compound, terminal, numberous flowered; petals ovate, white, pinkish in bud. Follicles suberect, spreading. Fl. May-Jun. Fr. Aug-Sep. Mountain valley forests or rocky slopes at 100-2400 m. Distributed in SW, C, N and E China. Also in Korean Peninsula.

陕西绣线菊
Spiraea wilsonii Duthie

灌木。叶长圆形、倒卵形或椭圆状长圆形，全缘，稀先端有少数锯齿。复伞房花序，生侧生小枝顶端；萼筒钟状，外面被长柔毛；花瓣白色；花盘圆环形，10裂；花柱短于雄蕊。蓇葖果5，开张，密被短柔毛。花期5-7月，果期8-9月。生海拔1000-3200米的山谷林下、石坡、开阔地或路边。产华中、华西和华北。

Shrubs. Leaves oblong, obovate or elliptic-oblong, margin entire or with a few teeth apically. Corymbs terminal on lateral branchlets, compound; hypanthium campanulate, villous abaxially; petals white; disk annular, 10-lobed; styles shorter than stamens. Follicles 5, spreading, densely pubescent. Fl. May-Jul. Fr. Aug-Sep. Mountain valley forests, rocky slopes, open places or roadsides at 1000-3200 m. Distributed in C, W and N China.

陕西绣线菊 *Spiraea wilsonii*

翠蓝绣线菊
Spiraea henryi Hemsl.

灌木。叶椭圆形、椭圆状长圆形或倒卵状长圆形，有时边缘自中部以上具少量糙齿，有时全缘。复伞房花序生侧生短枝顶端，多花；花瓣白色；花盘圆环形，具10个肥厚的圆球形裂片。蓇葖果开展。花期4-5月，果期7-8月。生海拔1300-3000米的山地林下、山麓或岩坡。产中国西南、华中和华西。

Shrub. Leaves elliptic, elliptic-oblong or obovate-oblong, margin sometimes with a few coarse teeth above middle, sometimes entire. Corymbs terminal on short lateral branchlets, compound, many flowered; petals white; disk annular, with 10 thick globular lobes. Follicles spreading. Fl. Apr-May. Fr. Jul-Aug. Mountain forests, foothills or rocky slopes at 1300-3000 m. Distributed in SW, C and W China.

滇中绣线菊
Spiraea schochiana Rehder

直立灌木。叶背面初密被绒毛，后为乳头状突起，上面疏被绒毛。复伞房花序生侧枝顶端，多花密集；花序轴和花梗密被暗黄色长绒毛。蓇葖果稍直立，外被柔毛。生海拔2000-2200米的山坡林中或山谷中。产云南中部。

Shrubs erect. Leaves abaxially densely tomentose initially, later papillose, adaxially sparsely tomentose. Corymbs terminal on lateral branchlets, compound, densely many flowered; rachises and pedicels densely dark yellow villous. Follicles somewhat erect, softly hairy. Forested slopes or mountain valleys at 2000-2200 m. Distributed in C Yunnan.

翠蓝绣线菊 *Spiraea henryi*

滇中绣线菊 *Spiraea schochiana*

新高山绣线菊 *Spiraea morrisonicola*

麻叶绣线菊

Spiraea cantoniensis Lour.

灌木。冬芽卵球形，小，具数枚鳞片。叶菱状披针形至菱状长圆形，具羽状脉，叶缘自中部以上具锐锯齿。花序与萼筒无毛；伞形花序具总梗，多花；萼筒钟状；花瓣白色。花期4-5月，果期7-9月。中国广泛栽培。日本亦有。

Shrubs. Winter Buds ovoid, small, with several scales. Leaves rhomboidal-lanceolate to rhomboidal-oblong, pinnately veined, margin incised serrate above middle.

新高山绣线菊

Spiraea morrisonicola Hayata

灌木或亚灌木。冬芽小，有2枚外露棕褐色鳞片。叶卵形或菱状卵形，边缘自中部以上有粗锐单锯齿，两面无毛，上面脉纹下陷，下面脉纹显著突起。复伞房花序顶生于侧生小枝，无毛，有多数花；萼筒钟状，内面有短柔毛，萼片果期直立；花瓣白色。蓇葖果仅沿腹缝线稍有短柔毛。花期7-8月，果期9-10月。生海拔4000米以下的山区。产台湾。

Shrubs or subshrubs. Winter buds minute, with 2 brownish exposed scales. Leaves ovate or rhombic-ovate, margin entire toward base, grossly sharply serrate distally, glabrous on both surfaces, depressed adaxially, lateral veins distinctly raised abaxially. Corymbs terminal on lateral branchlets, compound, glabrous, densely many flowered; hypanthium campanulate, pubescent adaxially, sepals erect in fruit; petals white. Follicles puberulous on adaxial suture. Fl. Jul-Aug. Fr. Sep-Oct. Mountain regions below 4000 m. Distributed in Taiwan.

楔叶绣线菊

Spiraea canescens D. Don

灌木，高2-4米。小枝有棱角；冬芽先端常具短尖头，具2枚褐色外露鳞片。叶卵形、倒卵形至倒卵状披针形，边缘自中部以上有3-5钝锯齿，下面具短柔毛。复伞房花序密具短柔毛；萼筒和萼片两面均被短柔毛；花白色或淡粉色。蓇葖果具短柔毛。花期7-8月，果期9-10月。生海拔2300-4000米的河岸、山坡和灌丛。产四川、云南和西藏。中南亚和印度亦有。

Shrubs, 2-4 m tall. Branchlets angled; winter buds usually with mucro at apex and 2 brown exposed scales. Leaves ovate, obovate to obovate-lanceolate, margin obtusely 3-5-dentate above middle, abaxially pubescent. Compound corymbs densely pubescent; hypanthium and sepals pubescent on both sides; petals white or pinkish. Follicles pubescent. Fl. Jul-Aug. Fr. Sep-Oct. River banks, slopes and thickets at 2300-4000 m. Distributed in Sichuan, Yunnan and Xizang. Also in SC Asia and India.

楔叶绣线菊 *Spiraea canescens*

麻叶绣线菊 *Spiraea cantoniensis*

Inflorescences and hypanthium glabrous; umbels pedunculate, many flowered; hypanthium campanulate; petals white. Fl. Apr-May. Fr. Jul-Sep. Widely cultivated in China. Also in Japan.

三裂绣线菊
Spiraea trilobata L.

灌木。叶近圆形，具显著3-5脉，边缘自中部以上有少数圆钝锯齿，常3裂。伞形花序有总梗，具15-30花；萼筒钟状；花瓣白色；花盘圆环形，裂片10。蓇葖果开张。花期5-6月，果期7-8月。生海拔400-2400米的灌丛中、向阳石坡或山区。产华北、华西、华东和东北。俄罗斯和朝鲜半岛亦有。

Shrubs. Leaves suborbicular, conspicuously 3-5-veined, margin slightly crenate above middle, usually 3-lobed. Umbels pedunculate, 15-30-flowered; hypanthium campanulate; petals white; disk annular, 10-lobed. Follicles spreading. Fl. May-Jun. Fr. Jul-Aug. Thickets, open rocky slopes or montane regions at 400-2400 m. Distributed in N, W, E and NE China. Also in Russia and Korean Peninsula.

绣球绣线菊
Spiraea blumei G. Don

灌木，高1-2米。叶菱状卵形至倒卵形或阔卵形，具不显明3脉或羽状脉。伞形花序，有总梗，具10-25花；花瓣白色；花盘环形，10裂。蓇葖果直立，无毛或被柔毛。花期4-6月，果期8-10月。生海拔500-2000米的混交林内、向阳山坡或路边。产华南、华中、华北、华西和东南。朝鲜半岛和日本亦有。

Shrubs, 1-2 m tall. Leaves rhombic-ovate to obovate or broadly ovate, inconspicuously 3-veined or pinnately veined. Umbels pedunculate, 10-25-flowered; petals white; disk annular, 10-lobed. Follicles erect, glabrous or pubescent. Fl. Apr-Jun. Fr. Aug-Oct. Mixed forests, sunny slopes or roadsides at 500-2000 m. Distributed in S, C, N, W and SE China. Also in Korean Peninsula and Japan.

三裂绣线菊 *Spiraea trilobata*

绣球绣线菊 *Spiraea blumei*

土庄绣线菊 *Spiraea pubescens*

丽江绣线菊 *Spiraea lichiangensis*

土庄绣线菊

Spiraea pubescens Turcz.

灌木。叶菱状卵形至椭圆形，下面具灰色柔毛，边缘具深锐锯齿，有时自中部以上3浅裂。伞形花序具总梗，有15-20花；萼筒钟状；花瓣白色；花盘圆环形，10裂。蓇葖果开张。花期4-6月，果期7-10月。生海拔200-2500米的混交林中、向阳或半阴处或干燥岩石坡地。产华南、华中、华北和东北。俄罗斯、蒙古和朝鲜半岛亦有。

Shrubs. Leaves rhombic-ovate to elliptic, abaxially gray-pubescent, margin deeply incised serrate, sometimes 3-lobed above middle. Umbels pedunculate, 15-20-flowered; hypanthium campanulate; petals white; disk annular, 10-lobed. Follicles spreading. Fl. Apr-Jun. Fr. Jul-Oct. Mixed forests, open or semishaded places or dry rocky slopes at 200-2500 m. Distributed in S, C, N and NE China. Also in Russia, Mongolia and Korean Peninsula.

毛花绣线菊 *Spiraea dasyantha*

毛花绣线菊

Spiraea dasyantha Bunge

灌木。小枝之字弯曲。叶菱状卵形，羽状脉显著。伞形花序具总梗，具10-20花，密集；萼筒钟状；花瓣白色；花盘圆环状，10裂。蓇葖果开张，全体被白色绒毛。花期5-6月，果期7-8月。生海拔400-1200米的林中、向阳干燥坡地或路边。产中国东南、华中、华北和华东。

Shrubs. Branchlets tortuous. Leaves rhombic-ovate, prominently pinnately veined. Umbels pedunculate, densely 10-20-flowered; hypanthium campanulate; petals white; disk annular, 10-lobed. Follicles spreading, white tomentose throughout. Fl. May-Jun. Fr. Jul-Aug. Forests, open dry slopes or roadsides at 400-1200 m. SE, C, N and E China.

丽江绣线菊

Spiraea lichiangensis W. W. Smith

灌木。叶阔卵形，两面无毛，边缘自中部以上具3-5个不明显的钝三角状齿。总状花序具总梗，伞状，具5-10花；花瓣圆形，白色；雄蕊40-50，长于花瓣；花盘环状，有裂。花期6-7月。生海拔3500-4000米的开阔松林或灌丛中。产云南西北部。

Shrubs. Leaves broadly ovate, glabrous on both surfaces, margin with 3-5 indistinct, obtusely triangular teeth on each side above middle. Racemes pedunculate, umbellate, 5-10-flowered; petals obicular, white; stamens 40-50, longer than petals; disk annular, lobed. Fl. Jun-Jul. Open *Pinus* forests or thickets at 3500-4000 m. Distributed in NW Yunnan.

高山绣线菊

Spiraea alpina Pall.

灌木，高0.5-1.2米。小枝有明显棱角，幼时被短柔毛；冬芽有数枚外露鳞片。叶线状披针形至长圆倒卵形，先端急尖或圆钝，全缘，两面无毛。伞形总状花序具短总梗；萼筒和萼片内面均具短柔毛。蓇葖果无毛或仅沿腹缝线具稀疏短柔毛。花期6-7月，果期8-9月。生海拔2000-4000米的向阳坡地或灌丛中。产华中、西藏和新疆。蒙古、俄罗斯和印度北部亦有。

高山绣线菊 *Spiraea alpina*

Shrubs, 0.5-1.2 m tall. Branchlets conspicuously angled, pubescent when young; winter buds with several exposed scales. Leaves linear-lanceolate to oblong-obovate, apex acute or obtuse, margin entire, glabrous on both surfaces. Racemes pedunculate, umbellate; hypanthium and sepals adaxially pubescent. Follicles glabrous or pilose on adaxial suture. Fl. Jun-Jul. Fr. Aug-Sep. Open slopes or thickets at 2000-4000 m. Distributed in C China, Xizang and Xinjiang. Also in Mongolia, Russia and N India.

绢毛绣线菊
Spiraea sericea Turcz.

灌木。叶卵状椭圆形或椭圆形，下面密被伏生长绢毛，全缘。总状花序具总梗，伞形，具15-30花；萼筒近钟状；花瓣白色；花盘圆环形，有10个明显的浅裂片；花柱顶生。蓇葖果分叉，具柔毛。花期5-6月，果期7-8月。生海拔500-1100米的混交林、草地或干燥山坡。产中国西南、华北和东北。俄罗斯、蒙古和日本亦有。

蒙古绣线菊 *Spiraea mongolica*

Shrubs. Leaves ovate-elliptic or elliptic, abaxially densely accumbent sericeous, margin entire. Racemes pedunculate, umbellate, 15-30-flowered; hypanthium subcampanulate; petals white; disk annular, distinctly 10-lobed; styles terminal. Follicles divergent, pubescent. Fl. May-Jun. Fr. Jul-Aug. Mixed forests, grasslands or dry slopes at 500-1100 m. Distributed in SW, N and NE China. Also in Russia, Mongolia and Japan.

绢毛绣线菊 *Spiraea sericea*

蒙古绣线菊
Spiraea mongolica Maxim.

灌木，高达3米。小枝有棱角；冬芽外被2枚棕褐色鳞片。叶长圆形或椭圆形，边缘全缘，上面无毛，下面无毛稀具短柔毛。伞形总状花序具总梗；花梗无毛；萼片在果期直立或反折；花瓣白色。蓇葖果具直立或反折萼片。花期5-7月，果期7-9月。生海拔1500-4700米的林地，山坡灌丛或山顶及山谷多石砾地和溪边。产华中、内蒙古、新疆和西藏。

Shrubs, up to 3 m tall. Branchlets angled; winter buds with 2 brownish scales. Leaves oblong or elliptic, margin entire, glabrous on both surfaces, rarely puberulous abaxially. Racemes pedunculate, umbellate; pedicels glabrous; sepals erect or slightly reflexed in fruit; petals white. Follicles with erect or reflexed sepals. Fl. May-Jul. Fr. Jul-Sep. Forests, slope thickets or mountain summits, rocky valleys and streamsides at 1500-4700 m. Distributed in C China, Neimenggu, Xinjiang and Xizang.

珍珠绣线菊 *Spiraea thunbergii*

珍珠绣线菊

Spiraea thunbergii Siebold ex Blume

灌木。叶条状披针形，两面无毛，具羽状脉。伞形花序无柄，基部簇生数枚小叶，具3-7花；萼筒钟状，无毛；花瓣白色；花盘圆环形，10裂。蓇葖果开张，无毛。花期4-5月，果期7-8月。原产华东，栽培于其他省区。日本亦有。

Shrubs. Leaves linear-lanceolate, glabrous on both surfaces, pinnately veined. Umbels sessile, with clustered leaves at base. 3-7-flowered; hypanthium campanulate, glabrous; petals white; disk annular, 10-lobed. Follicles spreading, glabrous. Fl. Apr-May. Fr. Jul-Aug. Native in E China, cultivated in other provinces. Also in Japan.

单瓣李叶绣线菊

Spiraea prunifolia Siebold et Zucc. var. **simpliciflora** (Nakai) Nakai

灌木。叶卵形至长圆状披针形，背面疏被散生短柔毛，老时近无毛，叶缘近基部或自中部至顶端有细锐单锯齿。花单生，直径小于1厘米；萼筒外被短柔毛。蓇葖果仅在腹缝线处被短柔毛。花期3-5月。生海拔500-1000米的灌丛、山坡或岩石上。产中国东南、华中和华东。

Shrubs. Leaves ovate to oblong-lanceolate, sparsely thinly pubescent abaxially, glabrescent when old, margin mostly minutely sharply serrate from near base or above middle to apex. Flowers single, less than 1 cm diam; hypanthium pubescent abaxially. Follicles pubescent along adaxial suture. Fl. Mar-May. Thickets, slopes or rocks at 500-1000 m. Distributed in SE, C and E China.

鲜卑花

Sibiraea laevigata (L.) Maxim.

灌木，高约1.5米。冬芽卵形，外被紫褐色鳞片。叶线状披针形、宽披针形或长圆倒披针形，全缘，上下两面无毛。顶生穗状圆锥花序；总花梗与花梗不具毛；花瓣倒卵形，无毛，白色。蓇葖果5，具直立稀开展的宿萼。花期7月，果期8-9月。生海拔2000-4000米的高山、溪边或草甸灌丛中。产青海、甘肃和西藏。俄罗斯(西伯利亚)亦有。

Shrubs, ca. 1.5 m tall. Winter buds ovoid, purplish brown scales. Leaves linear-lanceolate, broadly lanceolate or oblong-oblanceolate, margin entire, glabrous on both surfaces. Panicles terminal, spicate; peduncles and pedicels glabrous; petals obovate, glabrous, white. Follicles 5, rarely spreading persistent sepals. Fl. Jul. Fr. Aug-Sep. Mountains, streamsides or scrub meadows at 2000-4000 m. Distributed in Qinghai, Gansu and Xizang. Also in Russia (Siberia).

单瓣李叶绣线菊 *Spiraea prunifolia* var. *simpliciflora*

鲜卑花 *Sibiraea laevigata*

窄叶鲜卑花 *Sibiraea angustata*

毛叶鲜卑花 *Sibiraea tomentosa*

窄叶鲜卑花

Sibiraea angustata (Rehder) Hand-Mazz.

灌木。叶在当年生枝上单生，往年生枝上丛生，狭披针形或倒披针形，叶两面无毛。圆锥花序；花梗密被短柔毛；萼片宿存、直立；花瓣白色，阔倒卵形。花期6月，果期8-9月。生海拔3000-4000米的疏林、山坡或山谷路边。产云南、四川、西藏、甘肃和青海。

Shrubs. Leaves solitary at axils on current year's growth, clustered on older branchlets, narrowly lanceolate or oblanceolate, glabrous on both surfaces. Panicles; peduncles densely pubescent; sepals persistent, erect; petals white, broadly obovate. Fl. Jun. Fr. Aug-Sep. Open forests, slopes or valley roadsides at 3000-4000 m. Distributed in Yunnan, Sichuan, Xizang, Gansu and Qinghai.

毛叶鲜卑花

Sibiraea tomentosa Diels

灌木，高约1米。冬芽长卵形，有2-4外露鳞片。叶长圆倒卵形至倒披针形，全缘，幼叶上下两面均密被白色绢状绒毛，逐渐脱落，老叶仅下面有稀疏绒毛。顶生密集穗状圆锥花序；总花梗和花梗均被稀疏长柔毛；花瓣匙形，浅黄白色。蓇葖果5，直立，腹缝有柔毛。花期6月,果期8月。生海拔3500-4000米的高山、溪岸和潮湿多石地。产云南。

Shrubs, ca. 1 m tall. Winter buds long-ovoid, with 2-4 exposed scales. Leaves oblong-obovate to oblanceolate, margin entire, densely white sericeous-tomentose on both surfaces when young, later sparsely tomentose only abaxially. Spicate panicles terminal, compact; peduncles and pedicels sparsely villous; petals spatulate, pale yellowish white. Follicles 5, erect, sparsely villous. Fl. Jun. Fr. Aug. Slopes, streamsides and moist places on rocks at 3500-4000 m. Distributed in Yunnan.

假升麻

Aruncus sylvester Kostel. ex Maxim.

多年生草本，高达3米。叶为二至三回羽状复叶；小叶菱状卵形、卵状披针形或长椭圆形。圆锥花序疏松；花单性；萼筒杯状，外面微被毛；心皮3-4。蓇葖果无毛。花期6月，果期8-9月。生海拔1800-3500米的山坡混交林或山谷。产中国西南、华中、华西、华东和东北。南亚、东南亚、西南亚和欧洲亦有。

Herbs perennial, to 3 m tall. Leaves 2- or 3-pinnate; leaflets rhombic-ovate, ovate-lanceolate or long elliptic. Panicles lax; flowers unisexual; hypanthium cup-shaped, abaxially slightly pubescent; carpels 3-4. Follicles glabrous. Fl. Jun. Fr. Aug-Sep. Mixed forests on mountain slopes or valleys at 1800-3500 m. Distributed in SW, C, W, E and NE China. Also in S, SE and SW Asia, and Europe.

假升麻 *Aruncus sylvester*

贡山假升麻 *Aruncus gombalanus*

华北珍珠梅 *Sorbaria kirilowii*

贡山假升麻

Aruncus gombalanus (Hand-Mazz.) Hand-Mazz.

多年生草本，高可达70厘米。基部具肥厚木质地下茎。一至二回羽状复叶或一至二回三出复叶，小叶近圆形或宽卵形，稀菱状卵形。圆锥花序紧密，长5-25厘米；雄花的花丝稍长于花瓣。蓇葖果并立。花期6月，果期8-9月。生海拔3000-4000米的山顶草坡。产云南西北部和西藏东部。

Herbs perennial, to 70 cm tall. Rhizomes thick, woody. Leaves 1- or 2-pinnate or 1- or 2-ternate, leaflets suborbicular or broadly ovate, rarely rhombic-ovate. Panicles dense, 5-25 cm long; male flowers with filaments longer than petals. Follicles parallel. Fl. Jun. Fr. Aug-Sep. Grassy slopes on mountain summits at 3000-4000 m. Distributed in NW Yunnan and E Xizang.

珍珠梅

Sorbaria sorbifolia (L.) A. Braun

灌木。羽状复叶，小叶11-17枚，侧脉12-16对。圆锥花序长10-12厘米；萼筒外面基部被柔毛；萼片果期宿存，反折；雄蕊40-50枚，长于花瓣；花柱顶生。蓇葖果圆筒状，果梗直立。花期7-8月，果期9月。生海拔200-1500米的向阳林中。产中国东北。蒙古、朝鲜半岛和日本亦有。

Shrubs. Leaves pinnate, leaflets 11-17, lateral veins in 12-16 pairs. Panicles 10-12 cm long; hypanthium pubescent abaxially at base; sepals persistent and reflexed in fruit; stamens 40-50, longer than petals; styles terminal. Follicles cylindric, fruiting pedicels erect. Fl. Jul-Aug. Fr. Sep. Open forests at 200-1500 m. Distributed in NE China. Also in Mongolia, Korean Peninsula and Japan.

珍珠梅 *Sorbaria sorbifolia*

华北珍珠梅

Sorbaria kirilowii (Regel) Maxim.

灌木，高达3米。羽状复叶具小叶13-21，连叶柄长21-25厘米，光滑无毛；小叶披针形至长圆披针形，羽状网脉，侧脉15-23对近平行。大型密集的圆锥花序顶生，分枝斜出或稍直立；雄蕊约20，与花瓣等长或稍短于花瓣；花柱稍侧生，向外弯曲；萼片宿存，反折。花期6-7月，果期9-10月。生海拔200-1300米的山坡阳处和杂木林中。产华中、山东和内蒙古。

Shrubs, up to 3 m tall. Pinnate leaves with 13-21 leaflets, together with petiole 21-25 cm long, glabrous; leaflets lanceolate to oblong-lanceolate, pinnate veins, lateral veins 15-23 pairs, subparallel. Large and dense panicles terminal, branches oblique or slightly erect; stamens ca. 20, equaling or slightly shorter than petals; style lateral, curved outward; sepals persistent, reflexed. Fl. Jun-Jul. Fr. Sep-Oct. Sunny slopes and open mixed forests at 200-1300 m. Distributed in C China, Shandong and Neimenggu.

高丛珍珠梅

Sorbaria arborea C. K. Schneid.

灌木。叶片、叶轴和花序稍具星状毛。小叶13-17，对生，无柄或近无柄，披针形至长圆状披针形。圆锥花序多花。蓇葖果圆柱状，无毛，宿存萼片反折，果序下垂。花期6-7月，果期9-10月。生海拔2500-3500米的林缘坡地或溪边。产中

高丛珍珠梅 *Sorbaria arborea*

国西南、华中、华西和西北。

Shrubs. Leaflets, rachis, and inflorescence slightly stellate hairy. Leaflets 13-17, opposite, sessile or nearly so, lanceolate to oblong-lanceolate. Panicles many flowered. Follicles cylindric, glabrous, with persistent, reflexed sepals, fruiting pedicels pendulous. Fl. Jun-Jul. Fr. Sep-Oct. Slopes near forest edges or by streams at 2500-3500 m. Distributed in SW, C, W and NW China.

光叶高丛珍珠梅

Sorbaria arborea C. K. Schneid. var. **glabrata** Rehder

灌木。叶片、叶轴和花序均平滑无毛。小叶13-17，对生，无柄或近无柄，披针形至长圆状披针形。圆锥花序多花。蓇葖果圆柱状，无毛，宿存萼片反折，果序下垂。生海拔2500-3500米的密林中、山坡或溪边。产云南、四川、湖北、陕西和甘肃。

Shrubs. Leaflets, rachis, and inflorescence glabrous. Leaflets 13-17, opposite, sessile or nearly so, lanceolate to oblong-lanceolate. Panicles many flowered. Follicles cylindric, glabrous, with persistent, reflexed, fruiting pedicels pendulous. Dense forests, mountain slopes or streamsides at 2500-3500 m. Distributed in Yunnan, Sichuan, Hubei, Shaanxi and Gansu.

风箱果

Physocarpus amurensis (Maxim.) Maxim.

灌木。托叶早落；单叶互生，三角状卵形或宽卵形，常为三出脉，背面被星状毛与短柔毛。伞房花序顶生，直径3-4厘米，多花；花瓣白色。蓇葖果卵球形，渐无毛。花期6月，果期7-8月。生阔叶林林缘或山谷。产河北和黑龙江。俄罗斯(远东地区)和朝鲜半岛亦有。

Shrubs. Stipules caduceus; leaves simple and alternate, triangular-ovate or broadly ovate, usually 3-veined, abaxially stellate hairy and pubescent. Corymbs terminal, 3-4 cm diam, many flowered; petals white. Follicles ovoid, glabrescent. Fl. Jun. Fr. Jul-Aug. Broad-leaved forest edges or valleys. Distributed in Hebei and Heilongjiang. Also in Russia (Far East) and Korean Peninsula.

光叶高丛珍珠梅 *Sorbaria arborea* var. *glabrata*

风箱果 *Physocarpus amurensis*

云南绣线梅

Neillia serratisepala H. L. Li

灌木。叶椭圆状卵形至三角状卵形。圆锥花序顶生或腋生，具多花；花直径约4毫米；萼筒钟状，外面密被柔毛；花瓣白色；雄蕊约20；子房内含3-5胚珠。蓇葖果长卵球形。花期7-8月，果期9-10月。生海拔约2000米的疏林中或山坡林缘。产云南。

Shrubs. Leaves elliptic-ovate to triangular-ovate. Panicles terminal or axillary, many flowered; flowers ca. 4 mm diam; hypanthium campanulate, abaxially densely pubescent; petals white; stamens ca. 20; ovules 3-5. Follicles long ovoid. Fl. Jul-Aug. Fr. Sep-Oct. Open forests or forest edges on slopes at ca. 2000 m. Distributed in Yunnan.

云南绣线梅 *Neillia serratisepala*

毛果绣线梅

Neillia thyrsiflora D. Don var. **tunkinensis** (J. E. Vidal) J. E. Vidal

灌木。高达2米。小枝棕红色，具棱。叶卵形至卵状椭圆形，边缘常3深裂，具尖锐重锯齿。圆锥花序分枝较少，常多花；子房被柔毛；胚珠(8-)10-12。蓇葖果长圆形。花期7月，果期9-10月。生海拔1000-3000米的山谷密林或山坡林边。产云南、四川、西藏、贵州和广西。印度东北部、越南北部和印度尼西亚(爪哇、苏门答腊)亦有。

Shrubs, to 2 m tall. Branchlets red-brown, angled. Leaves ovate to ovate-elliptic, margin usually 3-parted, sharply doubly serrate. Panicles laxly branched, usually many flowered; ovary pubescent;

毛果绣线梅 *Neillia thyrsiflora* var. *tunkinensis*

矮生绣线梅 *Neillia gracilis*

中华绣线梅 *Neillia sinensis*

ovules (8-)10-12. Follicles cylindric. Fl. Jul. Fr. Sep-Oct. Dense forests or forest edges in valleys at 1000-3000 m. Distributed in Yunnan, Sichuan, Xizang, Guizhou and Guangxi. Also in NE India, N Vietnam and Indonesia (Java, Pulau Sumatera).

矮生绣线梅
Neillia gracilis Franch.

矮生亚灌木，高达0.5米。小枝弯曲有棱。叶卵形至三角卵形，稀近肾形，基部心形，边缘不规则3-5裂，具尖锐重锯齿。总状花序腋生；花瓣白色或粉红色；子房狭卵形，密被毛。蓇葖果被短柔毛。种子卵球形。花期5-7月，果期8-9月。生海拔2800-3000米的高山草甸或湿山坡上。产云南北部和四川西部。

Subshrubs, to 0.5 m tall. Branchlets recurved, angled. Leaves ovate to triangular-ovate, rarely subreniform, base cordate, margin irregularly 3-5-lobed, sharply doubly serrate. Racemes axillary; petals white or pinkish; ovary narrowly ovoid, densely villous. Follicles pubescent. Seeds ovoid. Fl. May-Jul. Fr. Aug-Sep. Alpine meadows or moist slopes at 2800-3000 m. Distributed in N Yunnan and W Sichuan.

中华绣线梅
Neillia sinensis Oliv.

灌木。小枝红褐色，无毛。叶卵形或卵状椭圆形，基部圆形或近心形，边缘常不规则分裂，具重锯齿。总状花序多花，长4-9厘米；花瓣粉色，倒卵形。蓇葖果长椭圆体形。种子卵球形。花期5-7月，果期8-9月。生海拔1000-2500米的沟谷、山坡或杂木林中。产中国西南、华南、华中、华北和华西。

Shrubs. Branchlets red-brown, glabrous. Leaves ovate or ovate-elliptic, base rounded or subcordate, margin often irregularly lobed, doubly serrate. Racemes 4-9 cm long, many flowered; petals pink, obovate. Follicles long ellipsoid. Seeds ovoid. Fl. May-Jul. Fr. Aug-Sep.Valleys, slopes or mixed forests at 1000-2500 m. Distributed in SW, S, C, N and W China.

华空木
Stephanandra chinensis Hance

灌木，高达 1.5米。叶卵形至长椭圆状卵形，长5-7厘米，宽2-3厘米，边缘具重锯齿，常浅裂。圆锥花序直径2-3厘米；总花梗和花梗无毛；萼筒无毛。蓇葖果直径约2毫米。种子1粒，卵球形。花期5月，果期7-8月。生海拔1000-1500米的阔叶林林缘。产中国西南、华南、东南、华中和华北。

Shrubs, to 1.5 m. Leaves ovate to long elliptic-ovate, 5-7 × 2-3 cm, margin doubly serrate, usually shallowly lobed. Panicles 2-3 cm diam; peduncles and pedicels glabrous; hypanthium glabrous. Follicles ca. 2 mm diam. Seeds 1, ovoid. Fl. May. Fr. Jul-Aug. Broad-leaved forest edges at 1000-1500 m. Distributed in SW, S, SE, C and N China.

华空木 *Stephanandra chinensis*

小米空木 *Stephanandra incisa*

齿叶白鹃梅 *Exochorda serratifolia*

小米空木

Stephanandra incisa (Thunb.) Zabel

灌木，高达2.5米。叶卵形至三角卵形，长2-4厘米，宽1.5-2.5厘米，边缘具重锯齿。圆锥花序疏松，顶生，具多花；总花梗与花梗均被柔毛。蓇葖果被柔毛；宿存萼片直立或开展。花期6-7月，果期8-9月。生海拔500-1000米的山坡或溪边。产台湾、山东东部和辽宁。朝鲜半岛和日本亦有。

Shrubs, to 2.5 m tall. Leaves ovate to triangular-ovate, 2-4 × 1.5-2.5 cm, margin doubly serrate. Panicles lax, terminal, many flowered; peduncle and pedicels pubescent. Follicles pubescent, persistent sepals erect or spreading. Fl. Jun-Jul. Fr. Aug-Sep. Slopes or by streams at 500-1000 m. Distributed in Taiwan, E Shandong and Liaoning. Also in Korean Peninsula and Japan.

白鹃梅

Exochorda racemosa (Lindl.) Rehder

灌木。叶椭圆形、狭椭圆形至长圆状倒卵形，全缘，稀中部以上有钝锯齿；叶柄长5-15毫米或近无柄。花梗长3-5毫米；花瓣基部有短爪；雄蕊15-20，3-4枚一束，与花瓣对生。蒴果无毛。花期5月，果期6-7月。生海拔200-500米的阴坡。产浙江、河南、江苏和江西。

Shrubs. Leaves elliptic, narrowly so to oblong-obovate, entire, rarely obtusely serrate above middle; petioles 5-15 mm long or nearly absent. Pedicels 3-5 mm long; petals base shortly clawed; stamens 15-20, 3- 4-fascicled, antepetalous. Capsules glabrous. Fl. May. Fr. Jun-Jul. Shady slopes at 200-500 m. Distributed in Zhejiang, Henan, Jiangsu and Jiangxi.

白鹃梅 *Exochorda racemosa*

齿叶白鹃梅

Exochorda serratifolia Rehder

灌木。叶缘中部以下全缘，中部以上有锐锯齿，叶柄长1-2厘米。总状花序有4-7花；萼筒浅钟状，外面无毛；花瓣长圆形至倒卵形，基部有长爪，顶端微凹；雄蕊约25枚，2-4枚一束。蒴果无毛。花期5-6月，果期7-8月。生灌丛、山坡或河边。产河北东北部和辽宁(千山)。朝鲜半岛亦有。

Shrubs. Leaf margin entire below middle, serrate above middle, petioles 1-2 cm long. Racemes 4-7-flowered; hypanthium shallowly campanulate, abaxially glabrous; petals oblong to obovate, base long clawed, apex emarginated; stamens ca. 25, 2-4-fascicled. Capsules glabrous. Fl. May-Jun. Fr. Jul-Aug. Thickets slopes or riversides. Distributed in NE Hebei and Liaoning (Qianshan Mountain). Also in Korean Peninsula.

牛筋条
Dichotomanthus tristaniaecarpa Kurz

灌木。叶椭圆形至长圆状披针形，背面密被白色柔毛；托叶丝状。花瓣白色；子房外被柔毛；花柱近顶生或侧生，无毛。果红色，长圆柱状，突出于肉质萼筒中。花期4-5月，果期8-11月。生海拔1300-2500米的杂木林、常绿林林缘和山坡开阔地。产云南和四川。

Shrubs. Leaves elliptic to oblong-lanceolate, abaxially densely white-villose; stipules filimentous. Petals white; ovary abaxially pubescent; styles subterminal or lateral, glabrous. Fruits red, long cylindrical , exserted from fleshy hypanthium. Fl. Apr-May. Fr. Aug-Nov. Mixed forests, evergreen forest edges and open slopes at 1300-2500 m. Distributed in Yunnan and Sichuan.

福建假稠李
Maddenia fujianensis Y. T. Chang

灌木。叶卵形，背面淡绿色，无毛或沿脉腋被柔毛。总状花序疏散，长3-5厘米，花序梗和花梗初被粉状柔毛，后脱落；萼筒钟状，无毛；雄蕊20-28枚；子房无毛；花柱短于雄蕊。花期4月。生海拔约1700米的疏林。产福建北部。

Shrubs. Leaves ovate, abaxially pale green, glabrous or sometimes pubescent along vein axils. Racemes lax, 3-5 cm long, peduncles and pedicels pulverulently puberulous, glabrescent; hypanthium campanulate, glabrous; stamens 20-28; ovary glabrous; styles shorter than stamens. Fl. Apr. Sparse forests at ca. 1700 m. Distributed in N Fujian.

牛筋条 *Dichotomanthus tristaniaecarpa*

福建假稠李 *Maddenia fujianensis*

柳叶栒子 *Cotoneaster salicifolius*

柳叶栒子
Cotoneaster salicifolius Franch.

灌木。叶椭圆状长圆形至卵状披针形或条状披针形，上面有浅皱，背面起初密被绒毛，先端急尖或渐尖。复伞房花序，花多数；总花梗和花梗密被灰白色绒毛；花瓣白色。果猩红色，小核2-3。花期6月，果期10月。生海拔400-3000米的山地或河边混交林中。产云南、四川、贵州、湖北和湖南。

Shrubs. Leaves elliptic-oblong to ovate-lanceolate or linear-lanceolate, adaxially shallowly rugose, abaxially initially densely tomentose, apex acute or acuminate. Compound corymbs, many flowered; rachises and pedicels densely gray-white tomentose; petals white. Fruits scarlet, pyrenes 2-3. Fl. Jun. Fr. Oct. Mountains or mixed forests along river banks at 400-3000 m. Distributed in Yunnan, Sichuan, Guizhou, Hubei and Hunan.

厚叶栒子
Cotoneaster coriaceus Franch.

灌木。叶倒卵形至椭圆形，厚革质，全缘，下面密被黄色绒毛。复伞房花序具多数小形而密集花；总花梗和花梗密被黄色绒毛；花瓣开展，白色。果红色，倒卵球形。花期5-6月，果期9-10月。生海拔1800-2700米的山谷疏林中。产云南、四川、西藏和贵州。

Shrubs. Leaves obovate to elliptic, thickly leathery, margin entire, abaxially densely yellow tomentose. Compound corymbs densely many small flowered; rachises and pedicels densely yellow tomentose; petals spreading, white. Fruits red, obovoid. Fl. May-Jun. Fr. Sep-Oct. Grassy slopes along riversides at 1800-2700 m. Distributed in Yunnan, Sichuan, Xizang and Guizhou.

水栒子
Cotoneaster multiflorus Bunge

灌木。叶卵形或宽卵形，长2-4厘米，下面具柔毛，后脱落。伞房花序疏松，具5-21花；花梗和萼筒无毛；花瓣白色，平展。梨果红色或淡紫红色，直径7-8毫米。花期5-6月，果期8-9月。生海拔1200-3500米的沟谷或杂木林中。产中国西南、华中、华北、华西、西北和东北。西亚、中亚、高加索地区和俄罗斯（西伯利亚）亦有。

Shrubs. Leaves ovate or broadly ovate, 2-4 cm long, abaxially pilose, glabrescent. Corymbs lax, 5-21-flowered; pedicels and hypanthium glabrous; petals white, spreading. Pomes red or purplish red, 7-8 mm diam. Fl.

厚叶栒子 *Cotoneaster coriaceus*

水栒子 *Cotoneaster multiflorus*

May-Jun. Fr. Aug-Sep. Valleys or mixed forests at 1200-3500 m. Distributed in SW, C, N, W, NW and NE China. Also in W and C Asia, Caucasus and Russia (Siberia).

紫果水栒子

Cotoneaster multiflorus Bunge var. **atopurpureus** T. T. Yü

灌木。叶卵形或宽卵形，长2-4厘米，下面具柔毛，后脱落。伞房花序疏松，具5-21花；花梗和萼筒无毛；花瓣白色，平展。梨果紫黑色，小，直径5-6毫米。花期5-6月，果期9-10月。生海拔2500-3100米的林缘、溪边或灌丛中。产云南西北部、四川西部和西藏。

Shrubs. Leaves ovate or broadly ovate, 2-4 cm long, abaxially pilose, glabrescent. Corymbs lax, 5-21-flowered; pedicels and hypanthium glabrous; petals white, spreading. Pomes purplish black, small, 5-6 mm diam. Fl. May-Jun. Fr. Sep-Oct. Forest edges, streamsides or thickets at 2500-3100 m. Distributed in NW Yunnan, W Sichuan and Xizang.

毛叶水栒子

Cotoneaster submultiflorus Popov

落叶灌木。叶卵形或菱状卵形至椭圆形，背面具柔毛。伞房花序具多数花；总花梗和花梗具长柔毛；萼筒钟状，外被柔毛；花瓣平展，白色。果亮红色，近球形。花期5-6月，果期8-9月。生海拔900-2000米的岩石缝或灌丛中。产中国西南、华中、华北和西北。中亚亦有。

Shrubs deciduous. Leaves ovate or rhombic-ovate to elliptic, abaxially pubescent. Corymbs many flowered; rachis and pedicels villous; hypanthium campanulate, abaxially pilose; petals spreading, white. Fruits bright red, subglobose. Fl. May-Jun. Fr. Aug-Sep. Rock crevices or thickets at 900-2000 m. Distributed in SW, C, N and NW China. Also in C Asia.

毛叶水栒子 *Cotoneaster submultiflorus*

华中栒子

Cotoneaster silvestrii Pamp.

落叶灌木。叶椭圆形至卵形，下面被薄层灰色绒毛。聚伞花序有3-7花；萼筒钟状，外被细长柔毛；花瓣平展，近圆形，白色；花药黄色。果实红色，近球形，通常2小核联合为1个。花期5-6月，果期8-9月。生海拔500-2600米的杂木林内。产四川、湖北、河南、甘肃、安徽、江苏和江西。

Shrubs deciduous. Leaves elliptic to ovate, abaxially thinly gray tomentose. Corymbs 3-7-flowered; hypanthium campanulate, abaxially thinly villous; petals spreading, subglobose, white; anthers yellow. Fruits red, subglobose, pyrenes 2, united into 1. Fl. May-Jun. Fr. Aug-Sep. Mixed forests at 500-2600 m. Distributed in Sichuan, Hubei, Henan, Gansu, Anhui, Jiangsu and Jiangxi.

紫果水栒子 *Cotoneaster multiflorus* var. *atopurpureus*

华中栒子 *Cotoneaster silvestrii*

麻核栒子 *Cotoneaster foveolatus*

西北栒子 *Cotoneaster zabelii*

麻核栒子
Cotoneaster foveolatus Rehder et E. H. Wils

落叶灌木。叶椭圆形、椭圆状卵形或椭圆状倒卵形，下面被短柔毛，老时近无毛。伞房花序具3-7花；总花梗与花梗被柔毛；萼筒钟状；花瓣直立，粉色，基部具短爪；雄蕊15-17；子房顶部密被柔毛。果实黑色，近球形。花期5-6月，果期9-10月。生海拔1400-3400米的灌丛、密林、林缘、河道边或弃荒地。产中国西南、华中和华西。

Shrubs deciduous. Leaves elliptic, elliptic-ovate or elliptic-obovate, abaxially initially pubescent, subglabrous when old. Corymbs 3-7-flowered; rachises and pedicels pubescent; hypanthium campanulate; petals erect, pink, base shortly clawed; stamens 15-17; ovary densely pilose apically. Fruits black, subglobose. Fl. May-Jun. Fr. Sep-Oct. Thickets, dense forests, forest edges, near water courses or waste fields at 1400-3400 m. Distributed in SW, C and W China.

灰栒子
Cotoneaster acutifolius Turcz.

落叶灌木。叶椭圆状卵形至长圆状卵形，叶脉在上面凹陷，在下面突起。伞房花序具2-5花；花瓣白色外带红晕，基部略带长爪。果实黑色，椭圆体形、倒卵球形或近球形。花期5-7月，果期8-9月。生海拔1000-3700米的山坡、山脚、森林、林缘、灌丛、山谷或裸露草地。产中国西南、华南、华中、华北和华西。俄罗斯和蒙古亦有。

Shrubs deciduous. Leaves elliptic-ovate to oblong-ovate, veins impressed adaxially and raised abaxially. Corymbs 2-5-flowered; petals white, tinged reddish, base somewhat long clawed. Fruits black, ellipsoid, obovoid or subglobose. Fl. May-Jul. Fr. Aug-Sep. Slopes, foothills, forests, forest edges, thickets, valleys or exposed grasslands at 1000-3700 m. Distributed in SW, S, C, N and W China. Also in Russia and Mongolia.

灰栒子 *Cotoneaster acutifolius*

西北栒子
Cotoneaster zabelii C. K. Schneid.

落叶灌木。叶椭圆形至卵形，先端多圆钝，背面密被绒毛。伞房花序具3-13花；萼筒外被绒毛；花瓣浅粉红色；雄蕊18-20。果亮红色，倒卵状或卵球形，常具2核。花期5-6月，果期8-9月。生海拔800-2500米的石灰岩山地、阴坡、河谷或灌丛中。产华中、华北、华东和西北。

Shrubs deciduous. Leaves elliptic to ovate, apex obtuse, abaxially densely tomentose. Corymbs 3-13-flowered; hypanthium abaxially tomentose; petals pinkish; stamens 18-20. Fruits bright red, obovoid or ovoid-globose, pyrenes often 2. Fl. May-Jun. Fr. Aug-Sep. Calcareous mountain regions, shaded slopes, river valleys or thickets at 800-2500 m. Distributed in C, N, E and NW China.

细枝栒子
Cotoneaster tenuipes Rehder et E. H. Wils

落叶灌木。小枝淡红褐色，细弱。叶卵形或椭圆状卵形至狭椭圆状卵形，全缘，先端急尖或稍钝。伞房花序，有2-4花；萼筒外面密被平贴长柔毛；花瓣白色。果紫黑色，1或2小核。花期5-6月，果期9-10月。生海拔1900-3100米的林下或多石地。产云南、四川、西藏、陕西、甘肃和青海。

Shrubs deciduous. Branchlets brownish red, slender. Leaves ovate or elliptic-ovate to narrowly elliptic-ovate, entire, apex acute or obtuse. Corymbs, 2-4-flowered;

细枝栒子 *Cotoneaster tenuipes*

hypanthium densely appressed villose abaxially; petals white. Fruits purplish black, pyrenes 1 or 2. Fl. May-Jun. Fr. Sep-Oct. Forests or rocky mountain areas at 1900-3100 m. Distributed in Yunnan, Sichuan, Xizang, Shaanxi, Gansu and Qinghai.

木帚栒子

Cotoneaster dielsianus Pritz.

落叶灌木。叶椭圆形至卵形，背面密被绒毛。伞房花序具3-7花；总花梗和花梗具长柔毛；花瓣粉色；雄蕊15-20。果深红色或珊瑚红色，近球形或倒卵球形，具3-5小核。花期6-7月，果期9-10月。生海拔1000-3600米的荒地、山谷、草地、灌丛或密林中。产中国西南和华中。

Shrubs deciduous. Leaves elliptic to ovate, abaxially densely tomentose. Corymbs 3-7-flowered; rachis and pedicels villous; petals pink; stamens 15-20. Fruits dark red or coral-red, subglobose or obovoid, pyrenes 3-5. Fl. Jun-Jul. Fr. Sep-Oct. Wastelands, valleys, grasslands, thickets or dense forests at 1000-3600 m. Distributed in SW and C China.

木帚栒子 *Cotoneaster dielsianus*

球花栒子

Cotoneaster glomerulatus W. W. Smith

落叶灌木。叶卵状披针形至长圆卵形，下面被稀疏黄色绒毛。聚伞花序1.5- 2.5厘米长，有3-11花；总花梗和花梗密被黄色长柔毛；花瓣直立，白色带红晕。果实亮红色，近球形，内具5小核。花期5-6月，果期9-10月。生海拔2000-2600米的山坡、河谷、疏林或稀疏灌丛。产云南。

Shrubs deciduous. Leaves ovate-lanceolate to oblong-ovate, abaxially sparsely yellow tomentose. Corymbs 1.5- 2.5 cm long, 3-11-flowered; rachises and pedicels densely yellow tomentose-villous; petals erect, white, stained reddish. Fruits bright red, subglobose, pyrenes 5. Fl. May-Jun. Fr. Sep-Oct. Slopes, river valleys, sparse forests or open scrubs at 2000-2600 m. Distributed in Yunnan.

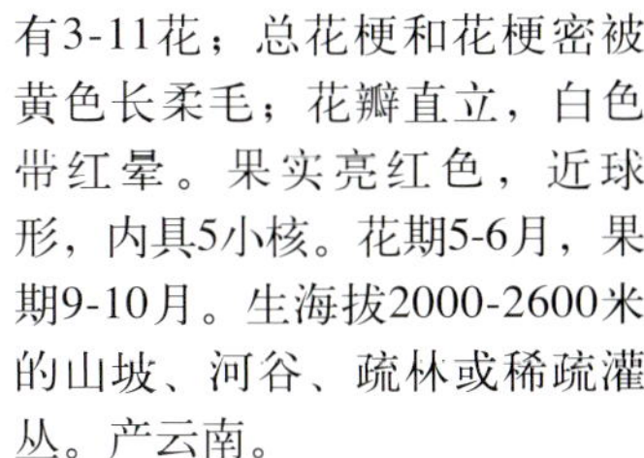

球花栒子 *Cotoneaster glomerulatus*

西南栒子 *Cotoneaster franchetii*

西南栒子
Cotoneaster franchetii Boiss.

半常绿灌木。叶椭圆形至卵形，下面密被黄色或白色绒毛。伞房花序，5-11花；总花梗和花梗密被绒毛；花瓣粉红色；雄蕊20。梨果球形，橘红色，具3-5枚小核。花期6-7月，果期9-10月。生海拔2000-2900米的向阳山坡或灌丛中。产中国西南。泰国亦有。

Shrubs semievergreen. Leaves elliptic to ovate, abaxially densely yellow or white tomentose. Corymbs, 5-11-flowered; rachis and pedicels densely tomentose; flowers pink; stamens 20. Pomes globose, orangish red, pyrenes 3-5. Fl. Jun-Jul. Fr. Sep-Oct. Sunny slopes or thickets at 2000-2900 m. Distributed in SW China. Also in Thailand.

矮生栒子
Cotoneaster dammeri C. K. Schneid.

半常绿矮小灌木。叶柄长2-3毫米，叶厚革质，椭圆形至椭圆长圆形。花常单生，稀2-3组成聚伞花序；花梗长4-6毫米；花瓣平展，白色，基部具短爪；子房顶端具柔毛。果实近球形。花期5-6月，果期9-10月。生海拔1300-4100米的多石山地或稀疏杂木林内。产云南、四川、西藏、贵州、湖北和甘肃。

Shrubs semievergreen, low. Petioles 2-3 mm long, leaves thickly leathery, elliptic to oblong-elliptic. Flowers usually solitary, rarely 2-3-flowered corymbs; pedicels 4-6 mm long; petals spreading, white, base shortly clawed; ovary pilose apically. Fruits subglobose. Fl. May-Jun. Fr. Sep-Oct. Rocky mountain areas or sparse mixed forests at 1300-4100 m. Distributed in Yunnan, Sichuan, Xizang, Guizhou, Hubei and Gansu.

小叶栒子
Cotoneaster microphyllus Wall. ex Lindl.

平卧灌木。叶倒卵形或长圆倒卵形，厚革质，背面被灰色短柔毛，叶缘反卷。花序具1(-3)花；花梗疏具柔毛；花瓣白色；雄蕊15-20。梨果球形，猩红色，2小核。花期5-6月，果期8-9月。生海拔2000-4200米的石砾山坡或灌丛中。产云南、四川和西藏。缅甸、不丹和印度亦有。

Shrubs prostrate. Leaves obovate or oblong-ovate, thickly leathery, abaxially gray pubescent, margin revolute. Inflorescences 1(-3)-flowered; pedicels sparsely pubescent; petals white; stamens 15-20. Pomes globose, scarlet-red, pyrenes 2. Fl. May-Jun. Fr. Aug-Sep. Gravel slopes or thickets at 2000-4200 m. Distributed in Yunnan, Sichuan and Xizang. Also in Myanmar, Bhutan and India.

矮生栒子 *Cotoneaster dammeri*

小叶栒子 *Cotoneaster microphyllus*

圆叶栒子 *Cotoneaster rotundifolius*

黄杨叶栒子 *Cotoneaster buxifolius*

圆叶栒子

Cotoneaster rotundifolius Wall. ex Lindl.

常绿灌木。叶近圆形或宽卵形，全缘，下面疏被毛，上面无毛。花序具1-3花；萼筒钟状，外面疏被毛；花瓣白色或带粉红色；雄蕊约20枚。果红色，具2-3小核。花期5-6月，果期8-9月。生海拔1200-4000米的草坡、多石地或山顶。产中国西南。印度、尼泊尔和不丹亦有。

Shrubs evergreen. Leaves suborbicular or broadly ovate, entire, abaxially pilose, adaxially glabrous. Inflorescences 1-3-flowered; hypanthium campanulate, abaxially pilose; petals white or tinged reddish; stamens 20. Fruits red, pyrenes 2-3. Fl. May-Jun. Fr. Aug-Sep. Grassy slopes, rocks or mountain summits at 1200-4000 m. Distributed in SW China. Also in India, Nepal and Bhutan.

黄杨叶栒子

Cotoneaster buxifolius Lindl.

常绿灌木。叶椭圆形至椭圆状倒卵形，背面密被灰色绒毛。花序具3-5(-9)花，稀单生；萼筒和萼片外面密被宿存糙毛；花瓣白色。果近球形，红色，具2小核。花期4-6月，果期9-10月。生海拔1000-3900米的多石山坡、灌丛或路边。产中国西南。不丹、印度、缅甸和尼泊尔亦有。

Shrubs evergreen. Leaves elliptic to elliptic-obovate, abaxially densely gray tomentose. Inflorescences 3-5(-9)-flowered, rarely solitary; hypanthium and sepals abaxially densely persistent strigose tomentose; petals white. Fruits subglobose, red, pyrenes 2. Fl. Apr-Jun. Fr. Sep-Oct. Rocky mountain slopes, thickets or roadside at 1000-3900 m. Distributed in SW China. Also in Bhutan, India, Myanmar and Nepal.

平枝栒子

Cotoneaster horizontalis Decne.

平卧小灌木。枝水平开张成整齐两列状。叶近圆形或阔椭圆形，背面被柔毛。花序具1-2花；萼筒外被柔毛；花瓣直立，粉红、红色或白色；子房顶端被毛；花柱(2-)3，离生。果近球形，亮红色。花期5-6月，果期9-10月。生海拔1500-3500米的灌丛、岩石地、多石山坡或干山地。产中国西南、华南、东南、华中、华西和华东。尼泊尔亦有。

Shrubs prostrate, low. Stems horizontally spreading. Leaves suborbicular or broadly elliptic, abaxially pilose. Inflorescences 1- or 2-flowered; hypanthium abaxially pubescent; petals erect, pink, reddish or whitish; ovary pilose apically; styles (2-)3, free. Fruits subglobose, bright red. Fl. May-Jun. Fr. Sep-Oct. Thickets, rocks, rocky slopes or dry mountain areas at 1500-3500 m. Distributed in SW, S, SE, C, W and E China. Also in Nepal.

平枝栒子 *Cotoneaster horizontalis*

火棘

Pyracantha fortuneana
(Maxim.) H. L. Li

灌木。叶背面绿色，两面无毛，边缘有圆锯齿，顶端最宽，先端钝或凹缺。复伞房花序极疏松，直径3-4厘米。梨果橘红色至暗红色，近球形，直径约5毫米。花期3-5月，果期8-11月。生海拔500-2800米的灌丛、溪边或路边。产中国西南、华南、东南、华中和华西。

Shrubs. Leaves abaxial surface green, both surfaces glabrous, margin crenate, broadest apically, apex obtuse or emarginate. Compound corymb rather loose, 3-4 cm diam. Pomes orangish red or dark red, subglobose, ca. 5 mm diam. Fl. Mar-May. Fr. Aug-Nov. Thickets, streamsides or roadsides at 500-2800 m. Distributed in SW, S, SE, C and W China.

火棘 *Pyracantha fortuneana*

全缘火棘

Pyracantha atalantioides
(Hance) Stapf

常绿灌木或小乔木。枝常有刺。叶椭圆形或长圆形，稀长圆倒卵形，先端微尖或圆钝，叶边通常全缘或有时具不显明

火棘 *Pyracantha fortuneana*

全缘火棘 *Pyracantha atalantioides*

的细锯齿，两面幼时有黄褐色柔毛，老时两面无毛，下面微带白霜。花梗和花萼外被黄褐色柔毛；花梗长5-10毫米。花期4-5月，果期9-11月。生海拔500-1700米的山坡、谷地或疏林中。产陕西、湖北、湖南、四川、贵州、广东和广西。

Evergreen shrubs or small trees. Branches usually thorny. Leaves elliptic or oblong, rarely oblong-obovate, apex apiculate or obtuse, margin usually entire, sometimes inconspicuously serrulate, both surfaces initially yellowish-brown pubescent, glabrescent, abaxially becoming slightly glaucescent. Pedicles and sepals yellowish brown pubescent; pedicels 5-10 mm long. Fl. Apr-May. Fr. Sep-Nov. Slopes, valleys and open forests at 500-1700 m. Distributed in Shaanxi, Hubei, Hunan, Sichuan, Guizhou, Guangdong and Guangxi.

细圆齿火棘

Pyracantha crenulata (D. Don) M. Roem.

灌木或小乔木。叶长圆形或倒披针形，稀卵状披针形，无毛，边缘有细圆锯齿，或具稀疏锯齿，先端急尖或钝。复伞房花序直径3-5厘米，多花；萼筒钟状，无毛。梨果橘黄色或橘红色，近球形，直径3-8毫米。花期3-5月，果期9-11月。生海拔700-2500米的山坡、路边、溪边、灌丛或草地。产中国西南、华南、华中、华西和华东。尼泊尔、不丹、印度和缅甸亦有。

Shrubs or small trees. Leaves oblong or oblanceolate, rarely ovate-lanceolate, glabrous, margin crenulate or sparsely so, apex acute or obtuse. Compound corymb 3-5 cm diam, many flowered; hypanthium campanulate, glabrous. Pomes orangish yellow or orangish red, subglobose, 3-8 mm diam. Fl. Mar-May. Fr. Sep-Nov. Slopes, roadsides, streamsides, shrubs or grasslands at 700-2500 m. Distributed in SW, S, C, W and E China. Also in Nepal, Bhutan, India and Myanmar.

细圆齿火棘 *Pyracantha crenulata*

窄叶火棘 *Pyracantha angustifolia*

窄叶火棘

Pyracantha angustifolia (Franch.) C. K. Schneid.

灌木或小乔木。叶狭长圆形至倒披针状长圆形，全缘，背面密被白色绒毛。复伞房花序，直径2-4厘米；萼筒外面密被绒毛。梨果淡红色，扁球形，直径5-6毫米。花期5-6月，果期10-12月。生海拔1600-3000米的山坡灌丛或路边。产云南、四川、西藏、贵州、浙江和湖北。

Shrubs or small trees. Leaves narrowly oblong to oblanceolate-oblong, margin entire, abaxially densely white tomentose. Compound corymbs 2-4 cm diam; hypanthium abaxially densely tomentose. Pomes reddish, depressed-globose, 5-6 mm diam. Fl. May-Jun. Fr. Oct-Dec. Thickets on slopes or roadsides at 1600-3000 m. Distributed in Yunnan, Sichuan, Xizang, Guizhou, Zhejiang and Hubei.

山楂

Crataegus pinnatifida Bunge

落叶乔木。叶宽卵形或三角状卵形，稀菱状卵形，两侧各有3-5羽状深裂片。伞房花序具多花；总花梗和花梗被柔毛；萼筒外被灰白色柔毛。梨果近球形或梨形，深红色，小核3-5。花期5-6月，果期8-9月。生海拔100-2000米的灌木丛中或山坡林边。产华北、华东和东北。朝鲜半岛亦有。

Trees deciduous. Leaves broadly ovate or triangular-ovate, rarely rhomboidal-ovate, with 3-5 pairs of lobes. Corymbs many flowered; pedicels and peduncles pubescent; hypanthium abaxially grayish white pubescent. Pomes subglobose or pyriform, dark red, pyrenes 3-5. Fl. May-Jun. Fr. Aug-Sep. Shrubs or forest edges on slopes at 100-2000 m. Distributed in N, E and NE China. Also in Korean Peninsula.

云南山楂

Crataegus scabrifolia (Franch.) Rehder

落叶乔木。枝常无刺。叶卵圆状披针形至卵圆状椭圆形，边缘有稀疏不整齐圆钝重锯齿，通常不分裂。伞房花序或复伞房花序，具多花；总花梗无毛。梨果黄色或带红晕。花期4-6月，果期8-10月。生海拔1500-3000米的松林边的灌丛

山楂 *Crataegus pinnatifida*

中、溪边杂木林中或栽培。产云南、四川、贵州和广西。

Trees deciduous. Branches usually unarmed. Leaves ovate-lanceolate to ovate-elliptic, margin sparsely irregularly and doubly obtusely serrate, usually not lobed. Corymbs or compound corymbs, many flowered; peduncles glabrous. Pomes yellow or tinged red. Fl. Apr-Jun. Fr. Aug-Oct. Shrubs of *Pinus* forest edges, mixed forests by rivers or cultivated at 1500-3000 m. Distributed in Yunnan, Sichuan, Guizhou and Guangxi.

云南山楂 *Crataegus scabrifolia*

湖北山楂

Crataegus hupehensis Sargent

乔木或灌木。叶缘具圆钝锯齿，上半部具2-4对浅裂片。伞房花序具7-9花；萼筒钟状，无毛；花瓣白色，卵形；雄蕊20；子房5室，每室2胚珠。梨果红色，近球形，直径约2.5厘米，萼片宿存，反折。花期5-7月，果期8-9月。生海拔500-2000米的山坡灌丛中。产中国西南、东南、华中、华北、华西和华东。

Trees or shrubs. Leaves margin crenate-serrate, with 2-4 pairs of shallow lobes at apical part. Corymbs 7-9-flowered; hypanthium campanulate, glabrous; petals white, ovate; stamens 20; ovary 5-loculed, with 2 ovules per locule. Pomes red, subglobose, ca. 2.5 cm diam, sepals persistent, reflexed. Fl. May-Jul. Fr. Aug-Sep. Thickets on slopes at 500-2000 m. Distributed in SW, SE, C, N, W and E China.

湖北山楂 *Crataegus hupehensis*

野山楂 *Crataegus cuneata*

野山楂

Crataegus cuneata Siebold et Zucc.

落叶灌木。叶阔倒卵形至倒卵状长圆形，基部楔形，先端3-7裂，背面被柔毛。伞房花序具5-7花；总花梗和花梗被柔毛。梨果红色或黄色，小核内面两侧平滑。花期5-6月，果期9-10月。生海拔200-2000米的山谷、多石地或灌丛中。产中国西南、华南、华东和华中。日本亦有。

Shrubs deciduous. Leaves broad-obovate to obovate-oblong, base cuneate, apex 3-7-lobed, abaxially pubescent. Corymbs 5-7-flowered; peduncles and pedicels pubescent. Pomes red or yellow, pyrenes smooth on both inner sides. Fl. May-Jun. Fr. Sep-Oct. Valleys, rocky places or thickets at 200-2000 m. Distributed in SW, S, E and C China. Also in Japan.

华中山楂

Crataegus wilsonii Sarg.

落叶灌木，常有刺。叶卵形或倒卵形，边缘具锐锯齿。伞房花序具多花；总花梗被白色绒毛；花瓣白色。梨果红色或紫红色，直径6-7毫米，有1-3小核，两边内侧有深凹痕。花期5月，果期8-9月。生海拔1000-2500米的阴湿山坡密林下或灌丛中。产中国西南、华中、华北、华西和华东。

Shrubs deciduous, usually thorny. Leaves ovate or obovate, margin sharply serrate. Corymbs many flowered; peduncles white tomentose; petals white. Pomes red or purplish red, 6-7 mm diam, pyrenes 1-3, with deep concave scars on both inner sides. Fl. May. Fr. Aug-Sep. Shaded dense forests on slopes or thickets at 1000-2500 m. Distributed in SW, C, N, W and E China.

毛山楂

Crataegus maximowiczii C. K. Schneid.

灌木或小乔木。叶边缘每侧各有3-5浅裂和疏生重锯齿，两面密被柔毛。复伞房花序具多花；总花梗被白色长柔毛。果红色或带紫褐色，小核3-5，两侧有凹痕。花期5-6月，果期8-9。生海拔200-1000米的混交

毛山楂 *Crataegus maximowiczii*

华中山楂 *Crataegus wilsonii*

光叶山楂 *Crataegus dahurica*

辽宁山楂 *Crataegus sanguinea*

林下、林缘，路边或河边等地。产中国东北和内蒙古。蒙古、俄罗斯(西伯利亚)、朝鲜半岛和日本亦有。

Shrubs or small trees. Leaves densely pubescent on both sides, margin remotely doubly serrate with 3-5 pairs of lobes. Compound corymbs many flowered; peduncles whitish villous. Fruits red or purplish brown, pyrenes 3-5, concave scars on both inner sides. Fl. May-Jun. Fl. Aug-Sep. Mixed forests, forest edges, roadsides or riversides at 200-1000 m. Distributed in NE China and Neimenggu. Also in Mongolia, Russia (Siberia), Korean Peninsula and Japan.

光叶山楂

Crataegus dahurica Koehne ex C. K. Schneid.

灌木或小乔木。刺细长。叶边缘具细锐重锯齿，在2/3部分有3-5对浅裂，两面无毛。复伞房花序具多花；总花梗无毛；花瓣白色。梨果橘红色或橘黄色，小核2-4。花期5月，果期8月。生海拔500-1500米的林下河边草地、沙坡或灌丛。产华北和东北。俄罗斯(东西伯利亚)和蒙古北部亦有。

Shrubs or small trees. Thorns slender. Leaves margin sharply doubly serrate, with 3-5 pairs at apical 2/3 of margin, glabrous on both sides. Compound corymbs many flowered; peduncle glabrous; petals white. Pomes orangish red or orangish yellow, pyrenes 2-4. Fl. May. Fr. Aug. Grassy places by rivers in forests, sandy slopes or thickets at 500-1500 m. Distributed in N and NE China. Also in Russia (E Siberia) and N Mongolia.

辽宁山楂

Crataegus sanguinea Pall.

灌木或小乔木，常具刺。叶宽卵形或菱状卵形，边缘具重锯齿。伞房花序多花；总花梗无毛或近无毛。梨果红色，近球形，直径约1厘米，小核3，稀5。花期5-6月，果期7-8月。生海拔900-3000米的山坡或河沟旁杂木林中。产河北、内蒙古、新疆、黑龙江和吉林。俄罗斯(西伯利亚)和蒙古北部亦有。

Shrubs or small trees, usually thorny. Leaves broadly ovate or rhombic-ovate, margin doubly serrate. Corymbs many flowered; peduncles glabrous or subglabrous. Pomes red, subglobose, ca. 1 cm diam; pyrenes 3, rarely 5. Fl. May-Jun. Fr. Jul-Aug. Slopes or mixed forests by rivers at 900-3000 m. Distributed in Hebei, Neimenggu, Xinjiang, Heilongjiang and Jilin. Also in Russia (Siberia) and N Mongolia.

中甸山楂

Crataegus chungtienensis W. W. Smith

灌木。叶宽卵形，先端圆钝，边缘具细锐重锯齿，有3-4对浅裂片，上面近无毛，下面疏生柔毛。伞房花序具多花；总花梗和花梗无毛或近无毛。梨果椭圆体形，红色，小核1-3。花期5月，果期9月。生海拔2500-3500米的山溪边杂木林或灌丛中。产云南西北部。

Shrubs. Leaves broadly ovate, apex obtuse, margin sharply doubly serrate, with 3-4 pairs of lobes, adaxially subglabrous, abaxially sparsely pubescent. Corymbs many flowered; rachises and peduncles glabrous or subglabrous. Pomes ellipsoid, red, pyrenes 1-3. Fl. May. Fr. Sep. Mixed forests by streams or thickets at 2500-3500 m. Distributed in NW Yunnan.

中甸山楂 *Crataegus chungtienensis*

甘肃山楂 *Crataegus kansuensis*

绿肉山楂 *Crataegus chlorosarca*

甘肃山楂
Crataegus kansuensis E. H. Wils

灌木或小乔木。叶宽卵形，边缘有尖锐重锯齿和5-7对浅裂片。伞房花序，具8-18花；花梗和总花梗无毛；子房顶端具柔毛。梨果红色或橙黄色，近球形，小核2-3。花期5月，果期7-9月。生海拔1000-1300米的林中、阴山坡或溪边。产四川、贵州、河北、山西、陕西和甘肃。

Shrubs or small trees. Leaves broadly ovate, margin sharply and doubly serrate, with 5-7 pairs of shallow lobes. Corymbs, 8-18-flowered; peduncles and pedicels glabrous; ovary apically tomentose. Pomes red or orange yellow, subglobose, pyrenes 2-3. Fl. May. Fr. Jul-Sep. Forests, shaded slopes or streamsides at 1000-3000 m. Distributed in Sichuan, Guizhou, Hebei, Shanxi, Shaanxi and Gansu.

绿肉山楂
Crataegus chlorosarca Maxim.

小乔木，刺少。叶边缘具3-5对浅裂片，两面被毛。萼筒钟状，无毛；数花呈伞房状；花瓣白色；子房顶端被毛。梨果黑色，果肉绿色，小核4-5，内面两侧有凹痕。花期6-7月，果期8-9月。栽培于辽宁。原产俄罗斯(远东地区)和日本。

Trees small, with few thorns. Leaves with 3-5 pairs of lobes, pubescent on both sides. Hypanthium campanulate, glabrous; corymb, several flowered; petals white; ovary pubescent apically. Pomes black, with green pulp, pyrenes 4-5, with concave scars on both inner sides. Fl. Jun-Jul. Fr. Aug-Sep. Cultivated in Liaoning. Native to Russia (Far East) and Japan.

华西小石积
Osteomeles schwerinae C. K. Schneid.

灌木或亚灌木，落叶或常绿。奇数羽状复叶；小叶椭圆形、椭圆状长圆形或倒卵状长圆形。伞房花序具3-5花；萼筒钟状，外面近无毛或散生柔毛；花瓣白色；花柱5，基部被长柔毛。花期4-5月，果期5-7月。生海拔1000-3000米的山坡灌丛

华西小石积 *Osteomeles schwerinae*

红果树 *Stranvaesia davidiana*

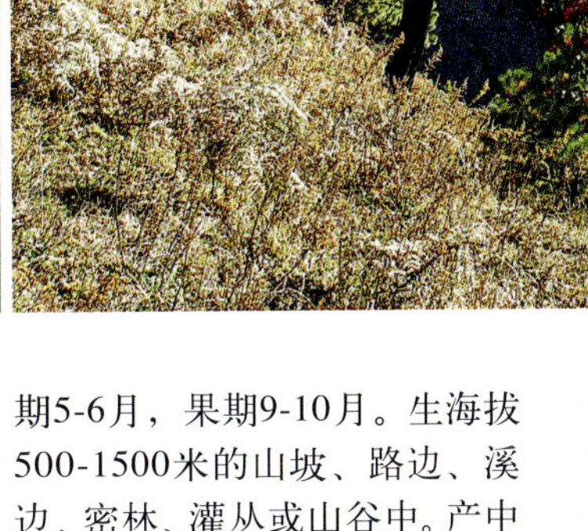

或路边。产中国西南、华西和台湾。

Shrubs or semishrubs, deciduous or evergreen. Leaves imparipinnate; leaflets elliptic, elliptic-oblong or obovate-oblong. Corymbs 3-5-flowered; hypanthium campanulate, abaxially subglabrous or sparsely pubescent; petals white; styles 5, villous basally. Fl. Apr-May. Fr. May-Jul. Scrubs on slopes or roadsides at 1000-3000 m. Distributed in SW and W China, and Taiwan.

红果树

Stranvaesia davidiana Decne.

灌木或小乔木。冬芽红褐色，狭卵球形。叶长圆形、长圆状披针形或倒披针形，全缘。花序直径5-10厘米，密集多花；总花梗和花梗密被长绒毛。果实橘红色，近球形。花期5-6月，果期9-10月。生海拔900-3000米的山坡、山顶、路边或灌丛中。产中国西南、东南、华中和华西。越南和马来西亚亦有。

Shrubs or small trees. Winter buds reddish brown, narrowly ovoid. Leaves oblong, oblong-lanceolate or oblanceolate, entire. Inflorescences 5-10 cm diam, densely flowered; rachises and pedicels densely villous. Fruits orangish red, subglobose. Fl. May-Jun. Fr. Sep-Oct. Slopes, mountain summits, roadsides or thickets at 900-3000 m. Distributed in SW, SE, C and W China. Also in Vietnam and Malaysia.

毛萼红果树

Stranvaesia amphidoxa C. K. Schneid.

灌木或小乔木。花序顶生，伞房状或近伞状，具数至10余花；总花梗与花梗密被棕黄色长绒毛或无毛。果实红黄色，卵球形，具浅色斑点；萼片直立或内弯，外面具长柔毛。花期5-6月，果期9-10月。生海拔500-1500米的山坡、路边、溪边、密林、灌丛或山谷中。产中国西南、东南、华中和华东。

Shrubs or small trees. Inflorescences terminal, corymbose or subumbellate, several to more than 10-flowered; rachises and pedicels densely brownish yellow tomentose-villous or glabrous. Fruits reddish yellow, ovoid, with small pale lenticels; sepals erect or incurved, abaxially tomentose-villous. Fl. May-Jun. Fr. Sep-Oct. Slopes, roadsides, streamsides, dense forests, thickets or mountain valleys at 500-1500 m. Distributed in SW, SE, C and E China.

毛萼红果树 *Stranvaesia amphidoxa*

石楠 *Photinia serratifolia*

石楠

Photinia serratifolia (Desf.) Kalkman

灌木或小乔木。叶柄长2-4厘米；叶缘明显锯齿状，仅近基部全缘。复伞房花序顶生；总花梗和花柄无毛、具长柔毛或绒毛。果实未成熟时红色，成熟后紫褐色，球形。花期4-5月，果期9-10月。生海拔700-2500米的杂木林、路边、山坡或田野。产华南大部分地区。印度南部、印度尼西亚、菲律宾和日本亦有。

Shrubs or small trees. Petiole 2-4 cm long; leaf margin prominently serrate, entire only basally. Compound corymbs terminal; rachis and pedicels glabrous, villous or tomentose. Fruits red when immature, brownish purple when mature, globose. Fl. Apr-May. Fr. Sep-Oct. Mixed forests, roadsides, slopes or fields at 700-2500 m. Distributed in most parts of S China. Also in S India, Indonesia, the Philippines and Japan.

贵州石楠

Photinia bodinieri H. Lévl.

常绿乔木。叶柄长(0.8-)1-1.5厘米。复伞房花序顶生，多花；总花梗和花梗被平伏柔毛；萼筒杯状，外面散生贴生柔毛；花瓣白色；花柱2或3，合生至中部，被白色长柔毛。梨果黄红色。花期4-5月，果期9-10月。生海拔300-1300米的林缘、灌丛、山谷或多石山坡。产中国西南、东南和华中。越南北部和印度尼西亚亦有。

Trees evergreen. Petiole (0.8-)1-1.5 cm long. Compound corymbs terminal, many flowered; rachis and pedicels appressed pubescent; hypanthium cupular, abaxially sparsely appressed pubescent; petals white; styles 2 or 3, connate from base to middle, white villous. Pomes yellowish red. Fl. Apr-May. Fr. Sep-Oct. Forest edges, thickets, valley or rocky slopes at 300-1300 m. Distributed in SW, SE and C China. Also in N Vietnam and Indonesia.

光叶石楠

Photinia glabra (Thunb.) Maxim.

常绿乔木。叶椭圆形、长圆形或长圆倒卵形，革质，两面无毛。复伞房花序顶生，具多

贵州石楠 *Photinia bodinieri*

光叶石楠 *Photinia glabra*

花；总花梗和花梗均无毛；花瓣内面近基部有白色绒毛。果实红色，倒卵球形或卵球形。花期4-5月，果期9-10月。生海拔500-800米的山坡杂木林中。产华南。缅甸、泰国和日本亦有。

Trees evergreen. Leaves elliptic, oblong or oblong-obovate, leathery, both surfaces glabrous. Compound corymbs terminal, numerous flowered; rachises and pedicels glabrous; petals adaxially white tomentose. Fruits red, obovoid or ovoid. Fl. Apr-May. Fr. Sep-Oct. Mixed forests on slopes at 500-800 m. Distributed in S China. Also in Myanmar, Thailand and Japan.

球花石楠

Photinia glomerata Rehd. et E. H. Wils.

灌木或小乔木。叶革质，侧脉12-20对，背面密被黄色绒毛。密集复伞房花序顶生；花芳香，萼筒杯状；花瓣白色；花柱2，合生达中部。梨果红色，卵球形。花期5-6月，果期8-9月。生海拔1500-2600米的混交林中、灌丛、路边或山坡。产湖北、云南和四川。

Shrubs or small trees. Leaves leathery, lateral veins 12-20 pairs, abaxially densely yellow tomentose. Compound corymbs terminal, densely glomerate; flowers fragrant, hypanthium cupular; petals white; styles 2, connate nearly to middle. Pomes red, ovoid. Fl. May-Jun. Fr. Aug-Sep. Mixed forests, thickets, roadsides or slopes at 1500-2600 m. Distributed in Hubei, Yunnan and Sichuan.

临桂石楠

Photinia chihsiniana Kuan

常绿小乔木。叶长圆披针形或倒披针形，先端急尖或短渐尖，基部渐狭，有疏生具腺细锐锯齿，初两面有灰色绒毛，以后脱落近无毛；侧脉12-15对；叶柄长5-20毫米。复伞房花序顶生；总花梗、花梗及萼筒均密生灰色绒毛。果实卵形，顶端有被白色绒毛的宿存萼片。花期4-5月，果期10-11月。生海拔300-1000米的山谷疏林或山坡。产广西东北部和湖南。

临桂石楠 *Photinia chihsiniana*

Evergreen small trees. Leaves oblong-lanceolate or oblanceolate, apex acute or shortly acuminate, base gradually attenuate, sparsely minutely sharply serrate with glands, both surfaces initially gray tomentose, glabrate; lateral veins 12-15-paired; petioles 5-20 mm long. Compound corymbs terminal; rachis, pedicels and hypanthium densely gray tomentose. Fruit ovoid, apex with white tomentose persistent sepals. Fl. Apr-May. Fr. Oct-Nov. Sparse forests in mountain valleys or slopes at 300-1000 m. Distributed in NE Guangxi and Hunan.

球花石楠 *Photinia glomerata*

桃叶石楠 *Photinia prunifolia*

厚叶石楠 *Photinia crassifolia*

桃叶石楠

Photinia prunifolia (Hook. et Arn.) Lindl.

常绿乔木。叶长圆形或长圆状披针形，背面密被黑色腺点，边缘密生锯齿。复伞房花序顶生，紧密，花多数；总花梗和花梗微具长柔毛；萼筒杯状，被长绒毛；花柱2(-3)，基部连合。梨果椭圆体形，红色。花期3-4月，果期10-11月。生海拔200-1700米的丘陵、山坡草地、河边、林中或路边竹林下。产中国西南、华中和东南。越南、马来西亚、印度尼西亚和日本亦有。

Trees evergreen. Leaves oblong or oblong-lanceolate, abaxially with black glands, margin densely serrate. Compound corymbs terminal, compact, many flowered; rachis and pedicels slightly villous; hypanthium cupular, villous; styles 2(-3), connate basally. Pomes ellipsoid, red. Fl. Mar-Apr. Fr. Oct-Nov. Hills, grassland on slopes, streamsides, forests, bamboo forests by roadsides at 200-1700 m. Distributed in SW, C and SE China. Also in Vietnam, Malaysia, Indonesia and Japan.

厚叶石楠

Photinia crassifolia Lévl.

常绿灌木，高4-5米。幼枝有锈色绒毛。叶厚革质，边缘稍外卷，全缘或有不显明锯齿，上面无毛，下面中脉和侧脉有绒毛；侧脉15-17对；叶柄长1.5-2毫米，有绒毛。复伞房花序顶生，花多数；总花梗和花梗密生绒毛；萼筒钟状，外面无毛；萼片三角形；花瓣白色。花期5月，果期9-11月。生海拔500-1700米的向阳山坡丛林中。产贵州、云南和广西。

Evergreen shrubs, 4-5 m tall. Branches rusty tomentose when young. Leaves thickly leathery, margin slightly revolute, entire or inconspicuously dentate, adaxially glabrous, abaxially tomentose along midvein and lateral veins; lateral veins 15-17 pairs; petioles 1.5-2 mm long, tomentose. Compound corymbs terminal, numerous flowered; rachis and pedicels densely tomentose; hypanthium campanulate, abaxially glabrous; sepals triangular; petals white. Fl. May. Fr. Sep-Nov. Sunny slopes, slope thickets at 500-1700 m. Distributed in Guizhou, Yunnan and Guangxi.

中华石楠

Photinia beauverdiana C. K. Schneid.

落叶灌木或小乔木。叶纸质，下面中脉疏生柔毛。复伞房花序顶生，花多数；总花梗和花梗无毛，上有多数长圆形皮孔；花瓣白色；花柱(2或)3。果紫红色，卵球形或近球形。花期4-5月，果期7-8月。生海拔200-3000米的山坡、山腰、林中、灌丛或沟谷。产中国西南、华南、东南、华中、华西和华东。不丹和越南北部亦有。

Shrubs or small trees, deciduous. Leaves papery, abaxially pilose along veins. Compound corymbs terminal, many flowered; rachis and pedicels glabrous, with numerous oblong lenticels; petals white; styles (2 or) 3. Fruits purplish red, ovoid or subglobose. Fl. Apr-May. Fr. Jul-Aug. Slopes, forests, thickets or valleys at 200-3000 m. Distributed in SW, S, SE, C, W and E China.

中华石楠 *Photinia beauverdiana*

小叶石楠 *Photinia parvifolia*

Also in Bhutan and N Vietnam.

小叶石楠

Photinia parvifolia (E. Pritz.) C. K. Schneid.

落叶灌木。叶长4-8厘米，宽1-3.5厘米，两面初疏生柔毛，以后无毛。花序伞形或单伞房状，数花，稀超过10；花瓣基部散生长绒毛；花梗和萼筒外无毛。梨果橙红色或紫色。花期4-5月，果期7-8月。生海拔300-2500米的丘陵、山谷、多石山坡、田边、林下或灌丛中。产中国西南、华南和华中。

Shrubs deciduous. Leaves 4-8 × 1-3.5 cm, both surfaces glabrous or slightly puberulous when young. Inflorescences umbellate or simple corymbose, several or rarely more than 10-flowered; petals sparsely villous at base; pedicels and hypanthium glabrous. Pomes orangish red or purple. Fl. Apr-May. Fr. Jul-Aug. Hills, mountain valley, rocky slopes, fields, forests or thickets at 300-2500 m. Distributed in SW, S and C China.

枇杷

Eriobotrya japonica (Thunb.) Lindl.

小乔木。叶披针形、倒披针形、倒卵形或椭圆状长圆形，背面密被灰锈色绒毛。花瓣白色；花柱5，离生。梨果黄色或橘黄色，球形或倒卵球形，直径1-1.5厘米，具锈色绒毛，后迅速无毛。花期6月，果期翌年7-8月。重庆和湖北有野生；栽培于中国西南、华南、华东和华中。印度、缅甸、泰国、越南、印度尼西亚和日本亦有。

Trees small. Leaves lanceolate, oblanceolate, obovate or elliptic-oblong, abaxially densely gray rusty tomentose. Petals white; styles 5, free. Pomes yellow or orangish yellow, globose or obovoid, 1-1.5 cm diam, rusty tomentose, soon glabrescent. Fl. Jun. Fr. next Jul-Aug. Native in Chongqing and Hubei; cultivated in SW, S, E and C China. Also in India, Myanmar, Thailand, Vietnam, Indonesia and Japan.

枇杷 *Eriobotrya japonica*

栎叶枇杷

Eriobotrya prinoides Rehder et E. H. Wils

小乔木。叶长圆形或椭圆形，边缘有波状锯齿，背面密被灰色绒毛。圆锥花序长6-10厘米，花多数；花瓣白色；花柱2，稀3，分离或基部联合。梨果卵球形，暗褐色，直径6-7毫米。花期6月，果期翌年7-8月。生海拔800-1700米的河边或湿润密林中。产四川西部和云南东南部。老挝亦有。

Trees small. Leaves oblong or elliptic, margin sinuate-toothed, abaxially densely gray-tomentose. Panicles 6-10 cm long, many flowered; petals white; styles 2, rarely 3, free or connate basally. Pomes ovoid, dark brown, 6-7 mm diam. Fl. Jun. Fr. next Jul-Aug. River banks or dense moist forest understories at 800-1700 m. Distributed in W Sichuan and SE Yunnan. Also in Laos.

栎叶枇杷 *Eriobotrya prinoides*

齿叶枇杷 *Eriobotrya serrata*

大花枇杷 *Eriobotrya cavaleriei*

齿叶枇杷
Eriobotrya serrata J. E. Vidal

乔木。叶倒卵形或倒披针形，边缘具锯齿。圆锥花序，花多数；花梗长2-3毫米；花柱3或4，基部有柔毛。梨果绿色，球形或梨形，直径1.5-1.8厘米，近无毛。花期11月，果期翌年5月。生海拔1000-1900米的山坡灌丛中。产广西和云南。老挝亦有。

Trees. Leaves obovate or oblanceolate, margin serrate. Panicles many flowered; pedicels 2-3 mm long; styles 3 or 4, base pubescent. Pomes green, globose or pyriform, 1.5-1.8 cm diam, subglabrous. Fl. Nov. Fr. next May. Thickets on slopes at 1000-1900 m. Distributed in Guangxi and Yunnan. Also in Laos.

大花枇杷
Eriobotrya cavaleriei (H. Lévl.) Rehder

乔木。叶簇生枝顶，叶长圆形、长圆状披针形或长圆状倒披针形，革质，边缘上部具锐锯齿，近基部全缘。圆锥花序，花多数；总花梗和花梗疏具黄褐色柔毛；花瓣白色。梨果黄红色，椭圆体形至近球形。花期4-5月，果期7-8月。生海拔500-2000米的山坡或河边杂木林。产中国西南、华南、华中和东南。越南北部亦有。

Trees. Leaves fascicle apically; leaves oblong, oblong-lanceolate or oblong-oblanceolate, leathery, margin sharply serrate, entire near base. Panicles many flowered; peduncles and pedicels sparsely yellowish-brown pubescent; petals white. Pomes yellowish red, ellipsoid to subglobose. Fl. Apr-May. Fr. Jul-Aug. Slopes or mixed river side forests at 500-2000 m. Distributed in SW, S, C and SE China. Also in N Vietnam.

小叶枇杷
Eriobotrya seguinii (Lévl.) Guillaumin

常绿灌木。叶革质，长圆形或倒披针形，长3-6厘米，宽1.2-2厘米，先端圆钝或急尖，边缘具内弯钝齿，下面幼时有长柔毛，以后脱落，中脉在上下两面隆起；叶柄长1-1.5厘米。圆锥花序或稀总状花序顶生，密生锈色绒毛；萼筒短钟状；雄蕊15；花柱3或4，离生。花期3-4月，果期6-7月。生海拔500-1500米的山坡灌丛中。产贵州西南部和云南东南部。

Evergreen shrubs. Leaves leathery, oblong or oblanceolate, 3-6 × 1.2-2 cm, apex obtuse or acute, margin incurved-crenate, abaxially villous when young, glabrescent, midvein prominent on both surfaces; petiole 1-1.5 cm long. Panicle or rarely raceme terminal, densely rusty tomentose; hypanthium shortly campanulate; stamens 15; styles 3 or 4, free. Fl. Mar-Apr. Fr. Jun-Jul. Thickets on slopes at 500-1500 m. Distributed in SW Guizhou and SE Yunnan.

小叶枇杷 *Eriobotrya seguinii*

石斑木
Rhaphiolepis indica (L.) Lindl.

灌木，高达4米。叶卵形或长圆形，稀倒卵形，长圆状披针形、狭椭圆形或披针状椭圆形，长(2-)4-8厘米，宽1.5-4厘米，革质。圆锥花序或总状花序顶生，多花或少花；总花梗和花梗被锈色绒毛；花瓣白色或淡粉色。梨果紫黑色，球形，直径5-8毫米，无毛。花期4月，

石斑木 *Rhaphiolepis indica*

果期7-8月。生海拔700-1600米山坡、路边或灌丛。产中国西南、华南、东南和华东。老挝、泰国、越南、柬埔寨和日本亦有。

Shrubs, to 4 m tall. Leaves ovate, oblong, rarely obovate, oblong-lanceolate, narrowly elliptic or lanceolate-elliptic, (2-)4-8 × 1.5-4 cm, leathery. Panicles or racemes terminal, many or few flowered; peduncles and pedicels rusty tomentose; petals white or pinkish. Pomes purplish black, globose, 5-8 mm diam, glabrous. Fl. Apr. Fr. Jul-Aug. Slopes, roadsides or thickets at 700-1600 m. Distributed in SW, S, SE and E China. Also in Laos, Thailand, Vietnam, Cambodia and Japan.

大叶石斑木

Rhaphiolepis major Cardot

灌木。叶窄椭圆形或倒卵状长圆形，边缘具浅圆齿，长7-15厘米。圆锥花序顶生，花多数；总花梗渐无毛。梨果黑色，球形，直径7-10毫米；萼片脱落后留有一圆环。花期4月，果期8月。生海拔200-300米的阴密林下或河边灌丛中。产福建、浙江、江苏南部和江西。

Shrubs. Leaves narrowly elliptic or obovate-oblong, margin shallowly crenulate, 7-15 cm long. Panicles terminal, many flowered; peduncles glabrescent. Pomes black, globose, 7-10 mm diam; sepals caducous, leaving an annular ring. Fl. Apr. Fr. Aug. Dense shady forests or among shrubs by streams at 200-300 m. Distributed in Fujian, Zhejiang, S Jiangsu and Jiangxi.

细叶石斑木

Rhaphiolepis lanceolata H. H. Hu

灌木。叶狭披针形，革质，下面无毛或近无毛，顶部钝或稍渐尖。圆锥花序顶生，花多数；花瓣白色或粉红色；雄蕊15；子房3室，每室具2胚珠；花柱3，基部合生。梨果黑色，球形，具柔毛。花期6月，果期10-11月。生海拔400-1500米的山坡开阔林中或山谷开阔灌丛中。产广西、广东和海南。

Shrubs. Leaves narrowly lanceolate, leathery, abaxially glabrous or nearly so, apex obtuse or shortly acuminate. Panicles terminal, many flowered; petals white or pinkish; stamens 15; ovary 3-loculed, with 2 ovules per locule; styles 3, connate at base. Pomes black, globose, pubescent. Fl. Jun. Fr. Oct-Nov. Open forests on slopes or open thickets in valleys at 400-1500 m. Distributed in Guangxi, Guangdong and Hainan.

大叶石斑木 *Rhaphiolepis major*

细叶石斑木 *Rhaphiolepis lanceolata*

大果花楸

Sorbus megalocarpa Rehder

灌木或小乔木。单叶，侧脉14-20对，直达叶边锯齿尖端。复伞房花序顶生小枝上，长4-7厘米，宽9-13厘米，花多数；萼筒钟状，外面被绒毛。果暗红褐色，卵球形。花期4-5月，果期7-8月。生海拔1200-2700米山谷、溪边或山坡林下多石处。产云南、四川、广西、贵州、湖北和湖南。

Shrubs or small trees. Leaves simple; lateral veins 14-20 pairs, nearly parallel and terminating in marginal teeth. Compound corymbs terminal on branchlets, 4-7 × 9-13 cm, many flowered; hypanthium campanulate, abaxially tomentose. Fruits dark reddish-brown, ovoid. Fl. Apr-May. Fr. Jul-Aug. Mountain valleys, streamsides or rocky forests on slopes at 1200-2700 m. Distributed in Yunnan, Sichuan, Guangxi, Guizhou, Hubei and Hunan.

白叶花楸 *Sorbus cuspidata*

大果花楸 *Sorbus megalocarpa*

康藏花楸

Sorbus thibetica (Cardot) Hand.-Mazz.

乔木。叶基部楔形。复伞房花序顶生，长3-6厘米，宽4-8厘米，有20-30花或更多；花柱基部无毛；花瓣白色。果实深红色，卵球形或倒卵球形，2室，具少数小皮孔。花期5-7月，果期9-10月。生海拔2400-3800米的山坡或沟谷密林中、石坡、溪边或灌丛中。产云南和西藏。不丹和缅甸北部亦有。

Trees. Leaf base cuneate. Compound corymbs terminal, 3-6 × 4-8 cm, 20-30-flowered or more; styles glabrous basally; petals white. Pomes scarlet, ovoid or obovoid, 2-locular, sparsely small lenticellate. Fl. May-Jul. Fr. Sep-Oct. Dense forests on slopes or in valleys, rocky slopes, stream banks or shrubby thickets at 2400-3800 m. Distributed in Yunnan and Xizang. Also in Bhutan and N Myanmar.

康藏花楸 *Sorbus thibetica*

白叶花楸

Sorbus cuspidata (Spach) Hedl.

乔木。叶椭圆形，长12-22厘米，先端急尖或钝，基部圆形或宽楔形，边缘有不规则的单锯齿或重锯齿，有时稍浅裂，下面密被白色绒毛；叶柄长1-2.5厘米。复伞房花序具多数花；花柱基部连合并具绒毛。果圆球形，具多数显明斑点，先端萼片宿存。花期6-7月，果期8-9月。生海拔2000-3500米的林内。产西藏。印度、缅甸、不丹和尼泊尔亦有。

Trees. Leaves elliptic, 12-22 cm long, apex acute or obtuse, base rounded or broadly cuneate, margin unevenly serrate or doubly serrate, sometimes slightly lobed, abaxially densely white tomentose; petioles 1-2.5 cm long. Compound corymbs with numberous flowers; style basally connate and tomentose. Fruits globose, with many conspicuous spots, apex with persistent sepals. Fl. Jun-Jul. Fr. Aug-Sep. Forests at 2000-3500 m. Distributed in Xizang. Also in India, Myanmar, Bhutan and Nepal.

水榆花楸 *Sorbus alnifolia*

水榆花楸

Sorbus alnifolia (Siebold et Zucc.) K. Koch.

乔木。叶柄长1.5-3厘米；单叶，叶边缘有不规则尖锐重锯齿。复伞房花序顶生，疏生6-25花；花柱2。果长圆形或卵状长圆形，稀近球形，不具皮孔或仅具少量皮孔。花期4-5月，果期8-9月。生海拔500-2300米的山坡、溪谷、混交林下或灌丛中。产中国东南、华中、华北、华东和东北。朝鲜半岛和日本亦有。

Trees. Petiole 1.5-3 cm long; leaves simple, margin irregularly sharply doubly serrate. Compound corymbs, terminal, loosely 6-25-flowered; styles 2. Fruits oblong or ovoid-oblong, rarely subglobose, without or with few minute lenticels. Fl. Apr-May. Fr. Aug-Sep. Slopes, gullies, mixed forests or thickets at 500-2300 m. Distributed in SE, C, N, E and NE China. Also in Korean Peninsula and Japan.

美脉花楸

Sorbus caloneura (Stapf) Rehder

乔木或灌木。叶柄长1-2(-3)厘米；叶狭椭圆形或狭卵形至倒卵状椭圆形，边缘具钝锯齿，侧脉10-18对。复伞房花序，花多数；花柱4或5。果球状或倒卵状，褐色。花期4-5月，果期8-10月。生海拔600-2100米的林中或山谷。产中国西南、华南、东南和华中。哈萨克斯坦、土库曼斯坦、乌兹别克斯坦、西南亚和欧洲亦有。

Trees or shrubs. Petioles 1-2(-3) cm long; leaves narrowly elliptic or narrowly ovate to obovate-elliptic, margin obtusely crenate, lateral veins 10-18 pairs. Compound corymbs, many flowered; styles 4 or 5. Fruits globose or obovoid, brown. Fl. Apr-May. Fr. Aug-Oct. Forests or valleys at 600-2100 m. Distributed in SW, S, SE and C China. Also in Kazakhstan, Turkmenistan, Uzbekistan, SW Asia and Europe.

石灰花楸

Sorbus folgneri (C. K. Schneid.) Rehder

乔木。叶柄长0.5-1.5(-2)厘米；侧脉8-15对，几平行。复伞房花序具20-30花；萼筒钟状；花瓣白色；花柱2或3，基部连合，被绒毛。梨果，萼片早落，留有圆穴。花期4-5月，果期7-8月。生海拔800-2000米的山谷、山坡或混交林下沟边。产华南至华中。

Trees. Petioles 0.5-1.5(-2) cm long; lateral veins 8-15 pairs, nearly parallel. Compound corymbs, 20-30-flowered; hypanthium campanulate; petals white; styles 2 or 3, basally connate and tomentose. Pomes, sepals caducous leaving an annular scar. Fl. Apr-May. Fr. Jul-Aug. Valleys, slopes or streamsides in mixed forests at 800-2000 m. Distributed from S to C China.

美脉花楸 *Sorbus caloneura*

石灰花楸 *Sorbus folgneri*

江南花楸 *Sorbus hemsleyi*

褐毛花楸 *Sorbus ochracea*

江南花楸

Sorbus hemsleyi (C. K. Schneid.) Rehder

乔木或灌木。单叶；叶卵形至狭椭圆状卵形，下面被灰白色绒毛，边缘具锯齿且稍反卷。复合伞房花序顶生，具(15-)20-30花；花柱基部合生且被灰白色绒毛。果实赤褐色，近球形，直径5-8毫米。花期5-6月，果期7-9月。生海拔900-3200米的山坡旱林中或常绿阔叶林中。产中国西南、东南、华中、华西和华东。

Trees or shrubs. Leaves simple; leaves ovate to narrowly elliptic-ovate, abaxially grayish white tomentose, margin serrulate and somewhat recurved. Compound corymbs terminal, (15-)20-30-flowered; styles basally connate and grayish white tomentose. Fruit russet, subglobose, 5-8 mm diam. Fl. May-Jun. Fr. Jul-Sep. Dry forests on slopes or evergreen broad-leaved forests at 900-3200 m. Distributed in SW, SE, C, W and E China.

褐毛花楸

Sorbus ochracea (Hand.-Mazz.) J. E. Vidal

乔木或灌木。叶柄长2-3厘米，密被锈棕色绒毛；单叶，叶卵圆形、椭圆状卵圆形或椭圆状倒卵形，侧脉10-12对。复伞房花序顶生，具20-30花至更多；花瓣黄白色。果实近球形，3或4室。花期3-4月，果期7-8月。生海拔1300-2700米的山坡杂木林中。产云南和西藏东部。

Trees or shrubs. Petioles 2-3 cm long, densely rust-brown tomentose; leaves simple, ovate, elliptic-ovate or elliptic-obovate, lateral veins 10-12 pairs. Compound corymbs terminal, 20-30-flowered or more; petals yellowish white. Fruits subglobose, 3- or 4-loculed. Fl. Mar-Apr. Fr. Jul-Aug. Mixed forests on slopes at 1300-2700 m. Distributed in Yunnan and E Xizang.

天山花楸

Sorbus tianschanica Rupr.

灌木或小乔木。奇数羽状复叶，有小叶(4-)6-7对，长5-7厘米，叶缘大部分具尖锐锯齿，仅基部全缘。大型复伞房花序，花多数，大而疏散；花柱(3-)5，基部密被白色柔毛。梨果球形，红色。花期5-6月，果期9-10月。生海拔2000-3200米的山谷、溪边或林缘。产甘肃、青海和新疆。阿富汗、巴基斯坦西部、俄罗斯和西南亚亦有。

Shrubs or small trees. Leaves imparipinnate; leaflets (4-)6-7-paired, 5-7 cm long, margin mostly sharply serrate, entire only basally. Compound corymbs, many flowered, large and lax; styles (3-)5, base densely white-pubescent. Pomes globose, red. Fl. May-Jun. Fr. Sep-Oct. Mountain valleys, streamsides or forest edges

天山花楸 *Sorbus tianschanica*

湖北花楸 *Sorbus hupehensis*

少齿花楸 *Sorbus oligodonta*

at 2000-3200 m. Distributed in Gansu, Qinghai and Xinjiang. Also in Afghanistan, W Pakistan, Russia and SW Asia.

湖北花楸

Sorbus hupehensis C. K. Schneid.

乔木。小叶4-8对，先端急尖，下面中脉上被白色绒毛。花序长(4-)5-8厘米，宽6-10厘米，花多数；花瓣白色，直径5-7毫米；花柱4或5。果实白色，有时带红晕，球形，无毛。花期5-7月，果期8-9月。生海拔300-3800米的林中、阴坡或灌丛。产中国西南、华中、华西和华东。

Trees. Leaflets 4-8 pairs, apex acute, abaxially white tomentose along midvein. Inflorescences (4-)5-8 × 6-10 cm, many flowered; petals white, 5-7 mm diam; styles 4 or 5. Fruits white, sometimes stained reddish, globose, glabrous. Fl. May-Jul. Fr. Aug-Sep. Forests, shaded slopes or thickets at 300-3800 m. Distributed in SW, C, W and E China.

少齿花楸

Sorbus oligodonta (Cardot) Hand.-Mazz.

乔木。奇数羽状复叶，小叶(4-)5-8对，基部阔楔形至圆形，顶端具少数锯齿。复伞房花序，密集多花；萼筒钟状，外面无毛；花柱4或5。果白色，有红晕，卵圆形；萼片宿存。花期5月，果期9月。生海拔2000-3600米的山坡、混交林下或沿河岸。产云南、四川和西藏。缅甸亦有。

Trees. Leaves imparipinnate; leaflets (4-)5-8-paired, basally broadly cuneate to rounded, with few teeth only at apex. Compound corymbs, densely flowered; hypanthium campanulate, abaxially glabrous; styles 4 or 5. Fruits white, stained-red, ovoid; sepals persistent. Fl. May. Fr. Sep. Mountain slopes, mixed forests or along river banks at 2000-3600 m. Distributed in Yunnan, Sichuan and Xizang. Also in Myanmar.

华西花楸

Sorbus wilsoniana C. K. Schneid.

乔木，高5-10米。奇数羽状复叶，小叶6-7对，上下两面无毛或仅在下面中脉附近有短柔毛，边缘每侧具8-20个细锯齿。花序多花；花瓣白色，卵形。果实橘红色，卵球形。花期5-6月，果期8-9月。生海拔1300-3300米的山区杂木林中。产中国西南和华中。

Trees, 5-10 m tall. Leaves imparipinnate, leaflets 6-7 pairs, both surfaces glabrous or abaxially pubescent along midvein, with 8-20 teeth per side. Inflorescences densely flowered; petals white, ovate. Fruits orangish red, ovoid. Fl. May-Jun. Fr. Aug-Sep. Mixed forests in mountain regions at 1300-3300 m. Distributed in SW and C China.

华西花楸 *Sorbus wilsoniana*

黄山花楸

Sorbus amabilis Cheng ex T. T. Yü et K. C. Kuan

乔木。奇数羽状复叶，叶轴和小叶轴下面具锈色柔毛；小叶边缘具粗锐锯齿，下面沿中脉具红褐色柔毛，渐无毛。复伞花序顶生，具多花。果近球状，红色。花期5月，果期9-10月。生海拔900-2000米的林中或山坡。产福建、浙江、湖北、安徽和江西。

Trees. Leaves imparipinnate, rachis and leaflets abaxially rust-brown pubescent; leaflets margin coarsely sharply serrate, abaxially reddish brown pubescent only along midvein, gradually glabrescent. Compound umbels terminal, many flowered. Fruits subglobose, red. Fl. May. Fr. Sep-Oct. Forests or mountain slopes at 900-2000 m. Distributed in Fujian, Zhejiang, Hubei, Anhui and Jiangxi.

黄山花楸 *Sorbus amabilis*

北京花楸

Sorbus discolor (Maxim.) Maxim.

乔木。小叶5-7对，两面无毛，边缘具细锐锯齿，每侧锯齿12-20，基部1/3以下全缘。复伞房花序疏散，花多数；花瓣白色；花柱3或4。果卵状，白色或黄色。花期5月，果期8-9月。生海拔1500-2000米的阳坡阔叶混交林中。产华北、华西和华东。

Trees. Leaflets 5-7 pairs, both surfaces glabrous, margin minutely sharply serrate, with 12-20 teeth on each margin, entire in basal 1/3. Compound corymbs loose, many flowered; petals white; styles 3 or 4. Fruits ovoid, white or yellow. Fl. May. Fr. Aug-Sep. Broad-leaved mixed forests on sunny slopes at 1500-2000 m. Distributed in N, W and E China.

北京花楸 *Sorbus discolor*

花楸树

Sorbus pohuashanensis (Hance) Hedl.

乔木。奇数羽状复叶，小叶5-7对，下面幼时被绒毛，渐无毛。花序密集多花；萼筒钟状，外面被绒毛状长柔毛或近无毛；花柱3，基部被柔毛。梨果红色或橙红色，近球形。花期6月，花期9-10月。生海拔900-2500米的山坡或山谷混交林下。产华北和东北。

Trees. Leaves imparipinnate, leaflets 5-7-paired, abaxially tomentose-villous when young, glabrescent. Inflorescences densely many flowered; hypanthium campanulate, abaxially tomentose-villous or subglabrous; styles 3, basally pubescent. Pomes red or orange-red, subglobose. Fl. Jun. Fr. Sep-Oct. Mountain slopes

花楸树 *Sorbus pohuashanensis*

or mixed forests in valleys at 900-2500 m. Distributed in N and NE China.

宾川花楸

Sorbus obsoletidentata (Card.) Yu

灌木或小乔木。奇数羽状复叶，连叶柄长6-9厘米；小叶11-14对，线状长圆形至长圆披针形，长1.2-2.3厘米，边缘全缘或少数在叶先端有2-3浅锯齿；叶轴下面被稀疏褐色柔毛，两侧微具窄翅，上面有沟，下面被稀疏褐色柔毛；托叶膜质，披针形。总花梗和花梗被褐色柔毛；萼筒钟状，外有褐色柔毛，内无毛；花柱常4。生山坡和路旁。产云南。

Shrubs or small trees. Leaves imparipinnate, together with petioles 6-9 cm long; leaflets 11-14-paired, linear-oblong to oblong-lanceolate, 1.2-2.3 cm long, margin entire or few apically shallowly 2-3-toothed; leaf rachis abaxially sparsely brown pubescent, both sides slightly narrowly winged, adaxially sulcate, abaxially slightly brown pubescent; stipules membranous, lanceolate. Rachis and pedicels densely brown pubescent; hypanthium campanulate, abaxially brown pubescent, adaxially glabrous; styles usually 4. Mountain slopes and roadsides. Distributed in Yunnan.

尼泊尔花楸

Sorbus foliolosa (Wallich) Spach

灌木或小乔木。奇数羽状复叶；小叶4-6(-9)对，长圆椭圆形或长圆形，长1.5-3.5(-4.5)厘米，边缘锯齿极细锐而不明显；叶轴下面被锈褐色柔毛；托叶草质，线状披针形，早落。总花梗和花梗密被锈褐色柔毛；萼筒外面无毛；花柱3。果具宿存闭合萼片。花期4-6月，果期9-10月。生海拔2500-3000米的杂木林、沟边。产云南和西藏。东南亚和印度亦有。

Shrubs or small trees. Leaves imparipinnate; leaflet 4-6(-9) pairs, oblong-elliptic or oblong, 1.5-3.5(-4.5) cm long, margin with sharp and unconspicuous tooth; leaf rachis abaxially densely rusty brown pubescent; stipules herbaceous, linear-lanceolate, caduceus. Rachis and pedicels with rusty brown hairs; hypanthium abaxially glabrous; styles 3. Fruit with persistent, closed sepals. Fl. Apr-Jun. Fr. Sep-Oct. Mixed forests, mizabe at 2500-3000 m. Distributed in Yunnan and Xizang. Also in SE Asia and India.

宾川花楸 *Sorbus obsoletidentata*

尼泊尔花楸 *Sorbus foliolosa*

球穗花楸 *Sorbus glomerulata*

陕甘花楸

Sorbus koehneana C. K. Schneid.

灌木或小乔木。奇数羽状复叶，小叶8-12对，稀更多，边缘除基部外具糙锯齿。复伞房花序顶生，多花；总花梗和花梗疏被白色短柔毛，老时近无毛。果白色，球形，萼片宿存。花期5-6月，果期8-9月。生海拔2300-4000米的山地杂木林或灌丛。产中国西南、华北和华西。

Shrubs or small trees. Leaves imparipinnate, leaflets 8-12-paired, rarely more, margin coarsely serrate except at base. Compound corymbs terminal, many flowered; rachises and pedicels sparsely white pubescent, nearly glabrous when old. Fruits white, globose, sepals persistent. Fl. May-Jun. Fr. Aug-Sep. Mixed forests in mountain regions or thickets at 2300-4000 m. Distributed in SW, N and W China.

球穗花楸

Sorbus glomerulata Koehne

灌木或小乔木。小枝无毛，具皮孔。奇数羽状复叶，小叶10-14(-18)对，顶生和基生小叶较其他叶小。复伞房花序具多数密集花；总花梗和花梗无毛或稍被柔毛；花瓣白色。果实白色，卵球形，无毛，萼片宿存。花期5-6月，果期9-10月。生海拔1600-4000米的灌丛、针阔混交林中。产云南、四川和湖北。

Shrubs or small trees. Branchlets glabrous, lenticellate. Leaves imparipinnate; leaflets 10-14(-18) pairs, terminal and basal ones smaller than others. Compound corymbs, densely flowered; rachises and pedicels glabrous or slightly pubescent; petals white. Fruits white, ovoid, glabrous, sepals persistent. Fl. May-Jun. Fr. Sep-Oct. Thickets and broad-leaved mixed forests at 1600-4000 m. Distributed in Yunnan, Sichuan and Hubei.

巨齿西南花楸

Sorbus rehderiana Koehne var. **grosseserrata** Koehne

灌木或小乔木。叶缘具深的重锯齿；冬芽、叶背面沿中脉和花序疏被锈色短柔毛，后渐无毛。复伞房花序顶生，稀腋生，密具花；花柱(4或)5。果粉红色。花期5-6月，果期8-9月。生海拔2600-3000米的林中或林缘。产四川。

Shrubs or small trees. Leaflet margin deeply coarsely serrate; winter buds, leaflets abaxially along midvein and inflorescences sparsely rust-brown pubescent, then glabrescent. Compound corymbs terminal, rarely axillary, densely flowered; styles (4 or) 5. Fruits pink. Fl. May-Jun. Fr. Aug-Sep. Forests or forest edges at 2600-3000 m. Distributed in Sichuan.

巨齿西南花楸 *Sorbus rehderiana* var. *grosseserrata*

陕甘花楸 *Sorbus koehneana*

云南移栎 *Docynia delavayi*

红毛花楸

Sorbus rufopilosa C. K. Schneid.

灌木或小乔木。奇数羽状复叶，小叶8-14(-17)对，幼时下面中脉被锈红色柔毛。伞房状或复伞房状花序，具3-8花或更多；花瓣粉红色。果红色，有直立而宿存萼片。花期5-6月，果期8-9月。生海拔2700-4000米的针叶林或杂木林，或山谷灌丛。产中国西南。印度、尼泊尔、不丹和缅甸北部亦有。

Shrubs or small trees. Leaves imparipinnate; leaflets 8-14(-17) pairs, abaxially rusty-red-pubescent on midvein when young. Inflorescences corymbose or compound-corymbose, 3-8(or more)-flowered; petals pink. Pomes red, with erect and persistent sepals. Fl. May-Jun. Fr. Aug-Sep. Coniferous or mixed forests, or thickets in valleys at 2700-4000 m. Distributed in SW China. Also in India, Nepal, Bhutan and N Myanmar.

榲桲

Cydonia oblonga Mill.

灌木或小乔木。叶卵形至长圆形，下面具明显叶脉和密被长柔毛，全缘。花单生；苞片膜质，早落；花瓣白色或粉红色；柱头基部密被长绒毛。果芳香，黄色，梨形，密被绒毛。花期4-5月，果期10月。栽培于贵州、福建、山西、陕西、江西和新疆。原产中亚。

Shrubs or small trees. Leaves ovate to oblong, abaxially with conspicuous veins and densely villous, margin entire. Flowers solitary; bracts membranous, caducous; petals white or pink; styles densely villous at base. Fruits fragrant, yellow, pear-shaped, densely tomentose. Fl. Apr-May. Fr. Oct. Cultivated in Guizhou, Fujian, Shanxi, Shaanxi, Jiangxi and Xinjiang. Native to C Asia.

云南移栎

Docynia delavayi (Franch.) C. K. Schneid.

常绿乔木。叶柄密被绒毛；叶披针形或卵状披针形，革质，背面密被微黄色绒毛。3-5花丛生；花瓣白色，基部具爪；雄蕊40-50。果实黄色，卵球形或长圆形，直径2-3厘米。花期3-4月，果期8-9月。生海拔1000-3000米的山谷溪旁、杂木林或山地。产云南、四川和贵州。

Trees evergreen. Petioles densely tomentose; leaves lanceolate or ovate-lanceolate, leathery, abaxially densely yellowish tomentose. Flowers 3-5-fascicled; petals white, base clawed; stamens 40-50. Pomes yellow, ovoid or oblong, 2-3 cm diam. Fl. Mar-Apr. Fr. Aug-Sep. Streamsides in valleys, mixed forests or mountains at 1000-3000 m. Distributed in Yunnan, Sichuan and Guizhou.

红毛花楸 *Sorbus rufopilosa*

榲桲 *Cydonia oblonga*

木瓜 *Chaenomeles sinensis*

木瓜
Chaenomeles sinensis (Thouin) Koehne

灌木或小乔木。小枝无刺。托叶卵状披针形，边缘有腺体，刺芒状尖锐锯齿，齿尖有腺。花单生，萼片反折。梨果芳香，暗黄色，狭椭圆体形，长10-15厘米，木质。花期4月，果期9-10月。生海拔约1000米的山坡。产华南、西南、华中和华东。

Shrubs or small trees. Branchlets unarmed. Stipules ovate-lanceolate, margin glandular, aristate and sharply serrate, teeth glandular at apices. Flowers solitary, sepals reflexed. Pomes fragrant, dark yellow, narrowly ellipsoid, 10-15 cm long, woody. Fl. Apr. Fr. Sep-Oct. Slopes at ca. 1000 m. Distributed in S, SW, C and E China.

皱皮木瓜
Chaenomeles speciosa (Sweet) Nakai

落叶灌木，具枝刺。叶卵圆形至椭圆形，边缘具短尖锯齿。花先叶开放，3-5花簇生，萼片直立；花瓣倒卵形或圆形，基部具短爪。梨果球形或卵球形，黄色或带黄绿色，直径4-6厘米。花期3-5月，果期9-10月。产中国西南、华南、东南和华西。缅甸亦有栽培。

Shrubs deciduous, with thorns. Leaves ovate to elliptic, margin shortly serrate. Flowers precocious, 3-5-fascicled, sepals erect; petals obovate or rotund, base shortly clawed. Pomes yellow or globose or ovoid, yellow or yellowish green, 4-6 cm diam. Fl. Mar-May. Fr. Sep-Oct. Distributed in SW, S, SE and W China. Also cultivated in Myanmar.

皱皮木瓜 *Chaenomeles speciosa*

毛叶木瓜
Chaenomeles cathayensis (Hemsl.) C. K. Schneid.

灌木或小乔木，具短枝刺。叶下幼时密被褐色绒毛，渐无毛，边缘有芒状细尖锯齿，上半部有时形成重锯齿，下半部锯齿较稀，有时近全缘。花先叶开放，2-3花簇生；花柱5。梨果黄红色，直径6-7厘米，卵球形或近柱状，芳香。花期3-5月，果期9-10月。生海拔900-2500米的山坡、林缘或路边，并广泛栽培。产中国西南、华南、东南和华中。

毛叶木瓜 *Chaenomeles cathayensis*

西藏木瓜 *Chaenomeles thibetica*

秋子梨 *Pyrus ussuriensis*

Shrubs or small trees, with short thorns. Leaves abaxially initially densely brown tomentose, glabrescent, margin minutely aristate-serrate, sparsely serrate or subentire basally, doubly serrate apically. Flowers precocious, 2-3-fascicled; styles 5. Pomes yellowish red, 6-7 cm diam, ovoid or subcylindric, fragrant. Fl. Mar-May. Fr. Sep-Oct. Slopes, forest edges or roadsides at 900-2500 m, widely cultivated. Distributed in SW, S, SE and C China.

西藏木瓜

Chaenomeles thibetica T. T. Yü

灌木或小乔木，常多枝刺。叶卵状披针形或长圆状披针形，革质，全缘，背面密被褐色绒毛。3-4花簇生；花柱5，基部合生且密被柔毛。梨果矩圆形或梨形，直径5-9厘米，黄色。花期7月，果期9-10月。生海拔2600-3800米的山坡或山谷灌丛中。产四川西部和西藏东部。

Shrubs or small trees, usually with many thorns. Leaves ovate-lanceolate or oblong-lanceolate, leathery, margin entire, abaxiallly densely brown tomentose. Flowers 3-4-fascicled; styles 5, connate and densely pubescent at base. Pomes oblong or pear-shaped, 5-9 cm diam, yellow. Fl. Jul. Fr. Sep-Oct. Shrubs on slopes or in valleys at 2600-3800 m. Distributed in W Sichuan and E Xizang.

秋子梨

Pyrus ussuriensis Maxim.

乔木。叶边缘具有带刺芒状长尖锐锯齿，长5-10厘米。伞房花序具5-7花，花柱5。梨果黄色，近球形，直径2-6厘米，5室；果梗长1-3厘米，无毛；萼片宿存。花期5月，果期8-10月。产华北和东北；华北广泛栽培。俄罗斯和朝鲜半岛亦有。

Trees. Leaf margin long spinulose-serrate, 5-10 cm long. Corymbs 5-7-flowered, styles 5. Pomes yellow, subglobose, 2-6 cm diam, 5-loculed; fruiting pedicels 1-3 cm long, glabrous; sepals persistent. Fl. May. Fr. Aug-Oct. Distributed in N and NE China; commonly cultivated in N China. Also in Russia and Korean Peninsula.

麻梨

Pyrus serrulata Rehder

乔木。叶缘具锯齿。伞形总状花序，有6-11花；萼筒外面被稀疏绒毛，子房3(-4)室，具2胚珠；花柱3(-4)，和雄蕊近等长。梨果暗褐色，直径1.5-2.2厘米，3或4室；萼片宿存或部分脱落；果梗长3-4厘米。花期4月，果期6-8月。生海拔100-1600米的灌木丛中或林缘。产华南、东南和华中。

Trees. Leaf margin serrulate. Racemes umbellike, 6-11-flowered; hypanthium sparsely tomentose abaxially; ovary 3(-4)-loculed, with 2 ovules per locule; styles 3(-4), as long as stamens. Pomes dark brown, 1.5-2.2 cm diam, 3- or 4-loculed; with persistent sepals or sometimes a few caducous; fruiting pedicels 3-4 cm long. Fl. Apr. Fr. Jun-Aug. Among shrubs or forest edges at 100-1600 m. Distributed in S, SE and C China.

麻梨 *Pyrus serrulata*

白梨
Pyrus bretschneideri Rehder

白梨 *Pyrus bretschneideri*

乔木。叶基部阔楔形，边缘具带芒锯齿。总状花序伞形，具7-10花；花柱5或4。梨果黄色，具细密斑点，卵球形或近球形，直径2-2.5厘米，4或5室；萼片早落；果柄长1.5-3厘米。花期4月，果期8-9月。生海拔100-2000米的山坡或干寒地区。产河南、河北、山西、陕西、甘肃、山东和新疆。

Trees. Leaves base broadly cuneate, margin spinulose-serrate. Racemes umbellike, 7-10-flowered; styles 5 or 4. Pomes yellow, with fine dots, ovoid or subglobose, 2-2.5 cm diam, 4- or 5-loculed; sepals caducous; fruiting pedicels 1.5-3 cm long. Fl. Apr. Fr. Aug-Sep. Slopes or dry cold regions at 100-2000 m. Distributed in Henan, Hebei, Shanxi, Shaanxi, Gansu, Shandong and Xinjiang.

沙梨
Pyrus pyrifolia (Burm. f.) Nakai

乔木。叶卵状披针形或卵形，边缘有带芒锯齿。总状花序伞形，有6-9花；花柱5或4。梨果浅褐色，具浅色斑点，近球形，直径2-2.5厘米，(4或)5室；萼片早落。花期4月，果期8月。生海拔100-1400米的多雨而温和的地区。产中国长江以南各省区。老挝和越南亦有。

Trees. Leaves ovate-elliptic or ovate, margin spinulose-serrate. Racemes umbellike, 6-9-flowered; styles 5 or 4. Pomes brownish, with pale dots, subglobose, 2-2.5 cm diam, (4- or)5-loculed; sepals caducous. Fl. Apr. Fr. Aug. Warm rainy regions at 100-1400 m. Distributed throughout the provinces on south of Yangtze River. Also in Laos and Vietnam.

沙梨 *Pyrus pyrifolia*

杜梨
Pyrus betulifolia Bunge

乔木。叶无毛或幼时具褐色绵毛，边缘具刺状锯齿。总状花序伞状，具6-9花。梨果褐色，有淡色斑点，近球形，直径2-2.5厘米，(4或)5室；萼片早落；果柄长3.5-5.5厘米。花期4月，果期8-9月。生海拔1800米以下的山坡或平原。产中国西南、华中、华北和华西。老挝亦有。

杜梨 *Pyrus betulifolia*

Trees. Leaves glabrous or brown lanate when young, margin spinulose-serrate. Racemes umbellike, 6-9-flowered. Pomes brown, with pale dots, subglobose, 2-2.5 cm diam, (4- or) 5-loculed; sepals caducous; fruiting pedicels 3.5-5.5 cm long. Fl. Apr. Fr. Aug-Sep. Open slopes or plains below 1800 m. Distributed in SW, C, N and W China. Also in Laos.

豆梨
Pyrus calleryana Decne.

乔木。叶无毛，边缘具重钝锯齿。总状花序伞形，有6-12花，花柱2(-3)。梨果球形，深褐色具淡色斑点，直径约1厘米，具2(或3)室；萼片早落。花期4月，果期8-9月。生海拔

豆梨 *Pyrus calleryana*

100-1800米的山坡、平原、山谷杂木林或灌丛中。产华南、华中和华东。越南和日本亦有。

Trees. Leaves glabrous, margin obtusely serrate. Racemes umbel-like, 6-12-flowered; styles 2(or 3). Pomes globose, blackish brown with pale dots, ca. 1 cm diam, 2(or 3)-loculed; sepals caducous. Fl. Apr. Fr. Aug-Sep. Slopes, plains, mixed forests in valleys or thickets at 100-1800 m. Distributed in S, C and E China. Also in Vietnam and Japan.

川梨

Pyrus pashia Buch.-Ham. ex D. Don

乔木。小枝和花序被绵毛。叶幼时具绒毛，渐无毛，边缘具钝锯齿。总状花序伞形，具7-13花；花瓣白色，雄蕊25-30，花柱3-5。梨果褐色，具浅色斑点，近球形，直径1-1.5厘米；萼片早落。花期3-4月，果期8-9月。生海拔600-3000米的山谷或灌丛中。产贵州、云南和四川。南亚和东南亚亦有。

Trees. Branchlets and inflorescences lanose. Leaves tomentose when young, glabrescent, margin obtusely serrate. Racemes umbel-like, 7-13-flowered; petals white, stamens 25-30, styles 3-5. Pomes brown, with pale dots, subglobose, 1-1.5 cm diam; sepals caducous. Fl. Mar-Apr. Fr. Aug-Sep. Valleys or among shrubs at 600-3000 m. Distributed in Guizhou, Yunnan and Sichuan. Also in S and SE Asia.

川梨 *Pyrus pashia*

山荆子 *Malus baccata*

山荆子

Malus baccata (L.) Borkh.

乔木。叶椭圆形或卵形，幼时无毛或被短柔毛。伞房花序具4-6花；萼片披针形，先端长渐尖；花柱5或4，基部密被长柔毛。梨果红色或黄色，近球形。花期4-6月，果期9-10月。生海拔1500米以下的林中或灌丛中。产华北和东北。俄罗斯(西伯利亚)、蒙古和朝鲜半岛亦有。

Trees. Leaves elliptic or ovate, glabrous or slightly puberulous when young. Corymbs 4-6-flowered; sepals lanceolate, apex long acuminate; styles 5 or 4, densely villous basally. Pomes red or yellow, subglobose. Fl. Apr-Jun. Fr. Sep-Oct. Forests or thickets below 1500 m. Distributed in N and NE China. Also in Russia (Siberia), Mongolia and Korean Peninsula.

毛山荆子

Malus manshurica (Maxim.) Kom. ex Juz.

乔木。小枝幼时密被柔毛，渐无毛。叶下面沿叶脉被毛或近无毛。伞房花序无总梗，伞状，具3-6花；萼片披针形；子房4或5室。梨果红色，椭圆体形或倒卵球形。花期5-7月，果期8-9月。生海拔100-2100米的山坡混交林中、山顶或峡谷。产华北和东北。俄罗斯亦有。

Trees. Branchlets densely puberulous when young, glabrescent. Leaves abaxially pubescent or subglabrous along veins. Corymbs sessile, umbellike, 3-6-flowered; sepals lanceolate; ovary 4- or 5-loculed. Pomes red, ellipsoid or obovoid. Fl. May-Jul. Fr. Aug-Sep. Mixed forests on slopes, mountain summits or valleys at 100-2100 m. Distributed in N and NE China. Also in Russia.

湖北海棠

Malus hupehensis (Pamp.) Rehder

乔木。叶边有细锐锯齿，幼时疏被柔毛，渐无毛。伞房花序具4-6花；萼筒和萼片背面无毛，花蕾粉红色，花瓣白色，花柱3，稀4。梨果黄绿色，带红色，椭圆体形或近球形。花期4-5月，果期8-9月。生海拔2900米以下的山坡或沟谷灌丛中。产华南、华东、华中和华北。

Trees. Leaves margin serrulate, sparsely puberulous when young, glabrescent. Corymbs 4-6-flowered; hypanthium and sepals abaxially glabrous, petals pink in bud, becoming white; styles 3, rarely 4. Pomes yellowish green, tinged red, ellipsoid or subglobose. Fl. Apr-May. Fr. Aug-Sep.

毛山荆子 *Malus manshurica*

湖北海棠 *Malus hupehensis*

Slopes or valley thickets below 2900 m. Distributed in S, E, C and N China.

垂丝海棠

Malus halliana Koehne

乔木。叶边有圆钝细锯齿。伞房花序具4-6花；萼筒和萼片背面无毛；花瓣粉红色；萼片先端钝；花柱4或5。梨果梨形或倒卵球形，淡紫色，直径6-8毫米，萼片早落。花期3-4月，果期9-10月。生海拔1200米以下的山坡灌丛或溪边。产中国西南、东南、华中、华西和华东。

Trees. Leaf margin obtusely serrulate. Corymbs 4-6-flowered; hypanthium and sepals abaxially glabrous; petals pink; sepals apically obtuse; styles 4 or 5. Pomes pyriform or obovoid, purplish, 6-8 mm diam, sepals caducous. Fl. Mar-Apr. Fr. Sep-Oct. Thickets on slopes or by streams below 1200 m. Distributed in SW, SE, C, W and E China.

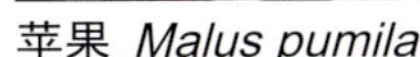

苹果 *Malus pumila*

新疆野苹果 *Malus sieversii*

苹果

Malus pumila Mill.

乔木。叶椭圆形、卵形至宽椭圆形，幼时两面具短柔毛，边缘具圆钝细锯齿。伞房花序，具3-7花；子房5室，每室具2个胚珠；花柱5。梨果红色或黄色，近扁球形，萼片宿存。花期5月，果期7-10月。栽培于中国西南、华北和西北。原产西南亚和欧洲；温带地区普遍栽培。

Trees. Leaves elliptic, ovate or broadly elliptic, both surfaces densely puberulous when young, margin obtusely serrate. Corymbs 3-7-flowered; ovary 5-loculed, with 2 ovules per locule; styles 5. Pomes red or yellow, almost depressed-globose, sepals persistent. Fl. May. Fr. Jul-Oct. Cultivated in SW, N and NW China. Native to SW Asia and Europe; commonly cultivated in temperate areas of the world.

垂丝海棠 *Malus halliana*

新疆野苹果

Malus sieversii (Ledeb.) Roem.

乔木。小枝嫩时具短柔毛；冬芽卵形，外被长柔毛。叶边缘具圆钝锯齿，幼叶下面密被长柔毛。花序近伞形；萼片先端渐尖，比萼筒稍长；花柱5。梨果大，球形或扁球形，直径3-4.5厘米，萼洼下陷，萼片宿存，反折；果梗长3.5-4厘米。花期5月，果期8-10月。生海拔1200-1300米的山顶、山坡或河谷地带有大面积野生林。产新疆西部。西亚和中亚亦有。

Trees. Branchlets puberulous when young; winter buds ovoid, villous abaxially. Leaves margin obtusely serrate, abaxially densely villous when young. Inflorescence umbellike; sepals apex acuminate, longer than hypanthium; styles 5. Pomes large, globose or depressed-globose, 3-4.5 cm diam, sepals with cavity at apex, persistent, reflexed; fruiting pedicels 3.5-4 cm long. Fl. May. Fr. Aug-Oct. Mountain summits, slopes, valleys, often the dominant tree of forests at 1200-1300 m. Distributed in W Xinjiang. Also in W and C Asia.

花红 *Malus asiatica*

花红
Malus asiatica Nakai

乔木。叶边缘有尖锐锯齿，背面密被短柔毛。伞房花序生在小枝顶端，伞状，具4-7(-10)花。梨果卵球形，直径4-5厘米；果梗长1.5-2.5厘米；萼片宿存。花期4-5月，果期8-9月。生海拔2800米以下的开阔山坡或平原沙地。产中国除华南以外大部分地区。

Trees. Leaf margin serrulate, abaxially densely pubescent. Corymbs at apices of branchlets, umbellike, 4-7(-10)-flowered. Pomes ovoid, 4-5 cm diam; pedicels 1.5-2 cm long; fruiting pedicels 1.5-2.5 cm long; sepals persistent. Fl. Apr-May. Fr. Aug-Sep. Open slopes or sandy soils of plains below 2800 m. Distributed in most parts of China, except S part.

西府海棠 *Malus × micromalus*

西府海棠
Malus × micromalus Makino

小乔木。叶柄长2-3.5厘米；叶狭椭圆形至椭圆形，幼时具柔毛，渐无毛，基部楔形，边缘具锯齿。伞房花序，具4-7花；花瓣粉色。梨果红色，近球形，直径1-1.5厘米，顶端凹陷，基部下陷；萼片早落或有少量宿存。花期4-5月，果期8-9月。产中国西南、东南、华北和华西。

Trees small. Petioles 2-3.5 cm long; leaves narrowly elliptic or elliptic, puberulous when young, glabrescent, base cuneate, margin serrate. Corymb umbellike, 4-7-flowered; petals pink. Pomes red, subglobose, 1-1.5 cm diam, impressed at apex, with cavity at base; sepals caducous or a few persistent. Fl. Apr-May. Fr. Aug-Sep. Distributed in SW, SE, N and W China.

山楂海棠
Malus komarovii (Sargent) Rehder

灌木或小乔木，高达3米。叶阔卵形，边缘具重锯齿，常在中部3裂，近基部两侧各具1浅裂。伞房花序具6-8花；花瓣白色。梨果红色，椭圆体形，直径0.8-1厘米，果心先端分离，萼片早落，果柄长1.2-1.5厘米。花期5月，果期9月。生海拔1100-1300米的灌丛中。产吉林(长白山)。朝鲜半岛亦有。

Shrubs or small trees, to 3 m tall. Leaves broadly ovate, margin doubly serrate, usually 3-parted at middle, 1-lobed on each side near base. Corymbs 6-8-flowered; petals white. Pomes red, ellipsoid, 0.8-1 cm diam, cores free at apex, sepals caduceus, fruiting pedicels 1.2-1.5 cm long. Fl. May. Fr. Sep. Among shrubs at 1100-1300 m. Distributed in Jilin (Changbai Mountain). Also in Korean Peninsula.

山楂海棠 *Malus komarovii*

变叶海棠 *Malus toringoides*

锡金海棠 *Malus sikkimensis*

变叶海棠

Malus toringoides (Rehder) Hughes

灌木或小乔木。叶形多变，背面沿中脉和侧脉密被短柔毛，腹面被稀疏柔毛。伞房花序伞状，具3-5花；花瓣白色，卵形或椭圆状卵形。梨果黄色，带红色，倒卵球形或狭椭圆体形，长1-1.3厘米。花期4-5月，果期9月。生海拔2000-3000米的山坡灌丛中。产甘肃、西藏和四川。

Shrubs or small trees. Leaves variable in shape, abaxially densely pubescent along midveins and lateral veins, adaxially sparsely pubescent. Corymbs umbellike, 3-5-flowered; petals white, ovate or elliptic-obovate. Pomes yellow, tinged red, obovoid or narrowly ellipsoid, 1-1.3 cm long. Fl. Apr-May. Fr. Sep. Thickets on slopes at 2000-3000 m. Distributed in Gansu, Xizang and Sichuan.

锡金海棠

Malus sikkimensis (Wenz.) Koehne

小乔木。叶椭圆形至椭圆状披针形，背面被短绒毛。伞房花序具6-10花；萼片卵状披针形，长于萼筒；花柱基部无毛。梨果倒卵球形或梨形，直径1-1.8厘米，暗红色，有白色点。花期5-7月，果期9-10月。生海拔2500-3000米的山坡疏林下和山谷混交林中。产四川、西藏和云南。印度东北部、尼泊尔和不丹亦有。

Trees small. Leaves elliptic to elliptic-lanceolate, abaxially tomentose. Corymbs 6-10-flowered; sepals ovate-lanceolate, longer than hypanthium; styles glabrous at base. Pomes obovoid or pyriform, 1-1.8 cm diam, dark red, white-punctate. Fl. May-Jul. Fr. Sichuan, Xizang and Yunnan. Sep-Oct. Open forests on slopes and in mixed forests in valleys at 2500-3000 m. Distributed in Sichuan, Xizang and Yunnan. Also in NE India, Nepal and Bhutan.

河南海棠

Malus honanensis Rehder

灌木或小乔木。叶边缘具重锯齿，托叶条状披针形，膜质，早落。伞形伞房状花序，具5-10花；花瓣基部近心形，有短爪，粉白色。梨果黄红色，近球形，直径约8毫米；萼片宿存。花期5月，果期8-9月。生海拔800-2600米的山坡或沟谷灌丛。产湖北、河南、河北、山西、陕西和甘肃。

Shrubs or small trees. Leaf margin doubly serrate, stipules linear-lanceolate, membranous, caduceus. Corymbs umbellike, 5-10-flowered; petals base subcordate and shortly clawed, pinkish white. Pomes yellowish red, subglobose, ca. 8 mm diam; sepals persistent. Fl. May. Fr. Aug-Sep. Slopes or thickets in valleys at 800-2600 m. Distributed in Hubei, Henan, Hebei, Shanxi, Shaanxi and Gansu.

河南海棠 *Malus honanensis*

滇池海棠 *Malus yunnanensis*

台湾林檎 *Malus doumeri*

滇池海棠
Malus yunnanensis (Franch.) C. K. Schneid.

乔木。叶两侧各3-5浅裂，背面密被绒毛。伞房花序伞状，具8-12花；萼筒和花梗密被绒毛，花柱5，萼片宿存。梨果红色，球形，直径1-1.5厘米，具白色斑点。花期5月，果期8-9月。生海拔1600-3800米的山坡杂木林中或沟谷溪边。产云南、四川、西藏、贵州、湖北和陕西。缅甸亦有。

Trees. Leaf margin each side 3-5-lobed, abaxially densely tomentose. Corymbs umbellike, 8-12-flowered; hypanthium and pedicles densely tomentose, styles 5, sepals persistent. Pomes red, globose, 1-1.5 cm diam, white punctate; Fl. May. Fr. Aug-Sep. Mixed forests on slopes or by streams in valleys at 1600-3800 m. Distributed in Yunnan, Sichuan, Xizang, Guizhou, Hubei and Shaanxi. Also in Myanmar.

台湾林檎
Malus doumeri (Bois) A. Chevalier

乔木，高达15米。叶狭椭圆状卵形或倒卵状披针形，两面初具绒毛，渐无毛。伞房花序伞状，具4-5花；花瓣黄白色。梨果黄红色，球形，直径2.5-5.5厘米，果柄长1-3厘米，萼片宿存。花期5月，果期8-9月。生海拔1000-2000米的林中。产中国西南、华南、华中和华东。老挝和越南亦有。

Trees, to 15 m tall. Leaves narrowly elliptic-ovate or obovate-lanceolate, both surfaces tomentose when young, glabrescent. Corymbs umbel-like, 4-5-flowered; petals yellowish white. Pomes yellowish red, globose, 2.5-5.5 cm diam, fruiting pedicels 1-3 cm long, sepals persistent. Fl. May. Fr. Aug-Sep. Forests at 1000-2000 m. Distributed in SW, S, C and E China. Also in Laos and Vietnam.

陇东海棠
Malus kansuensis (Batal.) Schneid.

灌木或小乔木，高3-5米。叶卵形或宽卵形，基部圆形或截形，边缘具重锯齿，常3裂。伞房花序具4-10花；花瓣白色。梨果黄红色，椭圆体形或倒卵球形，直径1-1.5厘米，果柄长2-3.5厘米，萼片早落。花期5-7月，果期7-8月。生海拔1500-3300米的杂木林或灌丛中。产中国西南、华中、华北、华西和西北。

Shrubs or small trees. 3-5 m tall. Leaves ovate or broadly ovate, base rounded or truncate, margin doubly serrulate, often 3-lobed. Corymbs 4-10-flowered; petals white. Pomes yellowish red, ellipsoid or obovoid, 1-1.5 cm diam; fruiting pedicels 2-3.5 cm long, sepals caducous. Fl. May-Jul. Fr. Jul-Aug. Mixed forests or scrubs at 1500-3300 m. Distributed in SW, C, N, W and NW China.

唐棣
Amelanchier sinica (Schneid.) Chun

小乔木。冬芽具浅褐色鳞片，鳞片边缘有柔毛。叶卵形或长椭圆形，通常在中部以上有细锐锯齿，基部全缘，幼时下面沿中脉和侧脉被绒毛或柔毛，老时脱落无毛。总状花序，多花；总花梗和花梗无毛或最初有毛，以后脱落。梨果近球形或扁圆形；萼片宿存，反折。花期5月，果期9-10月。生海拔1000-2000米的山坡、灌木丛中。产华中。

陇东海棠 *Malus kansuensis*

唐棣 *Amelanchier sinica*

Small trees. Winter buds with pale brown scales, margin of scales pubescent. Leaves ovate or oblong, usually serrulate above middle, entire basally, abaxially pubescent along midvein when young, glabrescent. Raceme many flowered; rachis and peduncles glabrous or initially sparsely pubescent, glabrescent. Pomes subglobose or depressed-globose; sepals persistent, reflexed. Fl. May. Fr. Sep-Oct. Slopes, among shrubs at 1000-2000 m. Distributed in C China.

光萼林檎
Malus leiocalyca S. Z. Huang

灌木。叶椭圆形或卵状椭圆形，幼时疏被柔毛，渐无毛。伞房状状伞形花序，5-7花；花瓣紫白色；萼片全缘，外面无毛，里面被绒毛。梨果黄红色，球形；萼片宿存。花期5月，果期8-9月。生海拔700-2400米的山地杂木林或沟谷溪边。产中国西南、东南、华中和华东。

Shrubs. Leaves elliptic or ovate-elliptic, sparsely pubescent when young, glabrescent. Corymbs umbellike, 5-7-flowered; petals purplish white; sepal margin entire, abaxially glabrous, adaxially tomentose. Pomes yellowish red, globose; sepals persistent. Fl. May. Fr. Aug-Sep. Mixed montane forests or streamsides in valleys at 700-2400 m. Distributed in SW, SE, C and E China.

棣棠花
Kerria japonica (L.) DC.

落叶灌木。叶互生，三角状卵形或卵形，边缘有重锯齿。花瓣黄色，先端具凹缺；副萼片无裂片；心皮5，离生，雌蕊5-8。瘦果黑褐色，倒卵球形或半球形，具脉纹。花期4-6月，果期6-8月。生海拔200-3000米的山坡灌丛中。产中国除东北以外大部分地区。日本亦有。

光萼林檎 *Malus leiocalyca*

Shrubs deciduous. Leaves alternate, triangular-ovate or ovate, margin doubly serrate. Petals yellow, apex emarginate; epicalyx segments absent; carpels 5, free, pistils 5-8. Achenes brownish black, obovoid or hemispheric, rugose. Fl. Apr-Jun. Fr. Jun-Aug. Thickets on slopes at 200-3000 m. Distributed in most parts of China, except NE China. Also in Japan.

棣棠花 *Kerria japonica*

鸡麻 *Rhodotypos scandens*

鸡麻

Rhodotypos scandens (Thunb.) Makino

灌木。叶对生，下面起初具绢毛，老时脱落仅沿脉被稀疏柔毛。萼筒杯碟状，萼片4，成2对，卵状椭圆形，副萼片条形；心皮4。核果1-4，褐黑色，斜卵球形，光滑。花期4-5月，果期6-9月。生海拔100-800米的山坡林下或山谷。产华中、华北和华东。朝鲜半岛和日本亦有。

Shrubs. Leaves opposite, abaxially sericeous when young, sparsely pilose on veins when old. Hypanthium saucer-shaped, sepals 4, in 2 pairs, ovate-elliptic, epicalyx segments linear, carpels 4. Drupes 1-4, brownish black, obliquely ellipsoid, glabrous. Fl. Apr-May. Fr. Jun-Sep. Forests on slopes or valleys at 100-800 m. Distributed in C, N and E China. Also in Korean Peninsula and Japan.

蚊子草

Filipendula palmata (Pall.) Maxim.

多年生草本。羽状复叶，具2对侧生小叶，顶生小叶5-9掌状深裂，裂片披针形至菱状披针形，背面密被白色绒毛。圆锥花序顶生，花多数。瘦果基部着生花托上，具短柄，直立。花果期7-9月。生海拔200-2300米的林缘、山坡、草地、山谷或阴湿处。产河北、内蒙古、陕西、黑龙江、吉林和辽宁。俄罗斯、蒙古和朝鲜半岛亦有。

Herbs perennial. Leaves pinnate, with 2 pairs of lateral leaflets, terminal leaflets palmately 5-9-parted, segments lanceolate to rhombic-lanceolate, abaxially densely white tomentose. Panicles terminal, many flowered. Achenes basally attached to receptacle, shortly stipitate, erect. Fl. and fr. Jul-Sep. Forest edges, slopes, grasslands, valleys, shady or moist places at 200-2300 m. Distributed in Hebei, Neimenggu, Shaanxi, Heilongjiang, Jilin and Liaoning. Also in Russia, Mongolia and Korean Peninsula.

槭叶蚊子草

Filipendula glaberrima Nakai

多年生草本。羽状复叶，具1-3对小叶。圆锥花序顶生，花直径4-5毫米；花瓣4或5，粉红色至白色，基部有短爪。瘦果直立，具柄，背腹两边有一行柔毛。花果期6-8月。生海拔700-1500米的林缘、林下或湿草地。产黑龙江、吉林和辽宁。俄罗斯、朝鲜半岛和日本亦有。

蚊子草 *Filipendula palmata*

槭叶蚊子草 *Filipendula glaberrima*

Herbs perennial. Leaves pinnate, with 1-3 pairs of leaflets. Panicles terminal; flowers 4-5 mm diam; petals 4 or 5, pink to white, base shortly clawed. Achenes erect, stipitate, long ciliate along abaxial and adaxial sides. Fl. and fr. Jun-Aug. Forest edges, under forests or moist grasslands at 700-1500 m. Distributed in Heilongjiang, Jilin and Liaoning. Also in Russia, Korean Peninsula and Japan.

白叶莓
Rubus innominatus S. Moore

灌木。奇数羽状复叶，小叶常3，稀于不孕枝上具5小叶；叶下密被灰色绒毛，边缘具不整齐糙锯齿或锐重锯齿。近总状或狭伞房状圆锥花序，顶生或腋生；萼片直立，卵形，先端急尖；花瓣紫色。聚合果橙红色，近球形。花期5-6月，果期7-8月。生海拔400-2500米的密或疏林、灌丛、山坡、路边、溪边或河边。产中国西南、华南、东南、华中、华西和华东。

Shrubs. Leaves imparipinnately compound, 3-foliolate, rarely 5-foliolate on sterile branches; abaxially densely gray tomentose, margin unevenly coarsely serrate or incised doubly serrate. Inflorescences subracemes or narrow cymose panicles, terminal or axillary; sepals erect, ovate, apex acute; petals purple. Aggregate fruits orange-red, subglobose. Fl. May-Jun. Fr. Jul-Aug. Dense or sparse forests, thickets, slopes, roadsides, streamsides or riversides at 400-2500 m. Distributed in SW, S, SE, C, W and E China.

红泡刺藤
Rubus niveus Thunb.

灌木。小叶常7-9，顶生小叶下面被灰白色绒毛，边缘常具不整齐粗锐锯齿；托叶线状披针形。花成伞房花序或短圆锥状花序；苞片披针形或线形；萼片顶端急尖或突尖；雌蕊55-70。聚合果半球形，密被灰白色绒毛。花果期5-9月。生海拔500-2800米的山坡灌丛、疏林、山谷、河滩和溪流旁。产华中、华南和西藏。东南亚、印度和克什米尔地区亦有。

白叶莓 *Rubus innominatus*

Shrubs. Leaflets 7-9, terminal ones abaxially gray-white tomentose, margin usually irregularly coarsely sharply serrate; stipules linear-lanceolate. Inflorescences corymbs, or short thyrses; bracts lanceolate or linear; sepals apex acute or abruptly pointed; pistils 55-70. Aggregate fruits semiglobose, densely gray-white tomentose. Fl. and fr. May-Sep. Thickets on slopes, sparse forests, montane valleys, flood plains and streamsides at 500-2800 m. Distributed in C and S China, and Xizang. Also in SE Asia, India and Kashmir.

红泡刺藤 *Rubus niveus*

桉叶悬钩子

Rubus eucalyptus Focke

灌木。奇数羽状复叶，具3-5(或7)小叶；叶柄和花梗疏具腺毛。花序顶生，稀腋生，具1-2花；花萼下面具疏腺毛，具针状刺。聚合果红色，近球形，直径1.2-2厘米，密被长绒毛。花期4-5月，果期6-7月。生海拔1000-3400米的杂木林、灌丛或草地。产四川、贵州、湖北、陕西和甘肃。

桉叶悬钩子 *Rubus eucalyptus*

Shrubs. Leaves imparipinnate, 3-5(or 7)-foliolate; petioles and pedicels with sparse, glandular hairs. Inflorescences terminal, rarely axillary, 1-2-flowered; calyx abaxially pubescent, with stalked glands and needlelike prickles. Aggregate fruits red, subglobose, 1.2-2 cm diam, densely long tomentose. Fl. Apr-May. Fr. Jun-Jul. Forests, thickets or grasslands at 1000-3400 m. Distributed in Sichuan, Guizhou, Hubei, Shaanxi and Gansu.

库页悬钩子 *Rubus sachalinensis*

库页悬钩子

Rubus sachalinensis H. Lévl.

灌木。奇数羽状复叶，通常3(-5)小叶。花序伞房状，有5-9花；总花梗、花梗和花萼外面密被柔毛、腺毛和针刺。聚合果红色，直径约1厘米，被绒毛。花期6-7月，果期8-9月。生海拔400-3100米的林下、林缘、灌丛、草地、山谷或石隙。产华北、西北和东北。俄罗斯、蒙古、朝鲜半岛、日本和欧洲亦有。

Shrubs. Leaves imparipinnate, usually 3(-5)-foliolate. Inflorescences corymbose, 5-9-flowered; rachises, pedicels and abaxial surface of calyx pubescent, and glandular hairy with needlelike prickles. Aggregate fruits red, ca. 1 cm diam, tomentose. Fl. Jun-Jul. Fr. Aug-Sep. Forests, forest edges, thickets, grasslands, valleys or rock crevices at 400-3100 m. Distributed in N, NW and NE China. Also in Russia, Mongolia, Korean Peninsula, Japan and Europe.

菰帽悬钩子

Rubus pileatus Focke

藤状灌木。奇数羽状复叶，常具5-7小叶。伞房花序顶生，具3-5花，稀花单生；花瓣白色；花直径1-1.5(-2)厘米；花萼紫色，下面无毛。聚合果红色，直径1-1.5(-2)厘米，卵球形，密被灰色长绒毛。花期6-7月，果期8-9月。生海拔1400-2800米的路边疏林、山谷密林、河边或河谷中。产中国西南、华中、华西和西北。

Shrubs scandent. Leaves imparipinnate, usually 5-7-foliolate. Corymbose terminal, 3-5-flowered, rarely flowers solitary; petals white; flowers 1-1.5(-2) cm diam; calyx purple, abaxially glabrous. Aggregate fruits red, 0.8-1.2 cm diam, ovoid, densely gray tomentose. Fl. Jun-Jul. Fr. Aug-Sep. Sparse forests on roadsides, dense forests in montane valleys,

菰帽悬钩子 *Rubus pileatus*

栽秧泡 *Rubus ellipticus* var. *obcordatus*

riversides or ravines at 1400-2800 m. Distributed in SW, C, W and NW China.

栽秧泡

Rubus ellipticus Sm var. **obcordatus** (Franch.) Focke

灌木。小叶3，倒卵形，顶端浅心形或截形。花序顶生，密集成顶生短总状花序，数花至10花至更多，或数花腋生成束；花梗和花萼上几无刺毛。聚合果近球形，金黄色。花期3-4月，果期4-5月。生海拔300-1000米的山坡、路边或灌丛中。产云南、西藏、广西、贵州和四川。印度、老挝、泰国和越南亦有。

Shrubs. Leaflets 3, obovate, apex shallowly cordate or subtruncate. Inflorescences terminal, dense glomerate racemes, flowers several to 10 or more, or flowers several in clusters in leaf axils; pedicels and calyx very sparsely bristly. Aggregate fruits subglobose, golden yellow. Fl. Mar-Apr. Fr. Apr-May. Slopes, roadsides or thickets at 300-1000 m. Distributed in Yunnan, Xizang, Guangxi, Guizhou and Sichuan. Also in India, Laos, Thailand and Vietnam.

茅莓

Rubus parvifolius L.

灌木。小枝、叶柄和花梗均被柔毛。奇数羽状复叶，小叶菱状卵形或倒卵形。花序顶生，伞房状，具数至多花，腋生花序伞房状；花瓣粉红色至紫红色。聚合果红色，卵球形，直径1-1.5厘米，无毛或具稀疏柔毛。花期5-6月，果期7-8月。生海拔400-2700米的疏林、山坡、路边或荒地。产中国大部分地区。越南、朝鲜半岛和日本亦有。

Shrubs. Branches, petioles and pedicels with soft hairs. Leaves imparipinnate, leaflets rhomboid-ovate or obovate. Inflorescences terminal, corymbose, several to many flowered, axillary inflorescences corymbose; petals pink to purplish red. Aggregate fruits red, ovoid, 1-1.5 cm diam, glabrous or somewhat pubescent. Fl. May-Jun. Fr. Jul-Aug. Open forests, slopes, roadsides or wastelands at 400-2700 m. Distributed in most parts of China. Also in Vietnam, Korean Peninsula and Japan.

茅莓 *Rubus parvifolius*

喜阴悬钩子 *Rubus mesogaeus*

黄色悬钩子 *Rubus lutescens*

喜阴悬钩子
Rubus mesogaeus Focke

藤状灌木。小枝具稀疏针状皮刺或近无刺。奇数羽状复叶，小叶常3，稀5。伞房花序顶生或腋生，具数花至多于20花；花瓣白色或粉红色。聚合果紫黑色，扁球形，无毛。花期4-5月，果期7-8月。生海拔600-3600米的林中、山坡、山谷、河岸或路旁。产中国西南、华南、华中、华北和华西。印度北部、尼泊尔、不丹、萨哈林岛(库页岛)和日本亦有。

Shrubs scandent. Branchlets with sparse, needlelike prickles or nearly unarmed. Leaves imparipinnate, often 3-foliolate, rarely 5-foliolate. Corymbose terminal or axillary, several to more than 20-flowered; petals white or pink. Aggregate fruits purplish black, compressed globose, glabrous. Fl. Apr-May. Fr. Jul-Aug. Forests, slopes, montane valleys, river banks or roadsides at 600-3600 m. Distributed in SW, S, C, N and W China. Also in N India, Nepal, Bhutan, Sakhalin and Japan.

红花悬钩子
Rubus inopertus (Focke) Focke

攀援灌木。小叶7-11，卵状披针形或卵形，边缘具粗锐重锯齿，侧生小叶几无柄，与叶轴均具稀疏小钩刺。花序顶生，伞房状，具数花，或数花簇生；花瓣卵形，粉红至紫红色。聚合果成熟后紫黑色，球形。花期5-6月，果期7-8月。生海拔800-2800米的山地密林、河谷和岩石坡地。产云南、四川、贵州、广西、台湾、湖北、湖南和陕西。越南亦有。

Shrubs climbing. Leaflets 7-11, ovate or ovate-lanceolate, margin coarsely sharp doubly serrate, lateral leaflets subsessile, petiolules and rachises with sparse, curved minute prickles. Inflorescences terminal, corymbs, several flowered, or flowers several in clusters; petals ovate, pink to purplish red. Aggregate fruits purplish black at maturity, globose. Fl. May-Jun. Fr. Jul-Aug. Dense forests in mountainous regions, river valleys and rocky slopes at 800-2800 m. Distributed in Yunnan, Sichuan, Guizhou, Guangxi, Taiwan, Hubei, Hunan and Shaanxi. Also in Vietnam.

红花悬钩子 *Rubus inopertus*

黄色悬钩子
Rubus lutescens Franch.

低矮亚灌木。奇数羽状复叶，小叶7-11，两面被柔毛，下面沿脉疏生小皮刺。1-2花，顶生或腋生，有时3-4花生枝顶；花瓣白色，转为淡黄色。果实球形，密被细柔毛。花期5-6月，果期7-8月。生海拔2500-4300米的山坡林缘、杂木林或多石地。产云南、四川和西藏。

Subshrubs low. Leaves imparipinnate, 7-11-foliolate, both surfaces soft hairy, with sparse, minute prickles along veins. Flowers 1-2, terminal or axillary, sometimes 3-4 flowers terminal; petals white, turning yellowish. Fruits globose, densely thinly pubescent. Fl. May-Jun. Fr. Jul-Aug. Forest edges on slopes, mixed forests or stony places at 2500-4300 m. Distributed in Yunnan, Sichuan and Xizang.

插田泡
Rubus coreanus Miq.

灌木。奇数羽状复叶具小叶5或7；小叶下面具柔毛或仅沿脉具

插田泡 *Rubus coreanus*

柔毛，或具短柔毛。花序顶生侧生短小枝上，伞房状，具数朵至30多朵花；萼片先端锐尖至尾尖；花瓣粉色至深红色。聚合果深红色或紫黑色，近球形。花期4-6月，果期6-8月。生海拔100-3100米的山坡灌丛、山地沟谷、峡谷、河岸或路边。产中国西南、东南、华中、华西、华东和西北。朝鲜半岛和日本亦有。

Shrubs. Leaves imparipinnate, 5- or 7-foliolate; leaflets abaxially pubescent or only along veins or shortly tomentose. Inflorescences terminal on lateral short branchlets, corymbose, several to more than 30-flowered; sepals apex acuminate to caudate; petals pink to dark red. Aggregate fruits dark red or purplish black, subglobose. Fl. Apr-Jun. Fr. Jun-Aug. Thickets on slopes, montane valleys, ravines, river banks or roadsides at 100-3100 m. Distributed in SW, SE, C, W, E and NW China. Also in Korean Peninsula and Japan.

黄果悬钩子

Rubus xanthocarpus Bureau et Franch.

低矮半灌木。茎疏生较长直立针刺。小叶常3，常长圆形或椭圆状披针形，侧生小叶长宽约为顶生小叶之半，顶生小叶叶柄长1-2.5厘米；托叶披针形或线状披针形。花1-4成伞房状；花萼外被较密直立针刺和柔毛；花瓣白色；雌蕊多数。聚合果扁球形，无毛。花期5-6月，果期8月。生海拔600-3200米的山坡、路旁、林中或石砾山沟。产华中。

Low subshrubs. Stems with sparse, long needlelike prickles. Leaflets usually 3, usually oblong or elliptic-lanceolate, length and width of lateral leaflets about half of terminal leaflets, petiolules of terminal leaflets 1-2.5 cm long; stipules lanceolate or linear-lanceolate. Flowers 1-4, corymbose; calyx abaxially with dense needlelike prickles and pubescent; petals white; pistils numerous. Aggregate fruits compressed globose, glabrous. Fl. May-Jun. Fr. Aug. Slopes, roadsides, forests or rocky ravines at 600-3200 m. Distributed in C China.

红腺悬钩子

Rubus sumatranus Miq.

直立或攀援灌木。小枝常具散生紫红色腺毛和弯皮刺。小叶(3-)5-7，卵状披针形至披针形。花序顶生，伞房状，3至数花；花瓣白色。聚合果橘红色，长圆形。花期4-6月，果期7-8月。生海拔700-2500米以下的林中、林缘、灌丛或草地。产中国西南、华南、华中和东南。南亚、东南亚和东北亚亦有。

Shrubs erect or scandent. Branchlets usually with scattered setose purplish red glandular hairs and curved prickles. Leaflets (3-)5-7, ovate-lanceolate to lanceolate. Inflorescences terminal, corymbose, 3-to several flowered; petals white. Aggregate orange-red, fruits oblong. Fl. Apr-Jun. Fr. Jul-Aug. Forests, forest edges, thickets or grasslands below 700-2500 m. Distributed in SW, S, C and SE China. Also in S, SE and NE Asia.

黄果悬钩子 *Rubus xanthocarpus*

红腺悬钩子 *Rubus sumatranus*

光滑悬钩子 *Rubus tsangii*

光滑悬钩子

Rubus tsangii Merr.

攀援灌木。小枝无毛，具腺毛和疏生皮刺。叶具7-11小叶，下面沿脉无腺毛。3-5花成顶生伞房花序，稀单生。子房和果实常无腺毛。聚合果红色，近球形，直径达1.5厘米。花期4-5月，果期6-7月。生海拔800-2500米的山谷林中、山坡、山麓或河岸。产云南、四川、贵州、广西、广东、浙江和福建。

Shrubs scandent. Branchlets glabrous, with glands and sparse prickles. Leaves 7-11-foliolate, abaxially without gland-tipped hairs along veins. Inflorescences terminal, corymbose, 3-5-flowered or rarely solitary. Ovary and aggregate fruits usually without glandular hairs. Aggregate fruits red, subglobose, to 1.5 cm diam. Fl. Apr-May. Fr. Jun-Jul. Forests in valleys, slopes, foothills or river banks at 800-2500 m. Distributed in Yunnan, Sichuan, Guizhou, Guangxi, Guangdong, Zhejiang and Fujian.

空心泡

Rubus rosifolius Sm.

直立或攀援灌木。奇数羽状复叶，小叶常5-7。花序顶生或生叶腋，常1-2花；花瓣白色；子房无毛。聚合果红色，卵球形或狭倒卵形至狭长圆形，无毛，具稀疏腺体。花期3-5月，果期6-7月。生低海拔至中海拔的杂木林、草坡或路边。产中国西南、华南、华中、华西和华东。东南亚、日本南部、澳大利亚和非洲亦有。

Shrubs erect or climbing. Leaves imparipinnate, usually 5-7-foliolate. Inflorescences terminal or in leaf axils, 1-2-flowered; petals white; ovary glabrous. Aggregate fruits red, ovoid or narrowly obovoid to oblong, glabrous, with few glands. Fl. Mar-May. Fr. Jun-Jul. Mixed forests, grassy slopes or roadsides at low to medium elevations. Distributed in SW, S, C, W and E China. Also in SE Asia, S Japan, Australia and Africa.

大红泡

Rubus eustephanos Focke

灌木。奇数羽状复叶，具3-5(-7)小叶；小叶卵形或椭圆形。花序单花，稀2-3花；苞片披针形；花梗长2.5-5厘米，无毛；萼片反折。聚合果红色，近球形，直径0.9-1.1厘米。花期4-5月，果期6-7月。生海拔500-2300米的山麓、山坡密林中、灌丛或河边等处。产四川、贵州、浙江、湖北和陕西。

Shrubs. Leaves imparipinnate, 3-5(-7)-foliolate; leaflets ovate or elliptic. Inflorescences 1-flowered, rarely 2-3-flowered; bracts lanceolate; pedicels 2.5-5 cm long, glabrous; sepals reflexed. Aggregate fruits red, subglobose, 0.9-1.1 cm diam. Fl. Apr-May. Fr. Jun-Jul. Foothills, dense forested

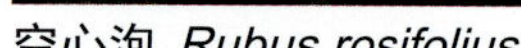

空心泡 *Rubus rosifolius*

大红泡 *Rubus eustephanos*

蓬蘽 *Rubus hirsutus*

slopes, thickets or river banks at 500-2300 m. Distributed in Sichuan, Guizhou, Zhejiang, Hubei and Shaanxi.

蓬蘽
Rubus hirsutus Thunb.

灌木。叶柄长2-3厘米；小叶3-5，顶生小叶叶柄长1厘米，先端渐尖，侧生小叶先端急尖。花序顶生或腋生，常具1花；花梗长2-6厘米。聚合果近球形，直径1-2厘米，无毛。花期4月，果期5-6月。生海拔900-3200米的山坡、路旁、荒地或灌丛。产华南、东南和华中。朝鲜半岛和日本亦有。

Shrubs. Petioles 2-3 cm long; leaflets 3-5, petiolules of terminal leaflets 1 cm long, apex of terminal leaflet acuminate, apex of lateral leaflets acute. Inflorescences terminal or axillary, often 1-flowered; pedicels 2-6 cm long. Aggregate fruits subglobose, 1-2 cm diam, glabrous. Fl. Apr. Fr. May-Jun. Slopes, roadsides, waste places or thickets at 900-3200 m. Distributed in S, SE and C China. Also in Korean Peninsula and Japan.

三叶悬钩子
Rubus delavayi Franch.

灌木。奇数羽状复叶，小叶3，披针形至狭披针形，边缘具不整齐粗锯齿。花单生或2-3成一束；花瓣白色，基部具爪。聚合果橘红色，球形，直径约1厘米。花期5-6月，果期6-7月。生海拔2000-3400米的山坡杂木林。产云南。

Shrubs. Leaves imparipinnate, 3-foliolate, leaflets lanceolate to narrowly lanceolate, margin irregularly coarsely serrate. Flowers solitary or with flowers in clusters of 2 or 3; petals white, base clawed. Aggregate fruits orange-red, globose, ca. 1 cm diam. Fl. May-Jun. Fr. Jun-Jul. Mixed forested slopes at 2000-3400 m. Distributed in Yunnan.

小柱悬钩子
Rubus columellaris Tutcher

灌木。奇数羽状数叶，通常3小叶。花序顶生或腋生，伞房状，具3-7花，在花序基部叶腋间常着生单花；雌蕊数300或更多，无毛。聚合果橙红色或褐黄色，近球形或微呈长圆形，直径5-1.7厘米，无毛。花期4-5月，果期6-7月。生海拔300-2200米的山坡、沟谷、杂木林或路边。产中国西南和华南。越南亦有。

Shrubs. Leaves imparipinnate, usually 3-foliolate. Inflorescences terminal or axillary, corymbose, 3-7-flowered, sometimes flowers solitary in leaf axils at bases; pistils 300 or more, glabrous. Aggregate fruits orange-red or brown-yellow, subglobose or slightly oblong, 1.5-1.7 cm diam, glabrous. Fl. Apr-May. Fr. Jun-Jul. Slopes, valleys, mixed forests or roadsides at 300-2200 m. Distributed in SW and S China. Also in Vietnam.

三叶悬钩子 *Rubus delavayi*

小柱悬钩子 *Rubus columellaris*

陷脉悬钩子 *Rubus impressinervus*

绵果悬钩子 *Rubus lasiostylus*

陷脉悬钩子

Rubus impressinervus F. P. Metcalf

灌木。单叶，脉9-12对，上面下陷；托叶条形，长8-10毫米，全缘。单花通常顶生或腋生；花瓣白色；雄蕊多数，短于花瓣。聚合果褐红色，核有较深洼孔。花期6-7月，果期8-9月。生海拔1300-1500米的山谷、密林下、草地或潮湿的荒地。产中国东南。

Herbs. Leaves simple, veins 9-12 pairs, impressed adaxially; stipules linear, 8-10 mm long, margin entire. Flowers solitary, usually terminal or axillary; petals white; stamens many, shorter than petals. Aggregate fruits brownish red, pyrenes deeply foveolate. Fl. Jun-Jul. Fr. Aug-Sep. Montane valley, dense forests, grasslands or wet waste places at 1300-1500 m. Distributed in SE China.

绵果悬钩子

Rubus lasiostylus Focke

灌木，高可达2米。奇数羽状复叶，具3(-7)小叶，小叶卵形、宽卵形或椭圆形。伞房花序顶生，多花；花瓣红色。聚合果红色，球形，密被长绒毛和柔毛，具宿存花柱。花期5-6月，果期7-8月。生海拔1000-3000米的山坡、灌丛、山谷林中或路边。产云南、四川、湖北和陕西。

Shrubs, to 2 m tall. Leaves imparipinnate; 3(-7)-foliolate, leaflets ovate, broadly ovate or elliptic. Corymbose terminal, several flowered; petals red. Aggregate fruits red, globose, densely long tomentose and soft hairy, with persistent styles. Fl. May-Jun. Fr. Jul-Aug. Slopes, thickets, forests in valleys, forest edges or roadsides at 1000-3000 m. Distributed in Yunnan, Sichuan, Hubei and Shaanxi.

山莓

Rubus corchorifolius L.

直立灌木。小枝圆柱状，具皮刺，被毛，渐无毛。单叶，卵形至卵状披针形。花单生或少数生短侧枝上；花瓣白色或淡粉色。聚合果红色，近球形或卵球形，直径1-1.2厘米，密被柔毛。花期2-4月，果期4-6月。生海拔200-2600米的向阳山坡、河边、山谷、灌丛或荒地。产中国大部分地区。缅甸、越南、朝鲜半岛和日本亦有。

Shrubs erect. Branchlets cylindric, prickly, finely hairy, glabrescent. Leaves simple, ovate to ovate-lanceolate. Inflorescences 1-flowered or few flowers terminal on short lateral branchlets; petals white or pinkish. Aggregate fruits red, subglobose or ovoid, 1-1.2 cm diam, densely finely pubescent. Fl. Feb-Apr. Fr. Apr-Jun. Sunny slopes, streamsides, mountain valleys, thickets or waste places at 200-2600 m. Distributed in most parts of China. Also in Myanmar, Vietnam, Korean Peninsula and Japan.

三花悬钩子

Rubus trianthus Focke

攀援灌木。单叶，两面无毛或微被柔毛，后无毛。花序顶生或腋生，3花簇生或少量花形成短总状花序；花瓣白色；雌蕊10-50。聚合果成熟后红色，近球形，无毛。花期4-5月，果期5-6月。生海拔500-2800米的山坡混交林下、林缘、草坡、路边、沟边或山谷。产中国西南、东南和华中。

山莓 *Rubus corchorifolius*

三花悬钩子 *Rubus trianthus*

Shrubs scandent. Leaves simple, glabrous on both surfaces or slightly pubescent, glabrescent. Inflorescences terminal or axillary, flowers in clusters of 3 or in short few flowered racemes; petals white; pistils 10-50. Aggregate fruits red at maturity, subglobose, glabrous. Fl. Apr-May. Fr. May-Jun. Mixed forests on slopes, forest edges, grassy slopes, roadsides, streamsides or montane valleys at 500-2800 m. Distributed in SW, SE and C China.

牛叠肚
Rubus crataegifolius Bunge

直立灌木。单叶，卵形至狭卵形，边缘3-5掌状浅裂。数花簇生或组成短总状花序；花瓣白色；花直径1-1.5厘米。果近球形，黄色至暗红色，直径约1厘米，无毛，有光泽。花期5-6月，果期7-9月。生海拔300-2500米的阳坡灌丛、林缘、山谷或路边。产华北和东北。俄罗斯东部、朝鲜半岛和日本亦有。

Shrubs erect. Leaves simple, ovate to narrowly ovate, margin palmately 3-5-lobed. Flowers several clustered or arranged to short racemes; petals white; flowers 1-1.5 cm diam. Fruits subglobose, yellow to dark red, ca. 1 cm diam, glabrous, lustrous. Fl. May-Jun. Fr. Jul-Sep. Thickets on sunny slopes, forest edges, valleys or roadsides at 300-2500 m. Distributed in N and NE China. Also in E Russia, Korean Peninsula and Japan.

掌叶覆盆子
Rubus chingii H. H. Hu

藤状灌木。单叶，近圆形，掌状(3-)5-7条脉，边缘掌状(3-)5(-8)深裂。花序顶生短小枝上，单花；花瓣白色。聚合果红色，近球形，直径1.5-2厘米，密被灰色柔毛。花期3-4月，果期5-6月。生海拔500-1000米的山坡、山地常绿阔叶林或针叶林下、灌丛或路边。产中国东南。日本亦有。

Shrubs lianoid. Leaves simple, suborbicular, palmately (3-)5-7-veined, margin usually palmately (3-)5(-8)-lobed. Inflorescences terminal on short branchlets, 1-flowered; petals white. Aggregate fruits red, subglobose, 1.5-2 cm diam, densely gray pubescent. Fl. Mar-Apr. Fr. May-Jun. Slopes, broad-leaved evergreen forests on hills, coniferous forests, thickets or roadsides at 500-1000 m. Distributed in SE China. Also in Japan.

牛叠肚 *Rubus crataegifolius*

掌叶覆盆子 *Rubus chingii*

绢毛悬钩子 *Rubus lineatus*

绢毛悬钩子
Rubus lineatus Reinw.

灌木。小叶羽状脉，侧脉30-50对，下面密被绢毛。花序顶生顶部叶腋，伞房状圆锥花序，具15-20花，有时花簇生叶腋；花瓣白色或绿白色。聚合果成熟时橘红色至红色，半球形或卵球形，幼时具绢毛，渐无毛。花期7-8月，果期9-10月。生海拔1400-3000米的山坡、沟谷、林下、林缘或废弃的耕地中。产云南和西藏。南亚和东南亚亦有。

Shrubs. Leaflets with pinnate veins, 30-50 pairs, densely sericeous abaxially. Inflorescences terminal and in axils of apical leaves, cymose panicles, 15-20-flowered, sometimes flowers in clusters in leaf axils; petals white or greenish white. Aggregate fruits orange to red at maturity, semiglobose or ovoid, sericeous when young, glabrescent. Fl. Jul-Aug. Fr. Sep-Oct. Slopes, valleys, forest understories, forest edges or fallow fields at 1400-3000 m. Distributed in Yunnan and Xizang. Also in S and SE Asia.

高粱泡
Rubus lambertianus Ser.

藤状灌木。小枝疏被短弯刺。单叶，边缘明显3-5裂或波状。顶生花序常聚伞圆锥状，腋生花序近总状；花瓣白色。聚合果成熟时红色，近球形，具多数小核果。花期7-8月，果期9-11月。生海拔200-2500米的山坡、路边、山谷、草地、灌丛、疏林或湿润地。产中国西南、华南、东南、华中、华北、华西和华东。泰国和日本亦有。

Shrubs lianoid. Branchlets with sparse, curved minute prickles. Leaves simple, margin distinctly 3-5-lobed or undulate. Inflorescences terminal ones usually cymose panicles, axillary ones often subracemes; petals white. Aggregate fruits red at maturity, subglobose, with many drupelets. Fl. Jul-Aug. Fr. Sep-Nov. Slopes, roadsides, montane valleys, grasslands, thickets, sparse forests or moist places at 200-2500 m. Distributed in SW, S, SE, C, N, W and E China. Also in Thailand and Japan.

高粱泡 *Rubus lambertianus*

宜昌悬钩子
Rubus ichangensis Hemsl. et Kuntze

藤状灌木。小枝常被疏腺毛和短刺。叶缘具疏锯齿。顶生花序狭聚伞圆锥状，腋生花序近总状，稍短；总花梗和花梗疏被柔毛、腺毛，有时混生短刺；花瓣白色。聚合果红色，近球形，无毛。花期7-8月，果期9-10月。生海拔800-2500米的山坡、弃荒地、山谷密林或灌丛中。产中国西南、华南、华中、华西和华东。

Shrubs scandent. Branchlets usually with sparse, glandular hairs and short prickles. Leaves margin sparsely serrulate. Inflorescences: terminal ones narrow cymose panicles, axillary ones subracemes, shorter; rachis and pedicels sparsely pubescent, with glandular hairs, sometimes intermixed with small prickles; petals white. Aggregate fruits red, subglobose, glabrous. Fl. Jul-Aug. Fr. Sep-Oct. Slopes, fallow fields, dense forests in valleys or thickets at 800-2500 m. Distributed in SW, S, C, W and E China.

宜昌悬钩子 *Rubus ichangensis*

毛萼莓 *Rubus chroosepalus*

灰白毛莓 *Rubus tephrodes*

毛萼莓
Rubus chroosepalus Focke

攀援灌木。单叶，掌状5脉；托叶脱落，离生，披针形，分裂或顶端浅裂。花序顶生，聚伞圆锥状；无花瓣；雌蕊约15，长于雄蕊。聚合果紫褐色或黑色，球形，无毛；核具脉纹。花期5-6月，果期7-8月。生海拔300-2000米的山坡灌丛或林缘。产中国西南、华南和华中。越南亦有。

Shrubs scandent. Leaves simple, palmately 5-veined; stipules caducous, free, lanceolate, undivided or apically lobed. Inflorescences terminal, cymose-panicles; petals absent; pistils ca. 15, longer than stamens. Aggregate fruits purple-brown or black, globose, glabrous; pyrenes rugose. Fl. May-Jun. Fl. Jul-Aug. Thickets on slopes or forest edges at 300-2000 m. Distributed in SW, S and C China. Also in Vietnam.

黄毛悬钩子
Rubus fuscorubens Focke

落叶灌木。单叶狭卵形，掌状5出脉，侧脉4或5对，厚纸质，下面密被灰色绒毛。花序常顶生，聚伞状圆锥花序；总花梗与花梗被黄色绸状柔毛。聚合果黑色，近球形，具少数小核果。花期5-7月，果期7-8月。生海拔400-1200米的山地区域。产湖北。

Shrubs deciduous. Leaves simple, narrowly ovate, palmately 5-veined, lateral veins 4 or 5 pairs, thickly papery, abaxially densely gray tomentose. Inflorescences often terminal, cymose panicles; rachises and pedicels yellowish sericeous-villous. Aggregate fruits black, subglobose, with few drupelets. Fl. May-Jul. Fr. Jul-Aug. Mountainous areas at 400-1200 m. Distributed in Hubei.

灰白毛莓
Rubus tephrodes Hance

藤状灌木。小枝疏被弯曲皮刺。单叶掌状5出脉，边缘具5-7裂片，具不规则锯齿，下面密被灰色绒毛。花序顶生，聚伞圆锥状，多花；总花梗和花梗具疏腺毛和硬毛；花瓣白色。聚合果紫黑色，球形，具多数小核。花期7-8月，果期8-10月。生海拔1500米以下的山区、山脚、山坡、路边、灌丛或山谷中。产中国西南、华南、东南、华中和华东。

Shrubs scandent. Branchlets with sparse, curved prickles. Leaves simple, palmately 5-veined, margin 5-7-lobed, irregularly serrate, abaxially densely gray tomentose. Inflorescences terminal, cymose panicles, many flowered; rachises and pedicels with sparse, glandular hairs and bristles; petals white. Aggregate fruits purplish black, globose, with many drupelets. Fl. Jul-Aug. Fr. Aug-Oct. Mountains, foothills, slopes, roadsides, thickets or valleys below 1500 m. Distributed in SW, S, SE, C and E China.

黄毛悬钩子 *Rubus fuscorubens*

粗叶悬钩子 *Rubus alceifolius*

川莓 *Rubus setchuenensis*

粗叶悬钩子

Rubus alceifolius Poir.

攀援灌木。单叶，腹面疏生长柔毛，脉间有囊泡状小突起，下面密被黄色至锈色绒毛，沿叶脉被长柔毛，边缘具不规则3-7浅裂。花序顶生，狭聚伞圆锥花序或近总状。聚合果近球形，红色。花期7-9月，果期10-11月。生海拔500-2000米的向阳山坡、山谷杂木林内、沼泽灌丛、路旁或岩隙。产中国西南、华南和东南。东南亚和日本亦有。

Shrubs scandent. Leaves simple, abaxially yellowish gray to rust colored tomentose, villous along veins, adaxially sparsely villous and distinctly bullate between veins, margin shallowly 3-7-lobed. Inflorescences terminal, narrow cymose panicles or subraccmes. Aggregate fruits subglobose, red. Fl. Jul-Sep. Fr. Oct-Nov. Sunny slopes, mixed forests in valleys, boggy thickets, roadsides or rock crevices at 500-2000 m. Distributed in SW, S and SE China. Also in SE Asia and Japan.

大乌泡

Rubus pluribracteatus L. T. Lu et Boufford

灌木。单叶，近圆形，下面密具灰黄色或黄色绒毛，沿脉具长柔毛，边缘具7-9浅裂，顶生裂片钝或近截形，不明显3裂，具不均等糙锯齿。花序顶生及腋生；花梗长1-1.5厘米；花直径1.5-2.5厘米；花瓣白色。聚合果红色，球形，直径达2厘米。花期3-6月，果期8-9月。生海拔300-2700米的山坡、河谷、路边、林中或林缘。产云南、贵州、广西和广东。老挝、泰国、越南和柬埔寨亦有。

Shrubs. Leaves simple, suborbicular, abaxially densely yellowish gray or yellow tomentose, villous along veins, margin 7-9-lobed, terminal lobe obtuse or subtruncate, inconspicuously 3-lobed, unevenly coarsely serrate. Inflorescences terminal and axillary; pedicels 1-1.5 cm long; flowers 1.5-2.5 cm diam; petals white. Aggregate fruits red, globose, to 2 cm diam. Fl. Mar-Jun. Fr. Aug-Sep. Slopes, river valleys, roadsides, forests or forest edges at 300-2700 m. Distributed in Yunnan, Guizhou, Guangxi and Guangdong. Also in Laos, Thailand, Vietnam and Cambodia.

川莓

Rubus setchuenensis Bureau et Franch.

灌木。单叶，叶近圆形或宽卵形，直径7-15厘米，掌状5条脉。花序窄聚伞圆锥状；轴和花梗密被黄色绒毛状柔毛。聚合果黑色，半球形，直径约1厘米，无毛，常包被于宿存花萼中。花期7-8月，果期9-10月。生海拔500-3000米山坡、路边、林下或灌丛边。产中国西南至华中。

Shrubs. Leaves simple, blade suborbicular or broadly ovate, 7-15 cm diam, palmately 5-veined. Inflorescences narrowly cymose-panicle, rachis and pedicels densely yellowish-tomentose-villose.

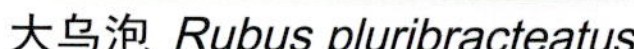

大乌泡 *Rubus pluribracteatus*

寒莓 *Rubus buergeri*

Aggregate fruits black, semiglobose, ca. 1 cm diam, glabrous, often enclosed in persistent calyx. Fl. Jul-Aug. Fr. Sep-Oct. Slopes, roadsides, forest edges or thickets at 500-3000 m. Distributed in SW to C China.

寒莓
Rubus buergeri Miq.

直立或匍匐小灌木。单叶，卵形至近圆形，边缘5-7浅裂。花序顶生及腋生，短近总状，具少花，或数花簇生叶腋。聚合果近球形，紫黑色，直径6-10毫米，无毛。花期7-8月，果期9-10月。生中低海拔的山地阔叶林或混交林中。产中国西南、华中、华南和东南。朝鲜半岛和日本亦有。

Shrubs erect to creeping. Leaves simple, ovate to suborbicular, margin 5-7-lobed. Inflorescences terminal and axillary, short subracemes, few flowered, or flowers several in clusters in leaf axils. Aggregate fruits subglobose, purplish black, 6-10 mm diam, glabrous. Fl. Jul-Aug. Fr. Sep-Oct. Broad-leaved forests or mixed forests in mountainous regions at low to medium elevations. Distributed in SW, C, S and SE China. Also in Korean Peninsula and Japan.

竹叶鸡爪茶
Rubus bambusarum Focke

常绿攀援灌木。小枝稍具弯皮刺。掌状复叶，具3-5小叶，小叶革质，边缘有不明显的稀疏小锯齿。花序近总状；总花梗与花梗具疏短皮刺；花瓣紫红色至红色。聚合果红色至红黑色，近球形。花期5-7月，果期7-8月。生海拔1000-3000米的山上空旷地或林下。产四川、贵州、湖北和陕西。

Shrubs scandent, evergreen. Branchlets with slightly curved prickles. Leaves palmately compound, 3-5-foliolate, leaflets leathery, margin sparsely inconspicuously serrulate. Inflorescences subracemes; rachises and pedicels with sparse, minute prickles; petals purplish red to reddish. Aggregate fruits red to reddish black, subglobose. Fl. May-Jul. Fr. Jul-Aug. Clearings on hills or forests at 1000-3000 m. Distributed in Sichuan, Guizhou, Hubei and Shaanxi.

鸡爪茶
Rubus henryi Hemsl. et Kuntze

常绿攀援灌木。小枝疏生稍弯短皮刺。叶单生，革质，掌状3-5裂，边缘具稀疏细锐锯齿。花序总状，具9-20花；花瓣红色。聚合果黑色，近球形，直径1.3-1.5厘米。花期5-6月，果期7-8月。生海拔2500米以下的山坡、多山区域、山谷、林下或灌丛中。产四川、贵州、湖北和湖南。

Shrubs climbing, evergreen. Branchlets with sparse, slightly curved small prickles. Leaves simple, leathery, palmately 3-5-lobed, margin with sparse, sharp serrations. Inflorescences racemes, 9-20-flowered; petals red. Aggregate fruits black, subglobose, 1.3-1.5 cm diam. Fl. May-Jun. Fr. Jul-Aug. Slopes, mountainous areas, valleys, forests or thickets below 2500 m. Distributed in Sichuan, Guizhou, Hubei and Hunan.

竹叶鸡爪茶 *Rubus bambusarum*

鸡爪茶 *Rubus henryi*

木莓 *Rubus swinhoei*

木莓
Rubus swinhoei Hance

攀援灌木。单叶，叶纸质。总花梗和花梗被绒毛状长柔毛，散被针状皮刺，亦混有紫褐色腺毛，长1-3毫米；花序顶生，短总状，具(1或)5-7花，或数花簇生。聚合果成熟后紫黑色，球形。花期5-6月，果期7-8月。生海拔300-1500米的混交林、灌丛、山坡或山谷。产华南和华中。琉球群岛亦有。

Shrubs scandent. Leaves simple, blades papery. Rachises and pedicels tomentose-villose, with sparse needlelike prickles, with intermixed purplish brown glandular hairs, 1-3 mm long. Inflorescences terminal, short racemes, (1 or)5-7-flowered, or flowers several in clusters. Aggregate fruits purple-black at maturity, globose. Fl. May-Jun. Fr. Jul-Aug. Mixed forests, thickets, slopes or valleys at 300-1500 m. Distributed in S and C China. Also in Ryukyu Islands.

网脉悬钩子
Rubus reticulatus Wall. ex Hook. f.

攀援灌木。单叶，宽卵形或近圆形，下面密被黄灰色绒毛，边缘常明显5浅裂，叶脉5-7对；叶柄长4-9厘米；托叶宽大，近扇形，梳齿状深裂，裂片再分裂。顶生花序为狭长圆锥花序或近总状，腋生花序近总状或团聚成束。果实球形。花期7-8月，果期9-10月。生海拔600-2100米的山沟谷地常绿林下或山坡灌丛中。产西藏。尼泊尔和印度亦有。

Scandent shrubs. Leaves simple, broadly ovate or suborbicular, abaxially densely yellowish gray tomentose, margin usually distinctly 5-lobed, veins 5-7 pairs; petioles 4-9 cm long; stipules broad, subflabellate, pectinately lobed, lobes divided again. Terminal inflorescences narrow panicles or subracemes, axillary ones subracemes or flowers in clusters. Aggregate fruits globose. Fl. Jul-Aug. Fr. Sep-Oct. Evergreen forests in montane valleys, ravines or thickets on slopes at 600-2100 m. Distributed in Xizang. Also in Nepal and India.

大花悬钩子
Rubus wardii Merr.

平卧矮小灌木或半灌木。枝具腺毛、柔毛、稀疏针刺或钩状小皮刺。小叶3，顶生小叶菱状卵形，侧生小叶近圆形或卵圆形，两面脉上有柔毛。花常单生，极稀2-3；花梗长3-4厘米；苞片常宿存；花直径3-4厘米；花瓣绿白色；雌蕊很多。花期6-7月，果期8-9月。生海拔1800-3000米的杂木林中、林缘或山谷石砾地。产云南和西藏。缅甸和印度北部亦有。

Prostrate small shrubs or subshrubs. Branchlets with glandular hairs, soft hairs and sparse, needlelike or small curved prickles. Leaflets 3, terminal ones rhombic-ovate, lateral ones suborbicular or ovate, both surfaces soft hairy along veins. Flowers usually solitary, rarely 2-3; pedicels 3-4 cm long; bracts usually persistent; flowers 3-4 cm diam; petals greenish white; pistils

网脉悬钩子 *Rubus reticulatus*

大花悬钩子 *Rubus wardii*

墨脱悬钩子 *Rubus metoensis*

周毛悬钩子 *Rubus amphidasys*

numerous. Fl. Jun-Jul. Fr. Aug-Sep. Mixed forests, forest margins, valleys, rocky slopes at 1800-3000 m. Distributed in Yunnan and Xizang. Also in Myanmar and N India.

墨脱悬钩子
Rubus metoensis Yü et Lu

小灌木。小叶3，顶生小叶常菱状卵形，侧生小叶卵形或椭圆形，上面疏生平贴柔毛，下面沿叶脉具柔毛和针刺，边缘有不整齐粗钝锯齿，侧脉5-7对；托叶离生，深裂，有柔毛。花常单生于叶腋；花梗长约1厘米，被柔毛和针刺；花直径1-1.5厘米；花萼紫褐色；萼片三角状披针形。花期7-8月。生海拔约2500米的混合林和灌丛。产西藏。

Small shrubs. Leaflet 3, terminal ones rhombic-ovate, lateral ones ovate or elliptic, adaxially sparsely appressed pubescent, abaxially pubescent and with needlelike prickles along veins, margin unevenly coarsely obtusely serrate, lateral veins 5-7 pairs; stipules free, deeply lobed, soft hairy. Flowers axillary, usually 1-flowered in leaf axils; pedicels ca. 1 cm long, pubescent, with needlelike prickles; flowers 1-1.5 cm diam; calyx purplish-brown; sepals triangular-lanceolate. Fl. Jul-Aug. Mixed forests and thickets at ca. 2500 m. Distributed in Xizang.

周毛悬钩子
Rubus amphidasys Focke

蔓性小灌木。小枝棕红色，密被棕红色长腺毛、软刺毛及黄色长毛，常无刺。单叶边缘3-5裂。花序近总状，具5-12花；花瓣白色。聚合果深红色，扁球形，包藏在宿萼内。花期5-7月，果期7-8月。生海拔400-1600米的山坡、路边、灌丛或竹林中。产中国西南、华南、东南、华中和华东。

Shrubs trailing, small. Branchlets reddish brown, with dense reddish brown long stipitate glands, soft bristles and long yellowish hairs, usually unarmed. Leaves simple, margin 3-5-lobed. Inflorescences subracemes, 5-12-flowered; petals white. Aggregate fruits dark red, compressed globose, enclosed in persistent calyx. Fl. May-Jul. Fr. Jul-Aug. Slopes, roadsides, shrubs or bamboo thickets at 400-1600 m. Distributed in SW, S, SE, C and E China.

齿萼悬钩子
Rubus calycinus Wall. ex D. Don

匍匐草本。单叶，托叶卵形；叶卵圆形或近圆形，基部深心形，边缘有浅齿，稀全缘。1或2花常顶生；花萼下面常被软毛和近钻状直皮刺。聚合果球形，深红色。花期5-6月，果期7-8月。生海拔1200-3000米的杂木林中、林缘或山坡。产云南、四川和西藏。印度北部、尼泊尔、不丹、缅甸北部和印度尼西亚(爪哇)亦有。

Herbs creeping. Leaves simple, stipules ovate; leaves orbicular-ovate or suborbicular, base deeply cordate, margin crenate or rarely entire. Inflorescences usually terminal, 1- or 2-flowered; calyx abaxially with soft hairs and straight subulate prickles. Aggregate fruits globose, dark red. Fl. May-Jun. Fr. Jul-Aug. Mixed forests, forest edges or slopes at 1200-3000 m. Distributed in Yunnan, Sichuan and Xizang. Also in N India, Nepal, Bhutan, N Myanmar and Indonesia (Java).

齿萼悬钩子 *Rubus calycinus*

匍匐悬钩子 *Rubus pectinarioides*

北悬钩子 *Rubus arcticus*

匍匐悬钩子
Rubus pectinarioides Hara

匍匐半灌木。茎、叶柄和花梗均被红褐色软刺毛和柔毛。单叶，心状圆卵形或近圆形，两面疏生柔毛；叶柄长2-5厘米；托叶椭圆形，全缘或顶端有锯齿。花单生或2-3，顶生，直径1.5-2.3厘米；萼片卵状披针形，顶端常浅条裂。果实近球形。花期7-8月，果期9-10月。生海拔2800-3300米的山溪边岩石上或石砾坡地林下。产云南和西藏。不丹和印度北部亦有。

Creeping subshrubs. Stems, petioles and pedicels with soft reddish brown bristles and hairs. Leaves simple, orbicular-ovate or suborbicular, both surfaces sparsely pilose; petioles 2-5 cm long; stipules ovate, margin entire or apex serrate. Flowers solitary or 2-3, terminal, 1.5-2.3 cm diam; sepals ovate-lanceolate, apex usually shallowly laciniate. Aggregate fruits subglobose. Fl. Jul-Aug. Fr. Sep-Oct. Rocky streamsides in mountains and forests on rocky slopes at 2800-3300 m. Distributed in Yunnan and Xizang. Also in Bhutan and N India.

北悬钩子
Rubus arcticus L.

草本，高10-30厘米。复叶具3小叶；小叶菱形至倒卵状菱形。花序常顶生，常为单花；花瓣紫红色。聚合果深红色，半球形，具少量小核果，宿存萼片反折；小核近光滑或稍具脉纹。花期6-7月，果期7-8月。生海拔约1200米的山坡、树木或峡谷中。产内蒙古、黑龙江、吉林和辽宁。俄罗斯、蒙古、朝鲜半岛和欧洲北部亦有。

Herbs, 10-30 cm tall. Leaves compound, 3-foliolate; leaflets rhombic to obovate-rhombic. Inflorescences usually terminal, usually 1-flowered; petals purplish red. Aggregate fruits dark red, semiglobose, with few drupelets, persistent sepals reflexed; pyrenes nearly smooth or slightly rugulose. Fl. Jun-Jul. Fr. Jul-Aug. Slopes, forests or ravines at ca. 1200 m. Distributed in Neimenggu, Heilongjiang, Jilin and Liaoning. Also in Russia, Mongolia, Korean Peninsula and N Europe.

凉山悬钩子
Rubus fockeanus Kurz

多年生草本。茎、叶柄和花梗仅具柔毛。叶为3小叶复叶。花序顶生，具1或2花；花瓣白色，倒卵状长圆形至带状长圆形；雌蕊4-20。聚合果红色，球形，无毛。花期5-6月，果期7-8月。生海拔2000-4000米的山坡草地或林中。产云南、四川、西藏和湖北。不丹、印度北部、尼泊尔和缅甸亦有。

Herbs perennial. Stems, petioles and pedicels only pubescent. Leaves compound, 3-foliolate. Inflorescences terminal, 1- or 2-flowered; petals white, obovate-oblong to linear-oblong; pistils 4-20. Aggregate fruits red, globose, glabrous. Fl. May-Jun. Fr. Jul-Aug. Grassy slopes or forests at 2000-4000 m. Distributed in Yunnan, Sichuan, Xizang and Hubei. Also in Bhutan, N India, Nepal and Myanmar.

红刺悬钩子
Rubus rubrisetulosus Card.

多年生草本。茎被细柔毛，常具刺毛或混生腺毛。复叶具3小叶，小叶近圆形；叶柄具细长柔毛，有时具稀疏刺毛和腺毛；托叶顶端梳齿状深裂，稀具3-5锯齿。花梗和花萼均被细柔毛、紫红色刺毛和腺毛。果实球状，红色，具红紫色宿萼。花期6-7月，果期8-9月。生海拔2000-3500米的山地树林、沟谷边或荒野阴湿处。产四川和云南。

凉山悬钩子 *Rubus fockeanus*

红刺悬钩子 *Rubus rubrisetulosus*

Perennial herbs. Stems thinly villous, intermixed bristly or glandular. Leaves compound, 3-foliolate, leaflets suborbicular; petioles thinly villous, sparsely bristly and glandular; stipules apex pectinately lobed, sometimes with 3-5 shallow teeth. Pedicels and calyx villous, with purplish red bristles and glands. Aggregate fruits globose, red, with reddish purple persistent calyx. Fl. Jun-Jul. Fr. Aug-Sep. Montane forests, ravines or wet places in waste fields at 2000-3500 m. Distributed in Sichuan and Yunnan.

东亚仙女木

Dryas octopetala L. var. **asiatica** (Nakai) Nakai

小灌木。叶柄密被白色绒毛和黄褐色分枝长柔毛；叶椭圆形、阔椭圆形或近圆形，近革质。萼筒疏生白色丛卷毛、密被暗紫色分枝长柔毛和亮黄腺毛；花柱被绢毛。瘦果褐色，长卵球形。花果期7-8月。生海拔2200-2800米的高山草甸。产新疆和吉林。俄罗斯东部、朝鲜半岛和日本亦有。

Shrublets. Petioles densely white-tomentose and yellow-brown branched villous; leaves elliptic, broadly so or suborbicular, subleathery. Hypanthium sparsely white floccose, densely dark purple branched villous and light yellow glandular-hairy; styles sericeous. Achenes brown, long ovoid. Fl. and fr. Jul-Aug. Alpine meadows at 2200-2800 m. Distributed in Xinjiang and Jilin. Also in E Russia, Korean Peninsula and Japan.

路边青

Geum aleppicum Jacq.

多年生草本。茎生叶羽状，具2-6小叶，有时复羽状分裂；小叶不等大。花序顶生，疏松；花直径1-1.7厘米；花瓣黄色；雌蕊多数，具长的钩状花柱，成熟时脱落。聚合瘦果倒卵球形。花果期7-10月。生海拔200-3500米的林中、林下开阔地、草坡、河边或田间。产中国西南、华北、西北和东北。北温带亦有。

Herbs perennial. Cauline leaves pinnate, 2-6-foliolate, sometimes repeatedly pinnatifid; leaflets unequal. Inflorescences terminal, lax; flowers 1-1.7 cm diam; petals yellow; pistils numerous, with long hooklike styles, deciduous at fruit maturity. Achenes aggregate obovoid. Fl. and fr. Jul-Oct. Forests, open places in forests, grasslands on slopes, river banks or fields at 200-3500 m. Distributed in SW, N, NW and NE China. Also in the north temperate zone.

东亚仙女木 *Dryas octopetala* var. *asiatica*

路边青 *Geum aleppicum*

柔毛路边青 *Geum japonicum* var. *chinense*

大萼羽叶花 *Acomastylis macrosepala*

柔毛路边青

Geum japonicum Thunb. var. **chinense** F. Bolle

多年生草本。茎生叶通常单叶，不裂或3浅裂。花序顶生，数花疏松排列；花直径1.5-1.8厘米；花瓣黄色；雌蕊多数，具长的钩状花柱，成熟时脱落。聚合瘦果卵球形或椭圆体形。花果期5-10月。生海拔200-2300米的疏林、灌丛、山坡草地、河边或田间。产中国西南、华南、华中、华北和西北。

Herbs perennial. Cauline leaves usually simple or 3-lobed. Inflorescences terminal, laxly several flowered; flowers 1.5-1.8 cm diam; petals yellow; pistils numerous, with long hooklike styles, deciduous at fruit maturity. Achenes aggregate ovoid or ellipsoid. Fl. and fr. May-Oct. Sparse forests, thickets, grasslands on slopes, river banks or fields at 200-2300 m. Distributed in SW, S, C, N and NW China.

大萼羽叶花

Acomastylis macrosepala (Ludlow) Yü et Li

多年生草本。基生叶为大头羽状复叶，有小叶5-10对；茎生叶单叶。花直径2.5-3.5厘米；副萼片卵形，是萼片1/4-1/3，常黄色带紫褐色，外面被硬毛；花瓣黄色，外面被疏柔毛；花柱顶生，上部无毛，下部被长硬毛；花柱短，宿存，直立。瘦果长椭圆形，被长硬毛。花果期8-9月。生海拔3800-4400米的山坡草地或灌丛中。产西藏。印度和不丹亦有。

Perennial herbs. Radical leaves lyrately pinnate, with 5-10 pairs of leaflets; cauline leaves simple. Flowers 2.5-3.5 cm diam; epicalyx segments ovate, 1/4-1/3 as long as sepals, usually purple-brownish yellow, abaxially hirsute; petals yellow, abaxially sparsely pilose; style terminal, upper part glabrous, lower part hirsute; style short, persistent, erect. Achenes long-ellipsoid, hirsute. Fl. and fr. Aug-Sep. Meadow on mountain slopes or thickets at 3800-4400 m. Distributed in Xizang. Also in India and Bhutan.

缘毛太行花

Taihangia rupestris T. T. Yü et C. L. Li var. **ciliata** T. T. Yü et C. L. Li

多年生草本。叶心状卵形，稀三角状卵形，边缘常具细密齿；叶柄被显著疏柔毛。花瓣

缘毛太行花 *Taihangia rupestris* var. *ciliata*

无尾果 *Coluria longifolia*

白色，倒卵状椭圆形，顶端圆钝；花托果期延长到10毫米。瘦果长3-4毫米，被毛。花期5-6月。生海拔1000-1200米的阴坡悬崖。产河北南部。

Herbs perennial. Leaves cordate-ovate, rarely triangular-ovate, margin usually densely serrate; petioles markedly pilose. Petals white, obovate-elliptic, apex rounded; fruiting receptacle slender, elongated to 10 mm long. Achenes 3-4 mm long, with hairs. Fl. May-Jun. Cliffs on shady slopes at 1000-1200 m. Distributed in S Hebei.

无尾果
Coluria longifolia Maxim.

多年生草本。叶间断状羽状分裂，具小叶9-20对，近轴小叶稍小，远轴者稍大。花直径1.5-2.5厘米；花瓣黄色；心皮多数。瘦果黑褐色，长球形，光滑无毛。花期6-7月，果期8-10月。生海拔2700-4600米的高山草甸。产云南、四川、西藏、甘肃和青海。

Herbs perennial. Leaves interrupted pinnate, with 9-20 pairs of leaflets, proximal ones smaller, distal ones larger. Flowers 1.5-2.5 cm diam; petals yellow; carpels numerous. Achenes black-brown, long globose, glabrous. Fl. Jun-Jul. Fr. Aug-Oct. Alpine meadows at 2700-4600 m. Distributed in Yunnan, Sichuan, Xizang, Gansu and Qinghai.

大头叶无尾果
Coluria henryi Batal.

多年生草本。基生叶大头羽状全裂；小叶4-10对，被棕黄色绒毛；茎生叶卵形；花茎自基部斜升，长6-30厘米，具1-4花。花瓣黄色或白色，基部具短爪；心皮多数；子房卵球形。瘦果棕色，卵球形或倒卵球形，有乳头状疣。花期4-6月，果期5-7月。生海拔1600-2400米的岩石。产四川、贵州和湖北。

Herbs perennial. Radical leaves lyrately pinnatisect; leaflets in 4-10 pairs, yellow-brown villous; cauline leaves ovate; flowering stems radical, ascending, 6-30 cm long, 1-4-flowered. Petals yellow or white, base shortly clawed; carpels numerous; ovary ovoid. Achenes brown, ovoid or obovoid, papillate. Fl. Apr-Jun. Fr. May-Jul. Rocks at 1600-2400 m. Distributed in Sichuan, Guizhou and Hubei.

金露梅
Potentilla fruticosa L.

直立灌木。羽状复叶，小叶2(-3)对；小叶长圆形、倒卵状长圆形或卵状披针形，全缘，边缘平坦。花序顶生，疏松总状或伞房状，小，具1至数花；花瓣黄色。瘦果近卵球形，外被长柔毛。花果期5-9月。生海拔400-5000米的山坡草地、砾石坡、灌丛或林缘。产华北、华西、西北和东北。亚洲、欧洲和北美洲亦有。

Shrubs erect. Leaves pinnate, with 2(-3) pairs of leaflets; leaflets oblong, obovate-oblong or ovate-lanceolate, margin entire, flat. Inflorescences terminal, laxly racemose or corymbiform, small, 1- to several flowered; petals yellow. Achenes subovate, abaxially villous. Fl. and fr. May-Sep. Grasslands on slopes, gravelly slopes, thickets or forest edges at 400-5000 m. Distributed in N, W, NW and NE China. Also in Asia, Europe and North America.

大头叶无尾果 *Coluria henryi*

金露梅 *Potentilla fruticosa*

银露梅

Potentilla glabra Lodd.

灌木。羽状复叶具2(或3)对小叶；小叶椭圆形至倒卵状椭圆形，两面疏具柔毛、绢毛或几无毛。单花或数花顶生不同的枝条上；花瓣白色。瘦果具毛。花果期5-11月。生海拔1400-4200米的草地、沟谷、灌丛或疏林中。产华中、华北和华西。俄罗斯、蒙古和朝鲜半岛亦有。

Shrubs. Leaves pinnate with 2(or 3) pairs of leaflets; leaflets elliptic to obovate-elliptic, both surfaces sparsely pilose, sericeous or glabrescent. Flowers terminal on separate branches, 1 to several; petals white. Achenes hairy. Fl. and fr. May-Nov. Grasslands, ravines, thickets or open forests at 1400-4200 m. Distributed in C, N and W China. Also in Russia, Mongolia and Korean Peninsula.

银露梅 *Potentilla glabra*

小叶金露梅

Potentilla parvifolia Fisch. ex Lehm.

矮灌木。羽状复叶，具2或3对小叶，基部两对小叶呈掌状或轮状排列。花序顶生，单花或数花总状排列；花瓣黄色；花柱侧生，顶端粗，呈棍棒状。瘦果平滑。花果期6-8月。生海拔900-5000米的林中、林缘、石隙或开阔草原。产云南、四川、西藏、内蒙古、青海、甘肃和黑龙江。俄罗斯和蒙古亦有。

Shrubs low. Leaves pinnate with 2 or 3 pairs of leaflets, basal 2 pairs usually palmately arranged or appearing whorled. Inflorescences terminal, a solitary flower or a few-flowered raceme; petals yellow; styles lateral, thickened and clavate at apex. Achenes smooth. Fl. and fr. Jun-Aug. Forests, forest edges, rock crevices or steppes at 900-5000 m. Distributed in Yunnan, Sichuan, Xizang, Neimenggu, Qinghai, Gansu and Heilongjiang. Also in Russia and Mongolia.

小叶金露梅 *Potentilla parvifolia*

矮生二裂委陵菜

Potentilla bifurca L. var. **humilior** Ost.-Sack. et Rupr.

多年生草本或矮小亚灌木。小叶常3-8对，顶端全缘，偶2裂；小叶对生，无柄，椭圆形或倒卵形。花茎短于7厘米；花常单生；花瓣黄色。瘦果光滑。花果期5-10月。生海拔1100-1400米的山坡草地、河滩沙地或干旱草原。产华北和华西。俄罗斯和蒙古亦有。

Herbs perennial or low subshrubs. Leaflets usually in 3-8 pairs, apex usually entire, occasionally 2-fid; leaflets opposite, sessile, elliptic or obovate.

矮生二裂委陵菜 *Potentilla bifurca* var. *humilior*

长叶二裂委陵菜 *Potentilla bifurca* var. *major*

楔叶委陵菜 *Potentilla cuneata*

Flowering stems less than 7 cm tall; inflorescences usually a solitary flower; petals yellow. Achenes smooth. Fl. and fr. May-Oct. Grassy mountain slopes, sandy river banks or steppe meadows at 1100-4000 m. Distributed in N and W China. Also in Russia and Mongolia.

长叶二裂委陵菜

Potentilla bifurca L. var. **major** Ledeb.

多年生草本或小灌木。羽状复叶，具3-8对小叶；小叶线形或长椭圆形。聚伞花序，具多花；花茎短于7厘米，下部与叶柄同具伏生柔毛或几无毛；花瓣黄色。瘦果光滑。花果期5-10月。生海拔400-3200米的山坡草地、河岸沙地或田边。产华北、西北和东北。亚洲温带、欧洲中部和东部亦有。

Herbs perennial or subshrubs. Leaves pinnate with 3-8 pairs of leaflets; leaflets linear or long elliptic. Inflorescences cymose, many flowered; flowering stems less than 7 cm tall, lower part together with petioles appressed pilose or glabrescent; petals yellow. Achenes smooth. Fl. and fr. May-Oct. Grassy slopes, sandy river banks or field edges at 400-3200 m. Distributed in N, NW and NE China. Also in other parts of temperate Asia, and C and E Europe.

楔叶委陵菜

Potentilla cuneata Wall. ex Lehm.

矮小亚灌木。叶为3小叶；小叶边缘或顶端具锯齿或浅裂。1-2花，顶生；萼片三角状；副萼片长圆状椭圆形，比萼片稍短；花柱近基生。瘦果稍长于宿萼。花果期6-10月。生海拔2700-3600米的林缘、灌丛、高山草甸或石隙。产云南、四川和西藏。印度北部、克什米尔地区和不丹亦有。

Subshrubs low. Leaves 3-foliolate; leaflets margin or apex serrate or lobed. Inflorescences terminal, 1-2-flowered; sepals triangular; epicalyx oblong-elliptic, slightly shorter than sepals; styles sub-basal. Achenes slightly longer than persistent sepals. Fl. and fr. Jun-Oct. Forest edges, thickets, alpine meadow or rock crevices at 2700-3600 m. Distributed in Yunnan, Sichuan and Xizang. Also in N India, Kashmir and Bhutan.

裂叶毛果委陵菜

Potentilla eriocarpa Wall. ex Lehm. var. **tsarongensis** W. E. Evans

亚灌木。小叶上下两面初时密被白色长柔毛，以后脱落减少，先端2-5裂，裂片宽带形或披针形，顶端渐尖或急尖。花序顶生，1-3花；花黄色；心皮密被扭曲长柔毛。花果期7-10月。生海拔2800-4300米的碎石坡或高山岩缝。产云南、四川和西藏。

Subshrubs. Leaflets on both surfaces usually densely white villous, later glabrescent, apically 2-5-lobed, lobes broad-linear or lanceolate, apex acute or acuminate. Inflorescences terminal, 1-3-flowered; petals yellow; carpels densely twisted villous. Fl. and fr. Jul-Oct. Talus slopes or alpine rock crevices at 2800-4300 m. Distributed in Yunnan, Sichuan and Xizang.

裂叶毛果委陵菜 *Potentilla eriocarpa* var. *tsarongensis*

薄叶皱叶委陵菜 *Potentilla ancistrifolia* var. *dickinsii*

关节委陵菜 *Potentilla articulata*

薄叶皱叶委陵菜

Potentilla ancistrifolia Bunge var. **dickinsii** (Franch. et Sav.) Koidz.

多年生草本。羽状复叶具2-4小叶，有时混生3小叶；小叶无柄，长圆状椭圆形或椭圆状卵形，近革质，两面被稀疏柔毛或脱落几无毛；基生叶有小叶2-3对。成熟瘦果光滑或脉纹不明显。花果期6-9月。生海拔200-2700米的山坡、林中、草甸、沟谷或岩石缝。产河南、河北、山西、陕西、甘肃、安徽和辽宁。朝鲜半岛和日本亦有。

Herbs perennial. Leaves pinnate with 2-4 pairs of leaflets, sometimes also 3-foliolate; leaflets sessile, oblong-elliptic or elliptic-ovate, subleathery, on both surfaces sparsely pilose or glabrescent; basal leaves with 2 or 3 pairs of leaflets. Achenes smooth or inconspicuously rugose at maturity. Fl. and fr. Jun-Sep. Slopes, forests, meadows, ravines or rock crevices at 200-2700 m. Distributed in Henan, Hebei, Shanxi, Shaanxi, Gansu, Anhui and Liaoning. Also in Korean Peninsula and Japan.

关节委陵菜

Potentilla articulata Franch.

多年生垫状草本。花茎丛生。基生叶为3小叶，小叶带状披针形或更狭窄，上面幼时密具长柔毛，边缘全缘而向下反卷，基部有明显或不明显关节。花单生；花瓣黄色。瘦果光滑。花果期6-9月。生海拔4200-4800米的高山雪线附近流石滩。产云南、四川和西藏。

Herbs perennial, pulvinate. Flowering stems tufted. Radical leaves 3-foliolate, leaflets linear-lanceolate or narrower, adaxially densely villous when young, margin entire and revolute, base markedly or inconspicuously articulate. Flower solitary; petals yellow. Achenes smooth. Fl. and fr. Jun-Sep. Alpine debris near snow line at 4200-4800 m. Distributed in Yunnan, Sichuan and Xizang.

多叶委陵菜

Potentilla polyphylla Wall. ex Lehm.

多年生草本。基生叶连叶柄长4-25厘米；小叶倒卵形至宽倒卵形，背面被硬毛。花序聚伞状，少花；花直径1-1.5厘米；花瓣黄色；心皮无毛。瘦果光滑。花果期7-9月。生海拔2500-4500米的林中、林缘或山坡草甸。产云南。印度、不丹、巴基斯坦、斯里兰卡、尼泊尔、缅甸和印度尼西亚亦有。

Herbs perennial. Radical leaves 4-25 cm long including petiole; leaflets obovate to broadly ovate, abaxially strigose. Inflorescences cymose, few flowered; flowers 1-1.5 cm diam; petals yellow; carpels glabrous. Achenes smooth.

多叶委陵菜 *Potentilla polyphylla*

Fl. and fr. Jul-Sep. Forests, forest edges or meadows on mountain slopes at 2500-4500 m. Distributed in Yunnan. Also in India, Nepal, Bhutan, Pakistan, Sri Lanka, Myanmar and Indonesia.

西南委陵菜
Potentilla lineata Treviranus

多年生草本。花茎密被开展长柔毛及短柔毛。基生叶为间断羽状复叶，小叶下面密被白色绢毛及绒毛。伞房状聚伞花序顶生；萼片及副萼片密生长柔毛或白色绢毛；花瓣黄色；花柱近基生，梭状；子房无毛。瘦果光滑。花果期6-10月。生海拔1500-3800米的山坡草地、灌丛和林中。产湖北、四川、贵州、云南和广西。广布南亚和东南亚。

Perennial herbs. Flowering stems densely spreading villous and pubescent. Radical leaves interrupted pinnate, leaflets abaxially densely white sericeous and tomentose. Inflorescence corymbose-cymose, terminal; sepals and epicalyx segments densely villous or white sericeous; petals yellow; style sub-basal, fusiformis; ovary glabrous. Achenes smooth. Fl. and fr. Jun-Oct. Grassy mountain slopes, thickets and forests at 1500-3800 m. Distributed in Hubei, Sichuan, Guizhou, Yunnan and Guangxi. Also widely in S and SE Asia.

银叶委陵菜
Potentilla leuconota D. Don

多年生草本。花茎直立或上升。基生叶常间断羽状复叶，小叶下面密被银白色绢毛；茎生叶1-2。花序集生在顶端，呈假伞形花序；花直径常8毫米；副萼片与萼片近等长；花柱侧生，小枝状，柱头扩大。瘦果光滑，无毛。花果期5-10月。生海拔1300-4600米的林下及山坡草地。产甘肃、湖北、四川、云南、贵州和台湾。不丹、尼泊尔和印度北部亦有。

Perennial herbs. Flowering stems erect or ascending. Radical leaves usually interrupted pinnate, leaflets abaxially densely silvery sericeous; cauline leaves 1-2. Inflorescence compactly terminal, pseudoumbellate; flowers 8 mm in diam; epicalyx segments as long as sepals; style lateral, virguliform, stigma enlarged. Achenes smooth, glabrous. Fl. and fr. May-Oct. Forests, meadows on mountain slopes at 1300-4600 m. Distributed in Gansu, Hubei, Sichuan, Yunnan, Guizhou and Taiwan. Also in Bhutan, Nepal and N India.

西南委陵菜 *Potentilla lineata*

银叶委陵菜 *Potentilla leuconota*

狭叶委陵菜 *Potentilla stenophylla*

多茎委陵菜 *Potentilla multicaulis*

狭叶委陵菜

Potentilla stenophylla (Franch.) Diels

多年生草本。羽状复叶具5-25对小叶；基生叶小叶长圆形，先端通常有3-5齿；茎生叶小叶状，全缘。聚伞花序；花瓣黄色，矩圆形至倒卵形。瘦果表面光滑或有皱纹。花果期6-9月。生海拔3200-5800米的林缘或高山草甸。产云南、四川和西藏。

Herbs perennial. Leaves pinnate with 5-25 pairs of leaflets; leaflets of radical leaves oblong, 3-5 teeth at apex; cauline leaves leafletlike, margin entire. Inflorescences cymose; petals yellow, oblong to obovate. Achenes smooth or rugose. Fl. and fr. Jun-Sep. Forest edges or alpine meadows at

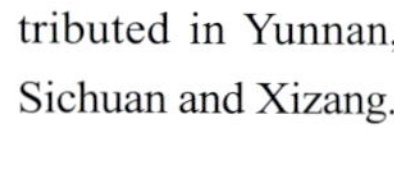

3200-5800 m. Distributed in Yunnan, Sichuan and Xizang.

蕨麻

Potentilla anserina L.

多年生草本。根有时具纺锤状或椭圆状块根。茎斜升、匍匐。间断羽状复叶，小叶5-11对，下面密被紧贴银白色绢毛。花单生；花瓣黄色，倒卵形，先端圆形；花柱侧生。花果期5-8月。生海拔500-4100米的山坡草地、草甸、河岸、湿地或路边。产中国大部分温带地区。世界各温带地区亦有。

Herbs perennial. Roots sometimes with fusiform or ellipsoid tubers. Stems prostrate, creeping. Leaves interrupted pinnate, leaflets 5-11 pairs, abaxially densely appressed silvery sericeous. Flower solitary; petals yellow, obovate, apex rounded; styles lateral. Fl. and fr. May-Aug. Grasslands on slopes, meadows, river banks, wet places or roadsides at 500-4100 m. Distributed in most temperate parts of China. Also in temperate parts of the world.

蕨麻 *Potentilla anserina*

多茎委陵菜

Potentilla multicaulis Bunge

多年生草本。花茎多而密集丛生，上升或铺散，常暗红色，被白色长柔毛或短柔毛。羽状复叶具4-6(-8)对小叶；小叶对生，椭圆形至倒卵形，下面被白色绒毛，脉上疏生白色长柔毛。聚伞花序，多花；花瓣黄色。瘦果卵球形，具皱纹。花果期4-9月。生海拔200-3800米的疏林、草甸、沟谷阴湿处、向阳碎石山坡或田边。产华北、华西、西北和东北。俄罗斯(西伯利亚)和蒙古亦有。

Herbs perennial. Flowering stems many, tufted, ascending or spreading, usually dark reddish, white villous or pubescent. Leaves pinnate with 4-6(-8) pairs of leaflets; leaflets opposite, elliptic to obovate, abaxially white tomentose, sparsely white villous on veins. Inflorescences cymose, many flowered; petals yellow. Achenes ovoid, rugose. Fl. and fr. Apr-Sep. Thinned forests, meadows, shady places in ravines, sunny gravelly slopes or field edges at 200-3800 m. Distributed in N, W, NW and NE China. Also in Russia (Siberia) and Mongolia.

西山委陵菜

Potentilla sischanensis Bunge ex Lehm.

多年生草本。羽状复叶，有3-5(-8)对小叶，近革质，下面密被白色绒毛。聚伞花序疏生；副萼条状披针形；花柱近顶生，基部稍微膨大，柱头稍扩大。瘦果卵球形。花果期4-8月。生海拔200-3600米的灌丛、草地、干旱山坡或黄土丘陵。产中国西南、华北、华西和西北。蒙古亦有。

Herbs perennial. Leaves pinnate with 3-5(-8) pairs of leaflets, subleathery, abaxially densely white tomentose. Inflorescences laxly cymose; epicalyx linear-lanceolate; styles subterminal, base

西山委陵菜 *Potentilla sischanensis*

slightly thickened, stigmas slightly dilated. Achenes ovoid. Fl. and fr. Apr-Aug. Thickets, grasslands, dry mountain slopes or loess hills at 200-3600 m. Distributed in SW, N, W and NW China. Also in Mongolia.

轮叶委陵菜

Potentilla verticillaris Stephan ex Willd.

多年生草本。基生叶3-5，叶边反卷，羽状深裂或掌状深裂几达叶轴；茎生叶掌状3-5裂。聚伞花序疏散少花；花瓣黄色，宽倒卵形；花柱近顶生，基部膨大。瘦果光滑。花果期5-8月。生海拔600-1900米的灌丛、草地、干山坡或河岸沙地。产河北、内蒙古、黑龙江和吉林。俄罗斯(西伯利亚)、蒙古、朝鲜半岛和日本亦有。

Herbs perennial. Radical leaves 3-5, leaflets revolute at margin, pinnately or palmately parted almost to midvein; cauline leaves palmately 3-5-sect. Inflorescences laxly cymose, few flowered; petals yellow, broadly obovate; styles subterminal, base thickened. Achenes smooth. Fl. and fr. May-Aug. Thickets, grasslands, dry mountain slopes or sandy river banks at 600-1900 m. Distributed in Hebei, Neimenggu, Heilongjiang and Jilin. Also in Russia (Siberia), Mongolia, Korean Peninsula and Japan.

委陵菜

Potentilla chinensis Ser.

多年生草本。羽状复叶具5-15对小叶；小叶下面被白色绒毛，沿脉被白色绢状长柔毛，边缘羽状深裂或全裂至中脉；裂片三角卵形、三角状披针形或长圆状披针形。花序为伞房状聚伞花序。瘦果卵球形，深褐色。花果期4-10月。生海拔400-3200米的疏林、林缘、灌丛、山坡草甸或沟谷。产中国大部分温带地区。俄罗斯、蒙古、朝鲜半岛和日本亦有。

Herbs perennial. Leaves pinnate with 5-15 pairs of leaflets; leaflets abaxially white tomentose, white sericeous-villous on veins, pinnatifid or parted to midvein; segments triangular-ovate, triangular-lanceolate or oblong-lanceolate. Inflorescences corymbose-cymose. Achenes ovoid, dark brown. Fl. and fr. Apr-Oct. Sparse forests, forest edges, thickets, meadows on slopes or ravines at 400-3200 m. Distributed in most parts of temperate China. Also in Russia, Mongolia, Korean Peninsula and Japan.

轮叶委陵菜 *Potentilla verticillaris*

委陵菜 *Potentilla chinensis*

细裂委陵菜 *Potentilla chinensis* var. *lineariloba*

大萼委陵菜 *Potentilla conferta*

细裂委陵菜

Potentilla chinensis Ser. var. **lineariloba** Franch. et Sav.

多年生草本。羽状复叶具5-15对小叶；小叶下面被白色绒毛，沿脉被白色绢状长柔毛，叶缘深裂至中脉或几达中脉；狭窄带形。花序为伞房状聚伞花序。瘦果深褐色，卵球形，有明显皱纹。花果期4-10月。生海拔800-1400米的草甸、草地或向阳山坡。产河南、河北、山东、江苏、黑龙江和辽宁。朝鲜半岛和日本亦有。

Herbs perennial. Leaves pinnate with 5-15 pairs of leaflets; leaflets abaxially white tomentose, white sericeous-villous on veins, leaflets parted to midveins or nearly so; segments linear. Inflorescences corymbose-cymose. Achenes dark brown, ovoid, markedly rugose. Fl. and fr. Apr-Oct. Meadows, grasslands or sunny slopes at 800-1400 m. Distributed in Henan, Hebei, Shandong, Jiangsu, Heilongjiang and Liaoning. Also in Korean Peninsula and Japan.

大萼委陵菜

Potentilla conferta Bunge

多年生草本。根木质化。羽状复叶具3-6对小叶；小叶披针形或长圆状椭圆形，下面被灰白色绒毛，羽状中裂或深裂。聚伞花序3至多花；副萼裂片披针形或长圆状披针形，果期显著膨大。瘦果卵状或半球形。花果期6-9月。生海拔3500米以下的灌丛、山坡草甸、峡谷或田边。产中国西南、华北和东北。俄罗斯和蒙古亦有。

Herbs perennial. Roots woody. Leaves pinnate with 3-6 pairs of leaflets; leaflets lanceolate or oblong-elliptic, abaxially canescent tomentose, pinnatifid or pinnately parted. Inflorescences cymose, 3-to many flowered; epicalyx segments lanceolate or oblong-lanceolate, markedly dilated in fruit. Achenes ovoid or hemispheric. Fl. and fr. Jun-Sep. Thickets, meadows on slopes, ravines or field edges below 3500 m. Distributed in SW, N and NE China. Also in Russia and Mongolia.

翻白草

Potentilla discolor Bunge

多年生草本。基生叶具2-4对小叶，背面密被白色或灰白色绵毛。聚伞花序有数花至多花，疏散；副萼片披针形，外面被白色绵毛；花瓣黄色，倒卵形。瘦果近肾形。花果期5-9月。生海拔100-1850米的山谷、沟边山坡、草甸或疏林下。产中国大部分温带地区。朝鲜半岛和日本亦有。

翻白草 *Potentilla discolor*

Herbs perennial. Radical leaves with 2-4 pairs of leaflets, abaxially densely white or gray lanate. Inflorescences cymose, laxly several to many flowered; epicalyx segments lanceolate, abaxially white lanate; petals yellow, obovate. Achenes subreniform. Fl. and fr. May-Sep. Valleys, slopes by ravines, meadows or sparse forests at 100-1850 m. Distributed in most temperate parts of China. Also in Korean Peninsula and Japan.

白萼委陵菜 *Potentilla betonicifolia*

白萼委陵菜

Potentilla betonicifolia Poiret

多年生草本。叶柄初被白色绒毛，以后脱落无毛。基生叶连叶柄长5-12厘米，具3小叶。聚伞花序圆锥状，疏散多花；花瓣黄色，倒卵形；副萼裂片外面被白色绒毛及柔毛。瘦果有脉纹。花果期5-6月。生海拔700-1600米的山坡草地或岩石缝间。产河北、内蒙古、黑龙江、吉林和辽宁。俄罗斯和蒙古亦有。

Herbs perennial. Petioles white tomentose when young, later glabrescent. Radical leaves 5-12 cm long including petioles, 3-foliolate. Inflorescences cymose-paniculate, laxly many flowered; petals yellow, obovate; epicalyx segments abaxially white tomentose and pilose. Achenes rugose. Fl. and fr. May-Jun. Meadows on mountain slopes or rock crevices at 700-1600 m. Distributed in Hebei, Neimenggu, Heilongjiang, Jilin and Liaoning. Also in Russia and Mongolia.

多齿雪白委陵菜

Potentilla nivea L. var. **elongata** Th. Wolf

多年生草本。叶柄被白色绒毛。基生叶连叶柄长1.5-8厘米，具3小叶；小叶每边具(6-)7-14锯齿，顶部圆钝。聚伞花序顶生；花瓣黄色，倒卵形；花柱基部不显著扩大。瘦果光滑。花果期5-9月。生海拔1600-3400米的草坡或岩石缝中。产河北和山西。俄罗斯(贝加尔地区)和蒙古亦有。

Herbs perennial. Petioles white tomentose. Radical leaves 1.5-8 cm long including petioles, 3-foliolate; leaflets margin (6-) 7-14-serrate on each side, apex obtuse. Inflorescences terminal, cymose; petals yellow, obovate; style base inconspicuously thickened. Achenes smooth. Fl. and fr. May-Sep. Grassy slopes or rocks crevices at 1600-3400 m. Distributed in Hebei and Shanxi. Also in Russia (Baikal region) and Mongolia.

多齿雪白委陵菜 *Potentilla nivea* var. *elongata*

窄裂委陵菜 *Potentilla angustiloba*

菊叶委陵菜 *Potentilla tanacetifolia*

窄裂委陵菜
Potentilla angustiloba Yu et Li

多年生草本。花茎铺散或上升。基生叶为五出掌状复叶；小叶倒卵长椭圆形或长圆状椭圆形，边缘深裂至中脉，每边有2-4个带形裂片，下面密被白色绒毛；茎生叶1-3，小叶3-5分裂。伞房状聚伞花序顶生，有花3-12朵；花直径0.8-1厘米；花柱近顶生，基部膨大，柱头稍扩大。花果期6-9月。生海拔2500-3200米的草原、山谷、河滩。产甘肃、青海和新疆。

Perennial herbs. Flowering stems spreading or ascending. Radical leaves palmately 5-foliolate; leaflets obovate-elliptic or oblong-elliptic, margin parted to midvein, 2-4 cestiform lobes each side, abaxially densely white tomentose; cauline leaves 1-3, leaflet margin 3-5-lobed. Inflorescence corymbose-cymose, terminal, 3-12-flowered; flowers 0.8-1 cm diam; style subterminal, base thickened, stigma slightly dilated. Fl. and fr. Jun-Sep. Grasslands, valleys, sandy river banks at 2500-3200 m. Distributed in Gansu, Qinghai and Xinjiang.

丛生荽叶委陵菜
Potentilla coriandrifolia D. Don var. **dumosa** Franch.

多年生草本。茎矮小且丛生。小叶2-4对，下面密被伏生长柔毛或以后脱落几无毛，仅沿中脉被伏生长柔毛。花序顶生，具1(-3)花；花黄色，基部不为紫红色。瘦果无毛。花果期7-12月。生海拔3300-4500米的高山草甸或岩石缝中。产云南、四川和西藏。缅甸亦有。

Herbs perennial. Stems low and tufted. Leaflets 2-4 pairs, abaxially densely appressed villous or glabrescent only along midvein. Inflorescences terminal, 1(-3)-flowered; flowers yellow, not purple-red at base. Achenes smooth. Fl. and fr. Jul-Dec. Alpine meadows or rock crevices at 3300-4500 m. Distributed in Yunnan, Sichuan and Xizang. Also in Myanmar.

菊叶委陵菜
Potentilla tanacetifolia Willd. ex Schltdl.

多年生草本。根强壮，圆柱形。羽状复叶具5-8对小叶，下面被柔毛，沿脉伏生柔毛，或被稀疏腺毛，边缘具缺刻状锯齿。伞房状聚伞花序，多花；花瓣黄色。瘦果卵状，具脉纹。花果期5-10月。生海拔400-2600米的林缘、草地、草甸、黄土高原或砾石地。产华北和东北。俄罗斯和蒙古亦有。

Herbs perennial. Roots robust, terete. Leaves pinnate with 5-8 pairs of leaflets, abaxially pube-

丛生荽叶委陵菜 *Potentilla coriandrifolia* var. *dumosa*

蛇含委陵菜 *Potentilla kleiniana*

scent, appressed pilose on veins, or sparsely glandular hairy, margin incised serrate. Inflorescence corymbose-cymose, many flowered; petals yellow. Achenes ovoid, rugose. Fl. and fr. May-Oct. Forest edges, grasslands, meadows, loess plateau lands or gravels at 400-2600 m. Distributed in N and NE China. Also in Russia and Mongolia.

蛇含委陵菜
Potentilla kleiniana Wight et Arn

一年生草本。根纤细。花茎上升或匍匐。基生叶为近乌足状5小叶，下面疏具长柔毛，全缘。聚伞花序密集枝顶如假伞形；花直径5-10毫米；花瓣黄色。瘦果近球形，一侧扁平，具皱纹。花果期4-9月。生海拔400-3000米的田边、草甸或山坡草地中。产中国除西北以外大部分地区。印度、尼泊尔、不丹、马来西亚、印度尼西亚、朝鲜半岛和日本亦有。

Herbs annual. Roots gracile. Flowering stems prostrate or ascending. Radical leaves subpedately 5-foliolate, abaxially sparsely villous, margin entire. Inflorescences terminal, cymose, congested, pseudoumbellate; flowers 5-10 mm diam; petals yellow. Achenes subglobose, flattened on 1 side, rugose. Fl. and fr. Apr-Sep. Field edges, meadows or grasslands on slopes at 400-3000 m. Distributed in most parts of China, except NW part. Also in India, Nepal, Bhutan, Malaysia, Indonesia, Korean Peninsula and Japan.

朝天委陵菜
Potentilla supina L.

一年生或二年生草本。基生叶羽状复叶，有小叶2-5对，绿色，两面被疏柔毛或近无毛。花序顶生，伞房状聚伞形，花茎下部具腋生花；花直径6-8毫米，花梗长0.8-1.5厘米，密被短柔毛。花果期3-10月。生海拔100-2000米的田边、河岸沙地、山坡湿地或草甸中。分布几遍全国。广布于北半球和亚热带地区。

Herbs annual or biennial. Radical leaves pinnate, leaflets 2-5 pairs, both surfaces green, pilose or glabrescent. Inflorescences terminal, corymbose-cymose, with axillary flowers on lower part of flowering stem; flowers 6-8 mm diam; pedicels 0.8-1.5 cm long, densely pubescent. Fl. and fr. Mar-Oct. By fields, sandy river banks, wet places on slopes or meadows at 100-2000 m. Distributed almost throughout China. Also widely in Northern Hemisphere and subtropical regions.

朝天委陵菜 *Potentilla supina*

蛇莓委陵菜
Potentilla centigrana Maxim.

一年生或二年生草本。花茎上升或匍匐。基生叶3小叶；小叶椭圆形或倒卵形，先端圆形，边缘有缺刻状圆钝或急尖锯齿，两面绿色；茎生叶托叶卵形。花单生，直径4-8毫米；花瓣比萼片短；花柱近顶生，基部膨大。花果期4-8月。生海拔400-2300米的荒地、河岸至林下湿地。产中国东北、内蒙古、陕西、甘肃、四川和云南。俄罗斯、朝鲜半岛和日本亦有。

Herbs annual or biennial. Flowering stems ascending or prostrate. Radical leaves 3-foliolate; leaflets elliptic or obovate, apex rounded, margin obtusely or acutely incised serrate, both surfaces green; cauline leaves stipules ovate. Flowers solitary, 4-8 mm diam; petals shorter than sepals; style subterminal, base thickened. Fl. and fr. Apr-Aug. Fields, riverside to damp places under forests at 400-2300 m. Distributed in NE China, Neimenggu, Shaanxi, Gansu, Sichuan and Yunnan. Also in Russia, Korean Peninsula and Japan.

蛇莓委陵菜 *Potentilla centigrana*

狼牙委陵菜 *Potentilla cryptotaeniae*

狼牙委陵菜

Potentilla cryptotaeniae Maxim.

一年生或二年生草本。花茎直立或斜升，高50-100厘米。基生叶三出复叶，于花期枯萎；茎生叶披针形，草质。伞房状聚伞花序顶生，多花；花瓣黄色，倒卵形。瘦果卵球形，光滑。花果期7-9月。生海拔1000-2500米的林缘、草地、草甸、山谷、峡谷中。产中国西南、华西和东北。俄罗斯(远东地区)、朝鲜半岛和日本亦有。

Herbs annual or biennial. Flowering stems erect or ascending, 50-100 cm tall. Radical leaves 3-foliolate, withered at anthesis; cauline leaves lanceolate, herbaceous. Inflorescences terminal, corymbose-cymose, many flowered; petals yellow, obovate. Achenes ovoid, smooth. Fl. and fr. Jul-Sep. Forest edges, grasslands, meadows, valleys or ravines at 1000-2500 m. Distributed in SW, W and NE China. Also in Russia (Far East), Korean Peninsula and Japan.

星毛委陵菜

Potentilla acaulis L.

多年生草本。叶柄密被星状毛和开展的微硬毛；基生叶连叶柄长15-70厘米；茎生叶具3小叶。聚伞花序顶生，含1-5花；花瓣黄色，倒卵形；花柱近顶生，基部有乳突。瘦果近肾形。花果期4-8月。生海拔600-3000米的山坡草地、黄土山坡或多砾石山坡。产中国东北、华北、华西和西北。俄罗斯和蒙古亦有。

Herbs perennial. Petioles densely stellate hairy and spreading hirtellous; radical leaves 15-70 cm long including petioles; cauline leaves 3-foliolate. Inflorescences terminal, cymose, 1-5-flowered; petals yellow, obovate; styles subterminal, base papillate. Achenes subreniform. Fl. and fr. Apr-Aug. Meadows on mountain slopes, loess slopes or gravelly slopes at 600-3000 m. Distributed in NE, N, W and NW China. Also in Russia and Mongolia.

星毛委陵菜 *Potentilla acaulis*

莓叶委陵菜 *Potentilla fragarioides*

莓叶委陵菜
Potentilla fragarioides L.

多年生草本。羽状复叶，小叶2-3对，两面绿色，两面疏被毛；茎生叶常为3小叶。伞房状聚伞花序顶生，松散，多花；花瓣黄色，倒卵形。成熟瘦果近肾形，表面有脉纹。花期4-6月，果期6-8月。生海拔300-2400米的疏林、草甸、灌丛、田边或沟渠中。产中国大部分地区。俄罗斯(西伯利亚)、蒙古、朝鲜半岛和日本亦有。

Herbs perennial. Leaves pinnate, leaflets 2-3 pairs, green on both surfaces, both surfaces pilose; cauline leaves usually 3-foliolate. Inflorescences terminal, laxly corymbose-cymose, many flowered; petals yellow, obovate. Mature achenes subreniform, rugose. Fl. Apr-Jun. Fr. Jun-Aug. Thinned forests, meadows, thickets, field banks or ditches at 300-2400 m. Distributed in most parts of China. Also in Russia (Siberia), Mongolia, Korean Peninsula and Japan.

三叶委陵菜
Potentilla freyniana Bornm.

多年生草本。根多分枝。基生叶为3小叶，边缘有多数急尖锯齿，两面疏生平铺柔毛。伞房状聚伞花序顶生，疏散，具多花；花瓣浅黄色，长圆状倒卵形。瘦果卵球形，表面有显著皱纹。花果期3-6月。生海拔300-2100米的林中草地和潮湿处或山坡草甸。产中国除西北以外大部分地区。俄罗斯、朝鲜半岛和日本亦有。

Herbs perennial. Roots much branched. Radical leaves 3-foliolate, margin acutely many serrate, both surfaces appressed to spreading pilose. Inflorescences terminal, laxly corymbose-cymose, many flowered; petals pale yellow, oblong-obovate. Achenes ovoid, markedly rugose. Fl. and fr. Mar-Jun. Grassy and damp places in thinned forests or meadows on slopes at 300-2100 m. Distributed in most parts of China, except NW China. Also in Russia, Korean Peninsula and Japan.

绢毛匍匐委陵菜
Potentilla reptans L. var. **sericophylla** Franch.

多年生草本。叶为三出掌状复叶；小叶背面和叶柄被伏生绢状柔毛，侧生小叶边缘浅裂至深裂，有时全缘。单花自叶腋生或与叶对生；副萼裂片果时显著增大；花瓣黄色，阔倒卵形。瘦果黄褐色，卵球形，外面被显著点纹。花果期4-9月。生海拔300-3500米的林缘、溪边灌丛、高山草甸、渠沟边或田边潮湿处。产中国西南、东南、华北、华西和华东。

Herbs perennial. Leaves palmately 3-foliolate; leaflets abaxially or petioles appressed sericeous, margin of lateral leaflets lobed to parted, sometimes entire. Flowers solitary, axillary or opposite leaves; epicalyx segments markedly enlarged in fruit; petals yellow, broadly obovate. Achenes yellow-brown, ovoid, markedly rugose. Fl. and fr. Apr-Sep. Forest edges, thickets by streams, meadows on mountain slopes, ditch banks or damp field margins at 300-3500 m. Distributed in SW, SE, N, W and E China.

三叶委陵菜 *Potentilla freyniana*

绢毛匍匐委陵菜 *Potentilla reptans* var. *sericophylla*

等齿委陵菜 *Potentilla simulatrix*

匐枝委陵菜 *Potentilla flagellaris*

等齿委陵菜

Potentilla simulatrix Th. Wolf

多年生草本。根纤细，多分枝。三出复叶；小叶下面密被平伏柔毛，脉上尤甚，边缘具粗圆齿，有时具深锐齿。单花腋生；花瓣黄色，倒卵形。瘦果有不明显皱纹。花果期4-10月。生海拔300-2200米的林下溪边阴湿处。产河北、山西、内蒙古、陕西、甘肃、青海和四川。

Herbs perennial. Roots slender, much branched. Leaves 3-foliolate; leaflets abaxially somewhat densely appressed pilose, especially on veins, margin coarsely crenate-dentate, sometimes somewhat deeply incised. Flowers solitary, axillary; petals yellow, obovate. Achenes inconspicuously rugose. Fl. and fr. Apr-Oct. Streamsides in damp forests at 300-2200 m. Distributed in Heibei, Shanxi, Neimenggu, Shaanxi, Gansu, Qinghai and Sichuan.

匐枝委陵菜

Potentilla flagellaris Willd. ex Schltdl.

多年生匍匐草本。根丛生，细弱。掌状五出复叶，两面疏被平伏柔毛，边缘有3-6缺刻状大小不等急尖锯齿。单花与叶对生；花瓣黄色。成熟瘦果长圆柱状卵球形。花果期5-9月。生海拔300-2100米的疏林、湿润草甸、湖边或河岸。产河北、山西、甘肃、山东、黑龙江、吉林和辽宁。俄罗斯、蒙古和朝鲜半岛亦有。

Herbs perennial, stoloniferous. Roots tufted, slender. Leaves palmately 5-foliolate, both surfaces sparsely appressed pubescent, margin irregularly acutely incised 3-6-serrate. Flowers solitary, opposite to leaves; petals yellow. Mature achenes cylindric-ovoid. Fl. and fr. May-Sep. Thinned forests, damp meadows, lakes shores or river banks at 300-2100 m. Distributed in Hebei, Shanxi, Gansu, Shandong, Heilongjiang, Jilin and Liaoning. Also in Russia, Mongolia and Korean Peninsula.

沼委陵菜

Comarum palustre L.

多年生草本。花茎斜升，自基部分枝。奇数羽状复叶，小叶5-7。聚伞花序顶生或腋生，有1至数花；萼筒盘形；花瓣暗紫色。瘦果多数，黄褐色，扁卵球形，长约1毫米，无毛。花期5-8月，果期7-10月。生沼泽或泥炭沼泽地。产河北、内蒙古、吉林、辽宁和黑龙江。俄罗斯、蒙古、朝鲜半岛、日本、欧洲和北美洲亦有。

Herbs perennial. Flowering stems ascending, branched near base. Leaves imparipinnate, 5-7-foliolate. Inflorescence terminal or axillary, cymose, 1-to several flowered; hypanthium saucer-shaped; petals dark purple. Achenes numerous, yellow brown, compressed ovoid, ca. 1 mm long, glabrous. Fl. May-Aug. Fr. Jul-Oct. Marshes or bogs. Distributed in Hebei, Neimenggu, Jilin, Liaoning and Heilongjiang. Also in Russia, Mongolia, Korean Peninsula, Japan, Europe and North America.

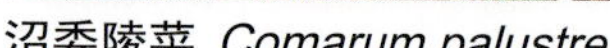

沼委陵菜 *Comarum palustre*

楔叶山莓草 *Sibbaldia cuneata*

四蕊山莓草 *Sibbaldia tetrandra*

楔叶山莓草

Sibbaldia cuneata Hornem. ex Ktze.

多年生草本。茎直立或上升，高5-14厘米，被伏生或斜展疏柔毛。基生叶为三出复叶，小叶广倒卵形至广椭圆形，先端截形，通常有3-5卵形急尖或圆钝锯齿，基部阔楔形。伞房状花序密集顶生；花瓣黄色，倒卵形，与萼片近等长或稍长。花果期5-10月。生海拔3400-4500米的高山草地和岩石缝。产云南、青海、西藏和台湾。俄罗斯和东南亚亦有。

Perennial herbs. Stems erect or ascending, 5-14 cm tall, appressed or subappressed pilose. Radical leaves 3-foliolate, leaflets broadly obovate to elliptic, apex truncate, margin usually 3-5-ovate-acute or dentate teeth, base broadly cuneate. Corymb compact, terminal; petals yellow, obovate, nearly equaling or slightly longer than sepals. Fl. and fr. May-Oct. Alpine meadows and rock crevices at 3400-4500 m. Distributed in Yunnan, Qinghai, Xizang and Taiwan. Also in Russia and SE Asia.

四蕊山莓草

Sibbaldia tetrandra Bunge

多年生草本。基生叶连叶柄长0.5-15厘米，三出复叶；小叶倒卵状长圆形，基部楔形。1(-3)花，常单性；花瓣4，淡黄色，倒卵状长圆形；雄蕊4。瘦果光滑。花果期5-8月。生海拔3000-5400米的森林、山坡草甸或岩石缝中。产西藏、青海和新疆。西南亚、中亚和东北亚亦有。

Herbs perennial. Radical leaves 0.5-15 cm long including petioles, 3-foliolate; leaflets obovate-oblong, base cuneate. Flowers 1(-3), usually unisexual; petals 4, pale yellow, obovate-oblong; stamens 4. Achenes glabrous. Fl. and fr. May-Aug. Forests, meadows on mountain slopes or rock crevices at 3000-5400 m. Distributed in Xizang, Qinghai and Xinjiang. Also in SW, C and NE Asia.

紫花山莓草

Sibbaldia purpurea Royle

多年生草本。基生叶为掌状五出复叶，小叶倒卵形或倒卵状长圆形，基部楔形或阔楔形，顶端通常具2-3齿。花序腋生，多花聚伞状或花单生；花瓣5，紫红色；雄蕊5。瘦果卵球形，紫褐色，无毛。花果期6-8月。生海拔3600-4700米的高山草甸、岩石缝中或雪线附近高山流石滩。产云南、四川、西藏和陕西。印度、尼泊尔和不丹亦有。

Herbs perennial. Radical leaves palmately 5-foliolate; leaflets obovate or obovate-oblong, base cuneate or broadly so, margin usually 2-3-serrate apically. Inflorescences axillary, corymbose and many flowered or a solitary flower; petals 5, purple-red; stamens 5. Achenes ovoid, purple-brown, glabrous. Fl. and fr. Jun-Aug. Alpine meadows, rock crevices or alpine debris near snow line at 3600-4700 m. Distributed in Yunnan, Sichuan, Xizang and Shaanxi. Also in India, Nepal and Bhutan.

紫花山莓草 *Sibbaldia purpurea*

伏毛山莓草 *Sibbaldia adpressa*

绢毛山莓草 *Sibbaldia sericea*

伏毛山莓草

Sibbaldia adpressa Bunge

多年生草本。基生叶连叶柄长1.5-7厘米，羽状复叶，具2对小叶，偶具3小叶，下面被绢状糙毛。总状花序具数花或单花顶生；花瓣黄色或白色，倒卵状椭圆形。瘦果明显具皱纹。花果期5-8月。生海拔600-4200米的山坡、沙生河岸、砾石地或田边。产中国西南、华北、华西、西北和东北。尼泊尔、俄罗斯和蒙古亦有。

Herbs perennial. Radical leaves 1.5-7 cm long including petioles, pinnate, with 2 pairs of leaflets, sometimes 3-foliolate, abaxially sericeous-strigose. Inflorescences cymose and several flowered or a solitary, terminal flower; petals yellow or white, obovate-oblong. Achenes markedly rugose. Fl. and fr. May-Aug. Mountain slopes, sandy river banks, gravels or field edges at 600-4200 m. Distributed in SW, N, W, NW and NE China. Also in Nepal, Russia and Mongolia.

绢毛山莓草

Sibbaldia sericea (Grubov) Soják

多年生草本。花茎丛生。基生叶连叶柄长1-4厘米，羽状复叶，具5小叶或3小叶，两面被平伏绢毛；茎生叶3。1-2花，4或5数；花瓣白色，倒卵形。生海拔600-1200米的山坡或荒漠草原。产内蒙古。蒙古亦有。

Herbs perennial. Flowering stems tufted. Radical leaves 1-4 cm long including petioles, pinnately 5-foliolate or 3-foliolate, both surfaces appressed sericeous; cauline leaves 3. Flowers 1-2, 4- or 5-merous; petals white, obovate. Mountain slopes or desert grasslands at 600-1200 m. Distributed in Neimenggu. Also in Mongolia.

地蔷薇

Chamaerhodos erecta (L.) Bunge

二年生或一年生草本。根木质。花茎单生。叶二回3深裂。聚伞花序顶生；萼片长1-2毫米；花瓣淡粉色或白色，长2-3毫米；心皮10-15，离生。瘦果深褐色，卵球形或圆柱状。花果期6-8月。生海拔约2500米的山坡、丘陵或河岸干旱沙地。产华中、华北、西北和东北。俄罗斯、蒙古和朝鲜半岛亦有。

Herbs biennial or annual. Roots woody. Flowering stems solitary. Leaves 2 times 3-parted. Cymoses terminal; sepals 1-2 mm long; petals pale pink or white, 2-3 mm long; carpels 10-15, free. Achenes dark brown, ovoid or cylindric. Fl. and fr. Jun-Aug. Mountain slopes, hills or dry sandy river banks at ca. 2500 m. Distributed in C, N, NW and NE China. Also in Russia, Mongolia and Korean Peninsula.

地蔷薇 *Chamaerhodos erecta*

三裂地蔷薇 *Chamaerhodos trifida*

三裂地蔷薇

Chamaerhodos trifida Ledeb.

多年生草本。花茎数个，簇生，高5-18厘米。茎生叶3-5裂。圆锥花序二歧分枝，多花；花瓣粉红色，基部渐狭成爪，先端钝，长于萼片；心皮6-10，离生。瘦果圆柱形。花期7月，果期8月。生山坡草甸。产黑龙江。俄罗斯和蒙古亦有。

Herbs perennial. Flowering stems many, tufted, 5-18 cm tall. Stem leaves 3-5-fid. Inflorescences paniculate, dichasially branched, many flowered; petals pink, base tapering into a cuneate claw, apex rounded, longer than sepals; carpels 6-10, free. Achenes cylindric. Fl. Jun. Fr. Aug. Meadows on mountain slopes. Distributed in Heilongjiang. Also in Russia and Mongolia.

野草莓

Fragaria vesca L.

多年生草本。茎与叶柄被开展柔毛。3小叶，稀羽状5小叶。花序聚伞状，有2-4(-5)花；花梗被贴伏柔毛；花瓣白色；心皮多数。聚合果卵球形，红色；瘦果卵球形。花期4-6月，果期6-9月。生林中、山坡或草甸。产云南、四川、贵州、陕西、甘肃、新疆和吉林。北温带亦有。

Herbs perennial. Stems together with petioles spreading hairy. Leaves 3-foliolate, rarely pinnately 5-foliolate. Inflorescences cymose, 2-4(-5)-flowered; pedicels appressed pilose; petals white; carpels numerous. Aggregate fruits ovoid, red; achenes ovoid. Fl. Apr-Jun. Fr. Jun-Sep. Forests, mountain slopes or meadows. Distributed in Yunnan, Sichuan, Guizhou, Shaanxi, Gansu, Xinjiang and Jilin. Also in the north temperate zone.

东方草莓

Fragaria orientalis Losinsk.

多年生草本。茎与叶柄被开展柔毛。小叶3，两面有毛。花序伞房状，具(1或)2-5(或6)花，叶状苞片位于基部。聚合果成熟时紫色，半球形；宿存萼片平展或稍反折；瘦果卵球形。花期5-7月，果期7-9月。生海拔600-4000米的林下、草地或山坡上。产华北和东北。俄罗斯东部、蒙古和朝鲜半岛亦有。

Herbs perennial. Stems together with petioles spreading pilose. Leaflets 3, both surfaces pilose. Inflorescences corymbiform, (1 or)2-5(or 6)-flowered, leafletlike bract at base. Aggregate fruits ripening purple, hemispheric; persistent sepals spreading or slightly reflexed; achenes ovoid. Fl. May-Jul. Fr. Jul-Sep. Under forests, grasslands or slopes at 600-4000 m. Distributed in N and NE China. Also in E Russia, Mongolia and Korean Peninsula.

野草莓 *Fragaria vesca*

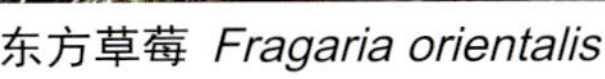

东方草莓 *Fragaria orientalis*

黄毛草莓 *Fragaria nilgerrensis*

西藏草莓 *Fragaria nubicola*

黄毛草莓

Fragaria nilgerrensis Schltdl. ex J. Gay

多年生草本。茎密被黄棕色绢状柔毛。小叶3，下面被黄棕色绢状柔毛。聚伞花序，具(1或)2-5(或6)花；花瓣白色。聚合果成熟时白色、淡白黄色或红色，球形；宿存萼片直立，紧贴聚合果；瘦果卵球形，无毛。花期4-7月，果期6-8月。生海拔700-3000米的山谷森林或山坡草地。产中国西南和华中。印度东北部、尼泊尔和越南北部亦有。

Herbs perennial. Stems densely fulvous sericeous. Leaflets 3, abaxially fulvous sericeous. Inflorescences cymose, (1 or)2-5(or 6)-flowered; petals white. Aggregate fruits ripening white, tinged yellow or red, globose; persistent sepals erect, appressed to aggregate fruit; achenes ovoid, glabrous. Fl. Apr-Jul. Fr. Jun-Aug. Valley forests or grasslands on slopes at 700-3000 m. Distributed in SW and C China. Also in NE India, Nepal and N Vietnam.

草莓

Fragaria × ananassa Duch.

多年生草本。三出复叶；小叶倒卵形或菱形。聚伞花序具5-15花，下面具一有短柄的叶状苞片；花两性，直径1.5-2厘米；副萼果时扩大；花瓣白色；心皮多数。聚合果成熟后红色，直径可达3厘米。花期4-5月，果期6-7月。中国广泛栽培。

Herbs perennial. Leaves 3-foliolate; leaflets obovate or rhombic. Inflorescences cymose, 5-15-flowered, proximally with a shortly petiolate, leafletlike bract; flowers bisexual, 1.5-2 cm diam; epicalyx segments enlarged in fruit; petals white; carpels numerous. Aggregate fruits ripening red, to 3 cm diam. Fl. Apr-May. Fr. Jun-Jul. Cultivated throughout China.

西藏草莓

Fragaria nubicola (Hook. f.) Lindl. ex Lacaita

多年生草本。茎被紧贴白色绢状柔毛。叶为3小叶，小叶顶端圆钝，边缘有缺刻状急尖锯齿；叶柄被白色紧贴绢状柔毛。花序有花1至数朵；副萼片披针形，全缘，稀有齿，比萼片小；花白色；雌蕊多数。聚合果卵球形，宿存萼片紧贴果实。花果期5-8月。生海拔2500-3900米的沟边林下、林缘及山坡草地。产西藏。广布喜马拉雅。

Perennial herbs. Stems appressed white sericeous. Leaves 3-foliolate, leaflets apex obtuse, margin sharply incised serrate; petioles appressed white sericeous. Inflorescence 1-to several flowered; epicalyx segments lanceolate, margin entire, rarely dentate, smaller than sepals; petals white; pistils numerous. Aggregate fruits ovoid, persistent sepals appressed to aggregate fruit. Fl. and fr. May-Aug. Valley forests, forest margins and meadows on mountain slopes at 2500-3900 m. Distributed in Xizang. Also widely in Himalaya.

草莓 *Fragaria × ananassa*

蛇莓

Duchesnea indica (Andrews) Focke

多年生草本。小叶倒卵形至菱状长圆形，长2.5-5厘米。花直径1-2.5厘米；花托果期鲜红色。聚合果成熟时红色，有光泽，直径1-2厘米，海绵质；瘦果新鲜时光亮，卵球形。花期6-8月，果期8-10月。生海拔1800米以下的山坡、草地、河

蛇莓 *Duchesnea indica*

长白蔷薇 *Rosa koreana*

岸或潮湿的地方。产中国大部分地区。亚洲、欧洲和美洲亦有。

Herbs perennial. Leaflets obovate to rhomboid-oblong, 2.5-5 cm long. Flowers 1-2.5 cm diam; receptacles scarlet when fruiting. Aggregate fruits ripening red, shining, 1-2 cm diam, spongy; achenes shining when fresh, ovoid. Fl. Jun-Aug. Fr. Aug-Oct. Slopes, grasslands, riverbanks or wet places below 1800 m. Distributed in most parts of China. Also in Asia, Europe and America.

长白蔷薇

Rosa koreana Kom.

小灌木。小叶7-11(-15)，带边缘腺尖锐锯齿，少部分为重锯齿；托叶大部贴生叶柄。花单生叶腋；花瓣带粉红色；花柱离生，比雄蕊短。蔷薇果橙红色，长圆形，萼片宿存，直立。花期5-6月，果期7-9月。生海拔600-1200米的林缘、灌丛或山坡多石地。产中国东北。朝鲜半岛亦有。

Shrubs small. Leaflets 7-11(-15), margin acutely glandular serrate, partly doubly serrate; stipules mostly adnate to petioles. Flowers solitary, axillary; petals tinged with pinkish; styles free, shorter than stamens. Hips orange-red, oblong, with persistent and erect sepals. Fl. May-Jun. Fr. Jul-Sep. Forest edges, scrubs or rocky places on slopes at 600-1200 m. Distributed in NE China. Also in Korean Peninsula.

黄蔷薇

Rosa hugonis Hemsl.

小灌木。小枝具散生皮刺、针刺和刚毛。托叶大部贴生叶柄；小叶5-13，全缘或具锐锯齿，无毛。花瓣浅黄色。蔷薇果紫红色或棕黑色，扁球形，无毛，光亮。花期5-6月，果期7-8月。生海拔600-2300米的林缘灌丛或开阔山坡。产四川、山西、陕西、甘肃和青海。

Shrubs small. Branches with scattered prickles, smaller prickles and bristles. Stipules mostly adnate to petioles; leaflets 5-13, margin entire or acutely serrate, glabrous. Petals light yellow. Hips purple-red or black-brown, depressed-globose, glabrous, shiny. Fl. May-Jun. Fr. Jul-Aug. Scrubs at forest edges or open slopes at 600-2300 m. Distributed in Sichuan, Shanxi, Shaanxi, Gansu and Qinghai.

黄蔷薇 *Rosa hugonis*

宽刺蔷薇 *Rosa platyacantha*

细梗蔷薇 *Rosa graciliflora*

宽刺蔷薇

Rosa platyacantha Schrenk

小灌木。皮刺多，黄色。小叶5-7(-9)。花单生或2-3簇生叶腋；花瓣5，黄色；萼筒球形或卵球形；花柱离生，比雄蕊短，微伸出。蔷薇果暗红色或紫褐色，球形或卵球形，有光泽；萼片宿存，直立。花期5-8月，果期8-11月。生海拔1100-1800米的林下、林缘、灌丛、溪边、干山坡或荒田。产新疆。哈萨克斯坦和蒙古亦有。

Shrubs small. Prickles abundant, yellow. Leaflets 5-7(-9). Flowers solitary or 2(-3)-fasciculated, axillary; petals 5, yellow; hypanthium globose or ovoid; styles free, shorter than stamens, slightly exserted. Hips dark red or purple-brown, globose or ovoid, shiny; sepals persistent, erect. Fl. May-Aug. Fr. Aug-Nov. Forests, forest edges, scrubs, streamsides, arid slopes or waste fields at 1100-1800 m. Distributed in Xinjiang. Also in Kazakhstan and Mongolia.

黄刺枚

Rosa xanthina Lindl.

直立灌木。刺于部分叶下成对。托叶大部贴生叶柄；小叶7-13。花单生叶腋；萼片披针形，全缘；花瓣5或重瓣，黄色。蔷薇果紫褐色或黑褐色，近球形或倒卵球形。花期4-6月，果期7-8月。生灌丛中或山坡开阔地。产中国东北和华北；也常栽培。

Shrubs erect. Prickles paired below some leaves. Stipules mostly adnate to petioles; leaflets 7-13. Flowers solitary, axillary; sepals lanceolate, margin entire; petals 5 or double, yellow. Hips purple-brown or black-brown, globose or obovoid. Fl. Apr-Jun. Fr. Jul-Aug. Scrubs or open slopes. Distributed in NE and N China; also commonly cultivated.

细梗蔷薇

Rosa graciliflora Rehder et E. H. Wils

直立灌木。叶连同叶柄5-8厘米长，叶轴和叶柄有短皮刺和腺毛。小叶9-11，稀7，卵形或椭圆形。花单生叶腋；花瓣5，粉红色或深红色。蔷薇果红色，倒卵球形或长圆状倒卵球形。花期7-8月，果期9-10月。生海拔3300-4500米的云杉林、林缘灌丛或山坡。产云南、四川和西藏。

Shrubs erect. Leaves including petioles 5-8 cm long, rachises and petioles shortly prickly and glandular-pubescent. Leaflets 9-11, rarely 7, ovate or elliptic. Flowers solitary, axillary; petals 5, pink or deep red. Hip red, obovoid or oblong-obovoid. Fl. Jul-Aug. Fr. Sep-Oct. *Picea* forests, scrubs at forest edges or slopes at 3300-

黄刺枚 *Rosa xanthina*

求江蔷薇 *Rosa taronensis*

峨眉蔷薇 *Rosa omeiensis*

4500 m. Distributed in Yunnan, Sichuan and Xizang.

求江蔷薇
Rosa taronensis T. T. Yu

灌木。小叶7-9(-13)，上部1/3-1/2边缘有锐锯齿，下部全缘，顶端截形。花单生，直径3.5-4厘米；萼筒倒圆锥形；萼片4；花瓣4，淡黄色；花柱离生，微伸出，短于雄蕊。蔷薇果橙黄色，倒圆锥形。生海拔2400-3300米的混交林中或草地。产云南西北部。

Shrubs. Leaflets 7-9(-13), margin acutely serrate at upper 1/3-1/2 part, entire at lower part, apex truncate. Flowers solitary, 3.5-4 cm diam; hypanthium obconic; sepals 4; petals 4, yellowish; styles free, slightly exserted, shorter than stamens. Hips orange-yellow, obconic. Mixed forests or grassy places at 2400-3300 m. Distributed in NW Yunnan.

峨眉蔷薇
Rosa omeiensis Rolfe f.

直立灌木。托叶大部贴生叶柄；小叶(5-)9-13(-17)，背面被柔毛或近无毛，无腺毛；边缘具锐锯齿。单花腋生；花瓣4，白色。蔷薇果深红色或黄色，倒卵球形或梨形。花期5-6月，果期7-9月。生海拔700-4000米的山坡、疏林或灌丛中。产中国西南、华中、华北和西北。

Shrubs erect. Stipules mostly adnate to petiole; leaflets (5-)9-13(-17), abaxially pubescent or subglabrous and non-glandular, margin acutely serrate. Flower solitary, axillary; petals 4, white. Hips deep red or yellow, obovoid or pyriform. Fl. May-Jun. Fr. Jul-Sep. Slopes, open forests or thickets at 700-4000 m. Distributed in SW, C, N and NW China.

绢毛蔷薇
Rosa sericea Lindl.

直立灌木。托叶大部贴生叶柄；小叶(5-)7-11(-13)，背面被绢状长柔毛。花单生叶腋；花瓣4，白色；花柱离生，短于雄蕊。蔷薇果红色或紫红色，倒卵球形或球形。花期5-6月，果期6-8月。生海拔2000-4400米的疏林、林缘、灌丛、山谷、山坡、山顶或向阳干燥的地方。产云南、四川、西藏和贵州。印度、不丹和缅甸亦有。

Shrubs erect. Stipules mostly adnate to petioles; leaflets (5-)7-11(-13), abaxially sericeous-villous. Flowers solitary, axially; petals 4, white; styles free, shorter than stamens. Hips red or purple-red, obovoid or globose. Fl. May-Jun. Fr. Jun-Aug. Sparse forests, forest edges, scrubs, valleys, slopes, mountain summits or dry sunny places at 2000-4400 m. Distributed in Yunnan, Sichuan, Xizang and Guizhou. Also in India, Bhutan and Myanmar.

绢毛蔷薇 *Rosa sericea*

川西蔷薇 *Rosa sikangensis*

川西蔷薇

Rosa sikangensis T. T. Yu et T. C. Ku

小灌木。托叶宽，大部贴生叶柄；小叶7-9(-13)，长圆形至倒卵形，背面被毛和腺体，边缘具细密重锯齿。单花腋生；花瓣4，白色。蔷薇果红色，近球形，直径约1厘米。花期5-6月，果期7-9月。生海拔2900-4200米的灌丛、河岸或路边。产云南、四川和西藏。

Shrubs small. Stipules broad, mostly adnate to petiole; leaflets 7-9(-13), oblong to obovate, abaxially pubescent and glandular, margin densely doubly serrate. Flower solitary, axillary; petals 4, white. Hips red, subglobose, ca. 1 cm diam. Fl. May-Jun. Fr. Jul-Sep. Scrubs, riversides or roadsides at 2900-4200 m. Distributed in Yunnan, Sichuan and Xizang.

弯刺蔷薇

Rosa beggeriana Schrenk

灌木。淡黄色弯皮刺。小叶5-9。数花至多数组成聚伞状或圆锥状；苞片1-3(-4)；萼筒近球形；花瓣5，白色。蔷薇果红色转为黑紫色，无毛，成熟后萼筒顶部和萼片一起脱落。花期5-7月，果期7-10月。生海拔900-2000米的山坡、山谷、河边或路边。产新疆和甘肃。阿富汗、哈萨克斯坦和蒙古亦有。

Shrubs. Prickles yellowish, hooked. Leaflets 5-9. Flowers several or numerous in corymb or panicle; bracts 1-3(-4); hypanthium subglobose; petals 5, white. Hips red, becoming black-purple, glabrous, after ripening apical part of hypanthium and sepals deciduous together. Fl. May-Jul. Fr. Jul-Oct. Slopes, valleys, riversides or roadsides at 900-2000 m. Distributed in Xinjiang and Gansu. Also in Afghanistan, Kazakhstan and Mongolia.

尾萼蔷薇

Rosa caudata Baker

灌木。小枝圆柱状，光滑，散生皮刺。小叶7-9，无毛或下面沿脉疏被柔毛。伞房花序花多数；萼片叶状；花瓣5，红色，背面无毛。果橙红色，矩圆状；萼片宿存，直立。花期6-7月，果期7-11月。生海拔1200-2500米的灌丛或山坡。产陕西、湖北和四川。

Shrubs. Branchlets terete, glabrous, with scattered prickles. Leaflets 7-9, glabrous or sparsely puberulous abaxially along veins. Corymbs many flowered; sepals leaflike; petals 5, red, abaxially glabrous. Fruits orange-red, oblong; sepals persistent, erect. Fl. Jun-Jul. Fr. Jul-Nov. Scrubs or slopes at 1200-2500 m. Distributed in Shaanxi, Hubei and Sichuan.

玫瑰

Rosa rugosa Thunb.

直立灌木。小枝密被绒毛。小叶5-9，椭圆形或椭圆状倒卵形，背面有毛。花单生或数花簇生；萼片常叶状；花瓣5，重瓣或半重瓣，紫红色、深粉色或白色。蔷薇果深红色，扁球形。花期5-6月，果期8-9月。生海拔100米以下的海岸丘陵、沙地或岸边岛屿。原产山东、吉林和辽宁，广泛栽培于中国其他地区。俄罗斯(远东地区)、朝鲜半岛和日本亦有；广泛栽培世界各地。

Shrubs erect. Branchlets tomentose. Leaflets 5-9, elliptic or elliptic-obovate, abaxially tomentose. Flower solitary or several and fasciculate; sepals often leaflike; petals 5, double or semi-

弯刺蔷薇 *Rosa beggeriana*

尾萼蔷薇 *Rosa caudata*

玫瑰 *Rosa rugosa*

double, purple-red, dark pink or white. Hips dark red, depressed-globose. Fl. May-Jun. Fr. Aug-Sep. Coastal hillsides, sandy soil or offshore islands below 100 m. Native to Shandong, Jilin and Liaoning; widely cultivated elsewhere in China. Also in Russia (Far East), Korean Peninsula and Japan; widely cultivated in the world.

山刺玫
Rosa davurica Pall.

直立灌木。皮刺散生。小叶7-9，背面有腺点，边缘具单锯齿及重细齿。花单生，或2-3花簇生腋处；萼筒近球形；花瓣5，粉色；花柱离生，比雄蕊短。蔷薇果红色，具颈部。花期6-7月，果期7-9月。生海拔400-2500米的林缘向阳地、丘陵草地或山坡。产中国东北和华北。俄罗斯(东西伯利亚)、蒙古南部、朝鲜半岛和日本亦有。

Shrubs erect. Branches sparsely prickly. Leaflets 7-9, abaxially glandular punctate, margin simple and doubly serrate. Flowers solitary, or 2-3 fasciculate, axillary; hypanthium subglobose; petals 5, pink; styles free, shorter than stamens. Hips red, with distinct neck. Fl. Jun-Jul. Fr. Jul-Sep. Sunny places at forest edges, grassy places on hills or slopes at 400-2500 m. Distributed in NE and N China. Also in Russia (E Siberia), S Mongolia, Korean Peninsula and Japan.

刺蔷薇
Rosa acicularis Lindl.

灌木。皮刺散生或较密。托叶大部贴生叶柄；小叶3-7，下面具柔毛。花单生，或2-3集生；花瓣5，粉色，芳香；花柱离生，短于雄蕊。蔷薇果红色，梨形，具有明显的颈部。花期6-7月，果期7-9月。生海拔400-1800米的桦木林下、矮灌丛、向阳山坡或路边。产中国西北、华北和东北。北亚、东北亚、欧洲北部和北美洲亦有。

Shrubs. Prickles sparse or dense. Stipules mostly adnate to petiole; leaflets 3-7, abaxially pubescent. Flowers solitary, or 2-3 and fasciculate; petals 5, pink, fragrant; styles free, shorter than stamens. Hips red, pyriform, with a distinct neck. Fl. Jun-Jul. Fr. Jul-Sep. *Betula* forests, brushland, sunny slopes or roadsides at 400-1800 m. Distributed in NW, N and NE China. Also in N and NE Asia, N Europe and North America.

山刺玫 *Rosa davurica*

刺蔷薇 *Rosa acicularis*

大红蔷薇 *Rosa saturata*

大红蔷薇
Rosa saturata Baker

灌木。叶连叶柄长7-16厘米；托叶约2/3部分贴生叶柄；总花梗与叶柄疏具刺；小叶7(-9)，边缘具单锯齿或部分具重锯齿。花常单生，直径3.5-5厘米；花瓣5，红色。果深红色，椭圆体形，具宿存直立的萼片。花期6月，果期7-10月。生海拔2200-2400米的灌丛或溪边。产四川、浙江和湖北。

Shrubs. Leaves including petioles 7-16 cm long; stipules adnate to petioles to 2/3 part; rachises and petioles sparsely small prickly; leaflets 7(-9), margin simply serrate or partly doubly serrate. Flowers usually solitary, 3.5-5 cm diam; petals 5, red. Hips deep red, ovoid, with persistent, erect sepals. Fl. Jun. Fr. Jul-Oct. Scrubs or streamsides at 2200-2400 m. Distributed in Sichuan, Zhejiang and Hubei.

美蔷薇
Rosa bella Rehder et E. H. Wils

灌木。小枝圆柱状，散生皮刺。小叶7-9，无毛或沿脉下面疏被柔毛和腺状柔毛，边缘具单锯齿。花单生或2-3簇生；萼片叶状，下部具带柄腺体；花瓣5，粉红色。果深红色，椭球形。花期5-7月，果期8-10月。生海拔约1700米的灌丛、山脚或溪边。产河南、河北、山西、内蒙古和吉林。

Shrubs. Branchlets terete, with scattered prickles. Leaflets 7-9, glabrous or abaxially along veins sparsely pubescent and glandular-pubescent, margin simply serrate. Flowers solitary or 2-3-clustered; sepals leaflike, abaxially stipitate glandular; petals 5, pink. Hips deep red, ellipsoid-ovoid. Fl. May-Jul. Fr. Aug-Oct. Thickets, bases of mountains or streamsides at ca. 1700 m. Distributed in Henan, Hebei, Shanxi, Neimenggu and Jilin.

华西蔷薇
Rosa moyesii Hemsl. et E. H. Wils

灌木。叶轴和小叶柄被短柔毛和腺毛，并散生皮刺。叶连同叶柄长7-13厘米；小叶7-13。花单生或2-3花簇生；花梗长1-3厘米，密被有柄腺点，稀光滑；花瓣5，深红色。蔷薇果紫红色或橘红色，卵球形。花期

美蔷薇 *Rosa bella*

华西蔷薇 *Rosa moyesii*

6-7月，果期8-10月。生海拔2700-3800米的灌丛或山坡。产云南、四川和陕西。

Shrubs. Rachises and petioles puberulous, glandular-pubescent, sparsely small prickly. Leaves including petioles 7-13 cm long; leaflets 7-13. Flowers solitary or 2-3 fasciculate; pedicels 1-3 cm long, usually densely stipitate glandular, rarely glabrous; petals 5, deep red. Hips purple-red or orange-red, ovoid. Fl. Jun-Jul. Fr. Aug-Oct. Scrubs or slopes at 2700-3800 m. Distributed in Yunnan, Sichuan and Shaanxi.

藏边蔷薇

Rosa webbiana Wall. ex Royle

灌木。小叶5-9，近圆形、倒卵形或宽椭圆形，长6-20毫米，上半部有单锯齿，近基部全缘，上面无毛，下面无毛或沿脉微被短柔毛。花单生，稀2-3；花梗长1-1.5厘米；花瓣淡红色或玫瑰色。果近球形或卵球形，萼片宿存开展。花果期6-9月。生海拔2000-4500米的山坡、林地、灌丛、河谷或田边。产西藏。中亚、印度北部和克什米尔地区亦有。

Shrubs. Leaflets 5-9, suborbicular, obovate or broadly elliptic, 6-20 mm long, margin serrate at upper part, near base entire, adaxially glabrous, abaxially glabrous or sparsely puberulous along veins. Flowers solitary, rarely 2-3; pedicels 1-1.5 cm long; petals pale red or rose. Hips subglobose or ovoid, with persistent, spreading sepals. Fl. and fr. Jun-Sep. Slopes, forests, scrubs, valleys or farmlands at 2000-4500 m. Distributed in Xizang. Also in C Asia, N India and Kashmir.

腺果蔷薇

Rosa fedtschenkoana Regel

灌木。托叶大部贴生叶柄；小叶通常7，稀5或9，近圆形或卵形，无毛，革质。花单生，有时2-4花簇生；花柱离生，短于雄蕊，被柔毛。蔷薇果深红色，密被腺柔毛，萼片宿存。花期7-8月，果期8-10月。生海拔2400-2700米的灌丛、山坡或山谷溪边。产新疆。哈萨克斯坦亦有。

藏边蔷薇 *Rosa webbiana*

Shrubs. Stipules mostly adnate to petioles; leaflets usually 7, rarely 5 or 9, suborbicular or ovate, glabrous, leathery. Flowers solitary, sometimes 2-4 fasciculate; styles free, shorter than stamens, pubescent. Hips deep red, densely glandular-pubescent, with persistent sepals. Fl. Jul-Aug. Fr. Aug-Oct. Scrubs, slopes or streamsides in valleys at 2400-2700 m. Distributed in Xinjiang. Also in Kazakhstan.

腺果蔷薇 *Rosa fedtschenkoana*

月季花 *Rosa chinensis*

月季花

Rosa chinensis Jacq.

灌木。小枝棕红色。小叶3-5，背面暗绿色，两面近无毛。花半重瓣或重瓣，4或5花簇生，稀单生；花瓣5，红色、粉色或白色；萼片全缘或少量羽状裂片。蔷薇果红色，卵球形或梨形。花期4-9月，果期6-11月。原产中国，各地普遍栽培。

Shrubs. Branchlets purple-brown. Leaflets 3-5, abaxially dark green, both surfaces subglabrous. Flowers semi-double or double, 4 or 5 fasciculate, rarely solitary; petals 5, red, pink or white; sepals margin entire or few pinnately lobed. Hips red, ovoid or pyriform. Fl. Apr-Sep. Fr. Jun-Nov. Native to China, widely cultivated.

桔黄香水月季

Rosa odorata (Andrews) Sweet var. **pseudoindica** (Lindl.) Rehder

灌木。叶连同叶柄长5-10厘米；小叶5-9，两面无毛。花重瓣，直径约8厘米，单生或2-3花簇生，极芳香；花瓣黄色或橙色；花柱离生，伸出，与雄蕊近等长，被毛。蔷薇果红色，扁球形，无毛。栽培于云南西北部。

Shrubs. Leaves including petiole 5-10 cm long; leaflets 5-9, both surfaces glabrous. Flowers double, ca. 8 cm diam, solitary or 2-3 fasciculate, very fragrant; petals yellow or orange; styles free, exserted, nearly equaling stamens, pubescent. Hips red, depressed-globose, glabrous. Cultivated in NW Yunnan.

桔黄香水月季 *Rosa odorata* var. *pseudoindica*

亮叶月季

Rosa lucidissima Levl.

常绿或半常绿攀援灌木。小枝有基部压扁的弯曲皮刺，有时密被刺毛。小叶常3，极稀5；托叶大部贴生，仅顶端分离，游离部分披针形，边缘有腺。花单生，直径3-3.5厘米；萼片全缘或稍有缺刻，花后反折；花瓣紫红色。果实梨形或倒卵球形。花期4-6月，果期5-8月。生海拔400-1400米的杂木林或灌丛中。产湖北、四川和贵州。

Evergreen or semi-evergreen climbing shrubs. Branchlets with curved, flat prickles gradually tapering to base, sometimes dense bristles. Leaflets usually 3, rarely 5; stipules mostly adnate to

亮叶月季 *Rosa lucidissima*

petiole, apex free only, free parts lanceolate, margin glandular. Flowers solitary, 3-3.5 cm diam; sepals margin entire or slightly incised, reflexed after anthesis; petals purple-red. Fruits pyriform or obovoid. Fl. Apr-Jun. Fr. May-Aug. Mixed forests or scrubs at 400-1400 m. Distributed in Hubei, Sichuan and Guizhou.

野蔷薇

Rosa multiflora Thunb.

攀援灌木。小枝常光滑。托叶篦齿状；小叶(3-)5-9，背面被柔毛。花多数排成伞房状，直径1.5-2厘米；花瓣白色，芳香，半重瓣或重瓣。蔷薇果红褐色或紫褐色，近球形。花期4-6月，果期7-10月。广泛栽培。产河南、江苏和山东。朝鲜半岛和日本亦有。

Shrubs climbing. Branchlets usually glabrous. Stipules pectinate; leaflets (3-)5-9, abaxially pubescent. Flowers numerous in corymb, 1.5-2 cm diam; petals white, fragrant, semi-double or double. Hips red-brown or purple-brown, subglobose, Fl. Apr-Jun. Fr. Jul-Oct. Widely planted. Distributed in Henan, Jiangsu and Shandong. Also in Korean Peninsula and Japan.

七姊妹

Rosa multiflora Thunb. var. **carnea** Thory

本变种与野蔷薇的区别在于本变种的花重瓣，粉红色。花期4-6月，果期7-10月。广泛栽培于中国。

This variety differs from the typical variety in its petals double, pink. Fl. Apr-Jun. Fr. Jul-Oct. Widely cultivated in China.

粉团蔷薇

Rosa multiflora Thunb. var. **cathayensis** Rehder et E. H. Wils

本变种与野蔷薇的区别在于本变种的花瓣粉红色，花直径达4厘米。生海拔300-2000米的山坡、灌丛或河边。产中国西南、华南、华中、华北、华西和华东。

This variety differs from the typical variety in its petals pink, flowers to 4 cm diam. Slopes, thickets or riversides at 300-2000 m. Distributed in SW, S, C, N, W and E China.

野蔷薇 *Rosa multiflora*

七姊妹 *Rosa multiflora* var. *carnea*

粉团蔷薇 *Rosa multiflora* var. *cathayensis*

伞花蔷薇 *Rosa maximowicziana*

伞花蔷薇
Rosa maximowicziana Regel

小灌木，具长匍枝。叶连同叶柄长4-11厘米；托叶大部贴生叶柄，离生部分披针形；小叶7-9，卵形、椭圆形或长圆形。数花成伞房状排列；花瓣白色或带粉红色。果实卵球形，棕黑色，有光泽。花期6-7月，果期9月。生灌丛、阳坡、溪旁或路旁。产山东和辽宁。俄罗斯(远东地区)和朝鲜半岛亦有。

Shrubs small, with long repent branches. Leaves including petioles 4-11 cm long; stipules mostly adnate to petioles, free parts lanceolate; leaflets 7-9, ovate, elliptic or oblong. Flowers several in corymb; petals white or tinged with pink. Hips ovoid, black-brown, shiny. Fl. Jun-Jul. Fr. Sep. Scrubs, open slopes, streamsides or roadsides. Distributed in Shandong and Liaoning. Also in Russia (Far East) and Korean Peninsula.

绣球蔷薇
Rosa glomerata Rehd. et Wils.

铺散灌木。小叶5-7，稀3或9，长圆形或长圆倒卵形，上面有显明褶皱，下面叶脉显明突起，密被长柔毛；托叶膜质，大部贴生于叶柄，离生部分耳状，全缘，有腺毛。伞房花序，密集多花，直径4-10厘米；花柱结合成束，伸出，比雄蕊稍长。花期7月，果期8-10月。生海拔1300-3000米的山坡、林缘和灌木丛中。产湖北、四川、云南和贵州。

Diffuse shrubs. Leaflets 5-7, rarely 3 or 9, oblong or oblong-obovate, adaxially conspicuously rugose, abaxially with prominent veins, densely villous; stipules membranous, mostly adnate to petiole, free parts auriculate, margin entire, glandular-pubescent. Corymbs densely numerous flowered, 4-10 cm diam; styles connate into column, exserted, slightly longer than stamens. Fl. Jul. Fr. Aug-Oct. Slopes, forest margins and thickets at 1300-3000 m. Distributed in Hubei, Sichuan, Yunnan and Guizhou.

悬钩子蔷薇 *Rosa rubus*

悬钩子蔷薇
Rosa rubus H. Lévl. et Vaniot

匍匐或攀援灌木。小叶(3-)5，背面被柔毛。10-25花组成圆锥状伞房花序，直径2.5-3厘米；花瓣5，白色，芳香；花柱合生成柱，比雄蕊稍长，被柔毛。蔷薇果亮红色、紫褐色或橙褐色，近球形。花期4-6月，果期

绣球蔷薇 *Rosa glomerata*

卵果蔷薇 *Rosa helenae*

长尖叶蔷薇 *Rosa longicuspis*

7-9月。生海拔500-1300米的灌丛、草地、山坡、山地、河边或路边。产中国西南、华南、东南、华西、华中和华东。

Shrubs creeping or scandent. Leaflets (3-)5, abaxially pubescent. Flowers 10-25 in a paniculate corymb, 2.5-3 cm diam; petals 5, white, fragrant, styles connate into column, slightly longer than stamens, pubescent. Hips bright-red, purple-brown or orange brown, subglobose. Fl. Apr-Jun. Fr. Jul-Sep. Scrubs, grassy places, slopes, montane regions, river banks or roadsides at 500-1300 m. Distributed in SW, S, SE, W, C and E China.

卵果蔷薇

Rosa helenae Rehder et E. H. Wils

铺散或攀援灌木。小叶5-9，连同叶柄长8-17厘米，叶缘有紧贴锐锯齿；叶柄有柔毛和小皮刺。顶生伞房花序，部分密集近伞形；花瓣5，芳香，白色；花柱合生成束。蔷薇果深红色，卵球形、椭圆体形或倒卵球形。花期5-7月，果期9-10月。生海拔1000-3000米的林缘、灌丛、山坡或溪边。产云南、四川、贵州、湖北、陕西和甘肃。泰国和越南亦有。

Shrubs diffuse or scandent. Leaflets 5-9, including petioles 8-17 cm long, margin appressed-serrate; petioles pubescent, with scattered small prickles. Terminal corymbs, partly umbellike, dense; petals 5, fragrant, white; styles connate into column. Hips deep red, ovoid, ellipsoid or obovoid, Fl. May-Jul. Fr. Sep-Oct. Forest edges, thickets, slopes or streamsides at 1000-3000 m. Distributed in Yunnan, Sichuan, Guizhou, Hubei, Shaanxi and Gansu. Also in Thailand and Vietnam.

长尖叶蔷薇

Rosa longicuspis Bertol.

常绿灌木。小叶(5-)7-9，革质，先端渐尖或长渐尖。花多数，排成伞房状，直径3-4(-5)厘米；花瓣5，白色或乳白色，芳香，外面有平铺绢毛。蔷薇果深红色，倒卵球形。花期5-7月，果期7-11月。生海拔400-2700米的常绿杂木林、灌木或干旱开阔地。产云南、四川和贵州。印度北部亦有。

软条七蔷薇 *Rosa henryi*

Shrubs usually evergreen. Leaflets (5-)7-9, leathery, apex acuminate or long acuminate. Flowers numerous, in corymb, 3-4(-5) cm diam; petals 5, white or creamy-white, fragrant, abaxially sericeous. Hips dark red, obovoid, Fl. May-Jul. Fr. Jul-Nov. Evergreen mixed forests, thickets or open dry lands at 400-2700 m. Distributed in Yunnan, Sichuan and Guizhou. Also in N India.

软条七蔷薇

Rosa henryi Boulenger

攀援灌木，具长匍匐枝。托叶大部贴生叶柄；小叶多5。花5-15，为伞状伞房花序；花柱合生成束，伸出，略长于雄蕊，被柔毛。蔷薇果褐红色，近球形。花期4-7月，果期7-9月。生海拔1700-2000米的林缘、灌丛、山谷或田野。产中国长江以南各省区。

Shrubs climbing, with long repent branches. Stipules mostly adnate to petioles; leaflets mostly 5. Flowers 5-15, in umbel-like corymb; styles connate in column, exserted, slightly longer than stamens, pubescent. Hips brown-red, subglobose. Fl. Apr-Jul. Fr. Jul-Sep. Forest edges, thickets, valleys or fields at 1700-2000 m. Distributed throughout the provinces on south of the Yangtze River.

木香花 *Rosa banksiae*

木香花
Rosa banksiae W. T. Aiton

攀援常绿灌木。小叶3-5，稀7。花4-5，成单伞形或伞房花序；萼筒球形或卵球形；萼片5，脱落；花瓣重瓣或半重瓣；心皮多数。蔷薇果橙黄色或黑褐色。花期4-5月，果期8-10月。生海拔500-2200米的灌丛、河边或路边。产云南和四川；各地广泛栽培。

Shrubs evergreen, climbing. Leaflets 3-5, rarely 7. Flowers 4-5 in simple umbels or corymbs; hypanthium globose or ovoid; sepals 5, deciduous; petals double or semi-double; carpels numerous. Hips orange-yellow or black-brown. Fl. Apr-May. Fr. Aug-Oct. Thickets, streamsides or roadsides at 500-2200 m. Distributed in Yunnan and Sichuan; also widely cultivated in China.

单瓣木香花 *Rosa banksiae* var. *normalis*

单瓣木香花
Rosa banksiae W. T. Aiton var. **normalis** Regel

本变种与木香花的区别在于本变种的花单瓣，有香味或无。生海拔500-1500米的山谷。产云南、四川、贵州、湖北、河南和甘肃。

This variety differs from the typical variety in its flowers solitary, fragrant or not. Valleys at 500-1500 m. Distributed in Yunnan, Sichuan, Guizhou, Hubei, Henan and Gansu.

小果蔷薇
Rosa cymosa Tratt.

常绿灌木。托叶膜质，早落，离生，条形；小叶3-5，两面无毛。花多数呈复伞房状花序；萼筒球形或卵球形，无毛。蔷薇果红色、黑色或黑褐色，球形，直径4-7毫米。花期5-6月，果期7-11月。生海拔200-1800米的丘陵、开旷山坡、河边或路边。产中国西南、华南、东南、华中、华西和东南。老挝和越南亦有。

Shrubs evergreen. Stipules membranous, caducous, free, linear; leaflets 3-5, both surfaces glabrous. Flowers numerous, in compound corymbs; hypanthium globose or ovoid, glabrous. Hips red, black, purple, or black-brown, globose, 4-7 mm diam. Fl. May-Jun. Fr. Jul-Nov. Hills,

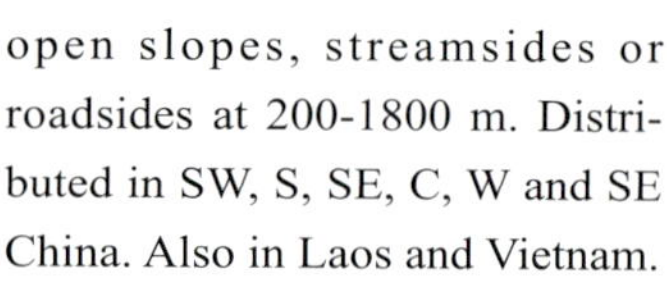
open slopes, streamsides or roadsides at 200-1800 m. Distributed in SW, S, SE, C, W and SE China. Also in Laos and Vietnam.

小果蔷薇 *Rosa cymosa*

金樱子
Rosa laevigata F. Michx.

常绿灌木。托叶离生或基部与叶柄合生；小叶3，革质，有时下面幼时沿中肋有腺毛，后渐脱落无毛。花单生；花瓣5，半重瓣或重瓣，白色。蔷薇果紫

金樱子 *Rosa laevigata*

褐色，梨形或倒卵球形，果和果梗外面密被刺毛。花期4-6月，果期7-11月。生海拔200-1600米的向阳田野、开阔山区或灌丛。产中国西南、东南和华中。

Shrubs evergreen. Stipules free or with base adnate to petioles; leaflets 3, leathery, sometimes abaxially minutely prickly and glandular bristly along midvein when young, glabrescent. Flowers solitary; petals 5, semi-double or double, white. Hips purple-brown, pyriform or obovoid, hips and fruiting pedicels densely setose. Fl. Apr-Jun. Fr. Jul-Nov. Sunny fields, open montane areas or thickets at 200-1600 m. Distributed in SW, SE and C China.

硕苞蔷薇

Rosa bracteata J. C. Wendl.

铺散常绿灌木，具长匍匐枝。小叶5-9，下面无毛或沿脉具柔毛。花单生或2-3束生，直径4.5-9厘米；花瓣5，白色或黄白色；花柱离生。蔷薇果球形，直径1.3-2.7厘米，密被茶褐色柔毛。花期5-7月，果期8-11月。生海拔300米以下的矮灌丛中、沙丘、河边、海岸或路边。产中国西南、华南、东南、华中和华东。日本南部亦有。

Shrubs evergreen, diffuse, with long repent branches. Leaflets 5-9, abaxially glabrous or pubescent along veins. Flowers solitary or 2-3 fasciculate, 4.5-9 cm diam; petals 5, white or yellowish white; styles free. Hips globose, 1.3-2.7 cm diam, densely tawny pubescent. Fl. May-Jul. Fr. Aug-Nov. Brushland, sandy hills, streamsides, seashores or roadsides below 300 m. Distributed in SW, S, SE, C and E China. Also in S Japan.

硕苞蔷薇 *Rosa bracteata*

缫丝花 *Rosa roxburghii*

缫丝花

Rosa roxburghii Tratt.

灌木。托叶大部贴生叶柄；小叶9-15，无毛，边缘具细锐单锯齿。花单生，或2-3花束生枝顶，直径4-6厘米；萼筒扁球形，密被刚毛；花柱离生，不外伸。蔷薇果扁球形，密被针刺。花期5-7月，果期8-10月。生海拔500-1400米的山林、灌丛、山坡或溪边。产中国西南、东南和华中。

Shrubs. Stipules mostly adnate to petioles; leaflets 9-15, glabrous, margin acutely simply serrulate. Flowers solitary, or 2-3 fasciculate apically on branches, 4-6 cm diam; hypanthium depressed-globose, densely bristly; styles free, not exerted. Hips depressed-globose, densely spinecent. Fl. May-Jul. Fr. Aug-Oct. Montane forests, thickets, slopes or stream-sides at 500-1400 m. Distributed in SW, SE and C China.

中甸刺玫

Rosa praelucens Byhouwer

灌木。托叶大部贴生叶柄；小叶7-13，两面密被柔毛。花单生，直径(5-)8-9厘米。花瓣5，红色；花柱离生；萼筒扁球形，外被柔毛和稀疏皮刺。蔷薇果扁球形，散生针刺。花期6-7月。生海拔2700-3000米的山坡林中。产云南西北部。

Shrubs. Stipules mostly adnate to petioles; leaflets 7-13, both surfaces densely puberulous. Flowers solitary, (5-)8-9 cm diam; petals 5, red; styles free; hypanthium depressed-globose, pubescent, glandular bristly. Hips depressed-globose, sparsely spinecent. Fl. Jun-Jul. Woods on slopes at 2700-3000 m. Distributed in NW Yunnan.

龙芽草

Agrimonia pilosa Ledeb.

多年生草本。茎疏被短柔毛及柔毛。奇数羽状复叶；小叶下面常沿脉具平伏短柔毛。花小，组成总状花序；萼筒顶端有数层钩刺；雌蕊2，包在萼筒内。瘦果。花果期5-12月。生海拔100-3800米的溪边、灌丛、草坡或疏林中。产中国大部分地区。越南北部、俄罗

中甸刺玫 *Rosa praelucens*

龙芽草 *Agrimonia pilosa*

黄龙尾 *Agrimonia pilosa* var. *nepalensis*

斯、蒙古、朝鲜半岛、日本和欧洲东部亦有。

Herbs perennial. Stems sparsely pilose and pubescent. Leaves imparipinnate; leaflets abaxially usually appressed pilose on veins, markedly glandular punctate. Flowers small, in racemes; hypanthium with a multiseriate crown of prickles apically; pistils 2, included in hypanthium. Achenes. Fl. and fr. May-Dec. Streamsides, thickets, grassy slopes or open forests at 100-3800 m. Distributed in most parts of China. Also in N Vietnam, Russia, Mongolia, Korean Peninsula, Japan and E Europe.

黄龙尾

Agrimonia pilosa Ledeb. var. **nepalensis** (D. Don) Nakai.

本变种与龙芽草的区别在于本变种茎下部密被粗硬毛。叶上面脉上被硬毛，脉间密被柔毛或绒毛状柔毛。花果期5-12月。生海拔100-3500米的水边、草地或疏林中。产中国大部分地区。南亚和东南亚亦有。

Stems densely rigidly hairy in lower part. Leaves abaxially densely pubescent or tomentose-pubescent between veins, adaxially hirsute or hirtellous on veins. Fl. and fr. May-Dec. By waters, grasslands or open forests at 100-3500 m. Distributed in most parts of China. Also in S and SE Asia.

马蹄黄

Spenceria ramalana Trimen

草本。小叶13-21。花序长5-20厘米，具12-15花；花瓣黄色，基部具短爪；萼片长4-6毫米，宽1.2-1.8毫米；副萼片长2-3.5毫米，宽1.5-3毫米。瘦果黄褐色。花期7-8月，果期9-10月。生海拔3000-5000米的高山草甸或石灰岩山坡上。产云南、四川和西藏。不丹亦有。

Herbs. Leaves with 13-21 leaflets. Inflorescences 5-20 cm long, 12-15-flowered; petals yellow, tapering into a short claw at base; sepals 4-6 × 1.2-1.8 mm; epicalyx segments 2-3.5 × 1.5-3 mm. Achenes yellow-brown. Fl. Jul-Aug. Fr. Sep-Oct. Alpine meadows or rocky slopes at 3000-5000 m. Distributed in Yunnan, Sichuan and Xizang. Also in Bhutan.

马蹄黄 *Spenceria ramalana*

地榆 *Sanguisorba officinalis*

地榆
Sanguisorba officinalis L.

多年生草本。根褐色或紫褐色，粗壮，常纺锤形。叶具4-6对小叶；小叶卵形或长圆状卵形，基部心形。穗状花序直立，椭圆体形、圆柱形或卵球形；萼片4，紫色、红色、粉色或白色。果托杯纵4裂。花果期7-10月。生海拔3000米以下的山坡草地、草甸、灌丛或疏林中。产中国大部分地区。亚洲和欧洲温带亦有。

Herbs perennial. Roots brown or purple-brown, robust, usually fusiform. Leaves with 4-6 pairs of leaflets; leaflets ovate or oblong-ovate, base cordate. Inflorescences erect, spicate, ellipsoid, cylindric or ovoid; sepals 4, purple, red, pink or white. Fruiting hypanthium longitudinally 4-ribbed. Fl. and fr. Jul-Oct. Grassy slopes, meadows, thickets or open forests below 3000 m. Distributed in most parts of China. Also in temperate regions of Asia and Europe.

细叶地榆 *Sanguisorba tenuifolia*

细叶地榆
Sanguisorba tenuifolia Fisch. ex Link

多年生草本，高达1.5米。基生叶具7-9对小叶；小叶有短柄，边缘具缺刻状急尖锯齿。穗状花序长圆柱形，下垂；花萼红色、淡红色或白色。果托杯纵4裂。花果期7-9月。生海拔200-1700米的树林、林缘、草甸、草坡或潮湿处。产内蒙古、黑龙江、吉林和辽宁。俄罗斯、蒙古、朝鲜半岛和日本亦有。

Herbs perennial, to 1.5 m tall. Radical leaves with 7-9 pairs of leaflets; leaflets petiolulate, margin acutely incised serrate. Inflorescences usually nodding,

宽蕊地榆 *Sanguisorba applanata*

spicate, long cylindric; sepals red, whitish red or white. Fruiting hypanthium longitudinally 4-ribbed. Fl. and fr. Jul-Sep. Forests, forest edges, meadows, grassy slopes or damp places at 200-1700 m. Distributed in Neimenggu, Heilongjiang, Jilin and Liaoning. Also in Russia, Mongolia, Korean Peninsula and Japan.

宽蕊地榆

Sanguisorba applanata T. T. Yü et C. L. Li

多年生草本。根粗壮，圆柱形。下部茎生叶边缘具锐锯齿；叶具3-5对小叶。穗状花序长圆柱形；花萼淡粉色或白色；花丝扁平，向上渐膨大，是萼片长的2倍以上，与花药同宽。花果期7-10月。生海拔100-500米的林中、山谷或溪岸。产河北、江苏和山东。

Herbs perennial. Rootstock robust, terete. Lower cauline leaves margin incised serrate; leaves with 3-5 pairs of leaflets. Inflorescences spicate, long cylindric; sepals pale pink or white; filaments gradually compressed-dilated distally, ca. 2 × as long as sepals, as broad as anthers. Fl. and fr. Jul-Oct. Forests, valleys or stream banks at 100-500 m. Distributed in Hebei, Jiangsu and Shandong.

矮地榆

Sanguisorba filiformis (Hook. f.) Hand.-Mazz.

多年生草本。茎高8-35厘米。基生叶为羽状复叶，有小叶3-5对，叶柄光滑；小叶宽卵形或近圆形。花序头状，花单性，几球形，周围为雄花，中央为雌花；苞片细小，卵形，边缘有稀疏睫毛；花柱长为萼片的1/2至等长，柱头呈乳头状扩大。果有4纵棱，成熟时萼片脱落。花果期6-9月。生海拔1200-4500米的山坡草地及沼泽。产四川、云南和西藏。印度北部亦有。

Perennial herbs. Stems 8-35 cm tall. Radical leaves, pinnate, with 3-5 pairs of leaflets, petioles glabrous; leaflets broadly ovate or suborbicular. Inflorescences capitate, flowers unisexual, subglobose, with male flowers surrounding females; bracts minute, ovate, margin sparsely ciliate; style 0.5-1 times as long as sepals, stigma dilated, papillate. Fruits longitudinally 4-ribbed, sepals deciduous when mature. Fl. and fr. Jun-Sep. Meadows on mountain slopes and marshes at 1200-4500 m. Distributed in Sichuan, Yunnan and Xizang. Also in N India.

矮地榆 *Sanguisorba filiformis*

疏花地榆 *Sanguisorba diandra*

高山地榆 *Sanguisorba alpina*

疏花地榆
Sanguisorba diandra Wall. ex Hordb.

多年生草本。茎高40-85厘米。羽状复叶，有小叶5-8对；小叶卵圆形、椭圆形或长椭圆形。头状花序组成圆锥花序，花两性；苞片披针形，有睫毛；萼筒外面疏被柔毛，有4棱；花柱细，与萼片近等长，柱头扩大，有很多分枝。果有4纵棱，棱宽阔呈翅状。花果期6-8月。生海拔3200-3900米的山坡草地、林缘及灌丛中。产西藏。不丹亦有。

Perennial herbs. Stems 40-85 cm tall. Pinnate leaves with 5-8 pairs of leaflets; leaflets ovate, elliptic or long elliptic. Panicles composed of heads, flowers bisexual; bracts lanceolate, with cilliate; hypanthium sparsely pubescent, 4-ribbed; style slender, as long as sepals, stigma dilated, much branched. Fruits with 4 longitudinal, broadly winged ribs. Fl. and fr. Jun-Aug. Meadows on mountain slopes, forest margins and thickets at 3200-3900 m. Distributed in Xizang. Also in Bhutan.

高山地榆
Sanguisorba alpina Bge.

多年生草本。小叶基部截形至微心形。穗状花序常圆柱形，从基部向上逐渐开放，下垂；苞片密被柔毛，未开花时显著比花蕾长，是萼片长的1-2倍；花丝从下部开始微扩大，中部最宽，到顶端渐狭明显比花药窄。果被疏柔毛，萼片宿存。花果期7-8月。生海拔1200-2700米的林缘、山坡、沟谷和沼地。产宁夏、甘肃和新疆。俄罗斯、蒙古和朝鲜半岛亦有。

Perennial herbs. Leaflets base truncate to subcordate. Spikes usually cylindric, flowering from base to apex, nodding; bracts densely pubescent, obviously longer than flower buds before anthesis, 1-2 × as long as sepals; filaments gradually dilated from base, broadest near middle, then gradually attenuate toward apex, obviously narrower than anther. Fruits sparsely pilose, with persistent sepals. Fl. and fr. Jul-Aug. Forest margins, mountain slopes, ravines and marshes at 1200-2700 m. Distributed in Ningxia, Gansu and Xinjiang. Also in Russia, Mongolia and Korean Peninsula.

大白花地榆
Sanguisorba stipulata Raf.

多年生草本。根粗壮，深长。基生叶具4-6对小叶；茎生叶2-4。穗状花序直立；萼片4；雄蕊4；花丝扁平，自中部至顶端膨大，顶部最阔，长为萼片的2-3倍。果托杯具短柔毛；萼片宿存。花果期6-9月。生海拔1400-2300米的林下、林缘、山

大白花地榆 *Sanguisorba stipulata*

羽衣草 *Alchemilla japonica*

地、深谷或沼泽地。产吉林和辽宁。俄罗斯、朝鲜半岛、日本和北美洲亦有。

Herbs perennial. Rootstocks robust, deep, long. Radical leaves with 4-6 pairs of leaflets; cauline leaves 2-4. Inflorescences erect, spicate; sepals 4; stamens 4; filaments compressed-dilated from middle to apex, broadest at apex, 2-3 × as long as sepals. Fruiting hypanthium pilose; sepals persistent. Fl. and fr. Jun-Sep. Thinned forests, forest edges, mountains, ravines or marshy places at 1400-2300 m. Distributed in Jilin and Liaoning. Also in Russia, Korean Peninsula, Japan and North America.

羽衣草

Alchemilla japonica Nakai et Hara

多年生草本，高10-13厘米。根状茎肥厚木质；茎密被白色长柔毛。茎生叶有长叶柄，心状圆形，基部深心形，顶端有7-9浅裂片，边缘有细锯齿，两面均被稀疏柔毛；叶柄密被开展长柔毛。伞房状聚伞花序较紧密；花黄绿色；萼筒外被稀疏柔毛。瘦果全部包在膜质花托内。生海拔2500-3500米的高山草原上。产华中、内蒙古和新疆。日本亦有。

Perennial herbs, 10-30 cm tall. Rhizome fleshy, thick, woody; stems densely white villous. Cauline leaves with long petioles, cordate-orbicular, base deeply cordate, apex slightly 7-9-lobed, margin serrulate, both surfaces sparsely pilose; petioles densely spreading villous. Inflorescences densely corymbose-cymose; petals yellow-green; hypanthium abaxially sparsely villous. Achenes included in membranous receptacle. Alpine grasslands at 2500-3500 m. Distributed in C China, Neimenggu and Xinjiang. Also in Japan.

纤细羽衣草

Alchemilla gracilis Opiz

多年生草本。基生叶肾状圆形，边缘7-9裂和细锐锯齿。伞房状聚伞花序较疏松；萼筒基部稍下延无毛。瘦果卵球形，长1-2毫米，无毛，先端略钝。生海拔1700-3500米的疏林或高山草地。产四川、山西、陕西、甘肃和新疆。俄罗斯(西西伯利亚)、蒙古和欧洲亦有。

Herbs perennial. Radical leaves reniform-orbicular, margin 7-9-lobed and serrulate. Inflorescences laxly corymbiform-cymose; hypanthium glabrous, base slightly decurrent. Achenes ovoid, 1-2 mm long, glabrous, apex subobtuse. Sparse forests or alpine grasslands at 1700-3500 m. Distributed in Sichuan, Shanxi, Shaanxi, Gansu and Xinjiang. Also in Russia (W Siberia), Mongolia and Europe.

纤细羽衣草 *Alchemilla gracilis*

扁核木 *Prinsepia utilis*

扁核木

Prinsepia utilis Royle

灌木。枝刺上生叶。叶长圆形至卵状披针形。总状花序具多花；花瓣白色，直径约1厘米，基部具短爪；雄蕊2-3轮。核果紫褐色或黑紫色，矩圆形或倒卵球状矩圆形。花期4-5月，果期8-9月。生海拔1000-2600米的山坡、荒地或山谷中。产云南、四川、西藏和贵州。尼泊尔、不丹、巴基斯坦和印度亦有。

Shrubs. Spines leafy. Leaves oblong to ovate-lanceolate. Racemes many flowered; petals white, ca. 1 cm diam, base shortly clawed; stamens in 2 or 3 whorls. Drupes purplish brown to blackish purple, oblong or obovoid-oblong. Fl. Apr-May. Fr. Aug-Sep. Slopes, wastelands or valleys at 1000-2600 m. Distributed in Yunnan, Sichuan, Xizang and Guizhou. Also in Nepal, Bhutan, Pakistan and India.

榆叶梅

Amygdalus triloba (Lindl.) Ricker

灌木，稀小乔木。短枝上叶簇生，一年生枝条上叶互生，先端常3裂。1或2花，先叶开放，直径2-3厘米，单瓣或重瓣；萼筒阔钟形；花瓣粉红色。核果近球形，直径1-1.6厘米。花期4-5月，果期5-7月。生海拔600-2500米的林下或灌丛中。产中国东北、华北和东南。朝鲜半岛和俄罗斯亦有。

Shrubs, rarely small trees. Leaves on short branchlets often fasciculate, those on previous year's branches alternate, apex usually 3-lobed. Flowers 1 or 2, opening before leaves, 2-3 cm diam, single or double; hypanthium broadly campanulate; petals pink. Drupes subglobose, 1-1.6 cm diam. Fl. Apr-May. Fr. May-Jul. Forests or thickets at 600-2500 m. Distributed in NE, N and SE China. Also in Korean Peninsula and Russia.

桃

Amygdalus persica L.

乔木。叶长圆状披针形，下面脉腋有少量毛或无毛，侧脉不

榆叶梅 *Amygdalus triloba*

桃 *Amygdalus persica*

直达叶缘，在叶边结合成网状。花单生，先叶开放；花萼被柔毛。核果颜色从浅绿色至橘黄色，卵球形、阔椭圆体形或扁球形；两侧扁平，顶端渐尖。花期3-4月，果期8-9月。逸生海拔1500-2000米的荒地或山坡。产中国大部分地区。世界各地亦广泛栽培。

Trees. Leaves oblong-lanceolate, abaxially with or without a few hairs in vein axils, lateral nerves anastomosis close to margin. Flowers solitary, opening before leaves; sepals pubescent. Drupes color varies from greenish white to orangish yellow, ovoid, broadly ellipsoid or compressed globose; endocarp compressed on both sides, apex acuminate. Fl. Mar-Apr. Fr. Aug-Sep. Escaped from cultivation in waste fields or on disturbed slopes at 1500-2200 m. Distributed in most parts of China. Also widely cultivated in the world.

光核桃 *Amygdalus mira*

山桃 *Amygdalus davidiana*

山桃

Amygdalus davidiana (Carrière) de Vos ex L. Henry

乔木。叶卵状披针形，基部楔形，边缘有细锐锯齿，两面无毛。花单生，先叶开放；花萼无毛。核果浅黄色，椭圆体形至矩圆形，直径2.5-3.5厘米，密具柔毛，表面具沟纹和空穴。花期3-4月，果期7-8月。生海拔800-3200米的林下、灌丛、山坡、山谷或荒田。产中国西南、华北、华西、华东、东北和西北。

Trees. Leaves ovate-lanceolate, base cuneate, margin acutely serrate, both surfaces glabrous. Flowers solitary, opening before leaves; sepals glabrous. Drupes yellowish, ellipsoid to oblong, 2.5-3.5 cm diam, densely pubescent; furrowed and pitted. Fl. Mar-Apr. Fr. Jul-Aug. Forests, thickets, slopes, mountain valleys or waste fields at 800-3200 m. Distributed in SW, N, W, E, NE and NW China.

光核桃

Amygdalus mira (Koehne) Ricker

乔木。叶披针形至卵状披针形，下面沿中脉具柔毛，边缘具浅圆齿，近顶端全缘，齿端常具小腺体。花单生，先叶开放；花瓣粉红色。中果皮肉质，不开裂，内果皮扁的卵球形，两侧稍压扁，表面光滑，顶端急尖。花期3-4月，果期8-9月。生海拔2000-4000米的山坡杂木林中、山谷或沟边，或栽培。产云南、四川和西藏。俄罗斯亦有。

Trees. Leaves lanceolate to ovate-lanceolate, abaxially pubescent along midvein, margin shallowly crenate but entire near apex and teeth usually gland-tipped. Flowers solitary, opening before leaves; petals pink. Mesocarp fleshy, not splitting endocarp ovoid, slightly flattened on both sides, surface smooth, apex acute. Fl. Mar-Apr. Fr. Aug-Sep. Slopes in mixed forests, mountain valleys or streamsides at 2000-4000 m, or cultivated. Distributed in Yunnan, Sichuan and Xizang. Also in Russia.

杏 *Armeniaca vulgaris*

杏

Armeniaca vulgaris Lam.

乔木。叶柄无毛；叶阔卵形至圆卵形，两面无毛；叶基部圆形至近圆形，上面光滑。花单生，先叶开放；花瓣白色、粉色或带红色。核果白色、黄色、橘黄色，常带红色，球形或卵球形，直径大于2.5厘米。花期3-4月，果期6-7。生海拔700-3000米的山坡疏林，也有栽培。产四川、河北、山西、陕西、甘肃、山东和新疆。中亚亦有。

Trees. Petioles glabrous; leaves broadly ovate to orbicular-ovate, both surfaces glabrous; base rounded to subcordate, adaxially glabrous. Flowers solitary, opening before leaves; petals white, pink or tinged with red. Drupes white, yellow, orange-yellow, often tinged red, globose or ovoid, more than 2.5 cm diam. Fl. Mar-Apr. Fr. Jun-Jul. Sparse forests on mountain slopes at 700-3000 m, also cultivated. Distributed in Sichuan, Hebei, Shanxi, Shaanxi, Gansu, Shandong and Xinjiang. Also in C Asia.

山杏

Armeniaca sibirica (L.) Lam.

灌木或小乔木。叶卵形至近圆形，先端长渐尖至尾尖。花单生，先叶开放；花瓣白色具粉色脉或淡粉色。中果皮干燥，苦、不可食，易与内果皮分开，成熟时沿腹缝线开裂。花期3-5月，果期6-7月。生海拔400-2500米的林中、灌丛、山区、丘陵草地、山坡、河谷或干燥向阳山坡。产华北、西北和东北。蒙古、俄罗斯和朝鲜半岛亦有。

Shrubs or small trees. Leaves ovate to suborbicular, apex long acuminate to caudate. Flowers solitary, opening before leaves; petals white with pink veins or pinkish. Mesocarp dry, bitter, inedible, easily separated from endocarp, splitting along ventral suture at maturity. Fl. Mar-May. Fr. Jun-Jul. Forests, thickets, mountainous areas, hill grasslands, slopes, river valleys or dry sunny slopes at 400-2500 m. Distributed in N, NW and NE China. Also in Mongolia, Russia and Korean Peninsula.

洪平杏

Armeniaca hongpingensis C. L. Li

乔木，高达10米。叶柄长1.5-2厘米，叶缘密被小锐锯齿。花未见。果柄长7-10毫米。核果近球形，长3.5-4厘米，密被黄褐色绒毛；两侧扁，腹棱钝，腹面具纵沟，表面具窝孔。果期6-7月。生海拔约1800米的道边或有时于村中栽培。产湖北西部。

Trees, to 10 m tall. Petioles 1.5-2 cm long; leaf margin densely acutely serrulate. Flowers unknown. Fruiting pedicels 7-10 mm long. Drupes subglobose, 3.5-4 cm long, densely yellowish brown pubescent; endocarp compressed on both sides, ventral rib

山杏 *Armeniaca sibirica*

洪平杏 *Armeniaca hongpingensis*

obtuse, longitudinally furrowed on ventral side, surface pitted. Fr. Jun-Jul. Along trails or sometimes cultivated in villages at ca. 1800 m. Distributed in W Hubei.

梅

Armeniaca mume Siebold

乔木。叶薄，卵形至椭圆形。花单生，或2朵簇生，先叶开放，香味浓郁；花梗短，1-3毫米；花瓣白色或粉色。核果黄色至绿白色，近球形，直径2-3厘米，具柔毛；椭圆形；基部楔形。花期冬季至翌年春季，果期5-6月(华北7-8月)。中国大部分地区有栽培。原产朝鲜半岛和日本。

Trees. Leaves thin, ovate to elliptic. Flowers solitary or 2 in a fascicle, opening before leaves, strongly fragrant; pedicels short, 1-3 mm; petals white or pink. Drupes yellow to greenish white, subglobose, 2-3 cm diam, pubescent; endocarps ellipsoid, base cuneate. Fl. winter to next spring. Fr. May-Jun (or Jul-Aug in N China). Commonly cultivated throughout most parts of China. Native to Korean Peninsula and Japan.

梅 *Armeniaca mume*

李

Prunus salicina Lindl.

乔木。叶长圆状倒卵形至长圆状卵圆形，背面沿主脉有稀疏柔毛或脉腋有髯毛。常3花簇生；花瓣白色。核果黄色或红色，有时绿色或紫色，球形、卵球形或圆锥形，直径3.5-5厘米，栽培品种可达7厘米，具白粉。花期4月，果期7-8月。生海拔200-2600米的山坡灌丛中、山谷疏林中或路边。产中国大部分地区，常见栽培。亦广泛栽培在亚洲其他地区、欧洲和北美洲。

Trees. Leaves oblong-obovate to oblong-ovate, abaxially sparsely pubescent on veins or barbate in vein axils. Flowers usually 3 in a fascicle; petals white. Drupes yellow or red, sometimes green or purple, globose, ovoid or conical, 3.5-5 cm diam to 7 cm diam in horticultural forms, glaucous. Fl. Apr. Fr. Jul-Aug. Thickets on slopes, open forests in valleys or roadsides at 200-2600 m. Distributed in most parts of China, commonly cultivated. Also widely cultivated in other regions of Asia, Europe and North America.

李 *Prunus salicina*

樱桃李 *Prunus cerasifera*

樱桃李
Prunus cerasifera Ehrh.

灌木或小乔木。枝深灰色，有时有棘刺；小枝暗红色，无毛。叶下面灰绿色，沿中脉具柔毛。花单生，稀2花成一束；花梗无毛或疏被柔毛。核果近球状至椭圆体形，表面平滑或粗糙，有时具沟。花期4月，果期8月。生海拔800-2000米的林中、溪边或砾石质山坡。产新疆。哈萨克斯坦、乌兹别克斯坦、塔吉克斯坦、西南亚和欧洲南部亦有。

Shrubs or small trees. Branches dark gray, sometimes spiny; branchlets dark red, glabrous. Leaves abaxially pale green and pubescent on midvein. Flowers solitary, rarely 2 in a fascicle; pedicels glabrous or sparsely pubescent. Drupes subglobose to ellipsoid, smooth or scabrous, sometimes pitted. Fl. Apr. Fr. Aug. Forests, streamsides or gravelly slopes at 800-2000 m. Distributed in Xinjiang. Also in Kazakhstan, Uzbekistan, Tajikistan, SW Asia and S Europe.

迎春樱桃
Cerasus discoidea Yü et Li

小乔木。叶倒卵状椭圆形或椭圆形，边有缺刻状急尖锯齿，齿端有小盘状腺体。伞形花序；花先叶开放；萼筒管状钟形；花瓣粉色，长椭圆形，顶端2裂。核果红色，直径约1厘米，内果皮上略有棱纹。花期3月，果期5月。生海拔200-1100米的山谷林中或溪边灌丛。产浙江、安徽和江西。

Small trees. Leaves obovate-oblong to elliptic, margin shallowly obtusely serrulate and teeth with a minute conical apical gland. Inflorescences umbellate; flowers opening before leaves; hypanthium tubular-campanulate; petals pink, long elliptic, apically 2-lobed. Drupes red, ca. 1 cm diam; endocarp sculptured. Fl. Mar. Fr. May. Forests in ravines or thickets by streams at 200-1100 m. Distributed in Zhejiang, Anhui and Jiangxi.

微毛樱桃
Cerasus clarofolia (C. K. Schneid.) T. T. Yü et C. L. Li

乔木或小灌木。冬芽卵球形，无毛。叶下面淡绿色，无毛或具短柔毛，边有齿，渐尖，齿端有小腺体或不明显。花序伞形或近伞形，具2-4花；苞片齿端有锥状或头状腺体；花叶同开。核果红色，长椭圆体形。花期4-6月，果期6-7月。生海拔800-3600米的林下或山

迎春樱桃 *Cerasus discoidea*

微毛樱桃 *Cerasus clarofolia*

浙闽樱桃 *Cerasus schneideriana*

坡灌丛中。产中国西南、华中和华北。

Shrubs or small trees. Winter buds ovoid, glabrous. Leaves abaxially pale green and glabrous or pilose, margin teeth with a minute or inconspicuous apical gland. Inflorescences umbellate or subumbellate, 2-4-flowered; bracts teeth with a conical to capitate apical gland; flowers opening at same time as leaves. Drupes red, long ellipsoid. Fl. Apr-Jun. Fr. Jun-Jul. Forests or thickets on slopes at 800-3600 m. Distributed in SW, C and N China.

浙闽樱桃
Cerasus schneideriana
(Koehne) T. T. Yü et C. L. Li

小乔木。小枝紫褐色。叶下被灰黄色粗毛，边缘齿端具一头状顶生腺体。伞形花序，具(1-)2(-3)花；萼筒管状，外面具平伏的褐色柔毛。核果紫红色，椭圆体形；内果皮具棱纹。花期3月，果期5月。生海拔600-1300米的林中。产广西、福建和浙江。

Small trees. Branchlets purplish brown. Leaves abaxially grayish yellow hirtellous, margin teeth with a capitate apical gland. Inflorescences umbellate, (1-)2(-3)-flowered; hypanthium tubular, outside appressed brown pubescent. Drupes purplish red, ellipsoid; endocarps sculptured. Fl. Mar. Fr. May. Forests at 600-1300 m. Distributed in Guangxi, Fujian and Zhejiang.

樱桃
Cerasus pseudocerasus
(Lindl.) Loudon

乔木。叶背面沿叶脉被疏柔毛，边缘齿端具一个微小的顶生腺体。花序伞房状或近伞形，具3-7花；萼筒钟状，萼片长为萼筒一半或近一半；花瓣白色，顶端微凹。核果红色，近球形，直径0.9-1.3厘米；内果皮稍具棱纹。花期3-4月，果期5-6月。产中国大部分地区。世界各地亦广泛栽培。

Trees. Leaves abaxially pilose along veins, margin teeth with a minute apical gland. Inflorescences corymbose or subumbellate, 3-7-flowered; hypanthium campanulate, sepals nearly 1/2 or more as long as hypanthium; petals white, apically emarginate. Drupes red, subglobose, 0.9-1.3 cm diam; endocarp sculptured. Fl. Mar-Apr. Fr. May-Jun. Distributed in most parts of China. Also widely planted in the world.

樱桃 *Cerasus pseudocerasus*

云南樱桃 *Cerasus yunnanensis*

云南樱桃

Cerasus yunnanensis (Franch.) T. T. Yü et C. L. Li

乔木。叶缘有尖锐锯齿，有时间有少数重锯齿，顶端有头状小腺体。花序近伞房总状，有3-5(-7)花；总花梗密被硬毛；花期苞片脱落；花柱基部疏被柔毛。核果紫红色，椭圆体形至卵球形，长7-10毫米。花期3-5月，果期5-6月。生海拔1900-2600米的林中或山坡上。产中国西南和华南。

Trees. Leaves margin acutely serrate and sometimes biserrate and teeth with a capitate apical gland. Inflorescences subcorymbose-racemose, 3-5(-7)-flowered; peduncles densely hirtellous; bracts deciduous after anthesis; styles basally pilose. Drupes purplish red, ellipsoid to ovoid, 7-10 mm diam. Fl. Mar-May. Fr. May-Jun. Forests or mountain slopes at 1900-2600 m. Distributed in SW and S China.

毛叶山樱花

Cerasus serrulata (Lindl.) Loudon var. **pubescens** (Makino) T. T. Yü et C. L. Li

乔木。叶柄、叶背面和花梗均被短柔毛。叶缘具渐尖细齿或重锯齿，齿端具一小的顶生腺体。花序伞房总状或近伞形，具2-3花；花瓣白色，先端凹缺。核果紫黑色，球形至卵球形，直径8-10毫米。花期4-5月，果期5-7月。生海拔400-1500米的山坡林中。产华北、华东和东北。

Trees. Petioles, abaxial surfaces of leaves and pedicels pubescent. Leaves margin acuminately serrate or biserrate and teeth with a minute apical gland. Inflorescences corymbose-racemose or subumbellate, 2-or 3-flowered; petals white, apex emarginate. Drupes purplish black, globose to ovoid, 8-10 mm diam. Fl. Apr-May. Fr. May-Jul. Forests on mountain slopes at 400-1500 m. Distributed in N, E and NE China.

华中樱桃

Cerasus conradinae (Koehne) T. T. Yü et C. L. Li

乔木。叶缘为重锯齿，两面无毛。伞形花序，3-5花；总苞片褐色；花直径1.5厘米，先叶开放或与叶同期开放；花瓣白色或粉色，先端2裂。核果红色，卵球形至近球形。花期3月，果

毛叶山樱花 *Cerasus serrulata* var. *pubescens*

华中樱桃 *Cerasus conradinae*

钟花樱桃 *Cerasus campanulata*

期4-5月。生海拔500-2100米的沟边或林下。产中国西南、华南、华中和华北。

Trees. Leaf margin biserrate, both surfaces glabrous. Inflorescences umbellate, 3-5-flowered; involucral bracts brown; flowers 1.5 cm diam, opening before or nearly at same time as leaves; petals white or pink, apically 2-lobed. Drupes red, ovoid to subglobose. Fl. Mar. Fr. Apr-May. By streams or forests at 500-2100 m. Distributed in SW, S, C and N China.

钟花樱桃
Cerasus campanulata
(Maxim.) A. N. Vassiljeva

乔木或灌木。叶下面脉腋有簇毛，叶缘有急尖锯齿，常稍不整齐。伞形花序，有2-4花，花先叶开放；花瓣先端颜色较深，下凹；萼筒钟形；花柱长于雄蕊或稀短于雄蕊，无毛。核果红色，卵球形。花期1-3月，果期4-5月。生海拔100-1300米的山谷林中或林缘。产广西、广东、海南、福建、浙江和湖南。越南和日本亦有。

Trees or shrubs. Leaves abaxially with tufts of hairs in vein axils, margin acutely and usually somewhat irregularly serrate. Inflorescences umbellate, 2-4-flowered, opening before leaves; petals apically darker and emarginate; hypanthium campanulate; styles longer or rarely shorter than stamens, glabrous. Drupes red, ovoid. Fl. Jan-Mar. Fr. Apr-May. Forests in valleys or forest edges at 100-1300 m. Distributed in Guangxi, Guangdong, Hainan, Fujian, Zhejiang and Hunan. Also in Vietnam and Japan.

高盆樱桃
Cerasus cerasoides (Buch.-Ham. ex D. Don) S. Y. Sokolov

乔木。伞形花序有1-4花；花叶同开或先叶开放；花梗长1-2.3厘米，果期长达3厘米，先端肥厚；萼筒红色至深红色，钟形至阔钟形；萼片常淡红色，三角形；花瓣白色或粉色。核果紫黑色；核顶端圆钝。花期10-12月。生海拔700-3700米的山沟林下。产云南西北部和西藏南部。印度北部、尼泊尔、不丹、克什米尔地区、缅甸、老挝北部和泰国亦有。

Trees. Inflorescences umbellate, 1-4-flowered; flowers opening at same time as or before leaves; pedicels 1-2.3 cm long, elongated to 3 cm long and apically thickened in fruit; hypanthium red to dark red, campanulate to broadly campanulate; sepals usually reddish, triangular; petals white or pink. Drupes purplish black; endocarp apex obtuse. Fl. Oct-Dec. Forests in ravines at 700-3700 m. Distributed in NW Yunnan and S Xizang. Also in N India, Nepal, Bhutan, Kashmir, Myanmar, N Laos and Thailand.

高盆樱桃 *Cerasus cerasoides*

细齿樱桃 *Cerasus serrula*

欧李

Cerasus humilis (Bunge) Sokolov.

灌木。叶近倒卵状长椭圆形或倒卵状披针形，下面无毛或疏具柔毛，先端急尖，基部楔形。花单生或3花束生，花叶同开；萼筒钟形；花瓣粉色或白色；花柱无毛。核果红色至紫红色，近球形。花期4-5月，果期6-10月。生海拔400-1800米的灌丛中或山坡。产中国西南、华北、华东和东北。

Shrubs. Leaves obovate-elliptic to obovate-oblanceolate, abaxially glabrous or sparsely pubescent, apex acute, base cuneate. Flowers solitary or to 3 in a fascicle, opening at same time as leaves; hypanthium campanulate; petals pink or white; styles glabrous. Drupes red to purplish red, subglobose. Fl. Apr-May. Fr. Jun-Oct. Thickets or slopes at 400-1800 m. Distributed in SW, N, E and NE China.

细齿樱桃

Cerasus serrula (Franch.) Yu et Li

乔木。嫩枝伏生疏柔毛，冬芽尖卵形。叶披针形至卵状披针形，长3.5-7厘米，边缘有尖锐单锯齿或重锯齿，齿端有小腺体，上面疏被柔毛，下面无毛或中脉下部两侧被疏柔毛。花单生或有2朵，与叶同开；萼筒管状钟形。核果紫红色，核有显著棱纹。花期5-6月，果期7-9月。生海拔1200-4000米的山坡、山谷林中、林缘。产四川、云南和西藏。

Trees. Young branchlets appressed pilose, winter buds acutely ovoid. Leaves lanceolate to ovate-lanceolate, 3.5-7 cm long, margin acutely serrate or biserrate and teeth with a minute gland, adaxially pilose, abaxially glabrous or pilose on lateral sides of midvein. Inflorescences 1- or 2-flowered; opening at same time as leaves; hypanthium tubular-campanulate. Drupes purplish-red, endocarp markedly sculptured. Fl. May-Jun. Fr. Jul-Sep. Mountain slopes, forests in ravines, forest margins at 1200-4000 m. Distributed in Sichuan, Yunnan and Xizang.

麦李

Cerasus glandulosa (Thunb.) Sokolov

灌木。叶柄长1.5-3毫米；叶缘有细钝重锯齿。花单生或2多簇生，与叶同时或近同时开放；萼筒钟形；花瓣粉色或白色。核果红色至淡紫色，近球形。花期3-4月，果期5-8月。生海拔800-2300米的山坡灌丛或沟谷边，亦栽培。产中国西南、东南、华南和华北。日本亦有。

Shrubs. Petioles 1.5-3 mm long; leaf margin obtusely finely biserrate. Flowers solitary or 2 in a fascicle, opening at same time as leaves or nearly so; hypanthium campanulate; petals pink or white. Drupes red to purplish, subglobose. Fl. Mar-Apr. Fr. May-Aug. Thickets of mountain slopes or sides of ravine at 800-2300 m, also cultivated. Distributed in SW, SE, S and N China. Also in Japan.

郁李

Cerasus japonica (Thunb.) Loisel.

灌木。叶卵形或卵状披针形，先端渐尖，基部圆形。花叶同开或先叶开放；萼筒陀螺状；

麦李 *Cerasus glandulosa*

欧李 *Cerasus humilis*

郁李 *Cerasus japonica*

花柱几等长于雄蕊，无毛。核果深红色或黑色，近球形，直径约1厘米；内果皮光滑。花期5月，果期7-8月。生海拔100-200米的山坡林下或灌丛，亦栽培。产华北、华东和东北。朝鲜半岛和日本亦有。

Shrubs. Leaves ovate or ovate-lanceolate, apex acuminate, base rounded. Flowers opening at same time as leaves or before; hypanthium turbinate; styles nearly as long as stamens, glabrous. Drupes dark red or black, subglobose, ca. 1 cm diam; endocarps smooth. Fl. May. Fr. Jul-Aug. Forests on mountain slopes or thickets at 100-200 m, also cultivated. Distributed in N, E and NE China. Also in Korean Peninsula and Japan.

毛樱桃

Cerasus tomentosa (Thunb.) Wall.

灌木。叶卵状椭圆形至倒卵状椭圆形，长2-7厘米，宽1-3.5厘米，下面灰绿色，密被灰色绒毛,之后脱落，上面深绿色，被疏柔毛,边有急尖或粗锐锯齿。花单生或2花簇生，花叶同开或先叶开放；花瓣粉色或白色。核果红色，近球形。花期4-5月，果期6-9月。生海拔100-3700米的山坡林中、林缘、灌丛或草甸，亦栽培。产中国西南、华北、华西、西北和东北。

Shrubs. Leaves ovate-elliptic to obovate-elliptic, 2-7 × 1-3.5 cm, abaxially grayish green or densely gray tomentose but glabrescent, adaxially dark green and pilose, margin coarsely or acutely serrate. Flowers solitary or 2 in a fascicle, opening at same time as or before leaves; petals pink or white. Drupes red, subglobose. Fl. Apr-May. Fr. Jun-Sep. Forests on mountain slopes, forest edges, thickets or meadows at 100-3700 m, also cultivated. Distributed in SW, N, W, NW and NE China.

毛樱桃 *Cerasus tomentosa*

橉木 *Padus buergeriana*

橉木

Padus buergeriana (Miq.) T. T. Yü et T. C. Ku

乔木。叶柄无腺体；叶椭圆形、长圆状椭圆形或稀倒卵状椭圆形，两面均无毛。总状花序长6-9厘米，常20-30花，基部无叶；花瓣白色，宽卵形。核果棕黑色，近球形或卵球形。花期4-5月，果期5-10月。生海拔1000-3400米的山坡密林、路旁或山坡阳处。产中国西南、华南、东南、华中、华北和华西。印度北部、不丹、朝鲜半岛和日本亦有。

Trees. Petioles without gland; leaves elliptic, oblong-elliptic or rarely obovate-elliptic, both surfaces glabrous. Racemes 6-9 cm long, usually 20-30-flowered, base leafless; petals white, broadly ovate. Drupes blackish brown, subglobose to ovoid. Fl. Apr-May. Fr. May-Oct. Dense forests on slopes, along trails or sunny places on slopes at 1000-3400 m. Distributed in SW, S, SE, C, N and W China. Also in N India, Bhutan, Korean Peninsula and Japan.

稠李

Padus avium Mill.

乔木。总状花序具7-8花，基部具1或2叶；萼片三角状卵形，迅速脱落；花瓣白色；雄蕊多数；子房无毛；花柱长为雄蕊的1/2。核果红褐色至黑色，卵球形，光滑。花期4-6月，果期5-10月。生海拔800-2600米的山坡、山谷或灌丛。产华北和东北。俄罗斯、朝鲜半岛和日本亦有。

Trees. Racemes 7-8-flowered, basally with 1 or 2 leaves; sepals triangular-ovate, soon caducous; petals white; stamens many; ovary glabrous; styles 1/2 as long as stamens. Drupes reddish brown to black, ovoid, smooth. Fl. Apr-Jun. Fr. May-Oct. Slopes, valleys or thickets at 800-2600 m. Distributed in N and NE China. Also in Russia, Korean Peninsula and Japan.

灰叶稠李

Padus grayana (Maxim.) C. K. Schneid.

乔木。叶柄长5-10毫米，常无毛；叶两面无毛或下面沿中脉有柔毛。总状花序5-8厘米，密集多花，基部具2-4(或5)叶；花瓣白色；雄蕊20-30；花柱长且通常伸出雄蕊和花瓣之外。核果黑褐色，卵球形，无毛。花期4-5月，果期6-9月。生海拔1000-3800米的山谷杂木林、山坡背阴处或路边。产中国西南、华南、东南和华中。日本亦有。

Trees. Petioles 5-10 mm long, usually glabrous; leaves glabrous or abaxially pubescent on midveins. Racemes 5-8 cm long, dense, many flowered, basally with 2-4(or 5) leaves; petals white; stamens 20-30; styles long and exserted. Drupes blackish

稠李 *Padus avium*

灰叶稠李 *Padus grayana*

brown, ovoid, glabrous. Fl. Apr-May. Fr. Jun-Sep. Mixed forests in valleys, shady places on slopes or along trails at 1000-3800 m. Distributed in SW, S, SE and C China. Also in Japan.

短梗稠李

Padus brachypoda (Batal.) C. K. Schneid.

乔木，高8-10米。叶柄顶部具2个腺体；叶长圆形，稀椭圆形或披针形，两面均无毛。总状花序长16-30厘米，多花，基部具1-3叶；花瓣白色，倒卵形。核果幼时紫红色，老时黑褐色，球形。花期4-5月，果期5-10月。生海拔1000-2500米的杂木林、山坡灌丛或沟谷中。产中国西南、华中、华北、华西和华东。

Trees, 8-10 m tall. Petioles apically with 2 nectaries; leaves oblong, rarely elliptic or lanceolate, both surfaces glabrous. Racemes 16-30 cm long, many flowered, basally with 1-3 leaves; petals white, obovate. Drupes purplish red but blackish brown with age, globose. Fl. Apr-May. Fr. May-Oct. Mixed forests, thickets on slopes or valleys at 1000-2500 m. Distributed in SW, C, N, W and E China.

细齿稠李

Padus obtusata (Koehne) T. T. Yü et T. C. Ku

乔木。叶柄通常顶端两侧各具1个腺体，叶狭长圆形、椭圆形或倒卵形，边缘具细齿，两面均无毛。总状花序长10-15厘米，多花，基部具2叶；花瓣白色，近圆形至长圆形。核果黑色，卵球形，无毛。花期5-7月，果期7-9月。生海拔800-3600米的山坡杂木林中、山谷或溪边。产中国西南、华南、华中、华北、华西和华东。

Trees. Petioles usually with a nectary on either side at apex, leaves narrowly oblong, elliptic or obovate, margin crenulate, both surfaces glabrous. Racemes 10-15 cm long, many flowered, basally with 2 leaves; petals white, suborbicular to oblong. Drupes black, ovoid, glabrous. Fl. May-Jul. Fr. Jul-Sep. Mixed forests on slopes, valleys or streamsides at 800-3600 m. Distributed in SW, S, C, N, W and E China.

短梗稠李 *Padus brachypoda*

细齿稠李 *Padus obtusata*

粗梗稠李 *Padus napaulensis*

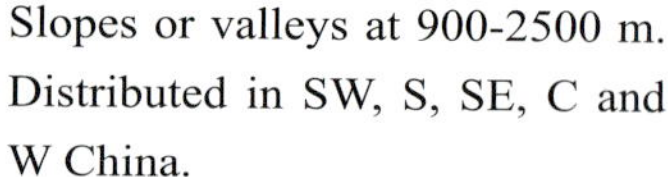

Slopes or valleys at 900-2500 m. Distributed in SW, S, SE, C and W China.

腺叶桂樱
Laurocerasus phaeosticta (Hance) C. K. Schneid.

灌木或乔木。叶草质至近革质，下面散生黑色腺点。总状花序具数花至10余花；萼筒杯状。核果紫黑色，近球形至横向椭圆体形，直径8-10毫米，无毛；内果皮薄，光滑。花期4-5月，果期7-10月。生海拔300-2500米的混交林下、山谷、高山草甸、沟边或路边。产中国西南、华南、东南至华中。印度、孟加拉国、缅甸北部、泰国北部和越南北部亦有。

Shrubs or trees. Leaves herbaceous to subcoriaceous, abaxially

粗梗稠李
Padus napaulensis (Ser.) C. K. Schneid.

乔木，高达27米。叶柄无毛，无蜜腺；叶缘具粗锯齿或有时波状。总状花序长7-14厘米，多花，基部具2或3叶；花瓣白色，基部具短爪。核果深紫色至黑色，卵球形，无毛。花期4月，果期7月。生海拔1200-2500米的常绿阔叶林或落叶杂木林中、或溪边开阔地。产中国西南、华中、华西和华东。印度北部、尼泊尔、不丹和缅甸北部亦有。

Trees, to 27 m tall. Petioles glabrous, without nectaries; leaf margin coarsely serrate or sometimes undulate. Racemes 7-14 cm long, many flowered, basally with 2 or 3 leaves; petals white, base shortly clawed. Drupes dark purple to black, ovoid, glabrous. Fl. Apr. Fr. Jul. Broad-leaved evergreen or deciduous mixed forests, or open places beside streams at 1200-2500 m. Distributed in SW, C, W and E China. Also in N India, Nepal, Bhutan and N Myanmar.

绢毛稠李
Padus wilsonii C. K. Schneid.

乔木。叶柄顶部具2个腺体，叶椭圆形、长圆形或长圆状倒卵形，叶背被绢毛。总状花序长7-14厘米，多花，基部具3-5叶；花瓣白色，倒卵状长圆形。核果幼果红褐色，老时黑紫色，球形或卵球形，无毛。花期4-5月，果期7-10月。生海拔900-2500米的山坡或山谷中。产中国西南、华南、东南、华中和华西。

Trees. Petioles apex with 2 nectaries, leaves elliptic, oblong or oblong-obovate, abaxially densely silky pubescent. Racemes 7-14 cm long, many flowered, basally with 3-5 leaves; petals white, obovate-oblong. Drupes reddish brown at first, becoming blackish purple, globose or ovoid, glabrous. Fl. Apr-May. Fr. Jul-Oct.

绢毛稠李 *Padus wilsonii*

腺叶桂樱 *Laurocerasus phaeosticta*

scattered black punctate. Racemes several to 10-flowered or more; hypanthium cup-shaped. Drupes purplish black, subglobose to transversely ellipsoid, 8-10 mm diam, glabrous; endocarps thin, smooth. Fl. Apr-May. Fr. Jul-Oct. Mixed forests, mountain valleys, mountain meadows, streamsides or along trails at 300-2500 m. Distributed in SW, S, SE to C China. Also in India, Bangladesh, N Myanmar, N Thailand and N Vietnam.

尖叶桂樱

Laurocerasus undulata (Buch.-Ham. ex D. Don) Roem.

灌木或乔木。叶草质或薄革质，两面无毛，椭圆形至长圆披针形，先端渐尖。总状花序单生或2-4个簇生；总花梗和花梗无毛；花瓣黄白色。核果紫黑色，卵球形至椭圆体形。花期8-10月，果期冬季至翌年春季。生海拔500-3600米的山坡混交林中。产中国西南、华南和华中。南亚和东南亚亦有。

Shrubs or trees. Leaves herbaceous or thinly leathery, both surfaces glabrous, elliptic to oblong-lanceolate, apex acuminate. Racemes solitary or 2-4 in a fascicle; rachises and pedicels glabrous; petals yellowish white. Drupes purplish black, ovoid to ellipsoid. Fl. Aug-Oct. Fr. from winter to next spring. Mixed forests on slopes at 500-3600 m. Distributed in SW, S and C China. Also in S and SE Asia.

大叶桂樱

Laurocerasus zippeliana (Miq.) Browicz

乔木。叶宽卵形至椭圆状长圆形或宽长圆形，长(5-)6-19厘米，叶边具稀疏或稍密粗锯齿，齿顶有黑色硬腺体；叶柄长1-2厘米。总状花序单生或4个簇生；花瓣白色。果实矩圆状或卵球状矩圆形。花期7-10月，果期冬季。生海拔400-2400米的杂木林、灌丛和石灰岩山区。产中国西南、华南、东南、华中和华西。越南北部和日本亦有。

Trees. Leaves broad-ovate to elliptic-oblong or broad-oblong, (5-)6-19 cm long, margin with sparse or dense coarse black glandular serration; petioles 1-2 cm long. Racemes solitary or to 4 in a fascicle; petals white. Drupes oblong or ovoid-oblong. Fl. Jul-Oct. Fr. winter. Mixed forests, thickets or calcareous mountains at 400-2400 m. Distributed in SW, S, SE, C and W China. Also in N Vietnam and Japan.

尖叶桂樱 *Laurocerasus undulata*

大叶桂樱 *Laurocerasus zippeliana*

刺叶桂樱 *Laurocerasus spinulosa*

大果臀果木 *Pygeum macrocarpum*

刺叶桂樱

Laurocerasus spinulosa (Sieb. et Zucc.) C. K. Schneid.

乔木。叶草质至薄革质，长圆形至倒卵状长圆形，边缘波状，中部以上或近顶端常具少数针状锐锯齿，顶端渐尖至尾尖。总花梗与花梗被短柔毛。核果无毛；内果皮薄而光滑。花期9-10月，果期11月至翌年3月。生海拔400-1500米的阳坡森林、沟边常绿阔叶林或山谷中。产中国西南、华南、东南和华中。日本亦有。

Trees. Leaves herbaceous to thinly leathery, oblong to obovate-oblong, margin undulate and with a few acicular teeth apically from middle to near apex, apex acuminate to caudate. Rachises and pedicels pubescent. Drupes glabrous; endocarp thin and smooth. Fl. Sep-Oct. Fr. Nov to next Mar. Sunny forest slopes, broad-leaved evergreen forests along rivers or mountain valleys at 400-1500 m. Distributed in SW, S, SE and C China. Also in Japan.

臀果木

Pygeum topengii Merr.

乔木。叶卵状椭圆形至椭圆形，侧脉5-8对，下面具褐色平伏柔毛，先端短渐尖而钝。总状花序长4-7厘米；子房无毛。核果深褐色，肾形，长0.8-1厘米，宽1-1.6厘米，顶端凹陷。花期6-9月，果期冬季。生海拔100-1600米的沟谷或林中。产中国西南和华南。

Trees. Leaves ovate-elliptic to elliptic, secondary veins 5-8 pairs, abaxially brown appressed pubescent, apex shortly acuminate and with an apical obtuse tip. Racemes 4-7 cm long; ovary glabrous. Drupes dark brown, reniform, 0.8-1 × 1-1.6 cm, apically depressed. Fl. Jun-Sep. Fr. winter. Valleys or forests at 100-1600 m. Distributed in SW and S China.

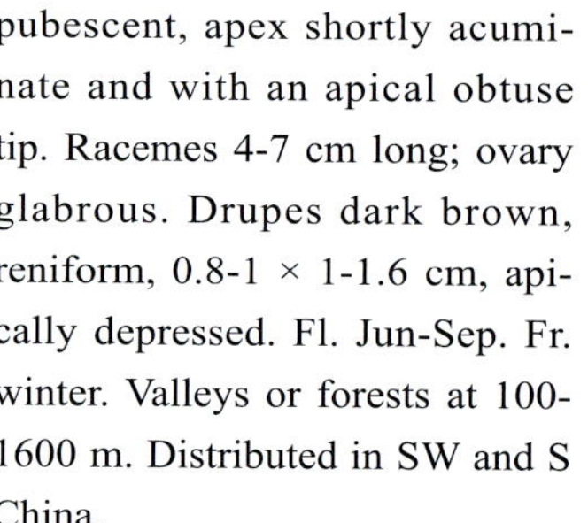

大果臀果木

Pygeum macrocarpum T. T. Yü et L. T. Lu

乔木。叶椭圆形或长圆状椭圆形，全缘，下面色浅，沿叶脉具稀疏褐色柔毛，渐无毛，基部圆形，先端突然狭缩成短尖头，侧脉6-8对。核果紫褐色，无毛，扁卵球形，长1.5-1.8厘米。花期8-10月，果期冬季至翌年春季。生海拔500-1000米的林中、林缘、深谷或溪边。产云南东南部。

Trees. Leaves elliptic or oblong-elliptic, margin entire, abaxially paler, sparsely brown pubescent along veins, glabrescent, base rounded, apex abruptly short pointed, secondary veins 6-8 pairs. Drupes pruplish brown, glabrous, compressed ovoid, 1.5-1.8 cm long. Fl. Aug-Oct. Fr. winter to next spring. Forests, forest edges, deep ravines or streamsides at 500-1000 m. Distributed in SE Yunnan.

臀果木 *Pygeum topengii*

花楸属一种 *Sorbus* sp.

牛栓藤科
Connaraceae

长尾红叶藤 *Rourea caudata*

小叶红叶藤 *Rourea microphylla*

长尾红叶藤
Rourea caudata Planch.

藤本或攀援灌木。奇数羽状复叶，小叶3或4对，顶端长尾尖，具钝头。圆锥花序腋生，具1-3轴；雄蕊约10；心皮5，离生。蓇葖果淡绿色。种子全部包以假种皮。花期4-10月，果期5月至翌年3月。生海拔800米以下的山地疏林。产云南、广西和广东。印度亦有。

Lianas or climbing shrubs. Leaves odd-pinnate; leaflets 3- or 4-paired, apex long caudate with obtuse tip. Inflorescences axillary, paniculate with 1-3 axes; stamens ca. 10; carpels 5, free. Follicles greenish. Seeds all covered by arils. Fl. Apr-Oct. Fr. May to next Mar. Sparse forests in the mountains below 800 m. Distributed in Yunnan, Guangxi and Guangdong. Also in India.

小叶红叶藤
Rourea microphylla (Hook. et Arn.) Planch.

藤本或攀援灌木。奇数羽状复叶，小叶3-8(-13)对，长1.5-4厘米，宽0.5-2厘米，小叶先端钝或渐尖。圆锥花序腋生于上部叶腋或假顶生，具1-5轴；花芳香。蓇葖果成熟时红色。种子橘黄色，基部被膜质假种皮包裹。花期3-9月，果期5月至翌年3月。生海拔100-600米的山坡或疏林。产云南、广西、广东和福建。印度、斯里兰卡、越南和印度尼西亚亦有。

Lianas or climbing shrubs. Leaves odd-pinnate, leaflets 3-8(-13)-paired, 1.5-4 × 0.5-2 cm, leaflets apex obtuse or acuminate. Inflorescences axillary in distal leaf axils or pseudoterminal, paniculate with 1-5 axes; flowers fragrant. Follicles red when mature. Seeds orange, with membranous arils at base. Fl. Mar-Sep. Fr. May to next Mar. Slopes or sparse forests at 100-600 m. Distributed in Yunnan, Guangxi, Guangdong and Fujian. Also in India, Sri Lanka, Vietnam and Indonesia.

红叶藤
Rourea minor (Gaertn.) Leenh.

藤本或攀援灌木。奇数羽状复叶，小叶1-3对，通常1对，先端锐尖至短渐尖。花序腋生或假顶生，疏松圆锥状或近总状。种子红色，基部被膜质假种皮包裹。花期4-10月，果期5月至翌年3月。生海拔800米以下的密林、竹林或灌丛。产云南、广东和台湾。南亚、东南亚和澳大利亚(昆士兰)亦有。

Lianas or climbing shrubs. Leaves odd-pinnate, leaflets 1-3-paired, usually 1-paired, apex acute to shortly acuminate. Inflorescences axillary or pseudoterminal, laxly paniculate or subracemose. Seeds red, base covered by membranous arils. Fl. Apr-Oct. Fr. May to next Mar. Dense forests, bamboo woods or shrubs below 800 m. Distributed in Yunnan, Guangdong and Taiwan. Also in S and SE Asia, and Australia (Queensland).

红叶藤 *Rourea minor*

豆科
Fabaceae

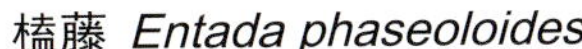

榼藤 *Entada phaseoloides*

海红豆
Adenanthera microsperma Teijsm. et Binnend.

乔木。二回羽状复叶，羽片3-5对；小叶互生；羽片具4-7对小叶。总状花序单生、腋生或在小枝顶端排成圆锥状，具柔毛；花白色或黄色，较小，芳香，具短梗。荚果狭长圆形，开裂后果瓣扭曲。花期4-7月，果期7-10月。生海拔1000米以下的山谷、林中、溪边，或栽培。产中国西南和华南。越南、老挝、缅甸、柬埔寨、马来西亚和印度尼西亚亦有。

Trees. Leaves bipinnate, pinnae 3-5 pairs; leaflets alternate; pinnae with leaflets in 4-7 pairs. Racemes simple, axillary or arranged in panicles at apices of branchlets, puberulent; flowers white or yellow, small, fragrant, shortly pedicellate. Legumes narrowly oblong, valves contorted after dehiscence. Fl. Apr-Jul. Fr. Jul-Oct. Valleys, forests, by streams, or planted. Distributed in SW and S China. Also in Myanmar, Laos, Vietnam, Cambodia, Malaysia and Indonesia.

榼藤
Entada phaseoloides (L.) Merr.

木质大藤本。羽片2对；每个羽片具1或2小叶；顶生一对羽片变为卷须。荚果大，长达1米，宽8-12厘米，弯曲，扁平，木质，内果皮羊皮纸状。种子大，圆形，直径4-6厘米，褐黑色。花期3-6月，果期8-11月。生海拔200-1300米的林中，攀援于大乔木上。产中国西南和华南。热带和亚热带亚洲、热带澳大利亚亦有。

Woody liana. Pinnae 2 pairs; leaflets 1 or 2 pairs per pinna; apical pair of pinnae transformed into a tendril. Legumes big, up to 1 m long, 8-12 cm broad, curved, flattened, woody, with a parchmentlike endocarp. Seeds large, orbicular, 4-6 cm diam, brown-black. Fl. Mar-Jun. Fr. Aug-Nov. Scandent to the trees in forests at 200-1300 m. Distributed in SW and S China. Also in tropical and subtropical Asia, and tropical Australia.

海红豆 *Adenanthera microsperma*

巴西含羞草 *Mimosa diplotricha*

巴西含羞草

Mimosa diplotricha C. Wright ex Sauvalle

直立、亚灌木状草本。茎攀援或平卧，五棱柱状，沿棱上密生钩刺。二回羽状复叶，长10-15厘米；总叶柄及叶轴有钩刺4-5列；羽片(4-)7-8对，长2-4厘米；小叶(12)20-30对，线状长圆形，被白色长柔毛。头状花序，1或2个生于叶腋，连花丝直径约1厘米；雄蕊8，花丝浅紫色。荚果边缘及荚节有刺毛。花果期3-9月。产广东。原产巴西。

Subshrubs or perennial herbs. Stems scandent or prostrate, 5-angulate, with prickles along angles. Leaves bipinnate, 10-15 cm long; petiole and rachis with 4-5 rows of recurved prickles; pinnae (4-)7-8 pairs, 2-4.5 cm long; leaflets (12-)20-30 pairs per pinna, linear-oblong, both surfaces white villous. Heads 1 or 2, axillary, ca. 1 cmn diam (including filaments); stamens 8; filaments pale purple-pink. Legumes margin with prickly bristles. Fl. and fr. Mar-Sep. Distributed in Guangdong. Native to Brazil.

含羞草

Mimosa pudica L.

灌木。羽片通常2对；小叶10-20对，线状披针形。花多数，粉色，小；花萼微小；花冠钟形；裂片外面具柔毛；雄蕊4。荚果排成星状，稍弯曲，扁平，长圆形，边缘具刺毛。花期3-10月，果期5-11月。生海拔1500米以下的荒地或灌丛，或栽培。产中国西南、华南和东南。原产热带美洲，其他热带地区亦有。

Shrubs. Pinnae usually 2 pairs; leaflets 10-20 pairs, linear-lanceolate. Flowers numerous, pink, small; calyx minute; corolla campanulate; lobes outside pubescent; stamens 4. Legumes arranged in a star, slightly recurved, flat, oblong, bristle along sutures. Fl. Mar-Oct. Fr. May-Nov. Wastelands or thickets below 1500 m, or cultivated. Distributed in SW, S and SE China. Native to tropical America, also in other tropical areas.

含羞草 *Mimosa pudica*

银合欢

Leucaena leucocephala (Lam.) De Wit

小乔木或灌木，高2-6米。羽片4-8对；小叶5-15对，线状长圆形。头状花序直径2-3厘米，常1-2生于叶腋，花梗长2-4厘米；苞片早落，被毛；花白色。荚果直伸，带状。花期4-7月，果期8-10月。栽培或归化于中国西南和华南。原产热带美洲，现广泛栽培于热带地区。

Shrubs or small trees, 2-6 m tall. Pinnae 4-8 pairs; leaflets 5-15 pairs, linear-oblong. Heads 2-3 cm diam, usually 1 or 2, axil-

银合欢 *Leucaena leucocephala*

台湾相思 *Acacia confusa*

lary; peduncles 2-4 cm long; bracts deciduous, pubescent; flowers white. Legumes straight, lorate. Fl. Apr-Jul. Fr. Aug-Oct. Cultivated or naturalized in SW and S China. Native to tropical America, now commonly cultivated in other tropical areas.

台湾相思

Acacia confusa Merr.

常绿乔木。叶退化，叶柄变成叶状柄。头状花序单生或2或3个簇生，腋生，球形；花金黄色，芳香；花瓣淡绿色；子房具黄褐色长柔毛。荚果于种子间有微缢缩。花期3-10月，果期8-12月。野生或栽培。产中国西南、华南和东南。印度尼西亚、菲律宾和斐济亦有；原产菲律宾。

Evergreen trees. Leaves reduced, petioles phyllodic. Heads solitary or 2-or 3-fasciculate, axillary, globose; flowers golden yellow, fragrant; petals greenish; ovary yellow-brown villous. Legumes slightly constricted between seeds. Fl. Mar-Oct. Fr. Aug-Dec. Wild or cultivated. Distributed in SW, S and SE China. Also in Indonesia, the Philippines and Fiji; native to the Philippines.

儿茶 *Acacia catechu*

儿茶

Acacia catechu (L. f.) Willd.

落叶乔木。小枝常于叶柄下具1对平、褐色的钩刺或无刺。二回羽状复叶；羽片10-30对。穗状花序1-4，腋生；花浅黄色或白色，离生；雄蕊多数。荚果褐色，直，带状，开裂。花期4-8月，果期9月至翌年1月。产华南和东南，除云南为野生外，其余均为引种。南亚亦有。

Deciduous trees. Branchlets often with a pair of flat, brown, hooked spines below stipules or without spines. Leaves bipinnate; pinnae 10-30 pairs. Spikes 1-4, axillary; flowers yellowish or white, free; stamens numerous. Legumes brown, straight, strap-shaped, nitid, dehiscent. Fl. Apr-Aug. Fr. Sep to next Jan. Distributed in S and SE China, only wild in Yunnan. Also in S Asia.

金合欢 *Acacia farnesiana*

金合欢

Acacia farnesiana (L.) Willd.

灌木或小乔木。枝多数；小枝之字形。托叶针状；羽片4-8对，腋生；小叶10-20对，线状长圆形。头状花序1-3，腋生；花黄色，有香味。荚果褐色，肿胀，近圆柱形。花期3-6月，果期7-12月。栽培于华南。原产热带美洲，现广布于热带地区。

Shrubs or small trees. Branches numerous; branchlets zigzag. Stipules spinelike; pinnae 4-8 pairs, axillary; leaflets 10-20 pairs, linear-oblong. Heads 1-3, axillary; flowers yellow, fragrant. Legumes brown, swollen, subcylindric. Fl. Mar-Jun. Fr. Jul-Dec. Cultivated in S China. Native to tropical America, cultivated throughout the tropics.

银荆

Acacia dealbata Link

灌木或小乔木。二回羽状复叶，有小叶10-20(-25)对。头状花序，直径6-7毫米，组成腋生总状花序或顶生圆锥花序；花淡黄色或橙黄色。荚果宽7-12毫米，无毛，具白霜。花期4月，果期7-8月。栽培于华南一带。原产澳大利亚。

Shrubs or small trees. Leaves bipinnate, with 10-20(-25) pairs of leaflets. capitulum small, 6-7 mm diam, forming an axillary raceme or terminal panicle; flowers yellowish or orange-yellow. Legumes 7-12 mm wide, glabrous, glaucous. Fl. Apr. Fr. Jul-Aug. Cultivated in S China. Native to Australia.

银荆 *Acacia dealbata*

粉被金合欢
Acacia pruinescens Kurz

灌木或小乔木。小枝具白霜，幼时具柔毛至绒毛。小叶纸质，无柄，线形，无毛或仅具缘毛；羽片9-11对；叶轴具柔毛，具少量弯刺。球形头状花序；花小，黄色，无柄。荚果纸质，光滑。花期4月，果期6-10月。生河谷中。产云南。缅甸和越南亦有。

Shrubs or small trees. Branchlets glaucous, puberulent to tomentose when young. Leaflets papery, sessile, linear, glabrous or ciliate only; pinnae 9-11 pairs; rachis pubescent, with few recurved spines. Heads globose; flowers small, yellow, sessile. Legumes papery, smooth. Fl. Apr. Fr. Jun-Oct. Grow in valleys. Distributed in Yunnan. Also in Myanmar and Vietnam.

粉被金合欢 *Acacia pruinescens*

棋子豆 *Archidendron robinsonii*

羽叶金合欢
Acacia pennata (L.) Willd.

攀援、多刺藤本。柄状腺体生叶柄下部，常仅位于叶枕上部；小叶30-54对，先端锐尖，常向前弯。头状花序单生或2或3个簇生，球形。荚果带状，缝线浅波状。花期3-10月，果期7月至翌年4月。生低海拔疏林中，常攀附于灌木或乔木的顶部。产云南、广东和福建。热带亚洲和热带非洲亦有。

Climbers, with copious prickles. Petiolar gland below middle of petiole, usually just above basal pulvinus; leaflets 30-54 pairs, apically sharply acute, often bent forward. Heads solitary or 2- or 3-fasciculate, globose. Legumes strap-shaped, sutures slightly sinuate. Fl. Mar-Oct. Fr. Jul to next Apr. Sparse forests at low elevations; scandent on tops of trees or shrubs. Distributed in Yunnan, Guangdong and Fujian. Also in tropical Asia and tropical Africa.

棋子豆
Archidendron robinsonii (Gagnep.) I. C. Nielsen

乔木。叶柄具圆形腺点；羽片1对；小叶3对，对生或近对生，椭圆形至披针形，两面无毛，侧脉3或4对。头状花序6-8花，排成总状生于叶腋；花萼坛状或杯状，外面无毛。荚果直伸，圆柱状。生海拔300-700米的山谷密林。产云南和广西。越南亦有。

Trees. Petiolar gland circular; pinnae 1 pair; leaflets 3 pairs, opposite or subopposite, elliptic to lanceolate, both surfaces glabrous, lateral veins 3 or 4 pairs. Heads 6-8-flowered, arranged in axillary panicles; calyx urceolate or cup-shaped, glabrous. Legumes straight, cylindric. Dense forests in valleys at 300-700 m. Distributed in Yunnan and Guangxi. Also in Vietnam.

羽叶金合欢 *Acacia pennata*

亮叶猴耳环 *Archidendron lucidum*

亮叶猴耳环

Archidendron lucidum (Benth.) I. C. Nielsen

无刺乔木。二回羽状复叶；羽片1-2对，小叶2-5对，除顶端一对生，余均互生，上面光亮且深绿色；叶轴腺体压扁。头状花序球形，具10-20花，排成圆锥状。荚果旋转成环状，宽2-3厘米。花期4-6月，果期7-12月。生疏林或密林中，或林缘灌丛中。产中国西南和华南。印度和越南亦有。

Trees, unarmed. Leaves bipinnate; pinnae 1-2 pairs, leaflets 2-5 pairs, alternate except terminal opposite ones, adaxially shiny and deep green; glands on leaf rachis depressed. Heads globose, 10-20-flowered, arranged in panicles. Legumes spiralled into rings, 2-3 cm broad. Fl. Apr-Jun. Fr. Jul-Dec. Open or dense forests, or thickets at forest edges. Distributed in SW and S China. Also in India and Vietnam.

猴耳环

Archidendron clypearia (Jack) I. C. Nielsen

乔木。小枝具棱，密具黄色绒毛。叶柄具4棱；羽片(3或)4或5(-8)对，密具黄色绒毛；小叶近无柄，倾斜，菱状梯形。伞房花序具数花，排成顶生或腋生圆锥状；花具柄；花冠白色或淡黄色。荚果扭曲。花期2-6月，果期4-8月。生海拔500-1800米的林中。产中国西南、华南和东南。

Trees. Branchlets angulate, densely yellow tomentose. Petioles 4-angulate; pinnae (3 or)4 or 5(-8) pairs, densely yellow tomentose; leaflets subsessile, oblique, rhombic-trapezoid. Corymbs several flowered, arranged in terminal or axillary panicles; flowers pedicellate; corolla white or yellowish. Legumes twisted. Fl. Feb-Jun. Fr. Apr-Aug. Forests at 500-1800 m. Distributed in SW, S and SE China.

薄叶猴耳环 *Archidendron utile*

薄叶猴耳环

Archidendron utile (Chun et F. C. How) I. C. Nielsen

灌木。小枝圆柱形，具褐色柔毛。羽片2或3对；叶4-7对，对生，长圆状菱形，2-9 × 1.5-4厘米。头状花序具约15花，排成顶生圆锥花序；花无柄，白

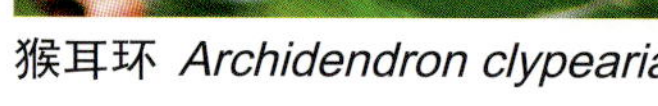

猴耳环 *Archidendron clypearia*

色，芳香。荚果红褐色，镰形。花期3-8月，果期4-12月。生海拔100-700米的密林中。产广西、广东、海南、福建和浙江。越南亦有。

Shrubs. Branchlets terete, brown pubescent. Pinnae 2 or 3 pairs; leaflets 4-7 pairs, opposite, oblong-rhombic, 2-9 × 1.5-4 cm. Heads ca. 15-flowered, arranged in terminal panicles; flowers sessile, white, fragrant. Legumes red-brown, falcate. Fl. Mar-Aug. Fr. Apr-Dec. Forests at 100-700 m. Distributed in Guangxi, Guangdong, Hainan, Fujian and Zhejiang. Also in Vietnam.

光叶合欢

Albizia lucidior (Steud.) I. C. Nielsen ex H. Hara

乔木。二回羽状复叶，羽片1-3对；小叶两面无毛或被柔毛。头状花序排成腋生的伞形圆锥花序；花白色，边花具柔毛。荚果开裂，浅褐色，直，舌状。种子轮廓圆形。花期4-6月，果期9-11月。生海拔600-1900米的河谷、河岸或山地林中。产云南、广西、贵州和台湾。南亚和东南亚亦有。

Trees. Leaves bipinnate, leaflets 1-3-paired; leaflets both surfaces glabrous or puberulent. Heads arranged in axillary umbel-panicles; flowers white; marginal flowers with puberulent. Legumes dehiscent, brownish, straight, ligulate. Seeds orbicular. Fl. Apr-Jun. Fr. Sep-Nov. Valleys, riverbanks or montane forests at 600-1900 m. Distributed in Yunnan, Guangxi, Guizhou and Taiwan. Also in S and SE Asia.

白花合欢

Albizia crassiramea Lace

乔木。小枝被锈色柔毛。叶柄近基部及叶轴近顶部具椭圆形腺体；小叶4-6对，椭圆形、卵形或倒卵形，长2-7厘米，背面被短柔毛，基部斜截形。头状花序具7-10花，再排成圆锥花序；花二型，无梗，白色；花萼杯状；花萼和花冠被淡黄色和白色绒毛。荚果开裂，舌状。花期8月，果期11月。生海拔500-1300米的林中。产云南和广西。东南亚亦有。

Trees. Branchlets ferruginous pubescent. Leaf glands elliptic, near base of petiole and near apex of rachis; leaflets 4-6 pairs, elliptic, ovate or obovate, 2-7 cm long, abaxially pubescent, base obliquely truncate. Heads 7-10-flowered, arranged into panicles; flowers dimorphic, sessile, white; calyx and corolla yellowish and white tomentose. Legume dehiscent, ligulate. Fl. Aug. Fr. Nov. Forests at 500-1300 m. Distributed in Yunnan and Guangxi. Also in SE Asia.

白花合欢 *Albizia crassiramea*

光叶合欢 *Albizia lucidior*

山槐 *Albizia kalkora*

阔荚合欢 *Albizia lebbeck*

山槐
Albizia kalkora (Roxb.) Prain

落叶小乔木或灌木。小叶5-14对，0.8-4.5 × 0.7-2厘米，两面具柔毛，中脉偏于上边缘。头状花序2-7，腋生或顶生，排列成圆锥状；花二型，起初白色，变为黄色，具明显的花梗。荚果带状。花期5-6月，果期8-10月。生海拔2600米以下的山坡灌丛或疏林中。产中国西南、华南、华中、华北和华东。印度、缅甸、越南和日本亦有。

Deciduous small trees or shrubs. Leaflets 5-14 pairs, 0.8-4.5 × 0.7-2 cm, both surfaces pubescent, mid-nerve of leaflets close to upper margin. Heads 2-7, axillary or terminal, arranged in panicles; flowers dimorphic, primarily white, turning yellow, with conspicuous pedicels. Legumes taeniform. Fl. May-Jun. Fr. Aug-Oct. Mountain slope or in sparse forests below 2600 m. Distributed in SW, S, C, N and E China. Also in India, Myanmar, Vietnam and Japan.

阔荚合欢
Albizia lebbeck (L.) Benth.

落叶乔木。二回羽状复叶；小叶4-8对，2-4.5 × (0.9-)1.3-2厘米，两面无毛或下面疏具细毛，顶端圆钝或凹缺。头状花序1至数个聚生于叶腋；花二型，芳香。荚果禾秆色，条形，扁平。花期5-9月，果期10月至翌年5月。华南有栽培。原产热带非洲，栽培于南亚。

Deciduous trees. Bipinnate coumpound leaves; leaflets 4-8 pairs, 2-4.5 × (0.9-)1.3-2 cm, both surfaces glabrous or abaxially sparsely finely pubescent, apically rounded-obtuse or emarginated. Capitulum 1- to several-clustered, axillary; flowers dimorphic, fragrant. Legumes straw-colored, strap-shaped, flat. Fl. May-Sep. Fr. Oct to next May. Cultivated in S China. Native to tropical Africa, cultivated in S Asia.

合欢
Albizia julibrissin Durazz.

落叶乔木。托叶脱落，线状披针形，小于小叶；二回羽状复

合欢 *Albizia julibrissin*

楹树 *Albizia chinensis*

叶；羽片4-12对。圆锥花序顶生；花萼管状；花序轴短且之字形；花及雄蕊粉红色。荚果带状，扁平，无毛。花期6-7月，果期8-10月。生山坡或栽培。中国各省区野生或栽培。中亚、东亚和非洲亦有；北美洲有栽培。

Deciduous trees. Stipules deciduous, linear-lanceolate, smaller than leaflets; leaves bipinnate; pinnae 4-12 pairs. Panicles terminal; calyx tubiform; rachis of inflorescence short and zigzag; flowers and stamens pink. Legumes strap-shaped, flat, glabrous. Fl. Jun-Jul. Fr. Aug-Oct. Slopes or cultivated. Widely distributed or cultivated in China. Also in C and E Asia, and Africa; also cultivated in North America.

楹树

Albizia chinensis (Osbeck) Merr.

乔木。托叶脱落，心形，大，膜质，先端尖；羽片6-12对；小叶20-35(-40)对。头状花序有10-20花，组成顶生圆锥花序；总花梗密被柔毛；花绿白色或淡黄色，密被黄褐色绒毛；雄蕊白绿色或黄色。荚果扁平。花期3-5月，果期6-12月。生海拔1000米以下的林中或开阔地。产中国西南、华南和东南。南亚至东南亚亦有。

Trees. Stipules deciduous, cordate, large, membranous, apex apiculate; pinnae 6-12 pairs; leaflets 20-35(-40) pairs. Heads 10-20-flowered, arranged in a terminal panicle; peduncles densely villous; corolla white-green or yellowish, densely yellow-brown tomentose; stamens green-white or yellow. Legumes compressed. Fl. Mar-May. Fr. Jun-Dec. Forests or open fields below 1000 m. Distributed in SW, S and SE China. Also in S to SE Asia.

毛叶合欢

Albizia mollis (Wall.) Boivin

乔木。小枝被短柔毛，有棱。小叶8-15对，镰状长圆形，密具长柔毛或上面老时近无毛。头状花序形成腋生的圆锥花序；花白色；花梗极短。荚果褐色，舌状，压扁。花期5-6月，果期8-12月。生海拔1500-2500米的山坡林中。产云南、西藏和贵州。印度和尼泊尔亦有。

Trees. Branchlets pubescent, angulate; leaflets 8-15 pairs, falcate-oblong, densely villous or adaxially glabrescent when old. Heads arranged in axillary panicles; flowers white; pedicels very short. Legumes brown, ligulate, compressed. Fl. May-Jun. Fr. Aug-Dec. Slopes in forests at 1500-2500 m. Distributed in Yunnan, Xizang and Guizhou. Also in India and Nepal.

毛叶合欢 *Albizia mollis*

肥皂荚 *Gymnocladus chinensis*

华南皂荚 *Gleditsia fera*

肥皂荚
Gymnocladus chinensis Baill.

落叶乔木，无刺。树皮灰褐色，具明显的白色皮孔。二回偶数羽状复叶；羽片5-10对；小叶互生，8-12对，近无柄，两面具缘毛状柔毛。总状花序具柔毛；花一雄多雌，下垂，白色或带紫色，具长柄。荚果长圆形，扁平或肿胀，无毛。果期8月。生海拔150-1500米的山坡、杂木林、村边或路旁。产中国西南、东南和华中。

Deciduous trees, prickleless. Bark grayish brown, with distinct white lentils. Bipinnate leaves; pinnae 5-10 pairs; leaflets alternate, 8-12 pairs, subsessile, both surfaces silky pubescent. Racemes puberulent; flowers polygamous, pendulous, whitish or tinged with purple, long pedicellate. Legumes oblong, compressed or turgid, glabrous. Fr. Aug. Slopes, mixed forests, near villages or roadsides at 150-1500 m. Distributed in SW, SE and C China.

野皂荚
Gleditsia microphylla D. Gordon ex Isely

灌木或小乔木。小叶长6-24毫米，全缘，上部小叶小于下部。花簇生，组成穗状花序或顶生的圆锥花序。荚果扁薄，斜椭圆状或斜矩圆形，红棕色，长3-6厘米，无毛。种子1-3粒。花期6-7月，果期7-10月。生海拔100-1300米的山坡阳处或路边。产河南、河北、山西、陕西、安徽、江苏和山东。

Shrubs or small trees. Leaflets 6-24 mm long, margin entire, upper leaflets smaller than those in lower part. Flowers clustered to spikes or terminal panicles. Legumes flat, thin, obliquely ellipsoid or obliquely oblong, reddish brown, 3-6 cm long, glabrous. Seeds 1-3. Fl. Jun-Jul. Fr. Jul-Oct. Sunny slopes or roadsides at 100-1300 m. Distributed in Henan, Hebei, Shanxi, Shaanxi, Anhui, Jiangsu and Shandong.

华南皂荚
Gleditsia fera (Lour.) Merr.

乔木。一回羽状复叶；小叶网脉明显凸起。花数个形成小聚伞花序，小聚伞花序再组成顶生或腋生的总状花序，长7-16厘米；花杂性，绿白色。荚果扁平，13.5-26(-41) × 2.5-3(-6.5)厘米，嫩果密被棕黄色短柔毛。花期4-5月，果期6-12月。生海

野皂荚 *Gleditsia microphylla*

山皂荚 *Gleditsia japonica*

拔300-1000米的山地缓坡、山谷林中、村旁或路边。产华南。越南亦有。

Trees. Leaves unipennate; leaflets with reticulate veinlets conspicuously raised. Flowers several in cymules, cymules in axillary or terminal racemes, 7-16 cm long; flowers ploygamous, greenish white. Legumes compressed, 13.5-26(-41) × 2.5-3(-6.5) cm, densely brownish yellow puberulous when young. Fl. Apr-May. Fr. Jun-Dec. Mountain slopes, valley forests, beside villages or roadsides at 300-1000 m. Distributed in S China. Also in Vietnam.

山皂荚

Gleditsia japonica Miq.

乔木或小乔木。小枝有散生的白色皮孔。小叶网脉不清晰，全缘或疏具浅锯齿。雌花直径5-6毫米。荚果扁，20-35 × 2-4厘米，不规则卷曲或镰刀状，密具黄绿色绒毛。生海拔100-1000米的山坡阳处、山谷、溪边或路旁。常见栽培。产华中、华北、华东和东北。朝鲜半岛和日本亦有。

Trees or small trees. Branchlets with scattered whitish lenticels. Leaflets with reticulate veinlets obscure, margin entire or sparsely shallowly crenate. Pistillate flowers 5-6 mm diam. Legumes compressed, 20-35 × 2-4 cm, irregularly twisted or falcate, densely yellowish green velutinous. Sunny slopes, valleys, streamsides or roadsides at 100-1000 m. Commonly cultivated. Distributed in C, N, E and NE China. Also in Korean Peninsula and Japan.

滇皂荚

Gleditsia japonica Miq. var. **delavayi** (Franch.) L. C. Li

本变种与山皂荚的区别在于本变种的雌花直径7-8(-9)毫米；荚果30-54 × 4.5-7厘米。花期4-6月，果期6-11月。生海拔1200-2500米的山林中或路边。产云南和贵州。

This variety differs from the typical variety in its pistillate flowers 7-9 mm diam; legumes 30-54 × 4.5-7 cm. Fl. Apr-Jun. Fr. Jun-Nov. Montane forests or roadsides at 1200-2500 m. Distributed in Yunnan and Guizhou.

滇皂荚 *Gleditsia japonica* var. *delavayi*

皂荚

Gleditsia sinensis Lam.

乔木或小乔木。刺粗壮，圆柱状、圆锥状，长达16厘米，常分枝。羽状复叶；小叶上面的网脉明显凸起，边缘具有细密的锯齿。荚果肥厚，直或扭转。花期3-5月，果期5-12月。生海拔200-2500米的山坡林中、谷地或路边。产中国除西北和东北以外大部分地区。

Trees or small trees. Spines robust, terete, conical, to 16 cm long, usually branched. Leaves pinnate; adaxial side of leaflet veins obviously convex, margin with fine close crenate. Legumes hypertrophic, straight or twisted. Fl. Mar-May. Fr. May-Dec. Mountain forests, valleys or roadsides at 200-2500 m. Distributed in most parts of China, except NW and NE China.

皂荚 *Gleditsia sinensis*

顶果树

Acrocarpus fraxinifolius Wight ex Arn.

大形乔木。二回羽状复叶；小叶4-8对，卵形或卵状长圆形，近革质。总状花序，花密集，较大，猩红色，雄蕊5。荚果淡紫褐色，扁平，沿腹缝线具窄翅。生海拔1000-1200米的疏林中。产广西和云南。南亚和东南亚亦有。

顶果树 *Acrocarpus fraxinifolius*

凤凰木 *Delonix regia*

Large trees. Leaves bipinnate; leaflets 4-8 pairs, ovate or ovate-oblong, subleathery. Inflorescences racemose, with scarlet aggregate flowers, flowers large; stamens 5. Legumes purplish brown, flattened, winged along ventral suture. Open forests at 1000-1200 m. Distributed in Guangxi and Yunnan. Also S and SE Asia.

凤凰木
Delonix regia (Bojer ex Hook.) Raf.

乔木。二回羽状复叶，小叶25对，密集，对生。花亮红色至橙红色，直径7-10厘米；花瓣5，鲜红色或橙红色，有黄色或白色斑。荚果带状，扁平，暗红褐色，成熟时黑褐色，顶端有宿存花柱。花期6-7月，果期8-10月。栽培于中国西南、华南和东南。广植于热带和亚热带地区；原产马达加斯加。

Trees. Leaves bipinnate. Leaflets 25 pairs, crowded, opposite, Flowers bright red to orange-red, 7-10 cm diam; petals 5, red or orange-red with yellow or white markings. Legumes lorate; compressed, dark reddish brown, blackish brown when mature, apex with persistent styles. Fl. Jun-Jul. Fr. Aug-Oct. Cultivated in SW, S and SE China. Also commonly cultivated in other tropical and subtropical areas; native to Madagascar.

刺果苏木
Caesalpinia bonduc (L.) Roxb.

有刺藤本，被黄色柔毛。叶长30-45厘米。总状花序腋生；花瓣黄色，雄蕊10，花丝离生，2轮排列。荚果革质，长圆形，顶端有喙，膨胀，外面有细长针刺。花期8-10月，果期10月至翌年3月。生海拔200米以下的灌丛、路边或海滨。产华南。世界热带地区亦有。

Climbers, prickly, yellow-pubescent. Leaves 30-45 cm long. Racemes axillary; petals yellow, stamens 10, filaments free, in 2 whorls. Legumes leathery, oblong, apex beaked, inflated, abaxially slender-spiny. Fl. Aug-Oct. Fr. Oct to next Mar. Thickets, roadsides or seashores below 200 m. Distributed in S China. Also in other tropical areas of the world.

刺果苏木 *Caesalpinia bonduc*

喙荚云实 *Caesalpinia minax*

喙荚云实
Caesalpinia minax Hance

具刺藤本。叶长达45厘米；小叶6-12对，中脉具柔毛。总状或圆锥花序顶生；花瓣白色，带紫色斑点。荚果长圆形，长10-15厘米，先端圆，具长喙。种子4-8粒，椭圆体形，铅灰色。花期3-11月，果期4-12月。生海拔100-1500米的山坡、山谷、灌丛或林缘。产中国西南和华南。印度、缅甸、老挝、泰国和越南亦有。

Climbers, prickly. Leaves to 45 cm long; leaflets 6-12 pairs, puberulent on midvein. Racemes or panicles terminal; petals whitish, tinged with purple spots. Legumes oblong, 10-15 cm long, apex round, with long beak. Seeds 4-8, ellipsoid, plumbeous. Fl. Mar-Nov. Fr. Apr-Dec. Slopes, valleys, thickets or forest edges at 100-1500 m. Distributed in SW and S China. Also in India, Myanmar, Laos, Thailand and Vietnam.

含羞云实
Caesalpinia mimosoides Lam.

木质藤本。小枝密被锈色腺毛和倒钩刺。二回羽状复叶长22-36厘米；羽片13-23对，长约3.5厘米；小叶7-14对，长圆形，长约9毫米，边缘和下面有刚毛。总状花序顶生，具50多朵花；花瓣亮黄色，近圆形，上面一片较小。荚果倒卵形，呈镰刀状，长4-5厘米，被刚毛。花期11-12月，果期翌年2-3月。生海拔600-700米的路旁灌丛。产云南。南亚和东南亚亦有。

Lianas woody. Branchlets densely ferruginous glandular hairy, with recurved prickles. Bipinnate leaves 22-36 cm long; pinnae 13-23 pairs, ca. 3.5 cm long; leaflets 7-14 pairs, oblong, ca. 9 mm long, marginally and abaxially with bristles. Racemes terminal, with more than 50 flowers; petals bright yellow, suborbicular, upper one smaller. Legume obovoid, falcate, 4-5 cm long, setose. Fl. Nov-Dec. Fr. Feb-Mar of next year. Among bushes near roads at 600-700 m. Distributed in Yunnan. Also in S and SE Asia.

鸡嘴簕
Caesalpinia sinensis (Hemsl.) J. E. Vidal

藤本。叶轴具刺；小叶2对，长圆形至卵形，6-9 × 2.5-3.5厘米，革质，下面沿脉具柔毛，

含羞云实 *Caesalpinia mimosoides*

鸡嘴簕 *Caesalpinia sinensis*

先端渐尖、锐尖或钝。圆锥花序腋生或顶生。荚果近球形，压扁，先端喙长约3毫米。花期3-5月，果期7-10月。生海拔100-900米的林中或灌丛中。产云南、贵州、广西、广东和湖北。缅甸、老挝和越南亦有。

Climbers. Rachises with prickles; leaflets 2 pairs, oblong to ovate, 6-9 × 2.5-3.5 cm, leathery, abaxially hairy on midvein, apex acuminate, acute or obtuse. Panicles axillary or terminal. Legumes subglobose, depressed, apex with beak ca. 3 mm long. Fl. Mar-May. Fr. Jul-Oct. Forests or thickets at 100-900 m. Distributed in Yunnan, Guizhou, Guangxi, Guangdong and Hubei. Also in Myanmar, Laos and Vietnam.

云实

Caesalpinia decapetala (Roth) Alston

藤本。枝、叶轴和花序具反折的刺及柔毛。小叶常长圆形。总状花序顶生，具大量的花；花瓣花期反折，黄色。荚果长圆形，长6-12厘米，宽2.5-3厘米。种子6-9粒。花果期4-10月。生海拔1800米以下的灌丛、山谷或河边。产中国除东北和西北以外大部分地区。南亚、东南亚和东亚亦有。

Climbers. Branches, rachis of leaves and inflorescence with recurved prickles and pubescent. Leaflets usually oblong. Racemes terminal, with abundant flowers; petals reflexed at anthesis, yellow. Legumes oblong, 6-12 cm long, 2.5-3 cm broad. Seeds 6-9. Fl. and fr. Apr-Oct. Thickets, valleys or riversides below 1800 m. Distributed in most parts of China, except NE and NW China. Also in S, SE and E Asia.

苏木 *Caesalpinia sappan*

苏木

Caesalpinia sappan L.

小乔木。小叶10-17对，长圆形至长圆状菱形，纸质，两面无毛或疏具毛。圆锥花序，顶生或腋生；花瓣黄色，最上面花瓣先端全缘，基部带粉色。荚果长圆形或倒卵球状长圆形。花期5-10月，果期7月至翌年3月。原产地不清。栽培于华南。亦栽培于世界热带地区。

Small trees. Leaflets 10-17 pairs, oblong to oblong-rhombic, papery, both surfaces glabrous or sparsely hairy. Panicles terminal or axillary; petals yellow, uppermost one entire at apex, tinged pink at base. Legumes oblong or obovoid-oblong. Fl. May-Oct. Fr. Jul to next Mar. Origin unknown. Cultivated in S China. Also planted in tropical areas of the world.

云实 *Caesalpinia decapetala*

金凤花 *Caesalpinia pulcherrima*

金凤花

Caesalpinia pulcherrima (L.) Swartz

灌木或小乔木。小枝具散生的疏刺。小叶7-11对，长圆形或倒卵形。伞房式总状花序，花瓣红色或橙红色。荚果倒披针状长圆形，不具翅，无毛，不开裂，先端圆形。花果期几年。广泛栽培于华南。原产于西印度群岛；现广植于热带各地。

Shrubs or small trees. Branches with scattered, sparse prickles. Leaflets 7-11 pairs, oblong or obovate. Racemes corymbose; petals red or orange-red. Legumes oblanceolate-oblong, not winged, glabrous, indehiscent, apex rounded. Fl. and fr. almost all year round. Widely planted in S China. Native to the West Indies; widely planted in tropical areas.

肉荚云实

Caesalpinia digyna Rottler

大藤本，具反折的刺。小叶7-9对，近无柄，紧密，长圆形，6-9 × 约3毫米，纸质，两面起初具毛；叶轴长17-23厘米，羽片长3-6厘米。总状花序顶生或腋生；花瓣黄色，近圆形。荚果长圆形，肉质。花期4-11月，果期5月至翌年3月。生海拔200-300米的山坡灌丛或海边。产云南和海南。南亚和东南亚亦有。

Climbers, large, with recurved prickles. Leaflets 7-9 pairs, subsessile, closely spaced, oblong, 6-9 × ca. 3 mm, papery, both surfaces pilose at first; leaf axes 17-23 cm long, pinnae 3-6 cm long. Racemes terminal or axillary; petals yellow, suborbicular.

肉荚云实 *Caesalpinia digyna*

老虎刺 *Pterolobium punctatum*

Legumes oblong, fleshy. Fl. Apr-Nov. Fr. May to next Mar. Thickets on slopes or seashores at 200-300 m. Distributed in Yunnan and Hainan. Also in S and SE Asia.

老虎刺
Pterolobium punctatum Hemsl.

木质藤本。小枝具散生或成对生叶柄基部的黑色、弯曲的短刺。羽片9-14对；小叶小且多数，常19-30对，脉不明显，下面具不明显或明显的黑色点。花密簇生；花瓣同形。荚果长4-6厘米。花期6-8月，果期9月至翌年4月。生海拔300-2000米的山坡疏林阳处、路旁干旱地或石灰岩山上。产中国西南、华南、东南和华中。

Climbers, woody. Branchlets with blackish, recurved, short prickles scattered or in pairs at bases of petioles. Pinnae 9-14 pairs; leaflets smaller and numerous, usually 19-30 pairs, veins obscure, abaxially with conspicuous or obscure blackish dots. Flowers densely fascicled; petals homomorphic. Legumes 4-6 cm long. Fl. Jun-Aug. Fr. Sep to next Apr. Sunny places of open forests on slopes, dry lands of roadsides or on limestone rocks at 300-2000 m. Distributed in SW, S, SE and C China.

格木
Erythrophleum fordii Oliv.

乔木。叶无毛；羽片常3对，对生或近对生；小叶8-12。圆锥花序15-20厘米；萼钟状，花瓣淡黄绿色，雄蕊长为花瓣的2倍；子房长圆形，密被黄白色柔毛，10-12室。荚果厚革质，有网脉。花期5-6月，果期8-10月。生山地密林或疏林中。产华南。越南亦有。

Trees. Leaves glabrous; pinnae usually 3 pairs, opposite or sub-opposite; leaflets 8-12. Panicles 15-20 cm long; calyx campaniform, petals yellowish green, stamens ca. 2 × as long as petals; ovary oblong, densely yellowish white pubescent, 10-12-ovuled. Legumes thick leathery, with reticulate veins. Fl. May-Jun. Fr. Aug-Oct. Dense or sparse forests in mountain slopes. Distributed in S China. Also in Vietnam.

格木 *Erythrophleum fordii*

任豆 *Zenia insignis*

任豆

Zenia insignis Chun

乔木。奇数羽状复叶；小叶互生，长圆状披针形，薄革质，下面具灰白色硬毛。圆锥花序顶生；花红色；子房7-9胚珠，边缘具贴伏毛。荚果矩圆形或椭圆状矩圆形，红褐色。花期5月，果期6-8月。生海拔200-1000米的林中或山坡。产广西和广东。越南亦有。

Trees. Leaves imparipinnate; leaflets alternate, oblong-lanceolate, thin leathery, abaxially grayish white strigose. Panicles terminal; flowers red; ovary 7-9-ovuled, margin adpressed pilose. Legumes oblong or ellipsoid-oblong, reddish brown. Fl. May. Fr. Jun-Aug. Forests or slopes at 200-1000 m. Distributed in Guangxi and Guangdong. Also in Vietnam.

腊肠树

Cassia fistula L.

落叶乔木。叶具3或4对小叶。雄蕊10，其中3枚具长而弯曲花丝，高出花瓣，4枚短而直，具有宽大花药，其余3枚很小，不育。荚果圆柱形，长30-60厘米，直径2-2.5厘米，黑褐色，不开裂。花期6-8月，果期10月。中国西南和华南均有栽培。原产印度、斯里兰卡和缅甸。

Deciduous trees. Leaves with 3 or 4 pairs of leaflets. Stamens 10, 3 of those with curved filaments, longer than petals, 4 short and erect, with big anthers, other 3 very small, sterile. Legumes terete, 30-60 cm long, 2-2.5 cm diam, black-brown, indehiscent. Fl. Jun-Aug. Fr. Oct. Cultivated in SW and S China. Native to India, Sri Lanka and Myanmar.

神黄豆

Cassia jawanica subsp. **agnes** (de Wit) K. Larsen

乔木。小叶6-10对，椭圆形或长圆状椭圆形，两面被柔毛。圆锥花序长6-9厘米，顶生于幼枝，由6-10个总状花序组成；花粉红色。荚果圆柱状，30-50 × 约0.2厘米，具环节。种子多数。花期3-4月，果期8-10月。生林中或山坡上。产云南和广西。老挝、越南和泰国亦有；广泛栽培于热带亚洲。

腊肠树 *Cassia fistula*

Trees. Leaflets 6-10 pairs, elliptic or oblong-elliptic, pilose on both surfaces. Panicles 6-9 cm long, terminal on young leafy shoots, composed of 6-10 racemes; flowers pink. Legumes terete, 30-50 × ca. 0.2 cm, with annular nodes. Seeds numerous. Fl. Mar-Apr. Fr. Aug-Oct. Forests or on mountain slopes. Distributed in Yunnan and Guangxi. Also in Laos, Vietnam and Thailand; widely planted in tropical Asia.

神黄豆 *Cassia jawanica* subsp. *agnes*

含羞草决明(山扁豆)

Chamaecrista mimosoides (L.) Greene

草本。小叶20-50对，长3-4毫米，1-1.5厘米宽，叶柄上端有一圆盘状腺体。花腋上生，多单花；花瓣亮黄色，不等大。荚果扁平，镰形。种子10-20粒，平滑。花果期6-11月。生山坡、荒地、灌丛或草地。产中国西南和东南。原产热带美洲，现遍布于热带和亚热带地区。

Herbs. Leaflets 20-50 pairs, 3-4 mm long, 1-1.5 mm broad, petioles with one disclike glands at top. Flowers supra-axillary, mostly solitary; petals bright yellow, unequal. Legumes flat, falcate. Seeds 10-20, smooth. Fl. and fr. Jun-Nov. Slopes, wastelands, bushes or grasslands. Distributed in SW and SE China. Native to tropical America, widely distributed in tropical and subtropical areas of the world.

望江南

Senna occidentalis (L.) Link

亚灌木或灌木。小叶卵状长圆形或椭圆形；叶柄近基部有一枚大而带褐色圆锥形腺体。聚伞总状花序具少花，腋生或顶生，长约5厘米；花瓣黄色，具紫色脉纹，外部2个稍大。荚果带状，扁平，10-13 × 约1厘米。花期4-8月，果期6-10月。生河边滩地、旷野或丘陵的灌木林或疏林中。产中国西南、华南和东南。热带和亚热带地区亦有。

Subshrubs or shrubs. Leaflets ovate-oblong or elliptic; petioles with one big and brownish conic glands near the base. Corymbose racemes few flowered, axillary or terminal, ca. 5 cm long; petals yellow, purplish veined, outer 2 slightly larger. Legumes lorate, compressed, 10-13 × ca. 1 cm. Fl. Apr-Aug. Fr. Jun-Oct. Shrubs or sparse forests on river banks, wilderness or hilly lands. Distributed in SW, S and SE China. Also in other tropical and subtropical areas.

含羞草决明(山扁豆) *Chamaecrista mimosoides*

望江南 *Senna occidentalis*

槐叶决明 *Senna sophera*

槐叶决明
Senna sophera (L.) Roxb.

灌木或亚灌木。小叶较小，4-10对，披针形或椭圆状披针形，先端锐尖至短渐尖；叶柄腺体狭，棍棒状至近钻形，基部节上生。聚伞花序腋生，具少花。荚果短，5-10 × 0.5-1厘米，起初扁平且稍加厚，近圆柱状，成熟后稍膨大。花期7-9月，果期10-12月。生山坡或路边。产中国西南、华南、东南和华中。原产亚洲热带。

Shrubs or subshrubs. Leaflets small, 4-10 pairs, lanceolate or elliptic-lanceolate, apex acute to shortly acuminate; petiolar gland narrow, clavate to subulate, above basal joint. Corymbs axillary, few flowered. Legumes short, 5-10 × 0.5-1 cm, flattened and slightly thick at first, subcylindric, swollen when ripe. Fl. Jul-Sep. Fr. Oct-Dec. Mountain slopes or roadsides. Distributed in SW, S, SE and C China. Native to tropical Asia.

决明
Senna tora (L.) Roxb.

一年生直立草本。小叶倒卵圆形或倒卵圆状长圆形，先端圆形钝而有小尖头，每两小叶间有1枚棒状腺体。总状花序腋生，短，具1或2(或3)花；花瓣黄色，下面2个稍长。荚果圆柱状，近四棱形，10-15 × 0.3-0.5厘米。花果期8-11月。生海拔250-2200米的草地、灌丛或山坡。产中国长江以南各省区。广布于热带和亚热带地区；原产美洲。

Herbs, annual, erect. Leaflets obovate or obovate-oblong, apex obtuse, mucronate, with 1 clavate gland between each pair of leaflets. Racemes axillary, short, 1- or 2(or 3)-flowered; petals yellow, lower 2 slightly longer. Legumes terete, subtetragonous,

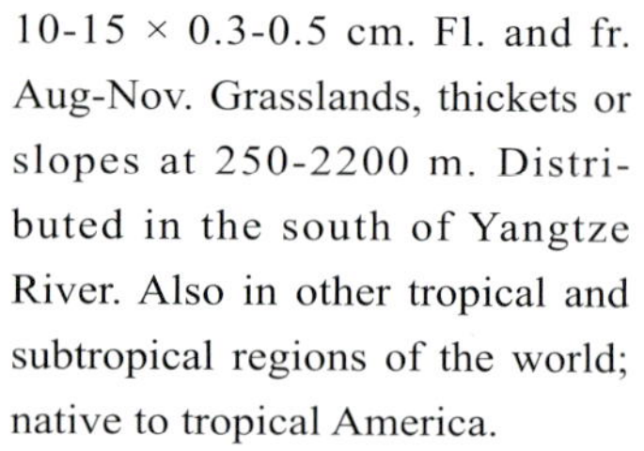

10-15 × 0.3-0.5 cm. Fl. and fr. Aug-Nov. Grasslands, thickets or slopes at 250-2200 m. Distributed in the south of Yangtze River. Also in other tropical and subtropical regions of the world; native to tropical America.

豆茶决明
Senna nomame (Makino) T. C. Chen

一年生草本。上部叶柄具一黑

豆茶决明 *Senna nomame*

决明 *Senna tora*

翅荚决明 *Senna alata*

褐色、碟形无柄的腺体；小叶8-28对，舌状披针形。花生叶腋，具柄，单生或2至数朵组成总状花序，腋生。荚果扁平，开裂。生山坡或草丛中。产中国大部分地区。朝鲜半岛和日本亦有。

Annual herbs. Leaves with a blackish brown, discoid, sessile gland on upper part of petioles; leaflets 8-28 pairs, ligulate-lanceolate. Flowers inserted in axils of leaves, with pedicels, solitary or 2 to several arranged to racemes, axillary. Legumes depressed, dehiscent. Slopes or grasses. Distributed in most parts of China. Also in Korean Peninsula and Japan.

翅荚决明

Senna alata (L.) Roxb.

灌木。偶数羽状复叶，有12-24小叶，薄革质，在近轴面的叶柄和叶轴上有狭翅。总状花序腋生，密集，具多花，或有时数个总状花序顶生圆锥状。荚果长带状，有翅。花期11月至翌年1月，果期12月至翌年2月。生疏林或较干旱山坡。产云南和广东。原产热带美洲；亦广布于全世界热带地区。

Shrubs. Leaves paripinnate, 12-24-foliolate, thinly coriaceous, adaxially with narrow wings on the petioles and leaf axes. Racemes axillary, dense, many flowered, or sometimes several racemes forming a terminal panicle. Legumes long lorate, winged. Fl. Nov to next Jan. Fr. Dec to next Feb. Open forests or dry slopes. Distributed in Yunnan and Guangdong. Native to tropical America; also widely distributed in tropical regions of the world.

毛荚决明

Senna hirsuta (L.) H. S. Irwin et Barneby

草本或灌木，密被长糙毛。叶有小叶8-12，叶柄基部的上面有呈黑褐色腺体。总状花序腋生或数个生顶端叶腋形成具叶圆锥花序；花瓣黄色。荚果条形，具棱，密被长糙毛。花期10-11月，果期11-12月。栽培或归化于海拔80-900米的林缘、村旁或路边。栽培于云南和广东。老挝、越南和印度尼西亚亦有；原产热带美洲。

Herbs or shrubs, dense hirsute. Leaves with 8-12 leaflets, petioles adaxially with one dark brown glands at base. Racemes axillary or several in axils of apical leaves forming a leafy panicle; petals yellow. Legumes linear, ribbed, dense hirsute. Fl. Oct-Nov. Fr. Nov-Dec. Cultivated or naturalized in forest edges, near villages or roadsides at 80-900 m. Cultivated in Yunnan and Guangdong. Also in Laos, Vietnam and Indonesia; native to tropical America.

毛荚决明 *Senna hirsuta*

光叶决明 *Senna septemtrionalis*

黄槐决明 *Senna surattensis*

光叶决明

Senna septemtrionalis (Viv.) H. S. Irwin et Barneby

灌木或小乔木。小叶3-4对，在每对小叶间和叶轴上有1个腺体；小叶卵形或卵状披针形，两面无毛。总状花序腋生或顶生，具4-10花；萼片黄绿色；花瓣黄色。荚果圆柱形，完全成熟后近四方形。花期5-7月，果期10-11月。栽培于云南、广西和广东。亦广布世界热带和亚热带地区。

Shrubs or small trees. Leaflets 3-4 pairs, with a gland between each pair of leaflets and on leaf axis; leaflets ovate or ovate-lanceolate, both surfaces glabrous. Racemes axillary or terminal, 4-10-flowered; sepals yellowish green; petals yellow. Legumes terete, slightly subquadrangular when fully mature. Fl. May-Jul. Fr. Oct-Nov. Cultivated in Yunnan, Guangxi and Guangdong. Also widely cultivated in tropical and subtropical regions of the world.

铁刀木 *Senna siamea*

黄槐决明

Senna surattensis (Burm. f.) H. S. Irwin et Barneby

灌木或小乔木。嫩枝、叶轴、叶柄被微柔毛。叶长10-15厘米；小叶6-9对。总状花序生于

广西紫荆 *Cercis chuniana*

枝条上部叶腋内；花瓣亮黄色至深黄色。荚果扁平，带状，开裂。栽培于华南。原产印度、斯里兰卡、印度尼西亚、菲律宾、澳大利亚和波利尼西亚等地。

Shrubs or small trees. Young branches, rachises of leaves and petioles puberulous. Leaves 10-15 cm long; leaflets 6-9 pairs. Racemes in axils of apical leaves; petals bright yellow to deep yellow. Legumes flat, strap-shaped, dehiscent. Cultivated in S China. Native to India, Sri Lanka, Indonesia, the Philippines, Australia and Polynesia.

铁刀木

Senna siamea (Lam.) H. S. Irwin et Barneby

乔木。小叶6-10对，下面无毛且具白粉；托叶线形，脱落。总状花序生顶端叶腋，常数个组成一个大的顶生圆锥花序。荚果扁平，边缘加厚，长15-30厘米，宽1-1.5厘米。花期10-11月，果期12月至翌年1月。产云南南部，栽培于华南。印度、缅甸和泰国亦有。

Trees. Leaflets 6-10 pairs, abaxially glabrous and farina-white; stipules linear, caducous. Racemes in axils of apical leaves, usually several forming a large terminal panicle. Legumes flattened, margin thickened, 15-30 cm long, 1-1.5 cm broad. Fl. Oct-Nov. Fr. Dec to next Jan. Distributed in S Yunnan, cultivated in S China. Also in India, Myanmar and Thailand.

广西紫荆

Cercis chuniana F. P. Metcalf

乔木。叶纸质；小叶(5-)7(-9)厘米，顶端锐尖至钝，稍钝，全缘，常波状，两面常被白粉。总状花序，7-15花；花瓣玫瑰粉色至浅粉色；雄蕊10，分离。荚果紫红色，干后呈红褐色，窄长圆形，极压扁。种子2-5粒，黑褐色。果期9-11月。生海拔600-1900米的山谷、溪边疏林或密林中。产广西、贵州、广东、湖南和江西。

Trees. Leaves papery; leaflets (5-)7(-9) cm, apex acute to obtuse, slightly retuse, margin entire, often sinuous, glaucous on both surfaces. Racemes, 7-15-flowered; petals rose-pink to whitish pink; stamens 10, free. Legumes purple-red, red-brown when dry, narrowly oblong, flattened. Seeds 2-5, blackish brown. Fr. Sep-Nov. Valleys, sparse and dense forests at 600-1900 m. Distributed in Guangxi, Guizhou, Guangdong, Hunan and Jiangxi.

湖北紫荆

Cercis glabra Pamp.

乔木。叶心脏形或三角状圆形，基部心形，无毛或脉腋疏具柔毛。总状花序短，花序轴短于2厘米；花淡紫色或粉红色。荚果紫红色，宽条形，缝不等长，背缝稍长。花期3-4月，果期9-11月。生海拔600-1900米的山地疏或密林中、山谷或岩石上。产中国西南、东南和华中。

Trees. Leaves cordate or triangular-orbicular, base cordate, glabrous or sparsely pubescent in axils of veins. Racemes short, rachis less than 2 cm long; flowers purplish or pink. Legumes purplish red, broadly linear, sutures unequal, dorsal suture slightly longer. Fl. Mar-Apr. Fr. Sep-Nov. Open or dense forests, valleys or on rocks at 600-1900 m. Distributed in SW, SE and C China.

湖北紫荆 *Cercis glabra*

紫荆 *Cercis chinensis*

紫荆
Cercis chinensis Bunge

灌木。叶近圆形或三角状圆形，纸质，两面常无毛。总状花序无梗；花簇生，通常先叶开放；花瓣紫红色或粉红色。荚果淡绿色，成熟后禾秆色，压扁，狭长圆形，背缝与腹缝等长或近等长。花期3-5月，果期8-10月。广为栽培。产中国西南、东南、华北、华中、华东和东北。

Shrubs. Leaves suborbicular or triangular-orbicular, papery, both surfaces usually glabrous. Racemes expedunculate; flowers fascicular, usually blooming before leaves; petals purplish red or pink. Legumes greenish, becoming stramineous at maturity, compressed, narrowly oblong, dorsal and ventral sutures equal or subequal. Fl. Mar-May. Fr. Aug-Oct. Widely planted. Distributed in SW, SE, N, C, E and NE China.

白花羊蹄甲
Bauhinia acuminata L.

灌木或小乔木。叶先端2裂，裂达叶长1/3-2/5，裂片先端急尖或稍渐尖。花序总状，具3-15花，腋生，似聚伞状；花瓣白色。荚果直或稍弯曲，条状倒披针形，压扁。花期4-6月，果期6-8月。生海拔280-800米的山坡阳处，或栽培。产云南、广西和广东。印度、斯里兰卡、越南、马来西亚和菲律宾亦有。

Shrubs or small trees. Leaf apex 2-lobed up to 1/3-2/5 of the leaves, lobe apex acute or slightly acuminate. Inflorescences a raceme, with 3-15 flowers, axillary, appearing cymose; petals white. Legumes straight or slightly curved, linear-oblanceolate, compressed. Fl. Apr-Jun. Fr. Jun-Aug. Sunny slopes or planted at 280-800 m. Distributed in Yunnan, Guangxi and Guangdong. Also in India, Sri Lanka, Vietnam, Malaysia and the Philippines.

红花羊蹄甲
Bauhinia × blakeana Dunn

乔木。叶革质，圆形或近圆形，先端两裂为叶全长的1/4-1/3，裂片顶端钝或狭圆，上面无毛，下面疏被短柔毛。总状花序顶生或腋生，有时复合为圆锥花序。通常不结果。花期11月至翌年3月。栽培于华南。世界热带地区广泛栽植。

白花羊蹄甲 *Bauhinia acuminata*

红花羊蹄甲 *Bauhinia × blakeana*

Trees. Leaves leathery, orbicular or suborbicular, apex bilobed to 1/4-1/3, lobes rounded or narrowly rounded, adaxially glabrous, abaxially puberulent. Inflorescences racemose, terminal or axillary, or several racemes together forming a panicle. Infertile. Fl. Nov to next Mar. Cultivated in S China. Also cultivated in tropical areas of the world.

洋紫荆
Bauhinia variegata L.

乔木。叶近圆形或阔卵形，近革质。花萼一侧深裂，顶端全缘；花瓣白色，或带粉红色或紫色，长4-5厘米。荚果条形，扁平，果瓣木质，15-25 × 1.5-2厘米。花期2-5月，果期3-7月。生海拔150-1500米的林中。原产云南南部，华南广泛栽培。缅甸、老挝、泰国和越南亦有。

Trees. Leaves suborbicular or broadly ovate, subleathery. Calyx deeply divided on one side, apex entire; petals white, or with pink or purplish spots, 4-5 cm long. Legumes linear, flat, valves woody, 15-25 × 1.5-2 cm. Fl. Feb-May. Fr. Mar-Jul. Forests at 150-1500 m. Native to S Yunnan, widely cultivated in S China. Also in Myanmar, Laos, Thailand and Vietnam.

洋紫荆 *Bauhinia variegata*

鞍叶羊蹄甲 *Bauhinia brachycarpa*

鞍叶羊蹄甲

Bauhinia brachycarpa Wall. ex Benth.

灌木。小叶小，长(5-)10-23毫米，先端深裂；基出脉7(-9)条。总状花序，长1.5-3厘米；花瓣小，长约5毫米；能育雄蕊通常10。荚果倒披针形，先端具长喙，果瓣黑褐色，有光泽。花期5-7月，果期8-10月。生海拔3200米以下的开阔森林或干山坡。产中国西南、华中和华西。缅甸、老挝和泰国亦有。

Shrubs. Leaflets small, (5-)10-23 mm long, apex parted, 7(-9)-nerved. Racemes 1.5-3 cm long; petals small, ca. 5 mm long; fertile stamens usually 10. Legumes oblanceolate, apex beaked, valves black-brown, lucid. Fl. May-Jul. Fr. Aug-Oct. Open forests or dry mountain slopes below 3200 m. Distributed in SW, C and W China. Also in Myanmar, Laos and Thailand.

阔裂叶羊蹄甲

Bauhinia apertilobata Merr. et Metc.

木质藤本，具卷须。叶卵形、阔椭圆形或近圆形，纸质，下面具淡褐色柔毛，浅2裂，具极宽的二叉裂片。伞房花序式的总状花序腋生或1-2个顶生。荚果倒披针形或长圆形。种子扁平。花期5-7月，果期8-11月。生海拔300-600米的山谷、山坡林中或灌丛中。产广西、贵州、广东、福建和江西。

Lianas, woody, with tendrils. Leaves ovate, broadly elliptic or suborbicular, papery, abaxially brownish puberulent, shallowly bifid with very broadly divergent lobes. Corymbiform racemes axillary or 1-2 terminal. Legumes oblanceolate or oblong. Seeds depressed. Fl. May-Jul. Fr. Aug-Nov. Valleys, forests on mountain slopes or bushes at 300-600 m. Distributed in Guangxi, Guizhou, Guangdong, Fujian and Jiangxi.

龙须藤

Bauhinia championii (Benth.) Benth.

藤本，具卷须。叶纸质，基出脉5-7条。总状花序腋生；花瓣白色；能育雄蕊3，花丝离生，无毛；退化雄蕊2-3。荚果倒卵球状长圆形，扁平，果瓣革质。种子2-5粒。花期6-10月，果期7-12月。生低海拔至中海拔的丘陵灌丛或山地疏林和密林中。产中国西南、东南和华中。越南亦有。

Lianas, with tendrils. Leaves papery, 5-7-nerved. Racemes axillary; petals white; fertile stamens 3; filaments free, glabrous; staminodes 2-3. Legumes obovoid-oblong, compressed,

阔裂叶羊蹄甲 *Bauhinia apertilobata*

龙须藤 *Bauhinia championii*

丽江羊蹄甲 *Bauhinia bohniana*

云南羊蹄甲 *Bauhinia yunnanensis*

valves leathery. Seeds 2-5. Fl. Jun-Oct. Fr. Jul-Dec. Thickets on hills or sparse and dense forests from low to middle elevation. Distributed in SW, SE and C China. Also in Vietnam.

丽江羊蹄甲

Bauhinia bohniana L. Chen

直立灌木，无卷须。叶扁圆形，宽大于长，近革质，下面幼时具锈色柔毛，老时柔毛渐少，先端2裂至1/4。花序聚伞状总状，顶生或侧生；花冠粉红色。荚果条形，压扁，缝膨大。花期4-8月，果期5-10月。生海拔1700-2000米的山坡阳处灌丛中。产云南。

Erect shrubs, tendrils absent. Leaves oblate, width longer than length, subleathery, abaxially rusty pubescent on young leaves, becoming less pubescent when old, apex bifid to 1/4. Inflorescences corymbose-racemose, terminal or lateral; corolla pink. Legumes strap-shaped, compressed, sutures swollen. Fl. Apr-Aug. Fr. May-Oct. Thickets on sunny hillsides at 1700-2000 m. Distributed in Yunnan.

云南羊蹄甲

Bauhinia yunnanensis Franch.

木质藤本，细弱，无毛，卷须成对。叶两面近无毛，先端2裂至近基部，裂片斜卵型，先端钝圆或圆。花序为总状，顶生或与叶对生，具10-20花；花托圆柱状，长7-8毫米；花瓣淡粉色，脉上具深红色条纹。荚果条状长圆形。花期8月，果期10月。生海拔400-2200米的山坡灌丛。产云南、四川和贵州。缅甸和泰国亦有。

Lianas, woody, slender, glabrous, tendrils in pairs. Leaves subglabrous on both surfaces, apex bifid almost up to the base, lobes oblique ovate, apex obtuse or rounded. Inflorescences a raceme, terminal or opposite to leaves, 10-20-flowered; receptacles cylindric, 7-8 mm long; petals pinkish, with dark red stripes along veins. Legumes linear-oblong. Fl. Aug. Fr. Oct. Thickets on hills at 400-2200 m. Distributed in Yunnan, Sichuan and Guizhou. Also in Myanmar and Thailand.

毛叶牛蹄麻

Bauhinia khasiana Baker var. **tomentella** T. C. Chen

木质藤本。叶革质至近革质，下面被锈色短柔毛，叶柄密被短柔毛。花序伞房状，或由数个伞房花序组成，顶生，具锈色柔毛；花瓣金黄色。荚果长圆状披针形，果瓣厚革质，无毛，有光泽。花期7月。生混交林中。产云南。

Lianas, woody. Leaves leathery to subleathery, pubescent with rusty hairs abaxially, and densely pubescent on petioles. Inflorescences corymbose, or consisting of several corymbs, terminal, rusty pubescent; petals golden yellowish. Legumes oblong-lanceolate, compressed, leathery, glabrous. Fl. Jul. Mixed forests. Distributed in Yunnan.

毛叶牛蹄麻 *Bauhinia khasiana* var. *tomentella*

缅甸羊蹄甲

Bauhinia nervosa (Wall. ex Benth.) Baker

木质攀援藤本，粗壮，具卷须。叶近圆形，革质，下面具薄绒毛，先端2裂至约1/3，裂片先端圆形。花蕾棒状，长4.5厘米；花瓣白色，基部近红色，近革质，倒卵圆状匙形；萼裂片长25-30毫米；花丝长约35毫米，无毛。花期9月。生海拔1500-1600米的疏林中。产云南。印度、缅甸和泰国亦有。

Lianas, woody, climbers, robust, with tendrils. Leaves suborbicular, subleathery, abaxially thinly pubescent, apex bifid to ca. 1/3, lobes rounded at apex. Flower buds clavate, 4.5 cm long; petals white, reddish at base, subleathery, obovate-spathulate; calyx lobes 25-30 mm long; filaments ca. 35 mm long, glabrous. Fl. Sep. Open forests at 1500-1600 m. Distributed in Yunnan. Also in India, Myanmar and Thailand.

粉叶羊蹄甲

Bauhinia glauca (Wall. ex Benth.) Benth.

木质藤本，具卷须。小枝幼时具浅红色柔毛。叶柄长2-4厘米；叶纸质，基出脉7-11条，先端2裂至1/3-1/2。花序相对较小；萼片外面被锈色绒毛；花瓣白色。荚果薄，长18-25厘米，无毛，不开裂。花期4-6月，果期7-9月。生山坡阳处疏林中或山谷荫蔽的密林中或灌丛中。产中国西南、华南和华中。柬埔寨、印度、中南半岛和印度尼西亚亦有。

Woody climbers, with tendrils. Shoots reddish pubescent when young. Petiole 2-4 cm long; leaves papery, 7-11-nerved, apex bifid to 1/3-1/2. Inflorescences relatively small; sepals abaxially rusty-tomentose; petals white. Legumes thin, 18-25 cm long, glabrous, indehiscent. Fl. Apr-Jun. Fr. Jul-Sep. Sparse forests on sunny slopes or dense forests in wet valleys or thickets. Distributed in SW, S and C China. Also in Cambodia, India, Indo-China Peninsula and Indonesia.

粉叶羊蹄甲 *Bauhinia glauca*

薄叶羊蹄甲

Bauhinia glauca (Wall. ex Benth.) Benth. subsp. **tenuiflora** (Watt ex C. B. Clarke) K. Larsen et S. S. Larsen

本亚种与粉叶羊蹄甲的区别在于本亚种的叶柄长1-2(-3)厘米，叶较大，长7-9厘米，主脉9-11，先端2裂至仅1/5。花序相对较大；花芽具毛；花托长25-30厘米。花期6-7月，果期9-12月。生海拔约1300米的沟谷密林中或灌丛中。产中国西南、华南和华中。缅甸、老挝、泰国和越南亦有。

This subspecies differs from the typical subspecies in its petioles

缅甸羊蹄甲 *Bauhinia nervosa*

薄叶羊蹄甲 *Bauhinia glauca* subsp. *tenuiflora*

首冠藤 *Bauhinia corymbosa*

1-2(-3) cm long, leaves relatively large, 7-9 cm long, primary veins 9-11, apex bifid to only 1/5. Inflorescences relatively large; flower buds hairy; receptacles 25-30 mm long. Fl. Jun-Jul. Fr. Sep-Dec. Dense forests in valleys or in thickets at ca. 1300 m. Distributed in SW, S and C China. Also in Myanmar, Laos, Thailand and Vietnam.

首冠藤

Bauhinia corymbosa Roxb. ex DC.

木质藤本。叶纸质，基出脉7条。花芳香；花瓣白色，有粉红色脉纹；能育雄蕊3；花丝淡红色；退化雄蕊2-5。荚果带状长圆形，长10-25厘米，果瓣厚革质。花期4-8月，果期6-12月。生山谷或阳坡疏林中。产广西、广东和海南。世界热带和亚热带地区有栽培。

Lianas, woody. Leaves papery, veins 7. Flowers aromatic; petals white, with pink stripes; fertile stamens 3; filaments pale red; staminodes 2-5. Legumes linear-oblong, 10-25 cm long, valves thick leathery. Fl. Apr-Aug. Fr. Jun-Dec. Sparse forests in valleys or sunny slopes. Distributed in Guangxi, Guangdong and Hainan. Also cultivated in tropical and subtropical regions of the world.

囊托羊蹄甲

Bauhinia touranensis Gagnepain

木质藤本。卷须扁平，一面被短柔毛。叶近圆形，先端2裂达叶长1/6-1/5；基出脉7-9条。伞房式总状花序被锈色短柔毛；花序梗基部具卷须；花萼裂片5，卵形；花瓣淡绿色或黄白色，具长爪，外面中部被毛。荚果带状，长12-16厘米，无毛，荚缝略增厚。花期3-6月，果期8-10月。生海拔500-1200米的疏林或山坡灌丛中。产云南、贵州和广西。东南亚亦有。

Lianas, woody. Tendrils compressed, pubescent on one side. Leaves suborbicular, apex bifid to 1/6-1/5, basal veins 7-9. Corymbose racemes rusty pubescent; peduncle with a tendril at base; calyx lobes 5, ovate; petals greenish or yellowish white, prominently clawed, abaxially hairy at middle. Legume strap-shaped, 12-16 cm long, glabrous, sutures slightly thickened. Fl. Mar-Jun. Fr. Aug-Oct. Open forests and thickets on slopes at 500-1200 m. Distributed in Yunnan, Guizhou and Guangxi. Also in SE Asia.

囊托羊蹄甲 *Bauhinia touranensis*

仪花 *Lysidice rhodostegia*

仪花
Lysidice rhodostegia Hance

灌木或小乔木。小叶长圆形或卵状披针形，纸质。圆锥花序长20-40厘米；花瓣紫红色，连爪长约1.2厘米；苞片和小苞片粉色；萼筒长1.2-1.5厘米，长于裂片；能育雄蕊2，花丝分离或基部稍连合。荚果倒卵球形状长圆形。花期6-8月，果期9-11月。生海拔500米以下的山坡、灌丛、路旁或沟谷河边。产云南、贵州、广西和广东。越南亦有。

Shrubs or small trees. Leaflets oblong or ovate-lanceolate, papery. Panicles 20-40 cm long; petals purplish red, ca. 1.2 cm with claws; bracts and bracteoles pink; calyx tubes 1.2-1.5 cm long, longer than lobes; fertile stamens 2, filaments free or slightly connate at base. Legumes obovoid-oblong. Fl. Jun-Aug. Fr. Sep-Nov. Mountain slopes, bushes, roadsides or along valleys by streams below 500 m. Distributed in Yunnan, Guizhou, Guangxi and Guangdong. Also in Vietnam.

中国无忧花
Saraca dives Pierre

乔木。小叶5或6对，幼时稍紫红色，下垂。花序腋生，较大；无花瓣；花萼裂片花瓣状；苞片较大，长10-50毫米；雄蕊8-10，花药长3-4毫米。荚果淡褐色，压扁，果瓣扭曲。花期4-5月，果期7-10月。生海拔200-1000米的密林或疏林、河岸、山谷旁或小溪边。产云南、广西和广东。

Trees. Leaflets 5 or 6 pairs, slightly purplish red when young, pendulous. Inflorescences axillary, larger; petals absent; calyx lobes petaloid; bracts large, 10-50 mm long; stamens 8-10, anthers 3-4 mm long. Legumes brownish, compressed, valves twisted. Fl. Apr-May. Fr. Jul-Oct. Dense or sparse forests, riversides, along valleys or by streams at 200-1000 m. Distributed in Yunnan, Guangxi and Guangdong.

中国无忧花 *Saraca dives*

酸豆
Tamarindus indica L.

乔木。小叶长圆形，小，无毛。花较少，浅黄色，带紫红色条纹；能育雄蕊3，花丝中部以下合生；小苞片花萼状或花瓣状，开花前包被花芽；子房柄与萼筒贴生。荚果圆柱形，直或弯曲。花期5-8月，果期12月至翌年5月。栽培或逸为野生。产华南。原产非洲；广泛栽培于热带地区。

Trees. Leaflets oblong, small, glabrous. Flowers few, yellowish tinged with purplish red stripes; fertile stamens 3, filaments connate below the middle; bracteoles sepaloid or petaloid, enclosing flower bud before flowering; stalk of ovary adnate to hypanthium. Legumes terete, erect or curved. Fl. May-Aug. Fr. Dec to next May. Cultivated or escaped. Distributed in S China. Native to Africa; widely cultivated in tropics.

肥荚红豆
Ormosia fordiana Oliv.

乔木。植株密被铁锈褐色柔毛。小叶(5-)7-9(-13)，倒卵状披针形或倒卵状椭圆形。花冠淡紫红色；旗瓣近基部中央具一黄色斑点。荚果半圆形或长圆形，压扁，外面浅褐色，内部象牙色。种子红褐色。花期6-7月，果期11月。生海拔100-1400米的山谷或疏林中。产华南。孟加拉国、缅甸、泰国和越南亦有。

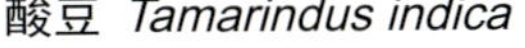

酸豆 *Tamarindus indica*

云南红豆 *Ormosia yunnanensis*

Trees. Plants densely ferruginous pubescent. Leaflets (5-)7-9(-13), obovate-lanceolate or obovate-elliptic. Corolla light purple; standards with a yellow spot at the middle near the base. Legumes semicircular or oblong, compressed, brownish outside, ivory inside. Seeds red-brown. Fl. Jun-Jul. Fr. Nov. Valleys or open forests at 100-1400 m. Distributed in S China. Also in Bangladesh, Myanmar, Thailand and Vietnam.

云南红豆

Ormosia yunnanensis Prain

常绿乔木。植株密被褐色绒毛。奇数羽状复叶；小叶3-6对，革质，上面无毛。圆锥花序顶生及腋生，密集。荚果倒卵球形，斜生，果瓣厚革质，种子间缢缩。花期3月，果期10月。生海拔500-1700米的平坝、山谷或疏林中。产云南南部。

Evergreen trees. Plants densely brown-tomentose. Leaves imparipinnate; leaflets 3-6 pairs, leathery, adaxially glabrous. Panicles terminal and axillary, congested. Legumes obovoid, oblique, valves thick leathery, constricted between seeds. Fl. Mar. Fr. Oct. Plains, valleys or open forests at 500-1700 m. Distributed in S Yunnan.

肥荚红豆 *Ormosia fordiana*

凹叶红豆 *Ormosia emarginata*

凹叶红豆
Ormosia emarginata (Hook. et Arn.) Benth.

常绿小乔木或灌木。奇数羽状复叶具(3-)5-7小叶。小叶倒卵形、倒卵状披针形、长倒卵形或长圆形，先端钝，具凹缺。圆锥花序顶生，少花；花冠白色或粉色。荚果深棕色或黑色，扁平，菱形或长圆形。种子1-4粒，红棕色。花期5-7月，果期8-10月。生山坡或山谷混交林中。产广西、广东和海南。越南亦有。

Small evergreen trees or sometimes shrubs. Leaves imparipinnate; leaflets (3-)5-7, obovate, obovate-elliptic, long-obovate or oblong, apex obtuse, emarginate. Panicles terminal, few-flowered; corolla white or pink. Legumes dark brown or black, compressed, rhombic or oblong. Seeds 1-4, red-brown. Fl. May-Jun. Fr. Aug-Oct. Mountain slopes or mixed valley forests. Distributed in Guangxi, Guangdong and Hainan. Also in Vietnam.

花榈木
Ormosia henryi Prain

常绿乔木。树皮灰绿色，平滑，有浅裂纹。奇数羽状复叶；小叶下面密具平伏绒毛。圆锥花序顶生或总状花序腋生。荚果长圆形，扁平，长5-12厘米。种子4-8粒。花期7-8月，果期10-11月。生海拔100-1300米的山坡或溪旁林中。产中国西南、华南和华中。泰国和越南亦有。

Evergreen trees. Bark gray-green, smooth, with shallow cracks. Leaves imparipinnate; leaflets densely appressed tomentose abaxially. Panicles terminal or racemes axillary. Legumes oblong, compressed, 5-12 cm long. Seeds 4-8. Fl. Jul-Aug. Fr. Oct-Nov. Forests on slopes or by streams at 100-1300 m. Distributed in SW, S and C China. Also in Thailand and Vietnam.

木荚红豆
Ormosia xylocarpa Chun ex Merr. et L. Chen

常绿乔木，高12-20米。奇数羽状复叶具小叶(3-)5-7，长圆形或至长圆状倒披针形，厚革质。圆锥花序顶生，具柔毛；花冠白色或粉色。荚果倒卵球形至长圆形或菱形，扁平，密具平伏茶色丝毛。种子1-5粒，红色。花期6-7月，果期10-11月。生海拔200-1600米的山坡、山谷、路边、溪边或林中。产中国西南、华南、东南、华中和华东。

Evergreen trees, 12-20 m tall. Leaves imparipinnate; leaflets (3-)5-7, oblong or oblong-oblanceolate, thick leathery. Panicles terminal, pubescent; corolla white or pink. Legumes obovate to oblong or rhombic, compressed, densely appressed fulvous sericeous. Seeds 1-5, red. Fl. Jun-Jul. Fr. Oct-Nov. Mountain slopes, valleys, roadsides, streamsides or forests at 200-1600 m. Distributed in SW, S, SE, C and E China.

海南红豆
Ormosia pinnata (Lour.) Merr.

常绿乔木或灌木。奇数羽状复叶；小叶7(-9)，披针形至倒披针形，薄革质，两面无毛。圆锥花序顶生，长20-30厘米。荚果镰刀状，果瓣厚木质，成熟后橘红色，干后褐色，具浅色斑点，无毛。花期7-8月，果期10月。生中低海拔的山坡、山谷或林中。产广东、广西和海南。泰国和越南亦有。

Trees or shrubs evergreen. Leaves imparipinnate; leaflets 7(-9), lanceolate to oblanceolate, thin leathery, both surfaces glabrous. Panicles terminal, 20-30 cm long. Legumes falciform, valves thick woody, orange-red at maturity, brown when dried, with light-colored spots, glabrous. Fl. Jul-Aug. Fr. Oct. Slopes, valleys or forests at middle and low elevations. Distributed in Guangdong,

花榈木 *Ormosia henryi*

木荚红豆 *Ormosia xylocarpa*

海南红豆 *Ormosia pinnata*

光叶马鞍树 *Maackia tenuifolia*

Guangxi and Hainan. Also in Thailand and Vietnam.

翅荚香槐
Cladrastis platycarpa (Maxim.) Makino

乔木，高达30米。奇数羽状复叶具小叶(7-)13(-15)。圆锥花序长9-30厘米；花冠白色，喉部具黄色斑点。荚果长椭圆体形或长圆形，两边具翼，不开裂。种子1-3粒，深棕色或黑色，长圆形。花期4-6月，果期7-10月。生海拔1000米以下的山谷林中或山坡。产中国西南、华南、东南、华中和华东。日本亦有。

Trees, to 30 m tall. Leaves imparipinnately compound; leaflets (7-)13(-15). Panicles 9-30 cm long; corolla white, with yellow spots in throat. Legumes long ellipsoid or oblong, winged on both sides, indehiscent. Seeds 1-3, dark brown or black, oblong. Fl. Apr-Jun. Fr. Jul-Oct. Forests in valleys or slopes below 1000 m. Distributed in SW, S, SE, C and E China. Also in Japan.

光叶马鞍树
Maackia tenuifolia (Hemsl.) Hand.-Mazz.

灌木或乔木，高2-7米。叶长12-16.5厘米；小叶3-5(-7)，主脉具毛；顶生小叶倒卵形或椭圆形。总状花序顶生；花冠绿白色。荚果棕色，条形，扁平，密被毛。种子2-4粒，淡红色至猩红色。花期7-9月，果期9-10月。生林中或山坡。产中国东南、华中、华北、华西和华东。

Shrubs or trees, 2-7 m tall. Leaves 12-16.5cm long; leaflets 3-5(-7), hairy along mid vein; terminal leaflets obovate or elliptic. Racemes terminal; corolla green-white. Legumes brown, linear, compressed, densely villous. Seeds 2-4, light red to scarlet. Fl. Jul-Sep. Fr. Sep-Oct Forests or slopes. Distributed in SE, C, N, W and E China.

翅荚香槐 *Cladrastis platycarpa*

短绒槐(灰毛槐) *Sophora velutina*

短绒槐(灰毛槐)

Sophora velutina Lindl.

灌木。幼枝、花序轴、花枝和叶轴等幼嫩部分密被绒毛。小叶13-41，对生或近对生，纸质。总状花序与叶对生或假顶生；花多数，疏松。荚果念珠状，稍压扁，具毛或近无毛。花果期4-8月。生海拔1000-2500米的山谷或溪边的灌木林中。产四川、贵州和云南。印度、孟加拉国和缅甸亦有。

Shrubs. Young branches, inflorescences axis, flower branches and leaf axis densely tomentose. Leaflets 13-41, opposite or nearly opposite, papery. Racemes opposite to leaves or false terminal; flowers many, widely spaced. Legumes moniliform, slightly compressed, hairy or nearly glabrous. Fl. and fr. Apr-Aug. Among bushes in valleys or by rivers at 1000-2500 m. Distributed in Sichuan, Guizhou and Yunnan. Also in India, Bangladesh and Myanmar.

柳叶槐

Sophora dunnii Prain

灌木。小枝被深黄色毛。小叶15-23，对生，纸质。总状花序与叶对生，花多数；花冠紫红色；子房密被棕黄色毛；胚珠4-6。荚果念珠形，具深棕色柔毛，沿背腹缝线开裂。种子2或3粒，长卵球形。花期3-6月，果期5-8月。生海拔500-2500米的沟谷或山坡林中。产云南、四川和贵州。缅甸和泰国亦有。

Shrubs. Branchlets dark yellow hairy; leaflets 15-23, opposite, papery. Racemes opposite a leaf, many-flowered; corolla purple-red; ovary densely yellow-brown hairy; ovules 4-6. Legumes moniliform, darkly brown pilose, dehiscent along sutures. Seeds 2 or 3, long ovoid. F1. Mar-Jun. Fr. May-Aug Valleys or forests on slopes at 500-2500 m. Distributed in Yunnan, Sichuan and Guizhou. Also in Myanmar and Thailand.

越南槐

Sophora tonkinensis Gagnep.

灌木。托叶极小；小叶5-9对，椭圆形、长圆形或卵状长圆形。顶生总状花序或基部具分枝形成圆锥花序；花黄色。荚果念珠状，疏被柔毛，沿两侧缝线开裂。花期5-7月，果期8-12月。生海拔1000-2000米的多石山地灌丛。产云南、贵州和广西。越南亦有。

Shrubs. Stipules very small; leaflets 5-9-paired, elliptic, oblong or ovate-oblong. Racemes or branched into panicles from base, terminal; flowers yellow. Legumes moniliform, sparsely pubescent, dehiscent along both sutures. Fl. May-Jul. Fr. Aug-Dec. Scrubs on stony mountains at 1000-2000 m. Distributed in Yunnan, Guizhou and Guangxi. Also in Vietnam.

砂生槐

Sophora moorcroftiana (Benth.) Baker

小灌木。小枝密被长柔毛。托叶钻状，后变成刺；小叶11-15，倒卵形，约10 × 6毫米，先端钝或微缺，常具芒尖，两

柳叶槐 *Sophora dunnii*

越南槐 *Sophora tonkinensis*

砂生槐 *Sophora moorcroftiana*

面被毛。总状花序生于小枝顶端；花较大；花萼蓝色，浅钟状，萼齿5，被长柔毛；花冠蓝紫色。荚果呈不明显串珠状，约6 × 0.7厘米。种子淡黄褐色。花期5-7月，果期7-10月。生海拔3000-4500米的山谷林中。产西藏。南亚亦有。

Small shrubs. Branchlets densely tomentose. Stipules subulate, spinescent; leaflets 11-15, obovate, ca. 10 × 6 mm, apex obtuse or retuse, usually mucronate, hairy on both surfaces. Racemes terminal at branchlets; flowers large; calyx blue, shortly campanulate, teeth 5, villose; corolla blue-purple. Legumes not obviously moniliform, slightly compressed, ca. 6 × 0.7 cm. Seeds light yellow-brown. Fl. May-Jul. Fr. Jul-Oct. Valley forests at 3000-4500 m. Distributed in Xizang. Also in S Asia.

白刺槐

Sophora davidii (Franch.) Skeels

灌木或小乔木。茎无毛。不育枝末端明显变成刺。小叶5-9对，长10-15毫米，椭圆状卵形或倒卵状长圆形。总状花序着生于小枝顶端；花白色或浅黄色，长约1.5厘米。荚果稍压扁。花期3-8月，果期6-10月。生海拔2500米以下的河谷沙丘或山坡灌丛中。产中国西南、华南、华中、华北和华东。

Shrubs or small trees. Stems glabrous. End of sterile branches prickled. Leaflets 5-9 pairs, 10-15 mm long, elliptic-ovate or obovate-oblong. Racemes on top of branchlets; flowers white or light yellow, ca. 1.5 cm long. Legumes slightly compressed. Fl. Mar-Aug. Fr. Jun-Oct. Valley dunes or bushes on mountain slopes below 2500 m. Distributed in SW, S, C, N and E China.

白刺槐 *Sophora davidii*

川西白刺槐 *Sophora davidii* var. *chuansiensis*

川西白刺槐
Sophora davidii (Franch.) Skeels var. **chuansiensis** C. Y. Ma

本变种与白刺槐的区别在于本变种的花冠蓝紫色；小叶极小，长5-6毫米，常阔卵形。花期3-8月，果期6-10月。生海拔2500-3400米的干旱山坡或河谷沙地。产云南、四川和西藏。

This variety differs from the typical variety in its blue-purple corolla; very small leaflets, which is 5-6 mm long, usually broadly ovate. Fl. Mar-Aug. Fr. Jun-Oct. Dry mountain slopes or valley sand dunes at 2500-3400 m. Distributed in Yunnan, Sichuan and Xizang.

苦豆子
Sophora alopecuroides L.

草本或亚灌木。小叶11-27，披针状长圆形或椭圆状长圆形。总状花序顶生，花多且密；花冠白色或奶白色；雄蕊10，基部稍合生，具短毛。荚果念珠状，伸直。花期5-7月，果期8-10月。生草地或沙地。产中国西南、东南、华北、华西和西北。南亚、中亚和西南亚亦有。

Herbs or subshrubs. Leaflets 11-27, lanceolate-oblong or elliptic-oblong. Racemes terminal; flowers many, dense; corolla white or creamy white; stamens 10, slightly fused at base, shortly hairy where fused. Legumes moniliform, straight. Fl. May-Jul. Fr. Aug-Oct. Grasslands or deserts. Distributed in SW, SE, N, W and NW China. Also in S, C and SW Asia.

苦参
Sophora flavescens Aiton

草本或半灌木。枝与小叶幼时无毛或具毛。羽状复叶，具13-25小叶。花形成疏松总状；花冠白色或浅黄色，龙骨瓣先端钝。荚果种子间稍缢缩，近四方形，疏或密具柔毛或渐无毛，4瓣裂。花期6-8月，果期7-10月。生海拔1500米以下的沙地草坡、灌丛中或田野。产中国大部分地区。印度、俄罗斯(西伯利亚)、朝鲜半岛和日本亦有。

Herbs or subshrubs. Branches and leaflets glabrous or pilose when young. Leaves pinnate, 13-25-foliolate. Flowers in lax

苦豆子 *Sophora alopecuroides*

苦参 *Sophora flavescens*

racemes; corolla white or light yellow, keel obtuse at apex. Legumes slightly constricted between seeds, slightly quadrangular, sparsely or densely pubescent or glabrescent, 4-valvate. Fl. Jun-Aug. Fr. Jul-Oct. Sandy or grassy slopes, thickets or fields below 1500 m. Distributed in most parts of China. Also in India, Russia (Siberia), Korean Peninsula and Japan.

绒毛槐
Sophora tomentosa L.

灌木或小乔木。枝被短茸毛。无托叶；小叶11-15(-19)，卵形或圆形，近革质，下面密具灰白色绒毛。通常总状花序，有时分枝呈圆锥状；花冠黄色或乳白色。荚果念珠状，具短毛。花期8-10月，果期9-12月。生海滨沙丘或附近的小灌木林中。产广东、海南和台湾。广布于热带海岸和岛屿。

Shrubs or small trees. Branches shortly tomentose. Without stipules; leaflets 11-15(-19), oval or rounded, nearly leathery, densely gray-white tomentose abaxially. Usually racemes, sometimes branched to paniculate; corolla yellow or creamy white. Legumes moniliform, shortly hairy. Fl. Aug-Oct. Fr. Sep-Dec. Coastal sandy areas or adjacent bushes. Distributed in Guangdong, Hainan and Taiwan. Widely distributed in tropical coasts and islands.

锈毛槐
Sophora prazeri Prain

灌木，被锈色茸毛。叶柄上面具凹槽；小叶7-15，卵形至长椭圆形，先端圆形或急尖，两面细脉明显。总状花序侧生或与叶互生，长5-20厘米；花萼5浅裂或近平截；花冠白色或淡黄色。荚果串珠状，长4-10厘米。种子2-4粒，深红色或鲜红色。花果期4-9月。生海拔2000米以下的山地林中、山谷或潮湿的山坡上。产云南、贵州和广西。缅甸亦有。

Shrubs, redbrown tomentose. Petiole sulcate adaxially; leaflets 7-15, usually ovate to long elliptic, apex rounded to acute, veinlets obvious on both surfaces. Racemes lateral, alternate with leaves, 5-20 cm long; calyx 5-lobed or subtruncate; corolla white or pale yellow. Legumes moniliform, 4-10 cm long. Seeds 2-4, dark red or light red. Fl. and fr. Apr-Sep. Mountain forests, valleys or wet slopes below 2000 m. Distributed in Yunnan, Guizhou and Guangxi. Also in Myanmar.

绒毛槐 *Sophora tomentosa*

锈毛槐 *Sophora prazeri*

槐 *Sophora japonica*

槐
Sophora japonica L.

乔木。当年生枝无毛。叶柄基部膨大；小叶9-15，卵状披针形或卵状长圆形。圆锥花序顶生；花冠白色或乳黄色，稀为紫红色；子房明显短于雄蕊。荚果厚，种子间明显缢缩。花期7-8月，果期8-10月。栽培于中国大部分地区。原产朝鲜半岛和日本，各地广栽培。

Trees. Branchlets of current year glabrous. Petioles inflated at base; leaflets 9-15, ovate-lanceolate or ovate-oblong. Panicles terminal; corolla white or creamy yellow, rarely purple-red; ovary obviously shorter than stamens in length. Legumes thick, obviously constricted between seeds. Fl. Jul-Aug. Fr. Aug-Oct. Cultivated in most parts of China. Native to Korean Peninsula and Japan, widely cultivated elsewhere.

藤槐
Bowringia callicarpa Champion ex Bentham

攀援灌木或木质藤本。托叶卵状三角形；叶长圆形或卵状长圆形，长6-13厘米，先端渐尖，无毛，叶脉两面隆起，细脉明显；叶柄两端稍膨大。总状花序腋生，长2-5厘米；花萼杯状，萼齿先端近截平；花冠白色。荚果卵球形，长2.5-3厘米，先端具喙，沿缝线开裂。花期4-6月，果期7-9月。生山谷林缘。产福建、广东、广西和海南。越南亦有。

Scandent shrubs or lianas. Stipules ovate-triangular; leaves oblong or ovate-oblong, 6-13 cm long, apex acuminate, glabrous; veins raised on both surfaces, veinlets obvious; petiole slightly inflated at base. Racemes axillary, 2-5 cm long; calyx cup-shaped, tooth subtruncate at apex; corolla white. Legumes ovoid, 2.5-3 cm long, beaked at apex, dehiscent along sutures. Fl. Apr-Jun. Fr. Jul-Sep. Forest margins in valleys. Distributed in Fujian, Guangdong, Guangxi and Hainan. Also in Vietnam.

斜叶黄檀
Dalbergia pinnata (Lour.) Prain

乔木或灌木。小叶斜长圆形，基部不对称，一侧圆形，另一侧楔形。圆锥花序腋生，密具锈色柔毛；花冠白色，具长爪；旗瓣卵形，反折。荚果干后褐色且光亮，长圆状舌形，薄，无毛。花期1-4月，果期5-7月。生海拔1400米以下的山

藤槐 *Bowringia callicarpa*

斜叶黄檀 *Dalbergia pinnata*

地密林中。产云南、西藏、广西和海南。缅甸、马来西亚、菲律宾和印度尼西亚亦有。

Trees or sometimes shrubby climbers. Leaflets oblique-oblong, base asymmetrical, one side rounded, other side cuneate. Panicles axillary, densely rusty puberulent; corolla white, long clawed; standard ovate, reflexed. Legumes brown and shiny when dry, oblong-ligulate, thin, glabrous. Fl. Jan-Apr. Fr. May-Jul. Dense forests on mountains below 1400 m. Distributed in Yunnan, Xizang, Guangxi and Hainan. Also in Myanmar, Malaysia, the Philippines and Indonesia.

香港黄檀

Dalbergia millettii Bentham

木质藤本。小叶23-35，线形或狭长圆形，先端截形，无毛。圆锥花序腋生，长1-1.5厘米；总花梗、花序轴和分枝疏被微柔毛；花长2.5-3毫米；花萼钟状，近无毛；花瓣白色，具柄，旗瓣圆形；雄蕊9，单体。荚果长圆形至带状，长4-6厘米，宽1.2-1.6厘米，无毛，全部有网纹。花期5月，果期8月。生海拔350-800米的山谷林中。产香港、广东、广西和浙江。

Woody climbers. Leaflets 23-35, linear or narrowly oblong, apex truncate, glabrous. Panicles axillary, 1-1.5 cm long; peduncles, rachis and branches sparsely puberulent; flowers 2.5-3 mm long; calyx campanulate, nearly glabrous; petals white, shortly clawed, standard orbicular; stamens 9, monadelphous. Legume oblong to linear, 4-6 × 1.2-1.6 cm, glabrous, reticulate veined throughout. Fl. May. Fr. Aug. Forests in ravines at 300-800 m. Distributed in HongKong, Guangdong, Guangxi and Zhejiang.

象鼻藤

Dalbergia mimosoides Franch.

灌木。小叶条状长圆形，幼时两面疏生褐色柔毛，老时无毛或近无毛。花冠白色或淡黄色；旗瓣长圆状倒卵形。荚果具柄，长圆形至条形，革质，无毛，对种子部分具网纹。花期4-5月，果期6-8月。生海拔800-2000米的疏林或山坡灌丛中。产云南、四川、浙江、西藏、湖北和陕西。印度亦有。

香港黄檀 *Dalbergia millettii*

Shrubs. Leaflets linear-oblong, both surfaces sparsely brown pubescent when young but glabrous or glabrescent when mature. Corolla white or pale yellow; standard oblong-obovate. Legumes stipitate, oblong to strap-shaped, leathery, glabrous, retic-ulate opposite 1 seed. Fl. Apr-May. Fr. Jun-Aug. Open forests or thickets at 800-2000 m. Distributed in Yunnan, Sichuan, Zhejiang, Xizang, Hubei and Shaanxi. Also in India.

象鼻藤 *Dalbergia mimosoides*

藤黄檀 *Dalbergia hancei*

大金刚藤 *Dalbergia dyeriana*

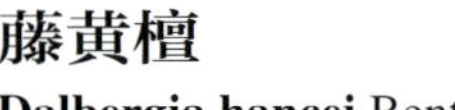
藤黄檀
Dalbergia hancei Benth.

木质藤本。托叶膜质，早落；小叶7-13，狭长圆形或倒卵状长圆形。花冠绿白色，芳香；旗瓣不反折，倒卵形或椭圆形；子房具短柄。荚果扁平，长圆形或带形。花期3-5月，果期6-11月。生山坡灌丛中或山谷溪旁。产中国西南、华南、华中和东南。

Woody climbers. Stipules membranous, caducous; leaflets 7-13, narrowly oblong or obovate-oblong. Corolla greenish white, fragrant; standard not reflexed, obovate or elliptic; ovary shortly stipitate. Legumes compressed, oblong or taeniform. Fl. Mar-May. Fr. Jun-Nov. Bushes on mountain slopes or streamsides in valleys. Distributed in SW, S, C and SE China.

大金刚藤
Dalbergia dyeriana Prain ex Harms

木质攀援植物。小叶(7-)9-15，薄革质，无毛，上面光亮，脉细密且紧密网结，在两面明显凸起。圆锥花序腋生；花冠黄白色；旗瓣长圆形，具凹缺。荚果长圆状或带状，扁平。花期5月。生海拔700-1500米的山坡灌丛中或山谷密林中。产中国西南、东南、华中和华西。

Woody climbers. Leaflets (7-)9-15, thin leathery, glabrous, shiny adaxially, veinlets finely and closely reticulate, conspicuously prominent on both surfaces. Panicles axillary; corolla yellowish white; standard oblong, emarginate. Legumes oblong or taeniform, depressed. Fl. May. Among shrubs on mountain slope or dense forests in valley at 700-1500 m. Distributed in SW, SE, C and W China.

滇黔黄檀
Dalbergia yunnanensis Franch.

木质攀援植物。小叶膜质，长常为宽的2倍，两面具平伏柔毛，下面中脉更密。聚伞状圆锥花序顶生或生于上部叶腋；花萼钟形，外面疏被柔毛，萼齿5，具缘毛，最下方1萼齿长于其余4片。花期4-6月，果期8-10月。生海拔1300-2200米的林中、灌丛或山坡。产云南、四川、贵州和广西。

Woody climbers. Leaflets membranous, length usually 2 × width, both surfaces appressed puberulent, more densely so on midvein abaxially. Cymose panicles terminal or axillary on upper part of stems; calyx campanulate, sparsely pubescent outside, lobes 5, ciliate, the lowest 1 longer than others. Fl. Apr-Jun. Fr. Aug-Oct. Forests, thickets or slopes at 1300-2200 m. Distributed in Yunnan, Sichuan, Guizhou and Guangxi.

降香黄檀(花梨)
Dalbergia odorifera T. Chen

乔木。小叶(7-)9-11(-13)，卵形或椭圆形，两面无毛。圆锥花序腋生，分枝呈伞房花序状；

滇黔黄檀 *Dalbergia yunnanensis*

降香黄檀(花梨) *Dalbergia odorifera*

花冠乳白色或浅黄色；花瓣近等长，具爪；旗瓣倒卵形。荚果舌状长圆形，网脉不明显，与种子对生处网脉明显凸起。花期4-6月，果期7-12月。生海拔100-500米的山坡疏林、林缘或空旷地。产海南。

Trees. Leaflets (7-)9-11(-13), ovate or elliptic, glabrous on both surfaces. Panicles axillary, branched to corymbiform; corolla creamy white or pale yellowish; petals subequal in length, clawed; standard obcordate. Legumes lorate-oblong, inconspicuously reticulate, reticulation distinctly prominent opposite seeds. Fl. Apr-Jun. Fr. Jul-Dec. Sparse forests, forest edges or open fields on slopes at 100-500 m. Distributed in Hainan.

多裂黄檀

Dalbergia rimosa Roxb.

木质攀援植物。小叶疏具毛。圆锥花序顶生或有时生于上部叶腋。花小；萼齿5，最下方1萼齿披针形，长于其余4片；花冠白色或黄绿色。荚果全部具网纹，与种子对生处网纹明显且稍凸起。花期4-6月，果期7-12月。生海拔800-1700米的山坡、山谷或疏林中。产云南南部和广西。喜马拉雅东部、缅甸、老挝、泰国和越南亦有。

Woody climbers. Leaflets sparsely hairy. Panicles terminal or sometimes extending into axils of uppermost leaves. Flowers minute; calyx lobes 5, lowest one lanceolate, longer than others; corolla white or yellowish green. Legumes reticulate throughout, conspicuously reticulate and slightly prominent opposite seeds. Fl. Apr-Jun. Fr. Jul-Dec. Slopes, valleys or open forests at 800-1700 m. Distributed in S Yunnan and Guangxi. Also in E Himalaya, Myanmar, Laos, Thailand and Vietnam.

黄檀

Dalbergia hupeana Hance

乔木。树皮呈薄片状剥落。小叶7-11，椭圆形至长圆状椭圆形，下面无明显的网脉。圆锥花序顶生或生上部叶腋，疏具锈色柔毛；花冠白色或淡紫色。荚果长圆形或阔舌状，果瓣薄革质。花期5-7月。生海拔600-1400米的山地林中、灌丛或山沟溪旁。产华南、华中和华东。

Trees. Bark peeled off by thin sections. Leaflets 7-11, elliptic to oblong-elliptic, abaxial surface without obvious reticulate veinlets. Panicles terminal or axillary in uppermost leaves, sparsely rusty puberulent; corolla white or light purple. Legumes oblong or wide lingulate, valves thin leathery. Fl. May-Jul. Mountain forests, shrubs or valleys by rivers at 600-1400 m. Distributed in S, C and E China.

多裂黄檀 *Dalbergia rimosa*

黄檀 *Dalbergia hupeana*

秧青 *Dalbergia assamica*

秧青

Dalbergia assamica Benth.

乔木。小叶6-7对，长为宽的1.5-2倍。圆锥花序腋生；花冠白色，旗瓣圆形，雄蕊10，二体；子房密被短柔毛。荚果阔舌形或长圆形至条形，较狭，先端锐尖，基部渐狭至楔形。花期4-7月，果期9-12月。生海拔300-1700米的山地杂木林中、水边或灌丛中。产中国西南、华南和东南。印度北部、缅甸、老挝、泰国和越南亦有。

Trees. Leaflets 6-7 pairs, length of leaflets 1.5-2 × width. Panicles axillary; corolla white, standards orbicular, stamens 10, diadelphous; ovary densely pubescent. Legumes broadly ligulate or oblong to strap-shaped, narrower, apex acute, base attenuate to cuneate. Fl. Apr-Jul. Fr. Sep-Dec. Mixed forests on mountains, riversides or thickets at 300-1700 m. Distributed in SW, S and SE China. Also in N India, Myanmar, Laos, Thailand and Vietnam.

紫檀

Pterocarpus indicus Willd.

乔木。树皮灰色。托叶卵形，早落；小叶5-7(-11)，卵状椭圆形，薄纸质，两面无毛。圆锥花序顶生或腋生，多花；花冠黄色。荚果球形，扁平，边缘具宽翅。花期3-4月，果期4-5月。生山坡疏林中，或栽培于庭院。产华南。印度、缅甸、印度尼西亚和菲律宾亦有。

Trees. Bark gray. Stipules ovate, caducous; leaflets 5-7(-11), ovate-elliptic, thin papery, both surfaces glabrous. Panicles terminal or axillary, many-flowered; corolla yellow. Legumes globose, depressed, broadly winged around margin. Fl. Mar-Apr. Fr. Apr-May. Sparse forests of mountain slopes or cultivated in gardens. Distributed in S China. Also in India, Myanmar, Indonesia and the Philippines.

紫檀 *Pterocarpus indicus*

猪腰豆

Afgekia filipes (Dunn) R. Geesink

攀援灌木。茎幼时被绢毛和深红色糙伏毛。托叶狭三角形；小叶(13-)17-19，长圆形，长6-10厘米，基部圆形，不对称。圆锥花序具苞片痕；花萼杯状，被毛，上部萼齿合生；花冠淡紫色，外面被短柔毛。荚果纺锤形，被短绒毛。种子肾形。花期7-8月，果期9-11月。生海拔200-1300米的稀疏灌丛或常绿阔叶林缘。产广西和云南。东南亚亦有。

Scandent shrubs. Stems sericeous and scarlet strigose when young. Stipules narrowly triangular; leaflets (13-)17-19, oblong, 6-10 cm long, base rounded, asymmetric. Panicles with scars of fallen bracts; calyx cup-shaped, with trichomes, adaxial pair of teeth connate; corolla lilac, abaxially puberulent. Legume spindle-shaped, velutinous. Seed reniform. Fl. Jul-Aug. Fr. Sep-Nov. In thickets or evergreen broad-leaved forest edges at 200-1300 m. Distributed in Guangxi and Yunnan. Also in SE Asia.

干花豆

Fordia cauliflora Hemsl.

灌木。托叶镰形，宿存；叶具(19-)23或25小叶；小叶长圆形，4-12 × 2.5-3厘米。假总状花序长15-40厘米；花冠红色至紫红色。荚果棍棒状，长7-10 × 2-2.5厘米，渐无毛。花期5-9月，果期6-12月。生海拔500米以下的山丘灌丛。产贵州、广西和广东。

Shrubs. Stipules falcate, persistent; leaves (19-)23-or 25-foliolate; leaflets oblong, 4-12 × 2.5-3 cm. Pseudoracemes 15-40 cm long; corolla red to purple-red. Legumes clavate, 7-10 × 2-2.5 cm, glabrescent. Fl. May-Sep. Fr. Jun-Dec. Thickets on hills below 500 m. Distributed in Guizhou, Guangxi and Guangdong.

干花豆 *Fordia cauliflora*

小叶干花豆

Fordia microphylla Dunn ex Z. Wei

灌木。托叶三角状披针形，脱落；叶具17-21小叶；小叶卵状披针形，被平伏细毛。总状花序长8-13厘米，生花节不隆起，2-5花簇生；苞片小，刺毛状；花长8-10毫米。花期4-6月，果期7-9月。生海拔800-2000米的石质山坡或灌丛中。产云南、贵州和广西。

Shrubs. Stipules triangular-lanceolate, caducous; leaves 17-21-foliolate; leaflets ovate-lanceolate, adpressed-pubescent. Racemes 8-13 cm long, nodes not swollen, flowers 2-5-clustered; bracts minute, setiform; flowers 8-10 mm long. Fl. Apr-Jun. Fr. Jul-Sep. Rocky slopes or thickets at 800-2000 m. Distributed in Yunnan, Guizhou and Guangxi.

猪腰豆 *Afgekia filipes*

小叶干花豆 *Fordia microphylla*

思茅崖豆 *Millettia leptobotrya*

思茅崖豆

Millettia leptobotrya Dunn

乔木。枝被褐色柔毛，易折断，后无毛。小叶长12-25厘米，侧脉11-13对。花冠白色；苞片和小苞片条形，密被褐色毛。荚果条状长圆形，扁平。种子厚棱镜状，光亮。花期4月，果期10月至翌年1月。生海拔300-1000米的山坡林中。产云南南部。老挝和越南亦有。

Trees. Branchlets brown pubescent, brittle, glabrescent. Leaflets 12-25 cm long, secondary veins 11-13 on each side of midvein. Corolla white; bracts and bracteoles linear, densely brown-hairy. Legumes linear-oblong, flat. Seeds thick lenticular, shiny. Fl. Apr. Fr. Oct to next Jan. Woodlands on slopes at 300-1000 m. Distributed in S Yunnan. Also in Laos and Vietnam.

厚果崖豆藤

Millettia pachycarpa Benth.

藤本。小叶13-17，下面具褐色丝毛，长圆状椭圆形至长圆状披针形，背面被平伏绢毛但中脉密被褐色绒毛。假总状花序在新枝下具2-6分枝，具褐色绒毛。荚果深褐色，膨大，密具浅黄色疣，渐无毛。花期4-6月，果期7-11月。生海拔100-2000米的山谷常绿阔叶林中。产中国西南、华南、东南和华中。南亚亦有。

Lianas. Leaflets 13-17, abaxially brown sericeous, oblong-elliptic to oblong-lanceolate, abaxially adpressed-sericeous but brown-tomentose on midrib. Pseudoracemes with 2-6 branches beneath new stems, brown tomentose. Legumes dark brown, inflated, densely covered with pale yellow warts, glabrescent. Fl. Apr-Jun. Fr. Jul-Nov. Evergreen broad-leaved forests in valleys at 100-2000 m. Distributed in SW, S, SE and C China. Also in S Asia.

美丽鸡血藤

Callerya speciosa (Champion ex Bentham) Schot

木质藤本。小叶13，长圆状披针形至椭圆状披针形，长4-8厘米，边缘微卷，上面无毛。总状花序腋生，常聚集于小枝近顶端形成圆锥花序，被褐色绒毛；花直径2.5-3.5厘米；花冠白色、乳白色或淡粉色。荚果条形，长10-15厘米，扁平，被

厚果崖豆藤 *Millettia pachycarpa*

美丽鸡血藤 *Callerya speciosa*

褐色绒毛，先端具喙。花期7-9月，果期8-10月。生海拔200-1700米的空地或疏林。产华南和湖南。越南亦有。

Lianas, woody. Leaflets 13, oblong-lanceolate to elliptic-lanceolate, 4-8 cm long, margins slightly revolute, adaxially glabrous. Racemes axillary, usually congested near apex of branchlets to form large panicles, brown tomentose; flowers 2.5-3.5 cm diam; corolla white, creamy white or pale pink. Legume linear, 10-15 cm long, flat, brown tomentose, apex beaked. Fl. Jul-Sep. Fr. Aug-Oct. Open places or sparse woodlands at 200-1700 m. Distributed in S China and Hunan. Also in Vietnam.

网络鸡血藤
Callerya reticulata (Benth.) Schot

藤本，长2-10米。叶具(5-)7-9小叶；叶轴长10-20厘米，带叶柄长2-5厘米；小叶卵状椭圆形、长圆形、条形或狭披针形。圆锥花序顶生或腋生于近顶端小枝，常下垂；花冠紫色。荚果干后黑色，条形；腹缝线不加厚。花期4-8月，果期6-11月。生海拔100-1200米的沟谷或山坡灌丛或溪边灌丛。产中国西南、华南、东南、华中和华东。越南北部亦有。

Lianas, 2-10 m long. Leaves (5-)7-9-foliolate; rachis 10-20 cm long, including petiole 2-5 cm long; leaflets ovate-elliptic, oblong, linear or narrowly lanceolate. Panicles terminal or axillary near apex of branchlets, often pendulous; corolla purple. Legumes black when dry, linear; ventral suture not thickened. Fl. Apr-Aug. Fr. Jun-Nov. Thickets on slopes or in valleys, or thickets by streams at 100-1200 m. Distributed in SW, S, SE, C and E China. Also in N Vietnam.

灰毛鸡血藤
Callerya cinerea (Benth.) Schot

攀援灌木。茎圆柱状，粗糙具脊，渐无毛。叶具5小叶；托叶长约4毫米；小叶柄长约4毫米；小叶倒卵状椭圆形。圆锥花序顶生；开花小枝开展至6厘米；花冠红色至紫色。荚果条状长圆形，密具灰色柔毛。花期2-7月，果期8-11月。生海拔500-1200米的沟谷畔次生阔叶林。产云南南部、四川西南部和西藏东南部。印度、尼泊尔、不丹、孟加拉国、缅甸和泰国亦有。

Shrubs, scandent. Stems terete, rough, ridged, glabrescent. Leaves 5-foliolate; stipules ca. 4 mm long; petiolules ca. 4 mm long; leaflets obovate-elliptic. Panicles terminal; flowering branchlets spreading to 6 cm long; corolla red to mauve. Legumes linear-oblong, densely gray pubescent. Fl. Feb-Jul. Fr. Aug-Nov. Secondary broad-leaved forests by ravines at 500-1200 m. Distributed in S Yunnan, SW Sichuan and SE Xizang. Also in India, Nepal, Bhutan, Bangladesh, Myanmar and Thailand.

网络鸡血藤 *Callerya reticulata*

灰毛鸡血藤 *Callerya cinerea*

球子鸡血藤 *Callerya sphaerosperma*

亮叶鸡血藤 *Callerya nitida*

球子鸡血藤
Callerya sphaerosperma (Z. Wei) Z. Wei et Pedley

攀援灌木。叶具3小叶；托叶长约2毫米；小叶椭圆状披针形，纸质。圆锥花序顶生，长12-15厘米；花枝开展，具柔毛；花冠红色至紫色。荚果球形，革质，具褐色绒毛，先端具喙；腹缝线明显。花期6-8月，果期10-11月。生海拔约1000米的阴湿沟谷。产贵州和广西。

Shrubs, scandent. Leaves 3-foliolate; stipels ca. 2 mm long; leaflets elliptic-lanceolate, papery. Panicles terminal, 12-15 cm long; flowering branchlets spreading, puberulent; corolla red to purple. Legumes globose, leathery, brown tomentose, apex beaked; ventral suture conspicuous. Fl. Jun-Aug. Fr. Oct-Nov. Shady ravines at ca. 1000 m. Distributed in Guizhou and Guangxi.

亮叶鸡血藤
Callerya nitida (Benth.) R. Geesink

攀援灌木。叶具5小叶；小叶卵状披针形或长圆形，近革质。圆锥花序顶生；花单生，淡紫色至紫色；旗瓣长圆形，具两个基生胼胝体。荚果条状长圆形，具柄，具褐色绒毛。花期5-9月，果期7-11月。生海拔1500米以下的海岸灌丛中或低山疏林中。产中国西南、华南和东南。

Climbing shrubs. Leaves 5-foliolate; leaflets ovate-lanceolate or oblong, subleathery. Panicles terminal; flowers solitary, violet to purple; standard oblong, with 2 basal calluses. Legumes linear-oblong, stipitate, brown tomentose. Fl. May-Sep. Fr. Jul-Nov. Thickets on coasts or lowland sparse forests below 1500 m. Distributed in SW, S and SE China.

香花鸡血藤
Callerya dielsiana (Harms) P. K. Lôc ex Z. Wei et Pedley

攀援灌木。羽状复叶，具5小叶；小叶下面散生柔毛，上面无毛，网状脉明显，先端锐尖。圆锥花序长12-20厘米，近

香花鸡血藤 *Callerya dielsiana*

水黄皮 *Pongamia pinnata*

无柄；花冠紫红色。荚果长圆形，果瓣近木质。花期5-9月，果期6-11月。生海拔300-2500米的山坡杂木林中或灌丛中，或溪边。产中国西南、华南、华中和东南。

Shrubs, scandent. Leaves pinnate, 5-foliolate; leaflets abaxially sparsely puberulous, adaxially glabrous and reticulate veins prominent, acute at apex. Panicles 12-20 cm long, subsessile; corolla purple-red. Legumes oblong, valves subwoody. Fl. May-Sep. Fr. Jun-Nov. Mixed forests or thickets on slopes, or streamsides at 300-2500 m. Distributed in SW, S, C and SE China.

水黄皮

Pongamia pinnata (L.) Pierre

乔木。老枝密生灰白色小皮孔。叶具5或7小叶。总状花序腋生，长15-20厘米，花序轴节处常具2花；子房具2胚珠。荚果不开裂，侧脉加厚而不达边缘，无毛。花期5-6月，果期8-10月。生溪边、池塘边或海边。产华南。热带亚洲和大洋洲亦有。

Trees. Old branches with many grayish lenticels. Leaves 5- or 7-foliolate. Racemes axillary, 15-20 cm long, rachis nodes usually with 2 flowers; ovary with 2 ovules. Legumes indehiscent, thick with secondary veins not reaching margins, glabrous. Fl. May-Jun. Fr. Aug-Oct. Streamsides, by ponds or by seas. Distributed in S China. Also in tropical Asia and Oceania.

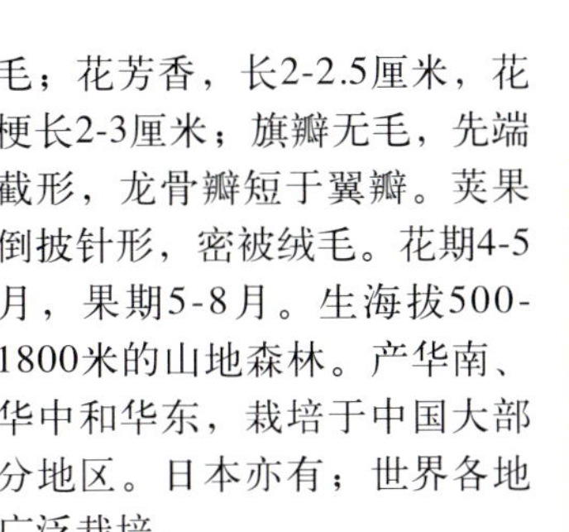

紫藤

Wisteria sinensis (Sims) Sweet

藤本。叶具7-13小叶。总状花序顶生或腋生于去年小枝，15-30 × 8-10厘米，具白色长柔毛；花芳香，长2-2.5厘米，花梗长2-3厘米；旗瓣无毛，先端截形，龙骨瓣短于翼瓣。荚果倒披针形，密被绒毛。花期4-5月，果期5-8月。生海拔500-1800米的山地森林。产华南、华中和华东，栽培于中国大部分地区。日本亦有；世界各地广泛栽培。

Lianas. Leaves 7-13-foliolate. Racemes terminal or axillary from branchlets of previous year, 15-30 × 8-10 cm, white villous; flowers fragrant, 2-2.5 cm long, pedicels 2-3 cm long; standard glabrous, apex truncate, keels shorter than wings. Legumes oblanceolate, densely tomentose. Fl. Apr-May. Fr. May-Aug. Mountain forests at 500-1800 m. Distributed in S, C and E China, cultivated in most parts of China. Also in Japan; widely cultivated worldwide.

紫藤 *Wisteria sinensis*

巴豆藤 *Craspedolobium unijugum*

巴豆藤

Craspedolobium unijugum (Gagnep.) Z. Wei et Pedley

攀援灌木。三出复叶，侧生小叶显著不对称。假总状花序常密集生于小枝顶端，开花小枝节处具3-5簇生的花；花冠红色；花瓣近等长；雄蕊2体。荚果条形，腹缝线具窄翅，被褐色绒毛。种子3-5(-7)粒。花期6-9月，果期9-10月。生海拔600-2000米的路边疏林。产云南、四川、贵州和广西。缅甸、老挝和泰国亦有。

Scandent shrubs. Leaves ternate, lateral leaflets markedly asymmetric. Pseudoracemes usually congested near apex of branchlets, flowering branchlet nodes with 3-5 fascicled flowers; corolla red; petals nearly equal in length; stamens diadelphous. Legumes linear, abaxial suture narrowly winged, brown tomentulose. Seeds 3-5(-7). Fl. Jun-Sep. Fr. Sep-Oct. Open forests by roads at 600-2000 m. Distributed in Yunnan, Sichuan, Guizhou and Guangxi. Also in Myanmar, Laos and Thailand.

毛果鱼藤

Derris eriocarpa F. C. How

木质藤本。幼枝、叶和花萼被淡黄色微柔毛。小叶13或15，侧生小叶长圆形至卵状长圆形，长5-7.5厘米。假总状花序腋生，长于叶；花萼杯状；花冠红白色；雄蕊单体。荚果线状长椭圆形，长6-11厘米，疏被柔毛，仅腹缝有宽约2毫米的翅，有1-8粒种子。花期6-7月，果期9月至翌年1月。生海拔800-1600米的山坡疏林中。产广西、贵州和云南。泰国亦有。

Lianas, woody. Young shoots, leaves and calyx yellowish puberulent. Leaflets 13 or 15, lateral ones oblong to ovate-oblong, 5-7.5 cm long. Pseudoracemes axillary, longer than leaves; calyx cup-shaped; corolla white flushed with red; stamens monadelphous. Legume linear-oblong, 6-11 cm long, sparsely villous, adaxial suture with a ca. 2 mm wide wing, 1-8-seeded. Fl. Jun-Jul. Fr. Sep to next Jan. Sparse forests on mountain slopes at 800-1600 m. Distributed in Guangxi, Guizhou and Yunnan. Also in Thailand.

鱼藤

Derris trifoliata Lour.

藤本。奇数羽状复叶，有小叶(3-)5(-7)。花3-15或更多常散生于沿结状或细弱的小枝上；花冠白色或粉红色；旗瓣近圆形；雄蕊单体。荚果斜卵形或圆形，扁平，无毛。种子1-2粒。花期4-8月，果期8-12月。生海拔1000米以下的沿海河岸灌木丛或林中。产华南。南亚、东南亚、日本、澳大利亚、太平洋岛屿和非洲东部亦有。

Lianas. Leaves odd-pinnate, with (3-)5(-7) leaflets. Flowers 3-15 or more usually scattered along knoblike or slender brachyblasts; corolla white or pink; standards suborbicular; stamens monadelphous. Legumes obliquely ovoid or globose, flat, glabrous. Seeds 1-2. Fl. Apr-Aug. Fr. Aug-Dec. Coastal areas along beaches or riverbanks, thickets or forests below 1000 m. Distributed in S China. Also in S and SE Asia, Japan, Australia, Pacific Islands and E Africa.

毛果鱼藤 *Derris eriocarpa*

鱼藤 *Derris trifoliata*

中南鱼藤 *Derris fordii*

中南鱼藤

Derris fordii Oliv.

藤本。叶具5或7小叶，小叶厚纸质至薄革质。假圆锥花序腋生，稍短于叶；短枝上的花序轴节上具数个簇生的花。荚果长4-10厘米，薄革质，边缘具翅。花期3-5月，果期6-11月。生山地路旁或山谷林中。产中国西南、华南、东南和华中。

Lianas. Leaves 5- or 7-foliolate, leaflets thickly papery to thinly leathery. Pseudopanicles axillary, slightly shorter than leaves; rachis nodes with several fascicled flowers on short branchlets. Legumes 4-10 cm long, thin leathery, margins winged. Fl.

Mar-May. Fr. Jun-Nov. Mountain roadsides or forests in valleys. Distributed in SW, S, SE and C China.

密锥花鱼藤

Aganope thyrsiflora (Bentham) Polhill

木质藤本或攀援灌木。叶长30-45厘米；小叶5-9，长圆形至长圆状披针形，两面无毛。假圆锥花序狭锥状，紧密；花萼钟形，疏被柔毛；花冠白色至紫红色。荚果长圆形，表面有明显网纹，腹背两缝均具宽3-8毫米的翅。花期5-6月，果期8-11月。生低海拔山中溪边灌丛，云南可达海拔2000米。产广东、广西、海南和云南。南亚和东南亚亦有。

Lianas or scandent shrubs. Leaves 30-45 cm long; leaflets 5-9, oblong to oblong-lanceolate, both surfaces glabrous. Pseudopanicles narrowly pyramidal, compact; calyx campanulate, sparsely pilose; corolla whitish to purplish red. Legume oblong, with conspicuous reticulate veins, both sutures with a 3-8 mm wide wing. Fl. May-Jun. Fr. Aug-Nov. Scrub by streams in mountains low elevations but to 2000 m in Yunnan. Distributed in Guangdong, Guangxi, Hainan and Yunnan. Also in S and SE Asia.

密锥花鱼藤 *Aganope thyrsiflora*

白灰毛豆

Tephrosia candida DC.

多年生草本。茎被灰白色茸毛。小叶8-12对，长圆形，3-6 × 0.6-1.4厘米，先端具凸尖，上面无毛，下面密被平伏绢毛。总状花序顶生或侧生；花长约2厘米；花梗长约1厘米；萼片近等长，圆头，长约1毫米。荚果长8-10厘米，密被褐色绒毛。种子10-15粒，橄榄绿色。花期10-11月，果期12月。生草地、旷野或山坡。产华南。原产印度；广泛栽培或逃逸。

Perennial herbs. Stems grayish white tomentose. Leaflet 8-12-paired, oblong, 3-6 × 0.6-1.4 cm, apex mucronate, adaxially glabrous, abaxially densely sericeous. Racemes terminal or lateral; flowers ca. 2 cm; pedicels ca. 1 cm; sepals subequal, apex rounded, ca. 1 mm long. Legume 8-10 cm long, densely brown tomentose. Seeds 10-15, olive-green. Fl. Oct-Nov. Fr. Dec. Grasslands, open places or slopes. Distributed in S China. Native to India; widely cultivated and escaped elsewhere.

白灰毛豆 *Tephrosia candida*

灰毛豆 *Tephrosia purpurea*

刺槐 *Robinia pseudoacacia*

灰毛豆

Tephrosia purpurea (L.) Persoon

多年生草本。茎近直立或伸展。叶具9-17(-21)小叶，侧脉多数，直，紧密平行。花冠淡紫色；旗瓣有白毛。荚果条形，具稀疏平伏毛；每个荚果中约具6粒种子，棕灰色，椭球形。花期5-10月，果期9-12月。生海拔700米以下的开阔地、山坡、海边沙质草地或河边草地。产中国西南、华南和东南。南亚和东南亚亦有。

Perennial herbs. Stems nearly erect or spreading. Leaves 9-17 (-21)-foliolate, secondary veins numerous, straight, closely parallel. Corolla light purple; standard white puberulent. Legumes linear, with sparse appressed trichomes. Seeds ca. 6 per legumes, grayish brown, ellipsoid. Fl. May-Oct. Fr. Sep-Dec. Open places, slopes, sandy grasslands by oceans or grasslands by rivers below 700 m. Distributed in SW, S and SE China. Also in S and SE Asia.

刺槐

Robinia pseudoacacia L.

乔木。小枝、轴和花梗具平的贴伏柔毛。托叶刺可达2厘米。小叶2-12对，长圆形、椭圆形或卵形。花冠白色；旗瓣内部具黄色斑点；雄蕊二体。荚果棕色或具棕红色条纹，条状长圆形，沿腹缝线具狭翅。花期4-6月，果期8-9月。中国除西藏和海南外广泛栽培。原产美洲东北部；现在世界许多地区栽培或归化。

Trees. Branchlets, rachis and pedicel with appressed adnate puberulence. Stipulate spines up to 2 cm. Leaflets 2-12 pairs, oblong, elliptic or ovate. Corolla white; standards with yellow spots inside; stamens diadelphous. Legumes brown or with reddish brown stripes, linear-oblong, narrow wings along ventral suture. Fl. Apr-Jun. Fr. Aug-Sep. Widely cultivated in China except Xizang and Hainan. Native to NE America; cultivated or naturalized in many parts of the world.

田菁

Sesbania cannabina (Retz.) Poir.

草本。小枝、叶轴和花轴无刺。小叶上面幼时具长柔毛，渐无毛。花冠黄色；旗瓣横椭圆形至近圆形，具紫黑色斑点和条纹。荚果长圆柱形，稍弯曲，开裂，外面具黑褐色条纹。花果期7-12月。栽培或归化于向阳的荒地或稻田边。产中国西南、华南、华中、华东和华北。亚洲、非洲、澳大利亚和太平洋岛屿亦有。可能原产澳大利亚和西南太平洋岛屿；广泛栽培。

Herbs. Branchlets, leaf rachises and flower rachises without prickles. Leaflets adaxially villous when young, glabrescent. Corolla yellow; standards transversely elliptic to subrotund, with purple-black spots and lines. Legumes long terete, slightly curved, dehiscent, outside with dark brown stripes. Fl. and fr. Jul-Dec. Cultivated or naturalized in open wastelands or paddy field edges. Distributed in SW, S, C, E and N China. Also in Asia, Africa, Australia and Pacific Islands. Probably native to Australia and SW Pacific Islands; widely cultivated.

田菁 *Sesbania cannabina*

大花田菁 *Sesbania grandiflora*

垂序木蓝 *Indigofera pendula*

大花田菁

Sesbania grandiflora (L.) Pers.

小乔木。叶痕及托叶痕明显。羽状复叶，具20-60小叶；小叶长圆形至长椭圆形，长2-5厘米，宽0.8-1.6厘米，先端圆钝至微凹，有小突尖。总状花序具2-4花；花长7-10厘米，在花蕾时显著呈镰状弯曲；花冠白色、粉红色至玫瑰红色。荚果线形，长20-60厘米，宽7-8毫米，先端渐狭成长3-4厘米的喙。花果期9月至翌年4月。栽培于华南。全世界热带地区广泛栽培。

Small trees. Leaf scars and stipule scars conspicuous. Leaves pinnate, 20-60-foliolate; leaflet oblong, 2-5 × 0.8-1.6 cm, apex obtuse to retuse and with a mucro. Racemes 2-4-flowered; flowers 7-10 cm long, conspicuously falcately curved in bud; corolla white, pink to rosy. Legume linear, 20-60 cm × 7-8 mm, apex tapering into a 3-4 cm beak. Fl. and fr. Sep to next Apr. Cultivated in S China. Cultivated throughout the tropics.

垂序木蓝

Indigofera pendula Franch.

灌木。羽状复叶，小叶(11或)13-23(-27)对，上面无毛。总状花序下垂，长达30厘米；花冠旗瓣蓝粉色，背面密具白色柔毛；翼瓣和龙骨瓣等长，边缘具缘毛。荚果褐色，圆柱状，内果皮具斑点。花期6-8月，果期9-10月。生海拔1900-3300米的山坡、沟谷或灌丛中。产云南和四川。

Shrubs. Leaves pinnate, leaflets (11 or)13-23(-27) pairs, adaxially glabrous. Racemes pendulous, to 30 cm long; corolla standard bluish pink, densely downy white puberulent dorsally; wings as long as keel, margin ciliate. Legumes brown, cylindric, endocarps blotched. Fl. Jun-Aug. Fr. Sep-Oct. Slopes, valleys or thickets at 1900-3300 m. Distributed in Yunnan and Sichuan.

滇木蓝

Indigofera delavayi Franch.

灌木。托叶钻形；叶长8-18厘米；小叶(13或)15-19，长圆形至稍倒卵形，长1.3-3厘米，下面被稀疏短丁字毛，中脉下面隆起，上面微凹。总状花序长达20厘米；花萼钟状，被贴伏细毛，萼齿三角形，先端渐尖；花冠白色或粉红色。荚果条状圆柱形，无毛，先端向上弯曲。花期7-9月，果期8-9月。生海拔1400-3400米的草地、灌丛、林中或河边。产云南和四川。

Shrubs. Stipules subulate; leaves 8-18 cm long; leaflets (13 or)15-19, oblong to slightly obovate, 1.3-3 cm long, abaxially with sparse appressed short medifixed trichomes, midvein abaxially prominent and adaxially impressed. Racemes ca. 20 cm long; calyx campanulate, with appressed minute trichomes, teeth triangular, apex acuminate; corolla white or pink. Legume linear-cylindric, glabrous, curved upward at apex. Fl. Jul-Sep. Fr. Aug-Sep. Grasslands, scrubs, forests or riverbanks at 1400-3400 m. Distributed in Yunnan and Sichuan.

滇木蓝 *Indigofera delavayi*

庭藤 *Indigofera decora*

花木蓝 *Indigofera kirilowii*

庭藤

Indigofera decora Lindl.

灌木。叶具小叶13-19；小叶上面无毛，卵状披针形、卵状长圆形或披针形，长2-7.5(-10)厘米；叶轴扁平或圆柱形。总状花序直立；花冠淡紫色、粉红色或稀为白色。荚果褐色，圆柱状，内果皮具斑点。花期4-6月，果期6-10月。生海拔200-1800米的溪边、沟谷、林中或灌丛中。产广东、福建、浙江、安徽和江苏。日本亦有。

Shrubs. Leaves 13-19-foliolate; leaflets adaxially glabrous, ovate-lanceolate, ovate-oblong or lanceolate, 2-7.5(-10) cm long; leaf axils depressed or cylindric. Racemes erect; corolla light purple, pink or rarely white. Legumes brown, cylindric, endocarps blotched. Fl. Apr-Jun. Fr. Jun-Oct. Streamsides, valleys, forests or bushes at 200-1800 m. Distributed in Guangdong, Fujian, Zhejiang, Anhui and Jiangsu. Also in Japan.

花木蓝

Indigofera kirilowii Maxim. ex Palibin

小灌木。叶具(5或)7-11小叶；小叶对生，阔卵形、卵状菱形或椭圆形。总状花序具柄，疏花，基部无宿存芽鳞；花瓣淡红色，稀白色，近等长。荚果圆柱状。花期5-7月，果期8-10月。生海拔100-400米的林中、灌丛或岩缝中。产河北、江苏、山东、吉林和辽宁。朝鲜半岛和日本亦有。

Shrublets. Leaves (5 or)7-11-foliolate; leaflets opposite, broadly ovate, ovate-rhombic, sparsely flowered or elliptic. Racemes pedunculate, without persistent bud scales at base; petals light red, rarely white, almost equal. Legumes cylindric. Fl. May-Jul. Fr. Aug-Oct. Forests, thickets or rock crevices at 100-400 m. Distributed in Hebei, Jiangsu, Shandong, Jilin and Liaoning. Also in Korean Peninsula and Japan.

华东木蓝

Indigofera fortunei Craib

灌木。小叶具7-15小叶，上面无毛。总状花序长8-18厘米；总花梗无毛，常短于叶柄；苞片卵形，长约1毫米，脱落；花冠紫色或粉色；雄蕊长7-9毫米。荚果褐色，条状圆柱形；开裂后果瓣旋卷。花期4-5月，果期5-9月。生海拔200-800米的山坡疏林或灌丛中。产浙江、湖北、安徽和江苏。

Shrubs. Leaves 7-15-foliolate, adaxially glabrous. Racemes 8-18 cm long; peduncles glabrous, usually shorter than petioles; bracts ovate, ca. 1 mm long, caducous; corolla purple or pink; stamens 7-9 mm long. Legumes brown, linear-cylindric; valvules twisted after dehiscent. Fl. Apr-May. Fr. May-Sep. Sparse forests or bushes of mountain slopes at 200-800 m. Distributed in Zhejiang, Hubei, Anhui and Jiangsu.

四川木蓝

Indigofera szechuensis Craib

灌木。茎红褐色至黑褐色，圆柱状，具散生黄色皮孔。托叶披针形；叶5-13；叶柄和叶轴上面扁平，贴伏白色丁字毛。总状花序长10-19厘米；花冠深红色。荚果线状圆柱形。花期5-6月，果期7-10月。生海拔2500-3800米的山坡、路旁或河边。产云南、四川、西藏和甘肃。

Shrubs. Stems rufous to blackish brown, terete, with sparse scattered yellow lenticels. Stipules lanceolate; leaves 5-13-foliolate; petioles and rachises adaxially flat, with appressed white medifixed trichomes. Racemes 10-19 cm long; corolla crimson. Legumes linear, cylindric. Fl. May-Jun. Fr. Jul-Oct. Slopes, trailsides or river banks at 2500-3800 m. Distributed in Yunnan, Sichuan, Xizang and Gansu.

华东木蓝 *Indigofera fortunei*

四川木蓝 *Indigofera szechuensis*

苍山木蓝 *Indigofera forrestii*

苍山木蓝
Indigofera forrestii Craib

灌木。托叶披针形；叶长6-9厘米；小叶(5-)9-19，狭椭圆形，长6-12(-18)毫米，两面疏被白色和棕色的贴伏丁字毛，中脉下面隆起，上面凹入。总状花序长6-12厘米；苞片卵形；花萼被贴伏的棕色丁字毛，萼齿三角状披针形，与萼筒等长；花冠红色。荚果线状圆柱形，疏被丁字毛。花期6-7月，果期8-10月。生海拔2400-2900米的阳坡或沟边。产云南和四川。

Shrubs. Stipules lanceolate; leaves 6-9 cm long; leaflets (5-)9-19, narrowly elliptic, 6-12 (-18) mm long, both surfaces with sparse white and brown appressed versatile hairs, midvein abaxially prominent and adaxially impressed. Racemes 6-12 cm long; bracts ovate; calyx with appressed brown versatile hairs, teeth triangular-lanceolate, as long as tube; corolla red. Legume linear-cylindric, with sparse appressed versatile hairs. Fl. Jun-Jul. Fr. Aug-Oct. Sunny slopes or near streams at 2400-2900 m. Distributed in Yunnan and Sichuan.

野青树
Indigofera suffruticosa Mill.

灌木或小灌木。叶具11-15(-19)小叶；小叶长圆形至倒披针形。总状花序呈穗状；苞片条形，长约2毫米，脱落；花冠红色，旗瓣倒卵形；雄蕊长3-4毫米。荚果镰刀状弯曲。花期3-5月，果期6-10月。生低海拔山地路旁、山谷疏林、空旷地、沟边或海滩沙地。栽培或野生于中国西南至东南。原产热带美洲；现广布于世界热带地区。

Shrubs or shrublets. Leaves 11-15(-19)-foliolate; leaflets oblong to oblanceolate. Racemes spicate; bracts linear, ca. 2 mm long, caducous; corolla red, standard obovate; stamens 3-4 mm long. Legumes sickleform. Fl. Mar-May. Fr. Jun-Oct. Mountain roadsides, sparse forests in valleys, open fields, ditches or coast sandy lands at low altitudes. Cultivated or naturalized in SW to SE China. Native to tropical America; widely cultivated in tropics of the world.

野青树 *Indigofera suffruticosa*

河北木蓝 *Indigofera bungeana*

河北木蓝
Indigofera bungeana Walp.

灌木。植株被白色丁字毛。叶具5-9(或11)小叶；小叶对生，椭圆形。总状花序具4-6(-10)厘米；苞片条形，长约1.5厘米；花冠紫色至紫红色；旗瓣阔倒卵形；翼瓣基部具耳状附属物；雄蕊长4-5毫米。荚果圆柱形，直。花期5-7月，果期8-10月。生海拔500-2300米的山坡林缘或灌丛中。产中国西南、华南、华中、华北、华西和华东。日本和朝鲜半岛亦有。

Shrubs; Plants with white medifixed trichomes. Leaves 5-9(or 11)-foliolate; leaflets opposite, elliptic. Racemes 4-6(-10) cm long; bracts linear, ca. 1.5 mm long; corolla purple to purplish red; standard broadly obovate; wings with earlike appendages at base; stamens 4-5 mm long. Legumes cylindric, straight. Fl. May-Jul. Fr. Aug-Oct. Forest edges on slopes or thickets at 500-2300 m. Distributed in SW, S, C, N, W and E China. Also in Japan and Korean Peninsula.

刺序木蓝
Indigofera silvestrii Pamp.

多枝灌木。叶长1-2厘米；小叶(5-)7-9，倒卵形至长圆形，长3-8毫米，下面有平贴丁字毛，中脉上面凹入。总状花序长2-5厘米，具刺；花萼钟状，长约2.5毫米，被丁字毛，萼齿线形；花冠紫红色。荚果线状圆柱形，被毛。花期6-7月，果期8-10月。生海拔100-2700米的干燥的山坡、向阳的岩石缝及河边。产甘肃、华中、贵州、云南和西藏。

Shrubs, many branched. Leaves 1-2 cm long; leaflets (5-)7-9, obovate to oblong, 3-8 mm long, abaxially with appressed medifixed trichomes, midvein adaxially impressed. Racemes 2-5 cm long, spinescent; calyx campanulate, ca. 2.5 mm long, with appressed medifixed trichomes, teeth linear; corolla purplish red. Legume linear-cylindric, hairy. Fl. Jun-Jul. Fr. Aug-Oct. Dry slopes, sunny rocks and riverbanks at 100-2700 m. Distributed in Gansu, C China, Guizhou, Yunnan and Xizang.

腺毛木蓝
Indigofera scabrida Dunn

直立灌木。枝、叶轴、叶缘、花序、苞片及萼片均有具柄的头状腺毛。托叶狭线形至狭三角形；叶长9-13厘米；小叶7-13，椭圆形至椭圆状长圆

刺序木蓝 *Indigofera silvestrii*

腺毛木蓝 *Indigofera scabrida*

形，长1.5-3厘米。总状花序长16-18厘米；萼齿线形，先端渐尖；花冠深红色。荚果线形，长1.8-3厘米。花期6-9月，果期8-10月。生海拔1400-2100米的疏林、松林或灌丛。产四川、云南和贵州。缅甸和老挝亦有。

Erect shrubs. Branches, leaf rachis, margin, inflorescences, bracts and sepals with stalked capitate glandular hairs. Stipules narrowly lanceolate to narrowly triangular; leaves 9-13 cm long; leaflets 7-13, elliptic to elliptic-oblong, 1.5-3 cm long. Racemes 16-18 cm long; calyx teeth linear, apex acuminate; corolla crimson. Legume linear, 1.8-3 cm long. Fl. Jun-Sep. Fr. Aug-Oct. Sparse forests, *Pinus* forests or thickets at 1400-2100 m. Distributed in Sichuan, Yunnan and Guizhou. Also in Myanmar and Laos.

穗序木蓝

Indigofera hendecaphylla Jacq.

草本。羽状复叶，有小叶2-5对。总状花序长5-16厘米；花序梗长达1.8厘米；苞片卵形至卵状披针形；花冠红色或紫红色；雄蕊长4-5毫米。荚果条形，4棱，贴伏丁字毛。花期9-11月，果期11月至翌年1月。生海拔800-1150米的空旷地或路边。产云南、广东和台湾，广西有栽培。南亚、东南亚和日本南部亦有。

Herbs. Leaves pinnate, leaflets 2-5 pairs. Racemes 5-16 cm long; peduncles to 1.8 cm long; bracts ovate to ovate-lanceolate; corolla red or violet-purple; stamens 4-5 mm long. Legumes linear, tetragonous, appressed medifixed trichomes. Fl. Sep-Nov to next Fr. Nov-Jan. Open lands or roadsides at 800-1150 m. Distributed in Yunnan, Guangdong and Taiwan, cultivated in Guangxi. Also in S and SE Asia, and S Japan.

单叶木蓝

Indigofera linifolia (L. f.) Retzius

小灌木或多年生草本。茎平卧或直立。单叶或复叶仅具一片小叶；叶或小叶条形。花序总状，长1-1.5厘米；花冠红色；子房仅具一个胚珠。荚果球形，密被平伏的丁字毛；每个荚果具一粒方形种子。花期4-9月，果期5-10月。生海拔1200米以下的旱山坡草地或阳地溪边。产云南、四川和台湾。西南亚、南亚、东南亚、澳大利亚和非洲亦有。

Shrublets or Perennial herbs. Prostrate or erect. Leaves simple or 1-foliolate. Leaves or leaflets usually linear. Racemes 1-1.5 cm long; corolla red; ovary with 1 ovule. Legumes ovoid, with dense appressed white medifixed trichomes; each legume with one cubic seed. Fl. Apr-Sep. Fr. May-Oct. Grasslands on dry slopes or sunny riversides below 1200 m. Distributed in Yunnan, Sichuan and Taiwan. Also in SW, S and SE Asia, Australia and Africa.

穗序木蓝 *Indigofera hendecaphylla*

单叶木蓝 *Indigofera linifolia*

单节假木豆 *Dendrolobium lanceolatum*

单节假木豆
Dendrolobium lanceolatum (Dunn) Schindl.

灌木。小叶3；顶生小叶长圆形或长圆状披针形，长2-5厘米，宽0.9-1.9厘米，侧生小叶较小，侧脉4-7对。花序近伞形，约有花10朵；花瓣白色或淡黄色，具柄，旗瓣椭圆形。荚果有1荚节，宽椭圆形或近圆形，长8-10毫米，无毛，有明显的网脉。花期5-8月，果期9-11月。生海拔100-800米的溪边草地、山坡灌丛或疏林中。产海南。东南亚亦有。

Shrubs. Leaflets 3; terminal leaflet oblong or oblong-lanceolate, 2-5 × 0.9-1.9 cm, lateral leaflets smaller, lateral veins 4-7 pairs. Inflorescences subumbellate, ca. 10-flowered; petals white or pale yellow, clawed, standard elliptic. Legume 1-jointed, broadly elliptic or nearly orbicular, 8-10 mm long, glabrous, distinctly reticulate veined. Fl. May-Aug. Fr. Sep-Nov. Grassy riverbanks, thickets or sparse forests on mountain slopes at 100-800 m. Distributed in Hainan. Also in SE Asia.

假木豆
Dendrolobium triangulare (Retz.) Schindl.

灌木。嫩枝三棱形。叶具3小叶，下面具长丝毛，沿脉尤甚。伞形花序或短总状花序腋生，具20-30花，白色或黄色。荚果长2-2.5厘米，稍弓形，具平伏丝毛，无柄，具3-6荚节。花期8-10月，果期10-12月。生海拔100-1400米的荒地或灌丛中。产中国西南和华南。印度、斯里兰卡、缅甸、老挝、越南、柬埔寨、马来西亚和非洲亦有。

Shrubs. Young branches 3-angled. Leaves 3-foliolate, abaxially long sericeous especially on veins. Umbels or short racemes axillary, 20-30-flowered, white or yellow. Legumes 2-2.5 cm long, slightly arcuate, appressed sericeous, sessile, 3-6-jointed. Fl. Aug-Oct. Fr. Oct-Dec. Wastelands or thickets at 100-1400 m. Distributed in SW and S China. Also in India, Sri Lanka, Myanmar, Laos, Vietnam, Cambodia, Malaysia and Africa.

长叶排钱树
Phyllodium longipes (Craib) Schindler

灌木，高约1米。小枝密被褐色短柔毛。叶柄长约3毫米；顶生小叶披针形或长圆形，长13-20厘米，侧生小叶斜卵形，长3-4厘米，下面密被褐色软毛，侧脉每边8-15条，网脉明显。伞形花序有花(5-)9-15朵，藏于叶状苞片内；花冠白色或淡黄色。花期8-9月，果期10-11月。生海拔900-1000米的山地灌丛或密林。产广东、广西和云南南部。东南亚亦有。

Shrubs, ca. 1 m tall. Branchlets densely brown pubescent. Petiole ca. 3 mm; terminal leaflet blade lanceolate or oblong, 13-20 cm

假木豆 *Dendrolobium triangulare*

长叶排钱树 *Phyllodium longipes*

毛排钱树 *Phyllodium elegans*

long, lateral ones obliquely ovate, 3-4 cm long, abaxially densely brown soft hairy, lateral veins 8-15 on each side, distinctly reticulate veined. Inflorescence umbellate, (5-)9-15-flowered, enclosed by pair of leaflike bracts. Corolla white or pale yellow. Fl. Aug-Sep. Fr. Oct-Nov. Mountain thickets or dense forests at 900-1000 m. Distributed in Guangdong, Guangxi and S Yunnan. Also in SE Asia.

排钱树

Phyllodium pulchellum (L.) Desv.

灌木。羽状3小叶，上面近无毛。花5或6，由一对叶状苞片所包被；叶状苞片圆形，稍具柔毛和缘毛；花冠白色或淡黄色。荚果具2荚节，无毛或稍具柔毛和缘毛。花期7-9月，果期10-11月。生海拔200-2000米的丘陵荒地、路边或山坡疏林中。产中国西南和华南。热带亚洲和澳大利亚北部亦有。

Shrubs. Leaves pinnately trifoliolate, adaxially nearly glabrous. Flowers 5 or 6, enclosed by pair of leaflike bracts; leaflike bracts orbicular, slightly pubescent and ciliate; corolla white or yellowish. Legumes 2-jointed, glabrous or slightly pubescent and ciliate. Fl. Jul-Sep. Fr. Oct-Nov. Wastelands on hills, roadsides or open forests on slopes at 200-2000 m. Distributed in SW and S China. Also in tropical Asia and N Australia.

毛排钱树

Phyllodium elegans (Lour.) Desv.

灌木。茎、枝和叶密被黄色绒毛。花通常4-9，由一对叶状苞片所包被；苞片阔椭圆形，密具黄色绒毛，基部倾斜，先端具凹缺。荚果密具银灰色绒毛，常具3或4荚节。花期7-8月，果期10-11月。生海拔1100米以下的平原、丘陵、山坡草地、疏林中或灌丛中。产福建、广东、广西、云南和海南。老挝、泰国、越南、柬埔寨和印度尼西亚亦有。

Shrubs. Stems, branches and leaves densely covered with yellow tomentum. Flowers 4-9, enclosed by a pair of leaflike bracts; bracts broadly elliptic, densely yellow tomentose, base oblique, apex emarginate. Legumes densely silver-gray tomentose, usually 3- or 4-jointed. Fl. Jul-Aug. Fr. Oct-Nov. Plains, hills, grasslands on slopes, sparse forests, or bushes below 1100 m. Distributed in Fujian, Guangdong, Guangxi, Yunnan and Hainan. Also in Laos, Thailand, Vietnam, Cambodia and Indonesia.

排钱树 *Phyllodium pulchellum*

单叶拿身草 *Desmodium zonatum*

单叶拿身草
Desmodium zonatum Miq.

小灌木。幼枝、花序轴和果密被小钩状毛。小叶1，卵状椭圆形或披针形，长5-12厘米，下面密被黄褐色柔毛；侧脉7-10对。总状花序长10-25厘米；花梗长4-10毫米；花冠白色或粉红色；雄蕊二体。荚果线形，长8-12厘米；荚节6-8，长圆状线形，长于12毫米。花期6-8月，果期8-9月。生海拔500-1300米的密林或林缘。产华南。南亚、东南亚和太平洋岛屿亦有。

Subshrubs. Young branches, inflorescences rachis and fruits with dense, minute, hooked hairs. Leaflet 1, ovate-elliptic or lanceolate, 5-12 cm long, abaxially densely yellow-brown pubescent; lateral veins 7-10-paired. Racemes 10-25 cm long; pedicel 4-10 mm long; corolla white or pink; stamens diadelphous. Legume linear, 8-12 cm long; 6-8-jointed, oblong-linear, longer than 12 mm. Fl. Jun-Aug. Fr. Aug-Sep. Dense forests or forest margins at 500-1300 m. Distributed in S China. Also in S and SE Asia, and Pacific Islands.

长圆叶山蚂蝗
Desmodium oblongum Wall. ex Benth.

直立灌木，疏生钩状毛。叶长圆形或长圆状披针形，长7-12厘米，上面仅脉上有毛，下面密被贴伏柔毛，脉上疏生小钩状毛。圆锥花序长10-30厘米，花稀疏；花梗长约1.2厘米；花冠紫色或堇色；雄蕊单体。荚果狭长圆形，长1.5-3厘米；荚节5-7，近圆形，几无毛。花果期8-11月。生海拔1000-1900米的阔叶林下或灌丛中。产广西和云南。南亚和东南亚亦有。

Erect shrubs, sparsely hooked hairy. Leaves oblong or oblong-lanceolate, 7-12 cm long, adaxially only veins hairy, abaxially densely adpressed pubescent, veins sparsely minutely hooked hairy. Panicles 10-30 cm long, laxly flowered; pedicel ca. 1.2 cm long; corolla purple or violet; stamens monadelphous. Legume narrowly oblong, 1.5-3 cm long; 5-7-jointed, suborbicular, subglabrous. Fl. and fr. Aug-Nov. Broad-leaved forests or thickets at 1000-1900 m. Distributed in Guangxi and Yunnan. Also in S and SE Asia.

疏果山蚂蝗
Desmodium griffithianum Benth.

亚灌木或草本。枝、叶柄、总花梗和花梗被伸展的黄褐色或锈色短柔毛。小叶3，倒三角状卵形或倒卵形，长1-2.5厘米，上面几无毛，下面被贴伏微柔毛。总状花序顶生；花冠紫红色。荚果反折，长1-1.5厘米；荚节3-4，近方形，被钩状毛和硬直毛。花果期8-9月。生海拔1500-2300米的草坡、路旁或松林下。产四川、贵州和云南。印度和东南亚亦有。

长圆叶山蚂蝗 *Desmodium oblongum*

Subshrubs or herbs. Branches, petiole, peduncle and pedicel spreading yellow-brown or rust-colored pubescent. Leaflets 3, obtriangular-ovate or obovate, 1-2.5 cm long, adaxially subglabrous, abaxially adpressed puberulent. Racemes terminal; corolla purple-red. Legume deflexed, 1-1.5 cm long, 3-4-jointed, nearly quadrate, with hooked and straight rigid hairs. Fl. and fr. Aug-Sep. Grassy slopes, roadsides or *Pinus* forests at 1500-2300 m. Distributed in Sichuan, Guizhou and Yunnan. Also in India and SE Asia.

疏果山蚂蝗 *Desmodium griffithianum*

假地豆

Desmodium heterocarpon (L.) DC.

灌木或亚灌木，直立或匍匐。叶具3小叶；顶生小叶背面背白色贴服短柔毛。总状花序顶生或腋生；花轴密具浅黄色开展的钩状毛；花冠紫色、紫红色或白色。荚果直立，狭长圆形。花期7-10月，果期10-12月。生海拔300-1800米的草地、草坡、水边、灌丛或林中。产中国西南、华中、华南和东南。南亚、东南亚、日本、澳大利亚、太平洋岛屿和非洲东部亦有。

Shrubs or subshrubs, erect or prostrate. Leaves 3-foliolate; terminal leaflets abaxially white adpressed pubescent. Racemes terminal or axillary; rachis densely yellowish spreading hooked hairy; corolla purple, purple-red or white. Legumes erect, narrowly oblong. Fl. Jul-Oct. Fr. Oct-Dec. Grasslands, grassy slopes, water sides, thickets or forests at 300-1800 m. Distributed in SW, C, S and SE China. Also in S and SE Asia, Japan, Australia, Pacific Islands and E Africa.

假地豆 *Desmodium heterocarpon*

肾叶山蚂蝗

Desmodium renifolium (L.) Schindl.

亚灌木，常无毛。小叶1，肾形或扁菱形，通常宽大于长，长1.5-3.5厘米；侧脉3-4对。圆锥花序顶生或总状花序腋生，长5-15厘米；花梗长2-8毫米；花冠白色至淡黄色或紫色；雄蕊单体。荚果狭长圆形，长2-3厘米，具2-5荚节。花果期9-11月。生海拔100-1000米的向阳草地、灌丛、林缘或阔叶林下。产海南、云南和台湾。南亚、东南亚和澳大利亚亦有。

Subshrubs, often glabrous. Leaflets 1, reniform or compressed rhombic, often broader than long, 1.5-3.5 cm long; lateral veins 3-4-paired. Panicles terminal or racemes axillary, 5-15 cm long; pedicel 2-8 mm long; corolla white to pale yellow or purple; stamens monadelphous. Legume narrowly oblong, 2-3 cm long, 2-5-jointed. Fl. and fr. Sep-Nov. Sunny grasslands, thickets, forest margins or broad-leaved forests at 100-1000 m. Distributed in Hainan, Yunnan and Taiwan. Also in S and SE Asia, and Australia.

肾叶山蚂蝗 *Desmodium renifolium*

小叶三点金 *Desmodium microphyllum*

小叶三点金

Desmodium microphyllum (Thunb.) DC.

多年生草本。小叶3或1，狭倒卵状椭圆形或狭椭圆形，长1-1.2厘米，或倒卵形或椭圆形，长2-6毫米。总状花序被黄褐色开展柔毛，有花6-10朵；花萼裂片是萼筒长的3-4倍；花冠粉红色，与花萼近等长；雄蕊二体。荚果长12毫米；荚节近圆形。花期5-9月，果期9-11月。生海拔150-2500米的荒地、草丛或灌丛中。产长江以南各地。亚洲和澳大利亚亦有。

Perennial herbs. Leaflets 3 or 1, narrowly obovate-elliptic or narrowly elliptic and 1-1.2 cm long, or obovate to elliptic and 2-6 mm long. Racemes yellow-brown spreading pubescent, 6-10-flowered; calyx lobes 3-4 × as long as tube; corolla pink, nearly as long as calyx; stamens diadelphous. Legume 12 mm long; articles suborbicular. Fl. May-Sep. Fr. Sep-Nov. Wastelands, grasslands or thickets at 150-2500 m. Distributed in the south of Yangtze River. Also in Asia and Australia.

三点金

Desmodium triflorum (L.) DC.

多年生草本，被开展柔毛。小叶3，顶生小叶倒心形，倒三角形或倒卵形，长2.5-10毫米。花单生或2-3朵腋生；花梗长3-8毫米；花冠紫红色；雄蕊二体。荚果狭长圆形，长5-12毫米；荚节3-5，近方形，长2-2.5毫米。花果期6-10月。生海拔200-600米的草地或河边沙地。产华南、浙江和江西。南亚、东南亚、热带非洲、美洲、澳大利亚和太平洋岛屿亦有。

Perennial herbs, spreading pubescent. Leaflets 3, terminal one obcordate, obtriangular or obovate, 2.5-10 mm long. Flowers solitary or 2-3 in leaf axils; pedicel 3-8 mm long; corolla purple-red; stamens diadelphous. Legume narrowly oblong, 5-12 mm long; 3-5-jointed, nearly quadrate, 2-2.5 mm long. Fl. and fr. Jun-Oct. Grasslands or riversides sandy soils at 200-600 m. Distributed in S China, Zhejiang and Jiangxi. Also in S and SE Asia, tropical Africa, Americas, Australia and Pacific Islands.

饿蚂蝗

Desmodium multiflorum DC.

直立灌木。3小叶复叶，下面具丝毛，中脉及侧脉明显。花序，顶生者多为圆锥花序，腋生者多为总状花序；花冠紫色。荚果密具平伏丝毛。花期7-9月，果期8-10月。生海拔500-2800米的山坡草地或林

三点金 *Desmodium triflorum*

饿蚂蝗 *Desmodium multiflorum*

圆锥山蚂蝗 *Desmodium elegans*

缘。产中国西南、华中、华南和东南。南亚亦有。

Erect shrubs. Leaves 3-foliolate, abaxially sericeous, midvein and lateral veins conspicuous. Inflorescences terminal ones mostly paniculate, axillary ones mostly racemose; corolla purple. Legumes densely appressed sericeous. Fl. Jul-Sep. Fr. Aug-Oct. Grasslands of mountain slopes or forest edges at 500-2800 m. Distributed in SW, C, S and SE China. Also in S Asia.

圆锥山蚂蝗
Desmodium elegans DC.

灌木。枝渐无毛。小叶3，纸质，下面被灰色短柔毛或几无毛。花序顶生者多圆锥状，腋生者多总状；花冠紫色或紫红色，旗瓣先端常凹缺；龙骨瓣先端不具棘尖。荚果扁平，条形，4-6(-9)荚节，疏被贴伏短柔毛。花果期6-10月。生海拔1000-4000米的松栎林缘或林下、山坡路边或水沟边。产华西。印度、尼泊尔、不丹和阿富汗亦有。

Shrubs. Branches glabrescent. Leaflets 3, papery, abaxially gray pubescent or nearly glabrous. Inflorescences terminal ones mostly paniculate, axillary ones mostly racemose; corolla purple or purple-red, apex of standard emarginate; keel not mucronate at apex. Legumes flattened, linear, 4-6(-9)-jointed, sparsely appressed pubescent. Fl. and fr. Jun-Oct. Pine-oak forest edges or under forests, roadsides or by streams at 1000-4000 m. Distributed in W China. Also in India, Nepal, Bhutan and Afghanistan.

云南山蚂蝗
Desmodium yunnanense Franch.

灌木。幼枝、叶柄和叶背密被白色或灰色绒毛。小叶3或1，顶生小叶近圆形、卵形或倒卵形，宽5-17厘米，上面疏被柔毛。圆锥花序顶生，长16-27厘米；花序轴和花梗被短绒毛；花梗长0.6-1厘米；花冠粉红色或紫色。荚果长4-6厘米；荚节4-7，长7-9毫米。花期8-9月，果期9-10月。生海拔1000-2200米的石砾地、荒草坡、灌丛及松栎林林缘。产云南和四川。

Shrubs. Young branches, petiole and leaf abaxial densely white or gray tomentose. Leaflets 3 or 1, terminal one nearly orbicular, ovate or obovate, 5-17 cm broad, adaxially sparsely pubescent. Panicles terminal, 16-27 cm long; rachis and pedicels shortly tomentose; pedicel 0.6-1 cm long; corolla pink or purple. Legume 4-6 cm long; 4-7-jointed, 7-9 mm long. Fl. Aug-Sep. Fr. Sep-Oct. Gravelly places, wastelands, grassy slopes, thickets and margins of *Pinus-Quercus* forests at 1000-2200 m. Distributed in Yunnan and Sichuan.

云南山蚂蝗 *Desmodium yunnanense*

长波叶山蚂蝗 *Desmodium sequax*

长波叶山蚂蝗

Desmodium sequax Wall.

直立灌木。托叶条形；叶具3小叶，顶生小叶4-10 × 4-6厘米。总状花序顶生和腋生或常为顶生的圆锥花序；花冠紫色；龙骨瓣等长于旗瓣。荚果念珠状，密被褐色钩状毛。花期7-9月，果期9-11月。生海拔1000-2800米的山坡草地或林缘。产中国西南、华南和华中。南亚和东南亚亦有。

Erect shrubs. Stipules linear; leaves 3-foliolate, terminal leaflets 4-10 × 4-6 cm. Racemes terminal and axillary or usually terminal panicles; corolla purple; keels equal to wings. Legumes moniliform, with densely brown hooked hairs. Fl. Jul-Sep. Fr. Sep-Nov. Grasslands or open forests at 1000-2800 m. Distributed in SW, S and C China. Also in S and SE Asia.

长柄山蚂蝗

Hylodesmum podocarpum (DC.) H. Ohashi et R. R. Mill

直立草本，高50-110厘米。三出羽状复叶；托叶条状披针形，长0.5-1毫米，基部最宽。总状或圆锥花序，顶生或顶生及腋生，每节常具2花；花冠紫红色。荚果常2荚节。生海拔100-2100米的路边、高山草甸、山坡、森林、灌丛、林缘或沟边。产中国西南、华南和华中。南亚和东南亚亦有。

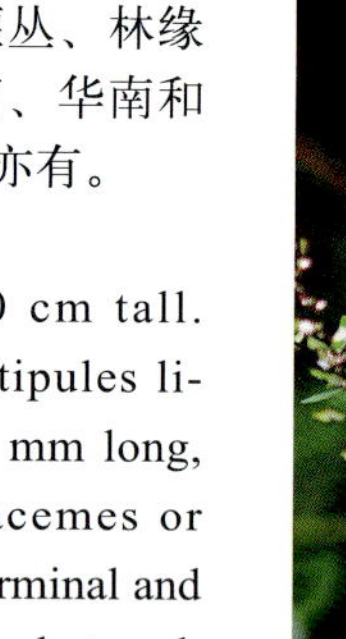

Erect herbs, 50-110 cm tall. Leaves 3-foliolate; stipules linear-lanceolate, 0.5-1 mm long, broadest at base. Racemes or panicles, terminal or terminal and axillary, often 2-flowered at each

长柄山蚂蝗 *Hylodesmum podocarpum*

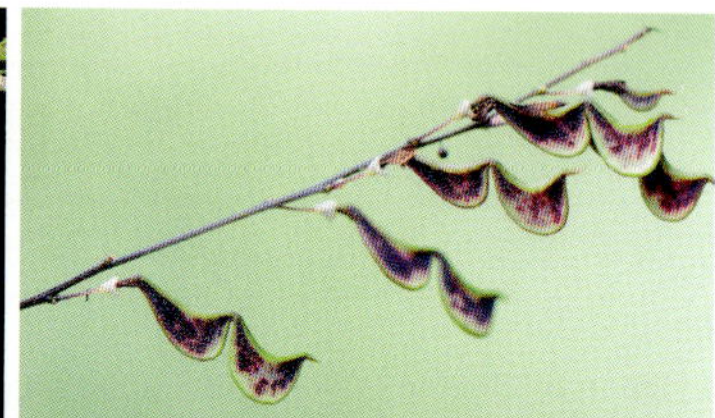

尖叶长柄山蚂蝗 *Hylodesmum podocarpum* subsp. *oxyphyllum*

node; corolla purplish red. Legumes often 2-jointed. Roadsides, grasslands on high mountains, mountain slopes, forests, thickets, forest edges or ditches at 100-2100 m. Distributed in SW, S, and C China. Also in S and SE Asia.

大苞长柄山蚂蝗 *Hylodesmum williamsii*

尖叶长柄山蚂蝗

Hylodesmum podocarpum subsp. **oxyphyllum** (Candolle) H. Ohashi et R. R. Mill Edinburgh

草本。托叶钻形；小叶3，顶生小叶椭圆状菱形至披针状菱形或卵形至菱状卵形，长4-10厘米，先端渐尖，全缘。苞片窄卵形；花萼钟形，裂片较萼筒短；花冠紫红色；雄蕊单体。荚节略呈宽半倒卵形。花期7-9月、果期8-11月。生海拔400-2100米的山坡、沟旁、林缘或阔叶林中。产秦岭淮河以南各省区。南亚、东南亚、东北亚和朝鲜半岛亦有。

Herbs. Stipules subulate; leaflets 3, terminal leaflet elliptic-rhombic to lanceolate-rhombic or ovate to rhombic-ovate, 4-10 cm long, apex acuminate, margin entire. Bracts broadly ovate; calyx campanulate, lobes slightly shorter than tube; corolla purple-red; stamen monadelphous. Articles slightly broadly semiobovate. Fl. Jul-Sep. Fr. Aug-Nov. Mountain slopes, ditches, forest margins or broad-leaved forests at 400-2100 m. Distributed in provinces in South of Qinling Mountain and Huaihe River. Also in S, SE and NE Asia, and Korean Peninsula.

大苞长柄山蚂蝗

Hylodesmum williamsii (H. Ohashi) H. Ohashi et R. R. Mill

多年生草本。羽状三出复叶，顶生小叶全缘，基部圆形；叶柄约长11厘米；小叶宽卵形或菱形。总状花序顶生，疏花，每节具2-4花；花冠粉红色或玫瑰紫色。荚果具1或2荚节，具钩状毛。花果期8-9月。生海拔1400-2700米的沟谷、草地、常绿林、石灰岩土壤或灌丛。产云南、四川和西藏。印度、尼泊尔和不丹亦有。

Perennial herbs. Leaves 3-foliolate, terminal leaflets entire base rounded; petioles ca. 11 cm long; leaflets broadly ovate or rhombic. Racemes terminal, laxly-flowered, 2-4-flowered at each node; corolla pink or roseate-purple. Legumes 1- or 2-jointed, uncinate pubescent. Fl. and fr. Aug-Sep. Ditches, grasslands, evergreen forests, limestone soils or thickets at 1400-2700 m. Distributed in Yunnan, Sichuan and Xizang. Also in India, Nepal and Bhutan.

舞草

Codariocalyx motorius (Houtt.) H. Ohashi

灌木。叶具3小叶，常两侧小叶退化而成1小叶；顶生小叶窄椭圆形或披针形。圆锥花序或总状花序；花序轴具反折的钩状毛和直伸的硬毛；花冠紫红色。荚果疏被短钩状毛。花期7-9月，果期10-11月。生海拔200-1500米的山坡或灌丛中。产中国西南和华南。南亚和东南亚亦有。

Shrubs. Leaves 3-foliolate, often 1-foliolate by reduction of lateral leaflets; terminal leaflets narrowly elliptic or lanceolate. Panicles or racemes; rachis with reflexed uncinate and straight rigid hairs; corolla purplish red. Legumes sparsely short hooklike hairy. Fl. Jul-Sep. Fr. Oct-Nov. Slopes or thickets at 200-1500 m. Distributed in SW and S China. Also in S and SE Asia.

舞草 *Codariocalyx motorius*

葫芦茶 *Tadehagi triquetrum*

葫芦茶

Tadehagi triquetrum (L.) H. Ohashi

灌木或亚灌木。叶仅具单小叶；成熟小叶5.8-13 × 1.1-3.5厘米，长常为宽的3倍。花序长15-30厘米，每节具2或3花；花冠粉色至浅蓝色或紫红色。荚果密被黄色或白色糙伏毛，不具网脉。花期6-10月，果期10-12月。生海拔1400米以下的荒地、山地林缘或路旁。产中国西南和华南。南亚、东南亚和太平洋岛屿亦有。

Shrubs or subshrubs. Leaves 1-foliolate; mature leaflets 5.8-13 × 1.1-3.5 cm, usually 3 × as long as wide. Inflorescences 15-30 cm long, 2- or 3-flowered at each node; corolla pink to bluish or reddish purple. Legumes densely yellow or white strigose, not reticulate veined. Fl. Jun-Oct. Fr. Oct-Dec. Wastelands, forest margins or roadsides below 1400 m. Distributed in SW and S China. Also in S and SE Asia, and Pacific Islands.

美花狸尾豆

Uraria picta (Jacq.) Desv. ex DC.

灌木或亚灌木。小叶5-7，条状长圆形或狭披针形，宽(0.8-)1-2厘米，上面沿中脉常具斑纹。总状花序顶生，长10-30厘米；花冠粉色或浅蓝色。荚果无毛，具3-5荚节。花果期4-10月。生海拔400-1500米的草坡。产中国西南。南亚、东南亚、澳大利亚和热带非洲亦有。

Subshrubs or shrubs. Leaflets 5-7, linear-oblong or narrowly lanceolate, (0.8-)1-2 cm broad, adaxially usually variegated on midvein. Racemes terminal, 10-30 cm long; corolla pink or pale blue. Legumes glabrous, 3-5-jointed. Fl. and fr. Apr-Oct. Grassy slopes at 400-1500 m. Distributed in SW China. Also in S and SE Asia, Australia and tropical Africa.

狸尾豆

Uraria lagopodioides (L.) Desv. ex DC.

草本。通常具3小叶，稀具1小叶，常圆形至阔卵形。总状花

美花狸尾豆 *Uraria picta*

狸尾豆 *Uraria lagopodioides*

中华狸尾豆 *Uraria sinensis*

序短，长3-6厘米，花密集；苞片宿存，在顶端开展；侧面及下面的萼裂片为上面的2倍。荚果无毛。花果期8-10月。生海拔1000米以下的荒地或灌丛中。产中国西南、华南、东南和华中。南亚、东南亚、日本南部、澳大利亚和太平洋岛屿亦有。

Herbs. Leaves mostly 3-foliolate, rarely 1-foliolate, usually orbicular to broadly ovate. Racemes short, 3-6 cm long, densely flowered; bracts persistent, spreading at apex; lateral and lowest calyx lobes 2 × as long as upper lobes. Legumes glabrous. Fl. and fr. Aug-Oct. Wastelands or thickets below 1000 m. Distributed in SW, S, SE and C China. Also in S and SE Asia, S Japan, Australia and Pacific Islands.

中华狸尾豆

Uraria sinensis (Hemsl.) Franch.

亚灌木。茎被灰黄色短粗硬毛。叶为羽状三出复叶；小叶长圆形、倒卵状长圆形、宽卵形。圆锥花序顶生，具灰黄色毛；每节有花1或2朵；花梗纤细；下部萼片与萼筒相等或较短；花冠紫色，较花萼长4倍。荚果与果梗几等长。花果期9-10月。生海拔500-2300米的干燥河谷、山坡、疏林、灌丛或高山草原。产华南、华中和西藏。不丹和印度亦有。

Subshrubs. Stems gray-yellow hispidulous. Leaves pinnately 3-foliolate; leaflets oblong, obovate-oblong or broadly ovate. Panicles terminal, gray-yellow hairy, 1- or 2-flowered at each node; pedicel slender; lower sepals ca. as long as tube or shorter; corolla purple, ca. 4 × as long as calyx. Legume ca. as long as pedicel. Fl. and fr. Sep-Oct. Dry river valleys, mountain slopes, sparse forests, thickets or alpine grasslands at 500-2300 m. Distributed in S and C China, and Xizang. Also in Bhutan and India.

蝙蝠草

Christia vespertilionis (L. f.) Bahn. f.

多年生草本。叶具1或3小叶；顶生小叶菱形或狭菱形，长0.8-1.5厘米，宽5-9厘米；侧生小叶倒心形或倒三角形。花序被短柔毛；花萼半透明，网脉明显，上部2裂片稍合生；花冠黄白色，不伸出萼外。荚果有荚节4-5，无毛，完全藏于萼内。花期3-5月，果期10-12月。生旷野草地、灌丛、路旁或海边。产广东、广西和海南。世界热带地区均有。

Perennial herbs. Leaves 1- or 3-foliolate; terminal leaflet rhombic or narrowly rhombic, 0.8-1.5 × 5-9 cm; lateral ones obcordate or obtriangular. Inflorescences pubescent; calyx half-hyaline, reticulate veined, upper 2 lobes slightly connate; corolla yellowish white, not exserted. Legume 4-5-jointed, glabrous, wholly enclosed by calyx. Fl. Mar-May. Fr. Oct-Dec. Open grasslands, thickets, roadsides or seasides. Distributed in Guangdong, Guangxi and Hainan. Widespread in all tropical regions.

蝙蝠草 *Christia vespertilionis*

链荚豆 *Alysicarpus vaginalis*

链荚豆
Alysicarpus vaginalis (L.) DC.

匍匐或开展的多年生草本。1小叶；叶卵状长圆形、披针形、心形、卵圆形或近圆形，网脉在两面凸起。花序疏具花；花冠红色、紫红色、蓝紫色或黄色，稍长于花萼。荚果扁圆柱形，具4-7荚节。花期9月，果期9-11月。生海拔100-700米的空旷草坡、旱地边、路边或海边沙地。产中国西南和华南。热带地区亦有。

Prostrate or spreading perennial herbs. 1 leaflet; leaves ovate-oblong, lanceolate, cordate, ovate or suborbicular, reticulate veinlets prominent on both surfaces. Inflorescences laxly flowered; corolla red, reddish purple, purplish blue or yellow, slightly longer than calyx. Legumes flattened-terete, 4-7-jointed. Fl. Sep. Fr. Sep-Nov. Open grasslands, by dry fields, roadsides or seaside sandy places at 100-700 m. Distributed in SW and S China. Also in other tropical areas.

毛杭子梢
Campylotropis hirtella (Franch.) Schindl.

小灌木，高0.5-1米。嫩枝多少密被开展锈色毛。小叶三角形至倒卵形。总状花序长4-17厘米，常呈圆锥状；花冠紫色。荚果斜倒卵球形。种子紫褐色，肾形。花期6-10月，果期10-11月。生海拔900-4100米的灌丛、溪边、疏林、山坡或草地。产云南、四川、西藏和贵州。印度亦有。

Shrublets, 0.5-1 m tall. Young branches densely ferruginous spreading hairy. Leaflets deltoid to ovate. Racemes 4-17 cm long, usually paniculate; corolla purple. Legumes obliquely obovoid. Seeds purplish brown, reniform. Fl. Jun-Oct. Fr. Oct-Nov. Thickets, streamsides, sparse forests, mountain slopes or grasslands at 900-4100 m. Distributed in Yunnan, Sichuan, Xizang and Guizhou. Also in India.

西南杭子梢
Campylotropis delavayi (Franch.) Schindl.

灌木。全株除小叶上面和花冠外均密被灰白色绢毛。小叶上面无毛。总状花序密具花，常圆锥状；苞片狭卵形，常短于3毫米；花冠深紫色。荚果偏斜，椭圆体形，顶端圆钝。花期10-12月，果期11-12月。生海拔400-2200米的山坡、灌丛或向阳草地。产云南和四川。

Shrubs. Whole plants densely covered with grayish white sericeous hairs, except adaxial sides of leaflelts and corolla. Leaflets adaxially glabrous. Racemes densely flowered, often paniculate; bracts narrowly ovate, usually shorter than 3 mm long; corolla dark purple. Legumes obliquely, elliptic, apex obtuse. Fl. Oct-Dec. Fr. Nov-Dec. Mountain slopes, among shrubs or sunny grassland at 400-2200 m. Distributed in Yunnan and Sichuan.

三棱枝杭子梢
Campylotropis trigonoclada (Franch.) Schindl.

灌木，通常高1-3米。幼枝三棱形，具稀疏短伏毛或无毛。小叶倒卵形、长圆形或卵形至狭卵形或椭圆形。总状花序长3-26厘米，有时呈圆锥状；花

毛杭子梢 *Campylotropis hirtella*

西南杭子梢 *Campylotropis delavayi*

三棱枝杭子梢 *Campylotropis trigonoclada*

冠黄色或紫色。荚果斜倒卵形。生海拔1000-3000米的山坡、灌丛、林缘、林中、草地或路边。产云南、四川、贵州和广西。

Shrubs, usually 1-3 m tall. Young branches triquetrous, with sparse appressed short hairs or glabrous. Leaflets obovate, oblong or ovate to narrowly ovate or elliptic. Racemes 3-26 cm long, sometimes paniculate; corolla yellow or purple. Legumes obliquely obovoid. Mountain slopes, thickets, forest edges, forests, grasslands or roadsides at 1000-3000 m. Distributed in Yunnan, Sichuan, Guizhou and Guangxi.

杭子梢

Campylotropis macrocarpa (Bunge) Rehder

灌木。小叶长圆形或卵形，顶生者长1.2-6.5厘米。总状花序，有时组成圆锥花序；苞片常在开花后脱落；花冠紫色至粉白色。荚果长圆形、近圆形或椭圆体形。花果期5-10月。生海拔100-2000米的山坡、林中、灌丛、林缘、溪边、山谷或开阔地。产中国大部分地区。朝鲜半岛亦有。

Shrubs. Leaflets oblong or ovate, terminal one 1.2-6.5 cm long. Racemes, sometimes paniculate; bracts usually caducous before flowering; corolla purple to pinkish white. Legumes oblong, suborbicular or ellipsoid. Fl. and fr. May-Oct. Slopes, forests, thickets, forest edges, stream sides, valleys or open places at 100-2000 m. Distributed in most parts of China. Also in Korean Peninsula.

小雀花

Campylotropis polyantha (Franch.) Schindl.

灌木。叶柄长0.5-4厘米；小叶长圆形、倒卵形或卵形至狭卵形，下面密具平伏或斜展的柔毛。总状花序长2-13厘米；花冠紫色至粉白色。荚果椭圆形或斜卵形，表面被毛。花果期3-11(-12)月。生海拔400-3200米的山坡、向阳灌丛、石山上、路边、草地或溪边。产云南、四川、西藏和贵州。

Shrubs. Petioles 0.5-4 cm long; leaflets oblong, obovate or ovate to narrowly ovate, abaxially densely appressed or ascending pubescent. Racemes 2-13 cm long; corolla purple to pinkish white. Legumes ellipsoid or oblique ovoid, with hairs on surface. Fl. and fr. Mar-Nov(-Dec). Mountain slopes, sunny thickets, rocky mountains, roadsides, grasslands or streamsides at 400-3200 m. Distributed in Yunnan, Sichuan, Xizang and Guizhou.

杭子梢 *Campylotropis macrocarpa*

小雀花 *Campylotropis polyantha*

滇杭子梢 *Campylotropis yunnanensis*

滇杭子梢

Campylotropis yunnanensis (Franch.) Schindl.

灌木。幼枝散生贴伏柔毛。小叶狭卵形至卵形或狭长圆形至长圆形，顶生小叶下面无毛或散生短伏毛，上面无毛。花序轴和花梗被上升或散生短柔毛；花梗长2.5-7毫米。花果期7-12月。生海拔1400-2800米的山坡、沟谷、灌丛或林缘。产云南。

Shrubs. Young branches sparsely shortly appressed hairy. Leaflets narrowly ovate to ovate or narrowly oblong to oblong, terminal one abaxially glabrous or sparsely shortly appressed hairy, adaxially glabrous. Inflorescences rachises and pedicels ascending or spreading shortly hairy; pedicels 2.5-7 mm long. Fl. and fr. Jul-Dec. Mountain slopes, valleys, thickets or forest edges at 1400-2800 m. Distributed in Yunnan.

丝莘子梢

Campylotropis yunnanensis subsp. **filipes** (Ricker) Iokawa et H. Ohashi

灌木。幼枝、花序轴、花梗和花萼疏被短伏毛。托叶条形；小叶3，狭卵形至卵形或狭长圆形至长圆形，顶生小叶长1.8-9厘米，宽0.3-3厘米，先端具小凸尖，两面无毛或下面疏被短伏毛。花梗长5-14毫米；花萼裂片狭三角形；花冠紫色。荚果斜长圆形，长8-12毫米，先端钝，侧面无毛。花果期8-12月。生海拔1900-2800米的山坡、林缘及灌丛中。产四川。

Shrubs. Young branches, inflorescence rachis, pedicels and calyx sparsely shortly appressed hairy. Stipels linear; leaflets 3, narrowly ovate to ovate or narrowly oblong to oblong, terminal one 1.8-9 × 0.3-3 cm, apex mucronulate, both surfaces glabrous or abaxially sparsely shortly appressed hairy. Pedicels 5-14 mm long; calyx lobes narrowly triangular; corolla purple. Legume obliquely oblong, 8-12 mm long, apex obtuse, lateral surface glabrous. Fl. and fr. Aug-Dec. Mountain slopes, forest margins and thickets at 1900-2800 m. Distributed in Sichuan.

绿胡枝子

Lespedeza buergeri Miq.

灌木。叶羽状3小叶。总状花序腋生；萼裂片卵状披针形或卵形，密具长柔毛；花冠淡黄绿色；旗瓣近圆形；雄蕊二体；子房1室，不裂，种子1粒。荚果长圆状披针形。生海拔1500米以下的山坡、林下、山沟和路边。产中国东南、华中和华北。朝鲜半岛和日本亦有。

Shrubs. Leaves pinnate, with 3 leaflets. Racemes axillary; calyx lobes ovate-lanceolate or ovate, densely villous; corolla pale yellow-green; standards suborbicular; stamens diadelphous; ovary 1-locular, not dehiscent, with 1 seed. Legumes oblong-lanceo-

绿胡枝子 *Lespedeza buergeri*

丝莘子梢 *Campylotropis yunnanensis* subsp. *filipes*

美丽胡枝子 *Lespedeza thunbergii* subsp. *formosa*

宽叶胡枝子 *Lespedeza maximowiczii*

late. Under forests, on slopes, by streams and roadsides below 1500 m. Distributed in SE, C and N China. Also in Korean Peninsula and Japan.

美丽胡枝子

Lespedeza thunbergii (DC.) Nakai subsp. **formosa** (Vogel) H. Ohashi

直立灌木。茎圆柱状或具条纹，具平伏的丝毛。小叶纸质，常卵状椭圆形，上面被毛或稀无毛。总状花序腋生，单一，比叶长。荚果倒卵状或倒卵状长圆形。花期7-9月，果期9-10月。生海拔2800米以下的山坡、林缘、路旁或灌丛中。产华南和东南。

Erect shrubs. Stems terete or striate, appressed sericeous. Leaflets papery, usually ovate-elliptic, adaxial surface of leaflets puberulent or rarely glabrescent. Racemes axillary, solitary, longer than leaves. Legumes obovoid or obovoid-oblong. Fl. Jul-Sep. Fr. Sep-Oct. Slopes, forest edges, roadsides or thickets below 2800 m. Distributed in S and SE China.

宽叶胡枝子

Lespedeza maximowiczii C. K. Schneid.

直立灌木。多分枝，被白色柔毛。羽状复叶具3小叶；小叶宽椭圆形或卵状椭圆形。总状花序腋生或为圆锥花序顶生；花冠紫红色。荚果卵状椭圆形，被柔毛，表面具网纹。花期7-8月，果期9-10月。生海拔1000米以下的山坡或林中。产浙江、河南和安徽。朝鲜半岛和日本亦有。

Erect shrubs. Much branched, white pilose. Leaves pinnately 3-foliolate; leaflets broadly elliptic or ovate-elliptic. Racemes axillary or in terminal panicles; corolla purplish red. Legumes ovoid-ellipsoid, pubescent, reticulate veined. Fl. Jul-Aug. Fr. Sep-Oct. Mountain slopes forests below 1000 m. Distributed in Zhejiang, Henan and Anhui. Also in Korean Peninsula and Japan.

胡枝子

Lespedeza bicolor Turcz.

直立灌木。叶具3小叶。总状花序腋生，比叶长；萼侧裂片卵形或三角形至狭卵形，顶端锐尖至短渐尖；花冠紫红色；龙骨瓣的距与其瓣片近等长。荚果斜倒卵形。花期7-9月，果期9-10月。生海拔100-1000米的山坡、林缘、路旁或灌丛中。产中国大部分地区。俄罗斯、蒙古、朝鲜半岛和日本亦有。

Erect shrubs. Leaves 3-foliolate. Racemes axillary, longer than leaves; lateral calyx lobes ovate or triangular to narrowly ovate, apically acute to shortly acuminate; corolla reddish purple; keels claw nearly as long as keel lamina. Legumes oblique obovoid. Fl. Jul-Sep. Fr. Sep-Oct. Slopes, forest edges, roadsides or thickets at 100-1000 m. Distributed in most parts of China. Also in Russia, Mongolia, Korean Peninsula and Japan.

胡枝子 *Lespedeza bicolor*

铁马鞭 *Lespedeza pilosa*

多花胡枝子 *Lespedeza floribunda*

铁马鞭
Lespedeza pilosa (Thunb.) Siebold et Zucc.

多年生草本。全株密被毛。茎匍匐。羽状复叶具3小叶；小叶宽倒卵形或倒卵形。总状花序腋生；花冠黄白色或黄色；常有闭锁花1-3。荚果宽卵形，凸镜状。花期7-9月，果期9-10月。生海拔1000米以下的荒山坡或草地。产中国西南、华南、华中、华西和华东。朝鲜半岛和日本亦有。

Perennial herbs. Densely villous throughout. Stems procumbent. Leaves pinnately 3-foliolate; leaflets broadly obovate or obovate. Racemes axillary; corolla yellowish white or yellow; cleistogamous flowers often 1-3. Legumes broadly ovoid, convex. Fl. Jul-Sep. Fr. Sep-Oct. Waste slopes or grasslands below 1000 m. Distributed in SW, S, C, W and E China. Also in Korean Peninsula and Japan.

横断山铁马鞭
Lespedeza fasciculiflora var. **hengduanshanensis** C. J. Chen

多年生草本。茎和小枝被紧贴的糙伏毛。小叶3，倒三角形，顶端小叶长4-7毫米，背面密被灰白毛，正面疏被糙伏毛。总状花序腋生，明显长于叶；花萼5深裂，裂片条状披针形；花冠白色或淡黄色，稍长于花萼。荚果狭卵球形，与宿存萼片近等长，密被硬毛，先端具长喙。花期8-9月。生海拔1800-2600米的干燥河谷灌丛中。产四川和西藏。

Perennial herbs. Stems and branchlets adpressed strigulose. Leaflets 3, obtriangular, terminal one 4-7 mm long, abaxially densely hoary, adaxially sparsely strigulose. Racemes axillary, distinctly overtopping leaves; calyx 5-parted, lobes linear-lanceolate; corolla white or yellowish, slightly overtopping calyx. Legume narrowly ovoid, subequal to persistent calyx, densely hirsute, apex long rostrate. Fl. Aug-Sep. Thickets in dry river valleys at 1800-2600 m. Distributed in Sichuan and Xizang.

多花胡枝子
Lespedeza floribunda Bunge

小灌木。叶具3小叶。总状花序腋生，显著较叶长；花多数；萼裂片披针形或卵状披针形；花冠紫色、紫红色或蓝紫色。荚果宽卵形，与宿存萼片近等长。花期6-9月，果期9-10月。生海拔1300米以下的石质山坡。产中国西南、华南、华中、华西、华北和东北。俄罗斯、朝鲜半岛和日本亦有。

Shrublets. Leaves 3-foliolate. Racemes axillary, conspicuously longer than leaves; flowers numerous; calyx lobes lanceolate or ovate-lanceolate; corolla purple, purplish red or bluish purple. Legumes wide ovoid, subequal to persistent calyx. Fl. Jun-Sep. Fr. Sep-Oct. Gravelly slopes below 1300 m. Distributed in SW, S, C, W, N and NE China. Also in Russia, Korean Peninsula and Japan.

细梗胡枝子
Lespedeza virgata (Thunb.) DC.

小灌木。小叶椭圆形或长圆形，背面密被伏毛。总状花序腋生，显著比叶长，疏松具少花；花序梗线形，无毛；花无柄或花梗短于1毫米；小苞片短于萼筒。荚果近圆形。花期7-9月，果期9-10月。生海拔800米以下的石质山坡、山林、路旁或灌丛。产中国除西北以外大部分地区。朝鲜半岛和日本亦有。

Shrublets. Leaflets often elliptic or oblong, abaxially densely adpressed pubescent. Racemes axillary, conspicuously longer than leaves, laxly few flowered; peduncles filiform, glabrous; flowers sessile or pedicel shorter than 1 mm long; bracteoles shorter

横断山铁马鞭 *Lespedeza fasciculiflora* var. *hengduanshanensis*

细梗胡枝子 *Lespedeza virgata*

长叶胡枝子 *Lespedeza caraganae*

than calyx tube. Legumes suborbicular. Fl. Jul-Sep. Fr. Sep-Oct. Rocky slopes, montane forests, roadsides or thickets below 800 m. Distributed in most parts of China, except NW China. Also in Korean Peninsula and Japan.

绒毛胡枝子

Lespedeza tomentosa (Thunb.) Siebold ex Maxim.

灌木。全株密被黄棕色绒毛。叶具3小叶；小叶椭圆形或卵状长圆形，3-6 × 1.5-3厘米，下面密具绒毛。总状花序顶生或腋生于茎上部；花冠黄色或黄白色。荚果倒卵球形。花期7-8月，果期9-10月。生海拔1000米以下的干旱山坡、草地或灌丛。产中国大部分地区。西南亚和东亚亦有。

Shrubs. Densely yellowish brown tomentose throughout. Leaves 3-foliolate; leaflets elliptic or ovate-oblong, 3-6 × 1.5-3 cm, abaxially densely tomentose. Racemes terminal or axillary at upper parts of stems; corolla yellow or yellowish white. Legumes obovoid. Fl. Jul-Aug. Fr. Sep-Oct. Dry slopes, grasslands or thickets below 1000 m. Distributed in most parts of China. Also in SW and E Asia.

兴安胡枝子

Lespedeza davurica (Laxm.) Schindl.

小灌木。小叶狭椭圆形至椭圆形。总状花序腋生，比叶短或与叶近等长，密花；花序梗具柔毛；花梗长1-3毫米；小苞片长于萼筒。荚果倒卵球形或长倒卵球形。花期7-8月，果期9-10月。生干旱山坡、草地、路旁或沙质地。产中国西南、华中、华北、华西和东北。俄罗斯、朝鲜半岛和日本亦有。

Shrublets. Leaflets narrowly elliptic to elliptic. Racemes axillary, shorter than or subequal to leaves, densely flowered; peduncles pubescent; pedicels 1-3 mm long; bracteoles longer than calyx tube. Legumes obovoid or long obovoid. Fl. Jul-Aug. Fr. Sep-Oct. Dry slopes, grasslands, roadsides or sandy sites. Distributed in SW, C, N, W and NE China. Also in Russia, Korean Peninsula and Japan.

长叶胡枝子

Lespedeza caraganae Bunge

灌木或多年生草本。茎直立。叶具3小叶；小叶长圆状条形。总状花序腋生，具3-5花；花冠白色或黄色，明显高出花萼。荚果长圆状卵形或倒卵形。花期6-9月，果期10月。生海拔1400米以下的山坡。产华北、华西、华东和东北。

Shrubs or perennial herbs. Stems erect. Leaves 3-foliolate; leaflets oblong-linear. Racemes axillary, 3-5-flowered; corolla white or yellow, distinctly overtopping calyx. Legumes oblong-ovoid or obovoid. Fl. Jun-Sep. Fr. Oct. Mountain slopes below 1400 m. Distributed in N, W, E and NE China.

绒毛胡枝子 *Lespedeza tomentosa*

兴安胡枝子 *Lespedeza davurica*

中华胡枝子 *Lespedeza chinensis*

中华胡枝子

Lespedeza chinensis G. Don

亚灌木。全株具白色平伏毛。直立或散生。具3小叶；小叶倒卵状长圆形、长圆形或倒卵形。总状花序腋生，少花；花冠白色或黄色；闭锁花簇生于茎下部叶腋。荚果卵形，密被白色伏毛，基部稍倾斜，顶端有喙。花期8-9月，果期10-11月。生海拔2500米以下的灌丛、林缘、路边、山坡、草地或林中。产中国西南、华南、东南、华中和华东。

Subshrubs. Adpressed white hairy throughout. Stems erect or diffuse. Leaves 3-foliolate; leaflets obovate-oblong, oblong or obovate. Racemes axillary, few-flowered; corolla white or yellow; cleistogamous flowers clustered in leaf axils of lower stems. Legumes ovoid, densely adpressed white hairy, base slightly oblique, apex rostrate. Fl. Aug-Sep. Fr. Oct-Nov. Thickets, forest edges, roadsides, mountain slopes, grasslands or forests below 2500 m. Distributed in SW, S, SE, C and E China.

截叶铁扫帚

Lespedeza cuneata (Dumont de Courset) G. Don

亚灌木或多年生草本。叶密集，具3小叶，小叶楔形或条状楔形，顶端截形或近截形，具小尖头，下面疏具平伏柔毛。总状花序腋生，具2-4花；花序梗短；花冠浅黄色或白色。荚果宽卵形，被伏毛。花期7-8月，果期9-10月。生海拔2500米以下的山坡或路旁。产中国大部分地区。南亚、东南亚、朝鲜半岛、日本和澳大利亚亦有。

Subshrubs or perennial herbs. Leaves congregate, 3-foliate, leaflets cuneate or linear-cuneate, apex truncate or subtruncate, mucronate, abaxial sparsely appressed pubescent. Racemes axillary, 2-4-flowered; peduncles short; corolla yellowish or white. Legumes broadly ovoid, adpressed hairy. Fl. Jul-Aug. Fr. Sep-Oct. Mountain slopes or roadsides below 2500 m. Distributed in most parts of China. Also in S and SE Asia, Korean Peninsula, Japan and Australia.

尖叶铁扫帚

Lespedeza juncea (L. f.) Pers.

亚灌木或多年生草本。小叶倒披针形或条状长圆形，先端具刺。总状花序腋生，稍超出叶，具3-7花；花冠白色或浅黄色；旗瓣基部具紫色斑点；闭锁花簇生于叶腋，近无柄。荚果阔卵形，两面具平伏的白色毛。花期7-9月，果期9-10月。生海拔1500米以下的灌丛中。产华北大部分地区。俄罗斯、蒙古、朝鲜半岛和日本亦有。

Subshrubs or perennial herbs. Leaflets oblanceolate or linear-oblong, apex spinescent. Racemes axillary, slightly overtopping leaves, 3-7-flowered; corolla white or yellowish; standard with purple spots at base; cleistogamous flowers clustered in leaf axils, subsessile. Legumes broadly ovoid, both surfaces adpressed white hairy. Fl. Jul-Sep. Fr. Sep-Oct. Thickets below 1500 m. Distributed in most parts of N China. Also in Russia, Mongolia, Korean Peninsula and Japan.

截叶铁扫帚 *Lespedeza cuneata*

尖叶铁扫帚 *Lespedeza juncea*

阴山胡枝子

Lespedeza inschanica (Maxim.) Schindl.

亚灌木或多年生草本。茎直立或上升，上部被短柔毛，下部近无毛。叶具3小叶；小叶长圆形或倒卵状长圆形，先端钝圆或凹缺。总状花序腋生，与叶近等长。荚果倒卵形。生于旱山坡。产华中、华北、华西、华东和东北。朝鲜半岛和日本亦有。

Subshrubs or perennial herbs. Stems erect or ascending, upper parts of stems pubescent, lower parts glabrous. Leaves 3-foliolate; leaflets oblong or obovate-oblong, apex obtuse-rounded or emarginate. Racemes axillary, subequal to petioles. Legumes obovoid. Dry mountain slopes. Distributed in C, N, W, E and NE China. Also in Korean Peninsula and Japan.

阴山胡枝子 *Lespedeza inschanica*

鸡眼草

Kummerowia striata (Thunb.) Schindl.

一年生草本。多分枝，小枝具向下的毛。三出羽状复叶，小叶先端常圆形；托叶大，膜质，具长缘毛，比叶柄长。花1-3生于上部叶腋；花冠长5-6毫米。荚果圆形或倒卵形，稍压扁。花期7-9月，果期8-10月。生海拔500米以下的路旁、沙地、田边、溪旁或山坡草地。产中国西南、华南、华中、华北、华东和东北。印度、越南、俄罗斯、朝鲜半岛和日本亦有。

Annual herbs. Many-branched, branchlets with downward-pointing hairs. Ternate-pinnate compound leaves, leaflets usually with rounded apex; stipules large, membranous, long ciliate, longer than petioles. Flowers 1-3 in upper axils of leaves; corolla 5-6 mm long. Legumes orbicular or obovoid, slightly compressed. Fl. Jul-Sep. Fr. Aug-Oct. Roadsides, sandy soils, fieldsides, streamsides, or grasslands on slopes below 500 m. Distributed in SW, S, C, N, E and NE China. Also in India, Vietnam, Russia, Korean Peninsula and Japan.

长萼鸡眼草

Kummerowia stipulacea (Maxim.) Makino

草本。茎与枝具疏生向上白毛。三出羽状复叶；小叶倒卵形或宽倒卵形，先端凹缺。花腋生，1或2朵；龙骨瓣上面具深紫色斑点。荚果卵形或椭圆形，稍扁平。花期7-8月，果期8-10月。生海拔100-1200米的路边、草地、山坡或沙丘。产中国大部分地区。俄罗斯、朝鲜半岛和日本亦有；归化于美国东南部。

Herbs. Stems and branches with sparse upward-pointing white hairs. Leaves 3-foliolate; leaflets obovate or broadly obovate, with apex emarginate. Flowers 1 or 2, axillary; keels with dark purple spots adaxially. Legumes ovoid or ellipsoid, slightly compressed. Fl. Jul-Aug. Fr. Aug-Oct. Roadsides, grasslands, mountain slopes or sand dunes at 100-1200 m. Distributed in most parts of China. Also in Russia, Korean Peninsula and Japan; naturalized in SE USA.

鸡眼草 *Kummerowia striata*

长萼鸡眼草 *Kummerowia stipulacea*

翅果刺桐 *Erythrina subumbrans*

翅果刺桐
Erythrina subumbrans (Hassk.) Merr.

乔木，皮刺粗壮。叶具3小叶，小叶卵状三角形，膜质，两面无毛。总状花序有褐色绒毛；花冠红色；龙骨瓣与翼瓣近等长。荚果长15厘米，仅顶端具种子，不开裂。生海拔300-600米的雨林中。产云南。老挝、越南、菲律宾和印度尼西亚(爪哇)亦有。

Trees, with robust prickles. Leaves 3-foliolate, leaflets ovate-deltoid, membranous, both surfaces glabrous. Racemes brown-tomentose; corolla red; keels subequal to wings. Legumes 15 cm long, with seeds only toward apex, indehiscent. Rain forests at 300-600 m. Distributed in Yunnan. Also in Laos, Vietnam, the Philippines and Indonesia (Java).

刺桐
Erythrina variegata L.

乔木，高可达20米。茎刺小，常黑色。三出羽状复叶，常簇生于枝顶；小叶宽卵形或菱状卵形，膜质，两面无毛。总状花序顶生，长10-16厘米；龙骨瓣和翼瓣近等大。荚果黑色，肾形。花期2-5月，果期4-8月。产华南。南亚、东南亚、日本南部、澳大利亚和太平洋岛屿。非洲、美洲中部和南部亦有引入。

Trees, up to 20 m tall. Stem prickles minute, usually black. Leaves pinnately 3-foliolate, usually clustered at branch tip; leaflets broadly ovate or rhomboid-ovate, membranous, both surfaces glabrous. Racemes terminal, 10-16 cm long; keels and wings subequal. Legumes black, reniform. Fl. Feb-May. Fr. Apr-Aug. Distributed in S China. Also in S and SE Asia, S Japan, Australia and Pacific Islands. Introduced to Africa, and Central and South America.

劲直刺桐
Erythrina stricta Roxb.

乔木。枝具短的白刺。叶具3小叶；托叶狭镰状。总状花序长约15厘米，花3或4个簇生；花萼佛焰苞状，不开裂或先端稍2裂；花冠红色。荚果长7-12厘米。生海拔约400米的河边疏林中或村旁。产云南、西藏和广西。南亚亦有。

Trees. Branches with short whitish prickles. Leaves 3-foliolate; stipules narrowly falcate. Racemes ca. 15 cm long, flowers in clusters of 3 or 4; calyx spathe-like, undivided or apex slightly 2-lobed; corolla red. Legumes 7-12 cm long. Open forests by rivers or near villages at ca. 400 m. Distributed in Yunnan, Xizang and Guangxi. Also in S Asia.

间序油麻藤
Mucuna interrupta Gagnep.

攀援藤本，近木质。羽状复叶，3小叶；小叶薄纸质，顶生小叶两面无毛或具稀疏的毛。花序腋生；花萼密具毛，裂片阔三角形；花冠白色。荚果卵

刺桐 *Erythrina variegata*

劲直刺桐 *Erythrina stricta*

形，被短毛和红褐色长刺毛。花期8月，果期10月。生海拔900-1100米的常绿阔叶林林缘。产云南。缅甸、老挝、泰国、越南和柬埔寨亦有。

Twining vines, woody. Leaves pinnate, 3-foliolate; leaflets thin papery, terminal one glabrous or sparsely puberulous on both surfaces. Inflorescences axillary; calyx densely hairy, lobes broadly triangular; corolla white. Legumes ovoid, pubescent and brown-red long spiny hairs. Fl. Aug. Fr. Oct. Evergreen broad-leaved forest edges at 900-1100 m. Distributed in Yunnan. Also in Myanmar, Laos, Thailand, Vietnam and Cambodia.

间序油麻藤 *Mucuna interrupta*

Climbing vines, to 5 m long. Leaves pinnately 3-foliolate; leaflets thin papery, terminal leaflets rhombic-ovate. Inflorescences axillary; corolla deep purple or reddish brown. Legumes narrowly oblong, asymmetric in outline. Seeds 2-5, deep reddish brown or black, smooth. Fr. Apr-May. Riversides, thickets, roadsides, mountain or valleys at 400-1500 m. Distributed in SW, S, SE, C and E China.

褶皮黧豆

Mucuna lamellata Wilmot-Dear

攀援藤本，长达5米。羽状复叶具3小叶；小叶薄纸质，顶生小叶菱状卵形。花序腋生；花冠深紫色或红棕色。荚果狭长圆形，轮廓不对称。种子2-5粒，深红棕色或黑色，光滑。果期4-5月。生海拔400-1500米的溪边、灌丛、路边、山地或山谷。产中国西南、华南、东南、华中和华东。

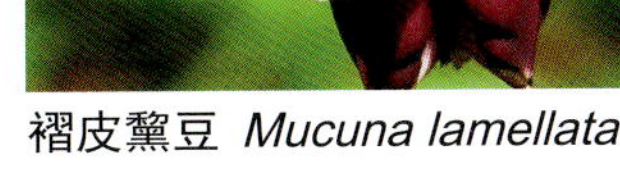

褶皮黧豆 *Mucuna lamellata*

白花油麻藤 *Mucuna birdwoodiana*

白花油麻藤

Mucuna birdwoodiana Tutch.

常绿大型木质藤本。小叶近革质，下面无毛或疏具毛；顶生小叶卵形、椭圆形或稍倒卵形。花序生老枝或腋生；花冠白色或绿白色。荚果边缘具一对厚木质翅，平均3-5毫米宽，具清晰的脊，明显念珠状。花期4-6月，果期6-11月。生海拔800-2500米的山地阳处、路旁或河边。产中国西南和华南。

Evergreen large woody vines. Leaflets almost leathery, glabrous or sparsely hairy abaxially; terminal leaflets ovate, elliptic or slightly obovate. Inflorescences on old branches or axillary; corolla white or greenish white. Legumes margins with a pair of thick woody wings, evenly 3-5 mm wide and with definite edges, markedly torulose. Fl. Apr-Jun. Fr. Jun-Nov. Mountain sunny sites, roadsides or by rivers at 800-2500 m. Distributed in SW and S China.

大果油麻藤

Mucuna macrocarpa Wall.

大型木质藤本。小叶纸质或革质，顶生小叶椭圆形或长披针

大果油麻藤 *Mucuna macrocarpa*

常春油麻藤 *Mucuna sempervirens*

形、卵状椭圆形、卵形或稍倒卵形，长10-19厘米，宽5-10厘米。花序常生自老茎；花冠二色。果木质，带形，26-48 × 3-5厘米。花期11月至翌年5月，果期翌年4-11月。生海拔800-2500米的山地或河边林中。产中国西南和华南。南亚和日本南部亦有。

Large woody vines. Leaflets papery or leathery, terminal leaflets elliptic, ovate-elliptic, ovate or subobovate, 10-19 cm long, 5-10 cm broad. Inflorescences usually arising from old stems; corolla bicolored. Legumes woody, taeniform, 26-48 × 3-5 cm. Fl. Nov to next May. Fr. next Apr-Nov. Mountains or forests by rivers at 800-2500 m. Distributed in SW and S China. Also in S Asia and S Japan.

常春油麻藤

Mucuna sempervirens Hemsl.

木质藤本。小叶纸质或质厚；顶生小叶椭圆形或椭圆状卵形。花序常生老茎；花冠深紫色，干后黑色。荚果木质，被红褐色短毛和刚毛。种子4-12粒，带红色、褐色或黑色。花期4-5月，果期8-10月。生海拔300-3000米的林中、灌丛或河边。产中国西南、华南、东南、华中和华东。南亚和日本亦有。

Woody vines. Leaflets papery or thicker textured; terminal leaflets elliptic or elliptic-ovate. Inflorescences usually on old stems; corolla deep purple, black after drying. Legumes woody, umber pubescent and hispid. Seeds 4-12, red, brown or black. Fl. Apr-May. Fr. Aug-Oct. Forests, thickets or riversides at 300-3000 m. Distributed in SW, S, SE, C and E China. Also in S Asia and Japan.

黧豆

Mucuna pruriens (L.) DC. var. **utilis** (Wall. ex Wight) Baker ex Burck

一年生缠绕藤本。茎疏具长的开展的毛。羽状复叶具3小叶。总状花序下垂，有花10-20朵；花萼密具长的浅色毛。荚果宽达2厘米，密具丝毛，具1或2个凸起的肋纹。花期10月，果期11月。栽培中国西南和华南，或逸生。亚洲热带和亚热带地区有栽培。

Annual twining vines. Stems with sparse long fine spreading hairs. Leaves pinnate compound with 3 leaflets. Racemes pendulous, 10-20-flowered; calyx with dense long pale hairs. Legumes to 2 cm wide, densely covered with silky hairs, with 1 or 2 prominent ribs. Fl. Oct. Fr. Nov. Cultivated in SW and S China, or escaped. Also cultivated in tropical and subtropical Asia.

黧豆 *Mucuna pruriens* var. *utilis*

紫矿 *Butea monosperma*

紫矿
Butea monosperma (Lam.) Taub.

乔木。小叶不等大，厚革质，上面无毛，网脉明显，下面沿叶脉被短柔毛。总状花序或圆锥花序腋生或生无叶枝的节处；花序轴、花梗和花萼外面密具褐色或黑褐色绒毛；花冠橙红色，后变为黄色。种子肾形，红褐色。花期3-4月。生林中或路旁潮湿处。产云南和广西。南亚亦有。

Trees. Leaflets unequal, thickly leathery, glabrous adaxially, reticulate veins distinct, abaxially pubescent along nerves. Racemes or panicles axillary or at nodes of leafless branches; rachis, pedicels and calyx outside densely brown or blackish brown velutinous; corolla orange-red, becoming yellow later. Seeds reniform, brown-red. Fl. Mar-Apr. Forests or wet places by roads. Distributed in Yunnan and Guangxi. Also in S Asia.

密花豆
Spatholobus suberectus Dunn

木质藤本。叶纸质或近革质。圆锥花序腋生或生小枝顶端，长达50厘米；花序轴和花梗具黄褐色柔毛；萼筒长为萼齿的2-3倍；花萼下面3齿先端圆或略钝，长不超过1毫米，上面2齿稍长，多少合生。种子扁长圆形。花期6月，果期11-12月。生海拔800-1700米的山坡林中或灌丛中。产福建、广东、广西和云南。

Woody climbers. Leaves papery or subleathery. Panicles axillary or at apex of branchlets, to 50 cm long; rachis and pedicels yellowish brown puberulent; calyx tubes 2-3 × longer than teeth; lower 3 sepals apex rotund or slightly obtuse, less than 1 mm long, upper 2 sepals slightly longer, connate. Seeds flat, oblong. Fl. Jun. Fr. Nov-Dec. Forests or thickets on slopes at 800-1700 m. Distributed in Fujian, Guangdong, Guangxi and Yunnan.

云南土圞儿
Apios delavayi Franchet

缠绕草本。托叶钻状；小叶5，狭卵状披针形，长2-5厘米，先端渐尖，基部圆形，边缘具短睫毛，两面疏被硬毛。总状花序具5-10朵排列稀疏的花，每节有1-3花；花萼宽钟形，长约为花冠的1/6，二唇形，上部的2萼片合生为三角形；花冠淡黄色。荚果长达15厘米，线形，无毛。花期7-8月，果期9月。生海拔1300-3500米的灌丛。产云南、四川和西藏。

Twining herbs. Stipules subulate; leaflets 5, lanceolate, 2-5 cm long, apex acuminate, base rounded, margin shortly ciliate, both surfaces sparsely hirsute. Raceme sparsely 5-10-flowered, nodes 1-3-flowered; calyx broadly campanulate, ca. 1/6 as long as corolla, 2-lipped, upper 2 lobes connate into triangle; corolla light yellow. Legume to 15 cm long, linear, glabrous. Fl. Jul-Aug. Fr. Sep. Shrublands at 1300-3500 m. Distributed in Yunnan, Sichuan and Xizang.

密花豆 *Spatholobus suberectus*

云南土圞儿 *Apios delavayi*

肉色土圞儿 *Apios carnea*

肉色土圞儿

Apios carnea (Wall.) Benth. ex Baker

缠绕藤本。叶具羽状5小叶；小叶长圆形至卵状长圆形，长3.5-13厘米，纸质，无毛。总状花序长15-40厘米；节上具2或3花；花冠淡红色、淡紫色或橙红色。荚果条形。花期7-9月，果期8-11月。生海拔600-2600米的沟边林中、溪边或路旁。产中国西南和华南。印度、尼泊尔、不丹、泰国和越南亦有。

Twining vines. Leaves pinnately 5-foliolate; leaflets oblong to ovate-oblong, 3.5-13 cm long, papery, glabrous. Racemes 15-40 cm long; nodes 2- or 3-flowered; corolla reddish, light purple or orange-red. Legumes linear. Fl. Jul-Sep. Fr. Aug-Nov. Forests by valleys, by rivers or roadsides at 600-2600 m. Distributed in SW and S China. Also in India, Nepal, Bhutan, Thailand and Vietnam.

刀豆

Canavalia gladiata (Jacq.) DC.

草质藤本。羽状复叶具3小叶，小叶卵形，两面被微柔毛或近无毛，侧生小叶偏狭。圆锥花序具数花，生花序轴中部以上；花瓣白色或粉红；子房线形，被毛。荚果条状长圆形，稍弯曲。花期7-9月，果期10月。中国长江以南各省有栽培。产亚洲；热带地区广栽培。

Climbing herbs. Leaves paripinnate, 3-foliolate; leaflets ovate, slightly pubescent or subglabrous on both surface. Inflorescenes panicles, several flowers above middle of rachises; petals white or pink; ovary linear, hairy. Legumes linear-oblong, slightly curved. Fl. Jul-Sep. Fr. Oct. Cultivated throughout the provinces in the south of the Yangtze River. Distributed in Asia; widely cultivated in the tropics.

刀豆 *Canavalia gladiata*

豆薯

Pachyrhizus erosus (Linnaeus) Urban

粗壮缠绕藤本。根块状，直径20-30厘米。小叶菱形或卵形，中部以上不规则浅裂，侧生小叶极偏斜，仅下面微被毛。总状花序长15-30厘米，每节有花3-5朵；花萼钟形，表面紧被毛；花冠浅紫色或淡红色。荚果带形，扁平，被细长糙伏毛。种子每荚8-10粒，近方形。花期8月，果期11月。产华中、华南、东南、西南和台湾。原产热带美洲。

Vines, robust, twining. Root tubers, 10-20 cm diam. Leaflets rhombic or ovate, upper margin often somewhat dentate or lobed, lateral ones very oblique, sparsely pubescent abaxially. Racemes, 15-30 cm long, flowers 3-5 at thickened nodes; calyx campanulate, adpressed pilose; corolla purplish or reddish. Legumes compressed, hirsute. Seeds 8-10 per article, subsquare. Fl. Aug. Fr. Nov. Distributed in C, S, SE and SW China, and Taiwan. Native to tropical America.

草葛 *Neustanthus phaseoloides*

豆薯 *Pachyrhizus erosus*

草葛

Neustanthus phaseoloides (Roxburgh) Bentham

草质藤本。茎纤细，长2-4米，被褐黄色长硬毛。托叶基着；侧生小叶被紧贴的长硬毛。总状花序单生；花冠浅蓝色或淡紫色；旗瓣近圆形，基部具2枚内弯的耳，龙骨瓣镰刀状，顶端具短喙，基部截形，具瓣柄。荚果近圆柱状。花期8-9月，果期10-11月。生山地、丘陵的灌丛中。产云南、广东、广西、台湾和浙江。南亚和东南亚亦有。

Herbaceous vines. Stem slender, 2-4 m long, brownish hirsute. Stipules basifixed; lateral leaflets abaxially densely hirsute. Racemes solitary. Corolla bluish or lilac; standard suborbicular, base with 2 incurved auricles; keel falcate, apex with short beak, base truncate, clawed. Legumes subcylindric. Fl. Aug-Sep. Fr. Oct-Nov. Thickets of mountainous and hilly areas. Distributed in Yunnan, Guangdong, Guangxi, Taiwan and Zhejiang. Also in S and SE Asia.

木琼豆 (紫花琼豆)

Teyleria stricta Kurz

灌木，偶蔓生。茎高1-2.5米。枝有条纹，嫩时被灰色短柔毛，老时无毛。托叶三角状卵形，被灰色短柔毛。总状花序通常不分枝；旗瓣倒卵形，长5-8毫米，顶端微凹，基部具瓣柄，耳内折；对旗瓣的一枚雄蕊连合至中部。荚果长圆形，扁平。种子5-10粒，种皮上有细疣点。花期5-6月，果期9-10月。生林中或草地。产云南。缅甸和泰国亦有。

Shrubs, erect or rarely climbing. Stem 1-2.5 m tall. Branches striate, gray pubescent when young, glabrous when old. Stipules triangular-ovate, gray pubescent. Racemes usually unbranched; standard obovate, apex emarginate, base clawed, auricles inflexed; vexillary stamen joined to middle. Legumes oblong, flattened. Seeds 5-10, tuberculate. Fl. May-Jun. Fr. Sep-Oct. Forests, or grasses. Distributed in Yunnan. Also in Myanmar and Thailand.

苦葛

Toxicopueraria peduncularis (Graham ex Benth.) Benth.

木质藤本。托叶基着；小叶卵

木琼豆（紫花琼豆） *Teyleria stricta*

形或斜卵形，两面具粗毛。总状花序细弱，长20-40厘米；花白色，3-5簇生于花序轴节上。荚果条形，果瓣近纸质。花期8月，果期10月。生荒地或林中。产中国西南。印度、尼泊尔、不丹、孟加拉国、巴基斯坦和缅甸亦有。

Woody lianas. Stipules basifixed; leaflets ovate or obliquely ovate, hirsute on both surfaces. Racemes slender, 20-40 cm long; flowers white, 3-5-clustered at nodes of rachis. Legumes linear, valvules almost papery. Fl. Aug. Fr. Oct. Wastelands or forests. Distributed in SW China. Also in India, Nepal, Bhutan, Bangladesh, Pakistan and Myanmar.

须弥葛
Pueraria wallichii DC.

灌木状缠线藤本。托叶披针形；顶生小叶倒卵形，长10-13厘米。总状花序长达15厘米，常簇生或排成圆锥花序式；花萼长约4毫米，近无毛，膜质；花冠淡红色，旗瓣倒卵形，长1.2厘米。荚果直，长7.5-12.5厘米，宽6-12毫米。花期9-10月。生海拔1700米的山坡灌丛中。产西藏和云南。泰国、缅甸、印度北部和东北部、不丹和尼泊尔亦有。

Shrubs, sometimes climbing. Stipules lanceolate; terminal leaflet obovate, 10-13 cm long. Racemes up to 15 cm, often fascicled or paniculate; calyx ca. 4 mm, subglabrous, membranous; corolla reddish; standard obovate, 1.2 cm long. Legumes straight, 7.5-12.5 cm × 6-12 mm. Fl. Sep-Oct. Hill slopes in thickets at 1700 m. Distributed in Xizang and Yunnan. Also in Thailand, Myanmar, N and NE India, Bhutan and Nepal.

苦葛
Toxicopueraria peduncularis

须弥葛 *Pueraria wallichii*

葛(野葛) *Pueraria montana*

葛(野葛)

Pueraria montana (Lour.) Merr.

粗壮藤本，具块根。羽状复叶，具3小叶；托叶基部不裂。花冠紫色；苞片短于小苞片；花萼长7-8毫米；旗瓣直径约8毫米。荚果长椭圆形，被褐色长硬毛，4-9厘米 × 6-8毫米。花期7-9月，果期10-12月。生海拔1800-2700米的山地疏林或密林中。产中国西南、华南、东南和华中。老挝、缅甸、菲律宾、泰国、越南和日本亦有。

Robust climbers, with tuberous roots. Leaves pinnate-ternate, leaflet 3; stipules undivided at base. Corolla purple; bracts shorter than bracteoles; calyx 7-8 mm long; standard ca. 8 mm diam. Legumes long ellipsoid, brown-hirsute, 4-9 cm × 6-8 mm long. Fl. Jul-Sep. Fr. Oct-Dec. Open or dense forests at 1800-2700 m. Distributed in SW, S, SE and C China. Also in Laos, Myanmar, the Philippines, Thailand, Vietnam and Japan.

密花葛(狐尾葛)

Pueraria alopecuroides Craib

木质藤本。托叶基部2裂，箭头形；小叶宽卵形，先端尾状渐尖，基部圆形，小叶膜质。花序圆锥状，花前极密，长约22厘米；苞片长于花芽，被锈色长硬毛；旗瓣长不逾1.5厘米。生海拔200-1300米的林中或草地。产云南南部。缅甸和泰国亦有。

Woody climbers. Stipules 2-lobed at base, sagittate; leaflets broad ovate, apex caudate-acuminate, base round, leaflets membranous. Inflorescences paniculate, very dense before flowering, ca. 22 cm long; bracts longer than flower buds, ferruginous-hirsute; standards less than 1.5 cm long. Forests or grasslands at 200-1300 m. Distributed in S Yunnan. Also in Myanmar and Thailand.

食用葛

Pueraria edulis Pampanini

缠绕草本，具块根。托叶背着，箭头形，基部2裂；顶生小

密花葛(狐尾葛) *Pueraria alopecuroides*

叶卵形，3裂，侧生的斜宽卵形，稍小，多少2裂。总状花序长达30厘米，不分枝或具1分枝；苞片卵形，无毛或具缘毛；萼裂片4，长4-7毫米；旗瓣近圆形。荚果带形，被极稀疏的黄色长硬毛。花期9月，果期10月。生海拔1000-3200米的山林。产四川和云南。不丹和印度亦有。

Twining herbs, with tuberous roots. Stipules dorsifixed, sagittate, basal 2 lobes; terminal leaflet ovate, 3-lobed to entire, lateral ones obliquely broadly ovate, smaller, ± 2-lobed. Racemes up to 30 cm long, simple or once branched; bracts ovate, glabrous or ciliate; calyx lobes 4, 4-7 mm long; standard suborbicular. Legumes linear-oblong, very sparsely hirsute with yellowish hairs. Fl. Sep. Fr. Oct. Montane forests at 1000-3200 m. Distributed in Sichuan and Yunnan. Also in Bhutan and India.

食用葛 *Pueraria edulis*

华扁豆

Sinodolichos lagopus (Dunn) Verdc.

多年生缠绕草本。茎及叶柄密被黄色短毛。羽状复叶，具3小叶；小叶卵形或菱形，4-10 × 2.5-7厘米，两面被粗柔毛，先端渐尖。总状花序腋生；花萼被灰色或黄色粗毛，裂片4，线状披针形；花冠紫色。荚果线形，长5.5-6.5厘米，宽约6毫米，被黄色粗长毛。花期11月。生海拔100-1700米的林中或灌丛中。产云南、广西和海南。马来西亚和泰国亦有。

Perennial twining herbs. Stems and petiole densely yellow pubescent. Leaves pinnate, leaflet 3; leaflet ovate or rhombic, 4-10 × 2.5-7 cm, hirsute on both surfaces, apex acuminate. Racemes axillary; calyx gray or yellow hirsute; lobes 4, linear-lanceolate; corolla purple. Legumes linear, 5.5-6.5 cm × ca. 6 mm, densely yellowish bristly pilose. Fl. Nov. Forests or thickets at 100-1700 m. Distributed in Yunnan, Guangxi and Hainan. Also in Malaysia and Thailand.

华扁豆 *Sinodolichos lagopus*

大豆 *Glycine max*

大豆

Glycine max (L.) Merr.

一年生草本。茎粗壮，直立或上部近攀援状。羽状3小叶。总状花序短或长；花少至多；花冠紫色、浅紫色或白色。荚果长圆形，稍弯曲，下垂，密被毛。种子2-5粒。花期6-7月，果期7-9月。中国大部分地区有栽培。全球广栽培。

Annual herbs. Stems robust, erect or upper almost climbing. Leaves pinnately 3-foliolate. Racemes short or long; flowers few to many; corolla purple, light purple or white. Legumes oblong, slightly curved, pendulous, densely hairy. Seeds 2-5. Fl. Jun-Jul. Fr. Jul-Sep. Cultivated in most parts of China. Aslo cultivated worldwide.

野大豆

Glycine soja Siebold et Zucc.

一年生缠绕草本。茎细弱，具粗毛。羽状3小叶，长达14厘米；顶生小叶卵状圆形至卵状披针形，两面具糙毛。总状花序通常短；花梗密生黄色长硬毛。荚果长圆形。花期7-8月，果期8-10月。生海拔2600米以下的潮湿田边、河岸、沼泽、草甸、沿海和岛屿向阳的矮灌木丛中。产中国大部分地区。

Annual twining herbs. Stem slender, hirsute. Leaves pinnately 3-foliolate, to 14 cm long; terminal leaflets ovate-circular to ovate-lanceolate, both surfaces strigose. Racemes usually short, pedicels densely yellow long hirsute. Legumes oblong. Fl. Jul-Aug. Fr. Aug-Oct. Damp fields, by rivers, swamps, marshlands, meadows or sunny low thickets by seas and islands below 2600 m. Distributed in most parts of China.

宿苞豆

Shuteria involucrata (Wall.) Wight et Arn.

草本。羽状3小叶；小叶膜质至薄纸质。总状花序腋生，下部轴上具小的、无柄的圆形或肾形的小叶；苞片和小苞片披针形，宿存；花冠红色、紫色或淡紫色。荚果条形，扁。花期9月至翌年3月，果期11月至翌年3月。生海拔900-2800米的山上、灌丛、林缘或路旁。产广西和云南。印度北部、越南、柬埔寨、泰国和印度尼西亚(爪哇)亦有。

Herbs. Leaves pinnately ternate; leaflets membranous to thin papery. Racemes axillary, lower part of inflorescence axis with small, sessile, rounded or reni-

野大豆 *Glycine soja*

宿苞豆 *Shuteria involucrata*

form leaflets; bracts and bracteoles lanceolate, persistent; corolla red, purple or light purple. Legumes linear, flattened. Fl. Sep to next Mar. Fr. Nov to next Mar. Mountains, thickets, forest edges or roadsides at 900-2800 m. Distributed in Guangxi and Yunnan. Also in N India, Vietnam, Cambodia, Thailand and Indonesia (Java).

西南宿苞豆

Shuteria involucrata (Wall.) Wight et Arn. var. **glabrata**

草质缠绕藤本，长1-3米。托叶披针形；顶生小叶椭圆形至近菱形，长1.5-4厘米。总状花序腋生；花冠紫色至淡紫红色，长约8毫米。荚果线形，长2.5-3.5厘米，宽4毫米，压扁。种子5-8。花期11月至翌年1月，果期1-3月。产云南、广西和海南。印度、斯里兰卡、尼泊尔、不丹、缅甸、泰国、越南、菲律宾和印度尼西亚亦有。

Twining vines, up to 1-3 m tall. Stipules lanceolate; terminal leaflet oval to nearly rhombus, 1.5-4 cm long. Racemes axillary; corolla purple to light purple-red, 8 mm long. Legumes linear, 2.5-3.5 cm long and 4 mm wide, flattened. 5-8-seeded. Fl. Nov to next Jan. Fr. Jan-Mar. Distributed in Yunnan, Guangxi and Hainan. Also in India, Sri Lanka, Nepal, Bhutan, Myanmar, Thailand, Vietnam, the Philippines and Indonesia.

心叶山黑豆

Dumasia cordifolia Benth. ex Baker

缠绕小藤本。茎细弱。小叶膜质，几心形或肾形，基部截形或浅心形。总状花序腋生，细弱，疏具柔毛或无毛，具2至多花；花冠亮黄色，具柄。荚果倒披针形至长圆形，裂片卷曲或平展。花期8-9月，果期10-12月。生海拔1200-2800米的山坡阳处灌丛中。产云南、四川和西藏。印度亦有。

Small twining vines. Stems slender. Leaflets membranous, almost cordate or reniform, base truncate to shallowly cordate. Racemes axillary, slender, sparsely hairy or glabrous, 2- to many-flowered; corolla light yellow, stipitate. Legumes oblanceolate to oblong, valves twisting or flat after dehiscence. Fl. Aug-Sep. Fr. Oct-Dec. Among bushes of mountain sunny slopes at 1200-2800 m. Distributed in Yunnan, Sichuan and Xizang. Also in India.

西南宿苞豆 *Shuteria involucrata* var. *glabrata*

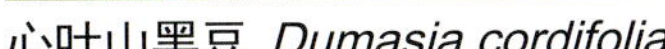

心叶山黑豆 *Dumasia cordifolia*

小鸡藤

Dumasia forrestii Diels

缠绕草本。叶具3小叶，小叶卵形、宽卵形或近圆形。总状花序腋生，长3-12厘米，无毛或疏具毛，花密集；苞片和小苞片披针形；花冠黄色。荚果条状长圆形。花期8-9月，果期10月以后。生海拔1800-3200米的山坡灌丛中。产云南、四川和西藏。

Herbs twining. Leaves 3-foliolate, leaflets ovate, broadly ovate or suborbicular. Racemose axillary, 3-12 cm long, glabrous or sparsely hairy, densely flowered; bracts and bracteoles lanceolate; corolla yellow. Legumes linear-oblong. Fl. Aug-Sep. Fr. after Oct. Thickets on slopes at 1800-3200 m. Distributed in Yunnan, Sichuan and Xizang.

小鸡藤 *Dumasia forrestii*

柔毛山黑豆

Dumasia villosa DC.

缠绕草本，全株被长柔毛。顶生小叶卵形至宽卵形，长3.5-5(-9)厘米，基部圆形、近截平或短楔形。总状花序腋生；苞片和小苞片刚毛状；花冠黄色；雄蕊二体。荚果长圆形，长2-3厘米，密被黄色柔毛。花期9-10月，果期11-12月。生海拔400-2500米的山谷或溪边。产西藏、云南、广西、贵州、四川和陕西。热带亚洲、非洲、大洋洲和马达加斯加亦有。

Herbs twining, villous throughout. Terminal leaflet ovate to broadly ovate, 3.5-5(-9) cm long, base rounded, almost truncate or broadly cuneate. Raceme axillary; bracts and bracteoles setiform; corolla yellow; stamens diadelphous. Legume oblong, 2-3 cm long, densely yellow villous. Fl. Sep-Oct. Fr. Nov-Dec. Mountain valleys or riversides at 400-2500 m. Distributed in Xizang, Yunnan, Guangxi, Guizhou, Sichuan and Shaanxi. Also in tropical Asia, Africa, Oceania and Madagascar.

柔毛山黑豆 *Dumasia villosa*

山黑豆 *Dumasia truncata*

山黑豆

Dumasia truncata Siebold et Zucc.

草质藤本。茎细弱，长1-3米，通常无毛。小叶膜质，三角形或卵状三角形。总状花序腋生，细弱，长1-4厘米；花冠黄色或淡黄色。荚果倒披针形至披针状椭圆形，稍膨大，基部尖，顶部具喙。花期8-9月，果期10-11月。生海拔300-1000(-2300)米的山地、路边或湿地。产华南、东南、华中、华北、华西和华东。日本亦有。

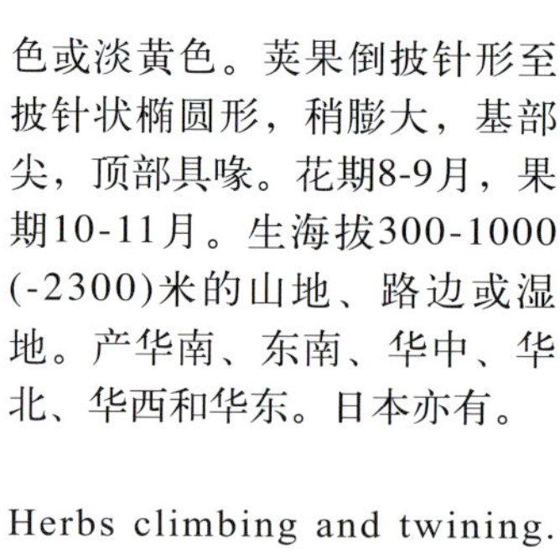

Herbs climbing and twining. Stems slender, 1-3 m long, usually glabrous. Leaflets membranous, triangular or ovate-triangular. Racemes axillary, slender, 1-4 cm long; corolla yellow or light yellow. Legumes oblanceolate to lanceolate-ellipsoid, slightly inflated, base acuminate, apex with beak. Fl. Aug-Sep. Fr. Oct-Nov. Mountains, roadsides or wet places at 300-1000(-2300) m. Distributed in S, SE, C, N, W and E China. Also in Japan.

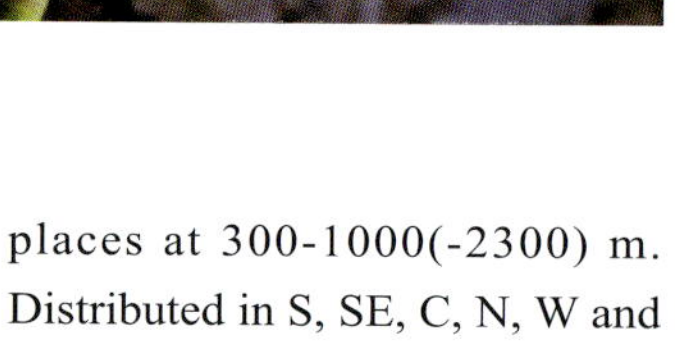

云南山黑豆

Dumasia yunnanensis Y. T. Wei et S. Lee

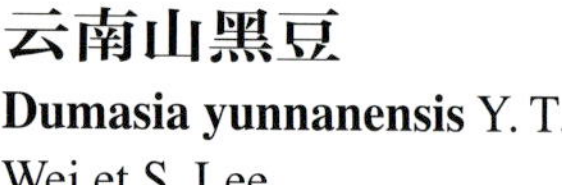

多年生缠绕草本。茎被短粗毛。小叶3，椭圆形至椭圆状卵形，长2-4厘米，上面近无毛，下面被短伏毛。总状花序腋生，长1-3.5厘米，被短粗毛，有花3-6朵；苞片和小苞片刚毛状；花梗长1-2毫米；花冠黄色。荚果狭镰形，长3-5厘米，宽3-6毫米，先端具喙，无毛，具3-4粒种子。花期8-10月。生海拔1300-2500米的高山、路旁

或沟谷。产云南和四川。

Perennial, twining herbs. Stems hirtellous. Leaflets 3, elliptic to elliptic-ovate, 2-4 cm long, adaxially almost glabrous, abaxially with short adpressed hairs. Raceme axillary, 1-3.5 cm long, hirtellous, 3-6-flowered; bracts and bracteoles setiform; pedicel 1-2 mm long; corolla yellow. Legume narrowly falcate, 3-5 × 0.3-0.6 cm, apex with beak, glabrous, 3-4-seeded. Fl. Aug-Oct. Mountains, roadsides or valleys at 1300-2500 m. Distributed in Yunnan and Sichuan.

云南山黑豆 *Dumasia yunnanensis*

锈毛两型豆 *Amphicarpaea ferruginea*

两型豆

Amphicarpaea edgeworthii Benth.

一年生草本。羽状复叶具3小叶；顶生小叶菱状卵形或扁卵形。具闭锁花；花冠紫色或白色。荚果多型：正常花所结实长圆形或倒卵状长圆形，扁平，具2-5粒种子；闭锁花所结实椭圆形或近球形，具1-4粒种子。花果期8-11月。生海拔300-3000米的山坡、路边、田地或草地。产中国西南、华南、华中、华北、华西、华东和东北。印度、越南、俄罗斯、朝鲜半岛和日本亦有。

Annual herbs. Leaves pinnately 3-foliolate; terminal leaflets rhomboid-ovate or oblate-ovate. Cleistogamous flowers often present; corolla purplish or white. Legumes dimorphic: those of normal flowers oblong or obovoid-oblong, compressed, 2-5-seeded; legumes of cleistogamous flowers ellipsoid or suborbicular, 1-4-seeded. Fl. and fr. Aug-Nov. Mountain slopes, roadsides, fields or grasslands at 300-3000 m. Distributed in SW, S, C, N, W, E and NE China. Also in India, Vietnam, Russia, Korean Peninsula and Japan.

锈毛两型豆

Amphicarpaea ferruginea Benth.

多年生草本。羽状复叶具3小叶；顶生小叶常卵形或卵状椭圆形至宽椭圆形，两面密具黄褐色长柔毛。无闭锁花；总状花序具柔毛；花常2-5聚生；花瓣红色或紫红色。荚果椭圆形，稍扁。种子(1-)2或3粒，黑灰色，肾形。花果期8-10月。生海拔2300-3000米的路边或开阔地。产云南和四川。

Perennial herbs. Leaves pinnately 3-foliolate; terminal leaflets usually ovate or ovate-elliptic to broadly elliptic, both surfaces densely yellowish brown villous. Cleistogamous flowers not recorded; racemes pubescent; flowers 2-5-clustered; corolla red or purple-red. Legumes ellipsoid, slightly inflated. Seeds (1-)2 or 3, blackish gray, reniform. Fl. and fr. Aug-Oct. Roadsides or open fields at 2300-3000 m. Distributed in Yunnan and Sichuan.

两型豆 *Amphicarpaea edgeworthii*

腺毛两型豆 *Amphicarpaea linearis*

腺毛两型豆

Amphicarpaea linearis Chun et T. C. Chen

缠绕草质藤本。茎纤细，密被长硬毛，后脱落。托叶条形至条状披针形；顶生小叶卵形至宽卵形，长4.5-6厘米，先端具急尖的小尖头，两面疏被紧贴早落硬毛，基出脉3；侧生小叶斜卵形，较小。总状花序腋生；花萼管状钟形，几无毛，5浅裂，裂片不等，最下部的最大；花冠蓝色；子房条形。花期1月。生路边或空地。产海南和云南。

Herbs twining. Stems slender, densely hirsute, glabrescent. Stipules linear to linear-lanceolate; terminal leaflet ovate to broadly ovate, 4.5-6 cm long, apex with acute mucro, both surfaces sparsely adpressed deciduous hirsute, basal veins 3; lateral leaflets obliquely ovate, much smaller. Racemes axillary; calyx tubular-campanulate, almost glabrous, 5-lobed, lobes unequal, lowest one longest; corolla blue; ovary linear. Fl. Jan. Roadsides or open fileds. Distributed in Hainan and Yunnan.

蝶豆

Clitoria ternatea L.

攀援状草质藤本。小叶5-7。花大，单生于叶腋；小苞片绿色，小，近圆形或倒卵形，膜质；花冠天蓝色、粉色或白色，长达5.5厘米；旗瓣中部粉白色或橙色。荚果褐色，条状长圆形，压扁，具长喙。花果期7-11月。产中国西南至东南。世界热带地区普遍栽培。

Herbaceous vines. Leaflets 5-7. Flowers large, solitary in axil; bracteoles green, small, suborbicular or obovate, membranous; corolla sky blue, pink or white, to 5.5 cm long; standard faintly pink-white or orange in middle. Legumes brown, linear-oblong, compressed, with long beak. Fl. and fr. Jul-Nov. Distributed from SW to SE China. Commonly cultivated in tropical areas.

蝶豆 *Clitoria ternatea*

三叶蝶豆

Clitoria mariana L.

草本。茎疏被长柔毛。小叶3，椭圆形至卵状椭圆形，长4-11厘米，上面无毛；侧脉7-11对，在下面明显凸起；叶柄长2.8-11.5厘米。花通常单朵腋生；小苞片卵形至卵状披针形；花萼筒状，膜质；花冠淡蓝色或紫色。荚果长2.5-10厘米，先端具喙。花期5-9月，果期9月至翌年1月。生海拔100-

三叶蝶豆 *Clitoria mariana*

镰瓣豆 *Dysolobium grande*

2000米的灌丛、路旁或林中。产云南和广西。南亚、东南亚和北美洲亦有。

Herbs. Stems sparsely villous. Leaflets 3, elliptic to ovate-elliptic, 4-11 cm long, adaxially glabrous; lateral veins 7-11 pairs, obviously convex abaxially; petiole 2.8-11.5 cm long. Flowers usually solitary, axillary; bracteoles ovate to ovate-lanceolate; calyx tubular, membranous; corolla light blue or purple. Legume 2.5-10 cm long, apex beaked. Fl. May-Sep. Fr. Sep to next Jan. Shrubs, roadsides or forests at 100-2000 m. Distributed in Yunnan and Guangxi. Also in S and SE Asia, and North America.

镰瓣豆

Dysolobium grande (Benth.) Prain

缠绕木质藤本，长5米。托叶披针形，约6毫米，密被绒毛。总状花序腋生，长达40厘米，被短绒毛；花萼钟形，密被短柔毛，5裂；花冠紫蓝色，宽卵形，约3 × 2.6厘米，顶端圆钝，稍弯；雄蕊近5厘米长。荚果12-16 × 2厘米。种子2-10粒，深棕色，椭圆体。花期7-10月，果期8-11月。生海拔300-500米的山地斜坡、山谷和河边。产贵州和云南。缅甸、尼泊尔和泰国亦有。

Vines, woody, twining, up to 5 m. Stipules lanceolate, ca. 6 mm, densely villous. Raceme axillary, up to 40 cm, shortly villous; calyx campanulate, exterior densely pubescent, 5-lobed; corolla purplish blue, standard broadly ovate, ca. 3 × 2.6 cm, apex emarginate, slightly reflexed; stamens subequal, ca. 5 cm. Legume 12-16 × 2 cm. Seeds 2-10, dark brown, oblong. Fl. Jul-Oct. Fr. Aug-Nov. Hill slopes, mountain valleys and river-sides at 300-500 m. Distributed in Guizhou and Yunnan. Also in Myanmar, Nepal and Thailand.

扁豆

Lablab purpureus (L.) Sweet

缠绕草本。羽状复叶，具3小叶；小叶宽三角状卵形。总状花序腋生，直立，长15-25厘米；花2-5簇生于每个节间；花冠白色或紫色。荚果长圆状镰形，压扁。花期4-12月。中国各地均有栽培。世界热带、亚热带和温带地区亦广为栽培。

Twinning herbs. Leaves pinnate, 3-foliolate; leaflets triangular-ovate. Racemes axillary, erect, 15-25 cm long; flowers 2-5-clustered at each node; corolla white or purple. Legumes oblong-falcate, compressed. Fl. Apr-Dec. Cultivated throughout whole China. Also widely cultivated in tropical, subtropical and temperate regions of the world.

扁豆 *Lablab purpureus*

镰扁豆 *Dolichos trilobus*

镰扁豆
Dolichos trilobus L.

缠绕草质藤本。托叶卵形；小叶菱形或卵状菱形，长2-6厘米，宽2-4.5厘米。总状花序腋生，花1-4；花萼阔钟状，裂齿三角形；花冠白色，长10-12毫米；子房无柄。荚果线状长椭圆形，稍弯，长6厘米，宽8毫米，扁平，无毛。种子6-7粒。花期10月至翌年3月。生旷野灌丛中。产台湾和海南。非洲和亚洲的热带地区亦有。

Twining herbs. Stipules ovate; leaflets rhombic or ovate-rhombic, 2-6 × 2-4.5 cm. Racemes axillary, 1-4-flowered; calyx broadly campanulate, teeth triangular; corolla white, 10-12 mm long; ovary sessile. Legumes linear-oblong, 6 cm long and 8 mm wide, slightly curved, compressed and glabrous. Seeds 6-7. Fl. Oct to next Mar. Thickets. Distributed in Taiwan and Hainan. Also in tropical Africa and Asia.

滇南镰扁豆
Dolichos junghuhnianus Benth.

缠绕草本。托叶宽披针形；小叶3，宽菱状卵形或卵形，长10-11厘米，宽9-9.5厘米，先端急尖或稍钝或具极短的尖头，疏被长柔毛。总状花序被短柔毛；花常成对生于花序轴肿胀的节上；花梗长4-7毫米；花萼长7-8毫米，齿短；花冠紫色，无毛；旗瓣近圆形，长约15毫米。荚果嫩时被绒毛，老时变无毛。产云南。泰国和印度尼西亚亦有。

Twining herbs. Stipules broadly lanceolate; leaflets 3, broadly rhombic-ovate or ovate, 10-11 × 9-9.5 cm, apex acute or slightly obtuse or with very short acumen, sparsely pilose. Racemes pubescent; flowers usually paired at swollen nodes of axis; peduncles 4-7 mm long; calyx 7-8 mm, teeth short; corolla purple, glabrous; standard suborbicular, ca. 15 mm long. Legumes tomentose when young, glabrescent when mature. Distributed in Yunnan. Also in Thailand and Indonesia.

野豇豆
Vigna vexillata (L.) A. Rich.

多年生攀援草本。根木质，纺锤形。小叶膜质，形状多变，卵形至披针形。总状花序腋生，具2-6花，近伞形。荚果伸直，条状圆柱形，具刚毛。种子10-18粒。花期7-9月。生灌丛、开阔林中。产中国西南、华南、华中、华北、华西和华东。广泛分布于热带和亚热带地区。

Perennial herbs, twining. Roots woody, fusiform. Leaflets membranous, variable in shape, ovate to lanceolate. Racemes axillary, 2-6-flowered, subumbellate. Legumes erect, linear-terete, bristly. Seeds 10-18. Fl. Jul-Sep. Thickets, open forests. Distributed in SW, S, C, N, W and E China. Also widely in tropical and subtropical regions.

滇南镰扁豆 *Dolichos junghuhnianus*

野豇豆 *Vigna vexillata*

贼小豆 *Vigna minima*

绿豆

Vigna radiata (L.) R. Wilczek

一年生直立草本。托叶盾状，卵形；小叶卵形。总状花序腋生，4至多花；小苞片条状披针形或长圆形，具条纹；旗瓣外部黄绿色；翼瓣黄色。荚果条状圆柱形，被淡褐色、散生的长硬毛。花期5-6月，果期6-9月。中国大部分地区有栽培。世界热带和亚热带地区广栽培。

Annual herbs, erect. Stipules peltate, ovate; leaflets ovate. Racemes axillary, with 4 to more flowers; bracteoles linear-lanceolate or oblong, striate; standard yellow-green outside; wings yellow. Legumes linear-oblong, sparsely pale brown hirsute. Fl. May-Jun. Fr. Jun-Sep. Cultivated in most parts of China. Also cultivated in tropical and subtropical areas of the world.

贼小豆

Vigna minima (Roxb.) J. Ohwi et H. Ohashi

一年生缠绕草本。托叶盾状，披针形；小叶大小与形状多变。总状花序腋生，柔弱，具花3-4；总花梗远长于叶柄；花冠黄色。荚果圆柱形，开裂后旋卷。花果期8-10月。生旷野、草丛或灌丛中。产华南、华北和华东。印度、菲律宾和日本亦有。

Annual twining herbs. Stipules peltate, lanceolate; leaflets variable in size and shape. Racemes axillary, weak, 3-4-flowered; peduncles longer than petioles; corolla yellow. Legumes cylindric, twisted after dehiscence. Fl. and fr. Aug-Oct. Fields, among grasses or bushes. Distributed in S, N and E China. Also in India, the Philippines and Japan.

绿豆 *Vigna radiata*

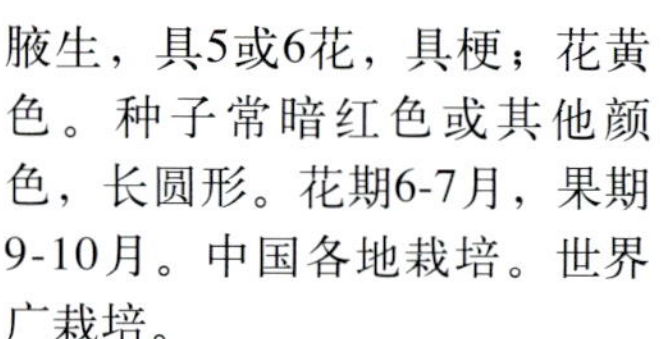

赤豆 *Vigna angularis*

赤豆

Vigna angularis (Willd.) J. Ohwi et H. Ohashi

一年生直立或缠绕草本。小叶卵形或菱状卵形，两面疏具柔毛，侧生小叶偏斜。总状花序腋生，具5或6花，具梗；花黄色。种子常暗红色或其他颜色，长圆形。花期6-7月，果期9-10月。中国各地栽培。世界广栽培。

Annual herbs, erect or twining. Leaflets ovate or rhomboid-ovate, sparsely pilose on both surfaces, lateral leaflets oblique. Racemes axillary, 5- or 6-flowered, pedunculate; flowers yellow. Seeds dark red or other color, oblong. Fl. Jun-Jul. Fr. Sep-Oct. Cultivated in most parts of China. Also cultivated worldwide.

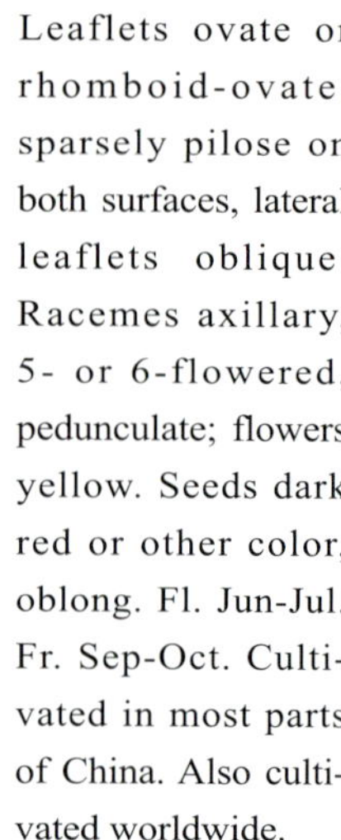

赤小豆

Vigna umbellata (Thunb.) Ohwi et Ohashi

一年生缠绕草本。茎幼时被黄色长柔毛，老时无毛。托叶盾状，披针形或卵状披针形，长10-15毫米；小叶3，卵形或披针形，长10-13厘米，宽5-7.5厘米，先端急尖，全缘或3浅裂。总状花序腋生，具2-3花；花冠黄色，长约1.8厘米。荚果长6-10厘米，宽约0.5厘米，无毛。种子6-10粒，通常暗红色。花期5-8月。产华南。朝鲜半岛、日本和东南亚亦有。

Annual twining herbs. Stems pilose with yellow hairs when young, later glabrescent. Stipules peltate, lanceolate or ovate-lanceolate, 10-15 mm long; leaflets 3, ovate or lanceolate, 10-13 × 5-7.5 cm, apex acute, entire or slightly 3-lobed. Racemes axillary, 2-3-flowered; corolla yellow, ca. 1.8 cm long. Legumes 6-10 × ca. 0.5 cm, glabrous. Seeds 6-10, usually dull red. Fl. May-Aug. Distributed in S China. Also in Korean Peninsula, Japan and SE Asia.

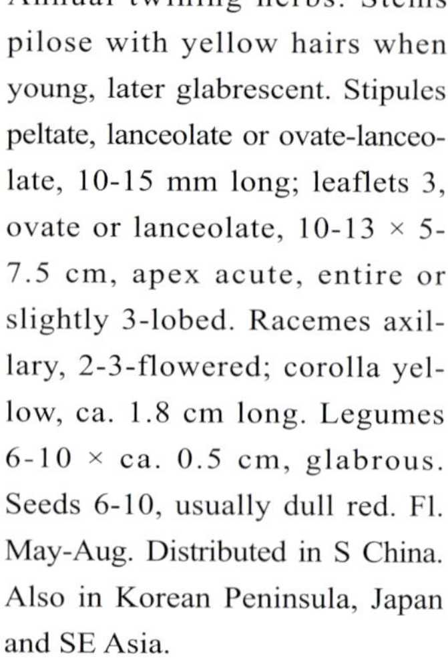

豇豆

Vigna unguiculata (L.) Walp.

一年生缠绕草本。茎无毛。托叶披针形，合生点下面具一狭距；小叶卵状菱形，两面具柔毛或无毛。总状花序腋生，具长梗。荚果下垂，有种子多颗。种子深红色或黑色，带黑色或褐色斑点。花期6-7月，果期8月。中国大部分地区有栽培。原产非洲；世界热带和亚热带地区广栽培。

Annual herbs, twining. Stems glabrous. Stipules lanceolate, with a narrow spur below point of attachment; leaflets ovate-rhomboid, puberulent or glabrous on both surfaces. Racemes axillary, long petiolate. Legumes pendulous, many-seeded. Seeds dark red or black, mottled with black or brown. Fl. Jun-Jul. Fr. Aug. Cultivated in most parts of China. Native to Africa; widely cultivated in tropical and subtropical areas of the world.

赤小豆 *Vigna umbellata*

紫花大翼豆

Macroptilium atropurpureum (DC.) Urban

多年生蔓生草本。茎节上生根。托叶卵形，长4-5毫米，具柔毛；小叶卵形至菱形。花序具10-25厘米长的花序梗；花冠深紫色；旗瓣具长瓣柄。荚果条形。种子杂有褐色及黑色斑

豇豆 *Vigna unguiculata*

纹。产广东和云南。原产热带美洲；热带地区广泛栽培或野生。

Perennial prostrate herbs. Stems rooted at nodes. Stipules ovate, 4-5 mm long, pilose; leaflets ovate to rhombic. Inflorescences with peduncles 10-25 cm long; corolla dark purple; standards with long claw. Legumes linear. Seeds marbled with brown and black striae. Distributed in Guangdong and Yunnan. Native to tropical America; widely cultivated or naturalized in the tropics.

紫花大翼豆 *Macroptilium atropurpureum*

菜豆

Phaseolus vulgaris L.

一年生攀援草本。托叶披针形，长约4毫米；小叶阔卵形或倒卵状菱形。总状花序短于叶，常数花生花序轴顶端；花冠白色、黄色、紫色或红色。荚果带状，稍弯曲，顶端不变宽，常无毛，顶有喙。花期4-7月。栽培于中国各地。广泛栽培于世界各地。

Annual herbs, climbing. Stipules lanceolate, ca. 4 mm long; leaflets broadly ovate or obovate-rhombic. Racemes shorter than leaves, usually several flowered at top of rachis; corolla white, yellow, violet or red. Legumes lorate, slightly curved, apex not widened, often glabrous, apex with beak. Fl. Apr-Jul. Cultivated in most parts of China. Also commonly planted worldwide.

菜豆 *Phaseolus vulgaris*

荷包豆 *Phaseolus coccineus*

荷包豆
Phaseolus coccineus L.

多年生缠绕草本，具块根。小叶两面被柔毛或无毛。总状花序长于叶，数花生花序轴顶端；花冠鲜红色，稀白色。荚果镰状长圆形。种子深紫色具红色或黑色斑点。中国西南、华中、华北和东北栽培。原产中美洲。

Perennial twining herbs, roots tubers. Leaflets pubescent or glabrous on both surfaces. Racemes longer than leaves, several flowered at top of rachis; corrola bright red, rarely white. Legumes sickleform-oblong. Seeds deep purple with red or black spots. Cultivated in SW, C, N and NE China. Native to Central America.

木豆
Cajanus cajan (L.) Millsp.

直立灌木。小叶披针形至椭圆形，长2.8-10厘米，先端渐尖或急尖。总状花序长3-7厘米；花冠黄色，长为花萼的约3倍。荚果条状长圆形，稍膨大，具褐色柔毛。种子3-6粒，灰色，有时具褐色斑点；无种阜。花果期1-11月。生海拔100-900米的路边或丘陵。产中国西南、华南和东南。广泛栽培于热带和亚热带地区。

Erect shrubs. Leaflets lanceolate to elliptic, 2.8-10 cm long, apex acuminate or acute. Racemes 3-7 cm long; corolla yellow, ca. 3 × calyx in length. Legumes linear-oblong, in flared, dun pubescent. Seeds 3-6, gray, sometimes with brown spots, strophiole absent. Fl. and fr. Jan-Nov. Roadsides or hills at 100-900 m. Distributed in SW, S and SE China. Also commonly cultivated in other tropical and subtropical areas.

木豆 *Cajanus cajan*

蔓草虫豆
Cajanus scarabaeoides (L.) Thouars

蔓生或缠绕藤本。羽状3小叶；小叶下面有腺状斑点。总状花序腋生，有花1-5；花冠黄色，长约1厘米。荚果长圆形，革质，密具长柔毛。种子2-7粒，暗褐色。花期9-10月，果期11-12月。生海拔150-1500米的旷野、路旁或山坡草丛。产中国西南和华南。南亚、东南亚、日本、大洋洲和非洲亦有。

Trailing or twining vines. Leaves pinnately 3-foliolate; abaxial sides of leaflets with glandular spots. Racemes axillary, 1-5-flowered; corolla yellow, ca. 1 cm long. Legumes oblong, leathery, densely villous. Seeds 2-7, dark brown. Fl. Sep-Oct. Fr. Nov-Dec. Fields, roadsides or grasslands on mountain slopes at 150-1500 m. Distributed in SW and S China. Also in S and SE Asia, Japan, Oceania and Africa.

蔓草虫豆 *Cajanus scarabaeoides*

墨江千斤拔
Flemingia chappar Buch.-Ham.

灌木。单叶，互生，纸质或近革质，圆心形，下面被棕色绒毛和腺点。花序为腋生或顶生的圆锥状聚伞花序；花序轴长3-7厘米，密具褐色毛。荚果椭圆形，密被褐色绒毛。花期12月至翌年3月，果期3-5月。生林下。产云南。印度、孟加拉国、缅甸、老挝、泰国和柬埔寨亦有。

Shrubs. Leaves simple, alternate, papery or subleathery, orbicular-cordate, abaxially brown-tomentose and glandular-punctate. Inflorescences an axillary or terminal thyrse; inflorescence axis

墨江千斤拔 *Flemingia chappar*

3-7 cm long, densely brown hairy. Legumes ellipsoid, densely brown-tomentose. Fl. Dec to next Mar. Fr. Mar-May. Forests. Distributed in Yunnan. Also in India, Bangladesh, Myanmar, Laos, Thailand and Cambodia.

球穗千斤拔
Flemingia strobilifera (L.) R. Brown

直立灌木。叶柄常长0.5-1.5厘米，密具毛；叶宽3-7厘米，薄革质，基部圆形，稍心形。聚伞圆锥花序，有时分枝；小聚伞花序包藏于贝状苞片内。荚果椭圆形，膨胀。花期2-8月，果期4-11月。生海拔200-1600米的山坡草丛或灌丛中。产中国西南和华南。南亚和东南亚亦有。

Erect shrubs. Petioles usually 0.5-1.5 cm long, densely hairy; leaves 3-7 cm wide, thinly leathery, base rounded, slightly cordate. Inflorescences a thyrse, sometimes branched; small cymes included in conchiform bracts. Legumes ellipsoid, inflated. Fl. Feb-Aug. Fr. Apr-Nov. Grasses or bushes of mountain slopes at 200-1600 m. Distributed in SW and S China. Also in S and SE Asia.

球穗千斤拔 *Flemingia strobilifera*

宽叶千斤拔
Flemingia latifolia Benth.

直立灌木。幼枝、花序和果密被锈色绒毛。小叶3，密具深褐色腺体，顶生小叶长8-14厘米，被短柔毛；侧生小叶偏斜。苞片椭圆形或椭圆状披针形；花冠紫色或粉红色，较花萼长；雄蕊二体。荚果椭圆形，长1.2-1.5厘米，先端具尖喙，具2粒种子。花果期几全年。生海拔560-2700米的旷野、山坡或林中。产云南和广西。印度、缅甸和老挝亦有。

Erect shrubs. Young branchlets, inflorescences and fruits densely rusty tomentose. Leaflets 3, with dense dark brown glands, terminal leaflet 8-14 cm long, pubescent; lateral leaflets oblique. Bracts elliptic or elliptic-lanceolate; corolla purple or pink, longer than calyx; stamens diadelphous. Legume elliptic, 1.2-1.5 cm long, apex with acute beak, 2-seeded. Fl. and fr. almost year-round. Fields, mountain slopes or forests at 560-2700 m. Distributed in Yunnan and Guangxi. Also in India, Myanmar and Laos.

宽叶千斤拔 *Flemingia latifolia*

大叶千斤拔 *Flemingia macrophylla*

大叶千斤拔
Flemingia macrophylla (Willd.) Prain

灌木。叶为掌状3小叶；叶柄长3-6厘米，具狭翅；小叶纸质至薄纸质，8-15 × 4-7厘米，除脉外无毛。总状花序常簇生于叶腋，具多个簇生的花；花萼被丝质柔毛；花冠紫色，稍长于花萼。荚果椭圆形。花期6-9月，果期10-12月。生海拔200-1700米的草地、灌丛、山谷路旁或疏林阳处。产中国西南、华南和东南。南亚和东南亚亦有。

Shrubs. Leaves digitately 3-foliolate; petioles 3-6 cm long, narrowly winged; leaflets papery to thinly papery, 8-15 × 4-7 cm, glabrous except along veins. Racemes usually clustered at axil, with many clustered flowers; calyx silky-pubescent; corolla purple, slightly longer than calyx. Legumes elliptic. Fl. Jun-Sep. Fr. Oct-Dec. Grasslands, thickets, roadsides in valleys or open forests at 200-1700 m. Distributed in SW, S and SE China. Also in S and SE Asia.

千斤拔
Flemingia prostrata Roxburgh

直立亚灌木。叶具指状3小叶；托叶线状披针形，被毛，早落；小叶长圆形或卵状披针形，偏斜，长4-7厘米，基部圆形，疏被短柔毛。总状花序腋生，总花梗短，密被灰白色柔毛；苞片狭卵状披针形；花密生；花冠紫色。荚果椭圆状，长6-8毫米，被短柔毛。花期3-6月，果期5-10月。产华南和华中。孟加拉国、印度、日本和缅甸亦有。

Erect subshrubs. Leaves digitately 3-foliolate; stipules linear-lanceolate, hairy, deciduous; leaflet oblong or ovate-lanceolate, oblique, 4-7 cm long, base rounded, sparsely pubescent. Raceme axillary, peduncle short, densely pale villous; bracts narrowly ovate-lanceolate; flowers clustered; corolla purple. Legume elliptic, 6-8 mm long, pubescent. Fl. Mar-Jun. Fr. May-Oct. Distributed in S and C China. Also in Bangladesh, India, Japan and Myanmar.

千斤拔 *Flemingia prostrata*

鹿藿
Rhynchosia volubilis Lour.

缠绕草纸藤本。茎具棱，被灰色至淡黄色绒毛。羽状复叶或有时几为指状3小叶，纸质，顶生小叶先端钝。总状花序，1-3腋生；花冠黄色。荚果紫红色，长圆形，极度压扁。花期5-8月，果期9-12月。生海拔200-1000米的山坡路旁草丛中。产中国长江以南各省。越南、朝鲜半岛和日本亦有。

Twining herbs. Stems ribbed, densely gray to light yellow villous. Leaves pinnately or sometimes almost digitately 3-foliolate, papery, terminal leaflet apex obtuse. Racemes 1-3 axillay; corolla yellow. Legumes reddish purple, oblong, extremely compressed. Fl. May-Aug. Fr. Sep-Dec. Grasslands of roadside on slopes at 200-1000 m. Distributed throughout the provinces in the south of Yangtze River. Also in Vietnam, Korean Peninsula and Japan.

鹿藿 *Rhynchosia volubilis*

菱叶鹿藿 *Rhynchosia dielsii*

菱叶鹿藿

Rhynchosia dielsii Harms

草本。茎缠绕。羽状复叶具3小叶，顶生小叶卵形、阔椭圆形或菱状卵形，两面密具柔毛，先端渐尖或尾状渐尖。总状花序腋生，花稀疏；花冠黄色。荚果长圆形或倒卵形，扁平，成熟后紫红色，具柔毛。花期6-7月，果期8-11月。生海拔600-2100米的山地或路边。产中国西南、华南、华中、华北和华西。

Herbs. Stems twining. Leaves pinnately 3-foliolate, terminal leaflet ovate, broadly elliptic or rhomboid-ovate, both sides densely pubescent, apex acuminate or caudate-acuminate. Racemes axillary, flowers sparse; corolla yellow. Legumes oblong or obovate, compressed, reddish purple when mature, pubescent. Fl. Jun-Jul. Fr. Aug-Nov. Mountains or roadsides at 600-2100 m. Distributed in SW, S, C, N and W China.

小鹿藿

Rhynchosia minima (Linnaeus) Candolle

一年生缠绕草本。茎纤细。叶具羽状3小叶；托叶披针形；顶生小叶菱状圆形，长、宽均1.5-3厘米，先端钝或圆，下面密被腺点。总状花序腋生；花长8毫米；萼片披针形；花冠黄色。荚果倒披针形至椭圆形，长1-2厘米，宽0.4-0.5厘米。花期5-10月，果期9-11月。生海拔900-2500米的林中。产云南、四川和台湾。日本、南亚、东南亚和东非热带亦有。

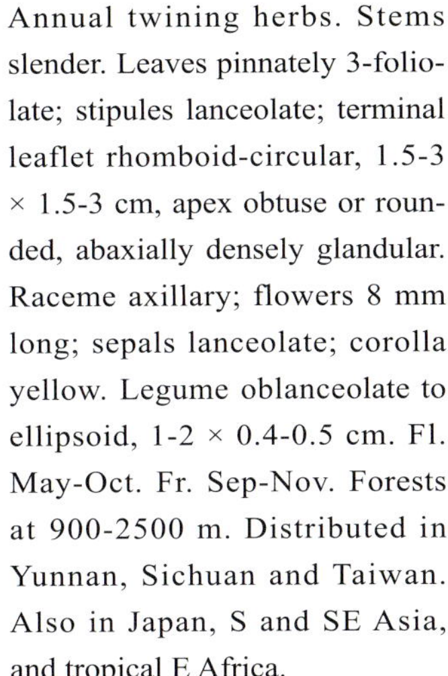

Annual twining herbs. Stems slender. Leaves pinnately 3-foliolate; stipules lanceolate; terminal leaflet rhomboid-circular, 1.5-3 × 1.5-3 cm, apex obtuse or rounded, abaxially densely glandular. Raceme axillary; flowers 8 mm long; sepals lanceolate; corolla yellow. Legume oblanceolate to ellipsoid, 1-2 × 0.4-0.5 cm. Fl. May-Oct. Fr. Sep-Nov. Forests at 900-2500 m. Distributed in Yunnan, Sichuan and Taiwan. Also in Japan, S and SE Asia, and tropical E Africa.

小鹿藿 *Rhynchosia minima*

喜马拉雅鹿藿 *Rhynchosia himalensis*

喜马拉雅鹿藿

Rhynchosia himalensis Benth. ex Baker

攀援草本。茎、叶两面、花序梗和果密被腺毛和短柔毛。叶具羽状3小叶，托叶狭卵形；顶生小叶宽卵形，先端渐尖；侧生小叶基部偏斜。总状花序腋生；花黄色；萼片5，外面密被毛和腺点，上部2裂片基部合生，最下面1裂片与花冠等长。荚果2.5-3 × 0.9厘米。生海拔1200-3300米的林中落叶层、河谷、高山和农田。产西藏和四川。缅甸和南亚亦有。

Climbing herbs. Stems, both surfaces of leaves, peduncles and fruits densely glandular-hairy and pubescent. Leaves pinnately 3-foliolate, stipule narrowly ovate; terminal leaflets broadly ovate, apex acuminate; lateral leaflets base oblique. Raceme axillary; flowers yellow; sepals 5, densely pubescent and glandular punctate, 2 upper lobes connate at base, 1 lowest lobe as long as corolla. Legumes 2.5-3 × 0.9 cm. Forest understories, river valleys, mountains and fields at 1200-3300 m. Distributed in Xizang and Sichuan. Also in Myanmar and S Asia.

紫脉花鹿藿

Rhynchosia himalensis Benth. ex Baker var. **craibiana** (Rehder) E. Peter

草本。茎缠绕。羽状复叶具3小叶；顶生小叶宽卵形或圆菱形，先端渐尖。总状花序腋生，具3-5花；花萼5裂，最下方裂片短于花冠；花冠黄色，具明显紫纹。荚果密被微柔毛和腺毛。花期7-9月，果期8-10月。生海拔1200-3300米的林下叶层、河谷、山地或田地。产云南、四川和西藏。

Herbs. Stems twining. Leaves pinnately 3-foliolate; terminal leaflets broadly ovate or circular-rhomboid, apex acuminate. Racemes axillary, 3-5-flowered; calyx 5-lobed, lowest calyx shorter than corolla; corolla yellow with obvious purple striations. Legumes densely microvillous and glandular hairy. Fl. Jul-Sep. Fr. Aug-Oct. Forest understories, river valleys, mountains or fields at 1200-3300 m. Distributed in Yunnan, Sichuan and Xizang.

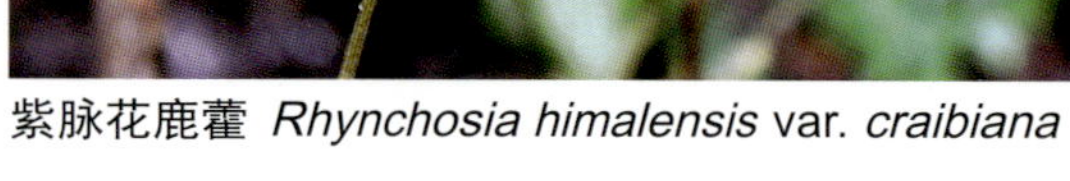

紫脉花鹿藿 *Rhynchosia himalensis* var. *craibiana*

鸡头薯

Eriosema chinense Vogel

多年生直立草本。块根纺锤形，肉质。叶为单叶。总状花序腋生，极短，通常有花1-2。荚果菱状椭圆形或长圆形，成熟时黑色，具粗毛。种柄着生于长线形种脐的一端。花期5-6月，果期7-10月。生海拔300-2000米的山间或草坡上。产中国西南和华南。南亚和东南亚亦有。

Perennial erect herbs. Root tuber fusiform, succulent. Leaves 1-foliolate. Racemes axillary, extremely short, usually with flowers 1-2. Legumes rhomboid-elliptic or oblong, black when mature, hirsute. Seeds with funicle attached at end of linear hilum. Fl. May-Jun. Fr. Jul-Oct. Mountains or grassy slopes at 300-2000 m. Distributed in SW and S China. Also in S and SE Asia.

鸡头薯 *Eriosema chinense*

补骨脂 *Cullen corylifolium*

补骨脂
Cullen corylifolium (L.) Medik.

一年生草本。茎具腺体。叶为单叶，宽卵形，两面有黑色腺点。花序腋生，密聚生，头状或短总状，具10-30花；花冠蓝色至浅黄色。荚果黑色，卵球形，不开裂。花期7-8月，果期9-10月。生山坡、溪边或田边。产云南、贵州和四川，华中和华东有栽培。南亚、东南亚和西南亚亦有。

Herbs, annual. Stems gland-dotted. Leaves simple, broadly ovate, both surfaces black glandular-punctate. Inflorescences axillary, densely congested, capitate or shortly racemose, 10-30-flowered; corolla blue to yellowish. Legumes black, ovoid, indehiscent. Fl. Jul-Aug. Fr. Sep-Oct. Slopes, by streams or field edges. Distributed in Yunnan, Guizhou and Sichuan, cultivated in C and E China. Also in S, SE and SW Asia.

紫穗槐
Amorpha fruticosa L.

落叶灌木，丛生。托叶针刺状；小叶11-25，卵形至椭圆形，下面具白色柔毛，具黑色腺点。总状花序1至数个，顶生或近顶生；旗瓣紫色；翼瓣和龙骨瓣缺无。荚果深褐色，长圆形。花期5-6月，果期7-9月。生河岸沙地，栽培或逸生。产中国大部分地区。原产北美洲；北亚和欧洲引种。

Deciduous shrubs, clustered. Stipules bristlelike; leaflets 11-25, ovate to elliptic, abaxially white puberulent, black glandular-dotted. Racemes 1 to many, terminal or subterminal; standard purple; wings and keel absent. Legumes dark brown, oblong. Fl. May-Jun. Fr. Jul-Sep. Sandy banks of ravines, cultivated or escaped. Distributed in most parts of China. Native to North America; introduced in N Asia and Europe.

紫穗槐 *Amorpha fruticosa*

合萌 *Aeschynomene indica*

合萌

Aeschynomene indica L.

小灌木或一年生草本。小叶小，20-30对或更多，长5-10(-15)毫米，上面密被腺点，下面有白霜。花序总状，疏松；苞片膜质，常宿存；花冠浅黄色，具淡紫色纵条纹。荚果条状长圆形。花期7-9月，果期8-10月。生林下或林缘。中国广布。热带亚洲、朝鲜半岛、日本、非洲和大洋洲亦有。

Shrublets or annual herbs. Leaflets small, 20-30 pairs or more, 5-10(-15) mm long, adaxially densely glandular-punctate, abaxially glaucous. Inflorescences racemose, lax; bracts membranous, usually persistent; corolla pale yellow with purplish longitudinal striations. Legumes linear-oblong. Fl. Jul-Sep. Fr. Aug-Oct. Forests or forest edges. Widely distributed in China. Also in tropical Asia, Korean Peninsula, Japan, Africa and Oceania.

坡油甘 *Smithia sensitiva*

坡油甘

Smithia sensitiva Aiton

一年生草本。茎纤细，多分枝，无毛。叶具6-20小叶，上面有毛。总状花序和叶部不密集在小枝顶部。总状花序于茎顶具1-6或更多花。荚果为褶状节荚，有4-6节。花期7-9月，果期9-11月。生海拔1000米以下的田边或湿地。产中国西南、华南和东南。南亚、东南亚、非洲和澳大利亚北部亦有。

Annual herbs. Stems slender, with many branches, glabrous. Leaves 6-20-foliolate, adaxially hairy. Racemes and leaves not aggregated on top of branchlets. Racemes with 1-6 or more flowers clustered near apex. Legumes a plicate loment, divided into 4-6 joints. Fl. Jul-Sep. Fr. Sep-Nov. Field edges or damp places below 1000 m. Distributed in SW, S and SE China. Also in S and SE Asia, Africa and N Australia.

缘毛合叶豆

Smithia ciliata Royle

一年生草本。茎和小枝无毛。叶具10-14小叶；小叶倒披针形或线状长圆形，6-12 × 2-4毫米，先端钝或圆形，边缘和中脉上具刺毛。花萼膜质，具网状脉纹，边缘密生刺毛；花冠稍长于萼。荚果有荚节6-8节，荚节密具乳头状凸起。花期8-9月，果期10-11月。生海拔100-2800米的路旁、高山和湿地。产华南、湖南、江西和浙江。南亚、东南亚和日本亦有。

Annual herbs. Stems and branchlets glabrous. Leaves 10-14-foliolate; leaflet oblanceolate or linear-oblong, 6-12 × 2-4 mm, apex obtuse to rounded, margin and midvein bristly. Calyx membranous, with reticulate veins, margin densely setose; slightly longer than calyx. Legume divided into 6-8 joints; joints densely papillate. Fl. Aug-Sep. Fr. Oct-Nov. Roadsides, mountains and wetlands at 100-2800 m. Distributed in S China, Hunan, Jiangxi and Zhejiang. Also in S and SE Asia, and Japan.

丁葵草

Zornia gibbosa Span.

草本。叶指状，小叶2；小叶具透明腺点；叶轴不为肉质。总状花序腋生；小苞片2枚，盾状着生，具缘毛；花冠黄色。荚果裂为2-7荚节；荚节近球形，具刺毛。花期4-7月，果期7-9

缘毛合叶豆 *Smithia ciliata*

丁葵草 *Zornia gibbosa*

月。生海拔100-1200米的田边、村边或干旱的旷野草地。产中国西南、华南和东南。南亚、东南亚、日本南部和澳大利亚亦有。

Herbs. Leaves digitate, with 2 leaflets; leaflets pellucid punctuate; leaf rachis not fleshy. Racemes axillary; bracteoles 2, peltate, ciliate; corolla yellow. Legumes divided into 2-7 articles; articles subglobose, echinate-setose. Fl. Apr-Jul. Fr. Jul-Sep. Field edges, by villages or dry and grassy wilderness at 100-1200 m. Distributed in SW, S and SE China. Also in S and SE Asia, S Japan and Australia.

落花生 *Arachis hypogaea*

落花生
Arachis hypogaea L.

一年生草本。茎直立或匍匐，被淡黄色长柔毛，后变无毛。叶通常具4小叶，卵状长圆形至倒卵形，基部贴生于叶柄。花冠黄色或金黄色。荚果膨胀，荚厚，具网纹，具1-4(-6)粒种子。花果期6-8月。生雨量适中的沙质地区。栽培于中国大部分地区。世界各地广泛栽培；原产南美洲。

Annual herbs. Stems erect or procumbent, yellowish pubescences, then glabrescent. Leaves usually with 4 leaflets, ovate-oblong to obovate, basally adnate to stipule. Corolla yellow or golden yellow. Legumes inflated, thick-walled, reticulate veined, with 1-4(-6) seeds. Fl. and Fr. Jun-Aug. Sandy places with moderate rainfall. Cultivated in most parts of China. Widely planted throughout the world; native to South America.

鱼鳔槐
Colutea arborescens L.

落叶灌木。幼枝被细小白毛。小叶7-13，长圆形至倒卵形，长1-3厘米，上面无毛，下面疏生短毛。总状花序生6-8花；萼齿三角形，外面疏被黑褐色及白色伏毛；花冠鲜黄色，翼瓣近基部最宽，上部常渐狭；子房密被短柔毛。荚果狭卵形，长6-8厘米，无毛至近无毛。花期5-7月，果期7-10月。辽宁、北京、山东、陕西和江苏有栽培。原产欧洲。

Deciduous shrubs. Young branches with fine white hairs. Leaflets 7-13, oblong to obovate, 1-3 cm long, adaxially glabrous, abaxially sparsely shortly hairy. Racemes 6-8-flowered; calyx teeth triangular, outside sparsely blackish brown and white sericeous; corolla light yellow, wings widest near base, upper part often attenuate; ovary densely pubescent. Legume narrowly ovate, 6-8 cm long, glabrous to subglabrate. Fl. May-Jul. Fr. Jul-Oct. Cultivated in Liaoning, Beijing, Shandong, Shaanxi and Jiangsu. Native to Europe.

鱼鳔槐 *Colutea arborescens*

苦马豆 *Sphaerophysa salsula*

苦马豆
Sphaerophysa salsula (Pall.) DC.

亚灌木或多年生草本。奇数羽状复叶，小叶11-21。总状花序，有花6-16；旗瓣近圆形；雄蕊二体；花柱弯曲。荚果椭圆形或卵形，果瓣膜质，膨胀，被白色柔毛。花期5-8月，果期6-9月。生海拔950-3200米的山坡、草原、荒地、沙滩或沟渠等地。产华北、西北和东北。蒙古和俄罗斯亦有。

Subshrubs or perennial herbs. Leaves odd-pinnate, with 11-21 leaflets. Racemes 6-16-flowered; standards suborbicular; stamens diadelphous; styles curved. Legumes ellipsoid or ovoid, valves membranous, turgid, white-pubescent. Fl. May-Aug. Fr. Jun-Sep. Mountain slopes, grasslands, wastelands, sandy beaches or ditches at 950-3200 m. Distributed in N, NW and NE China. Also in Mongolia and Russia.

铃铛刺
Halimodendron halodendron (Pall.) Voss

灌木。叶柄和叶轴宿存，刺状；小叶倒披针形，起初密具银白色毛，渐无毛。总状花序具花2-5，总花梗长1.5-3厘米；花瓣紫色或白色。荚果扁平，背缝线和腹缝线下凹，无纵隔膜。花期7月，果期8月。生盐化土壤上。产内蒙古、甘肃和新疆。俄罗斯和蒙古亦有。

Shrubs. Rachis and petiole persistent, spine-shaped; leaflets oblanceolate, with dense silvery white trichomes at first, glabrescent. Racemes 2-5-flowered with a 1.5-3 cm long peduncle; petals purple or white. Legumes applanate, dorsal and ventral line impressed, without septa. Fl. Jul. Fr. Aug. Salinized sands. Distributed in Neimenggu, Gansu and Xinjiang. Also in Russia and Mongolia.

锦鸡儿
Caragana sinica (Buc'hoz) Rehder

灌木。羽叶复叶或有时掌状，具4小叶，顶端一对小叶常最大。花单生；萼筒钟形，长12-14毫米；花冠黄色，长2.8-3厘米。荚果圆柱状，长3-3.5厘米。花期4-5月，果期7月。生海拔400-2000米的山坡上。产中国西南、华中、华北和华东。

Shrubs. Leaves pinnate or sometimes digitate, 4-foliolate, apical pairs often largest. Flowers solitary; calyx tubes campanulate, 12-14 mm long; corolla yellow, 2.8-3 cm long. Legumes cylindric,

铃铛刺 *Halimodendron halodendron*

锦鸡儿 *Caragana sinica*

川西锦鸡儿 *Caragana erinacea*

二色锦鸡儿 *Caragana bicolor*

3-3.5 cm long. Fl. Apr-May. Fr. Jul. Slopes at 400-2000 m. Distributed in SW, C, N and E China.

川西锦鸡儿

Caragana erinacea Kom.

灌木。羽状复叶，有小叶4-8；托叶成针刺状，脱落或宿存。花单生或1-4簇生叶腋；萼筒长0.8-1厘米；花冠旗瓣阔卵形至长圆状倒卵形。荚果圆柱形，被毛。花期5-6月，果期7-9月。生海拔2000-4600米的山坡、草地、林缘、灌丛、河岸或沙丘。产云南、四川、西藏、甘肃和青海。

Shrubs. Leaves pinnate, 4-8-foliolate; stipules spinescent, caducous or persistent. Flowers solitary or 1-4 in a fascicle in leaf axils; calyx tubes 0.8-1 cm long; corolla standard broadly ovate to oblong-obovate. Legumes cylindric, pubescent. Fl. May-Jun. Fr. Jul-Sep. Slopes, grasslands, forest edges, thickets, river banks or sand dunes at 2000-4600 m. Distributed in Yunnan, Sichuan, Xizang, Gansu and Qinghai.

二色锦鸡儿

Caragana bicolor Kom.

灌木。偶数羽状复叶，有小叶8-16，小叶顶端钝至锐尖。花单生或成对生于总花梗；萼裂片深褐色；花冠黄色，但是旗瓣干后紫色；旗瓣先端凹缺，翼瓣具1耳。荚果圆筒状，被白色柔毛。花期6-7月，果期9-10月。生海拔2400-3600米的山坡灌丛或杂木林中。产云南、四川和西藏。

Shrubs. Leaves paripinnate, 8-16-foliolate, leaflets apices obtuse to acute. Flowers solitary or in pairs on a peduncle; calyx tubes dark brown; corolla yellow but standard violet-purple when dry; standard emarginate at apex, wing with 1 auricle. Legumes cylindric, sparsely white pubescent. Fl. Jun-Jul. Fr. Sep-Oct. Thickets or mixed forests on slopes at 2400-3600 m. Distributed in Yunnan, Sichuan and Xizang.

云南锦鸡儿

Caragana franchetiana Kom.

灌木。小叶5-9对，顶端钝。每花序梗具一花；花黄色，单生；萼筒黄褐色；花冠黄色，旗瓣有时紫色；旗瓣近圆形，顶端不凹陷，翼瓣具2耳。荚果圆柱形。花期5-6月，果期7-8月。生海拔2800-4000米的山坡灌丛、林中或林缘。产云南、四川和西藏。

Shrubs. Leaflets 5-9 pairs, apices obtuse. Each peduncles with one flower; flowers yellow, solitary; calyx tubes yellowish brown; corolla yellow but standard sometimes purple; standards suborbicular, apex not emarginate, wings with 2 auricles. Legumes cylindric. Fl. May-Jun. Fr. Jul-Aug. Slopes, forests or forest edges at 2800-4000 m. Distributed in Yunnan, Sichuan and Xizang.

云南锦鸡儿 *Caragana franchetiana*

鬼箭锦鸡儿 *Caragana jubata*

北京锦鸡儿 *Caragana pekinensis*

鬼箭锦鸡儿
Caragana jubata (Pall.) Poir.

灌木。羽状复叶，小叶8-12；叶柄和叶轴宿存。花单生；萼筒管状，长1.4-1.7厘米；花冠玫瑰色、紫红色、亮紫色、粉色或白色；翼瓣具单耳，生于瓣片上。荚果长达3厘米，密具长柔毛。花期6-7月，果期8-9月。生海拔2400-4700米的山坡或林缘。产华北、华西和西北。印度、尼泊尔、不丹、俄罗斯和蒙古亦有。

Shrubs. Leaves pinnate, with 8-12 leaflets; petioles and rachis persistent. Flowers solitary; calyx tubes tubular, 1.4-1.7 cm long; corolla rosy, reddish purple, bright purple, pink or white; wings with 1 auricle on limb. Legumes to 3 cm long, densely villous. Fl. Jun-Jul. Fr. Aug-Sep. Slopes or forest edges at 2400-4700 m. Distributed in N, W and NW China. Also in India, Nepal, Bhutan, Russia and Mongolia.

树锦鸡儿
Caragana arborescens Lam.

小乔木或大灌木，高2-6米。托叶针刺状，长5-10毫米；羽状复叶具8-16小叶；小叶长圆状倒卵形至椭圆形。花单生或2-5簇生；花冠黄色，长1.6-2厘米。荚果圆柱状。花期5-6月，果期8-9月。生海拔1000-1900米的林缘或林地。产华北大部分地区。俄罗斯、哈萨克斯坦和蒙古亦有。

Small trees or large shrubs, 2-6 m tall. Stipules spinelike, 5-10 mm long; leaves pinnate, 8-16-foliolate; leaflets oblong-obovate to elliptic. Flowers solitary or 2-5-clustered; corolla yellow, 1.6-2 cm long. Legumes cylindric. Fl. May-Jun. Fr. Aug-Sep. Forest edges or woodlands at 1000-1900 m. Distributed in most parts of N China. Also in Russia, Kazakhstan and Mongolia.

北京锦鸡儿
Caragana pekinensis Kom.

灌木，高达2米。小枝褐色至深棕色。羽状复叶具12-16小叶；叶柄和叶轴长2-6厘米，脱落；小叶椭圆形至倒卵状披针形。花单生或2(-4)成一束；花冠黄色。荚果扁平，密被毛。花期5月，果期7月。生海拔400-1000米的山腰或黄土丘陵。产河北和山西。

Shrubs, to 2 m tall. Branches brown to dark brown. Leaves pinnate, 12-16-foliolate; petioles and rachis 2-6 cm long, caducous; leaflets elliptic to obovate-elliptic. Flowers solitary or 2(-4) in a fascicle; corolla yellow. Legumes compressed, densely pubescent. Fl. May. Fr. Jul. Hillsides or loess hills at 400-

树锦鸡儿 *Caragana arborescens*

小叶锦鸡儿 *Caragana microphylla*

柠条锦鸡儿 *Caragana korshinskii*

1000 m. Distributed in Hebei and Shanxi.

小叶锦鸡儿

Caragana microphylla Lam.

灌木。托叶针刺状，长1.5-5毫米；羽状复叶有10-20小叶，倒卵形至倒卵状长圆形，先端钝至截形。花单生；花冠黄色；子房常无毛。荚果圆筒形，具锐尖头。花期5-6月，果期7-8月。生海拔1000-2000米的固定、半固定沙地或岩石山坡。产内蒙古、吉林和辽宁。俄罗斯(西伯利亚)和蒙古亦有。

Shrubs. Stipules spinelike, 1.5-5 mm long; leaves pinnate, 10-20-foliolate, obovate to obovate-oblong, apex obtuse to truncate. Flowers solitary; corolla yellow; ovary often glabrous. Legumes cylindric, with acute prong. Fl. May-Jun. Fr. Jul-Aug. Fixed or semi-fixed dunes, or rocky mountain slopes at 1000-2000 m. Distributed in Neimenggu, Jilin and Liaoning. Also in Russia (Siberia) and Mongolia.

柠条锦鸡儿

Caragana korshinskii Kom.

灌木或小乔木。老枝金黄色，小枝被短柔毛。托叶刺状，长3-7毫米；羽状复叶具12-16小叶；叶轴早落；小叶披针形至狭长圆形，长7-8毫米，两面密被伏贴柔毛。花单生；花萼管状钟形，长8-9毫米；花冠长2-2.3厘米；子房无毛。荚果披针形，长2-2.5厘米。花期5-6月，果期6-7月。生海拔900-2400米的半固定沙丘。产中国西北、内蒙古和山西。蒙古亦有。

Shrubs or small trees. Branches golden yellow, branchlets pubescent. Stipules spinelike, 3-7 mm long; leaves pinnate, 12-16-foliolate; rachis caducous; leaflets lanceolate to narrowly oblong, 7-8 mm long, both surfaces densely appressed sericeous. Flowers solitary; calyx tubular-campanulate, 8-9 mm long; corolla 2-2.3 cm long; ovary glabrous. Legume lanceolate, 2-2.5 cm long. Fl. May-Jun. Fr. Jun-Jul. Semiconsolidated sand dunes at 900-2400 m. Distributed in NW China, Neimenggu and Shanxi. Also in Mongolia.

白皮锦鸡儿

Caragana leucophloea Pojark.

灌木，高达1.5米。小枝淡黄白色至黄色，当年生小枝紫红色。掌状复叶具4小叶；小叶狭倒披针形。花单生或2朵成一束；花冠黄色；子房无毛。荚果圆柱状，无毛。花期5-6月，果期7-8月。生海拔900-2700米的干旱山坡或沙漠谷地。产内蒙古、甘肃和新疆。哈萨克斯坦和蒙古亦有。

Shrubs, to 1.5 m tall. Branches yellowish white to yellow, current-year branchlets purplish red. Leaves digitate, 4-foliolate; leaflets narrowly oblanceolate. Flowers solitary or 2 in a fascicle; corolla yellow; ovary glabrous. Legumes cylindric, glabrous. Fl. May-Jun. Fr. Jul-Aug. Dry slopes or desert valleys at 900-2700 m. Distributed in Neimenggu, Gansu and Xinjiang. Also in Kazakhstan and Mongolia.

白皮锦鸡儿 *Caragana leucophloea*

毛掌叶锦鸡儿 *Caragana leveillei*

毛掌叶锦鸡儿

Caragana leveillei Kom.

灌木。树皮暗棕色。小枝灰褐色。掌状4小叶，密具柔毛。花单生；花瓣长2.5-2.8厘米，黄色至淡粉色；旗瓣倒卵状长圆形；子房密被长绒毛。荚果圆柱状，密具长柔毛。花期4-5月，果期6月。生海拔500-1300米的山坡。产河南、河北、山西、陕西和山东。

Shrubs. Bark dark brown. Branchlets grayish brown. Leaves digitate, 4-foliolate, densely pubescent. Flowers solitary; corolla 2.5-2.8 cm long, yellow to pinkish; standard obovate-oblong; ovary densely villous. Legumes cylindric, densely villous. Fl. Apr-May. Fr. Jun. Slopes at 500-1300 m. Distributed in Henan, Hebei, Shanxi, Shaanxi and Shandong.

红花锦鸡儿

Caragana rosea Turcz. ex Maxim.

灌木。叶柄脱落或宿存成针刺；叶掌状，有4小叶，无毛。花单生；花冠至少于枯萎时浅粉色至浅红色；旗瓣长圆状倒卵形；子房无毛。荚果圆柱状，具渐尖头。花期5-6月，果期6-7月。生海拔200-2100米的山谷或山坡。产华中、华北、华东和东北。

Shrubs. Petioles caduceus or persistent to acupuncture; leaves digitate, with 4 leaflets, glabrous. Flowers solitary; corolla pinkish to reddish at least when wilted; standard oblong-obovate; ovary glabrous. Legumes cylindric, apex acuminate. Fl. May-Jun. Fr. Jun-Jul. Mountain valleys or slopes at 200-2100 m. Distributed in C, N, E and NE China.

猫尾草 *Uraria crinita*

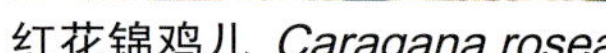

红花锦鸡儿 *Caragana rosea*

猫尾草

Uraria crinita (L.) Desv. ex DC.

亚灌木。茎直立，被灰色短柔毛。奇数羽状复叶具小叶3-7。总状花序顶生，长15-30厘米，粗壮，密被灰白色长硬毛；花萼浅杯状；花冠紫色。荚果微被短柔毛，具网状脉。花果期4-9月。生海拔900米以下的干燥坡地、路旁或灌丛。产中国西南和华南。南亚、东南亚、日本南部和澳大利亚北部亦有。

云雾雀儿豆 *Chesneya nubigena*

川滇雀儿豆 *Chesneya polystichoides*

Subshrubs. Stems erect, gray pubescent. Leaves imparipinnate, with 3-7 leaflets. Racemes terminal, 15-30 cm long, stout, densely gray-white hirsute; calyx shallowly cup-shaped, long white hirsute; corolla purple. Legumes slightly pubescent, reticulate veined. Fl. and fr. Apr-Sep. Dry waste slopes, roadsides or thickets below 900 m. Distributed in SW and S China. Also in S and SE Asia, S Japan and N Australia.

云雾雀儿豆

Chesneya nubigena (D. Don) Ali

垫状草本。托叶条形，长约10毫米，与叶一半合生；羽状复叶，具15-21小叶，两面密被开展的长柔毛，基部圆形或稍圆形。花单生；花冠黄色或紫色。荚果长圆形，疏被白色长柔毛。花期7-8月，果期8-9月。生海拔3600-5300米的山坡和碎石坡。产云南和西藏。印度和尼泊尔亦有。

Herbs cushionlike. Stipules linear, ca. 10 mm long, adnate to petiole for ca. 1/2 its length; leaves pinnate, 15-21-foliolate, both surfaces with dense spreading appressed hairs, base rounded or slightly so. Flowers solitary; corolla yellow or purple. Legumes oblong, sparsely white villose. Fl. Jul-Aug. Fr. Aug-Sep. Slopes and gravel slopes at 3600-5300 m. Distributed in Yunnan and Xizang. Also in India and Nepal.

川滇雀儿豆

Chesneya polystichoides (Hand.-Mazz.) Ali

垫状植物。小叶19-41，无柄，密生，下面灰白色，两面具毛，基部偏斜；托叶线形，长约15毫米。花单生；花冠黄色；子房无柄，无毛。荚果狭椭圆形，革质。花期7月，果期8月。生海拔3400-4400米的高山灌丛、石质山坡或石缝中。产云南、四川和西藏。

Plants cushionlike. Leaflets 19-41, sessile, dense, abaxially gray-white, both surfaces hairy, base oblique; stipules linear, ca. 15 mm long. Flowers solitary; corolla yellow; ovary sessile, glabrous. Legumes narrowly ellipsoid, leathery. Fl. Jul. Fr. Aug. Alpine thickets, rocky slopes or rock crevices at 3400-4400 m. Distributed in Yunnan, Sichuan and Xizang.

笔直黄耆

Astragalus strictus Graham ex Benth.

多年生草本。叶柄长1-2厘米；羽状复叶，小叶8-12对，狭椭圆形至椭圆形。总状花序短，具多花；花瓣浅粉色至紫色；旗瓣椭圆形，7-9 × 4-5.5(-6)毫米。荚果微弯曲，1室。花期7-8月，果期8-9月。生海拔3000-5600米的山坡草地、溪边、田野或石砾地。产云南、四川、西藏、青海和新疆。印度、尼泊尔、克什米尔和巴基斯坦亦有。

Perennial herbs. Petioles 1-2 cm long; leaves pinnate, in 8-12 pairs, narrowly elliptic to elliptic. Racemes short, flowers numerous; petals pale pinkish to purple; standard elliptic, 7-9 × 4-5.5(-6) mm. Legumes slightly curved, 1-locular. Fl. Jul-Aug. Fr. Aug-Sep. Grassy slopes, streamsides, fields or gravelly places at 3000-5600 m. Distributed in Yunnan, Sichuan, Xizang, Qinghai and Xinjiang. Also in India, Nepal, Kashmir and Pakistan.

笔直黄耆 *Astragalus strictus*

边向花黄耆 *Astragalus moellendorffii*

边向花黄耆

Astragalus moellendorffii
Bunge ex Maxim.

多年生草本，高35-45厘米，具长0.5-0.8(-1)毫米的柔毛。叶长6-10厘米；小叶2-4对，狭椭圆形至狭卵形。总状花序具10至多花，不断伸长；花瓣黄色或紫色。荚果狭椭圆体形。生海拔1800-2600米。产河北、山西、甘肃和宁夏。

Perennial herbs, 35-45 cm tall, vegetative parts with hairs 0.5-0.8(-1) mm long. Leaves 6-10 cm long; leaflets 2-4 pairs, narrowly elliptic to narrowly ovate. Racemes densely 10- to many-flowered, elongating with age; petals yellow or purplish. Legumes narrowly ellipsoid. At 1800-2600 m. Distributed in Hebei, Shanxi, Gansu and Ningxia.

蒙古黄耆

Astragalus mongholicus Bunge

小灌木。叶近无柄；小叶8-12对，狭卵形至狭椭圆形，先端阔圆形至狭凹缺。总状花序疏具多花；花瓣黄色；旗瓣宽椭圆形，长13-20 × 7-9毫米。荚果侧面观斜椭圆体形至近半圆形。生海拔800-2000米的典型草原、草甸、干旱灌丛、针叶林或山坡区域。产中国西南、华北、华西、西北和东北。哈萨克斯坦、俄罗斯和蒙古亦有。

Shrublets. Leaves nearly sessile; leaflets in 8-12 pairs, narrowly ovate to narrowly elliptic, apex widely rounded to shallowly emarginate. Racemes loosely many flowered; petals yellow; standard widely elliptic, 13-20 × 7-9 mm. Legumes obliquely ellipsoid to nearly semicircular as seen from side. Steppes, meadows, xerophytic shrubs, coniferous forests or montane zone at 800-2000 m. Distributed in SW, N, W, NW and NE China. Also in Kazakhstan, Russia and Mongolia.

金翼黄耆

Astragalus chrysopterus
Bunge ex Maxim.

多年生草本，高60(-100)厘米，具伏贴的白色柔毛，花序上的毛黑色。叶近无柄；具小叶5-11对。总状花序6-10厘米，具6-20花；所有花瓣近等长。荚果斜椭圆形。生海拔1600-3700米。产中国西南、华中、华北、华西和西北。

蒙古黄耆 *Astragalus mongholicus*

金翼黄耆 *Astragalus chrysopterus*

Perennial herbs, 60(-100) cm tall, hairs appressed to subappressed, hairs on inflorescence black. Leaves subsessile; leaflets in 5-11 pairs. Racemes 6-10 cm long, 6-20-flowered; petals all of nearly equal length. Legumes obliquely ellipsoid. At 1600-3700 m. Distributed in SW, C, N, W and NW China.

云南黄耆

Astragalus yunnanensis Franch.

多年生草本。叶上面无毛，下面被白色长柔毛。总状花序长1-2.5厘米，密生2-13花，果期花序轴稍伸出至5厘米长；花瓣浅黄色，有时浅紫色；子房被长柔毛。荚果膜质，狭卵形，长12-23毫米。花期7月。生海拔3000-4300米的山坡或草甸。产中国西南、华北和华西。印度和尼泊尔亦有。

Perennial herbs. Leaflets adaxially glabrous, abaxially white villose; densely hairy. Racemes 1-2.5 cm long, rather densely 2-13-flowered, at fruiting time somewhat elongated up to 5 cm long; petals pale yellow, sometimes slightly fading violet; ovary villose. Legumes membranous, narrowly ovoid, 12-23 mm long. Fl. Jul. Slopes or meadows at 3000-4300 m. Distributed in SW, N and W China. Also in India and Nepal.

大托叶黄耆 *Astragalus stipulatus*

大托叶黄耆

Astragalus stipulatus D. Don

多年生草本。托叶叶状，长(2-)3-7厘米；小叶(8-)11-20对，狭卵形至狭长圆形，长(15-)20-55厘米，上面无毛，下面被白毛，后变无毛，边缘具明显缘毛。总状花序多达100朵花；花冠白黄色至黄色。荚果狭椭圆体形至狭长圆形，长1.8-3.2厘米，2室。花期7-8月，果期9月。生海拔1500-3700米的山坡林缘。产西藏和云南。不丹、印度和尼泊尔亦有。

Perennial herbs. Stipules leaflike, (2-)3-7 cm long; leaflets (8-)11-20-paired, narrowly ovate to narrowly oblong, (15-)20-55 cm long, adaxially glabrous, abaxially white hairy, glabrescent, margin distinctly ciliate. Racemes up to 100-flowered; corolla whitish yellow to yellow. Legumes narrowly ellipsoid to narrowly oblong, 1.8-3.2 cm long, 2-locular. Fl. Jul-Aug. Fr. Sep. Forest margin on slopes at 1500-3700 m. Distributed in Xizang and Yunnan. Also in Bhutan, India and Nepal.

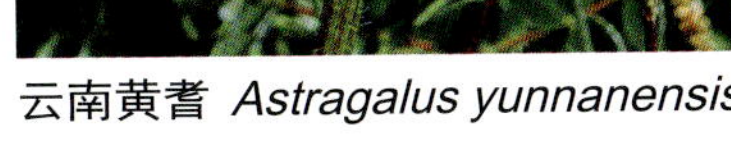

云南黄耆 *Astragalus yunnanensis*

草珠黄耆 *Astragalus capillipes*

草珠黄耆
Astragalus capillipes Bunge

多年生草本。托叶膜质；小叶(1或)2或3对，椭圆形，5-20 × 2.5-8厘米。总状花序疏生多花；花瓣白色、乳白色、橙色或红色；萼齿长0.1-0.5毫米，外侧及内侧具黑毛。荚果卵球形。花期7-9月，果期9-10月。生海拔300-2000米的河谷沙地、山坡或路旁。产河北、山西、内蒙古、陕西、甘肃和宁夏。

Perennial herbs. Stipules membranous; leaflets in (1 or)2 or 3 pairs, elliptic, 5-20 × 2.5-8 mm long. Racemes loosely many-flowered; petals white, cream white, orange or red; calyx teeth 0.1-0.5 mm long, black hairy on outer and inner sides. Legumes ovoid. Fl. Jul-Sep. Fr. Sep-Oct. Sandy places in valleys, slopes or roadsides at 300-2000 m. Distributed in Hebei, Shanxi, Neimenggu, Shaanxi, Gansu and Ningxia.

草木樨状黄耆
Astragalus melilotoides Pall.

多年生草本，高45-90厘米，具伏贴的白色柔毛。叶近无柄；小叶(1-)2或3对。总状花序，松散多花，果期长达16厘米；花瓣白色，乳白色、黄色、紫色或紫色带黑色斑点。荚果无柄，椭圆体形。生海拔200-2600米的草原，或沙质或石质山坡。产中国西南、华中、华北、华西和东北。俄罗斯(西伯利亚)、蒙古和日本亦有。

Perennial herbs, 45-90 cm tall, furnished with appressed white hairs. Leaves subsessile; leaflets in (1-)2 or 3 pairs. Racemes loosely many-flowered, elongating in fruit up to 16 cm long; petals cream white, yellow, purple or violet with dark spots. Legumes sessile, ellipsoid. Steppes, or stony or sandy slopes at 200-2600 m. Distributed in SW, C, N, W and NE China. Also in Russia (Siberia), Mongolia and Japan.

草木樨状黄耆 *Astragalus melilotoides*

中国黄耆
Astragalus chinensis Linnaeus f.

多年生草本。茎单一，无毛。托叶条形，基部具弯曲的短耳；叶长7-15厘米；小叶10-15对，长16-25毫米，狭椭圆形，下面疏被贴伏白毛。总状花序松散，具多花；花萼无毛；花冠黄色或黄白色。荚果具长6-8毫米的果茎，球形至倒卵球形，先端具细短喙，无毛，果瓣具横向皱纹。花期6-7月，果期7-8月。产中国东北、华北和青海。蒙古和俄罗斯亦有。

Perennial herbs. Stem solitary, glabrous. Stipules linear, with a curved short auricle at base; leaves 7-15 cm long; leaflets 10-15-paired, 16-25 mm long, narrowly elliptic, abaxially sparsely appressed white hairy. Racemes loosely many flowered; calyx glabrous; petals yellow or whitish yellow. Legumes with a slender stipe 6-8 mm, globose to obovoid, with a very short slender beak, glabrous, valves transversely wrinkled-nerved. Fl. Jun-Jul. Fr. Jul-Aug. Distributed in NE and N China, and Qinghai. Also in Mongolia and Russia.

多枝黄耆
Astragalus polycladus Bureau et Franch.

多年生草本。茎平卧或上升，被灰白色伏贴柔毛或混有黑毛。奇数羽状复叶，无柄至近无柄。总状花序密集成头状；

中国黄耆 *Astragalus chinensis*

多枝黄耆 *Astragalus polycladus*

花瓣浅紫色，龙骨瓣色深；旗瓣长6-7毫米。荚果长圆形，微弯曲，不完全2室。花期7-8月，果期9月。生海拔2000-4500米的山坡或路旁。产云南、四川、西藏、甘肃、青海和新疆。

Perennial herbs. Stems ascending to erect, sparsely to loosely covered with white appressed hairs also with black hairs. Leaves odd-pinnate, sessile to subsessile. Racemes capitate; petals pale violet, with darker keel; standard 6-7 mm long. Legumes oblong, slightly curved, incompletely 2-locular. Fl. Jul-Aug. Fr. Sep. Slopes or roadsides at 2000-4500 m. Distributed in Yunnan, Sichuan, Xizang, Gansu, Qinghai and Xinjiang.

紫云英

Astragalus sinicus L.

一年生植物或为短命多年生。茎于节上生根。叶柄疏具毛；小叶3-5对，下面疏具平伏毛。花组成稍紧密呈伞形的总状花序；花瓣白色、粉色、浅红色或紫色。荚果条状长圆形，稍弯曲。花期2-6月，果期3-7月。生海拔100-3000米的山坡、溪边或湿地。产中国西南、华南、东南和华中。

Plants annual or short-lived perennial. Stems rooting at nodes. Petioles sparsely hairy; leaflets in 3-5 pairs, abaxially sparsely appressed hairy. Flowers in a close umbel-raceme; petals white, pink, light red, or purple. Legumes linear-oblong, slight curved. Fl. Feb-Jun. Fr. Mar-Jul. Slopes, by streams or wet places at 100-3000 m. Distributed in SW, S, SE and C China.

紫云英 *Astragalus sinicus*

四川黄耆 *Astragalus sutchuenensis*

无茎黄耆 *Astragalus acaulis*

四川黄耆
Astragalus sutchuenensis Franch.

多年生草本。茎散生，匍匐或直立，25-40厘米，节间长2-5厘米，疏被柔毛。小叶4-8对，椭圆形。总状花序初密集，后伸长而疏散，具4-7花。荚果直伸，狭条形。花期5-6月，果期7月。生海拔400-3300米的山坡林下。产甘肃和四川。

Herbs perennial. Stems several, prostrate or erect, 25-40 cm long, internodes 2-5 cm long, sparsely to loosely hairy. Leaflets in 4-8 pairs, elliptic. Racemes at first densely, later elongating and rather loosely, 4-7-flowered. Legumes erect, narrowly linear. Fl. May-Jun. Fr. Jul. Forests on slopes at 400-3300 m. Distributed in Gansu and Sichuan.

无茎黄耆
Astragalus acaulis Baker

多年生草本。茎短缩，高3-5厘米。小叶7-12对，狭卵形，下面无毛或于中脉边缘疏具毛。总状花序近无柄，生2-4花。荚果近无柄，直立，腹部和背部狭圆形，无喙。花果期6-8月。生海拔3300-4500米的高山草地或沙石滩中。产云南、四川和西藏。印度北部和不丹亦有。

Perennial herbs. Stems short, 3-5 cm tall. Leaflets in 7-12 pairs, narrowly ovate, abaxially glabrous or sparsely hairy at margins and midvein. Racemes subsessile, flowers 2-4. Legumes subsessile, erect, narrowly rounded ventrally and dorsally, without beak. Fl. and fr. Jun-Aug. Alpine grasslands or screes at 3300-4500 m. Distributed in Yunnan, Sichuan and Xizang. Also in N India and Bhutan.

达乌里黄耆
Astragalus dahuricus (Pall.) DC.

一年生或二年生草本，被白色薄毛，花期亦有时黑色。小叶4-9对，狭椭圆形。总状花序，10-25花密生；花瓣紫色；旗瓣椭圆形，急缩成短爪，顶端具深V形齿。荚果条形。花期7-9月，果期8-10月。生海拔400-2500米的河边、草甸、山坡或田中。产中国西南、华中、华北、华东、西北和东北。俄罗斯、蒙古和朝鲜半岛亦有。

Annual or biennial herbs, hairs thin, white, in inflorescence also black. Leaflets in 4-9 pairs, narrowly elliptic. Racemes, densely up to 10-25-flowered; petals violet; standard elliptic, abruptly contracted into short claw, apex deeply V-like incised. Legumes linear. Fl. Jul-Sep. Fr. Aug-Oct. Riversides, meadows, slopes or fields at 400-2500 m. Distributed in SW, C, N, E, NW and NE China. Also in Russia, Mongolia and Korean Peninsula.

狐尾黄耆
Astragalus alopecurus Pallas

多年生草本。茎密被上向柔毛。托叶长10-18毫米，被毛；叶长20-30厘米；小叶17-27对，卵形至椭圆形，先端钝，下面疏被柔毛，上面无毛。总状花序呈卵球形或圆柱状，长5-9.5厘米；花萼稍膨大，钟状，被开展柔毛；花瓣黄色，无毛。荚果卵球形，密被柔毛。花期6-8月，果期8-9月。产新疆。哈萨克斯坦、俄罗斯、东南亚和欧洲亦有。

Perennial herbs. Stem with ascending hairs. Stipules 10-18 mm long, hairy; leaves 20-30 cm long; leaflets 17-27 pairs, ovate to elliptic, apex obtuse, abaxially sparsely hairy, adaxially gla-

达乌里黄耆 *Astragalus dahuricus*

狐尾黄耆 *Astragalus alopecurus*

斜茎黄耆 *Astragalus laxmannii*

brous. Racemes ovoid or cylindric, 5-9.5 cm long; calyx slightly inflated, spreading hairy; petals yellow, glabrous. Legumes ovoid, densely hairy. Fl. Jun-Aug. Fr. Aug-Sep. Distributed in Xinjiang. Also in Kazakhstan, Russia, SW Asia and Europe.

斜茎黄耆

Astragalus laxmannii Jacq.

多年生草本。茎多数或数个丛生。小叶6-16对，5-25 × 2-7毫米，疏具毛。总状花序椭圆体形，具多花；花瓣紫色至蓝紫色，干后多为黄色；旗瓣具凹缺。荚果长圆形，背缝凹入成沟槽。花期6-8月，果期8-10月。生海拔1100-4200米的向阳山坡灌丛或林缘地带。产中国西南、华北、西北和东北。俄罗斯、蒙古、朝鲜半岛、日本和北美洲亦有。

Perennial herbs. Stems several or many-clustered. Leaflets in 6-16 pairs, 5-25 × 2-7 mm, sparsely hairy. Racemes ellipsoid, many flowered; petals purple to bluish violet, mostly yellow when dry; standard emarginate. Legumes oblong, dorsal sutures concave to grooved. Fl. Jun-Aug. Fr. Aug-Oct. Shrubs on sunny mountain slopes or forest edges at 1100-4200 m. Distributed in SW, N, NW and NE China. Also in Russia, Mongolia, Korean Peninsula, Japan and North America.

地八角

Astragalus bhotanensis Baker in J. D. Hooker

多年生草本。茎中空。叶长4-20厘米；小叶10-17对，狭椭圆形，长5-23毫米，下面被紧贴的丁字毛。总状花序伞状球形，具8-20朵花；花萼长5-8毫米，被黑色极短或有时长达0.5毫米的柔毛，萼齿长2-3.5毫米；花瓣深紫色至蓝紫色。荚果条形，具短直喙。花期3-8月，果期8-10月。生海拔600-2800米。产贵州、云南、西藏、四川、陕西和甘肃。不丹和朝鲜半岛亦有。

Perennial herbs. Stems hollow. Leaves 4-20 cm long; leaflets 10-17-paired, narrowly elliptic, 5-23 mm long, abaxially covered with appressed, medifixed hairs. Racemes umbellate-globose, 8-20-flowered; calyx 5-8 mm, covered with predominantly black, very short, sometimes also longer hairs up to 0.5 mm, teeth 2-3.5 mm long; petals dark purple to blue-purple. Legumes linear, with a short straight beak. Fl. Mar-Aug. Fr. Aug-Oct. Distributed in Guizhou, Yunnan, Xizang, Sichuan, Shaanxi and Gansu. Also in Bhutan and Korean Peninsula.

地八角 *Astragalus bhotanensis*

毛叶黄耆

Astragalus pallasii Sprengel

多年生草本。小叶4-7对，长4-8毫米，上面无毛或近叶缘处被稀疏毛，下面被较密半开展白色毛。总状花序2-5花；总花梗密被毛；花萼管状，长13-16毫米，被白色和黑色半开展的毛；花冠淡紫色。荚果膨大，卵球形，长18-24毫米，被稀疏开展的白色毛，具弯曲的喙。花期4-5月，果期5-6月。生于旱山坡。产新疆。哈萨克斯坦和乌兹别克斯坦亦有。

Perennial herbs. Leaflets 4-7-paired, 4-8 mm long, adaxially glabrous or with sparse hairs along margins, abaxially densely white semi-appressed hairy. Racemes 2-5-flowered; peduncle densely hairy; calyx tubular, 13-16 mm long, black and white semi-appressed hairy; corolla purplish. Legumes inflated, ovoid, 18-24 mm long, sparsely white spreading hairy, with a curved beak. Fl. Apr-May. Fr May-Jun. Dry slopes. Distributed in Xinjiang. Also in Kazakhstan and Uzbekistan.

毛叶黄耆 *Astragalus pallasii*

糙叶黄耆 *Astragalus scaberrimus*

单叶黄耆

Astragalus efoliolatus Han.-Mazz.

多年生草本，无茎。托叶边缘具丁字毛；单叶，无柄，线形，长1-5厘米，边缘平展或反折，两面被伏贴毛。总状花序松散，多达8花；花萼长约5毫米，密被贴伏毛，萼齿长约2.5毫米；花瓣红紫色或紫色。荚果条状长圆形，先端有短喙，无柄，密被白色伏贴毛。花期6-9月，果期9-10月。生沙质冲积土中。产内蒙古、陕西、宁夏和甘肃。

Perennial herbs, acaulescent. Stipules at margins often with basifixed hairs; leaves simple, sessile, linear, 1-5 cm long, margin flat or folded, both surfaces appressed hairy. Racemes loosely up to 8-flowered; calyx ca. 5 mm long, densely appressed hairy, teeth ca. 2.5 mm long; petals red-purple or violet. Legumes linear-oblong, with a short beak, sessile, densely white hairy. Fl. Jun-Sep. Fr. Sep-Oct. Sandy alluvial soil. Distributed in Neimenggu, Shaanxi, Ningxia and Gansu.

单叶黄耆 *Astragalus efoliolatus*

糙叶黄耆

Astragalus scaberrimus Bunge

多年生草本，密被白色伏贴毛。根状茎缩短，多分枝。小叶3-6(-8)对，狭椭圆形至椭圆形，两面具疏至极密的毛。总状花序疏具3-5花；总花梗长0.3-1.5(-3.5)厘米。荚果披针状长圆形。花期4-8月，果期5-9月。生山坡石砾质草地、沙丘或沿河岸沙地。产中国西北、华北和东北。俄罗斯和蒙古亦有。

Perennial herbs, densely covered with white adnate hairs. Rhizomes shortened, many-branched. Leaflets in 3-6(-8) pairs, narrowly elliptic to elliptic, both surfaces sparsely to rather densely hairy. Racemes loosely 3-5-flowered; peduncles 0.3-1.5 (-3.5) cm long. Legumes lanceolate-oblong. Fl. Apr-Aug. Fr. May-Sep. Stony grasslands of mountain slopes, dunes or sandy banks. Distributed in NW, N and NE China. Also in Russia and Mongolia.

乳白花黄耆

Astragalus galactites Pallas

植株高3-12厘米，常具紧密的白色毡毛。具小叶4-8对，狭椭圆形。总状花序多生，在叶基部形成密的头状花序，(1-)2-4花；花瓣白色，旗瓣和龙骨瓣的顶端带紫色。荚果具白毛。生草原。产内蒙古、陕西和甘肃。俄罗斯(西伯利亚)和蒙古亦有。

Plants 3-12 cm tall, often forming compact mats, white hairy. Leaflets in 4-8 pairs, narrowly elliptic. Racemes many, forming dense, capitate synflorescences around base of leaves, (1-)2-4-flowered; petals whitish, tip of standard and keel often violet. Legumes white hairy. Steppes. Distributed in Neimenggu, Shaanxi and Gansu. Also in Russia (Siberia)

乳白花黄耆 *Astragalus galactites*

and Mongolia.

费尔干岩黄耆

Hedysarum ferganense Korsh.

多年生草本。托叶三角状披针形；叶长8-10厘米；小叶7-11(-13)，卵状长圆形或狭圆形，上面被疏柔毛，下面密被贴伏柔毛。总状花序长圆形，具8-20花；花长12-15毫米；花萼短钟状，被疏柔毛，萼齿披针状钻形，长为萼筒2-3倍；翼瓣长约为旗瓣的1/4。荚果被透明鳞片。花期6-7月，果期7-8月。生海拔800-1700米的干旱草原。产新疆。中亚和南亚亦有。

Perennial herbs. Stipules triangular-lanceolate; leaves 8-10 cm long; leaflets 7-11(-13), ovate-oblong or narrowly elliptic, adaxially sparsely pilose, abaxially densely appressed pilose. Racemes oblong, 8-20-flowered; flowers 12-15 mm long; calyx shortly campanulate, sparsely pilose, teeth lanceolate-subulate, 2-3 × as long as tube; wings ca. 1/4 as long as standard. Legume with hyaline scales. Fl. Jun-Jul. Fr. Jul-Aug. Steppes at 800-1700 m. Distributed in Xinjiang. Also in C and S Asia.

费尔干岩黄耆 *Hedysarum ferganense*

背扁膨果豆

Phyllolobium chinense Fisch.

多年生草本，被平伏的、极短的、鳞片状、泡状毛，长达0.1毫米。小叶4-10对。总状花序具2-7疏散排列的花，较叶长；花冠白色带红紫色。荚果狭长圆形，1室。花期7-9月，果期8-10月。生海拔1000-1700米的山坡、草地或路边。产华中、华北、华西和东北。

Perennial herbs, covered with appressed, very short, scalelike, bladderlike hairs up to 0.1 mm long. Leaflets in 4-10 pairs. Racemes remotely 2-7-flowered, longer than leaves; corolla white with reddish purple. Legumes narrowly oblong, 1-locular. Fl. Jul-Sep. Fr. Aug-Oct. Slopes, grasslands and roadsides at 1000-1700 m. Distributed in C, N, W and NE China.

背扁膨果豆 *Phyllolobium chinense*

硬毛棘豆 *Oxytropis hirta*

硬毛棘豆
Oxytropis hirta Bunge

草本，无茎。奇数羽状复叶，具9-21小叶，小叶椭圆形、披针形或卵形。总状花序伸长，多花；花冠蓝紫色、紫色、蓝色、红色、浅黄色、黄绿色、紫红色、粉色、黄色或白色。荚果无柄，卵球形，密被毛。花期5-9月，果期7-9月。生海拔1000-4100米的山区、路边、草地、荒沙地、灌丛、旱山坡或疏林下。产华北、华西、华东和东北。俄罗斯和蒙古亦有。

Herbs, acaulescent. Leaves imparipinnate, leaflets 9-21-foliolate, elliptic, lanceolate or ovate. Racemes elongate, many-flowered; corolla bluish purple, purple, blue, red, light yellow yellowish green, reddish violet, pink, yellow, or white. Legumes sessile, ovoid, with dense trichomes. Fl. May-Sep. Fr. Jul-Sep. Mountains, roadsides, grasslands, sandy tracts, shrubs, dry slopes, or beneath sparse forests at 1000-4100 m. Distributed in N, W, E and NE China. Also in Russia and Mongolia.

黄花棘豆
Oxytropis ochrocephala Bunge

多年生草本。叶具(11或)13-27(-39)小叶；小叶卵状披针形，长2.5-3厘米，两面具糙毛杂生疏短或长柔毛。花序轴被白褐色腺点和黄色长丁字毛；总状花序紧密，多花；花冠黄色。荚果长圆状，膨胀。花期6-8月，果期7-9月。生海拔1800-4500米的草地、山坡或草甸。产华北、华西和西北。

Perennial herbs. Leaves (11 or)13-27(-39)-foliolate; leaflets ovate-lanceolate, 2.5-3 cm long, both surfaces strigose with sparse short or long trichomes. Rachises with pale brown glands and yellow long trichomes; racemes compact, many-flowered; corolla yellow. Legumes oblong, inflated.

黄花棘豆 *Oxytropis ochrocephala*

小花棘豆 *Oxytropis glabra*

黑萼棘豆 *Oxytropis melanocalyx*

Fl. Jun-Aug. Fr. Jul-Sep. Grasslands, slopes and meadows at 1800-4500 m. Distributed in N, W and NW China.

小花棘豆
Oxytropis glabra DC.

草本。茎直立。奇数羽状复叶，具11-29小叶，小叶披针形、卵状披针形或椭圆形。总状花序稀疏；花冠紫色至蓝紫色；萼裂片长1.5-2毫米；旗瓣长(5-)7-8(-10)毫米。荚果近圆柱状，稍膨胀，近无毛。花期6-9月，果期7-9月。生海拔400-4400米的山腰、路边、沙地、灌丛、冲积平原、田地或荒漠草原。产中国西南、华北、华西、西北和东北。哈萨克斯坦、俄罗斯和蒙古亦有。

Herbs. Stems erect. Leaves imparipinnate, leaflets 11-29-foliolate, lanceolate, ovate-lanceolate or elliptic. Racemes, lax; corolla purple to bluish purple; calyx lobes 1.5-2 mm long; standard (5-)7-8 (-10) mm long. Legumes subcylindric, slightly inflated, glabrescent. Fl. Jun-Sep. Fr. Jul-Sep. Hillsides, roadsides, sandy areas, scrubs, floodplains, fields or desert meadows at 400-4400 m. Distributed in SW, N, W, NW and NE China. Also in Kazakhstan, Russia and Mongolia.

黑萼棘豆
Oxytropis melanocalyx Bunge

草本。叶具小叶9-25；叶轴细，疏生黄色长柔毛。总状花序密集，具花3-15朵；萼具黑色短柔毛杂以黄色和白色柔毛，萼裂片长2.5-4.7毫米；旗瓣长(10-)11-14毫米。荚果阔长圆形，下垂，膨大，纸质，1室。花期7-8月，果期8-9月。生海拔2100-5100米的山坡、灌丛中、草地、沙砾地、高山草甸或落叶松林中。产中国西南和西北。

Herbs. Leaves with 9-25 leaflets; rachises thin, with sparse yellow long trichomes. Racemes compact, 3-15-flowered; calyx with black short trichomes intermixed with yellow and white trichomes, calyx lobes 2.5-4.7 mm long; standard (10-)11-14 mm long. Legumes broadly oblong, pendulous, inflated, papery, 1-locular. Fl. Jul-Aug. Fr. Aug-Sep. Hillsides, among scrubs, grasslands, gravel areas, alpine meadows or *Larix* forests at 2100-5100 m. Distributed in SW and NW China.

云南棘豆
Oxytropis yunnanensis Franch.

草本，无茎或具短茎。羽状复叶长3-5厘米。总状花序具3-10花；花萼具黑色和白色的长柔毛；花冠紫色或紫红色；旗瓣长1.2-1.3厘米，先端2裂；翼瓣长1-1.2厘米。荚果大，长2-3厘米。花果期7-9月。生海拔1800-4900米的草甸、石灰岩山地、开阔多石草地、山坡草地或冻原。产云南、四川、西藏、甘肃和青海。

Herbs, acaulescent or shortly caulescent. Leaves pinnate, 3-5 cm long. Racemes 3-10-flowered; calyx with black and white long trichomes; corolla purple or purplish red; standard 1.2-1.3 cm long, apex 2-lobed; wings 1-1.2 cm long. Legumes large, 2-3 cm long. Fl. and fr. Jul-Sep. Meadows, limestone screes, open stony pastures, grassy slopes, rocky slopes or frozen steppes at 1800-4900 m. Distributed in Yunnan, Sichuan, Xizang, Gansu and Qinghai.

云南棘豆 *Oxytropis yunnanensis*

蓝花棘豆 *Oxytropis caerulea*

蓝花棘豆

Oxytropis caerulea (Pall.) DC.

多年生草本。叶具15-41小叶；小叶卵形至披针状椭圆形，宿存，近无毛。总状花序疏散，具花12-20朵；花冠紫色、淡紫色、蓝色、红色或白色；龙骨瓣喙长2-3毫米；子房无柄，无毛。荚果矩圆状卵球形，膨胀。花期6-9月，果期8-9月。生海拔1000-3000米的山坡、林下、草甸或路边。产河北、山西、内蒙古、甘肃和黑龙江。俄罗斯和蒙古亦有。

Perennial herbs. Leaves 15-41-foliolate; leaflets ovate to lanceolate-elliptic, persistent, subglabrous. Racemes lax, 12-20-flowered; corolla purple, violet, blue, red or white; keel beak 2-3 mm long; ovary sessile, glabrous. Legumes oblong-ovoid, inflated. Fl. Jun-Sep. Fr. Aug-Sep. Slopes, under forests, meadows or roadsides at 1000-3000 m. Distributed in Hebei, Shanxi, Neimenggu, Gansu and Heilongjiang. Also in Russia and Mongolia.

宽苞棘豆

Oxytropis latibracteata Jurtzev

草本。无茎。羽状复叶具(13-)15-23小叶，对生或有时轮生；小叶椭圆形、狭卵形或披针形。总状花序极密，具5-13花或更多；苞片椭圆形；花冠蓝紫色至淡紫色。荚果无柄，卵球状椭圆体形，具柔毛。花果期7-8月。生海拔1700-3800米的向阳山坡、白桦林中、山腰、草地、冲积平原或灌丛中。产中国西南、华北、华西和西北。

Herbs. Acaulescent. Leaves imparipinnate; leaflets (13-)15-23-foliolate, opposite or sometimes whirled; leaflets elliptic, narrowly ovate or lanceolate. Racemes fairly dense, 5-13-flowered or more; bracts elliptic; corolla bluish purple to pale purple. Legumes sessile, ovoid-ellipsoid, pubescent. Fl. and fr. Jul-Aug. Sunny slopes, *Betula* forests, hillsides, grasslands, floodplains or among shrubs at 1700-3800 m. Distributed in SW, N, W and NW China.

窄膜棘豆

Oxytropis moellendorffii Bunge ex Maxim.

草本。无茎。奇数羽状复叶具13-21小叶；小叶披针形至狭披针形。总状花序紧缩，具(2-)3-5花；花冠紫色；翼瓣长约1.5厘米。荚果无柄，长圆形，约2 × 0.6厘米，膜质，密具黑色硬直的和白色的长柔毛。花期6-7月，果期7-8月。生海拔2400-3400米的路边、山腰或多石山顶。产河北和山西。

宽苞棘豆 *Oxytropis latibracteata*

窄膜棘豆 *Oxytropis moellendorffii*

缘毛棘豆 *Oxytropis ciliata*

Herbs. Acaulescent. Leaves imparipinnate, with 13-21-foliolate; leaflets lanceolate to narrowly lanceolate. Racemes compact, (2-)3-5-flowered; corolla purple; wings ca. 1.5 cm long. Legumes sessile, oblong, ca. 2 × 0.6 cm long, membranous, with dense black stout and white long trichomes. Fl. Jun-Jul. Fr. Jul-Aug. Roadsides, hillsides or gravelly hilltops at 2400-3400 m. Distributed in Hebei and Shanxi.

长白棘豆

Oxytropis anertii Nakai ex Kitag.

草本。茎极缩短，丛生。叶具17-35小叶；小叶卵形，近无毛。总状花序极疏松，具2-8花；苞片三角形；花冠蓝色、蓝紫色或浅蓝色。荚果卵球形至卵球状长圆形，膨胀。花期6-8月，果期8-9月。生海拔1800-2600米的高山冻原、高山草甸、高山石缝、林缘或向阳山坡。产吉林。朝鲜半岛亦有。

Herbs. Stems extremely shortened, clustered. Leaves 17-35-foliolate; leaflets ovate, subglabrous. Racemes rather lax, 2-8-flowered; bracts triangular; corolla blue, bluish purple or light purple. Legumes ovoid to ovoid-oblong, inflated. Fl. Jun-Aug. Fr. Aug-Sep. Alpine tundras, meadows, stone cracks, forest edges or sunny slopes at 1800-2600 m. Distributed in Jilin. Also in Korean Peninsula.

缘毛棘豆

Oxytropis ciliata Turcz.

草本。无茎。奇数羽状复叶，具9-17小叶，条状长圆形、长圆形、条状披针形或倒披针形，边缘具长缘毛。总状花序极疏松，具2-8花；花冠淡黄色。荚果无柄，深棕色至黄棕色，卵球形，膨胀，无毛，顶端具喙。花期5-7月，果期6-7月。生海拔1800-1900米的路边、向阳山坡、山谷、灌丛或白桦林中。产河北、内蒙古和宁夏。蒙古亦有。

Herbs. Acaulescent. Leaves imparipinnate; leaflets 9-17-foliolate, linear-oblong, oblong, linear-lanceolate or oblanceolate, margin long ciliate. Racemes rather lax, 2-8-flowered; corolla pale yellow. Legumes sessile, dark brown to yellowish brown, ovoid, inflated, glabrous, apex beaked. Fl. May-Jul. Fr. Jun-Jul. Roadsides, sunny slopes, valleys, scrubs or in *Betula* forests at 1800-1900 m. Distributed in Hebei, Neimenggu and Ningxia. Also in Mongolia.

长白棘豆 *Oxytropis anertii*

山泡泡 *Oxytropis leptophylla*

山泡泡

Oxytropis leptophylla (Pallas) DC.

草本。无茎，具带分枝的假茎。羽状复叶具9-13小叶；小叶条形。总状花序紧缩至疏松，具2-5花；花冠紫色；旗瓣长1.8-2.3厘米；翼瓣长1.9-2厘米。荚果无柄，卵球形，膨胀，膜质，1室。花期5-6月，果期6-7月。生海拔800-1900米的山腰、多石地区、沙丘或草地。产河北、山西、内蒙古和吉林。俄罗斯和蒙古亦有。

Herbs. Acaulescent, from a branching superficial caudex. Leaves imparipinnate; leaflets 9-13-foliolate; leaflets linear. Racemes compact to rather lax, 2-5-flowered; corolla purple; standard 1.8-2.3 cm long; wings 1.9-2 cm long. Legumes sessile, ovoid, inflated, membranous, 1-locular. Fl. May-Jun. Fr. Jun-Jul. Hillsides, gravelly areas, sandy dunes or grasslands at 800-1900 m. Distributed in Hebei, Shanxi, Neimenggu and Jilin. Also in Russia and Mongolia.

黄毛棘豆

Oxytropis ochrantha Turcz.

多年生草本，具平伏的黄色长柔毛。小叶轮生，1-9轮；小叶成熟后下面具长柔毛。总状花序伸长，圆筒状，具多花；花瓣白色或淡黄色。荚果卵球形，膨胀，膜质。花期6-7月，果期7-8月。生海拔500-4800米的草甸、山坡、山谷或沙地。产华北、华西和西北。蒙古亦有。

Perennial herbs, with appressed yellow long trichomes. Leaflets verticillate, in 1-9 whorls; leaflets abaxially with long trichomes when mature. Racemes elongate, cylindric, many-flowered; petals white or light yellow. Legumes ovoid, inflated, membranous. Fl. Jun-Jul. Fr. Jul-Aug. Meadows, slopes, valleys or sandy areas at 500-4800 m. Distributed in N, W and NW China. Also in Mongolia.

地角儿苗

Oxytropis bicolor Bunge

多年生草本，具平伏的浅黄色或腺柔毛。小叶至少部分轮生，每叶具3-10轮。总状花序密集或疏生，具花7-25；花萼圆柱形，长(10-)12-17毫米。荚果无柄，卵球状长圆形，膨胀。花期4-7月，果期6-8月。生海拔400-2700米的山坡、灌丛、草地、路边或沙地。产河北、山西、内蒙古、甘肃、陕西、宁夏、河南和山东。蒙古亦有。

Perennial herbs, with appressed yellowish or glandular trichomes. Leaflets at least some verticillate, in 3-10 whorls per leaf. Racemes dense or lax, 7-25-flowered; calyx cylindric, (10-)12-17 mm long. Legumes sessile, ovoid-oblong, inflated. Fl. Apr-Jul. Fr. Jun-Aug. Slopes, thickets, grasslands, roadsides or sandy sites at

黄毛棘豆 *Oxytropis ochrantha*

地角儿苗 *Oxytropis bicolor*

多叶棘豆 *Oxytropis myriophylla*

400-2700 m. Distributed in Hebei, Shanxi, Neimenggu, Gansu, Shaanxi, Ningxia, Henan and Shandong. Also in Mongolia.

多叶棘豆

Oxytropis myriophylla (Pallas) DC.

草本。无茎。小叶12-16(-50)轮，每轮4-8片，条形、长圆形或披针形。总状花序紧缩，多花；花冠淡紫色。荚果直升，略有柄，革质，不完全两室，密被长柔毛。花期5-6月，果期7-8月。生海拔200-2600米的山腰、草地、多石山坡、沙地或白桦林缘。产华北、华西和东北。俄罗斯和蒙古亦有。

Herbs. Acaulescent. Leaflets verticillate, in 12-16(-50) whorls, 4-8 leaflets per whorl, linear, oblong or lanceolate. Racemes compact, many-flowered; corolla pale purple. Legumes erect-ascending, substipitate, leathery, ± 2-locular, with dense long trichomes. Fl. May-Jun. Fr. Jul-Aug. Hillsides, grasslands, rocky mountain slopes, sandy areas or *Betula* forest edges at 200-2600 m. Distributed in N, W and NE China. Also in Russia and Mongolia.

砂珍棘豆

Oxytropis racemosa Turcz.

多年生草本，高5-30厘米。托叶膜质，卵形；小叶轮生，6-12轮，长圆形、线形或披针形，长5-10毫米，宽1-2毫米。总状花序，长4厘米；花冠紫红色或淡紫红色。荚果膜质，卵状球形，膨胀，长约10毫米。种子肾状圆形，长约1毫米，暗褐色。花期5-7月，果期6-10月。生海拔600-1900米的沙滩、沙荒地、沙丘、沙质坡地及丘陵地区阳坡。产中国东北、内蒙古、河北、山西、陕西和宁夏。蒙古和朝鲜半岛亦有。

Perennial herbs, 5-30 cm tall. Stipules membranous, ovate; leaflets verticillate, in 6-12 whorls, leaflet blades oblong, linear, or lanceolate, 5-10 × 1-2 mm. Racemes to 4 cm; corolla purple-red or pale purple-red. Legume membranous, ovoid, inflated, 10 mm long. Seed kidney-round, ca. 1 mm long, crineous. Fl. May-Jul. Fr. Jun-Oct. Sandy places on hillsides, dry valleys, grasslands, damp places, sandy or gravelly floodplains and riverbanks at 600-1900 m. Distributed in NE China, Neimenggu, Hebei, Shanxi, Shaanxi and Ningxia. Also in Mongolia and Korean Peninsula.

砂珍棘豆 *Oxytropis racemosa*

镰荚棘豆 *Oxytropis falcata*

镰荚棘豆
Oxytropis falcata Bunge

多年生草本。无茎。托叶狭三角形，膜质；小叶对生或互生，11-19、25-31或27-47小叶，具白色长柔毛。总状花序具6-10花。荚果幼时浅红色，具柄，镰状长圆形，稍膨胀。花期5-8月，果期7-9月。生海拔2700-5200米的阳坡、河滩、河边草甸、沙石地、山谷或林下。产四川、西藏、甘肃、青海和新疆。蒙古亦有。

Perennial herbs. Acaulescent. Stipules narrowly triangular, membranous; leaflets opposite or alternate, either 11-19, 25-31 or 27-47 blades, with whitish long trichomes. Racemes 6-10-flowered. Legumes reddish when young, stipitate, falcate-oblong, slightly inflated. Fl. May-Aug. Fr. Jul-Sep. Sunny slopes, river flood plains, riverside meadows, sandy and stony areas, valleys or under forests at 2700-5200 m. Distributed in Sichuan, Xizang, Gansu, Qinghai and Xinjiang. Also in Mongolia.

少花米口袋 *Gueldenstaedtia verna*

少花米口袋
Gueldenstaedtia verna (Georgi) Boriss.

多年生草本。主根直而粗壮。茎短。奇数羽状复叶，具(5或)7-19小叶。伞形花序有2-6花；花冠紫红色、紫色、粉红色、玫瑰色或白色。荚果圆筒状，被长柔毛。花期4-5月，果期6-7月。生海拔2600米以下的山坡、草地、路旁或田边。产华中、华西、华东、华北和东北。印度、巴基斯坦、缅甸、老挝、俄罗斯和蒙古亦有。

Perennial herbs. Taproots straight and robust. Stems short. Leaves imparipinnate, (5 or)7-19-foliolate. Umbels with flowers 2-6; corolla purplish red, purple, pink, rose or white. Legumes cylindric, villose. Fl. Apr-May. Fr. Jun-Jul. Mountain slopes, grasslands, roadsides or by fields below 2600 m. Distributed in C, W, E, N and NE China. Also in India, Pakistan, Myanmar, Laos, Russia and Mongolia.

太行米口袋
Gueldenstaedtia taihangensis H. P. Tsui

多年生草本。叶具5-13小叶；小叶宽卵形至近圆形，长6-9毫米，先端截形、深缺、钝或急尖并具小尖头，基部圆形，两面疏被柔毛。伞形花序，具2-3花，总花梗纤细，开花时与叶近等长；花萼密被伏贴绒毛，5齿裂，上面两裂片较大；花冠紫堇色。荚果圆锥状，疏被柔毛，长约1.5厘米。花期5月，果期8月。产河北、山西和广西。

Perennial herbs. Leaves 5-13-foliolate; leaflets broadly ovate to suborbicular, 6-9 mm long, apex truncate, notched, obtuse or acute and with mucro; base rounded, both surfaces sparsely pilose. Umbel 2-3-flowered; rachis slender, subequal to leaves at anthesis; calyx densely appressed villous, 5-toothed; upper 2 teeth larger; corolla purple. Legume conical, pilose, ca. 1.5 cm long. Fl. May. Fr. Aug. Distributed in Hebei, Shanxi and Guangxi.

太行米口袋 *Gueldenstaedtia taihangensis*

黄花高山豆 *Tibetia tongolensis*

亚东高山豆 *Tibetia yadongensis*

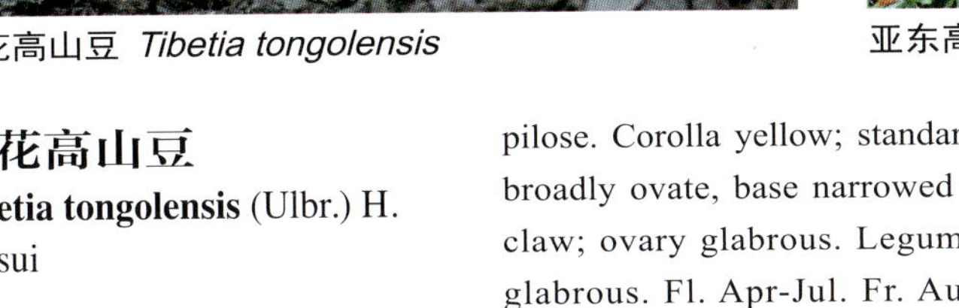

黄花高山豆

Tibetia tongolensis (Ulbr.) H. P. Tsui

多年生草本。茎纤细。托叶具紫褐色斑点；小叶5-9，倒卵形或宽椭圆形，叶上面常有小黑点、无毛，下面被疏柔毛。花黄色；旗瓣宽卵形，基部狭窄成爪；子房无毛。荚果无毛。花期4-7月，果期8-9月。生海拔3000米以上的丘陵。产云南和四川。

Perennial herbs. Stems slender. Stipules with brown-purple marks; leaflets 5-9, obovate or broadly elliptic, adaxially black-punctate, glabrous, abaxially pilose. Corolla yellow; standards broadly ovate, base narrowed to claw; ovary glabrous. Legumes glabrous. Fl. Apr-Jul. Fr. Aug-Sep. Hills above 3000 m. Distributed in Yunnan and Sichuan.

高山豆

Tibetia himalaica (Baker) H. P. Tsui

多年生草本，植株被毛密。茎明显。小叶9-12(-19)，下面被贴生长柔毛，先端圆或微裂。伞形花序具1-3花；花冠蓝紫色、紫色、蓝色、淡紫色、紫罗兰色或红色；子房被长柔毛。荚果具毛或渐无毛。花期5-6月，果期7-8月。生海拔3000-5000米的丘陵、高山草甸、石山坡或林中。产云南、四川、西藏、甘肃和青海。印度、尼泊尔、不丹和巴基斯坦亦有。

Perennial herbs, whole plants dense villose. Stems obvious. Leaflets 9-12(-19), abaxially apressed-villose, apex rotund or slightly lobed. Umbels 1-3-flowered; corolla bluish purple, purple, blue, mauve, violet or red; ovary villose Legumes pilose or glabrescent. Fl. May-Jun. Fr. Jul-Aug. Hilly areas, alpine meadows, rocky slopes or forests at 3000-5000 m. Distributed in Yunnan, Sichuan, Xizang, Gansu and Qinghai. Also in India, Nepal, Bhutan and Pakistan.

亚东高山豆

Tibetia yadongensis H. P. Tsui

多年生草本。茎长而分枝。托叶先端圆或渐尖；小叶7-15，椭圆形至倒心形，长7-10毫米，先端深裂直至几二裂状，两面被稀疏柔毛。伞形花序具1-2花；花萼狭钟状，棕色，疏被贴伏黄色柔毛，萼齿5，上2萼齿几合生至顶部，花冠紫色；子房密被长柔毛。荚果圆柱状，长2厘米。花期5-6月，果期7-9月。生海拔3000-4100米的山坡草地或灌丛。产西藏。

Perennial herbs. Stems long, branched. Stipules, apex rounded or acuminate; leaflets 7-15, elliptic to obcordate, 7-10 mm long, apex parted to bifid, both surfaces pilose. Umbel 1-2-flowered; calyx narrowly campanulate, brown, sparsely appressed yellow villose, teeth 5, 2 upper teeth almost connate to apex; corolla purple; ovary densely villous. Legume terete, 2 cm long. Fl. May-Jun. Fr. Jul-Sep. Grasslands on hills or thickets at 3000-4100 m. Distributed in Xizang.

高山豆 *Tibetia himalaica*

云南高山豆

Tibetia yunnanensis (Franch.) H. P. Tsui

多年生草本。茎发达，纤细。托叶抱茎，倒卵形，先端渐尖，边缘具牙齿状腺体；小叶3-7(-9)，倒卵形至倒心形，顶端截形至微缺，被贴伏柔毛。伞形花序具1-2(-3)花；花萼钟状，长4-5毫米，萼齿披针形，上2萼齿合生至2/3处，被长柔毛；花冠紫色；子房被长柔毛。花期6-7月，果期9月。生海拔2500米以上的山区。产云南、四川和西藏。

Perennial herbs. Stems very developed, slender. Stipules amplexicaul, obovate, apex acuminate, margin glandular toothed; leaflets 3-7(-9), obovate to obcordate, apex truncate to retuse, appressed pilose. Umbel 1-2(-3)-flowered; calyx campanulate, 4-5 mm long, teeth lanceolate, upper 2 teeth joined for up to 2/3 length, villous; corolla purple; ovary villous. Fl. Jun-Jul. Fr. Sep. Hilly areas at above 2500 m. Distributed in Yunnan, Sichuan and Xizang.

骆驼刺

Alhagi sparsifolia Shap.

亚灌木。茎直立，丛生。托叶钻状；叶全缘，顶端圆，有短尖头。总状花序腋生；花萼钟状，萼片三角形或钻状三角形；花冠深紫红色；旗瓣倒长卵形。荚果条形，常弯曲。花期6-8月，果期7-10月。生荒漠地区的沙地、河岸或农田。产中国西北。中亚亦有。

Subshrubs. Stems erect, caespitose. Stipules subulate; leaves entire, apex rounded and shortly acute. Racemes axillary; calyx campanulate, sepals triangular or subulate-triangular; corolla deeply purple-red; standard long obovate. Legumes linear, often curved. Fl. Jun-Aug. Fr. Jul-Oct. Sandy places on wastelands, river banks or by fields. Distributed in NW China. Also in C Asia.

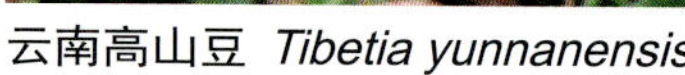

云南高山豆 *Tibetia yunnanensis*

骆驼刺 *Alhagi sparsifolia*

甘草

Glycyrrhiza uralensis Fisch. ex DC.

多年生草本。茎密被鳞片状腺点和白色或棕色绒毛。小叶5-17，密被腺点和短柔毛，叶全缘或微波状。总状花序多花；花冠紫色、白色或黄色。荚果镰刀状或弯曲呈环状，密生瘤状突起和刺毛状腺体。花期6-8月，果期7-10月。生海拔400-2700米的沙地、干旱河岸或山丘草地。产华北、西北和东北。中亚和东北亚亦有。

Perennial herbs. Stems densely scaly glandular punctate and white or brown tomentose; leaflets 5-17, densely glandular punctate and pubescent, margin entire or repand. Racemes much flowered; corolla purple, white or yellow. Legumes falcate or curved into a ring, densely tuberculate and glandular hairy. Fl. Jun-Aug. Fr. Jul-Oct. Sandy lands, dry river banks or grasslands on hills at 400-2700 m. Distributed in N, NW and NE China. Also in C and NE Asia.

甘草 *Glycyrrhiza uralensis*

胀果甘草 *Glycyrrhiza inflata*

胀果甘草
Glycyrrhiza inflata Batal.

多年生草本。根和根状茎强壮。小叶3-7(-9)。总状花序腋生，具多数疏生的花；上部2萼齿自基部合生至1/2；花冠紫色或淡紫色。荚果椭圆体形或矩圆形，直伸或稍弯曲。花期5-7月，果期6-10月。生海拔约1100米的河岸、田边或荒地。产内蒙古、甘肃和新疆。哈萨克斯坦、乌兹别克斯坦、吉尔吉斯斯坦、塔吉克斯坦、土库曼斯坦和蒙古亦有。

Perennial herbs. Roots and rhizomes robust. Leaflets 3-7(-9). Racemes axillary, with many sparsely arranged flowers; calyx with upper 2 teeth joined to 1/2 from base; corolla purple or light purple. Legumes ellipsoid or oblong, straight or slightly curved. Fl. May-Jul. Fr. Jun-Oct. River banks, fields or wastelands at ca. 1100 m. Distributed in Neimenggu, Gansu and Xinjiang. Also in Kazakhstan, Uzbekistan, Kyrgyzstan, Tajikistan, Turkmenistan and Mongolia.

刺果甘草
Glycyrrhiza pallidiflora Maxim.

多年生草本。羽状复叶，有小叶9-15。总状花序腋生，密集成头状；苞片膜质，具腺点；花萼钟状，密被腺点；花冠淡紫色、紫色或淡紫红色；旗瓣卵圆形。荚果卵球形，顶端尖，具刚硬的刺。种子2粒。花期6-7月，果期7-9月。生海拔2600-3100米的河滩地、旷野或路旁。产华北、华东和东北。俄罗斯(远东地区)和蒙古亦有。

Perennial herbs. Leaves pinnate, with 9-15 leaflets. Racemes axillary, aggregated in heads; bracts membranous, glandular; calyx campanulate, densely glandular; corolla pale purple, purple or pale purple-red; standards ovate. Legumes ovoid, apex acute, thorny. Seeds 2. Fl. Jun-Jul. Fr. Jul-Sep. Flood lands, wilderness or roadsides at 2600-3100 m. Distributed in N, E and NE China. Also in Russia (Far East) and Mongolia.

刺果甘草 *Glycyrrhiza pallidiflora*

云南甘草

Glycyrrhiza yunnanensis S. H. Cheng et L. K. Dai ex P. C. Li

多年生草本。小叶披针形或卵状披针形，密具鳞片状腺点，疏具柔毛。总状花序球形或近球形；花冠紫色；旗瓣狭卵形或椭圆形。荚果长卵球形，先端尖凸，密具硬刺。花期5-6月，果期7-9月。生海拔约2700米的林边、开阔草地或路旁。产云南。

Herbs, perennial. Leaflets lanceolate or ovate-lanceolate, densely scaly glandular punctate, sparsely pubescent. Racemes globose or subglobose; corolla purple; standard narrowly ovate or elliptic. Legumes long-ovoid, cuspidate at apex, densely rigidly spiny. Fl. May-Jun. Fr. Jul-Sep. Forest edges, open grassy slopes or roadsides at ca. 2700 m. Distributed in Yunnan.

细枝山竹子

Corethrodendron scoparium (Fischer et C. A. Meyer) Fischer et Basiner

亚灌木。茎被纤维状条纹。基部的叶具7-11小叶，上部的具3-5小叶，有时无；小叶条状长圆形至狭披针形，长1.5-3厘米，下面被长柔毛。总状花序具多花；总花梗长于叶；花萼钟形，被短柔毛，萼齿不等大；花冠紫色。荚果具2-4果荚，密被白色绒毛。花期6-9月，果期8-10月。生海拔600-1100米的沙漠或半沙漠。产中国西北。哈萨克斯坦和蒙古亦有。

Shrublets. Stems with fibrous stripping. Basal leaves 7-11-foliolate, apical ones 3-5-foliolate or sometimes none; leaflets linear-oblong to narrowly lanceolate, 1.5-3 cm long, abaxially pilose. Racemes with many flowers; peduncle longer than leaves; calyx campanulate, pubescent, teeth unequal; corolla purple. Legume with 2-4 articles, densely white lanate. Fl. Jun-Sep. Fr. Aug-Oct. Deserts or semideserts at 600-1100 m. Distributed in NW China. Also in Kazakhstan and Mongolia.

红花山竹子

Corethrodendron multijugum (Maximowicz) B. H. Choi et H. Ohashi

亚灌木。托叶卵状披针形，基部合生；小叶15-29，宽卵形至近圆形，长5-8(-15)毫米，上面无毛，下面被短柔毛。总状花序具多花；总花梗明显长于叶；花萼斜钟形，长5-6毫米，被短柔毛；花冠紫色、粉紫色，稀白色。荚果具2-3果荚，被短柔毛和皮刺。花期6-8月，果期8-9月。生海拔500-3800米的砾石地或多石山坡。产华北、西北、华中和西藏。

Shrublets. Stipules ovate-lanceolate, basally connate; leaflets 15-29, broadly ovate to suborbicular, 5-8(-15) mm long, adaxially glabrous, abaxially pubescent. Racemes with many flowers; peduncle conspicuously longer than leaves; calyx obliquely campanulate, 5-6 mm long, pubescent; corolla purple, pinkish purple, rarely white. Legume with 2-3 articles, pubescent and prickly. Fl. Jun-Aug. Fr. Aug-Sep. Gravelly areas or stony slopes at 500-3800 m. Distributed in N,

云南甘草 *Glycyrrhiza yunnanensis*

细枝山竹子 *Corethrodendron scoparium*

红花山竹子 *Corethrodendron multijugum*

NW and C China, and Xizang.

山竹子

Corethrodendron fruticosum

(Pallas) B. H. Choi et H. Ohashi

小灌木。茎丛生，直立。叶具11-19小叶；小叶椭圆形至长圆形，1.4-2.2 × 0.3-0.6厘米。总状花序疏松，具多花；花萼钟形；萼齿三角形；花冠紫色。荚果开裂为2或3节；膨大，具柔毛，有或无刺。花期7-8月，果期8-9月。生海拔600-1100米的典型草原沙地。产内蒙古东北部、辽宁西部、吉林和黑龙江。俄罗斯东部和蒙古东部亦有。

Shrublets. Stems caespitose, erect. Leaves 11-19-foliolate; leaflets elliptic to oblong, 1.4-2.2 × 0.3-0.6 cm. Racemes lax, with many flowers; calyx campanulate; teeth triangular; corolla purple. Legumes divided into 2 or 3 articles, inflated, pubescent, prickly or not. Fl. Jul-Aug. Fr. Aug-Sep. Sandy areas in steppes at 600-1100 m. Distributed in NE Neimenggu, W Liaoning, Jilin and Heilongjiang. Also in E Russia and E Mongolia.

山竹子 *Corethrodendron fruticosum*

多序岩黄耆

Hedysarum polybotrys

Hand.-Mazz.

多年生草本。托叶披针形；叶长5-9厘米；小叶11-19，卵状披针形至卵状长圆形，仅下面被贴伏柔毛。总状花序具多数花；花萼斜宽钟状，被短柔毛，萼齿三角状钻形，下萼齿长为上萼齿的2倍；花冠淡黄色，翼瓣等于或稍长于旗瓣。荚果疏被短柔毛，节荚近圆形或宽卵形，边缘狭且全缘。花期7-8月，果期8-9月。生沙砾石质山坡。产甘肃和四川。

Perennial herbs. Stipules lanceolate; leaves 5-9 cm long, leaflets 11-19, ovate-lanceolate to ovate-oblong, only abaxially with appressed trichomes. Racemes with many flowers; calyx obliquely campanulate, pubescent; teeth triangular-subulate, most abaxial tooth ca. 2 × as long as others; corolla yellow, wings as long as to slightly longer than standard. Legume sparsely pubescent, articles subglobose or broadly ovoid, margin narrow and entire. Fl. Jul-Aug. Fr. Aug-Sep. Gravelly and stony slopes. Distributed in Gansu and Sichuan.

多序岩黄耆 *Hedysarum polybotrys*

拟蚕豆岩黄耆 *Hedysarum ussuriense*

拟蚕豆岩黄耆

Hedysarum ussuriense Schischk. et Kom.

多年生草本。茎簇生，直立。小叶11-19，下面沿叶脉被短柔毛，上面无毛。总状花序密具多花；萼上面齿长为萼筒的0.7-1.5倍，其他4个三角形，长为萼筒的1/4-1/3；花冠黄色。荚果开裂为2或4节。花期7-8月，果期8-9月。生海拔2500-3200米的林中或亚高山草甸。产四川、河北、吉林和辽宁。俄罗斯和朝鲜半岛北部亦有。

Perennial herbs. Stems caespitose, erect. Leaflets 11-19, abaxially pubescent along veins, adaxially glabrous. Racemes dense, with many flowers; calyx most abaxial tooth 0.7-1.5 × as long as tube, other 4 teeth triangular and 1/4-1/3 as long as tube; corolla yellow. Legumes divided into 2 or 4 articles. Fl. Jul-Aug. Fr. Aug-Sep. Forests or subalpine meadows at 2500-3200 m. Distributed in Sichuan, Hebei, Jilin and Liaoning. Also in Russia and N Korean Peninsula.

湿地岩黄耆

Hedysarum inundatum Turcz.

多年生草本。托叶披针形；叶长7-12厘米，小叶11-17，卵形至长圆形，下面被疏柔毛，上面无毛。总状花序具多数花；花长14-18毫米；花萼钟形，被柔毛，萼齿三角形，长为萼筒的1半；花冠紫红色。荚果3-4节，节间明显缢缩，节荚椭圆形，边缘全缘。花期6-7月，果期7-8月。生海拔2500-3000米的亚高山草甸。产河北和山西。俄罗斯(西伯利亚)亦有。

Perennial herbs. Stipules lanceolate; leaves 7-12 cm long, leaflets 11-17, ovate to oblong, abaxially sparsely pubescent, adaxially glabrous. Racemes with many flowers; flowers 14-18 mm long; calyx campanulate, pubescent, teeth triangular, ca. 1/2 as long as tube; corolla purple. Legume divided into 3-4 loments, inconspicuously constricted between articles, articles ellipsoid, margin entire. Fl. Jun-Jul. Fr. Jul-Aug. Subalpine meadows at 2500-3000 m. Distributed in Hebei and Shanxi. Also in Russia (Siberia).

湿地岩黄耆 *Hedysarum inundatum*

山岩黄耆 *Hedysarum alpinum*

疏忽岩黄耆 *Hedysarum neglectum*

山岩黄耆

Hedysarum alpinum L.

多年生草本，高50-100厘米。叶长8-12厘米，具9-21小叶；小叶卵状椭圆形、狭椭圆形或倒卵状披针形。总状花序紧密、多花；花冠紫红色。荚果分为2或3(-5)个节荚。花果期7-8月。生沼泽草甸、北方针叶林、多石山坡或灌丛中。产中国西南、华北、华西和东北。西南亚、东北亚、欧洲东北部和北美洲亦有。

Perennial herbs, 50-100 cm tall. Leaves 8-12 cm long, 9-21-foliolate; leaflets ovate-oblong, narrowly elliptic or oblong-lanceolate. Racemes dense, with many flowers; corolla purple-red. Legumes divided into 2 or 3(-5) loments. Fl. and fr. Jul-Aug. Swampy meadows, taiga forests, stony slopes or scrubs. Distributed in SW, N, W and NE China. Also in SW and NE Asia, NE Europe and North America.

疏忽岩黄耆

Hedysarum neglectum Ledeb.

多年生草本。托叶披针形；叶长8-12厘米；小叶11-15，长卵形至卵状长圆形，宽8-14毫米，仅下面被疏柔毛。总状花序具多花；花萼钟状，被短柔毛，萼齿三角状披针形，与萼筒近等长；花冠紫色。节荚圆形或卵形，被贴伏疏柔毛，边缘全缘。花期7-8月，果期8-9月。生海拔1200-2600米的高山草甸、灌丛和林中。产新疆。哈萨克斯坦、蒙古和俄罗斯亦有。

Perennial herbs. Stipules lanceolate; leaves 8-12 cm long; leaflets 11-15, long ovate to ovate-oblong, 8-14 mm broad, only abaxially sparsely pubescent. Racemes with many flowers; calyx campanulate, pubescent, teeth triangular-lanceolate, as long as tube; corolla purple. Articles globose or ovoid, appressed pubescent, margin entire. Fl. Jul-Aug. Fr. Aug-Sep. Alpine meadows, scrubs or forests at 1200-2600 m. Distributed in Xinjiang. Also in Kazakhstan, Mongolia and Russia.

短翼岩黄耆

Hedysarum brachypterum Bunge

多年生草本。茎丛生，匍匐。叶具11-19小叶；小叶下面具平伏柔毛。总状花序密集，多花；花冠紫色；翼瓣长约为旗瓣的2/5。荚果2-4节荚，密被柔毛，具刺。花期5-8月，果期7-9月。生海拔600-800米的草原。产河北、内蒙古和宁夏。

Perennial herbs. Stems caespitose, decumbent. Leaves 11-19-foliolate; leaflets abaxially appressed pubescent. Racemes dense, with many flowers; corolla purple; wings ca. 2/5 as long as standard. Legumes divided into 2-4 articles, densely pubescent, prickly. Fl. May-Aug. Fr. Jul-Sep. Steppes at 600-800 m. Distributed in Hebei, Neimenggu and Ningxia.

短翼岩黄耆 *Hedysarum brachypterum*

藏豆 *Stracheya tibetica*

藏豆

Stracheya tibetica Benth.

多年生草本。茎短缩，不明显。托叶卵形；叶长4-7厘米；小叶11-15，长卵形至椭圆形，下面被贴伏毛，上面无毛。总状花序伞房状，具3-6朵花；花长17-20毫米；花萼斜钟状，被长柔毛，萼齿披针形；花冠淡红色；子房线形。荚果边缘和沿两侧中线具1-1.5毫米长的皮刺。花期7-8月，果期8-9月。生海拔4000-4600米的高山草甸。产西藏和青海。南亚亦有。

Perennial herbs. Stems abbreviated, inconspicuous. Stipules ovate; leaves 4-7 cm long; leaflets 11-15, long ovate to elliptic, abaxially appressed pubescent, adaxially glabrous. Racemes corymbose, with 3-6 flowers; flowers 17-20 mm long; calyx obliquely campanulate, pilose, teeth lanceolate; corolla reddish; ovary linear. Legume with 1-1.5 mm prickles along margin and lateral midline. Fl. Jul-Aug. Fr. Aug-Sep. Alpine meadows at 4000-4600 m. Distributed in Xizang and Qinghai. Also in S Asia.

顿河红豆草

Onobrychis tanaitica Spreng.

多年生草本。茎多数，直立。小叶9-13。总状花序紧密排列成穗状，多花；花冠粉紫色；旗瓣倒卵形；翼瓣长约为旗瓣的1/4。荚果有一近圆形荚节，具乳突。花期6-7月，果期7-8月。生海拔1400-1800米的山地草甸或苔原灌丛。产新疆。中亚、西南亚和欧洲东南部亦有。

Herbs. perennial. Stems numerous, erect. Leaves with 9-13 leaflets. Racemes densely spicate, with many flowers; corolla pinkish purple; standard obovate; wings ca. 1/4 as long as standard. Legumes with one subglobose loment, papillose-prickly. Fl. Jun-Jul. Fr. Jul-Aug. Meadows or scrubs on steppes at 1400-1800 m. Distributed in Xinjiang. Also in C and SW Asia, and SE Europe.

驴食草

Onobrychis viciifolia Scop.

多年生草本。茎直立。复叶具13-19小叶；小叶倒卵状披针形至披针形。总状花序密穗状，多花；花冠粉紫色；旗瓣倒卵形；翼瓣长约为旗瓣的1/4。荚果1个节荚；节荚半圆形，上部具刺。花期6-7月，果期7-8月。栽培于华北、华西和西北。可能原产欧洲中部。

Perennial herbs. Stems erect; 13-19-foliolate; leaflets oblong-lanceolate to lanceolate. Racemes densely spicate, with many flowers; corolla pinkish purple; standard obovate; wings ca. 1/4 as

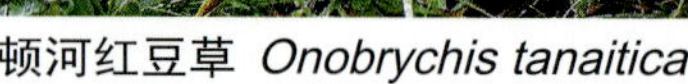

顿河红豆草 *Onobrychis tanaitica*

驴食草 *Onobrychis viciifolia*

百脉根 *Lotus corniculatus*

long as standard. Legumes with 1 loment; loment subglobose, adaxially with prickles. Fl. Jun-Jul. Fr. Jul-Aug. Cultivated in N, W and NW China. Probably native to C Europe.

百脉根

Lotus corniculatus L.

多年生草本，全株散生稀疏白色柔毛或无毛。伞形花序，具3-7花；花长(9-)10-18毫米；萼齿几等长；花冠黄色或部分或全部橙红色，干后常蓝黑色。荚果直，褐色。花期5-9月，果期7-10月。生山坡、草地、旷野或河岸边。产中国西南、华中和西北。亚洲、澳大利亚、欧洲和北美洲亦有。

Perennial herbs, scattered sparsely white puberulous or glabrescent. Umbels 3-7-flowered; flowers (9-)10-18 mm long; calyx teeth almost equal in length; corolla yellow or partly or wholly orange-red, often bluish black when dry. Legumes straight, brown. Fl. May-Sep. Fr. Jul-Oct. Slopes, grasslands, fields or river banks. Distributed in SW, C and NW China. Also in Asia, Australia, Europe and North America.

绣球小冠花

Securigera varia (L.) Lassen

多年生草本。托叶披针形；小叶11-17(-25)，椭圆形或长圆形，先端具短尖头，基部近圆形，两面无毛。伞形花序腋生，比叶短；总梗疏生小刺；花5-10(-20)，排列成绣球状；花冠紫色、淡红色或白色，有紫色条纹。荚果狭圆柱形，具4棱，先端有宿存的喙状花柱。花期6-7月，果期8-9月。中国东北南部有栽培。原产地中海地区。

Perennial herbs. Stipules lanceolate; leaflets 11-17(-25), elliptic or oblong, apex mucronate, base subrounded, glabrous on both surfaces. Umbels axillary, shorter than leaves; rachis with sparse prickles; flowers 5-10(-20), arranged into Hydrangea-shaped; corolla purple, reddish or white, with purple stripes. Legume narrowly cylindrical, quardrangulate, with persistent beaked style at apex. Fl. Jun-Jul. Fr. Aug-Sep. Cultivated in the south of NE China. Native to Mediterranean.

绣球小冠花 *Securigera varia*

广布野豌豆
Vicia cracca L.

广布野豌豆 *Vicia cracca*

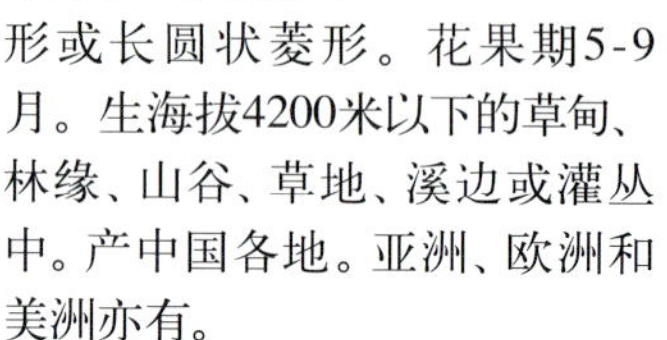

多年生草本。偶数羽状复叶；叶轴顶端具2-3分叉的卷须；小叶条形、条状披针形或长圆形，11-30 × 2-4毫米。总状花序与叶近等长，具10-40花；花冠紫色、蓝紫色或紫红色；旗瓣提琴形。荚果长圆形或长圆状菱形。花果期5-9月。生海拔4200米以下的草甸、林缘、山谷、草地、溪边或灌丛中。产中国各地。亚洲、欧洲和美洲亦有。

Perennial herbs. Leaves paripinnate; tendrils 2-3 branched at rhachis apex; leaflets linear, linear-lanceolate or oblong, 11-30 × 2-4 mm. Racemes subequal leaves, 10-40-flowered; corolla purple, blue-purple or purple-red; standard violin-shaped. Legumes oblong or oblong-rhomboid. Fl. and fr. May-Sep. Meadows, forest edges, valleys, grasslands, streamsides or thickets below 4200 m. Widely distributed in China. Also in Asia, Europe and America.

大叶野豌豆 *Vicia pseudorobus*

长柔毛野豌豆
Vicia villosa Roth

一年生草本，植株被长柔毛。偶数羽状复叶；小叶4-12对，条形至长圆形或披针形。总状花序短于或稍长于叶，具10-30花；旗瓣长圆形，中间缢缩；翼瓣短于旗瓣和龙骨瓣。荚果长圆状菱形，侧扁，先端具喙。花果期4-10月。中国大部分地区有栽培。原产伊朗、中亚和欧洲。

Annual herbs, whole plant villose. Leaves paripinnate; leaflets 4-12-paired, linear to oblong or lanceolate. Racemes shorter than or slightly longer than leaf, 10-30-flowered; standard oblong, constricted at middle; wings shorter than standards and keels. Legumes oblong-rhomboid, depressed, apex beaked. Fl. and fr. Apr-Oct. Cultivated in most parts of China. Native to Iran, C Asia and Europe.

大叶野豌豆
Vicia pseudorobus Fisch. ex C. A. Meyer

多年生草本。偶数羽状复叶；卷须发达，托叶长0.8-1.5厘米；小叶2-5对，卵形、椭圆形或狭披针形，长(2-)3-6(-10)厘米，宽1-3.5厘米，先端圆或渐尖，有短尖头；侧脉清晰。总状花序长于叶，具15-30花；花萼斜钟状；花长1-2厘米。花期6-9月，果期7-10。生海拔400-3000米的山坡、灌丛或林中。产中国东北、华北、西北、华东、华中和西南。俄罗斯、蒙古、朝鲜半岛和日本亦有。

Perennial herbs. Leaves paripinnate; tendril well-developed, stipules 0.8-1.5 cm long; leaflets 2-5-paired, ovate, elliptic or narrowly lanceolate, (2-)3-6(-10) × 1-3.5 cm, apex obtuse or acuminate, mucronate; lateral veins obvious. Raceme longer than leaf, 15-30-flowered; calyx obliquely campanulate; flowers 1-2 cm long. Fl. Jun-Sep. Fr. Jul-Oct. Hill slopes, thickets or forests at 400-3000 m. Distributed in NE, N, NW, E, C and SW China. Also in Russia, Mongolia,

长柔毛野豌豆 *Vicia villosa*

山野豌豆 *Vicia amoena*

西藏野豌豆 *Vicia tibetica*

Korean Peninsula and Japan.

山野豌豆
Vicia amoena Fisch. ex Ser.

多年生草本。叶无柄，偶数羽状复叶；小叶4-7对，椭圆形至卵状披针形，上面具平伏长柔毛。总状花序常长于叶，密具10-30花；花冠蓝色、蓝紫色、紫红色或白色。荚果长圆形。花期4-9月，果期7-10月。生海拔4000米以下的草地、山坡、灌丛或林中。产华南以外中国大部分地区。俄罗斯、蒙古、朝鲜半岛和日本亦有。

Perennial herbs. Leaves subsessile, paripinnate; leaflets 4-7-paired, elliptic to ovate-lanceolate, adaxially appressed villous. Racemes usually longer than leaf, densely 10-30-flowered; corolla blue, blue-purple, red-purple or white. Legumes oblong. Fl. Apr-Sep. Fr. Jul-Oct. Grasslands, slopes, thickets or forests below 4000 m. Distributed in most parts of China, except S China. Also in Russia, Mongolia, Korean Peninsula and Japan.

西藏野豌豆
Vicia tibetica C. A. C. Fisch.

多年生草本。偶数羽状复叶，卷须有2-3分枝；托叶三角形，具3-5齿；小叶3-6对，长圆形，长(0.4-)1-2厘米，宽(0.15-)0.3-0.7厘米，先端圆，具短尖头；叶脉密致，两面凸出。总状花序长于叶，具4-13花；花萼斜钟状；花冠红色、紫红色或淡蓝色。花果期4-9月。生海拔1300-4300米的高山松林、山坡草地或灌丛。产青海、西藏、四川和云南。不丹和印度亦有。

Perennial herbs. Leaves paripinnate; tendril 2-3-branched; stipules triangular, 3-5-toothed; leaflets 3-6-paired, oblong, (0.4-)1-2 × (0.15-)0.3-0.7 cm, apex obtuse and mucronate; veins dense, raised on both surfaces. Raceme longer than leaf, 4-13-flowered; calyx obliquely campanulate; corolla red, purple-red or bluish. Fl. and fr. Apr-Sep. Alpine pines, hill slopes, or scrubs at 1300-4300 m. Distributed in Qinghai, Xizang, Sichuan and Yunnan. Also in Bhutan and India.

确山野豌豆
Vicia kioshanica L. H. Bailey

多年生草本。偶数羽状复叶；小叶3-7对。总状花序柔软而弯曲，具6-20朵疏散排列的花；花冠长7-8(-14)毫米，强烈向前弯曲，中部形成90°角。荚果菱形或长圆形。花期4-6月，果期6-9月。生海拔100-1000米的山坡、山谷、田边或路边。产华中、华北、华西和华东。

Perennial herbs. Leaves paripinnate; leaflets 3-7-paired. Racemes weak and curved, sparsely 6-20-flowered; corolla 7-8(-14) mm long, strongly bent upward at middle forming a 90° angle. Legumes rhomboid or oblong. Fl. Apr-Jun. Fr. Jun-Sep. Slopes, valleys, fieldsides or roadsides at 100-1000 m. Distributed in C, N, W and E China.

确山野豌豆 *Vicia kioshanica*

大野豌豆 *Vicia gigantea*

大野豌豆
Vicia gigantea Bge.

多年生草本，被白色柔毛。偶数羽状复叶；卷须2-3分枝或单一；托叶长约0.6厘米；小叶3-6对，椭圆形或倒卵状椭圆形，长1.5-3(-3.5)厘米，宽0.6-1.7(-2)厘米，先端钝，具短尖头。总状花序长于叶；具花6-16朵；花萼钟状；花冠白色、粉红色或紫色，长约0.7厘米。花期6-8月，果期7-10月。生海拔600-3000米的林中、灌丛、草地及山坡。产华北、华中和西南。

Perennial herbs, white pilose. Leaves paripinnate; tendril 2-3-branched or unbranched; stipules ca. 0.6 cm long; leaflets 3-6-paired, elliptic or obovate-elliptic, 1.5-3(-3.5) × 0.6-1.7(-2) cm, apex obtuse, mucronate. Raceme longer than leaf, 6-16-flowered; calyx campanulate; corolla white, pink or purple, ca. 0.7 cm long. Fl. Jun-Aug. Fr. Jul-Oct. Forests, scrubs, grasslands and slopes at 600-3000 m. Distributed in N, C and SW China.

新疆野豌豆
Vicia costata Ledeb.

多年生草本。偶数羽状复叶；小叶3-8对，长圆状披针形或椭圆形。总状花序明显长于叶，具3-11花；花冠黄色、淡黄色或白色；旗瓣倒卵形，中部缢缩。荚果条形。种子1-4粒。花果期6-8月。生海拔500-3700米的山坡、砾石地、沙地或沙漠。产中国西南、华北、西北和东北。哈萨克斯坦、俄罗斯(阿尔泰、图瓦)和蒙古亦有。

Perennial herbs. Leaves paripinnate; leaflets 3-8-paired, oblong-lanceolate or elliptic. Racemes obviously longer than leaf, 3-11-flowered; corolla yellow, light yellow or white; standard obovate, constricted at middle. Legumes linear. Seeds 1-4. Fl. and fr. Jun-Aug. Hill slopes, gravels, sandy lands or deserts at 500-3700 m. Distributed in SW, N, NW and NE China. Also in Kazakhstan, Russia (Altay, Tuva) and Mongolia.

北野豌豆
Vicia ramuliflora (Maxim.) Ohwi

多年生草本。偶数羽状复叶；小叶常3对，披针形或卵状披针形。总状花序腋生，2-3分枝，短于叶，具4-9朵疏散排列的花；花冠蓝色、蓝紫色或紫红色。荚果长圆状菱形。花果期6-9月。生海拔200-1600米的山坡、草地或林中。产华北和东北。俄罗斯、蒙古、朝鲜半岛和日本亦有。

Perennial herbs. Leaves paripinnate; leaflets usually 3-paired, lanceolate or ovate-lanceolate. Racemes axillary, 2-3-branched, shorter than leaf, sparsely 4-9-flowered; corolla blue, blue-purple or purple-red. Legumes oblong-rhomboid. Fl. and fr. Jun-Sep. Slopes, grasslands or forests at 200-1600 m. Distributed in N and NE China. Also in Russia, Mongolia, Korean Peninsula and Japan.

新疆野豌豆 *Vicia costata*

北野豌豆 *Vicia ramuliflora*

歪头菜 *Vicia unijuga*

歪头菜

Vicia unijuga A. Braun

多年生草本。偶数羽状复叶；小叶1对，卵状披针形，革质或厚纸质。总状花序单生，稀具分枝，近圆柱形，常明显长于叶；花萼渐无毛；花冠深或浅蓝色、紫色至红色。荚果长圆形，扁平，无毛。花期6-9月，果期7-10月。生海拔4000米以下的疏林、山坡、草地、溪边或灌丛中。产中国大部分地区。俄罗斯、蒙古、朝鲜半岛和日本亦有。

Perennial herbs. Leaves paripinnate; leaflets 1-paired, ovate-lanceolate, leathery or thick papery. Racemes single, rarely branched, ± cylindric, usually obviously longer than leaf; calyx glabrescent; corolla deep or light blue, purple to red. Legumes oblong, flat, glabrous. Fl. Jun-Sep. Fr. Jul-Oct. Open forests, slopes, grasslands, stream banks or thickets below 4000 m. Distributed in most parts of China. Also in Russia, Mongolia, Korean Peninsula and Japan.

大花野豌豆

Vicia bungei Ohwi

一年生或二年生草本。偶数羽状复叶；托叶半戟形，长3-7毫米，边缘具齿；小叶3-5对，条状长圆形、长圆形或狭长圆状倒披针形至长圆状倒卵形。总状花序具花2-5朵；花冠红紫色、蓝紫色或淡紫罗兰色。荚果扁长圆形。花期4-7月，果期5-8月。生海拔4200米以下的山坡、山谷、草地或路边。产中国西南、华北、华西、华东和东北。朝鲜半岛亦有。

Annual or biennial herbs. Leaves pari pinnate; stipules semihastate, 3-7 mm long, margin toothed; leaflets 3-5-paired, linear-oblong, oblong or narrowly oblong-oblanceolate to oblong-obovate. Racemes 2-5-flowered; corolla red-purple, blue-purple or pale violet. Legumes depressed oblong. Fl. Apr-Jul. Fr. May-Aug. Slopes, valleys, grasslands or roadsides below 4200 m. Distributed in SW, N, W, E and NE China. Also in Korean Peninsula.

大花野豌豆 *Vicia bungei*

四籽野豌豆 *Vicia tetrasperma*

四籽野豌豆
Vicia tetrasperma (L.) Schreb.

一年生草本。茎纤细，缠绕，多分枝。小叶2-6对。总状花序，具小花1-2；花冠淡蓝色、淡紫色、淡玫瑰色或白色，长4-8毫米。荚果长圆形，无毛。种子4。花期2-8月，果期3-8月。生海拔2900米以下的山坡、山谷、田野、荒地、路边或草地。产中国除东北以外大部分地区。亚洲、欧洲、北美洲和北非亦有。

Annual herbs. Stems slender, twining, much-branched. Leaflets 2-6-paired. Racemes with 1-2 very small flowers; corolla pale blue, pale violet, pale rose or white, 4-8 mm long. Legumes oblong, glabrous. Seeds 4. Fl. Feb-Aug. Fr. Mar-Aug. Slopes, valleys, fields, wastelands, roadsides or grasslands below 2900 m. Distributed in most parts of China, except NE China. Also in Asia, Europe, North America and N Africa.

小巢菜
Vicia hirsuta (L.) S. F. Gray

一年生草本。茎细柔，有棱，近无毛。偶数羽状复叶；托叶半戟形至披针形；小叶4-8对，条形或狭长圆形，无毛；卷须有分枝。总状花序明显短于叶，密具2-4(-7)花；花冠白色至浅紫色。荚果长圆状菱形，具粗毛。种子扁圆形。花期2-6月，果期2-8月。生海拔2900米以下的山谷、河滩、田边或路旁草丛中。产中国西南、华南、华西、华东和西北。西南亚、中亚、东亚、北欧和北美洲亦有。

Annual herbs. Stems gracilis, ribbed, almost glabrous. Leaves paripinnate; stipules semisagittate or lanceolate; leaflets 4-8-paired, linear or narrowly oblong, glabrous; tendril branched. Racemes obviously shorter than leaf, densely 2-4(-7)-flowered; corolla white to light purple. Legumes oblong-rhomboid, hirsute. Seeds oblate. Fl. Feb-Jun. Fr. Feb-Aug. Valleys, flood lands, fieldsides or grasslands along roads below 2900 m. Distributed in SW, S, W, E and NW China. Also in SW, C and E Asia, N Europe and North America.

小巢菜 *Vicia hirsuta*

救荒野豌豆
Vicia sativa L.

一年生或二年生草本。茎被微柔毛。小叶长圆状楔形至倒心形，先端截形或凹缺。花紫红色或红色，长18-30毫米；萼齿等长或稍长于萼筒。荚果条状长圆形，褐色或黄褐色，种子间收缩，常具毛。花期1-8月，果期2-9月。栽培或野生海拔3000米以下的林中、山坡、草地、农地、田野、荒地或路边。产中国大部分地区。世界广布。

Annual or biennial herbs. Stems slightly pubescent. Leaflets oblong-cuneate to obcordate, apex truncate or emarginate. Flowers purple-red or red, 18-30 mm long; calyx teeth equaling or lon-

救荒野豌豆 *Vicia sativa*

窄叶野豌豆 *Vicia sativa* subsp. *nigra*

蚕豆 *Vicia faba*

ger than calyx tube. Legumes linear-oblong, brown or yellow-brown, contracted between seeds, usually hairy. Fl. Jan-Aug. Fr. Feb-Sep. Cultivated or naturalized in forests, slopes, grasslands, farms, fields, wastelands or roadsides below 3000 m. Distributed in most parts of China. Also in Worldwide.

窄叶野豌豆

Vicia sativa L. subsp. **nigra** Ehrh.

本亚种与救荒野豌豆的区别在于本亚种的小叶条形至长圆状楔形，先端锐尖、钝或截形。花冠长(8-)10-18毫米；萼齿短于萼筒。荚果黑色或黑褐色，种子间不收缩，常无毛。花期3-7月，果期3-9月。生海拔200-3700米的河岸、湿草地、山谷或田边。产中国西南、华南和西北。中亚、西南亚、欧洲和非洲亦有。各地温带区域均有引种和归化。

This subspecies differs from the typical variety in its leaflets linear to oblong-cuneate, apex acute, obtuse or truncate. Corolla (8-)10-18 mm long; calyx teeth shorter than calyx tubes. Legumes black or brownish black, not contracted between seeds, usually glabrous. Fl. Mar-Jul. Fr. Mar-Sep. River banks, damp grasslands, valleys or fieldsides at 200-3700 m. Distributed in S, SW and NW China. Also in C and SW Asia, Europe and Africa. Introduced and naturalized in temperate regions elsewhere.

蚕豆

Vicia faba L.

一年生草本。茎粗壮，直立，无毛。偶数羽状复叶；小叶通常1-5对，长圆形、椭圆形或倒卵形，互生。花常2-4(-6)腋生成簇；花冠白色，带紫色纹理。荚果粗壮，50-100 × 20-30毫米，具绒毛。花果期3-9月。中国大部分地区有栽培。原产地中海沿岸、西南亚和北非；世界广栽培。

Annual herbs. Stems robust, erect, glabrous. Leaves paripinnate; leaflets usually 1-5 pairs, oblong, elliptic or obovate, alternate. Flowers 2-4(-6) in axillary fascicles; corolla white with purple veins. Legumes stout, 50-100 × 20-30 mm, tomentose. Fl. and fr. Mar-Sep. Cultivated in most parts of China. Native to Mediterranean coastal, SW Asia and N Africa; widely cultivated in the world.

大山黧豆

Lathyrus davidii Hance

多年生草本。叶具粗壮的具分枝的卷须；托叶大，与小叶近等长或等长；小叶(2或)3或4(或5)对，常卵形，40-70 × 50-110毫米，无毛。总状花序腋生，具花10-40。荚果条形，橙褐色。花期5-7月，果期8-10月。生海拔1800米以下的林缘、灌丛中或山坡。产华中、华北、华东和东北。俄罗斯、朝鲜半岛和日本亦有。

Perennial herbs. Leaves with strong, branched tendrils; stipules large, often similar to leaflets; leaflets (2 or)3-or 4(or 5)-paired, usually ovate, 40-70 × 50-110 mm, glabrous. Racemes axillary, 10-40-flowered. Legumes linear, orange-brown. Fl. May-Jul. Fr. Aug-Oct. Forest edges, thickets or slopes below 1800 m. Distributed in C, N, E and NE China. Also in Russia, Korean Peninsula and Japan.

大山黧豆 *Lathyrus davidii*

大托叶山黧豆 *Lathyrus pisiformis*

毛山黧豆 *Lathyrus palustris* var. *pilosus*

大托叶山黧豆

Lathyrus pisiformis L.

多年生草本。茎具翅，无毛。托叶卵形或椭圆形，长3.5-6.5厘米；小叶3-5对，狭卵形、卵状披针形或椭圆状披针形，长5.5-9厘米，两面无毛，先端具分枝的卷须。总状花序腋生，有花8-14；花萼钟形，无毛，萼齿不等长；花红紫色。荚果长4.5厘米。花期5-6月，果期7-8月。生海拔1100-1500米的林中、河谷或河边。产新疆。俄罗斯、欧洲中部和东部亦有。

Perennial herbs. Stem winged, glabrous. Stipules ovate or elliptic, 3.5-6.5 cm; leaflets 3-5-paired, narrowly ovate, ovate-lanceolate or elliptic-lanceolate, 5.5-9 cm long, glabrous, with branched tendril at apex. Raceme axillary, 8-14-flowered; calyx campanulate, glabrous, unequally toothed; corolla red-purple. Legume 4.5 cm long. Fl. May-Jun. Fr. Jul-Aug. Forests, valleys or riverbanks at 1100-1500 m. Distributed in Xinjiang. Also in Russia, C and E Europe.

矮山黧豆

Lathyrus humilis (Ser.) Spreng.

多年生草本，高20-30厘米。茎细弱，直立，具柔毛，无翅。偶数羽状复叶；小叶(2-)3或4对，卵形或椭圆形。总状花序短于叶，具2-5花；花冠紫红色。荚果条形。种子棕红色，椭圆体形，光滑。花期5-7月，果期8-9月。生海拔2500米以下的林缘、灌丛、林中或山坡草地。产华北、西北和东北。俄罗斯、蒙古和朝鲜半岛亦有。

Perennial herbs, 20-30 cm tall. Stems slender, erect, puberulent, wingless. Leaves paripinnate; leaflets (2-)3- or 4-paired, ovate or elliptic. Racemes shorter than leaf, 2-5-flowered; corolla purple-red. Legumes linear. Seeds red-brown, ellipsoid, smooth. Fl. May-Jul. Fr. Aug-Sep. Forest edges, scrubs, forests or grassy slopes below 2500 m. Distributed in N, NW and NE China. Also in Russia, Mongolia and Korean Peninsula.

毛山黧豆

Lathyrus palustris var. **pilosus** (Chamisso) Ledeb.

多年生草本，被短柔毛。茎具翅。托叶半箭形，长12-15毫

矮山黧豆 *Lathyrus humilis*

山黧豆 *Lathyrus quinquenervius*

米；小叶2-4对，线形或线状披针形，长3.5-4厘米，先端具有分歧的卷须。总状花序腋生；花长13-15(-20)毫米；萼钟状，萼齿狭三角形；花冠紫色。荚果线形，长3-4厘米，先端具喙。花期6-7月，果期7-9月。产中国东北、华北、华东、湖北、青海、云南和四川。俄罗斯、日本、朝鲜半岛和蒙古亦有。

Perennial herbs, pubescent. Stem winged. Stipules semisagittate, 12-15 mm long; leaflets 2-4-paired, linear or linear-lanceolate, 3.5-4 cm long, apex with branched tendril. Raceme axillary; flowers 13-15(-20) mm long; calyx campanulate, tooth narrowly triangular; corolla purple. Legume linear, 3-4 cm long, beaked at apex. Fl. Jun-Jul. Fr. Jul-Sep. Distributed in NE, N and E China, Hubei, Qinghai, Yunnan and Sichuan. Also in Russia, Japan, Korean Peninsula and Mongolia.

山黧豆

Lathyrus quinquenervius (Miq.) Litv.

多年生草本。小叶具5个凸起的平行脉；托叶基部具反折的距。总状花序腋生，具5-8花；花萼钟状，具短柔毛，最下一萼齿约与萼筒等长；花紫蓝色或紫色。荚果条形。花期5-7月，果期8-9月。生海拔2500米以下的林中、山坡或路边。产华中、华北、华东和东北。俄罗斯、朝鲜半岛和日本亦有。

Perennial herbs. Leaflets with 5 prominent parallel veins; stipules with reflexed basal spur. Racemes axillary with 5-8 flowers; calyx campanulate, pubescent, lowest tooth equal to tube in length; flowers purple-blue or purple. Legumes linear. Fl. May-Jul. Fr. Aug-Sep. Forests, hill slopes or roadsides below 2500 m. Distributed in C N, E and NE China. Also in Russia, Korean Peninsula and Japan.

玫红山黧豆

Lathyrus tuberosus L.

多年生草本。茎无翅。托叶半箭形，长5-20毫米；小叶1对，椭圆形、长圆形或倒卵形，先端具单一或分枝的卷须，两面无毛。总状花序具2-7花；花萼钟状，最长萼齿稍短于萼筒；花冠玫瑰红色；花柱扭转。荚果线形，长2-4厘米，无毛。花期6-8月，果期8-9月。生海拔500-2400米的山地阴坡及河谷旁。产新疆。哈萨克斯坦、俄罗斯和欧洲亦有。

Perennial herbs. Stem wingless. Stipules semisagittate, 5-20 mm long; leaflets 1-paired, elliptic, oblong or obovate, with simple or branched tendril at apex, both surfaces glabrous. Raceme 2-7-flowered; calyx campanulate, longest tooth shorter than tube; corolla purple-red; style twisted. Legume linear, 2-4 cm long, glabrous. Fl. Jun-Aug. Fr. Aug-Sep. Water meadows and riverbanks at 500-2400 m. Distributed in Xinjiang. Also in Kazakhstan, Russia and Europe.

玫红山黧豆 *Lathyrus tuberosus*

宽叶山黧豆 *Lathyrus latifolius*

宽叶山黧豆
Lathyrus latifolius L.

多年生草本。茎四棱形，具翅。托叶半箭形，披针形至卵形；叶具1对小叶，椭圆形至椭圆状圆形或卵形至线形，长(3-)4-15厘米，先端具有分枝的卷须；具平行脉。总状花序具5-15花；花萼钟状，长6毫米，5萼齿近相等；花冠紫色至粉红色；花柱扭转。荚果长5-11厘米，无毛。陕西有栽培。原产欧洲中部和南部；现世界温带地区广泛栽培。

Perennial herbs. Stem quadrangular, winged. Stipules semisagittate, lanceolate to ovate; leaflets 1-paired, elliptic to elliptic-orbicular or ovate to linear, (3-)4-15 cm long, with branched tendril at apex; with parallel veins. Raceme 5-15-flowered; calyx campanulate, ca. 6 mm long, equally 5-toothed; corolla purple to pink; style twisted. Legume 5-11 cm long, glabrous. Cultivated in Shaanxi. Native to C and S Europe; cultivated worldwide in temperate areas.

牧地山黧豆
Lathyrus pratensis L.

多年生草本。茎无翅。托叶箭形；小叶1对，椭圆形、披针形或线状披针形，长1-3(-5)厘米，被微柔毛，先端具单一或分枝的卷须。总状花序长为叶的3-6倍，具5-12花；花萼钟形，被短柔毛，最长萼齿长于萼筒；花冠黄色。荚果线形，长2.5-4.5厘米。花期6-8月，果期8-10月。生海拔1000-3000米的林中、山坡或路边。产华西、黑龙江和湖北。欧洲和亚洲亦有。

Perennial herbs. Stem wingless. Stipules sagittate; leaflets 1-paired, elliptic, lanceolate or linear-lanceolate, 1-3(-5) cm long, puberulent, with simple or branched tendril at apex. Raceme 3-6 × as long as leaf, 5-12-flowered; calyx campanulate, pubescent, longest tooth longer than tube; corolla yellow. Legume linear, 2.5-4.5 cm long. Fl. Jun-Aug. Fr. Aug-Oct. Forests, hill slopes or roadsides at 1000-3000 m. Distributed in W China, Heilongjiang and Hubei. Also in Europe and Asia.

豌豆
Pisum sativum L.

一年生攀援草本。叶具4-6小叶；小叶卵形，2-7 × 1-4厘米。总状花序具1-3花；花冠颜色多变，常白色和/或紫色；子房无毛；花柱扁平。荚果2.5-12 × 1-2.5厘米。花果期2-9月。中国广泛栽培。世界各地普遍栽培。

牧地山黧豆 *Lathyrus pratensis*

豌豆 *Pisum sativum*

小叶鹰嘴豆 *Cicer microphyllum*

紫雀花 *Parochetus communis*

Annual climbing herbs. Leaflets 4-6; leaflets ovate, 2-7 × 1-4 cm. Racemes 1-3-flowered; corolla variable in color, usually white and/or purple; ovary glabrous; styles flat. Legumes 2.5-12 × 1-2.5 cm. Fl. and fr. Feb-Sep. Widely cultivated in China. Also cultivated worldwide.

小叶鹰嘴豆

Cicer microphyllum Royle ex Bentham

一年生草本。茎、托叶、叶、花梗和花萼均被腺毛。托叶叶状，边缘5-7裂；羽状复叶顶端具卷须；小叶6-15对，倒卵状楔形，长4-12毫米，先端具小短尖，上半部边缘具深锯齿。花单生；花冠蓝紫色或淡蓝色，长约2.5厘米。荚果椭圆形，长2.5-3.5厘米，密被白色短柔毛。生海拔1600-4600米的山坡、阳坡草地、河滩或沙砾地。产西藏和新疆。南亚亦有。

Annual herbs. Stems, stipules, leaves, pedicels and calyx glandular hairy. Stipules leaflike, margin 5-7-toothed; leaves paripinnate, with a terminal tendril; leaflets 6-15-paired, obovate-cuneate, 4-12 mm long, apex mucronate, margin dentate only in distal half. Flower solitary; corolla blue-purple or light blue, ca. 2.5 cm long. Legume elliptic, 2.5-3.5 cm long, densely white pubescent. Hill slopes, meadows on sunny slopes, riverbanks or gravels at 1600-4600 m. Distributed in Xizang and Xinjiang. Also in S Asia.

紫雀花

Parochetus communis Buch.-Ham. ex D. Don

匍匐草本，高10-20厘米，被稀疏柔毛。掌状三出复叶；托叶长4-5毫米，膜质，无毛，全缘；小叶倒心形，侧脉4-5对；小叶柄短于1毫米。伞状花序生于叶腋，花1-3朵；总花梗长于叶柄或等长；苞片2-4；萼钟形，密被褐色细毛。花果期4-11月。生海拔2000-3000米的林缘草地、山坡、路旁荒地。产四川、西藏和云南。南亚和东南亚亦有。

Stems prostrate to ascending, 10-20 cm, tubers lacking. Ternate palmate leaf; stipules 4-5 mm long, membranous, glabrous, entire; lateral veins 4-5 pairs; petiolule less than 1 mm. Inflorescence umbellate axillary; 1-3-flowered; peduncle longer than or equal to petiole; bract 2-4; calyx campanulate, densely brown villous. Fl. and fr. Apr-Nov. Along woodland margins grasslands, mountain slopes, roadsides at 2000-3000 m. Distributed in Sichuan, Xizang and Yunnan. Also in S and SE Asia.

白花草木犀

Melilotus albus Medikus

一年生或二年生草本。茎直立，圆柱形，中空，多分枝。羽状三出复叶；托叶尖刺状锥形，全缘；叶柄比小叶短，纤细；小叶边缘具浅锯齿。总状花序具40-100花；花冠白色，长4-5毫米。荚果椭圆形至长圆形，长3-3.5毫米，先端锐尖，具尖喙。花期5-7月，果期7-9月。生湿润沙地、路旁和荒地。产华北、东北、西北和西南。中亚、西南亚和欧洲亦有。

Annual or biennial herbs. Stems erect, terete, hollow, much branched. Leaves pinnately 3-foliolate; stipules subulate, entire; petiole slender, shorter than leaflet; leaflets margins shallowly serrate. Racemes 40-100-flowered; corolla white, 4-5 mm long. Legume elliptic to oblong, 3-3.5 mm long, apex acute, beaked. Fl. May-Jul. Fr. Jul-Sep. Moist soil in fields, roadsides wastelands. Distributed in N, NE, NW and SW China. Also in C and SW Asia, and Europe.

白花草木犀 *Melilotus albus*

草木犀 *Melilotus officinalis*

草木犀

Melilotus officinalis (L.) Pall.

二年生草本。叶三出复叶；托叶镰刀状条形，基部全缘或具1小齿；小叶8-12对，边缘具浅齿。总状花序具30-70花，起初密集，花期疏松；花黄色。荚果卵球形，脉为横向网纹状，深褐色。花期5-9月，果期6-10月。生山坡、河岸、路边、沙地或林缘。产中国大部分地区。亚洲和欧洲亦有。

Biennial herbs. Leaves ternately compound; stipules falcate-linear, base entire or with 1 minute tooth; leaflets 8-12 pairs, margins shallowly serrate. Racemes 30-70-flowered, dense at first, becoming lax in anthesis; corolla yellow. Legumes ovoid, veins transversely reticulate, dark brown. Fl. May-Sep. Fr. Jun-Oct. Slopes, river banks, roadsides, sandy places or forest edges. Distributed in most parts of China. Also in Asia and Europe.

印度草木犀

Melilotus indicus (L.) All.

一年生草本。托叶披针形，基部具耳。总状花序纤细，花序轴长，具15-25花；花冠黄色，长达3毫米；花梗短于1毫米。荚果球形，微伸出于花萼，表面具网纹。花期3-5月，果期5-6月。生开阔地、草甸碱地或路边。产中国西南、华南和东南。孟加拉国、印度、近东地区、巴基斯坦和欧洲亦有。

Annual herbs. Stipules lanceolate, base auriculate. Racemes slender, peduncles long, with 15-25 flowers; corolla yellow, up to 3 mm long; pedicels less than 1 mm long. Legumes globose, slight exserted from calyx, surface veins reticulate. Fl. Mar-May. Fr. May-Jun. Open places, alkaline soil of meadows or roadsides. Distributed in SW, S and SE China. Also in Bangladesh, India, Near East, Pakistan and Europe.

印度草木犀 *Melilotus indicus*

胡卢巴

Trigonella foenum-graecum L.

一年生草本。羽状三出复叶；小叶狭倒卵形、卵形至长圆状披针形，长15-40毫米，先端钝，边缘上半部具三角形尖齿。花无梗，1-2朵着生叶腋；花萼筒状，被长柔毛；花冠乳白色或淡黄色，基部堇青色。荚果圆筒状，长7-12厘米，具纵长网纹，先端具长喙。花期

胡卢巴 *Trigonella foenum-graecum*

4-7月，果期7-9月。生田间或路旁。中国各地均有栽培。喜马拉雅和东南亚亦有。

Annual herbs. Leaves pinnately 3-foliolate; leaflets narrowly obovate, ovate to oblong-lanceolate, 15-40 mm long, apex obtuse, margin with sharply triangular teeth in upper 1/2. Flowers sessile, axillary, solitary or in pairs; calyx tubular, villous; corolla creamy or pale yellow, base violet. Legume conical, 7-12 cm long, veins longitudinally reticulate, apex beaked. Fl. Apr-Jul. Fr. Jul-Sep. Cultivated throughout China. Also in Himalaya and SE Asia.

天蓝苜蓿
Medicago lupulina L.

一年生或短命多年生草本植物。小叶3，基部楔形，纸质，具柔毛。花10-20成小头状；苞片刺毛状，较小；花冠黄色。荚果肾形，散生柔毛，约3 × 2毫米。花期4-9月，果期6-10月。生林缘、田边、路边或河边。产中国各地。亚洲和欧洲亦有；归化于世界各地。

Annual or short-lived perennial herbs. Leaflets 3, cuneate at base, papery, pubescent. Flowers 10-20 in small heads; bracts bristlelike, minute; corolla yellow. Legumes reniform, sparsely pubescent, ca. 3 × 2 mm. Fl. Apr-Sep. Fr. Jun-Oct. Forest edges, field edges, roadsides or along rivers. Distributed in most parts of China. Also in Asia and Europe; naturalized in most regions of the world.

阔荚苜蓿
Medicago platycarpos (Linnaeus) Trautvetter

多年生草本。茎直立或上升，四棱形，无毛或微被柔毛，基部常带紫色。羽状三出复叶；小叶纸质，边缘具不整齐尖齿。花序伞形，具花(4-)5-8(-15)朵；花冠黄色带紫色条纹，干后变蓝色条纹。荚果长圆状镰形至近半圆形，扁平。果期7-8月。生海拔1200-2000米的河谷、针叶林缘草地。产新疆。哈萨克斯坦、吉尔吉斯斯坦、蒙古和俄罗斯亦有。

Perennial herbs. Stems erect, quadrangular, glabrous or puberulent, base often purplish. Leaves pinnately 3-foliolate; leaflets papery, margin irregularly dentate. Flowers (4-)5-8(-15) in umbels; corolla yellow with purple stripes, bluish when dry. Legume oblong-falcate to semilunar, flat. Fr. Jul-Aug. Ravines, meadows by margins of coniferous forests at 1200-2000 m. Distributed in Xinjiang. Also in Kazakhstan, Kyrgyzstan, Mongolia and Russia.

花苜蓿 *Medicago ruthenica*

花苜蓿
Medicago ruthenica (L.) Trautv.

多年生草本。托叶披针形，具尖头；小叶在不同生境中形状各异。(4-)6-9(-15)花组成伞形花序；花冠黄褐色，中心具绯红色至紫色条纹。荚果矩圆形或矩圆状卵球形。花期6-9月，果期8-10月。生沙地、河岸、山坡或草地。产华北、华西和东北。俄罗斯和蒙古亦有。

Perennial herbs. Stipules lanceolate, cuspidate; leaflets varied in shape with different habitats. Flowers (4-)6-9(-15) in umbels; corolla yellow-brown, with scarlet to purple stripes in center. Legumes oblong or oblong-ovoid. Fl. Jun-Sep. Fr. Aug-Oct. Sandy soil, river banks, slopes or grasslands. Distributed in N, W and NE China. Also in Russia and Mongolia.

天蓝苜蓿 *Medicago lupulina*

阔荚苜蓿 *Medicago platycarpos*

野苜蓿(黄花苜蓿)

Medicago falcata L.

多年生草本植物。托叶披针形至条状披针形；羽状三出复叶。花序短总状，腋生，有6-20花，稠密；花冠黄色；花柱短。荚果镰形，2.5-3.5(-4)毫米，被毛。种子2-4粒。花期6-8月，果期7-9月。生沙地、山坡或草丛中。产中国东北、华北和西北。俄罗斯、蒙古、中亚和欧洲亦有。世界各国有引种栽培。

Perennial herbs. Stipules lanceolate to linear-lanceolate; leaves ternate. Racemes short, axillary, 6-12-flowered, very dense; corolla yellow; styles short. Legumes falcate, 2.5-3.5(-4) mm, hairy. Seeds 2-4. Fl. Jun-Aug. Fr. Jul-Sep. Sandy places, slopes or grasslands. Distributed in NE, N and NW China. Also in Russia, Mongolia, C Asia and Europe. Introduced to most regions of the world.

紫苜蓿

Medicago sativa L.

多年生草本植物。小叶3，长卵形、倒卵形至条状卵形。头状或总状花序，5-30花；花梗直立，长于叶。荚果紧密卷成2-4(-6)个螺旋状，中间实心或近实心。花期5-7月，果期6-10月。中国各地栽培，常逸生于路边、田野、草地或溪边。并广泛栽培世界各地。

Perennial herbs. Leaflets 3, long ovate, obovate to linear-ovate. Heads or racemes with 5-30 flowers; peduncles straight, longer than leaves. Legumes tightly coiled in 2-4(-6) spirals, center solid or nearly so. Fl. May-Jul. Fr. Jun-Oct. Cultivated throughout China, often escaped to roadsides, fields, grasslands or stream banks. Also widely cultivated in most regions of the world.

南苜蓿

Medicago polymorpha Linnaeus

一年生或二年生草本。茎平卧或直立，近四棱形，基部分枝，近无毛。托叶边缘具不整齐条裂，成丝状细条或深齿状缺刻。荚果盘形，直径4-6(-10)毫米，顺时针方向紧旋1.5-2.5(-6)圈，有多条辐射状脉纹，近边缘处环结，每圈具棘刺或瘤突15枚。花期3-5月，果期5-6月。产甘肃、陕西和长江以南。非洲北部、欧洲南部和西南亚亦有。

Annual or biennial herbs. Stems prostrate or erect, subquadrangular, branched at base, glabrescent. Stipules margin irregularly laciniate or deeply incised. Legume discoid, 4-6(-10) mm diam, tightly coiled in 1.5-2.5(-6) spirals, turning clockwise, radial veins connected near edge on coil face, spines or tubercles 15 in each row. Fl. Mar-May. Fr. May-Jun. Distributed in Gansu, Shaanxi and the south of Yangtze River. Also in N Africa, S Europe and SW Asia.

野苜蓿(黄花苜蓿) *Medicago falcata*

紫苜蓿 *Medicago sativa*

南苜蓿 *Medicago polymorpha*

小苜蓿 *Medicago minima*

小苜蓿

Medicago minima (Linnaeus) Bartalini

一年生草本，被伸展柔毛。茎铺散，平卧并上升，基部多分枝。羽状三出复叶；托叶卵形，全缘或不明浅齿；小叶边缘1/3以上具锯齿，两面均被毛。花序头状，具2-10花；总花梗细，挺直；花冠淡黄色。荚果球形，旋转3-5圈，直径2.5-4.5毫米。花期3-4月，果期4-5月。生荒坡、沙地、河岸。产黄河流域和长江以北。非洲、亚洲和欧洲亦有。

Annual herbs, spreading hairy. Stems diffuse, prostrate or ascending, branched at base. Leaves pinnately 3-foliolate; stipules ovate, margin entire or obscurely shallowly serrate; leaflets margin serrulate in apical 1/3, villous on both surfaces. Flowers 2-10 in capitate racemes; peduncles slender and straight; corolla pale yellow. Legume globose, tightly coiled in 3-5 spirals, 2.5-4.5 mm diam. Fl. Mar-Apr. Fr. Apr-May. Waste fields, sandy slopes, stream banks. Distributed in Yellow River basin and the north of Yangtze River. Also in Africa, Asia and Europe.

野火球

Trifolium lupinaster L.

多年生草本。掌状复叶，有小叶(3-)5(-7)，侧脉多达50对以上。头状花序顶生或生于叶腋，有20-35花；花冠淡红色或紫红色；旗瓣椭圆形；雄蕊10。荚果长圆形，膜质。种子2-6粒。花果期6-10月。生湿草地、林缘或山坡。产华北、西北和东北。俄罗斯、蒙古、朝鲜半岛和日本亦有。

Perennial herbs. Leaves palmately compound with (3-)5(-7) leaflets, lateral veins 50 pairs or more. Capitulum 20-35-flowered; corolla reddish or purplish red; standards elliptic; stamens 10. Legumes oblong, membranous. Seeds 2-6. Fl. and fr. Jun-Oct. Wet grasslands, forest edges or slopes. Distributed in N, NW and NE China. Also in Russia, Mongolia, Korean Peninsula and Japan.

野火球 *Trifolium lupinaster*

白车轴草
Trifolium repens L.

多年生短命植物。叶具长柄，掌状3小叶。总花梗长6-20厘米，花萼脉纹10条；花冠白色、奶油色或带浅粉色，长7-12毫米，有香味；旗瓣椭圆形，比翼瓣和龙骨瓣长近1倍。荚果长圆形，常具3粒种子。花果期5-10月。中国广泛栽培或野化。世界各地广泛栽培。

Short-lived perennials. Leaves long petiolate, palmately 3-foliolate. Peduncles 6-20 cm long; calyx with 10 nerves; corolla white, cream or tinged with pink, 7-12 mm long, fragrant; standards elliptic, ca. twice longer than wings and keel. Legumes oblong, usually with 3 seeds. Fl. and fr. May-Oct. Cultivated or naturalized in most parts of China. Widely planted in the world.

草莓车轴草 *Trifolium fragiferum*

草莓车轴草
Trifolium fragiferum L.

多年生草本，全株除花萼外几无毛。掌状三出复叶；叶柄长5-10厘米；小叶倒卵形或倒卵状椭圆形，先端钝圆，微凹，基部苍白色。花序半球形至卵形，具10-30花；花长6-8毫米；花萼二唇形，在果期极膨大，上方2萼齿稍长；花冠白色或粉红色。花果期5-8月。生盐碱地、沼泽、沟边或路旁。产中国东北、华北和西北。北非、中亚、东南亚和欧洲。

Perennial herbs, almost glabrous except calyx. Leaves palmately 3-foliolate; petiole 5-10 cm long; leaflets obovate to obovate-elliptic, apex rounded, retuse, base broadly cuneate, pale. Heads hemispheric to ovoid, 10-30-flowered; flowers 6-8 mm long; calyx bilabiate, strongly inflated in fruit; 2 upper longer than 3 lower; corolla white or pink. Fl. and fr. May-Aug. Alkaline soils, swamps, ditches or roadsides. Distributed in NE, N and NW China. Also in N Africa, C and SE Asia, and Europe.

白车轴草 *Trifolium repens*

红车轴草 *Trifolium pratense*

红车轴草
Trifolium pratense L.

多年生短命植物。托叶卵形，膜质；羽状3小叶；小叶两面疏被棕色长绒毛，中央有“V”字形斑。花30-70朵组成顶生球形或卵球形的头状花序；花冠

猪屎豆 *Crotalaria pallida*

紫色；旗瓣匙形。荚果卵球形。花果期5-9月。栽培或逸生于林缘、湿润草地或路边。产中国大部分地区。北非、西南亚、欧洲和北美洲亦有。

Perennial herbs. Stipules ovate, membranous; leaves pinnate with 3 leaflets; leaflets sparsely brown villous on both surfaces, with a V-shaped white spot. Flowers 30-70 crowded in globoid or ovoid heads, terminal; corolla purple; standard spatulate. Legumes ovoid. Fl. and fr. May-Sep. Cultivated or escaped in forest edges, wet meadows or roadsides. Distributed in most parts of China. Also in N Africa, SW Asia, Europe and North America.

猪屎豆

Crotalaria pallida Aiton

多年生草本。小叶长圆形或椭圆形，长3-6厘米，宽1.5-3厘米。总状花序顶生，具10-40花；花萼密被短柔毛；花冠黄色，伸出于萼外。荚果长圆形，具20-30粒种子。花期9-10月，果期11-12月。生海拔100-1100米的草地或荒地中。产中国西南、华南、华中和华东。热带和亚热带地区亦有。

Herbs, perennial. Leaflets oblong or elliptic, 3-6 cm long, 1.5-3 cm broad. Racemes terminal, 10-40-flowered; calyx densely pubescent; corolla yellow, exserted beyond calyx. Legumes oblong, 20-30-seeded. Fl. Sep-Oct. Fr. Nov-Dec. Grasslands or wastelands at 100-1100 m. Distributed in SW, S, C and E China. Also in other tropical and subtropical areas.

光萼猪屎豆

Crotalaria trichotoma Bojer

草本或小灌木。叶具3小叶；小叶狭椭圆形，全缘。总状花序顶生，具10-20花；花萼无毛至疏具柔毛；花冠黄色，伸出于萼外；旗瓣基部具2附属物，先端具芒。荚果圆柱形，具50-70粒种子。花期4-8月，果期9-12月。生海拔100-2000米的草地或路边。栽培或逸生于华南和华中。原产东非；热带亚洲和澳大利亚引进。

Herbs or shrublets. Leaves 3-foliolate; leaflets narrowly elliptic, entire. Racemes terminal, 10-20-flowered; calyx glabrous to sparsely pubescent; corolla yellow, exserted beyond calyx; standard base with 2 appendages, apex awned. Legumes cylindric, 50-70-seeded. Fl. Apr-Aug. Fr. Sep-Dec. Grasslands or roadsides at 100-2000 m. Cultivated or escaped in S and C China. Native to E Africa; introduced to tropical Asia and Australia.

光萼猪屎豆 *Crotalaria trichotoma*

大猪屎豆
Crotalaria assamica Benth.

直立草本。叶单生；叶倒披针形至狭椭圆形，下面具绢毛，基部楔形，先端钝且具短尖头。总状花序顶生或与叶对生，具20-30花；花萼2唇形；花冠深金黄色。荚果长圆形，具20-30粒种子。花期5-9月，果期8-12月。生海拔100-3000米的山地草地或路边。产云南、贵州、广西、广东、海南和台湾。印度、缅甸、老挝、泰国、越南和菲律宾亦有。

Herbs, erect. Leaves simple; leaves oblanceolate to narrowly elliptic, abaxially sericeous, base cuneate, apex obtuse and mucronate. Racemes terminal or leaf-opposed, 20-30-flowered; calyx 2-lipped; corolla deep golden yellow. Legumes oblong, 20-30-seeded. Fl. May-Sep. Fr. Aug-Dec. Montane grasslands or along trails at 100-3000 m. Distributed in Yunnan, Guizhou, Guangxi, Guangdong, Hainan and Taiwan. Also in India, Myanmar, Laos, Thailand, Vietnam and the Philippines.

四棱猪屎豆
Crotalaria tetragona Roxb. ex Andr.

草本。茎四棱形，被丝质短柔毛。托叶线形或线状披针形；叶长圆状椭圆形或线状披针形，长10-20(-25)厘米，两面被细短伏毛，中脉在叶背发白且凸起；花萼二唇形，长1.5-2.5厘米，密被褐色长柔毛；花冠黄色。荚果长圆形，长4-5厘米，种子10-20粒。花期9-11月，果期12月至翌年2月。生海拔500-1600米的路旁疏林中。产广东、广西、四川和云南。南亚和东南亚亦有。

Herbs. Stems 4-angled, silky pubescent. Stipules linear or linear-lanceolate; leaves oblong-elliptic or linear-lanceolate, 10-20 (-25) cm long, both surfaces finely appressed pubescent, midvein abaxially pale and prominent; calyx 2-lipped, 1.5-2.5 cm long, densely brown pilose; corolla yellow. Legume oblong, 4-5 cm long, 10-20-seeded. Fl. Sep-Nov. Fr. Dec to next Feb. Sparse forests along trails at 500-1600 m. Distributed in Guangdong, Guangxi, Sichuan and Yunnan. Also in S and SE Asia.

假地蓝
Crotalaria ferruginea Graham ex Benth.

草本，直立或斜升。托叶披针形至三角状披针形，长5-8毫米。叶单生，椭圆形，2-6 × 1-3厘米，两面具柔毛，但下面稍密。总状花序顶生，具2-6花；花冠黄色。荚果长圆形。花期6-10月，果期9-12月。生海拔400-2200米的开阔林中或山坡草地。产中国西南、华南、东南、华中和华东。南亚、东南亚和巴布亚新几内亚亦有。

大猪屎豆 *Crotalaria assamica*

四棱猪屎豆 *Crotalaria tetragona*

假地蓝 *Crotalaria ferruginea*

紫花猪屎豆 *Crotalaria occulta*

Herbs, erect to ascending. Stipules lanceolate to triangular-lanceolate, 5-8 mm long. Leaves simple, elliptic, 2-6 × 1-3 cm, both surfaces pilose but abaxially more densely so. Racemes terminal, 2-6-flowered; corolla yellow. Legumes oblong. Fl. Jun-Oct. Fr. Sep-Dec. Open forests or montane grasslands at 400-2200 m. Distributed in SW, S, SE, C and E China. Also in S and SE Asia, and Papua New Guinea.

紫花猪屎豆

Crotalaria occulta Grah. ex Benth.

草本。枝、叶背和花萼密被长柔毛。托叶线形至丝状，长5-8毫米；叶近无柄，线状长圆形或椭圆状倒披针形，长5-8厘米，基部渐狭，上面无毛。花萼二唇形，长1.5-1.8厘米；花冠黄色或紫蓝色，包被萼内，龙骨瓣具扭曲长喙。荚果圆柱形，长约1.5厘米，无毛，种子10-15粒。花期8-10月，果期11月至翌年2月。生海拔800-1000米的路旁疏林。产云南。南亚和老挝亦有。

Herbs. Branches, leaf abaxials and calyx densely pilose. Stipules linear to filiform, 5-8 mm long; leaves subsessile, linear-oblong or elliptic-oblanceolate, 5-8 cm long, base attenuate, adaxially glabrous. Calyx 2-lipped, 1.5-1.8 cm long; corolla yellow or purplish blue, included in calyx, keel with a long twisted beak. Legume cylindric, ca. 1.5 cm long, glabrous, 10-15-seeded. Fl. Aug-Oct. Fr. Nov to next Feb. Sparse forests along trails at 800-1000 m. Distributed in Yunnan. Also in S Asia and Laos.

长萼猪屎豆

Crotalaria calycina Schrank

一年生或短命多年生草本。茎、叶背和花萼被褐色长柔毛。托叶丝状；叶长圆状线形或线状披针形，长3-12厘米，上面沿中脉有毛。花萼二唇形，长2-3厘米；花冠黄色，包被于萼内。荚果近圆柱形，长2-2.5厘米，无毛，具种子20-30粒。花期6-9月，果期10-12月。生海拔100-2400米的路旁林中。产华南、四川和西藏。非洲、大洋洲、南亚和东南亚亦有。

Annual or short-lived perennial herbs. Stems, leaf abaxials and calyx brown pilose. Stipules filiform; leaves oblong-linear or linear-lanceolate, 3-12 cm long, adaxially pilose on midvein. Calyx 2-lipped, 2-3 cm long; corolla yellow, included in calyx. Legume subcylindric, 2-2.5 cm long, glabrous, 20-30-seeded. Fl. Jun-Sep. Fr. Oct-Dec. Open forests along trails at 100-2400 m. Distributed in S China, Sichuan and Xizang. Also in Africa, Oceania, S and SE Asia.

长萼猪屎豆 *Crotalaria calycina*

野百合 *Crotalaria sessiliflora*

响铃豆 *Crotalaria albida*

野百合
Crotalaria sessiliflora L.

直立草本，体高30-100厘米。托叶线形，长2-3毫米；单叶，通常为线形或线状披针形，长3-8厘米，宽0.5-1厘米。总状花序顶生、腋生或密生枝顶形似头状，花1至多数；花梗短，长约2毫米；花冠蓝色或紫蓝色；旗瓣长圆形，翼瓣长圆形或披针状长圆形。荚果短圆柱形，长约10毫米。种子10-15粒。花果期5月至翌年2月。生海拔70-1500米的荒地路旁及山谷草地。产辽宁、河北、山东、江苏、安徽、浙江、江西、福建、台湾、湖南、湖北、广东、海南、广西、四川、贵州、云南和西藏。

Herbs, erect, to 30-100 cm tall. Stipules linear, 2-3 mm; leaves simple, usually linear or linear-lanceolate, 3-8 × 0.5-1 cm, base attenuate. Racemes terminal or leaf-opposed or densely congested and headlike on branch apices, 1 to many flowered; pedicel ca. 2 mm; corolla blue or purplish blue; wings oblong to linear-oblong, shorter than standard. Legume cylindric, ca. 10 mm. 10-15-seeded. Fl. and fr. May to next Feb. Valley grasslands, along trails at 70-1500 m. Distributed in Liaoning, Hebei, Shandong, Jiangsu, Anhui, Zhejiang, Jiangxi, Fujian, Taiwan, Hunan, Hubei, Guangdong, Hainan, Guangxi, Sichuan, Guizhou, Yunnan and Xizang.

响铃豆
Crotalaria albida Heyne ex Roth

草本。托叶针状，小，脱落，有时明显缺无；单叶，近无柄，下面具糙毛。总状花序顶生，常生侧枝，具20-30花；花冠浅黄色。荚果短，圆柱形，长约1厘米，无毛。花期5-9月，果期9-12月。生海拔200-2800米的路边或山坡疏林中。产中国西南、华南和东南。南亚、东南亚和太平洋岛屿亦有。

Herbs. Stipules acicular, minute, caducous, sometimes apparently absent; leaves simple, subsessile, abaxially strigose. Racemes terminal, often on lateral branches, 20-30-flowered; corolla pale yellow. Legumes shortly cylindric, ca. 1 cm long, glabrous. Fl. May-Sep. Fr. Sep-Dec. Roadsides or open forests on slopes at 200-2800 m. Distributed in SW, S and SE China. Also in S and SE Asia, and Pacific Islands.

云南猪屎豆
Crotalaria yunnanensis Franch.

多年生草本。茎被粗糙开展长柔毛。无托叶；叶长圆形至椭圆形，长2-6厘米，两面疏被褐色长柔毛，在下面中脉更密；叶脉在下凸起。总状花序有花5-30；花萼二唇形，长5-10毫米，密被褐色长柔毛；花冠黄色，龙骨瓣具扭曲的长喙。荚果短圆柱形，长约1厘米，无毛。花期5-8月，果期8-10月。生海拔100-3000米的草地或灌丛。产四川和云南。

Perennial herbs. Stems coarsely spreading pilose. Stipules absent; leaves oblong to elliptic, 2-6 cm long, both surfaces sparsely brown pilose but abaxially more densely so on midvein, veins abaxially raised. Racemes 5-30-flowered; calyx 2-lipped, 5-10 mm long, densely brown pilose; corolla yellow, keel with a long twisted beak. Legume cylindric, ca. 1 cm long, glabrous. Fl. May-Aug. Fr. Aug-Oct. Open grasslands or thickets at 100-3000 m. Distributed in Sichuan and Yunnan.

线叶猪屎豆
Crotalaria linifolia L. f.

一年生或短命多年生草本。茎密被丝质短柔毛。叶倒披针形至长圆形，长2-5厘米，两面被丝质柔毛，上面有时仅沿中脉被毛。花萼二唇形，长6-7毫米，密被锈色柔毛；花冠黄色，具深色脉。荚果菱形，长5-6毫米，无毛，种子8-10粒。

云南猪屎豆 *Crotalaria yunnanensis*

线叶猪屎豆 *Crotalaria linifolia*

花期5-10月，果期8-12月。生海拔500-2500米的山坡路旁。产华南、四川和西藏。印度、日本、缅甸和斯里兰卡亦有。

Annual or short-lived perennial herbs. Stems densely silky pubescent. Leaves oblanceolate to oblong, 2-5 cm long, both surfaces silky pilose, adaxially sometimes only along midvein. Calyx 2-lipped, 6-7 mm long, densely rusty pilose; corolla yellow with darker veins. Legume rhombic, 5-6 mm long, glabrous, 8-10-seeded. Fl. May-Oct. Fr. Aug-Dec. Slopes or along trails at 500-2500 m. Distributed in S China, Sichuan and Xizang. Also in India, Japan, Myanmar and Sri Lanka.

假苜蓿

Crotalaria medicaginea Lamk.

草本。茎和叶背被丝质短柔毛。叶柄0.2-1厘米；小叶3，倒披针形至倒卵形，长1-2厘米，宽3-6毫米，上面无毛。花萼近钟形，长2-4毫米；花冠黄色，龙骨瓣长喙扭转。荚果球形，直径3-5毫米，先端具短喙，被微柔毛。种子2粒。花期8-10月，果期11-12月。生海拔100-1400米的沙滩海滨及路边。产台湾、四川、云南、广东和广西。南亚和东南亚亦有。

Herbs. Stems and leaf abaxials silky pubescent. Petiole 0.2-1 cm long; leaflets 3, oblanceolate to obovate, 1-2 cm long, 3-6 mm wide, adaxially glabrous. Calyx subcampanulate, 2-4 mm long; corolla yellow, keel beak long and twisted. Legume globose, 3-5 mm diam, apex shortly beaked, pubescent. 2-seeded. Fl. Aug-Oct. Fr. Nov-Dec. Seashore sandy areas and along trails at 100-1400 m. Distributed in Taiwan, Sichuan, Yunnan, Guangdong and Guangxi. Also in S and SE Asia.

球果猪屎豆

Crotalaria uncinella subsp. **elliptica** (Roxburgh) Polhill

草本。托叶卵状三角形；叶柄长1-2厘米；小叶3，椭圆形，长1-3厘米，上面无毛，下面被短柔毛；中脉在下面凸起。花萼近钟形，长3-4毫米，裂片宽披针形，密被短柔毛；花冠黄色，龙骨瓣具短直喙。荚果卵球形，长6-7毫米，被短柔毛。种子2粒。花期8-10月，果期11-12月。生海拔100-1100米的山地路旁。产广东、广西和海南。东南亚亦有。

Herbs. Stipules ovate-triangular; petioles 1-2 cm long; leaflets 3, elliptic, 1-3 cm long, adaxially glabrous, abaxially pubescent; midvein abaxially raised. Calyx subcampanulate, 3-4 mm long, lobes broadly lanceolate, densely pubescent; corolla yellow, keel with a short straight beak. Legume ovoid, 6-7 mm long, pubescent. 2-seeded. Fl. Aug-Oct. Fr. Nov-Dec. Mountains along trails at 100-1100 m. Distributed in Guangdong, Guangxi and Hainan. Also in SE Asia.

假苜蓿 *Crotalaria medicaginea*

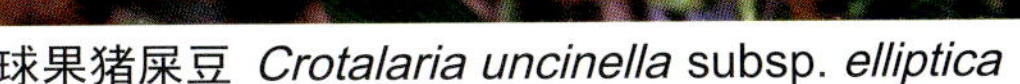
球果猪屎豆 *Crotalaria uncinella* subsp. *elliptica*

黄雀儿

Crotalaria psoraleoides D. Don

灌木。小叶3，椭圆形至长圆形。总状花序顶生或与叶对生，具10-30花；花萼具柔毛；花冠淡黄色，后变为红色，超出花萼。荚果扁，椭圆形，幼时被毛，成熟后无毛。花果期4-12月。生海拔800-1500米的山坡路边。产云南和西藏。印度和尼泊尔亦有。

Shrubs. Leaflets 3, elliptic to oblong. Racemes terminal or leaf-opposed, 10-30-flowered; calyx pubescent; corolla yellowish, turning red when old, exserted from calyx. Legumes flattened, ellipsoid, hairy when young, glabrescent when mature. Fl. and fr. Apr-Dec. Roadsides on slopes at 800-1500 m. Distributed in Yunnan and Xizang. Also in India and Nepal.

黄花木

Piptanthus nepalensis (Hook.) D. Don ex Sweet

灌木。茎具白色绵毛至平伏柔毛，渐无毛。小叶3，下面被黄色丝状及白色贴伏柔毛，渐无毛。总状花序顶生，具2-7轮花；花冠亮黄色。荚果带形，扁，被白色或淡锈色短柔毛。花期4-7月，果期6-9月。生海拔1600-4000米的针叶林、林缘、灌丛或草甸中。产云南、四川、陕西、西藏和甘肃。印度西北部、尼泊尔和缅甸亦有。

Shrubs. Stems white woolly to appressed pubescent, glabrescent. Leaflets 3, yellow silky and white appressed pubescent abaxially, gradually glabrescent. Racemes terminal, with flowers in 2-7 whorls; corolla bright yellow. Legumes flattened, lorate, white or slightly rusty pubescent. Fl. Apr-Jul. Fr. Jun-Sep. Coniferous forests, forest edges, thickets or meadows at 1600-4000 m. Distributed in Yunnan, Sichuan, Shaanxi, Xizang and Gansu. Also in NW India, Nepal and Myanmar.

沙冬青

Ammopiptanthus mongolicus (Maxim. ex Kom.) S. H. Cheng

常绿灌木。叶具1或3小叶；小叶菱状椭圆形或阔椭圆形至阔卵形，两面密被银色绒毛。花4-15，形成短的密集的顶生总状花序；花冠黄色，花瓣具长爪。荚果条状长圆形。花期4-6

黄雀儿 *Crotalaria psoraleoides*

黄花木 *Piptanthus nepalensis*

沙冬青 *Ammopiptanthus mongolicus*

月，果期5-8月。生沙丘或河滩边台地。产内蒙古、甘肃、宁夏和新疆。蒙古亦有。

Everygreen shrubs. Leaves 1- or 3-foliolate; leaflets rhombic-elliptic or broadly elliptic to broadly ovate, densely silvery tomentose on both surfaces. Flowers 4-15, in short dense terminal racemes; corolla yellow, petals long clawed. Legumes linear-oblong. Fl. Apr-Jun. Fr. May-Aug. Dunes or terraces by rivers. Distributed in Neimenggu, Gansu, Ningxia and Xinjiang. Also in Mongolia.

霍州油菜

Thermopsis chinensis Benth. ex S. Moore

多年生草本。茎直立，具沟棱。3小叶，倒卵形或线状披针形，2-4.5 × 0.8-2厘米，先端钝圆，具细尖。总状花序顶生；花互生；萼片5，疏被短柔毛；花冠黄色。荚果贴茎上指，披针状线形，薄木质，被淡黄色贴伏长硬毛。种子15-20粒，红褐色，密布腺点。花期4-5月，果期6-7月。生溪边、荒野和路旁。产河北、华中、华东和福建。日本亦有。

Perennial herbs. Stems erect, ridged. Leaflets 3, obovate or linear-lanceolate, 2-4.5 × 0.8-2 cm, apex rounded, mucronate. Racemes terminal; flowers alternate; sepals 5, sparsely puberulent; corolla yellow. Legume held erect and close to stem, linear-lanceolate, thinly woody, sparsely appressed yellowish hirsute. Seeds 15-20, reddish brown, densely glandular. Fl. Apr-May. Fr. Jun-Jul. Stream banks, wastelands and roadsides. Distributed in Hebei, C and E China, and Fujian. Also in Japan.

披针叶野决明

Thermopsis lanceolata R. Br.

多年生草本。茎直立，具乳白色柔毛。小叶条状长圆形或倒披针形至条形，下面具贴伏柔毛，长达7.5厘米。总状花序顶生，花2-3轮生；花冠黄色，花瓣具长爪。荚果条形或长圆形。花期5-7月，果期6-10月。生草地、河岸或荒地。产华北、西北和东北。吉尔吉斯斯坦、俄罗斯和蒙古亦有。

Perennial herbs. Stems erect, creamy pubescent. Leaflets linear-oblong or oblanceolate to linear, appressed puberulent abaxially, to 7.5 cm long. Racemes terminal, flowers 2-3-verticillate; corolla yellow, petals long clawed. Legumes linear or oblong. Fl. May-Jul. Fr. Jun-Oct. Grasslands, river banks or wastelands. Distributed in N, NW and NE China. Also in Kyrgyzstan, Russia and Mongolia.

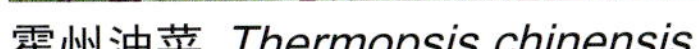

霍州油菜 *Thermopsis chinensis*

披针叶野决明 *Thermopsis lanceolata*

紫花野决明(紫花黄华)

Thermopsis barbata Benth.

多年生草本。基部叶4-7轮生，合生成鞘；小叶长圆形或披针形至倒披针形，两面密被白色长柔毛。总状花序疏松；花深紫色，有时干后变蓝；花瓣近等长。荚果褐色，狭椭圆形，锐尖。花期6-7月，果期8-9月。生海拔2700-4500米的山谷或山坡。产青海、四川西部、新疆(天山)、西藏和云南西南部。印度、尼泊尔、巴基斯坦和克什米尔地区亦有。

Perennial herbs. Leaves at base 4-7-verticillate and connate into a sheath; leaflets oblong or lanceolate to oblanceolate, densely white-villose on both surfaces. Racemes lax; flowers deep purple, sometimes blue when dry; petals subequal. Legumes brown, narrowly elliptic, acute. Fl. Jun-Jul. Fr. Aug-Sep. Valleys or slopes at 2700-4500 m. Distributed in Qinghai, W Sichuan, Xinjiang (Tianshan Mountain), Xizang and SW Yunnan. Also in India, Nepal, Pakistan and Kashmir.

紫花野决明(紫花黄华) *Thermopsis barbata*

矮生野决明

Thermopsis smithiana E. Peter

多年生草本。小叶倒卵形至狭椭圆形，先端钝圆或截形，偶有细尖，上面无毛，下面被白色长绒毛。总状花序短，3-5厘米，3花轮生；花冠亮黄色，花瓣具长爪。荚果椭圆形、长圆形或倒卵形。花期6-7月，果期7-8月。生海拔3500-4500米的山坡上。产云南、四川和西藏。

Perennial herbs. Leaflets obovate to narrow-elliptic, apex obtuse or truncate, occasionally mucronate, glabrous adaxially, white-villous abaxially. Racemes short, 3-5 cm long, flowers 3-verticillate in whorls; corolla bright yellow, petals long-clawed. Legumes elliptic, oblong or obovate. Fl. Jun-Jul. Fr. Jul-Aug. Mountain slopes at 3500-4500 m. Distributed in Yunnan, Sichuan and Xizang.

矮生野决明 *Thermopsis smithiana*

酢浆草科 Oxalidaceae

阳桃

Averrhoa carambola L.

乔木，高3-12(-15)米。叶互生，小叶(3-)5-13对；小叶卵形至椭圆形。圆锥或聚伞花序腋生或生枝干上；小枝和花芽绯红色；花多数，较小。浆果黄色或黄褐色，长圆形，具(3-)5(或6)棱，横切面星形，极肉质。花期4-12月，果期7-12月。栽培于中国西南、华南和东南，有时逸生至海拔1000米以下的路边和次生开阔林地。其他热带地区亦有。

Trees, 3-12(-15) m tall. Leaves alternate, leaflets (3-)5-13 pairs; leaflets ovate to elliptic. Inflorescences axillary or rameal, panicles or cymes; branches and flower buds crimson; flowers numerous, small. Berries yellow to yellow-brown, oblong, (3-)5(or 6)-ribbed, stellate in cross section, very fleshy. Fl. Apr-Dec. Fr. Jul-Dec. Cultivated in SW, S and SE China, sometimes escaping to roadsides and secondary open forests below 1000 m. Also in other tropics.

阳桃 *Averrhoa carambola*

三敛 *Averrhoa bilimbi*

山酢浆草 *Oxalis griffithii*

三敛

Averrhoa bilimbi L.

小乔木，高5-6米。叶聚生枝顶，小叶10-20对，长圆形，长3-5厘米，先端斜渐尖，基部圆形，全缘，两面多少被毛；小叶柄长2-4毫米，被柔毛。圆锥花序生分枝或树干上；萼片卵状披针形，被柔毛，长约4毫米；花瓣长圆状匙形，长为萼片2倍以上。浆果长圆体形，具钝棱。花期4-12月，果期7-12月。广东、广西和台湾栽培。原产亚洲热带。

Small trees, 5-6 m tall. Leaves aggregated at apex of branches; leaflets 10-20 pairs, oblong, 3-5 cm long, apex acuminate, base obliquely rounded, margin entire, both surfaces somewhat pubescent; petiolutes 2-4 mm long, pubescent. Panicles on branches or trunk; sepals ovate-lanceolate, pubescent, ca. 4 mm long; petals oblong-spatulate, over two times as long as sepals. Berries oblong, obscurely angle. Fl. Apr-Dec. Fr. Jul-Dec. Cultivated in Guangdong, Guangxi and Taiwan. Native to tropical Asia.

山酢浆草

Oxalis griffithii Edgew. et Hook. f.

多年生草本。无茎。叶基生；小叶倒三角形。花单生，下垂；花梗等长或长于叶；花萼披针形，宿存；花瓣白色，具紫色脉纹，稀粉红色，狭倒卵形。蒴果长圆状圆锥形。花期5-9月，果期5-10月。生海拔800-3400米的林下、灌丛、草甸或干燥阴处。产中国西南、东南、华中、华北、华东和华西。南亚和东南亚亦有。

Perennials herbs. Stemless. Leaves basal; leaflets obtriangular. Flowers solitary, nodding; peduncles equal to or longer than leaves; sepals lanceolate, persistent; petals white, with lilac veins, rarely pink, narrowly obovate. Capsules oblong-conic. Fl. May-Sep. Fr. May-Oct. Forests, thickets, meadows or dry shady places at 800-3400 m. Distributed in SW, SE, C, N, E and W China. Also in S and SE Asia.

三角叶酢浆草

Oxalis obtriangulata Maxim.

多年生草本，高5-12厘米。无地上茎，无鳞茎，根状茎匍匐，直径5-8毫米，密被残存的叶柄基。叶基生；小叶先端截形或近截形，裂片先端稍尖。花单生，俯垂；苞片生于花基部；萼片椭圆形；花瓣白色，狭椭圆状倒卵形。蒴果狭圆锥形，长3-4厘米。花果期5-6月。生海拔700-1500米的树林和灌丛中。产中国东北。东北亚亦有。

Perennials herbs, 5-12 cm tall. Stemless, bulbs absent, rhizome creeping, 5-8 mm diam, covered densely by remains of leaf bases. Leaves basal; leaflets truncate or subtruncate at apex, lobes apices subacute. Flowers solitary, nodding; bracts near base of flower; sepals elliptic; petals white, narrowly elliptic-obovate. Capsules narrowly conical, 3-4 cm long. Fl. and fr. May-Jun. Forests and thickets at 700-1500 m. Distributed in NE China. Also in NE Asia.

三角叶酢浆草 *Oxalis obtriangulata*

紫心酢浆草 *Oxalis articulate*

紫心酢浆草
Oxalis articulate Savigny

多年生草本，高10-25厘米。无地上茎，具木质的块茎。叶基生；小叶3，倒心形，长1.5-4厘米，被短柔毛。伞形花序具4-10花；苞片披针形，被短柔毛；萼片披针形，无毛；花瓣紫色，倒卵形；花柱3裂，子房被柔毛。花期4-9月，果期6-10月。山东、江苏、湖南和湖北栽培和归化。原产美洲。

Perennials herbs, 10-25 cm tall. Stemless, tubers woody. Leaves basal; leaflets 3, obcordate, 1.5-4 cm long, pubescent. Inflorescence umbellate 4-10-flowered; bracts lanceolate, pubescent; sepals lanceolate, glabrous; petals purple, obovate; styles 3-lobes, ovary pubescent. Fl. Apr-Sep. Fr. Jun-Oct. Cultivated and naturalized in Shandong, Jiangsu, Hunan and Hubei. Native to America.

宽叶酢浆草
Oxalis latifolia Kunth

多年生草本。根倒圆锥形，肉质。无地上茎，鳞茎近球形，具鳞片。叶基生；小叶3，宽倒三角形，长2-3厘米，无毛。伞形花序具6-10花；萼片长圆状披针形，先端有2枚长圆形、橙红色的腺体；花冠漏斗状，紫色，筒内绿色；花柱5，子房无毛。花期8-10月。福建、广东、广西和云南归化。原产美洲。

Perennials herbs. Root obconical, fleshy. Stemless, bulb subglobular, covered by scales. Leaves basal; leaflets 3, obtriangle, 2-3 cm long, glabrous. Inflorescence umbellate 6-10-flowered; sepals oblong-lanceolate, at apex with 2 oblong and orange gland; corolla funnel shaped, purple, inner tube green; styles 5, ovary glabrous. Fl. Aug-Oct. Cultivated in Fujian, Guangdong, Guangxi and Yunnan. Native to America.

红花酢浆草
Oxalis corymbosa DC.

多年生草本。无茎。叶基生；小叶倒心形。聚伞状圆锥花序，不规则分枝，具8-15花；萼片披针形，先端具2个棕红色胼胝体；花瓣紫红色，具深色脉纹，基部绿色。蒴果极稀发育。花期3-12月。海拔2300米以下栽培或逸为野生，常见杂草。产华北、华西、华东、华中和华南。原产美洲热带地区；归化于世界各地。

Perennials herbs. Stemless. Leaves basal; leaflets obcordate. Inflorescences corymbose cymes, irregularly branched, 8-15-flowered; sepals lanceolate, apex with 2 reddish brown calli; petals purplish red with darker veins, base green. Capsules rarely formed. Fl. Mar-Dec. Cultivated or escaped in wild below 2300 m, common weeds. Distributed in N, W, E, C and S China. Native to tropical America; naturalized in many parts of the world.

白花酢浆草
Oxalis acetosella L.

多年生草本，高8-10厘米。无地上茎，无鳞茎，根状茎直径

宽叶酢浆草 *Oxalis latifolia*

红花酢浆草 *Oxalis corymbosa*

白花酢浆草 *Oxalis acetosella*

约3毫米。叶基生；小叶3，倒心形。花单生；花梗长2-3厘米，被短柔毛；花瓣白色或具淡紫红色脉纹，倒卵形，长(1.2-)1.5-2.2厘米，花柱5。蒴果卵球形。花期7-8月，果期8-9月。生海拔800-3700米的林中。产中国东北、宁夏和新疆。欧洲温带和亚洲温带亦有。

Perennials herbs, 8-10 cm tall. Stemless, bulbs absent, rhizome ca. 3 mm diam. Leaves basal; leaflets 3, obcordate. Flowers solitary; pedicles 2-3 cm long, pubescent; petals white or lilac to pinkish veined, obovate, (1.2-) 1.5-2.2 cm long, stigma 5. Capsules angular-ovoid. Fl. Jul-Aug. Fr. Aug-Sep. Forests at 800-3700 m. Distributed in NE China, Ningxia and Xinjiang. Also in temperate Europe and temperate Asia.

酢浆草

Oxalis corniculata L.

一年生或多年生短命植物。小叶表面无斑点，倒心形，绿色或带紫红色。花序伞形，具(2-)1-5(-7)花；花瓣亮黄色，长圆状倒卵形。蒴果长圆柱状，具5棱。花果期2-10月。生海拔3400米以下的路边、山坡或疏林。广布中国各地。全球大部分地区均有。

Annuals or short-lived perennials. Leaflets without spots adaxially, obcordate, green or suffused purplish red. Inflorescences umbellate, (2-)1-5(-7)-flowered; petals bright yellow, oblong-obovate. Capsules long cylindric, 5-edge. Fl. and fr. Feb-Oct. Roadsides, slopes or open forests below 3400 m. Widespread in China. Almost cosmopolitan.

感应草

Biophytum sensitivum (L.) DC.

一年生草本。茎单一，被糙硬毛。叶集生茎顶，小叶6-14对，长圆形至倒卵状长圆形；花数朵，呈伞形花序；花瓣黄色；雄蕊10；蒴果椭圆状倒卵形，具柔毛。花果期7-12月。生海拔100-700米的路旁、山坡草地或林下阴湿地。产云南、贵州、广西、海南和台湾。热带亚洲和热带非洲亦有。

Annual herbs. Stems simple, hirsute. Leaves clustered at stem apex, leaflets 6-14-paired, oblong to obovate-oblong; flowers a few, arranged in umbels; petals yellow; stamens 10; capsules ellipsoid-obovoid, pubescent. Fl. and fr. Jul-Dec. Roadsides, grassy slopes or moist places under forests at 100-700 m. Distributed in Yunnan, Guizhou, Guangxi, Hainan and Taiwan. Also in tropical Asia and tropical Africa.

酢浆草 *Oxalis corniculata*

感应草 *Biophytum sensitivum*

牻牛儿苗科 Geraniaceae

汉荭鱼腥草 *Geranium robertianum*

野老鹳草 *Geranium carolinianum*

牻牛儿苗
Erodium stephanianum Willd.

多年生草本，高20-50(-120)厘米。茎多数，斜升至匍匐。叶对生，卵形至三角状卵形，羽状分裂，基部裂片显著。假伞形花序明显长于叶，具2-5两性花；花瓣紫色，基部无斑点。芒不为羽毛状。花期7-8月，果期8-9月。生海拔400-4000米的草甸、草原、洪积平原或农田。产中国西北、西南、华中、华北和东北。南亚、东北亚、阿富汗和吉尔吉斯斯坦亦有。

Perennial herbs, 20-50(-120) cm tall. Stems numerous, ascending to decumbent. Leaves opposite, ovate to triangular-ovate, pinnately parted with basal pair of lobes distinct. Pseudoumbels conspicuously longer than leaves, with 2-5 hermaphrodite flowers; petals purple, without a basal spot. Awns not plumose. Fl. Jul-Aug. Fr. Aug-Sep. Meadows, steppes, flood plains or farmlands at 400-4000 m. Distributed in NW, SW, C, N and NE China. Also in S and NE Asia, Afghanistan and Kyrgyzstan.

汉荭鱼腥草
Geranium robertianum L.

多年生草本。植株有鱼腥味。叶对生；叶掌状裂，二至三回三出羽状。聚伞花序单生，具2花；花瓣淡紫色；雄蕊花丝淡粉色。果实未成熟时直立；果皮网状，疏具脊且基部疏网结，顶端稍密。花期4-6月，果期5-8月。生海拔900-3300米的疏林或路边。产中国西南、东南和华中。南亚、中亚和东北亚亦有。

Perennial herbs. Plants with fishy smell. Leaves opposite; blades palmately divided, bi- to tri-ternate-pinnate. Cymules solitary, 2-flowered; petals purplish; filaments pinkish. Fruits erect when immature; mericarps reticulate, ridges sparse and scarcely anastomosing in basal half but denser apically. Fl. Apr-Jun. Fr. May-Aug. Open forests or roadsides at 900-3300 m. Distributed in SW, SE and C China. Also in S, C and NE Asia.

野老鹳草
Geranium carolinianum L.

一年生草本。叶1-3互生，但在花序上对生，圆肾形，掌状5-7裂至基部。聚伞花序伞形，密聚集于每个枝顶端；花瓣淡粉色。蒴果约长2厘米，果皮密被0.5-1.8毫米的无腺或有时为具腺绒毛。花期4-7月，果期5-9月。生海拔800米以下的平原或杂草中。逸生于中国中部至南部。原产北美洲。

Herbs annual. Leaves 1-3 alternate but opposite at inflorescence, orbicular-reniform, palmately 5-7 divided to base. Cymules umbelliform, in dense aggregates at apex of each branch; petals pale-pink. Capsules ca. 2 cm long, mericarps densely covered with 0.5-1.8 mm long nonglandular and sometimes glandular trichomes. Fl. Apr-Jul. Fr. May-Sep. Plains or weedy places below 800 m. Escaped in Centrol to South of China. Native to North America.

牻牛儿苗 *Erodium stephanianum*

老鹳草
Geranium wilfordii Maxim.

多年生草本。根状茎横走。基生叶圆肾形，5深裂达1/2处；裂片3(或5)，菱形。小聚伞花序单生或聚集在枝顶，

老鹳草 *Geranium wilfordii*

具1或2朵花；花瓣淡粉色或白色，长4-6.2(-6.9)毫米。果实长2-2.2厘米，未成熟时直立；果皮光滑，基部具胼胝体。花期6-7月，果期7-8月。生海拔100-1800米的灌丛、草甸或杂草中。产中国西南、东南、华中、华北、华西、华东和东北。俄罗斯、朝鲜半岛和日本亦有。

Perennial herbs. Rootstocks horizontal. Basal leaves orbicular-reniform, 5-parted to 1/2 part; segments 3(or 5), rhombic. Cymules solitary, 1- or 2-flowered; petals pale pink or white, 4-6.2(-6.9) mm long. Fruits 2-2.2 cm long, erect when immature; mericarps smooth, with a basal callus. Fl. Jun-Jul. Fr. Jul-Aug. Scrubby areas, meadows or weedy areas at 100-1800 m. Distributed in SW, SE, C, N, W, E and NE China. Also in Russia, Korean Peninsula and Japan.

鼠掌老鹳草
Geranium sibiricum L.

多年生草本。叶对生，掌状5深裂，中央裂片狭菱形；托叶披针形，棕褐色。聚伞花序单生，具1(或2)花；苞片对生，钻形，膜质；花瓣白色或粉红色带紫色脉纹，长4-5.4(-5.8)毫米。蒴果下垂，疏被毛。花期6-7月，果期8-9月。生林缘、疏灌丛、河谷草甸，或为杂草。产中国西南、华中、华北、华西、华东、西北和东北。中亚、东北亚、巴基斯坦和阿富汗亦有。

Perennial herbs. Leaves opposite, palmately 5-parted, middle segments narrowly rhombic; stipules lanceolate, rusty-brown. Cymules solitary, 1(or 2)-flowered; bracts opposite, subulate, membranous; petals white or pink with purplish veins, 4-5.4(-5.8) mm long. Capsules nodding, hairy. Fl. Jun-Jul. Fr. Aug-Sep. Forest edges, sparse thickets, meadows in valleys, or as weeds. Distributed in SW, C, N, W, E, NW and NE China. Also in C and NE Asia, Pakistan and Afghanistan.

尼泊尔老鹳草
Geranium nepalense Sweet

多年生草本。叶对生，掌状开裂，裂片先端锐尖或钝卵圆，中央裂片阔菱形。聚伞花序单生，具(1或)2花；花瓣白色、浅粉色或稀深粉红色，长5.1-5.9(-6.3)毫米。果实长1.4-1.8厘米；果皮光滑，顶端具1横脉。花果全年。生海拔1000-3600米的林下、灌丛、山坡、草地、路边、水边或废弃地。产中国西南、华中、华北、华东和华西。南亚、东南亚和阿富汗亦有。

Perennial herbs. Leaves opposite, palmately cleft, lobes of leaf apices acute or obtuse, middle segments broadly rhombic. Cymules solitary, (1 or)2-flowered; petals white, pale pink or rarely deep pink, 5.1-5.9 (-6.3) mm long, Fruits 1.4-1.8 cm long; mericarps smooth with 1 transversal vein at apex. Fl. and fr. all year. Forests, thickets, slopes, grasslands, roadsides, by waters or wastelands at 1000-3600 m. Distributed in SW, C, N, E and W China. Also in S and SE Asia, and Afghanistan.

鼠掌老鹳草 *Geranium sibiricum*

尼泊尔老鹳草 *Geranium nepalense*

陕西老鹳草 *Geranium shensianum*

陕西老鹳草
Geranium shensianum R. Knuth

多年生草本，高40-70厘米。托叶披针形，长10-12毫米，宽3-4毫米，被疏柔毛；叶五角状肾圆形，长7-10厘米，宽8-15厘米。花序腋生和顶生；萼片长卵形或卵状椭圆形，长6-8毫米，宽约3毫米；花瓣黄色或白色，倒卵形。花期6-7月，果期7-8月。生海拔1800-2800米的山地草甸和疏林下。产陕西南部、甘肃和四川西北部。

Perennials, stem 40-70 cm tall. Stipules lanceolate, 10-12 × 3-4 mm, puberulous; leaves 7-10 cm long, 8-15 cm wide. Inflorescence axillary and basidixed; sepal long-ovate or ovate-elliptic, 6-8 × ca. 3mm; petals yellow or white, obovate. Fl. Jun-Jul. Fr. Jul-Aug. Secondary forests, meadows at 1800-2800 m. Distributed in S Shaanxi, Gansu and NW Sichuan.

五叶老鹳草
Geranium delavayi Franch.

多年生草本。叶掌状深裂，裂片5，菱形，小裂片全缘；托叶干时棕色，膜质；苞片钻形。聚伞花序单生或聚集于每个枝顶端，具2花；花瓣深红色至粉色，基部白色，反折；子房被毛。蒴果熟时下垂。花期6-8月，果期8-10月。生海拔2300-4100米的山地草甸、林缘或灌丛中。产云南和四川南部。

Perennial herbs. Leaves palmately parted, segments 5, rhombic, lobes entire; stipules brown when dry, membranous; bracts subulate. Cymules solitary or in aggregates at apex of each branch, 2-flowered; petals blackish red to pink with a whitish base, reflexed; ovary hairy. Capsules nodding when ripe. Fl. Jun-Aug. Fr. Aug-Oct. Mountain meadows, forest edges or thickets at 2300-4100 m. Distributed in Yunnan and S Sichuan.

毛蕊老鹳草
Geranium platyanthum Duthie

多年生草本。根状茎横走。茎具腺毛或无腺毛。叶1(或2)互生，于花序对生，掌状分裂。小聚伞花序单生或聚集在枝顶，具2花；花瓣淡紫色或白色；雄蕊花丝淡紫色，披针形，下面具柔毛，上半部具纤毛。蒴果被糙毛和腺毛。花期6-7月，果期8-9月。生海拔1000-2700米的林中、灌丛中或草甸。产中国西南、华北、华西和东北。俄罗斯东部、蒙古和朝鲜半岛亦有。

Perennial herbs. Rootstocks horizontal. Stem with nonglandular trichomes and usually glandular trichomes. Leaves 1(or 2) alternate but opposite at inflorescence, palmately cleft. Cymules solitary or in aggregates at apex of each branch, 2-flowered; petals pur-

五叶老鹳草 *Geranium delavayi*

plish or white; staminal filaments purplish, lanceolate, abaxially pilose and proximal half ciliate. Capsules strigose and glandular-hairy. Fl. Jun-Jul. Fr. Aug-Sep. Forests, scrubby areas or meadows at 1000-2700 m. Distributed in SW, N, W and NE China. Also in E Russia, Mongolia and Korean Peninsula.

反瓣老鹳草
Geranium refractum
Edgeworth et J. D. Hooker

多年生草本。叶对生，具长柄；叶圆形或肾形，5深裂达基部。聚伞花序单生，具2花；花梗具钩状无腺毛及淡紫色腺毛；花瓣白色或浅粉红色，反折。果梗下垂；蒴果长约2.5厘米，被短柔毛。花期7-8月，果期8-9月。生海拔3000-4300米的山地草甸、林缘或灌丛中。产云南、四川和西藏南部。印度北部、尼泊尔、不丹和缅甸北部亦有。

Perennial herbs. Leaves opposite, with long petioles; blades orbicular or reniform, 5-parted to base. Cymules solitary, 2-flowered; pedicels with uncinate nonglandular trichomes and purplish glandular trichomes; petals white or pale pink, reflexed. Fruiting pedicels pendulous; capsules ca. 2.5 cm long, pubescent. Fl. Jul-Aug. Fr. Aug-Sep. Meadows on mountains, forest edges or thickets at 3000-4300 m. Distributed in Yunnan, Sichuan and S Xizang. Also in N India, Nepal, Bhutan and N Myanmar.

刚毛紫地榆
Geranium hispidissimum
(Franch.) R. Knuth

多年生草本。茎具棱。叶对生；叶掌状分裂，具柔毛及平伏无腺毛及腺毛。小聚伞花序单生，具花2朵；花瓣白色或淡粉色，长8.6-9.8(-10.6)毫米，直立至反折。果实未成熟时直立；果皮网结。花期6-7月，果期7-8月。生海拔1500-3400米的林下、灌丛或草坡。产云南西北部。

Perennial herbs. Stems angulate. Leaves opposite; leaves palmately cleft, pilose with appressed nonglandular and glandular trichomes. Cymules solitary, 2-flowered; petals white or pinkish, 8.6-9.8 (-10.6) mm long, erect to patent. Fruits erect when immature; mericarps reticulate. Fl. Jun-Jul. Fr. Jul-Aug. Forests, scrubby areas or grassy slopes at 1500-3400 m. Distributed in NW Yunnan.

毛蕊老鹳草 *Geranium platyanthum*

刚毛紫地榆 *Geranium hispidissimum*

反瓣老鹳草 *Geranium refractum*

大花老鹳草 *Geranium himalayense*

白花老鹳草 *Geranium albiflorum*

大花老鹳草

Geranium himalayense Klotzsch

多年生草本。茎直立。叶对生，长2.2-3.8(-5.5)厘米，掌状分裂。小聚伞花序单生，具2花；花瓣深蓝色至白色；萼片具一长0.7-1.3(-1.9)毫米的凸尖。果成熟之前反卷；果皮光滑，但顶端常具1或2个横脉。花期6-7月，果期8-9月。生海拔3700-4400米的高山草甸。产西藏南部至西部。印度北部、尼泊尔、巴基斯坦和阿富汗亦有。

Perennial herbs. Stems erect. Leaves opposite, 2.2-3.8(-5.5) cm long, palmately cleft. Cymules solitary, 2-flowered; petals deep blue to whitish; sepals with a mucro 0.7-1.3(-1.9) mm long. Fruits reflexed when immature; mericarps smooth but usually with 1 or 2 transversal veins at apex. Fl. Jun-Jul. Fr. Aug-Sep. Alpine meadows at 3700-4400 m. Distributed in S to W Xizang. Also in N India, Nepal, Pakistan and Afghanistan.

草地老鹳草

Geranium pratense L.

多年生草本。叶长(4.2-)6.2-11.5厘米，掌状7-9深裂。聚伞花序顶生；苞片披针形；花瓣淡蓝色或有时淡紫红色或白色；萼片具一长(1.7-)2.2-3.9毫米的凸尖；雌蕊被短柔毛。蒴果被短柔毛和腺毛。花期6-7月，果期7-9月。生海拔1400-4000米的山地草甸或亚高山草甸。产中国西南、华北和西北。南亚、中亚和东北亚亦有。

Perennial herbs. Leaves (4.2-)6.2-11.5 cm long, palmately 7-9-parted. Cymes terminal; bracts narrowly lanceolate; petals bluish or sometimes purplish or white, sepals with a mucro (1.7-)2.2-3.9 mm long; pistils pubescent. Capsules pubescent and glandular. Fl. Jun-Jul. Fr. Jul-Sep. Meadows on mountains or subalpine meadows at 1400-4000 m. Distributed in SW, N and NW China. Also in S, C and NE Asia.

白花老鹳草

Geranium albiflorum Ledeb.

多年生草本。叶圆肾形，掌状深裂达3/4，下部全缘。花序腋生或顶生；花瓣白色或有时淡紫色，长0.8-1.3厘米，顶端

草地老鹳草 *Geranium pratense*

甘青老鹳草 *Geranium pylzowianum*

钝，具一凹缺；雌蕊被长柔毛。蒴果长约3厘米，被柔毛。花期6-7月，果期7-8月。生海拔800-1800米的林中、河谷或亚高山草甸。产新疆北部。哈萨克斯坦、吉尔吉斯斯坦、俄罗斯和蒙古亦有。

Perennial herbs. Leaves orbicular-reniform, palmately parted to 3/4 part, below entire. Inflorescences axillary or terminal; petals white or sometimes pale-purple, 0.8-1.3 cm long, apex retuse with anotch; pistils villose. Capsules ca. 3 cm long, pubescent. Fl. Jun-Jul. Fr. Jul-Aug. Forests, valleys or subalpine meadows at 800-1800 m. Distributed in N Xinjiang. Also in Kazakhstan, Kyrgyzstan, Russia and Mongolia.

甘青老鹳草

Geranium pylzowianum Maxim.

多年生草本。根茎瘤近球状。叶互生，但于花序对生。小聚伞花序单生，具2花；萼片长7.1-10.9毫米；花瓣深玫瑰粉色，基部白色，长1.6-1.8厘米。果实长2.3-2.9厘米，未成熟时直立。花期7-8月，果期9-10月。生海拔2500-5000米的林缘、高山或亚高山草甸。产中国西南和华西。

Perennial herbs. Rhizome tubercles subglobose. Leaves alternate but opposite at inflorescence. Cymules solitary, 2-flowered; sepals 7.1-10.9 mm long; petals deep rose pink with a whitish base, 1.6-1.8 cm long. Fruits 2.3-2.9 cm long, erect when immature. Fl. Jul-Aug. Fr. Sep-Oct. Forest edges, alpine or subalpine meadows at 2500-5000 m. Distributed in SW and W China.

粗根老鹳草

Geranium dahuricum DC.

多年生草本。根垂直，具一束长且加粗的根。叶对生，掌状分裂。小聚伞花序单生，具2朵花；萼片长5.5-6.9(-9)毫米；花瓣淡粉色，长(0.8-)0.9-1.3(-1.4)厘米。果实长1.7-2.6厘米，未成熟时直立。花期7-8月，果期8-9月。生海拔1500-3500米的草甸、山坡或林缘。产中国西南、华北、华西、西北和东北。俄罗斯、蒙古和朝鲜半岛亦有。

Perennial herbs. Rootstocks vertical, with a fascicle of long and thickened roots. Leaves opposite, palmately cleft. Cymules solitary, 2-flowered; sepals 5.5-6.9 (-9) mm long; petals light pink, (0.8-)0.9-1.3(-1.4) cm long. Fruits 1.7-2.6 cm long, erect when immature. Fl. Jul-Aug. Fr. Aug-Sep. Meadows, slopes or forest edges at 1500-3500 m. Distributed in SW, N, W, NW and NE China. Also in Russia, Mongolia and Korean Peninsula.

朝鲜老鹳草

Geranium koreanum Kom.

多年生草本。根状茎垂直。茎直立。叶对生，掌状分裂。小聚伞花序单生，具2花；花瓣亮粉色，外部基部具毛，内部基部约1/3具毛，基部边缘具纤毛，先端圆形。果实未成熟时直立。花期7-8月，果期8-9月。生海拔500-800米的林中或草甸。产山东东北部和辽宁东部。朝鲜半岛亦有。

Perennial herbs. Rootstocks vertical. Stem erect. Leaves opposite, palmately cleft. Cymules solitary, 2-flowered; petals bright pink, outside basally with trichomes, inside basal ca. 1/3 with trichomes, margin basally ciliate, apex rounded. Fruits erect when immature. Fl. Jul-Aug. Fr. Aug-Sep. Forests or meadows at 500-800 m. Distributed in NE Shandong and E Liaoning. Also in Korean Peninsula.

粗根老鹳草 *Geranium dahuricum*

朝鲜老鹳草 *Geranium koreanum*

灰背老鹳草 *Geranium wlassovianum*

灰背老鹳草

Geranium wlassovianum Fisch. ex Link

多年生草本。根茎直立。茎直立。叶对生，掌状分裂，具柔毛及杂以贴伏无腺毛。小聚伞花序单生，具2花；萼片长(0.8-)0.9-1.1厘米；花瓣长1.6-2.1厘米，全缘，深紫红色；蜜腺顶端具一簇毛；柱头深红色至粉色。果成熟之前直立。花期7-8月，果期8-9月。生海拔1800-3400米的林中或草甸。产华北和东北。俄罗斯、蒙古和朝鲜半岛亦有。

Perennial herbs. Rhizomes erect. Stems erect. Leaves opposite, palmately cleft, pilose with appressed nonglandular trichomes. Cymules solitary, 2-flowered; sepals (0.8-)0.9-1.1 cm long; petals 1.6-2.1 cm long, entire, deep magenta; nectariesapex with a tuft of trichomes; stigma deep red to pink. Fruits erect when immature. Fl. Jul-Aug. Fr. Aug-Sep. Forests or meadows at 1800-3400 m. Distributed in N and NE China. Also in Russia, Mongolia and Korean Peninsula.

湖北老鹳草

Geranium rosthornii R. Knuth

多年生草本。叶对生，掌状分裂，具柔毛杂以无腺毛。花序腋生或顶生；花瓣粉色至紫色，长(1-)1.2-1.4厘米，外面基部具毛，内部基部1/3-1/2具毛；边缘有糙毛；雌蕊密被柔毛；花柱分枝，枝长2-3毫米。蒴果被毛。花期6-7月，果期8-9月。生海拔1600-2400米的山地林下或山坡草地。产中国西南、华西、华中至华北。

Perennial herbs. Leaves opposite, palmately cleft, pilose with appressed nonglandular trichomes. Inflorescences axillary or terminal; petals pink to purplish, (1-)1.2-1.4 cm long, outside basally with trichomes, inside basal 1/3-1/2 with trichomes; pistils densely pubescent; styles branched, branches 2-3 mm long. Capsules pubescent. Fl. Jun-Jul. Fr. Aug-Sep. Forests on mountain regions or grassy slopes at 1600-2400 m. Distributed in SW, W, C to N China.

天竺葵

Pelargonium hortorum Bailey

多年生草本，高30-60厘米。叶互生；托叶宽三角形或卵形，长7-15毫米；叶圆形或肾形。伞形花序腋生，具多花；花梗3-4厘米，被柔毛和腺毛；萼片狭披针形，长8-10毫米；花瓣红色、橙红、粉红或白色，宽倒卵形，长12-15毫米，宽6-8毫米。蒴果长约3厘米，被柔毛。花期5-7月，果期6-9月。我国各地普遍栽培。原产非洲南部。

Perennials, stem 30-60 cm tall. Leaves alternate; stipules wide triangular or ovate, 7-15 mm long; leaves roundness or reniform. Umbel axillary, many flowers; peduncle 3-4 cm long, villous and glandular hairy; sepal narrow lanceolate, 8-10 mm long; petals red, orange, pink or white, broad obovate, 12-15 × 6-8 mm. Capsule ca. 3 cm long, villous. Fl. May-Jul. Fr. Jun-Sep. Cultivated in various regions of China. Native to S Africa.

香叶天竺葵

Pelargonium graveolens L'Hér

多年生草本或灌木状，高达1米。叶近圆形，直径2-10厘米，掌状5-7深裂，两面被长糙毛。花5-12朵，排成伞形花

湖北老鹳草 *Geranium rosthornii*

天竺葵 *Pelargonium hortorum*

香叶天竺葵 *Pelargonium graveolens*

序；萼片狭披针形，外面被腺毛和长硬毛；花瓣玫瑰色或粉红色。花期5-7月，果期8-9月。全国各地庭院栽培。

Perennial herbs or shrublike, to 1 m tall. Leaves suborbicular, 2-10 cm diam, palmately 5-7-parted, hirsute on both surfaces. Flowers 5-12 arranged in umbels; sepals narrowly lanceolate, abaxially glandular-hairy and hirsute; petals rose or pink. Fl. May-Jul. Fr. Aug-Sep. Cultivated in gardens throughout China.

熏倒牛

Biebersteinia heterostemon Maxim.

一年生草本，高30-90厘米。叶狭条形或齿状；托叶半卵形，长约1厘米。圆锥聚伞花序；苞片披针形，长2-3毫米；萼片宽卵形，长6-7毫米；花瓣黄色，倒卵形。蒴果肾形。种子肾形，长约1.5毫米，宽约1毫米。花期7-8月，果期8-9月。生海拔1000-3200米的黄土山坡、河滩地和杂草坡地。产甘肃、宁夏、青海东部和南部、四川西北部。

熏倒牛 *Biebersteinia heterostemon*

Annual herbs, stem 30-90 cm tall. Leaves narrow linetype or dentation; stipules semiovate, ca. 1 cm long. Inflorescences long, many flowered, bract lanceolate, 2-3 mm long; sepal broad ovate, 6-7 mm long; petals yellow, obovate. Capsule reniform. Seed reniform and 1.5 × 1 mm. Fl. Jul-Aug. Fr. Aug-Sep. Loess slopes, meadows, gravelly areas along rivers at 1000-3200 m. Distributed in Gansu, Ningxia, E and S Qinghai, and NW Sichuan.

金莲花科 Tropaeolaceae

Annual herbs, fleshy. Leaves orbicular to somewhat reniform, peltate and with 9 main nerves radiating from petiole, abaxial surface usually papillose. Flowers axillary, yellow, orange or varicolored; torus cup-shaped; petals 5; stamens 8, distinct, unequal; ovary 3-loculed. Fruits oblate, separating into 3 1-seeded at maturity. Fl. Jun-Oct. Fr. Jul-Oct. Commonly cultivated or escaped in most parts of China. Introduced from South America.

旱金莲
Tropaeolum majus L.

一年生肉质草本。叶圆形至稍肾形，盾状，具9个主脉自叶柄着生处向四面辐射，叶背面常具乳突。花腋生，黄色、橘黄色或多色；花托杯状；花瓣5；雄蕊8，离生，不等长。果实扁球形，成熟后分离为3个具1粒种子的瘦果。花期6-10月，果期7-10月。广泛栽培或逸生于中国大部分地区。引自南美洲。

旱金莲 *Tropaeolum majus*

亚麻科 Linaceae

石海椒
Reinwardtia indica Dum.

常绿直立灌木。叶椭圆形至倒卵状披针形，纸质，边缘全缘或具锯齿。花萼离生；花瓣黄色，离生，但是基部合生，花瓣具爪；退化雄蕊近钻形。蒴果球形，开裂为6或8瓣。花果期4月到翌年1月。生海拔550-2300米的林中、灌丛或山坡，常生于石灰性土壤。产中国西南、华南和华中。南亚亦有。

Evergreen erect shrubs. Leaves elliptic to obovate-elliptic, papery, margin entire or crenate. Sepals distinct; petals yellow, distinct but basally confluent, petals with claws; staminodes subulate. Capsules globose, splitting into 6 or 8 mericarps. Fl. and fr. Apr to next Jan. Forests, thickets or slopes at 550-2300 m, often in calcareous soil. Distributed in SW, S and C China. Also in S Asia.

石海椒 *Reinwardtia indica*

青篱柴
Tirpitzia sinensis (Hemsl.) Hallier

灌木或小乔木。叶椭圆形、倒卵状椭圆形或卵形，纸质至厚

青篱柴 *Tirpitzia sinensis*

野亚麻 *Linum stelleroides*

亚麻 *Linum usitatissimum*

纸质。花序为顶生或腋生的聚伞状；花瓣白色；爪长2-3.8厘米；子房4室。蒴果长椭圆体形至卵球形，4瓣裂。种子每室2枚，具膜质翅。花期5-8月，果期8-12月。生海拔300-2000米的石灰岩山坡或向阳处。产云南、贵州和广西。越南亦有。

Shrubs or small trees. Leaves elliptic, obovate-elliptic or ovate, papery to thickly papery. Inflorescences terminal or axillary cymes; petals white; claws 2-3.8 cm long; ovary 4-loculed. Capsules long ellipsoid to ovoid, 4-valvate. Seeds usually 2 per locule, with a membranous wing. Fl. May-Aug. Fr. Aug-Dec. Limestone slopes or sunny places at 300-2000 m. Distributed in Yunnan, Guizhou and Guangxi. Also in Vietnam.

野亚麻
Linum stelleroides Planch.

一年生或二年生草本。茎圆柱状，无毛。叶互生，线形、线状披针形或狭披针形，两面无毛。花萼边缘膜质，具黑色头状腺点；花瓣蓝色、蓝紫色或粉色；花单生或多个聚集成宽聚伞花序。蒴果近球形，开裂。花期6-9月，果期8-10月。生海拔600-2800米的山坡或荒地。产中国大部分地区。乌兹别克斯坦、吉尔吉斯斯坦、土库曼斯坦、塔吉克斯坦、俄罗斯、朝鲜半岛和日本亦有。

Annual or biennial herbs. Stems cylindric, glabrous. Leaves alternate, linear, linear-lanceolate or narrowly lanceolate, both surfaces glabrous. Sepals margin membranous and with black stipitate glands; petals blue, bluish purple or pink; flowers solitary or numerous in broad cymes. Capsules subglobose, septicidal. Fl. Jun-Sep. Fr. Aug-Oct. Slopes or wastelands at 600-2800 m. Distributed in most parts of China. Also in Uzbekistan, Kyrgyzstan, Turkmenistan, Tajikistan, Russia, Korean Peninsula and Japan.

亚麻
Linum usitatissimum L.

一年生草本。叶互生，线形、线状披针形或披针形。花腋生，单生或成阔聚伞状；花大，直径约在2.5厘米；萼片无腺体；花瓣蓝色，稀白色或红色。蒴果球形。花期6-8月，果期7-10月。生海拔1800-2600米的田野或荒地。栽培于中国西南至华北。全世界亦有。

Annual herbs. Leaves alternate, linear, linear-lanceolate or lanceolate. Flowers axillary, solitary or in broad cymes; flowers large, ca. 2.5 cm diam; sepals nonglandular; petals blue, rarely white or red. Capsules globose. Fl. Jun-Aug. Fr. Jul-Oct. Fields or wastelands at 1800-2600 m. Cultivated in SW to N China. Also in worldwide.

宿根亚麻
Linum perenne L.

多年生草本。叶互生，狭线形或线状披针形。花萼卵形，具5或7脉，全缘；花多数，排成总状聚伞花序，蓝色、蓝紫色或淡蓝色。蒴果近球形，室间开裂。花期6-8月，果期7-9月。生海拔4100米以下的沙质干河滩、干燥山坡、疏林、砾石冲积平原或草地上。产中国西南、华北、华西和西北。俄罗斯(西伯利亚)、蒙古和西亚亦有。

Perennial herbs. Leaves alternate, narrowly-linear or linear-lanceolate. Sepals ovate, 5- or 7-veined, margin entire; flowers numerous, arranged in racemose cymes, blue, blue-purple or pale-blue. Capsules subglobose, septicidal. Fl. Jun-Aug. Fr. Jul-Sep. Sandy and dry flood lands, dry slopes, open forests, gravelly floodplains or grasslands below 4100 m. Distributed in SW, N, W and NW China. Also in Russia (Siberia), Mongolia and W Asia.

宿根亚麻 *Linum perenne*

黑水亚麻 *Linum amurense*

黑水亚麻
Linum amurense Alef.

多年生草本。茎数个，具不开花的长枝。叶互生，线形至线状披针形。花组成稀疏的聚伞花序，多数；花梗果期反折；花萼顶端具一凹尖；花瓣蓝紫色。蒴果近球形，开裂。花期6-7月，果期8月。生海拔600-4000米的草地、洪积平原或干旱山坡。产华西和东北。俄罗斯(远东地区)亦有。

Perennial herbs. Stems several, with long nonflowering shoots. Leaves alternate, linear to linear-lanceolate. Flowers in sparse cymes, numerous; pedicels excurved in fruit; sepals apex with a mucro; petals bluish purple. Capsules subglobose, septicidal. Fl. Jun-Jul. Fr. Aug. Grasslands, flood plains or dry slopes at 600-4000 m. Distributed in W and NE China. Also in Russia (Far East).

异腺草
Anisadenia pubescens Griff.

多年生草本。茎纤细，直立，被微柔毛。叶散生于茎但下部较大且密集；叶椭圆形至卵形，两面具柔毛。花序为穗状总状；花梗密具柔毛；外轮3花萼具开展的腺状刚毛；花瓣5，白色至浅紫色。蒴果长圆形，膜质。花期6-9月。生海拔1200-3200米的山坡、灌丛或林下。产西藏东南部和云南。印度和不丹亦有。

Perennial herbs. Leaves scattered along stem but more concentrated and larger basally; leaves elliptic to ovate, both surfaces pilose. Inflorescences a spikelike raceme; peduncles densely pubescent; outer 3 sepals with spreading gland-tipped bristles; petals 5, white to whitish mauve. Capsules oblong, membranous. Fl. Jun-Sep. Slopes, thickets or forests at 1200-3200 m. Distributed in SE Xizang and Yunnan. Also in India and Bhutan.

黏木
Ixonanthes reticulata Jack

乔木，高30米，基部具板状根。单叶，椭圆形、椭圆状长圆形或稍倒卵形，纸质至革质，侧脉5-17对。花序疏松；花瓣白色；雄蕊10，长于花瓣。蒴果长椭圆体形，具宿存的扩大的花萼和花瓣。花期8月至翌年6月，果期6-10月。生海拔1000米以下的路边、山谷、河边或林中。产中国西南、华南和东南。南亚和东南亚亦有。

Trees, to 30 m tall, buttressed at base. Leaves simple, elliptic, elliptic-oblong or slightly obovate, papery to leathery, with lateral veins 5-17 pairs. Inflorescences lax; petals white; stamens 10, longer than petals. Capsules long ellipsoid, with persistent enlarged sepals and petals. Fl. Aug to next Jun. Fr. Jun-Oct. Roadsides, valleys, riverbanks or forests below 1000 m. Distributed in SW, S and SE China. Also in S and SE Asia.

异腺草 *Anisadenia pubescens*

黏木 *Ixonanthes reticulata*

古柯科 Erythroxylaceae

东方古柯
Erythroxylum sinense C. Y. Wu

灌木或小乔木。单叶互生，狭椭圆形、倒披针形或倒卵形，纸质。花腋生，单生或2-7簇生于极短的总花梗；花瓣粉红色；雄蕊10；花丝基部合生成筒状；子房长圆形，3室。核果长圆形，稍弯曲，有3条纵棱。花期8月至翌年5月，果期翌年5-10月。生海拔200-2200米的山坡、林中、山谷或路旁。产中国西南、华南和东南。印度、缅甸和越南亦有。

Shrubs or small trees. Leaves alternate, simple, narrowly elliptic, oblanceolate or obovate, papery. Flowers axillary, solitary or 2-7-fascicled on a very short peduncle; petals pink; stamens 10; filament bases connate into a tube; ovary oblong, 3-locular. Drupes oblong, slightly curved, with 3 longitudinal ribs. Fl. Aug. to next May. Fr. next May-Oct. Slopes, forests, valleys or roadsides at 200-2200 m. Distributed in SW, S and SE China. Also in India, Myanmar and Vietnam.

东方古柯 *Erythroxylum sinense*

蒺藜科 Zygophyllaceae

泡泡刺
Nitraria sphaerocarpa Maxim.

灌木。枝平卧，幼枝白色，不育枝先端具刺。叶簇生，条形或倒披针状条形，全缘，宽2-5毫米。萼片5，绿色；花瓣白色。果未成熟时被黄褐色柔毛，成熟时膨大成球状，外果皮干膜质；果核窄圆锥体形。花期5-6月，果期6-7月。生戈壁、山麓或砾质沙地。产内蒙古、甘肃和新疆。哈萨克斯坦和蒙古亦有。

Shrubs. Branches prostrate, young branches white, sterile branches spiny at apex. Leaves clustered, linear or oblanceolate-linear, entire, 2-5 mm width. Sepals 5, green; petals white. Unripe fruits densely yellow-brown-pubescent, ripe fruits expanding into a ball, exocarps dry and membranous; stone spindle-shaped. Fl. May-Jun. Fr. Jun-Jul. In deserts, foothills or gravelly and sandy areas. Distributed in Neimenggu, Gansu and Xinjiang. Also in Kazakhstan and Mongolia.

泡泡刺 *Nitraria sphaerocarpa*

小果白刺 *Nitraria sibirica*

白刺 *Nitraria tangutorum*

小果白刺

Nitraria sibirica Pallas

灌木。枝铺散，幼枝白色。叶4-6簇生，倒披针形。花瓣黄绿色至近白色。果椭圆体形或球形，熟时暗红色；果汁暗蓝紫色；果核卵球形。花期5-6月，果期7-8月。生绿洲沙地或湖畔盐化沙地。产华北、华西、西北和东北。中亚、俄罗斯和蒙古亦有。

Shrubs. Branches prostrate, young branches white. Leaves in fascicles of 4-6, oblanceolate. Petals yellowish green to nearly white. Fruits dark red, ellipsoid to spherical, ripe fruits dark red; mesocarps dark blue to purplish; stone ovoid. Fl. May-Jun. Fr. Jul-Aug. Distributed in sandy areas in oases or saline sandy areas along lakeshores. Distributed in N, W, NW and NE China. Also in C Asia, Russia and Mongolia.

白刺

Nitraria tangutorum Bobrov

灌木，高1-2米。幼枝白色，不育枝先端具刺。叶2-3簇生于当年枝上，阔倒披针形。聚伞花序密集；花瓣5，白色。果熟时深红色，卵球形或椭圆体形；果汁紫红色。花期5-6月，果期7-8月。生荒漠和半荒漠的湖盆沙地、河流阶地。产中国西南、华北、华西和西北。

Shrubs, 1-2 m tall. Young branches white, sterile branches spiny at apex. Leaves in fascicles of 2-3 on current year branchlets, broadly oblanceolate. Cymes dense; petals 5, white. Ripe fruits dark red, ovoid or sometimes ellipsoid; mesocarps rosy. Fl. May-Jun. Fr. Jul-Aug. Distributed in deserts or semidesert sandy areas around lakes, river terraces. Distributed in SW, N, W and NW China.

骆驼蓬

Peganum harmala L.

多年生草本，高30-70厘米。茎直立或开展。叶全裂为3-5条形至披针状条形裂片。萼片5，裂片条形；花瓣黄白色。蒴果近球形。花期5-6月，果期7-10月。生荒漠地带干旱草地、低山坡、绿洲边缘沙地或河谷沙丘。产中国西南、华北、华西和西北。西亚、中亚、北非和欧洲南部亦有。

Perennial herbs, 30-70 cm tall. Stem erect or spreading. Leaves divided into 3-5 linear to lanceo-

骆驼蓬 *Peganum harmala*

骆驼蒿 *Peganum nigellastrum*

长梗霸王 *Zygophyllum obliquum*

late-linear lobes. Sepals 5, divided into linear lobes; petals yellow-white. Capsules subglobose. Fl. May-Jun. Fr. Jul-Oct. Dry grasslands in desert areas, low slopes, sandy places in oasis edges or dunes in valleys . Distributed in SW, N, W and NW China. Also in W and C Asia, N Africa and S Europe.

骆驼蒿
Peganum nigellastrum Bunge

多年生草本，高10-25厘米，密被短硬毛。茎直立或开展。叶二至三回深裂，先端渐尖。花单生于茎端或叶腋；萼片5，宿存；花瓣淡黄色，倒披针形。蒴果近球形，黄褐色。花期5-7月，果期7-9月。生沙质或砾质地、山前平原、丘间低地、固定或半固定沙地。产内蒙古、陕西、宁夏、甘肃和新疆。蒙古亦有。

Perennial herbs, 10-25 cm tall, hispid. Stems erect or spreading. Leaf blade 2-3 × divided into lobes; lobes apex acuminate. Flowers terminal or axillary; sepals 5, persistent; petals pale yellow, oblanceolate. Capsule subglobular, yellowish brown. Fl. May-Jul. Fr. Jul-Sep. Sandy or gravelly areas, dry grasslands, hilly slopes, semidesert and steppe areas. Distributed in Neimenggu, Shaanxi, Ningxia, Gansu and Xinjiang. Also in Mongolia.

长梗霸王
Zygophyllum obliquum Popov

多年生草本，高30-80厘米。茎多分枝，光滑。小叶1对，灰蓝色，先端锐尖，基部楔形。花梗长10-18毫米；萼片5；花瓣下部橘红色，上部色淡；雄蕊短于花瓣。蒴果竖立，圆柱体形，两端钝，具5棱。花期6-8月，果期7-9月。生低山山坡、河滩沙砾地或河谷。产甘肃(河西走廊)和新疆南部。中亚亦有。

Perennial herbs, 30-80 cm tall. Stems branches numerous, glossy. Leaves with 2 leaflets, glaucous, apex acute, base cuneate. Pedicel 10-18 mm; sepals 5; petals orange at base; stamens shorter than petals. Capsule erect, cylindric, both ends obtuse, with 5 ridges. Fl. Jun-Aug. Fr. Jul-Sep. Hillsides, sandy gravel beach or river valleys. Distributed in Gansu (Hexi) Corridor and S Xinjiang. Also in C Asia.

豆型霸王
Zygophyllum fabago L.

多年生草本，高30-80厘米。根粗壮。茎多分枝。小叶1对，倒卵形至长圆状倒卵形，质厚，先端圆形。花腋生；花瓣基部橘红色，顶部白色。蒴果矩圆柱形至圆柱状，具5肋。花期5-6月，果期7-9月。生冲积平原、绿洲、湿润沙地或荒地。产内蒙古、甘肃、青海和新疆。西亚、中亚、北非和欧洲东南部亦有。

Perennial herbs, 30-80 cm tall. Roots thick. Stems much branched. Leaves with 2 leaflets, leaflets obovate to oblong-obovate, thick, apex rounded. Flowers axillary; petals basally orangish red and apically white. Capsules oblong to cylindric, with 5 ridges. Fl. May-Jun. Fr. Jul-Sep. Alluvial plains, oases, wet sands or wastelands. Distributed in Neimenggu, Gansu, Qinghai and Xinjiang. Also in W and C Asia, N Africa and SE Europe.

豆型霸王 *Zygophyllum fabago*

蝎虎霸王 *Zygophyllum mucronatum*

大翅霸王 *Zygophyllum macropterum*

蝎虎霸王
Zygophyllum mucronatum
Maximowicz

多年生草本，高15-25厘米。茎多数，平卧或开展。小叶2-3对，条形或条状矩圆形。花1-2腋生；萼片5；花瓣5，上部近白色，下部橘红色；雄蕊长于花瓣。蒴果圆柱体形，稍具5棱，先端渐尖。花期6-8月，果期7-9月。生低山山坡、冲积扇、河流阶地或黄土山坡。产内蒙古、宁夏和青海。

Perennial herbs, 15-25 cm tall. Stems numerous, prostrate or spreading. Leaves with 4 or 6 leaflets, leaflet blades linear to linear-oblong. Flowers axillary, solitary or paired; sepals 5; petals 5, white at apex and orangish red at base; stamens longer than petals. Capsule cylindric, with slightly 5 ridges, apex acuminate. Fl. Jun-Aug. Fr. Jul-Sep. Hilly slopes, alluvial fans, terraces or loess hills. Distributed in Neimenggu, Ningxia and Qinghai.

大花霸王
Zygophyllum potaninii
Maximowicz

多年生草本，高10-25厘米。茎基部多分枝。叶1-2对；叶斜倒卵形、椭圆形或圆形。花腋生；萼片淡黄；花瓣白色，基部橙色，短于萼片；雄蕊长于萼片。蒴果卵球形至球形，具5翅。花期5-6月，果期6-9月。生砾质荒漠、石质低山坡。产内蒙古、甘肃和新疆。哈萨克斯坦和蒙古亦有。

Perennial herbs, 10-25 cm tall. Stems basally much branched. Leaves with 2 or 4 leaflets; leaflet blades obliquely obovate, elliptic or rotund. Flowers axillary; sepals yellowish; petals white but orange at base, shorter than sepals; stamens longer than sepals. Capsule ovoid-globose to globose, with 5 wings. Fl. May-Jun. Fr. Jun-Sep. Deserts, gravel hills. Distributed in Neimenggu, Gansu and Xinjiang. Also in Kazakhstan and Mongolia.

大翅霸王
Zygophyllum macropterum
C. A. Meyer

多年生草本，高5-25厘米。根木质。茎平展至直立。叶3-5对；叶柄1-2厘米；叶倒卵形到长圆形。花腋生；萼片椭圆形；花瓣橙色，倒卵形，长于萼片；雄蕊10，5等长于花瓣，5短于花瓣。蒴果球状至卵球形，翅膜质。花期4-5月，果期

大花霸王 *Zygophyllum potaninii*

霸王 *Zygophyllum xanthoxylon*

5-8月。生低山、河流阶地。产新疆。俄罗斯和中亚亦有。

Perennial herbs, 5-25 cm tall. Root woody. Stems spreading to erect. Leaves with 6-10 leaflets; petiole 1-2 cm; leaflet blades obovate to oblong. Flowers axillary; sepals elliptic; petals orange, obovate, longer than sepals; stamens 10, 5 equal to petal length, 5 shorter than petal length. Capsule globose to ovoid-globose, wings membranous. Fl. Apr-May. Fr. May-Aug. Hills, river terraces. Distributed in Xinjiang. Also in Russia and C Asia.

霸王

Zygophyllum xanthoxylon (Bunge) Maxim.

灌木，高50-100厘米。枝弯曲，顶端成刺尖。叶在老枝上簇生，幼枝上对生；小叶1对，长匙形，先端钝。花生于老枝叶腋；花瓣4，淡黄色；雄蕊长于花瓣。蒴果近球形，长18-40毫米，有翅。花期4-5月，果期7-8月。生沙砾质河流阶地或低山山坡。产河北、内蒙古、甘肃、宁夏、青海和新疆。蒙古亦有。

Shrubs, 50-100 cm tall. Branches curved, spreading spiny-pointed. Leaves clustered at old branches and opposite at young ones; with 2 leaflets, leaflets long spatulate, linear-oblong. Flowers axillary on old branches; petals 4, light yellow; stamens longer than petals. Capsules subglobose, 18-40 mm long, with wings. Fl. Apr-May. Fr. Jul-Aug. Sandy gravel rocks of terraces along streams or low hilly slopes. Distributed in Hebei, Neimenggu, Gansu, Ningxia, Qinghai and Xinjiang. Also in Mongolia.

大花蒺藜 *Tribulus cistoides*

蒺藜

Tribulus terrestris L.

一年生草本。茎平卧。偶数羽状复叶；小叶对生，3-8对，长圆形至斜长圆形，基部稍偏斜，全缘。花瓣5，黄色，直径约1厘米；花梗短于叶。分果瓣5，硬，中部边缘具2个加厚的刺。花期5-8月，果期6-9月。生沙地、荒地、山坡或居民点附近。产中国大部分地区。全世界几乎均分布。

Annual herbs. Stems prostrate. Even-pinnate; leaves opposite, leaflets 3-8 pairs, leaflets oblong to obliquely oblong, base slightly oblique, margin entire. Petals 5, yellow, ca. 1 cm diam; pedicels shorter than leaves. Schizocarps with 5 carpels with 2 hardened spines at mid margin. Fl. May-Aug. Fr. Jun-Sep. Sands, wastelands, slopes or residential areas. Distributed in most parts of China. Almost worldwide.

大花蒺藜

Tribulus cistoides L.

多年生草本。枝平卧地面或上升。小叶4-7对，长圆形或倒卵状长圆形。花单生于叶腋，直径约3厘米；花梗与叶近等长；萼片披针形，外被长柔毛；花瓣倒卵状长圆形。分果具4刺或凸尖。花期5-6月。生海拔350-500米的海滨疏林或干热河谷。产云南、海南和台湾。广布热带地区。

Perennial herbs. Branches prostrate to ascending. Leaflets 4-7 pairs, leaflets oblong or obovate-oblong. Flowers solitary, axillary, ca. 3 cm diam; pedicel equal to leaves; sepals lanceolate, abaxially villose; petals obovate-oblong. Schizocarps with 4 spines or murications. Fl. May-Jun. Open forests by seashores or dry and hot valleys at 350-500 m. Distributed in Yunnan, Hainan and Taiwan. Also widespread in tropical regions.

蒺藜 *Tribulus terrestris*

芸香科
Rutaceae

大叶臭花椒
Zanthoxylum myriacanthum Wall. ex Hook. f.

落叶乔木。叶无刺，7-17小叶；小叶红褐色至黑褐色，10-20 × 4-10厘米，下面不具白粉，两面被毛，油点多。花序顶生，多花；花瓣白色。蓇葖果红棕色，油点多。花期6-8月，果期9-11月。生海拔200-1500米的山坡林中。产中国西南、华南和东南。南亚和东南亚亦有。

Deciduous trees. Leaves without prickles, 7-17-foliolate; leaflets reddish brown to blackish brown, 10-20 × 4-10 cm long, abaxially not glaucous, hairy at both sides, with many oil spots. Inflorescences terminal, many-flowered; petals white. Follicles reddish brown, oil glands numerous. Fl. Jun-Aug. Fr. Sep-Nov. Forests on mountain slopes at 200-1500 m. Distributed in SW, S and SE China. Also in S and SE Asia.

花椒簕
Zanthoxylum scandens Blume

灌木或木质藤本。小叶5-25，4-10 × 1.5-4厘米，两面干后黑色或黑褐色，油点不明显。花序顶生或腋生；萼片4，淡紫绿色；花瓣4，淡黄绿色。果和果序轴柄无毛或疏具柔毛。花期3-5月，果期7-8月。生海拔1500米以下的灌丛中或疏林下。产中国西南、华南、华中和东南。印度、缅甸、马来西亚、印度尼西亚和日本南部亦有。

Shrubs or woody climbers. Leaflets 5-25, 4-10 × 1.5-4 cm long, both surfaces black or blackish brown when dry, oil glands inconspicuous. Inflorescences terminal or axillary; sepals 4, purplish green; petals 4, yellowish green; pedicels of fruit and rachis of infructescences glabrous or sparsely puberulent. Fl. Mar-May. Fr. Jul-Aug. Thickets or sparse forests below 1500 m. Distributed in SW, S, C and SE China. Also in India, Myanmar, Malaysia, Indonesia and S Japan.

石山花椒
Zanthoxylum calcicola C. C. Huang

灌木或木质攀援藤本。小枝和叶轴具刺。小枝具皮孔。叶具9-31小叶，小叶披针形或斜长圆形，2-5 × 0.7-2.5厘米，先端锐尖或短渐尖，油点不明显。花序腋生。果实圆锥形，长3-6厘米。花期3-4月，果期9-11月。生海拔500-1600米的山地开阔林中。产云南东南部、贵州和广西。

Shrubs or woody climbers. Branchlets and leaf rachises with prickles. Branchlets lenticellate. Leaves

大叶臭花椒 *Zanthoxylum myriacanthum*

花椒簕 *Zanthoxylum scandens*

石山花椒 *Zanthoxylum calcicola*

砚壳花椒 *Zanthoxylum dissitum*

9-31-foliolate, leaflets lanceolate or obliquely oblong, 2-5 × 0.7-2.5 cm long, apex acute or shortly acuminate, oil glands inconspicuous. Inflorescences axillary. Infructescences paniculate, 3-6 cm long. Fl. Mar-Apr. Fr. Sep-Nov. Montane open forests at 500-1600 m. Distributed in SE Yunnan, Guizhou and Guangxi.

砚壳花椒
Zanthoxylum dissitum Hemsl.

木质藤本。茎灰白色。叶轴和小叶中脉具棕红色刺。叶具(3-)5-9小叶，油点不明显。花序腋生，长达10厘米；花瓣淡黄绿色。蓇葖果紧实集于果序中；在干后开裂的果实中，外果皮和中果皮在裂隙每边均大于内果皮。花期4-5月，果期9-10月。生海拔300-2600米的山地灌丛中。产中国西南、华南、华中和华西。

Woody climbers. Stems grayish white. Leaf rachises and midvein of leaflets with brownish red prickles. Leaves (3-)5-9-foliolate, oil glands inconspicuous. Inflorescences axillary, to 10 cm long; petals pale yellowish green. Follicles densely pressed together in infructescences; exocarp and mesocarp extended beyond endocarps on each side of suture in dehisced fruits. Fl. Apr-May. Fr. Sep-Oct. Montane thickets at 300-2600 m. Distributed in SW, S, C and W China.

刺壳花椒
Zanthoxylum echinocarpum Hemsl.

木质藤本。小枝和叶具刺。叶具(3-)5-11小叶，小叶互生或对生，卵形、卵状椭圆形或长椭圆形，厚革质。花序腋生或顶生。果梗几无或长3毫米，蓇葖果具刺，连刺长1厘米。花期4-5月，果期10-12月。生海拔200-1800米的林中或山坡开阔林地和灌丛。产中国西南和华中。

Woody climbers. Branchlets and leaves with prickles. Leaves (3-)5-11-foliolate, leaflets alternate or opposite, ovate, ovate-elliptic or long elliptic, thickly leathery. Inflorescences axillary or terminal. Fruits pedicels obsolete or to 3 mm long, follicles with prickles to 1 cm long. Fl. Apr-May. Fr. Oct-Dec. Forests or hillside open forests and thickets at 200-1800 m. Distributed in SW and C China.

刺壳花椒 *Zanthoxylum echinocarpum*

狭叶花椒
Zanthoxylum stenophyllum Hemsl.

灌木或小乔木。小枝和小叶背面中脉具刺。叶具9-23小叶；小叶互生，披针形、狭披针形或卵形，油点不明显。花序顶生，伞房状聚伞花序具多达30花。蓇葖果淡紫红色至深红色。花期5-6月，果期8-9月。生海拔700-2400米的山地灌丛。产四川、湖北、湖南、河南、陕西和甘肃。

Shrubs or small trees. Branchlets and midvein of leaflets abaxially with prickles. Leaves 9-23-foliolate; leaflets alternate, lanceolate, narrowly lanceolate or ovate, oil glands inconspicuous. Inflorescences terminal, cymose-corymbiform, to 30-flowered. Follicles pale purplish red to dark red. Fl. May-Jun. Fr. Aug-Sep. Mountain thickets at 700-2400 m. Distributed in Sichuan, Hubei, Hunan, Henan, Shaanxi and Gansu.

狭叶花椒 *Zanthoxylum stenophyllum*

贵州花椒
Zanthoxylum esquirolii H. Lévl.

小乔木或灌木。小枝干时淡紫红色及绿灰色。叶具5-13小叶；小叶柄无毛；小叶互生，卵形或披针形，中脉在上面凹陷。伞房状聚伞花序顶生；雌花3或4心皮。蓇葖果紫红色。花期5-6月，果期9-11月。生海拔700-3200米的山地疏林中或灌木丛中。产云南、四川和贵州。

Small trees or shrubs. Branchlets pale purplish red and glaucous when dry. Leaves 5-13-foliolate; petiolules glabrous; leaflets alternate, ovate or lanceolate, midvein on adaxial surface impressed. Corymbose cymes terminal; pistillate flowers 3- or 4-carpelled. Follicles purplish red. Fl. May-Jun. Fr. Sep-Nov. Mountain sparse forests or among thickets at 700-3200 m. Distributed in Yunnan, Sichuan and Guizhou.

贵州花椒 *Zanthoxylum esquirolii*

簕欓花椒
Zanthoxylum avicennae (Lam.) DC.

落叶乔木。小枝和叶无毛，具刺。叶具11-21小叶；轴具翅；小叶对生，不对称，基部倾斜。花序顶生，多花；花序轴紫红色；萼片绿色，宽卵形；花瓣黄白色；雌花的雌蕊具2(或3)心皮。蓇葖果浅紫红色。花期6-8月，果期10-12月。生海拔400-700米的平坦低地次生林、山丘或山谷。产云南、广西、广东、海南和福建。南亚和东南亚亦有。

Deciduous trees. Branchlets and leaves glabrous, with prickles. Leaves 11-21-foliolate; rachis winged; leaflets opposite, asymmetric, oblique at base. Inflorescences terminal, many-flowered;

簕欓花椒 *Zanthoxylum avicennae*

rachis purplish red; sepals green, broadly ovate; petals yellowish white; gynoecium in pistillate flowers 2(or 3)-carpelled. Follicles pale purplish red. Fl. Jun-Aug. Fr. Oct-Dec. Secondary forests in lowland flat areas, hillsides or valleys at 400-700 m. Distributed in Yunnan, Guangxi, Guangdong, Hainan and Fujian. Also in S and SE Asia.

椿叶花椒 (食茱萸)

Zanthoxylum ailanthoides Siebold et Zucc.

落叶乔木，高达15米。小枝和花序轴具刺。叶具11-27小叶；小叶对生，背面灰绿色或灰蓝色，油点多。花序顶生，多花；花瓣淡黄白色。蓇葖果浅红棕色，油点多。花期8-9月，果期10-12月。生海拔300-1500米的山地灌丛。产中国西南、华南和东南。菲律宾、朝鲜半岛和日本南部亦有。

Trees, to 15 m tall, deciduous. Branchlets and inflorescence rachises with prickles. Leaves 11-27-foliolate; leaflets opposite, abaxially grayish green or glaucous, oil glands numerous. Inflorescences terminal, many-flowered; petals pale yellowish white. Follicles pale reddish brown, oil glands numerous. Fl. Aug-Sep. Fr. Oct-Dec. Montane thickets at 300-1500 m. Distributed in SW, S and SE China. Also in the Philippines, Korean Peninsula and S Japan.

椿叶花椒（食茱萸）*Zanthoxylum ailanthoides*

朵花椒 *Zanthoxylum molle*

朵花椒
Zanthoxylum molle Rehder

落叶乔木。小枝和花序轴具刺。叶具(5-)13-19小叶；小叶对生，背面具浅灰色至灰黄色的厚绒毛毛被，油点不明显。花序顶生，多花；花瓣白色。蓇葖果浅紫红色，油点多。花期6-8月，果期10-11月。生海拔100-900米的开阔的山地树林和灌丛中。产中国西南、中南和东南。

Deciduous trees. Young branches and rachises of inflorescences with prickles. Leaves (5-)13-19-foliolate; leaflets opposite, abaxially with soft grayish white to grayish yellow woolly-villous indumentum, oil glands inconspicuous. Inflorescences terminal, many-flowered; petals white. Follicles pale purplish red, oil glands numerous. Fl. Jun-Aug. Fr. Oct-Nov. Mountain open forests and thickets at 100-900 m. Distributed in SW, SC and SE China.

青花椒
Zanthoxylum schinifolium Siebold et Zucc.

灌木。茎枝有短刺。叶具7-19小叶；小叶宽4-6(-25)厘米，纸质，几无柄，上面具毛，下面无毛。花序顶生；雄花的退化雄蕊甚短。蓇葖果红褐色，干后墨绿色至黑褐色。花期7-9月，果期9-12月。生海拔800米以下的疏林或灌丛。产中国除西北以外大部分地区。朝鲜半岛和日本亦有。

青花椒 *Zanthoxylum schinifolium*

异叶花椒 *Zanthoxylum dimorphophyllum*

竹叶花椒 *Zanthoxylum armatum*

Shrubs. Stem branches with small prickles. Leaves 7-19-foliolate; leaflets 4-6(-25) mm wide, papery, almost sessile, adaxially with trichomes and abaxially glabrous. Inflorescences terminal; infertile stamens of staminate flowers very short. Follicles reddish brown but dark green to brownish black when dry. Fl. Jul-Sep. Fr. Sep-Dec. Sparse forests or thickets below 800 m. Distributed in most parts of China, except NW China. Also in Korean Peninsula and Japan.

异叶花椒
Zanthoxylum dimorphophyllum Hemsl.

落叶乔木。叶具3-5(-11)小叶或为单小叶；小叶卵形、椭圆形或有时倒卵形，油点多。花序顶生；雌花具2-3心皮，花柱常反折。蓇葖果紫红色，疏被油点，具柄。花期4-6月，果期9-11月。生海拔300-2400米的山地林中湿润处，或山腰开阔的树林或灌丛中。产中国西南、华南、华中和华西。泰国和越南亦有。

Deciduous trees. Leaves 3-5(-11)-foliate or simple; leaflets ovate, elliptic or sometimes obovate, oil glands numerous. Inflorescences terminal; gynoecium in pistillate flowers 2- or 3-carpelled and styles usually recurved. Follicles purplish red, with sparse oil glands, stipitate. Fl. Apr-Jun. Fr. Sep-Nov. Moist areas in montane forests, or hillside open forests or thickets at 300-2400 m. Distributed in SW, S, C and W China. Also in Thailand and Vietnam.

刺花椒
Zanthoxylum acanthopodium DC.

灌木、木质藤本或乔木，高达6米。小枝常具刺。小叶3-9，偶有单小叶，纸质，侧脉明显，每侧10-28条。雄花花期前花药紫红色；雌花具2-5心皮；花被6-8，淡黄绿色。果紫红色。花期4-5月，果期9-10月。生海拔1400-3200米的山地灌丛中或疏林中。产中国西南。南亚和东南亚亦有。

Shrubs, woody climbers or trees, to 6 m tall. Branchlets usually with prickles. Leaflets 3-9, occasionally 1, papery, secondary veins evident, 10-28 on each side. Anthers in staminate flowers reddish purple prior to anthesis; gynoecium in pistillate flowers 2-5-carpelled; tepals 6-8, yellowish green. Fruits purple-red. Fl. Apr-May. Fr. Sep-Oct. Montane thickets or open forests at 1400-3200 m. Distributed in SW China. Also in S and SE Asia.

竹叶花椒
Zanthoxylum armatum DC.

灌木或木质藤本。小枝和小叶下面中脉常具刺。叶具3-9(或11)小叶；轴翅每侧宽达6毫米；小叶无柄，侧脉(尤其上面)常不明显，每侧具7-15中脉。圆锥花序长 2.5-6厘米；花被片6-8。果实直径5毫米，腺点粗大而凸起。花期4-5月，果期8-10月。生海拔3100米以下的灌丛中。产中国西南、华南、华东和华中。南亚、东南亚和东亚亦有。

Shrubs or woody climbers. Branchlets and leaflets abaxially on midvein usually with prickles. Leaves 3-9(or 11)-foliolate; rachis wings to 6 mm wide on each side; leaflets subsessile, secondary veins of leaflets generally faint especially adaxially, 7-15 on each side of midvein. Panicles 2.5-6 cm long; tepals 6-8. Fruits 5 mm diam, glands big and convex. Fl. Apr-May. Fr. Aug-Oct. Thickets at 3100 m. Distributed in SW, S, E and C China. Also in S, SE and E Asia.

刺花椒 *Zanthoxylum acanthopodium*

花椒 *Zanthoxylum bungeanum*

花椒

Zanthoxylum bungeanum Maxim.

乔木，高3-7米。茎和小枝具刺。茎刺基部平。小叶5-13，叶背基部中脉两侧有丛毛或小叶两面均被短柔毛，仅叶缘具油腺点。花序腋生，但顶生于侧生小枝；雌花具2-5心皮。蓇葖果紫红色，直径4-5毫米。花期4-5月，果期7-10月。生海拔3200米以下的河边、山坡灌丛或栽培。产中国大部分地区。不丹亦有。

Trees, 3-7 m tall. Stems and branchlets with prickles. Stem prickles with a flat base. Leaflets 5-13, abaxially with tufted hairs on both sides of midrib or pubescent on both surfaces, oil glands only at margin. Inflorescences axillary but terminal on lateral branchlets; pistillate flowers 2-5-carpelled. Follicles purplish red, 4-5 mm diam. Fl. Apr-May. Fr. Jul-Oct. Riversides, thickets on slopes or planted below 3200 m. Distributed in most parts of China. Also in Bhutan.

浪叶花椒

Zanthoxylum undulatifolium Hemsl.

乔木，具少量枝刺或无刺。叶具3-5(-7)小叶；小叶下面无毛，上面具粗毛至疏被柔毛，边缘波状且具圆齿。花序顶生，伞房状。蓇葖果棕红色，油腺大且内陷。花期4-5月，果期8-10月。生海拔1600-3200米的林地或灌丛中。产云南东北部、四川、湖南和陕西。

Trees, with few prickles or unarmed. Leaves 3-5(-7)-foliolate; leaflets abaxially glabrous and adaxially hirsutulous to sparsely puberulent, margin undulate and crenulate. Inflorescences terminal, corymbose. Follicles reddish brown, oil glands large and impressed. Fl. Apr-May. Fr. Aug-Oct. Forests or thickets at 1600-3200 m. Distributed in NE Yunnan, Sichuan, Hunan and Shaanxi.

野花椒

Zanthoxylum simulans Hance

灌木或小乔木。茎与枝具刺。叶具5-15小叶，叶轴具翅；小叶腹面具刺，油点多，边缘具圆齿。花序顶生；花被片淡黄绿色。蓇葖果棕红色，油点多且稍突出，顶端不具喙。花期3-5月，果期7-9月。生平原或高原林地。产中国西南、华南、华中、华北、华西和华东。

Shrubs or small trees. Stems and branchlets with prickles. Leaves 5-15-foliolate; rachises winged; leaflets adaxially with spines, oil

浪叶花椒 *Zanthoxylum undulatifolium*

野花椒 *Zanthoxylum simulans*

glands numerous, margin crenate. Inflorescences terminal, tepals pale yellowish green. Follicles reddish brown, oil glands numerous and slightly protruding, not apically beaked. Fl. Mar-May. Fr. Jul-Sep. Plains or montane forests. Distributed in SW, S, C, N, W and E China.

梗花椒

Zanthoxylum stipitatum C. C. Huang

灌木或乔木。枝刺长 1.5厘米。叶具7-17小叶；小叶油腺稀疏，中脉下面具锈色簇生毛，叶缘具锯齿。花序顶生。蓇葖果紫红色，干后油腺稍突出，基部渐狭为长1-3毫米的柄。花期4-5月，果期7-8月。生海拔100-800米的林地。产广西、广东、福建和湖南。

Shrubs or trees. Prickles to 1.5 cm long. Leaves 7-17-foliolate; leaflets with sparse oil glands, midvein abaxially rust-colored flocculent, margin serrulate. Inflorescences terminal. Follicles purplish red, oil glands slightly protruding when dry, base attenuate into a 1-3 mm long stipe. Fl. Apr-May. Fr. Jul-Aug. Forests at 100-800 m. Distributed in Guangxi, Guangdong, Fujian and Hunan.

梗花椒 *Zanthoxylum stipitatum*

三椏苦

Melicope pteleifolia (Champ. ex Benth.) T. G. Hartley

灌木或乔木。叶为3小叶，偶有2小叶或单叶；小叶卵状椭圆形、椭圆形或椭圆状倒卵形，或狭椭圆状倒卵形。花序腋生，有时基生于叶，较大主枝常开展；雌雄异花；花瓣4，具腺点。蓇葖果近球形至椭圆体形至倒卵球形。花期4-6月，果期7-10月。生海拔2300米以下的林中或灌丛。产中国西南、华南和东南。缅甸、老挝、泰国、越南和柬埔寨亦有。

Shrubs or trees. Leaves with 3 leaflets, occasionally with 2 leaflets or as simple one; leaflets ovate-elliptic, elliptic, or elliptic-obovate or narrowly so. Inflorescences axillary and sometimes basal to leaves, larger primary branches usually spreading; flowers unisexual; petals 4, glandular punctate. Follicles subglobose to ellipsoid to obovoid. Fl. Apr-Jun. Fr. Jul-Oct. Forests or thickets below 2300 m. Distributed in SW, S and SE China. Also in Myanmar, Laos, Thailand, Vietnam and Cambodia.

三椏苦 *Melicope pteleifolia*

华南吴萸
Tetradium austrosinense (Hand.-Mazz.) T. G. Hartley

乔木，高6-20米。叶具小叶5-13；小叶下面密具乳突。花序11-18厘米，顶生，多花；花4或5基数；花瓣绿色或黄绿色，干后变棕色，外无毛或散生柔毛；每室1胚珠。花期6-7月，果期9-11月。生海拔300-1500米的山地疏林或山谷。产云南南部、广西和广东。越南北部亦有。

Trees, 6-20 m tall. Leaves with 5-13 leaflets; leaflets abaxially finely papillate. Inflorescences 11-18 cm long, terminal, many-flowered; flowers 4 or 5-merous; petals green to greenish yellow but drying brown, outside glabrous or sparsely puberulent; ovules 1 per carpel. Fl. Jun-Jul. Fr. Sep-Nov. Sparse forests on mountains or valleys at 300-1500 m. Distributed in S Yunnan, Guangxi and Guangdong. Also in N Vietnam.

华南吴萸 *Tetradium austrosinense*

牛科吴萸
Tetradium trichotomum Lour.

灌木或乔木，高达8米。小叶(3-)5-11，每侧侧脉11-14条，全缘。花4(或5)数；萼片和花瓣均为4；花瓣绿色、黄色或白色，干后棕色或近白色；每心皮具2胚珠，并生。花期4-8月，果期9-11月。生海拔300-1900米林中或灌丛。产中国西南、华南和华中。老挝、泰国北部和越南亦有。

Shrubs or trees, to 8 m tall. Leaflets (3-)5-11, secondary veins 11-14 on each side, entire. Flowers 4(or 5)-merous; sepals 4, petals 4; petals green, yellow or white but drying brown or whitish; ovules 2 per carpel, collateral. Fl. Apr-Aug. Fr. Sep-Nov. Forests or thickets at 300-1900 m. Distributed in SW, S and C China. Also in Laos, N Thailand and Vietnam.

吴茱萸
Tetradium ruticarpum (A. Juss.) T. G. Hartley

灌木或乔木。叶具(3-)5-13(-15)小叶；小叶侧脉9-17对。聚伞状圆锥花序；花瓣绿色、黄色或白色。蓇葖果近球形，常5心皮。每个蓇葖果具1粒发育的种子和1粒败育种子。花期4-6月，果期8-11月。生海拔100-3000米的林中、灌丛或开阔地。产中国西南、华南、东南和华中。印度东北部、尼泊尔、不丹和缅甸亦有。

Shrubs or trees. Leaves (3-)5-13(-15)-foliolate; leaflets secondary veins 9-17 on each side. Cymous panicle; petals green, yellow or white. Fruit follicles subglobose, usually 5-carpelled. Seed 1 per follicle but paired with an abortive seed. Fl. Apr-Jun. Fr. Aug-Nov. Forests, thickets or open places at 100-3000 m. Distributed in SW, S, SE and C China. Also in NE India, Nepal, Bhutan and Myanmar.

牛科吴萸 *Tetradium trichotomum*

吴茱萸 *Tetradium ruticarpum*

楝叶吴萸

Tetradium glabrifolium
(Champ. ex Benth.) T. G. Hartley

灌木或乔木。叶具(3-)5-19小叶；小叶下面常具白粉，网脉密集。花序长9-19厘米；花萼长约0.5毫米；每心皮具2胚珠，并生或近并生。果实常具5心皮；蓇葖果三棱形。花期6-9月，果期9-12月。生海拔1200米以下的林中、灌丛或开阔地。产中国西南、华南、东南、华中和华西。南亚、东南亚和日本南部亦有。

Shrubs or trees. Leaves (3-)5-19-foliolate; leaflet blades abaxially usually glaucous, reticulate veinlets dense. Inflorescences 9-19 cm long; sepals ca. 0.5 mm long; ovules 2 per carpel, collateral or subcollateral. Fruits usually 5-carpelled; follicles trigonous. Fl. Jun-Sep. Fr. Sep-Dec. Forests, thickets or open places at 1200 m. Distributed in SW, S, SE, C and W China. Also in S and SE Asia, and S Japan.

楝叶吴萸 *Tetradium glabrifolium*

石山吴萸 *Tetradium calcicola*

无腺吴萸 *Tetradium fraxinifolium*

石山吴萸

Tetradium calcicola (Chun ex C. C. Huang) T. G. Hartley

灌木或乔木，高达15米。小叶3-7，全缘，近革质。花序长5.5-13厘米，花甚多；萼片和花瓣均为5；花瓣紫色，干后暗紫红色，外面疏至密具平伏柔毛；每心皮具2个胚珠，叠生。花期6-9月，果期9-12月。生海拔600-800米的林中或灌丛。产云南东南部、贵州南部、广西北部和西部。

Shrubs or trees, to 15 m tall. Leaflets 3-7, margins entire, subleathery. Inflorescences 5.5-13 cm long; flowers numerous; sepals 5, petals 5; petals purple, drying dull purplish red, outside sparsely to densely appressed pubescent; ovules 2 per carpel, superposed. Fl. Jun-Sep. Fr. Sep-Dec. Forests or thickets at 600-800 m. Distributed in SE Yunnan, S Guizhou, and N and W Guangxi.

无腺吴萸

Tetradium fraxinifolium (Hook. f.) T. G. Hartley

乔木，高达12米。叶具5-15小叶；小叶侧脉13-22对，边缘具圆齿或稀全缘，叶背的油点干后黑色，稀疏分布于中脉两侧。花4(或5)数；每心皮2胚珠，近并生。分果爿干后略有皱纹，油点大。花期5月和11月，果期7-11月。生海拔700-3000米的山地林中、山谷或山坡灌丛中。产云南和西藏东南部。南亚和东南亚亦有。

Trees, to 12 m tall. Leaves 5-15-foliolate; leaflets secondary veins 13-22 on each side, margin crenulate or rarely entire, oil spots at abaxial side of leaflets, black after drying, sparsely distributed near mid nerve. Flowers 4(or 5)-merous; ovules 2 per carpel, subcollateral. Mericarps slightly corrugate after drying, with big oil spots. Fl. May and Nov. Fr. Jul-Nov. Mountain forests, valleys or among bushes on slopes at 700-3000 m. Distributed in Yunnan and SE Xizang. Also in S and SE Asia.

臭檀吴萸

Tetradium daniellii (Benn.) T. G. Hartley

灌木或乔木，高达20米。小叶纸质，5-9(-11)，边缘近全缘至具圆齿。花序为伞房状聚伞花序；花瓣白色或近白色，干后近白色或淡棕色，外面无毛；

臭檀吴萸 *Tetradium daniellii*

臭节草 *Boenninghausenia albiflora*

每心皮具2胚珠，叠生。花期6-8月，果期8-11月。生海拔3200米以下的林中、林缘和开阔山坡。产中国大部分地区。朝鲜半岛亦有。

Shrubs or trees, to 20 m tall. Leaflets 5-9(-11), papery, margins subentire to crenulate. Inflorescences corymbose-cymose; petals white or whitish, drying whitish or pale brown, outside glabrous; ovules 2 per carpel, superposed. Fl. Jun-Aug. Fr. Aug-Nov. Forests, forest edges or open slopes below 3200 m. Distributed in most parts of China. Also in Korean Peninsula.

臭节草

Boenninghausenia albiflora (Hook.) Rchb. ex Meisn.

多年生常绿草本，高达 1.2米。叶互生，二至三回三出复叶。花序长60厘米；花蕾球形至卵球形至椭圆体形至长圆形。果实为顶端4开裂的蓇葖果。种子肾形，种皮革质，瘤状。花果期5-11月。生海拔500-2800米的开阔林或草坡。产中国西南、华南、东南、华中、华西和华东。南亚、东南亚和日本亦有。

Perennial evergreen herbs, to 1.2 m tall. Leaves alternate, pinnately to ternately decompounds. Inflorescences 60 cm long; flowers globose to ovoid to ellipsoid to oblong in bud. Fruits of 4 distinct apically dehiscent follicles. Seeds reniform, seed coat leathery, tuberculate. Fl. and fr. May-Nov. Open forests or grassy slopes at 500-2800 m. Distributed in SW, S, SE, C, W and E China. Also in S and SE Asia, and Japan.

北芸香

Haplophyllum dauricum (L.) G. Don

多年生草本，有香气。叶厚纸质，披针形至线形，灰绿色，油腺点甚多。伞房状聚伞花序顶生，多花；苞片线形；花瓣黄色至淡黄白色，长圆形，散生半透明油腺点；雄蕊10。每果瓣2种子。种子肾形，褐黑色。花期6-7月，果期8-9月。生低海拔的山坡、草原。产甘肃、河北、黑龙江、吉林、内蒙古、宁夏、陕西和新疆。蒙古和俄罗斯亦有。

Perennial herbs, fragrant. Leaves thick papery, lanceolate to linear, grayish green, with many oil glands. Corymbose cyme terminal, many flowered; bracts linear; petals yellow to pale yellowish white, oblong, with scattered semipellucid oil glands; stamens 10. Fruiting carpels 2-seeded. Seeds reniform, black brown. Fl. Jun-Jul. Fr. Aug-Sep. Mountain slopes or grasslands at low elevations. Distributed in Gansu, Hebei, Heilongjiang, Jilin, Neimenggu, Ningxia, Shaanxi and Xinjiang. Also in Mongolia and Russia.

北芸香 *Haplophyllum dauricum*

裸芸香

Psilopeganum sinense Hemsl.

多年生草本。叶互生，具油腺斑点，有清香柑橘气味，具3小叶，小叶椭圆形或倒卵状椭圆形，先端钝或圆，微凹缺。萼片卵形，长约1毫米；花小，淡黄色，单生；花瓣4，卵状椭圆形；雄蕊8；子房常倒心形。蓇葖果椭圆体形，顶部开裂，2室。花果期5-8月。生海拔约800米的山坡。产贵州、湖北西北部和四川东北部。

Perennial herbs. Leaves alternate, with oil glands spots and citrus scent throughout, with 3 leaflets, leaflets elliptic or obovate-elliptic, apex obtuse or rounded, slightly emarginate. Sepals ovate, ca. 1 mm long; flowers small, yellowish, solitary; petals 4, ovate-elliptic; stamens 8; ovary usually obcordate. Fruit follicles ellipsoid, cracking at the top, 2-loculed. Fl. and fr. May-Aug. Mountain slopes at ca. 800 m. Distributed in Guizhou, NW Hubei and NE Sichuan.

九里香

Murraya paniculata (L.) Jack

小乔木。小叶倒卵形或倒卵状椭圆形，中部以上最宽，顶端圆或钝，基部短尖。伞状花序通常顶生或顶生兼腋生；花白色；萼片卵形；花瓣5片，长10-15毫米；雄蕊10；花丝白色；花药背部油点2颗；花柱较子房长。果肉有黏液。种子被棉毛。花期4-8月，果期9-12月。生海拔3000米以下的林中。产云南、贵州、湖南、广东、广西、福建、海南和台湾。

四数九里香 *Murraya tetramera*

Small trees. Leaflet blades mostly suborbicular to ovate to elliptic, margin entire or crenulate, apex rounded to acuminate. Inflorescences terminal or terminal and axillary; flower white; sepals ovate; petals 5, 10-15 mm long; stamens 10; filament white; anthers oil glands 2 at back; stigma long than ovary. Fruit frlsh slime. Seeds villous. Fl. Apr-Aug. Fr. Sep-Dec. Montane forests below 3000 m. Distributed in Yunnan, Guizhou, Hunan, Guangdong, Guangxi, Fujian, Hainan and Taiwan.

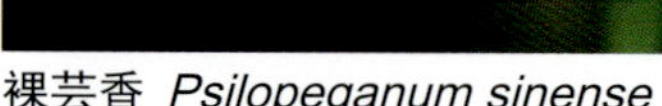

裸芸香 *Psilopeganum sinense*

四数九里香

Murraya tetramera Huang

小乔木。小叶狭长披针形，干后暗褐黑色，叶缘的上半段常有细裂齿，齿缝有油点。伞房状聚伞花序；萼片及花瓣均4片；萼片卵形；花瓣白色，长4-5毫米；雄蕊8；子房椭圆形，长约1毫米；花柱长约2毫米。果圆球形，油点甚多。种子1-3粒，种皮平滑。花期3-4月，果期7-8月。生石灰岩山地的山顶部。产广西西部和云南东南部。

Small trees. Leaflet blades lanceolate, dark brownish black when dry, serrate from middle to apex, oil glands. Inflorescences paniculate; flowers 4-merous; sepals ovate; petals white, 4-5 mm long; stamens 8; ovary ellipsoid, ca. 1 mm long; style ca. 2 mm long. Fruit globose, with many oil glands. 1-3-seeded; seed coat smooth. Fl. Mar-Apr. Fr. Jul-Aug. Often on limestone mountains. Distributed in W Guangxi and SE Yunnan.

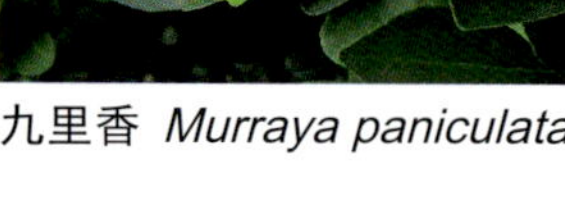

九里香 *Murraya paniculata*

霸王金橘 *Fortunella bawangica*

芸香 *Ruta graveolens*

霸王金橘

Fortunella bawangica Huang

小乔木，刺长达4厘米。小叶椭圆形或卵形，长4-7厘米，萌发枝的叶长达10厘米，先端圆，边缘中部以下具钝齿；叶柄长3-5毫米，萌发枝上的叶柄长达17毫米。花单朵腋生；花萼裂片长约1毫米；花瓣长圆形或披针形，长约7毫米。果梨形，基部狭窄且下延呈短柄状，长22-25毫米。种子单胚。生海拔约1200米的山坡杂木林中。产海南。

Small trees, spines up to 4 cm long. Leaflets elliptic or ovate, 4-7 cm long, leaves on germinated branches up to 10 cm long, apex rounded, margin obtusely serrate below middle; petiole 3-5 mm long, petiole on germinated branches up to 17 mm long. Flowers solitary, axillary; calyx lobes ca. 1 mm long; petals oblong or lanceolate, ca. 7 mm long. Fruits pyriform, base narrow and decurrent, short stalklike, 22-25 mm long. Seeds unigerminal. Weed-tree forests on slopes at ca. 1200 m. Distributed in Hainan.

芸香

Ruta graveolens L.

多年生木质草本，全草具浓烈特殊气味。叶为二至三回羽状复叶，长6-12厘米，灰绿色。花金黄色，花瓣4片；子房4室，多胚珠。果皮有明显的凸起的油点。种子多数，肾形。花期3-6月和冬季末期，果期7-9月。栽培于中国大部分地区。原产地中海沿岸地区。

Perennial woody herbs, whole plant with strong special flavor. Bipinnate or tripinnate compound leaves, 6-12 cm long, glaucous. Flowers golden yellow, petal 4; ovary 4 locules, many ovules. Pericarps covered with obvious convex oil spots. Seeds many, reniform. Fl. Mar-Jun and end of winter. Fr. Jul-Sep. Cultivated in most parts of China. Native to Mediterranean Region.

白鲜

Dictamnus dasycarpus Turcz.

多年生宿根草本。茎基部木质化。7-13小叶；叶轴边缘具狭翅；小叶卵状椭圆形、椭圆形或长圆形。总状花序长可达30厘米；花瓣粉白色至粉色，具淡紫色条纹，倒披针形。种子黑色，近球形，光亮。花期5月，果期8-9月。生丘陵、平地灌木丛、草地、疏林下或石灰岩山地。产中国东北、西北、华西、华北、华东和华中。俄罗斯(远东地区)、蒙古和朝鲜半岛亦有。

Perennial herbs with persistent roots. Stems woody at base. Leaves 7-13-foliolate; rachis marginate to narrowly winged; leaflets ovate-elliptic, elliptic or oblong. Racemes up to 30 cm long; petals pinkish white to pink, with purplish stripes, oblanceolate. Seeds black, subglobose, shiny. Fl. May. Fr. Aug-Sep. Hills, plain bushes, grasslands, under parse forests or limestone mountains. Distributed in NE, NW, W, N, E and C China. Also in Russia (Far East), Mongolia and Korean Peninsula.

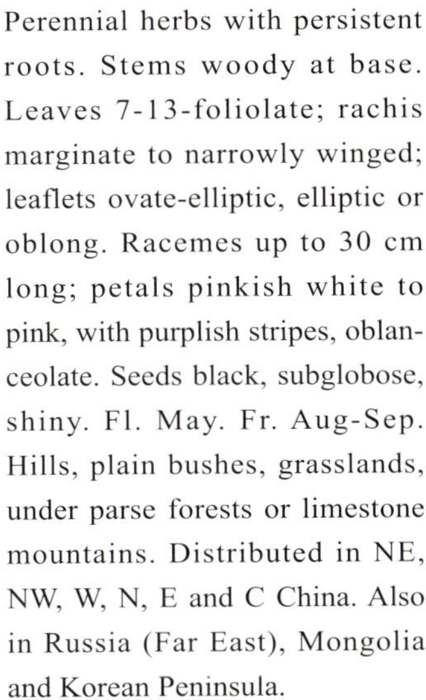

白鲜 *Dictamnus dasycarpus*

飞龙掌血 *Toddalia asiatica*

飞龙掌血

Toddalia asiatica (L.) Lam.

灌木(常平卧)或木质藤本，常具刺。小叶常无柄或近无柄，椭圆形或狭椭圆形至倒卵形至倒披针形。花序长达17厘米；花瓣乳白色，卵形至椭圆形。雌蕊柄于雌花中卵球形至椭圆体形，雄蕊中近圆柱形。果实直径5-10毫米。花期全年但多于春夏季，果期秋冬季。生海拔2000米以下的次生林或灌丛。产中国西南、华南、东南、华中和华西。南亚、东南亚和日本亦有。

Shrubs (usually sprawling) or woody climbers, usually armed. Leaflets usually sessile or subsessile, elliptic or narrowly elliptic to obovate to oblanceolate. Inflorescences to 17 cm long; petals cream-white, ovate to elliptic. Gynoecium in female flowers ovoid to ellipsoid, in male flowers subcylindric. Fruits 5-10 mm diam. Fl. year-round but mostly in spring and summer. Fr. autumn and winter. Secondary forests or thickets below 2000 m. Distributed in SW, S, SE, C and W China. Also in S and SE Asia, and Japan.

黄檗

Phellodendron amurense Rupr.

乔木。成年树的树皮有厚木栓层，具深沟状或不规则网状开裂。叶具7-13小叶；小叶卵形至卵状披针形，纸质至薄纸质。花序和果序稍疏松，分枝和花梗细弱；退化雌蕊短小。果圆球形。花期5-6月，果期9-10月。生山地林中或河谷沿岸。产中国东南、华北、华东和东北。俄罗斯、朝鲜半岛和日本亦有。

Trees. Mature bark with cork layer, deep sulciform or irregular dehiscent. Leaves 7-13-foliolate; leaflets ovate to ovate-lanceolate, papery to thinly papery. Inflorescences and infructescences lax, branches and pedicels slender; reduced pistils short. Fruits globose. Fl. May-Jun. Fr. Sep-Oct. Mountain forests or bank of river valleys. Distributed in SE, N, E and NE China. Also in Russia, Korean Peninsula and Japan.

川黄檗

Phellodendron chinense C. K. Schneid.

乔木，高达15米。叶具7-15小叶；叶轴及叶柄粗壮，通常密被褐锈色或棕色柔毛。花序和果序近紧密，花轴、分枝和花梗粗壮。果多数密集成团；果实近球形至椭圆体形，蓝黑色。花期5-6月，果期9-11月。生海拔800-1500(-3000)米的林中。产中国西南、华南、华中、华西和华东。

Trees, to 15 m tall. Leaves 7-15-foliolate; leaf axes and petioles robust, usually densely clothed with rubiginous or brown pubescence. Inflorescences and infructescences compact, rachis, branches and pedicels robust. Fruits usually many-clustered; fruits subglobose to ellipsoid, bluish black. Fl. May-Jun. Fr. Sep-Nov. Forests at 800-1500 (-3000) m. Distributed in SW, S, C, W and E China.

山油柑 (降真香)

Acronychia pedunculata (L.) Miq.

灌木或大乔木，高达28米。小叶常椭圆形至椭圆状长圆形。花序长2-25厘米，具少至多花；花瓣4，边缘内卷；雄蕊8。浆果绿色至黄色，外果皮和中果皮干后厚0.5-3毫米；中果皮木质或近木质。花期4-8月，果期8-12月。生海拔900米以下的次生林、灌丛或山谷林缘中。产中国西南和华南。南亚、东南亚和新几内亚岛亦有。

Shrubs or large trees, to 28 m

黄檗 *Phellodendron amurense*

川黄檗 *Phellodendron chinense*

tall. Leaflets usually elliptic to elliptic-oblong. Inflorescences 2-25 cm long, few to many flowered; petals 4, margins involute; stamens 8. Berries green to yellow, exocarps and mesocarps drying 0.5-3 mm thick; mesocarps woody or subwoody. Fl. Apr-Aug. Fr. Aug-Dec. Secondary forests, thickets or forest edges in valleys below 900 m. Distributed in SW and S China. Also in S and SE Asia, and New Guinea.

山油柑(降真香) *Acronychia pedunculata*

茵芋 *Skimmia reevesiana*

茵芋

Skimmia reevesiana (Fortune) Fortune

灌木，高1-2米。叶椭圆形、披针形、卵形或倒披针形，革质，中脉上面具柔毛。圆锥花序顶生，具密集的花；花有雄性、雌性或两性，芳香。果球形或椭圆状倒卵球形，红色。花期3-5月，果期9-11月。生海拔1200-2600米的山地林中。产中国西南、华南、东南、华中和华东。缅甸、越南和菲律宾亦有。

Shrubs, 1-2 m tall. Leaves elliptic, lanceolate, ovate or oblanceolate, leathery, midvein adaxially puberulent. Panicles terminal, densely flowered; flowers male, female or bisexual, fragrant. Fruits globose to ellipsoid-obovoid, red. Fl. Mar-May. Fr. Sep-Nov. Montane forests at 1200-2600 m. Distributed in SW, S, SE, C and E China. Also in Myanmar, Vietnam and the Philippines.

乔木茵芋

Skimmia arborescens T. Anders. ex Gamble

乔木，高达8米。叶纸质，无毛，中脉在上面凸起，侧脉7-10对。花序长2-5厘米，花轴具柔毛或无毛；花单性或杂性；花瓣5，淡黄色；雄花中退化雌蕊3或4室。果近圆球形，蓝黑色。花期4-6月，果期7-9月。生海拔1000-2800米的阴湿山区。产中国西南和华南。南亚亦有。

Trees, to 8 m tall. Leaves papery, glabrous, midvein adaxially prominent, secondary veins 7-10 on each side of midvein. Inflorescences 2-5 cm long, rachis puberulent or glabrous; flowers unisexual or polygamous; petals 5, yellowish; rudimentary gynoecium in staminate flowers 3- or 4-lobed. Fruits subglobose, blue-black. Fl. Apr-Jun. Fr. Jul-Sep. Shady moist montane areas at 1000-2800 m. Distributed in SW and S China. Also in S Asia.

黑果茵芋

Skimmia melanocarpa Rehder et E. H. Wisl.

灌木。枝具空髓。叶椭圆形至椭圆状披针形，中脉上面具柔毛。花序长达4厘米，密具花；花序轴具柔毛；花近无柄；花瓣5，黄白色，倒披针形至长圆形，雄花中反折。果实蓝黑色，近球形。花期3-5月，果期9-11月。生海拔2000-3000米的密林或开阔林中。产云南、四川、西藏东南部、湖北西部、陕西南部和甘肃南部。

乔木茵芋 *Skimmia arborescens*

Shrubs. Branches with hollow pith. Leaves elliptic to elliptic-lanceolate, midvein adaxially pubescent. Inflorescences to 4 cm long, with condensed flowers; rachis puberulent; flowers subsessile; petals 5, yellowish white, oblanceolate to oblong, reflexed in male flowers. Fruits bluish black, subglobose. Fl. Mar-May. Fr. Sep-Nov. Dense or open forests at 2000-3000 m. Distributed in Yunnan, Sichuan, SE Xizang, W Hubei, S Shaanxi and S Gansu.

黑果茵芋 *Skimmia melanocarpa*

小芸木
Micromelum integerrimum (Buch.-Ham.) M. Roem.

乔木。小叶7-15，基部不对称，边缘具圆齿、波状或全缘。花芽长圆形；花瓣5，浅黄色，长5-10毫米，外面密被柔毛。浆果椭圆体形至倒卵球形，表面具腺点。花期2-4月，果期7-9月。生海拔2000米以下的热带沟谷密林或石灰岩丘陵混交林中。产中国西南至华南。南亚亦有。

Trees. Leaflets 7-15, base asymmetrical, margin crenate, repand or entire. Flower buds oblong; petals 5, pale yellow, 5-10 mm long, densely pubescent outside. Berries ellipsoid to obovoid, glandular-punctate. Fl. Feb-Apr. Fr. Jul-Sep. Tropical dense forests in valleys or mixed forests on limestone hills below 2000 m. Distributed in SW to S China. Also in S Asia.

小芸木 *Micromelum integerrimum*

山小橘
Glycosmis pentaphylla (Retz.) DC.

乔木。羽状复叶，通常5小叶或1-3小叶，小叶边缘具疏齿。花序腋生或顶生，圆锥状，长于10厘米；花瓣白色或淡黄色；花丝在其上半部最宽；子房无毛。果浅红色，近球形，直径8-10毫米。花期7-10月，果期翌年1-3月。生海拔600-1200米的热带丛林中。产云南西南部和南部。南亚至东南亚亦有。

Trees. Leaves pinnate, leaflets 5 or 1-3, margins remotely serrate. Inflorescences axillary or terminal, paniculate, longer than 10 cm long; petals white or pale yellow; staminal filaments widest in their apical half; ovary glabrous. Fruits reddish, subglobose, 8-10 mm diam. Fl. Jul-Oct. Fr. next Jan-Mar. Tropical forests at 600-1200 m. Distributed in SW and S Yunnan. Also in S to SE Asia.

小花山小橘
Glycosmis parviflora (Sims) Little

灌木或小乔木。叶具(1或) 2-4(或5)小叶；小叶全缘。圆锥花序腋生或顶生，腋生者长3-5厘米，顶生者长达14厘米；花瓣白色，长约4毫米，长圆形。果球形或椭圆体形，油点明显，直径1-1.5厘米。花期3-5月，果期7-9月。生海拔200-1000米的山坡、林中、灌丛或路边。产中国西南和华南。越南、缅甸和日本南部亦有。

Shrubs or small trees. Leaves (1 or)2-4(or 5)-foliolate; leaflets margin entire. Panicles axillary or terminal, 3-5 cm long when axillary, to 14 cm long when terminal; petals white, ca. 4 mm long, oblong. Fruits globose or ellipsoid, oil spots obvious, 1-1.5 cm diam. Fl. Mar-May. Fr. Jul-Sep. Slopes, forests, bushes or roadsides at 200-1000 m. Distributed in SW and S China. Also in Vietnam, Myanmar and S Japan.

山小橘 *Glycosmis pentaphylla*

小花山小橘 *Glycosmis parviflora*

假黄皮 *Clausena excavata*

齿叶黄皮 *Clausena dunniana*

假黄皮
Clausena excavata Burm. f.

灌木，高1-2米，全株被毛，有刺激臭气。叶具21-27小叶，2-9 × 1-3厘米。花序顶生；苞片对生；花蕾球形；花白色或浅黄白色；雄蕊8，不等长。浆果椭圆球形，淡红或朱红色。花期4-5月和7-8月(海南至10月)，果期8-10月。生海拔1000米的林中、林缘或灌丛中。产中国西南和华南。南亚和东南亚亦有。

Shrubs, 1-2 m tall, whole plant with hairs, stinky smell. Leaves 21-27-foliolated, 2-9 × 1-3 cm. Inflorescences terminal; bracts opposite; flowers globose in bud; petals white or pale yellowish white; stamens 8, unequal in length. Berries ellipsoid, pink or vermilion. Fl. Apr-May and Jul-Aug(-Oct in Hannan). Fr. Aug-Oct. Forests, forest edges or thickets at 1000 m. Distributed in SW and S China. Also in S and SE Asia.

齿叶黄皮
Clausena dunniana H. Lévl.

乔木。羽状复叶，小叶5-15，基部两侧不对称，边缘有圆齿。花序顶生；花4(或5)数，花蕾球形；雄蕊8(或10)。浆果近球形，黑色至蓝黑色，直径1-1.5厘米，具1或2粒种子。花期6-7月，果期10-11月。生海拔300-1500米的混交林或石灰岩灌丛。产云南、四川、贵州、广西、广东和湖南。越南东北部亦有。

Trees. Leaves pinnate, leaflets 5-15, base asymmetrical, margins crenate. Inflorescences terminal; flowers 4(or 5)-merous, globose in bud; stamens 8(or 10). Berries subglobose, black to blue-black, 1-1.5 cm diam, 1- or 2-seeded. Fl. Jun-Jul. Fr. Oct-Nov. Mixed forests or thickets on limestone hills at 300-1500 m. Distributed in Yunnan, Sichuan, Guizhou, Guangxi, Guangdong and Hunan. Also in NE Vietnam.

黄皮
Clausena lansium (Lour.) Skeels

小乔木，高12米。叶具5-11小叶；小叶常一侧偏斜，边缘波浪状或具浅的圆裂齿。圆锥花序顶生；花蕾球形；雄蕊10。果实浅黄色，球形、椭圆体形或阔卵球形，1.5-3 × 1-2厘米，具1-4粒种子。花期4-5月，果期7-8月。产中国西南和华南。越南亦有。

Small trees, 12 m tall. Leaves 5-11-foliolate; leaflets oblique, margins sinuate or sinuous with round teeth. Panicles terminal; flowers globose in bud; stamens 10. Fruits pale yellow, globose, ellipsoid or broadly ovoid, 1.5-3 × 1-2 cm, 1-4-seeded. Fl. Apr-May. Fr. Jul-Aug. Distributed in SW and S China. Also in Vietnam.

黄皮 *Clausena lansium*

云南黄皮 *Clausena yunnanensis*

云南黄皮

Clausena yunnanensis C. C. Huang

乔木。小枝无毛。叶轴微被柔毛，叶柄光滑或微被柔毛；小叶5-11，大，长圆形或卵状椭圆形，10-40厘米，背面叶脉上被短柔毛。圆锥花序顶生，长达40厘米；花蕾球形；雄蕊10。果椭圆体形，橙黄色。花期6月，果期9-10月。生海拔500-1300米的山地密林中。产云南和广西。

Trees. Branchlets glabrous. Leaf rachises minutely pubescent, petiolules glabrous or minutely pubescent; leaves 5-11, large, oblong or ovate-elliptic, 10-40 cm long, abaxially shortly pubescent on veins. Inflorescences terminal, paniculate, to 40 cm long; flowers globose in bud; stamens 10. Fruits ellipsoid, orange. Fl. Jun. Fr. Sep-Oct. Dense forests at 500-1300 m. Distributed in Yunnan and Guangxi.

千里香

Murraya paniculata (L.) Jack

灌木或乔木，高1.8-12米。叶具2-5小叶；小叶卵形或卵状披针形，宽1.5-6厘米，背面沿中脉被白色绵毛。花序顶生或顶生及腋生；花五瓣，芳香。果橘红色至朱红色，狭椭圆体形或稀卵球形。种皮有绵毛。花期4-9月，果期4-12月。生海拔1300米以下的灌丛、山区林中或石灰岩地区。产中国西南和华南。南亚、东南亚、日本南部、澳大利亚和西南太平洋岛屿亦有。

Shrubs or trees, 1.8-12 m tall. Leaves 2-5-foliolate; leaflets ovate or ovate-lanceolate, 1.5-6 cm wide, white-lanate along midrib abaxially. Inflorescences terminal or terminal and axillary; flowers 5-merous, fragrant. Fruits orange to vermilion, narrowly ellipsoid or rarely ovoid. Seed coats lanate. Fl. Apr-Sep. Fr. Apr-Dec. Thickets, montane forests or limestone areas below 1300 m. Distributed in SW and S China. Also in S and SE Asia, S Japan, Australia and SW Pacific Islands.

调料九里香 *Murraya koenigii*

调料九里香

Murraya koenigii (L.) Spreng.

灌木或乔木。小叶17-31，斜卵形或斜卵状披针形，宽0.5-2厘米。圆锥花序顶生，多花；花蕾椭圆体形；花瓣5，白色；雄蕊10。果蓝黑色，卵球形至长圆形。花期3-4月，果期7-8月。生海拔500-1600米的湿润阔叶林中。产云南、广东和海南。南亚至东南亚亦有。

Shrubs or trees. Leaflets 17-31, oblique-ovate or ovate-lanceolate, 0.5-2 cm wide. Inflorescences terminal, paniculate, many flowered; petals ellipsoid in bud; petals 5, white; stamens 10. Fruits blue-black, ovoid to oblong. Fl. Mar-Apr. Fr. Jul-Aug. Moist broad-leaved forests at 500-1600 m. Distributed in Yunnan, Guangdong and Hainan. Also in S to SE Asia.

千里香 *Murraya paniculata*

枳 *Citrus trifoliata*

枳

Citrus trifoliata (L.) Raf.

落叶小乔木，高1-5米。树冠伞形或圆头状。叶具掌状3(-5)小叶。花单生或成对；萼片5-7，基部联合；雄蕊常20枚；花丝不等长；雌蕊萎缩。果深黄色，近圆形至梨形，表面粗糙，具环纹。花期5-6月，果期10-11月。产中国东北和西北以外中国大部分地区。

Small deciduous trees, 1-5 m tall. Crown umbellate or round cephaloid. Leaves palmately 3(-5)-foliolate. Flowers solitary or paired; calyx lobes 5-7, basally connate; stamens usually 20; filaments of different lengths; pistil reduced. Fruits dark yellow, subglobose to pyriform, with coarse ring-shaped furrows. Fl. May-Jun. Fr. Oct-Nov. Distributed in most parts of China, except NE and NW China.

宜昌橙

Citrus cavaleriei H. Lévl. ex Cavalerie

乔木或灌木。小枝近无毛；刺直伸，粗硬。叶状柄长为叶片的1-3倍，狭椭圆形，边缘具细圆齿；叶卵状披针形。花单生或多达9个成束；蕾浅紫红色；花瓣4或5，白色或粉色。果实浅黄色，扁球形、球形或梨形；果肉7-13瓣，黄白色，极酸。花期3-6月，果期10-12月。生海拔2500米以下的山坡、丘陵或沟谷。产中国西南、华中和华西。

Trees or shrubs. Branchlets subglabrous; spines straight, stout. Leafy petioles 1-3 × as long as blade, narrowly elliptic, margin finely crenulate; leaves ovate-lanceolate. Flowers solitary or to 9 in fascicles; buds pale purplish red; petals 4 or 5, white or pink. Fruits pale yellow, oblate, globose or pyriform; sarcocarps in 7-13 segments, yellowish white, very sour. Fl. Mar-Jun. Fr. Oct-Dec. Mountains, hills or valleys below 2500 m. Distributed in SW, C and W China.

香橼

Citrus medica L.

灌木或小乔木。分枝不规则；幼枝、叶芽、花蕾略带紫色。叶椭圆形至卵状椭圆形，边缘具锯齿。总状花序或有时兼有腋生单花，约有12花或有时花单生。果浅黄色，椭圆体形至近球形，达2千克。花期4-5月，果期10-11月。栽培或有时归化于云南、四川、西藏东部、贵州西南部、广西和海南。印度东北部和缅甸亦有。

Shrubs or small trees. Branches irregular; branchlets, leaf buds and flower buds purplish. Leaves elliptic to ovate-elliptic, margin serrate. Racemes or sometimes with solitary flower axillary, ca. 12-flowered or sometimes flowers solitary. Fruits pale yellow, ellip-

宜昌橙 *Citrus cavaleriei*

香橼 *Citrus medica*

soid to subglobose, to 2 kg. Fl. Apr-May. Fr. Oct-Nov. Cultivated or sometimes naturalized in Yunnan, Sichuan, E Xizang, SW Guizhou, Guangxi and Hainan. Also in NE India and Myanmar.

佛手

Citrus medica var. **sarcodactylis** (Noot.) Swingle

本变种与香橼的区别在于本变种的子房在花柱脱落后即行分裂，在果的发育过程中成手指状。果皮甚厚，通常无种子。栽培或有时归化于中国长江以南各地。原产印度东北部，缅甸可能亦有。

This variety differs from the typical variety in its ovary disintegrated after style caduceus, becoming fingerlike in the process of fruiting. Pericarp thick, usually without seed. Cultivated or sometimes naturalized throughout the provinces to the south of Yangtze River. Native to NE India, possibly also in Myanmar.

佛手 *Citrus medica* var. *sarcodactylis*

柚 *Citrus maxima*

柚

Citrus maxima (Burm.) Merr.

乔木。嫩枝、叶背、花梗、花萼及子房均被柔毛。叶宽卵形至椭圆形，9-16 × 4-8厘米或更大，厚，深绿色。花单生或呈总状花序；花蕾紫红色稀乳白色。果实浅黄色及黄绿色，球形、扁球形、梨形或阔倒圆锥形，常直径多于10厘米。花期4-5月，果期9-12月。华南栽培或归化。可能原产东南亚。

Trees. Young branches, abaxial surface of leaves, peduncles, calyx and ovaries pilose. Leaves broadly ovate to elliptic, 9-16 × 4-8 cm or larger, thick, dark green. Flowers solitary or in racemes; flower buds purplish or rarely milky white. Fruits pale yellow and yellowish green, globose, oblate, pyriform or broadly obconic, usually more than 10 cm diam. Fl. Apr-May. Fr. Sep-Dec. Cultivated and naturalized in S China. Probably native to SE Asia.

酸橙 *Citrus × aurantium*

酸橙

Citrus × aurantium (Christm.) Swingle

小乔木。枝条具长达约8厘米的刺。叶柄倒卵形；叶深绿色，厚。花序总状，具少花或花单生；完全花或雄花雌蕊近完全退化。果实橙色至淡红色，球形至扁球形，表面粗糙；果肉具10-13裂片，酸或甜或有时稍苦。花期4-5月，果期9-12月。栽培或有时归化于中国秦岭以南。

Small trees. Branches with spines up to ca. 8 cm long. Petioles obovate; leaves dark green, thick. Inflorescences racemes, with few flowers or flowers solitary; flowers perfect or male by complete abortion of pistil. Fruits orange to reddish, globose to oblate, surface coarse; sarcocarp with 10-13 segments, acidic and sweet or sometimes bitter. Fl. Apr-May. Fr. Sep-Dec. Cultivated and sometimes naturalized in South of Qinling Mountains in China.

黎檬

Citrus limonia Osb

小乔木。枝不规则，多锐刺。单身复叶，夏梢上的叶有明显的翼叶。少花簇生或单花腋生；花瓣背面淡紫色；雄蕊25-30；子房卵状，花柱比子房长3倍。果皮薄，难剥离，味颇酸，略有柠檬香味。种子长卵形，子叶绿色，常多胚。花期4-5月，果期9-10月。生较干燥坡地或河谷两岸坡地。产台湾、福建、广东、广西、湖南、贵州西南部和云南南部。

Small trees. Irregular branches, many sharp spines. Unifoliate compound leaves, in the summer the leaves have more apparent on the tip of wing leaves. Spend less tufted or single flowers altar; petals on the back of the lilac; stamens 25-30; ovary ovoid, styles about 3 times longer than ovary. Fruit skin thin, difficult to shed, taste acid, lemon flavor. Seeds long ovate, leaf green, often polyembryony. Fl. Apr-May. Fr. Sep-Oct. Dry slopes or river valley slope on both sides. Distributed in Taiwan, Fujian, Guangdong, Guangxi, Hunan, SW Guizhou and S Yunnan.

柑橘

Citrus reticulata Blanco

小乔木。单身复叶。花单生或2-3簇生；雄蕊20-25，花柱细长，柱头头状。果淡黄色、橙色、红色或洋红色，扁圆形至近圆形，表明光滑或粗糙；外果皮易剥落；果肉甜至酸，有时稍苦。花期4-5月，果期10-12月。生低海拔的山坡丛林。广泛栽培于中国秦岭南坡以南地区。

Small trees. Unifoliate compound leaves. Flowers solitary or 2-3 clustered; stamens 20-25; styles long, slender; stigma clavate. Fruits pale yellow, orange, red or carmine, oblate to subglobose, smooth or coarse; pericarps easily removed; sarcocarps sweet to acidic and sometimes bitter. Fl. Apr-May. Fr. Oct-Dec. Hillside forests at low elevations. Widely cultivated in south of Qinling Mountains in China.

黎檬 *Citrus limonia*

柑橘 *Citrus reticulata*

苦木科
Simarubaceae

刺臭椿
Ailanthus vilmoriniana Dode

乔木。小枝幼时具软刺。叶为奇数羽状复叶，叶柄紫红色，具刺；小叶8-17对，披针状长圆形，每边各具2-4个粗锯齿，锯齿背面具腺体。圆锥花序长约30厘米。翅果长约5厘米。生海拔500-2800米的山坡疏林或山谷中。产云南、四川和湖北。

Trees. Branches with soft thorns when young. Leaves odd-pinnate, with violet-red petioles and spiny; leaflets 8-17 pairs, lanceolate-oblong, each side 2-4-dentate, teeth abaxially glandular. Panicles ca. 30 cm long. Samaras ca. 5 cm long. Sparse woods of mountainous slopes or valleys at 500-2800 m. Distributed in Yunnan, Sichuan and Hubei.

臭椿 *Ailanthus altissima*

臭椿
Ailanthus altissima (Mill.) Swingle

乔木。嫩枝被黄色或黄褐色柔毛，渐无毛。叶腹面深绿色，背面灰绿色，揉碎后有臭味；小叶基部每侧具1或2齿，叶柄无刺。圆锥花序长10-30厘米；花淡绿色。翅果长圆形。种子生翅中部，扁球形。花期4-5月，果期8-10月。生海拔100-2500米的诸多生境。中国大部分地区有分布及栽培。

Trees. Young branches with yellow or yellow brown pubescence, then glabrescent. Leaflets dark green at adaxial side, grayish green at abaxial side, smelly when rubbed; leaflet blades base with 1 or 2 teeth on each side, petioles without thorns. Panicles 10-30 cm long; Flowers light green. Samarium oblong. Seed in middle of wing, flat-globose. Fl. Apr-May. Fr. Aug-Oct. Many habitats at 100-2500 m. Naturalized and widely cultivated in most parts of China.

刺臭椿 *Ailanthus vilmoriniana*

苦树

Picrasma quassioides (D. Don) Benn.

落叶乔木，雌雄异株。叶互生，奇数羽状复叶，长15-30厘米；小叶9-15，卵状披针形或宽卵形；叶痕明显，半圆形或圆形。花组成腋生聚伞花序；心皮4或5，离生。分核果成熟时蓝绿色，球形。花期4-5月，果期6-9月。生海拔1400-2400米的山地混生林中。产中国除西北以外大部分地区。印度、尼泊尔、不丹、斯里兰卡、朝鲜半岛和日本亦有。

Trees, deciduous. Leaves alternate, odd-pinnate, 15-30 cm long; leaflets 9-15, ovate-lanceolate or broadly ovate; leaf scars conspicuous, semi-rounded or rounded. Flowers in axillary cymes; carpels 4 or 5, free. Druparium blue-green when ripe, globose. Fl. Apr-May. Fr. Jun-Sep. Mountainous mixed forests at 1400-2400 m. Distributed in most parts of China, except NW China. Also in India, Nepal, Bhutan, Sri Lanka, Korean Peninsula and Japan.

苦树 *Picrasma quassioides*

中国苦树

Picrasma chinensis P. Y. Chen

乔木。嫩枝黄绿色，无毛。叶互生，小叶2-4对，全缘或有时波状或皱波状。圆锥花序腋生；萼片4。核果圆球形，包藏于宿存的花瓣内。分核果成熟时红褐色，球形。花期4-5月，果期6-8月。生海拔600-1400米的疏林或密林中。产云南、西藏和广西。

Trees. Branches yellow-green when young, glabrous. Leaves alternate, leaflets 2-4 pairs, entire or sometimes sinuate or wrinkled-sinuate. Panicles axillary; sepals 4. Drupes globose, contained in the persistent petals. Druparium red-brown when ripe, globose. Fl. Apr-May. Fr. Jun-Aug. Sparse or dense forests at 600-1400 m. Distributed in Yunnan, Xizang and Guangxi.

中国苦树 *Picrasma chinensis*

鸦胆子 *Brucea javanica*

鸦胆子
Brucea javanica (L.) Merr.

灌木或小乔木。小叶3-15，卵状披针形，边缘有粗齿。圆锥花序，雄花序长15-25(-40)厘米，雌花序长为雄花序的一半；花小，深紫色。分核果1-4，离生，长圆状卵球形，6-8 × 4-6毫米，成熟时灰黑色；外果皮干后网状皱缩。花期6-7月，果期8-10月。生海拔100-1000米的旷野或疏林中。产中国西南和华南。南亚、东南亚和澳大利亚亦有。

Shrubs or small trees. Leaflets 3-15, ovate-lanceolate, margin coarsely serrate. Panicles, 15-25 (-40) cm long in males, ca. half as long in females; flowers small, dark purple. Druparia 1-4, free, oblong-ovoid, 6-8 × 4-6 mm, gray-black when ripe; exocarps reticulately wrinkled when dry. Fl. Jun-Jul. Fr. Aug-Oct. Fields or sparse forestry at 100-1000 m. Distributed in SW and S China. Also in S and SE Asia, and Australia.

柔毛鸦胆子
Brucea mollis Wall ex Kurz

灌木或小乔木。枝条红紫色，密布皮孔。奇数羽状复叶，长20-45厘米；小叶披针形，全缘。圆锥花序，花序轴密被黄色柔毛；花萼外被短柔毛；花瓣匙形；雄花盘扁球形，雌花盘浅盘形；子房密被柔毛。核果卵圆形，8-12 × 6-8毫米，干后有浅网纹。花期3-4月，果期8-10月。生海拔750-1200米的山地疏林、密林中或路边灌丛中。产广西、广东和云南南部。

Shrubs or small trees. Branches red-purple, densely lenticellate. Leaves odd-pinnate, 20-45 cm long; blades lanceolate, margin entire. Rachis, densely yellow tomentose; sepals densely pubescent; petals spoon-shaped; disk flat and globose in males, shallowly disk-shaped in females; ovary densely pubescent. Druparium ovoid, 8-12 × 6-8 mm, shallowly reticulately wrinkled when dry. Fl. Mar-Apr. Fr. Aug-Oct. Mountainous sparse forests, thickets or roadside shrubs at 750-1200 m. Distributed in Guangxi, Guangdong and S Yunnan.

柔毛鸦胆子 *Brucea mollis*

橄榄科 Burseraceae

羽叶白头树
Garuga pinnata Roxb.

乔木。复叶、叶轴及小叶被长毛。圆锥花序腋生和侧生，被长柔毛；花开展时长0.7-1.0厘米；萼片三角形，两面被毛；花瓣长圆形，被毛；花丝基部有长毛；子房长卵形，被疏柔毛；花柱被长毛，柱头5裂。果近球形，长1.1-1.5厘米，直径1.1-1.8厘米。花期3-4月，果期4-10月。生海拔400-1400米的河谷灌丛、山坡疏林或杂木林中。产广西、四川和云南。

Trees. Compound leaves, rachis and leaflets pubescent with long hairs. Panicles lateral or axillary, densely pubescent with long hairs; flowers 0.7-1.0 cm; sepals deltoid, pubescent; petals oblong, pubescent; filaments with long hairs at base; ovary oblong, with short stipe; style pilose, stigma 5-lobed. Fruit globose, 1.1-1.5 cm long, 1.1-1.8 cm diam. Fl. Mar-Apr, fr. Apr-Oct. Mixed forests, sparse montane forests, valley scrub at 400-1400 m. Distributed in Guangxi, Sichuan and Yunnan.

橄榄
Canarium album (Lour.) Rauesch.

乔木。叶具托叶；小叶3-6对，纸质至革质，长6-14厘米，宽2-5.5厘米，全缘。花序腋生，雄花序为聚伞圆锥花序，雌花序为总状花序。核果卵球形或纺锤形，黄绿色。花期4-5月，果期10-12月。生海拔1300米以下的山谷或山坡林中。产中国西南和华南。越南亦有。

橄榄 *Canarium album*

Trees. Leaves stipulate; leaflets 3-6 pairs, papery to leathery, 6-14 × 2-5.5 cm, margin entire. Inflorescences axillary, male inflorescences cymose panicles, female inflorescences racemes. Drupes ovoid or spindle-shaped, yellow-green. Fl. Apr-May. Fr. Oct-Dec. Valleys or forests on slopes below 1300 m. Distributed in SW and S China. Also in Vietnam.

方榄
Canarium bengalense Roxb.

乔木，高达25米。叶具托叶；小叶5-6(-10)对，长圆形至倒卵状披针形，边缘波状或全缘。花序腋生，雄花序为聚伞花序。果序腋上生或腋生，具1-3果；核果绿色，纺锤形，具3凸肋。果期7-10月。生海拔400-1300米的混生林中。产广西和云南。印度、缅甸、老挝和泰国亦有。

Trees, up to 25 m tall. Leaves stipulate; leaflets 5-6 (-10) pairs, oblong to obovate-lanceolate, margin sinuate or entire. Inflorescences axillary, cymose panicles in male plants. Infructescences extra-axillary or axillary, 1-3-fruited; drupes green, spindle-shaped, 3-ribbed. Fr. Jul-Oct. Mixed forests at 400-1300 m. Distributed in Guangxi and Yunnan. Also in India, Myanmar, Laos and Thailand.

羽叶白头树 *Garuga pinnata*

方榄 *Canarium bengalense*

楝科 Meliaceae

香椿

Toona sinensis (A. Juss.) M. Roem.

乔木。树皮粗糙，切后具强烈的葱蒜味与胡椒味。小叶常8-20对，狭披针形至条状披针形，叶边缘有疏离锯齿或浅锯齿。花序下垂；花瓣白色或带粉色；花瓣边缘、子房和花盘无毛。种子在一端具翅。花期5-10月，果期8月至翌年1月。生海拔100-2900米的林中、山谷、溪边、杂木林或次生林。产中国除西北和东北以外大部分地区。南亚和东南亚亦有。

Trees. Bark fissured, smelling strongly of garlic and pepper when cut. Leaflets usually 8-20 pairs, narrowly lanceolate to linear-lanceolate, margin serrate or dentate. Inflorescences pendent; petals white or flushed pink; petals margin, ovary and disk glabrous. Seeds winged at one end. Fl. May-Oct. Fr. Aug to next Jan. Forests, valleys, along streams at 100-2900 m. Distributed in most parts of China, except NW and NE China. Also in S and SE Asia.

红椿

Toona ciliata M. Roem.

乔木。树皮内侧棕色至红色；边材切割时具强烈雪松香味。小叶(5-)9-15对，披针形至卵状披针形。花序长达55厘米，下垂；花芳香；花瓣白色至乳白色；子房每室具8胚珠。种子两端具翅。花期1-6月，果期2-11月。生海拔400-2800米的各类生境。产云南、四川、广东和海南。南亚、东南亚、澳大利亚东部和太平洋岛屿西部亦有。

红椿 *Toona ciliata*

Trees. Inner bark brown to reddish; sap-wood smelling strongly of cedar when cut. Leaflets (5-)9-15 pairs, lanceolate to ovate-lanceolate. Inflorescences to 55 cm long, pendent; flowers sweetly scented; petals white to creamy white; ovary with 8 ovules per locule. Seeds winged at both ends. Fl. Jan-Jun. Fr. Feb-Nov. Many kinds of habitats at 400-2800 m. Distributed in Yunnan, Sichuan, Guangdong and Hainan. Also in S and SE Asia, E Australia and W Pacific Islands.

香椿 *Toona sinensis*

红花香椿 *Toona fargesii*

麻楝

Chukrasia tabularis A. Juss.

乔木。小叶10-16。聚伞状圆锥花序疏松；花瓣奶油色或略带紫色，外面中部以上稀疏的短柔毛；花药生于雄蕊管顶部。蒴果灰黄色至褐色，近球形至长圆形。种子扁平，长圆形，翅宽。花期4-5月，果期7月至翌年1月。生海拔300-1600米的常绿阔叶林和落叶混交林或丘陵疏林中。产中国西南、华南和东南。南亚和东南亚亦有。

Trees. Leaflets 10-16. Thyrses lax; petals cream-colored or lavender, above middle part of outside sparsely puberulent; anthers inserted on apical margin of staminal tube. Capsules yellowish gray to brown, subglobose to oblong. Seeds flat, oblong, broadly winged. Fl. Apr-May. Fr. Jul to next Jan. Mixed evergreen broad-leaved and deciduous forests or sparse forests in hilly regions at 300-1600 m. Distributed in SW, S and SE China. Also in S and SE Asia.

红花香椿

Toona fargesii A. Chev.

乔木。叶长26-66厘米或更长；常具5-11对小叶；小叶卵状披针形至披针形。花序长达60厘米或稍长，下垂；花芽明显圆锥形；花瓣粉色、红色或紫色，倒梨形；雄花花药伸出花瓣；子房深灰褐色，花盘密具长柔毛。蒴果椭圆形。花期6-7月，果期9-12月。生海拔300-1900米的密林、杂木林、沟谷或溪边，多生潮湿处。产云南、四川、广西、广东、福建和湖北西部。

Trees. Leaves 26-66 cm long or more; leaflets usually 5-11-paired; leaflets ovate-lanceolate to lanceolate. Inflorescences to 60 cm long or more, pendent; flower buds distinctly conical; petals pink, red or purple, obpyriform; anthers exserted beyond petals of male flowers; ovary dark grayish brown, densely villous as disk. Capsules elliptic. Fl. Jun-Jul. Fr. Sep-Dec. Dense forests, mixed forests , valleys or streamsides, often in moist habitats at 300-1900 m. Distributed in Yunnan, Sichuan, Guangxi, Guangdong, Fujian and W Hubei.

麻楝 *Chukrasia tabularis*

单叶地黄连 *Munronia unifoliolata*

羽状地黄连 *Munronia pinnata*

单叶地黄连
Munronia unifoliolata Oliv.

小灌木。叶簇生于茎顶端，单生；叶膜质至坚纸质。聚伞圆锥花序近顶生或腋生于茎顶端，具1-3花；花冠白色；雄蕊管伸出，无毛。蒴果球形，稍具柔毛。种子黑色，半球形，背面凹陷。花期6-12月。生海拔200-600米的山地林中或崖边阴处。产中国西南、华南和华中。越南亦有。

Shrublets. Leaves clustered near stem apices, simple; leaves membranous to thickly papery. Thyrses subterminal or axillary on apical part of stems, with 1-3 flowers; corolla white; staminal tube exserted, glabrous. Capsules globose, puberulent. Seeds black, hemispheric, abaxially concave. Fl. Jun-Dec. Forests in mountainous regions or shady places near cliffs at 200-600 m. Distributed in SW, S and C China. Also in Vietnam.

羽状地黄连
Munronia pinnata (Wallich) W. Theobald

直立灌木。奇数羽状复叶，常聚生于茎上部，有5(3-7)小叶，椭圆形至卵状椭圆形，长5-7.5厘米，基部偏斜，全缘或具疏钝齿。总状花序腋生，少花；花萼5裂，裂片线状披针形，渐尖且外弯；花冠白色，裂片倒披针形，长1-2厘米。花期5月。生海拔200-1800米的林中、路旁灌丛、岩石缝或山坡草地。产重庆和华南。南亚和东南亚亦有。

Erect shrubs. Leaves odd-pinnate, usually aggregated apically on stem, leaflets 5(3-7), elliptic to ovate-elliptic, 5-7.5 cm long, base oblique, margin entire or with sparse obtuse teeth. Racemes axillary, few flowered; calyx 5-lobed, lobes linear-lanceolate, apex acuminate and incurved; corolla white, lobes oblanceolate, 1-2 cm long. Fl. May. Forests, thickets near roads, rock crevices or grassland on slopes at 200-1800 m. Distributed in Chongqing and S China. Also in S and SE Asia.

浆果楝
Cipadessa baccifera (Roth) Miq.

灌木或乔木。叶轴和花序无毛或近无毛。小叶常9-13，对生；小叶卵形至卵状长圆形。聚伞状圆锥花序长8-15厘米，分枝伞房状；花丝仅基部合生；子房常5室。核果成熟时紫色至黑色，球形，直径4-5毫米。花期4-10月，果期8月至翌年2月。生海拔200-2100米的山坡疏林或灌木林中。产云南、四川、贵州和广西。南亚和东南亚亦有。

Shrubs or trees. Leaf-axils and panicles glabrous or subglabrous. Leaflets usually 9-13, opposite; leaflets ovate to ovate-oblong. Thyrses 8-15 cm long, branches corymbose; filaments connate only at base; ovary usually 5-locular. Drupes purple to black when mature, globose, 4-5 mm diam. Fl. Apr-Oct. Fr. Aug to next Feb. Sparse forests or shrubs on slopes at 200-2100 m. Distributed in Yunnan, Sichuan, Guizhou and Guangxi. Also in S and SE Asia.

浆果楝 *Cipadessa baccifera*

割舌树 *Walsura robusta*

割舌树

Walsura robusta Roxb.

乔木，高10-25米。枝具皮孔，无毛。小叶3-5，对生。聚伞圆锥花序 8-17厘米，疏被短柔毛；花瓣白色；花丝基部或中部以下合生成管；子房2室，扁球形。浆果球形或卵球形，密被黄褐色柔毛。花期2-3月，果期4-6月。生海拔950-1200米的山丘密林或疏林中。产云南、广西和海南。南亚和东南亚亦有。

Trees, 10-25 m tall. Branches with lenticels, glabrous. Leaflets 3-5, opposite. Thyrses 8-17 cm long, sparsely pubescent; petals white; filaments base or basal to middle part connate into a tube; ovary 2-locular, oblate. Berries globose or ovoid, densely covered with yellowish-brown trichomes . Fl. Feb-Mar. Fr. Apr-Jun. Dense or sparse forests in hilly regions at 950-1200 m. Distributed in Yunnan, Guangxi and Hainan. Also in S and SE Asia.

鹧鸪花

Heynea trijuga Roxb.

乔木。花枝无毛，幼枝被黄色短柔毛。叶互生，小叶7-9，对生。聚伞圆锥花序腋生，花序轴稍短于叶。蒴果椭圆形，有柄，开裂成两瓣。种子1粒，以白色的假种皮包被。花期4-6月，果期5-6月和11-12月。生海拔200-1300米的丘陵林中。产云南、贵州、广西、广东和海南。南亚和东南亚亦有。

Trees. Old branches glabrous, young parts yellow pubescent. Leaves alternate, leaflets 7-9, opposite. Thyrses axillary, slightly shorter than leaves. Capsules ellipsoid and with a carpopodium, dehiscing into 2 segments. Seed 1, with a white aril. Fl. Apr-Jun. Fr. May-Jun and Nov-Dec. Forests on hills at 200-1300 m. Distributed in Yunnan, Guizhou, Guangxi, Guangdong and Hainan. Also in S and SE Asia.

米仔兰

Aglaia odorata Lour.

灌木或小乔木。幼枝顶部被星状锈色鳞片。叶轴和叶柄具狭翅；小叶3-7(或9)，对生，两面无毛。聚伞圆锥花序腋生，稍疏散，无毛；花芳香；花黄色。果不开裂，卵球形至近球形。花期5-12月，果期7月至翌年3月。生低海拔疏林或灌丛。产广西、广东和海南。老挝、泰国、越南和柬埔寨亦有。

Shrubs or small trees. Young branches apically with stellate lepidote trichomes. Petioles and rachises narrowly winged; leaflets 3-7(or 9), opposite, both surfaces glabrous. Thyrses axillary, lax, glabrous; flowers fragrant; petals yellow. Fruits indehiscent, ovoid to subglobose. Fl. May-Dec. Fr. Jul to next Mar. Sparse

鹧鸪花 *Heynea trijuga*

米仔兰 *Aglaia odorata*

forests or thickets at low elevations. Distributed in Guangxi, Guangdong and Hainan. Also in Laos, Thailand, Vietnam and Cambodia.

碧绿米仔兰

Aglaia perviridis Hiern

乔木，高达12米。小叶9-13，互生至近对生，长5-15(-18)厘米，侧脉每边12-16条。聚伞状圆锥花序腋生，具深灰色鳞片；花白色，极小；萼片5，圆形；花瓣5，圆形至卵形，白色。花期3-5月，果期9-12月。生海拔100-1400米的雨林、季雨林或常绿阔叶林中。产云南南部和东南部。印度洋群岛、南亚和东南亚亦有。

Trees, up to 12 m tall. Leaflets 9-13, alternate to subopposite, 5-15(-18) cm long, secondary veins 12-16 on each side. Thyrses axillary, dark gray squamate; flowers white, small; calyx 5, orbicular; petals 5, orbicular to ovate, white. Fl. Mar-May. Fr. Sep-Dec. Rain forests, seasonal rain forests or evergreen broad-leaved forests at 100-1400 m. Distributed in S and SE Yunnan. Also in Indian Ocean Islands, and S and SE Asia.

马肾果

Aglaia edulis (Roxb.) Wall.

乔木，高5-9米。树皮红褐色。叶柄及叶轴无毛，幼时被稀疏的棕色鳞片；小叶7(-11)，互生至近对生，中脉在上面明显凹陷。聚伞圆锥花序腋生，具褐色鳞片；花无柄，球形。浆果椭圆体形，具淡黄色皮孔，密具褐色鳞片。花期11月至翌年1月，果期翌年11月至第三年1月。生海拔1200-1800米的石灰岩山常绿阔叶林中。产云南东南部。南亚和东南亚亦有。

Trees, 5-9 m tall. Bark rufous. Petiole and rachis glabrous but sparsely brown squamate when young; leaflets 7(-11), alternate to subopposite, midveins adaxially conspicuously depressed. Thyrses axillary, brown squamate; flowers subsessile, globose. Berries elliptic, yellowish lenticellate, densely brown squamate. Fl. Nov to next Jan. Fr. next Nov to third Jan. Evergreen broad-leaved forests on limestone areas at 1200-1800 m. Distributed in SE Yunnan. Also in S and SE Asia.

马肾果 *Aglaia edulis*

山楝

Aphanamixis polystachya (Wall.) R. Parker

乔木。小叶11-21，对生。圆锥花序腋生；雄花聚伞圆锥状，雌花或两性花穗状；雄蕊管球形，无毛；花药5或6；子房3室，具密毛。蒴果，球状梨形至卵球形，成熟后带橙色。花期5-9月，果期10月至翌年4月。生海拔500-1000米的密林或疏林。产云南、广西和广东。南亚、东南亚和太平洋岛屿亦有。

Trees. Leaflets 11-21, opposite. Inflorescences axillary; staminate flowers in thyrses, pistillate flowers or bisexual flowers in spikes; staminal tubes globose, glabrous; anthers 5 or 6; ovary 3-locular, with thick trichomes. Capsules, spherical-pyriform to nearly ovoid, orangish when mature. Fl. May-Sep. Fr. Oct to next Apr. Dense or sparse forests at 500-1000 m. Distributed in Yunnan, Guangxi and Guangdong. Also in S and SE Asia, and Pacific Islands.

碧绿米仔兰 *Aglaia perviridis*

山楝 *Aphanamixis polystachya*

望谟崖摩 *Aglaia lawii*

望谟崖摩

Aglaia lawii (Wight) C. J. Saldanha et Ramamorthy

乔木或灌木。叶互生；小叶3-9，互生至近对生；小叶椭圆形、长圆形、卵状披针形或披针形，纸质至革质。圆锥状聚伞花序腋生，葡萄串状，密具鳞片毛或星状鳞片毛，少花或有时仅具1花；花瓣3或4。果实开裂。花期5-12月，果期几全年。生海拔1600米以下的多山林中、石灰岩区域、雨林沟谷、常绿阔叶林或灌丛。产中国西南和华南。南亚、东南亚和巴布亚新几内亚亦有。

Trees or shrubs. Leaves alternate; leaflets 3-9, alternate to subopposite; leaflets elliptic, oblong, ovate-lanceolate or lanceolate, papery to leathery. Thyrses axillary, botryose, densely lepidote or stellately lepidote, few flowered or sometimes with just 1 flower; petals 3 or 4. Fruits dehiscent. Fl. May-Dec. Fr. almost year-round. Forests in hilly regions, limestone regions, ravine rain forests, evergreen broad-leaved forests or thickets below 1600 m. Distributed in SW and S China. Also in S and SE Asia, and Papua New Guinea.

皮孔樫木 *Dysoxylum lenticellatum*

皮孔樫木

Dysoxylum lenticellatum C. Y. Wu ex H. Li

乔木，高10-30米。叶轴和叶柄显著有皮孔；叶互生，奇数羽状复叶。聚伞圆锥花序自老枝及二年生枝上生出，3-5束生，具褐色柔毛；花白色；萼碟形；花瓣5；雄蕊管坛状；子房与花柱被微柔毛；花盘环状，肉质。花期2-4月。生海拔900-1400米的沟谷雨林或石灰石溪旁杂木林中。产云南南部和东南部。缅甸和泰国亦有。

Trees, 10-30 m tall. Rachis and petiole conspicuously lenticellate; leaves alternate, odd-pinnate. Thyrses arising from old and second-year branches, 3-5-fascicled, brown pubescent; flowers white; calyx cup-shaped; petals 5; staminal tube urceolate; ovary and style puberulent; disc annular, fleshy. Fl. Feb-Apr. Ravine rain forests or deciduous forests near streams in limestone regions at 900-1400 m. Distributed in S and SE Yunnan. Also in Myanmar and Thailand.

海南樫木 *Dysoxylum mollissimum*

海南樫木

Dysoxylum mollissimum Blume

乔木，高7-10(-20)米，有浓厚葱臭味。小叶20-23，对生至近对生，背面被毛或被广展的长柔毛，中脉及侧脉上尤密。聚伞圆锥花序腋生，疏散，少花；花4基数；花萼碟状，裂片圆形，被柔毛；花瓣黄色。蒴果干后黄色，球形。花期1-2月或5-9月，果期10-11月或翌年3-4月。生中、低海拔的山坡疏

溪杪 *Chisocheton cumingianus* subsp. *balansae*

林或密林中。产云南、广西、广东和海南。印度、不丹、缅甸、马来西亚、印度尼西亚和菲律宾亦有。

Trees, 7-10(-20) m tall, smelled strongly like onion. Leaflets 20-23, opposite to subopposite; abaxially glabrous or sparsely villous but densely villous on midvein and secondary veins. Thyrses axillary, lax and with a few flowers; flowers 4-merous; calyx disciform, lobes round, pubescent; petals yellow. Capsules yellow when dry, globose. Fl. Jan-Feb or May-Sep. Fr. Oct-Nov or next Mar-Apr. Sparse or dense forests on slopes at low to middle elevations. Distributed in Yunnan, Guangxi, Guangdong and Hainan. Also in India, Bhutan, Myanmar, Malaysia, Indonesia and the Philippines.

溪杪

Chisocheton cumingianus (DC.) Harms subsp. **balansae** (DC.) Mabb.

乔木。幼枝和花序被黄褐色粗硬毛。偶数羽状复叶；小叶纸质或薄革质，全缘。圆锥花序腋生，常长于叶；花瓣4，线形至匙形；雄蕊管长柱状，外面上部被微柔毛。蒴果梨状球形。花期6-7月，果期10月。生山地沟谷密林中。产云南、广西和广东。印度、不丹、缅甸、老挝、泰国和越南亦有。

Trees. Young branches and inflorescences covered with brown trichomes. Leaves even-pinnate; leaflets papery or thin leathery, margin entire. Thyrses axillary, usually longer than leaves; petals 4, linear to spatulate; staminal tube long cylindric, outside apically densely puberulent. Capsule pyriform-globose. Fl. Jun-Jul. Fr. Oct. Dense forests in ravines and on hills. Distributed in Yunnan, Guangxi and Guangdong. Also in India, Bhutan, Myanmar, Laos, Thailand and Vietnam.

楝

Melia azedarach L.

落叶乔木，高达10米。二回羽状复叶；小叶有锯齿、浅锯齿或具缺刻，稀全缘。聚伞花序无毛或具短鳞片状柔毛；花淡紫色，芳香；雄蕊管紫色，具纵细脉。核果球形至椭球形；内果皮木质。花期3-5月，果期10-12月。生海拔500-2100米的常绿阔叶落叶混交林、疏林、旷野或路边。产中国除西北和东北以外大部分地区。南亚、东南亚、澳大利亚和太平洋岛屿亦有。

Trees, to 10 m tall, deciduous. Leaves bipinnate; leaflet blades serrate, crenulate or incised, rarely entire. Thyrses glabrous or covered with short lepidote pubescence; flowers lilac-colored, fragrant; staminal tubes purple, with longitudinal stripes. Drupe globose to ellipsoid; endocarps ligneous. Fl. Mar-May. Fr. Oct-Dec. Mixed evergreen broad-leaved and deciduous forests, sparse forests, field margins or roadsides at 500-2100 m. Distributed in most parts of China, except NW and NE China. Also in S and SE Asia, Australia and Pacific Islands.

楝 *Melia azedarach*

金虎尾科 Malpighiaceae

多花盾翅藤

Aspidopterys floribunda Hutch.

藤本灌木。叶椭圆形或椭圆形卵状，薄纸质，两面无毛。圆锥花序腋生或顶生，常达30 × 15厘米，具锈色柔毛；花梗长8毫米，关节上部和花萼外面无毛。翅果长圆形或长圆状披针形，先端渐狭，膜质，具明显网状脉纹，基部截状圆形。花期7-8月，果期9-10月。生海拔1400-1700米的山地灌丛或疏林中。产云南。

Lianoid shrubs. Leaves elliptic or elliptic-ovate, thinly papery, both surfaces glabrous. Panicles axillary or terminal, typically up to 30 × 15 cm long, ferruginous pubescent; pedicels 8 mm long, above articulation and abaxial surface of calyx glabrous. Samaras oblong or oblong-lanceolate, apex gradually attenuate, membranous, with distinct reticulate veins, base truncate-rounded. Fl. Jul-Aug. Fr. Sep-Oct. Sparse hill forests, shrub forests at 1400-1700 m. Distributed in Yunnan.

倒心盾翅藤

Aspidopterys obcordata Hemsl.

木质藤本。叶圆形至倒心形，背面密被黄褐色短柔毛。聚伞花序腋生，短于或与叶等长，密具黄褐色柔毛；花瓣白或淡黄色，倒卵形至长圆形。翅果长圆形或近圆形，先端微凹。花期2-3月，果期4-5月。生海拔600-1600米的高海拔林地、疏林或灌丛中。产云南南部和海南。

Woody lianas. Leaves rotund to obcordate, abaxially densely fulvous pubescent. Panicles axillary, shorter than or equaling leaves, densely yellow-brown pubescent; petals white or yellowish, obovate to oblong. Samaras oblong or suborbicular, apex retuse. Fl. Feb-Mar. Fr. Apr-May. Forests of higher-elevation hills, open forests or thickets at 600-1600 m. Distributed in S Yunnan and Hainan.

风筝果 *Hiptage benghalensis*

风筝果

Hiptage benghalensis (L.) Kurz

灌木或藤本。小枝与花序密具黄褐色或银灰色柔毛。叶革质，幼叶浅红色，具柔毛；老叶绿色，无毛，下面常具2腺体。花极芳香；花萼外被长圆形腺体。翅果之坚果背面具冠状附属物。花期2-4月，果期4-5月。生海拔(100-)200-1900米的沟谷、河岸、林缘、竹林或灌丛中。产中国西南和华南。南亚和东南亚亦有。

Shrubs or lianas. Branchlets and inflorescences densely yellowish brown or silver-gray pubescent. Leaves leathery, young leaves light red, pubescent; old leaves green, glabrous, abaxially often with 2 glands. Flowers very fragrant; sepals outside with oblong glands. Samara nuts with crested appendages on back. Fl. Feb-Apr. Fr. Apr-May. Valleys, riverbanks, forest edges, bamboo forests or thickets at (100-)200-1900 m. Distributed in SW and S China. Also in S and SE Asia.

多花盾翅藤 *Aspidopterys floribunda*

倒心盾翅藤 *Aspidopterys obcordata*

远志科 Polygalaceae

泰国黄叶树

Xanthophyllum flavescens Roxb.

乔木。腋芽1枚。叶黄绿色，光亮，披针形或长圆状披针形。总状花序或圆锥花序顶生或腋生，常多分枝；花白色，花瓣5。核果幼时绿色，无毛。花期3-4月，果期5-7月。生海拔500-2000米的潮湿林中。产云南南部。缅甸、老挝、泰国和柬埔寨亦有。

Trees. Axillary buds 1. Leaves greenish, shiny, lanceolate or oblong-lanceolate. Racemes or panicles terminal or axillary, usually much branched; flowers white, petals 5. Drupes green when young, glabrous. Fl. Mar-Apr. Fr. May-Jul. Damp forests at 500-2000 m. Distributed in S Yunnan. Also in Myanmar, Laos, Thailand and Cambodia.

云南黄叶树

Xanthophyllum yunnanense C. Y. Wu

乔木。腋芽2枚，重叠。叶线状披针形，革质，两面无毛，中脉黄色。总状花序腋生，密具柔毛。果序密具黄褐色绒毛；果梗短，粗壮；核果绿色，无毛。果期7-9月。生海拔1800-2000米的林中。产云南南部。

Trees. Axillary buds 2, superposed. Leaves linear-lanceolate, leathery, both surfaces glabrous, midvein yellow. Racemes axillary, densely yellow pubescent. Infructescences densely yellow-brown tomentose; fruit stalks short, stout; drupes green, glabrous. Fr. Jul-Sep. Forests at 1800-2000 m. Distributed in S Yunnan.

蝉翼藤

Securidaca inappendiculata Hassk.

攀援灌木。单叶互生；叶绿色，椭圆形或倒卵状长圆形，纸质或近革质，密具短伏毛。花序大，长13-15厘米；花瓣3，淡紫红色。核果较小，具长翅。花期5-8月，果期10-12月。生海拔500-1100米的林中。产云南、广西、广东和海南。南亚和东南亚亦有。

Climbing shrubs. Leaves simple, alternate; leaves green, elliptic or obovate-oblong, papery or subleathery, densely with short strigose hairs. Inflorescences large, 13-15 cm long; petals 3, pale purple-red. Drupes small, long winged. Fl. May-Aug. Fr. Oct-Dec. Forests at 500-1100 m. Distributed in Yunnan, Guangxi, Guangdong and Hainan. Also in S and SE Asia.

泰国黄叶树 *Xanthophyllum flavescens*

云南黄叶树 *Xanthophyllum yunnanense*

蝉翼藤 *Securidaca inappendiculata*

荷包山桂花 *Polygala arillata*

荷包山桂花

Polygala arillata Buch.-Ham. ex D. Don

灌木或小乔木。叶绿色，椭圆形或长圆状椭圆形至长圆状披针形，纸质。总状花序与叶对生，下垂，密具柔毛；萼片5，花后脱落，外面3枚小，内萼片2枚，花瓣状，红紫色，与花瓣几成直角着生；花瓣3，黄色。蒴果成熟后紫红色，浆果状，阔肾形或稍心形，边缘具翅。花期5-10月，果期6-11月。生海拔(700-)1000-2800(-3000)米的林下。产中国西南、东南、华中和华东。南亚和东南亚亦有。

Shrubs or small trees. Leaves green, elliptic or oblong-elliptic to oblong-lanceolate, papery. Racemes opposite to leaves, drooping, densely pubescent; sepals 5, caducous after anthesis, outer sepals 3, small, inner sepals 2, petaloid, red-purple, nearly at right angle between them and petals; petals 3, yellow. Capsules purple-red at maturity, baccate, broadly reniform or slightly cordate, margin winged. Fl. May-Oct. Fr. Jun-Nov. Forests at (700-)1000-2800(-3000) m. Distributed in SW, SE, C and E China. Also in S and SE Asia.

黄花倒水莲 *Polygala fallax*

黄花倒水莲

Polygala fallax Hemsl.

灌木或小乔木。叶披针形至椭圆状披针形。总状花序顶生或腋生，具柔毛；萼片5，早落，外面3枚小，里面2枚大，花瓣状；花瓣3，黄色。蒴果黄绿色，阔倒心形至圆形或倒卵球形，不具翅且无缘毛。花期5-8月，果期8-10月。生海拔(400-)1200-1700米的林下潮湿地或溪边。产中国西南和东南。

Shrubs or small trees. Leaves lanceolate to elliptic-lanceolate. Racemes terminal or axillary, pubescent; sepals 5, caducous, outer sepals 3, small, inner sepals 2, large, petaloid; petals 3, yellow. Capsules green-yellow, broadly obcordate to orbicular or obovoid, not winged and eciliate. Fl. May-Aug. Fr. Aug-Oct. Damp places in forests or streamsides at (400-)1200-1700 m. Distributed in SW and SE China.

尾叶远志

Polygala caudata Rehd. et Wils.

灌木。单叶互生，绝大部分螺旋状紧密地排列于小枝顶部，叶近革质，长圆形或倒披针形，先端具尾状渐尖或细尖，基部渐狭至楔形，全缘，略反卷，且波状。总状花序顶生或生于顶部数个叶腋内，数个密集成伞房状花序或圆锥状花序；萼片5，果时早落，外面3枚小，卵形，里面2枚大，花瓣状，倒卵形至斜倒卵形；花瓣3，白色、黄色或紫色，侧生花瓣较龙骨瓣短，龙骨瓣顶端背部具1盾状鸡冠状附属物。蒴果长圆状倒卵形，先端微凹，基部渐狭，具杯状环，边缘具狭翅。花期11月至翌年5月，果期翌年5-12月。生海拔1000-1800(-2100)米的石灰山林下。产广东、广西、湖北、四川、贵州和云南。

Shrubs. Leaves alternate, usually congested at apices of branchlets; leaf subleathery, blade oblong or oblanceolate, base attenuate to cuneate, margin entire, slightly recurved, undulate. Racemes terminal or axillary at apices of branchlets, several together corymbose or paniculate; sepals 5, caducous at fructescence; outer sepals 3, small, ovate, inner sepals 2, petaloid, obovate to obliquely obovate, large; petals 3, white, yellow or purple; keel longer than lateral petals; apex with peltate appendages. Capsule oblong-obovoid, apex retuse, base attenuate, with annular disk, margin narrowly winged. Fl. Nov to next May. Fr. next May-Dec. Forests on limestone mountains at 1000-1800(-2100) m. Distributed in Guangdong, Guangxi, Hubei, Sichuan, Guizhou and Yunnan.

长毛籽远志

Polygala wattersii Hance

灌木或小乔木。叶聚集于小枝顶端；叶椭圆形、椭圆状披针形或倒披针形。总状花序2-5，数个腋生于小枝顶端；萼片5，

尾叶远志 *Polygala caudata*

长毛籽远志 *Polygala wattersii*

早落，外面3枚卵形，内萼片花瓣状；花瓣3，黄色、稀白色或紫色，侧生花瓣略短于龙骨瓣，3/4以下与龙骨瓣合生，龙骨瓣具2兜状、先端圆形或2浅裂的鸡冠状附属物。蒴果倒卵形或楔形，长10-14毫米，边缘具由下而上逐渐加宽的狭翅。种子被长毛，无种阜。花期4-6月，果期5-7月。生海拔1000-1500米的林下或灌丛。产中国西南、东南和华中。越南北部亦有。

Shrubs or small trees. Leaves congested at apices of branchlets; leaves elliptic, elliptic-lanceolate or oblanceolate. Racemes 2-5, several axillary at apices of branchlets; sepals 5, caducous; outer sepals 3, ovate, inner sepals 2, petaloid; petals 3, yellow, rarely white or purple, lateral petals shorter than keel, connate in lower 3/4, keel apex with 2 cucullate appendages. Capsules obovoid or cuneate, 10-14 mm long. Seeds with long hairs, estrophiolate. Fl. Apr-Jun. Fr. May-Jul. Forests or thickets at 1000-1500 m. Distributed in SW, SE and C China. Also in N Vietnam.

密花远志

Polygala karensium Kurz.

灌木。叶线状披针形至椭圆状披针形，膜质至薄纸质。总状花序多数，生于枝条顶端，顶生或腋生；花密集；小苞片三角形；萼片5，不等大，常花瓣状，2轮排列，外面3枚小，内面2枚大，花后脱落；花瓣3，白色带紫色。花期12月至翌年4月，果期翌年3-7月。生海拔1000-2500米的山坡疏林下灌丛中。产云南、西藏和广西。不丹、缅甸、泰国和越南亦有。

Shrubs. Leaves linear-lanceolate to elliptic-lanceolate, membranous to thinly papery. Racemes numerous, terminal or axillary; flowers dense; bracteoles triangular; sepals 5, unequal, often petaloid, 2-whorled, the outer 3 small, inner 2 large, caducous aft er anthesis; petals 3, purplish white. Fl. Dec to next Apr. Fr. Mar-Jul. Sparse forests on slopes or thickets at 1000-2500 m. Distributed in Yunnan, Xizang and Guangxi. Also in Bhutan, Myanmar, Thailand and Vietnam.

密花远志 *Polygala karensium*

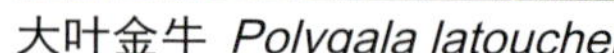
大叶金牛 *Polygala latouchei*

小扁豆 *Polygala tatarinowii*

大叶金牛
Polygala latouchei Franch.

亚灌木。叶密集于枝上部；叶背面淡红色或深紫色，长3.5-8厘米，纸质，侧脉4-5对。总状花序密具花；花瓣3，膜质，粉红色至紫红色。蒴果近球形，具翅。花期3-4月，果期4-5月。生海拔(100-)700-1300米的林下岩石上或山坡草地。产广西、广东、福建、浙江和江西。

Subshrubs. Leaves crowded on upper part of branches; leaves abaxially reddish or dark purple, 3.5-8 cm long, papery, lateral veins 4-5 pairs. Racemes densely flowered; petals 3, membranous, pink to purple-red. Capsules subglobose with wings. Fl. Mar-Apr. Fr. Apr-May. Rocks under forests or grassy slopes at (100-)700-1300 m. Distributed in Guangxi, Guangdong, Fujian, Zhejiang and Jiangxi.

小扁豆
Polygala tatarinowii Regel

一年生直立草本。茎单生或多分枝。叶绿色，卵形或椭圆形至阔椭圆形，纸质，疏具柔毛，羽状脉。总状花序顶生；小苞片2，早落；萼片5，绿色；花瓣3，红色或紫红色。蒴果扁圆形，疏具柔毛，具狭翅。花期8-9月，果期9-11月。生海拔(500-)1300-3000(-3900)米的草坡、灌丛、路边或杂木林下。产中国大部分地区。印度、不丹、缅甸北部、菲律宾、日本和马来半岛亦有。

Annual herbs, erect. Stems simple or much branched. Leaves green, ovate or elliptic to broadly elliptic, papery, sparsely pubescent, penniveined. Racemes termina; bracteoles 2, caducous; sepals 5, green; petals 3, red or purple-red. Capsules compressed-orbicular, sparsely pubescent, narrowly winged. Fl. Aug-Sep. Fr. Sep-Nov. Grassy slopes, thickets, roadsides or mixed forests at (500-)1300-3000(-3900) m. Distributed in most parts of China. Also in India, Bhutan, N Myanmar, the Philippines, Japan and Malay Peninsula.

心果小扁豆
Polygala isocarpa Chodat

一年生草本。茎圆柱状，坚硬，光滑。叶绿色，卵形或卵状三角形。总状花序顶生；苞片和小苞片早落；萼片5；花瓣3，黄色。蒴果宽倒心形或近球状，具狭翅。花果期9-10月。生海拔1200-1400米的林中、草地或路边。产云南、四川和贵州。

Annual herbs. Stems terete, ridged, glabrous. Leaves green, ovate or ovate-triangular. Racemes terminal; bracts and bracteoles caducous; sepals 5; petals 3, yellow. Capsules broadly obcordate or suborbicular, narrowly winged. Fl. and fr. Sep-Oct. Forests, grasslands or roadsides at 1200-1400 m. Distributed in Yunnan, Sichuan and Guizhou.

瓜子金
Polygala japonica Houtt.

多年生草本，高15-20厘米。叶全缘，卵形或卵状披针形，侧脉3-5对。总状花序与叶对生或

心果小扁豆 *Polygala isocarpa*

shaped. Fl. Mar-Apr. Fr. Apr-May. Damp rocks in forests at 1200-2100 m. Distributed in Guangxi and Yunnan. Also in N Vietnam.

瓜子金 *Polygala japonica*

腋外生，花序低于茎顶；萼片5，宿存；花瓣3，白色至紫色；雄蕊8。蒴果圆形，边缘具宽翅。花期4-5月，果期5-8月。生海拔800-2100米的山坡草地或田埂。产中国除华南以外各地。南亚、东南亚、东北亚和新几内亚岛亦有。

Perennial herbs, 15-20 cm tall. Leaves entire, ovate or ovate-lanceolate, lateral veins 3-5 pairs. Racemes opposite to leaves or extra-axillary, top raceme lower than apices of stems; sepals 5, persistent; petals 3, white to purple; stamens 8. Capsules orbicular, margin broadly winged. Fl. Apr-May. Fr. May-Aug. Grassy slopes or field ridges at 800-2100 m. Distributed in most parts of China, except S China. Also in S, SE and NE Asia, and New Guinea.

岩生远志

Polygala saxicola Dunn

直立草本或亚灌木。茎和枝被柔毛。叶倒卵形或椭圆形。总状花序顶生，具1-3花，被柔毛；花瓣3，粉红至淡紫色，先端具4枚片状附属物，黄色；花丝3/4以下合生。蒴果球形或近方形，具翅，顶端凹缺具短尖头。种子卵形，被白色柔毛，种阜翅状。花期3-4月，果期4-5月。生海拔1200-2100米的林下岩缝潮湿的岩石上。产广西和云南。越南北部亦有。

Herbs erect or subshrubs. Stems and branches pubescent. Leaves obovate or elliptic. Racemes terminal, 1-3-flowered, pubescent; petals 3, pink to purplish, apex with appendages 4-lamellate, yellow; filaments lower 3/4 united. Capsule orbicular or subsquare, winged, apex retuse and mucronate. Seeds ovoid, with white pubescent; strophiole wing-

坝王远志

Polygala bawanglingensis Xing et Z. X. Li

亚灌木。茎红褐色。叶倒卵形或椭圆形，长1-8厘米，先端钝，具短尖头，基部楔形，全缘，两面无毛，主脉背面隆起，上面凹陷。总状花序顶生，腋生或腋外生，长约5厘米，无毛；萼片5，早落；花瓣3，黄色，龙骨瓣具舟状鸡冠状附属物。种子椭圆形，密被白色柔毛，具种阜和附属物。花果期5-6月。生海拔900-1000米的石灰岩山上。产海南。

Subshrubs. Stems red brown. Leaves obovate or elliptic, 1-8 cm long, apex obtuse, mucronate, base cuneate, margin entire, both surfaces glabrous, midvein raised abaxially, impressed adaxially. Racemes terminal, axillary or extra-axillary, ca. 5 cm long, glabrous; sepals 5, caducous; petals 3, yellow, keel with boatlike appendages. Seeds ellipsoidal, densely white pilose, with strophiole and appendage. Fl. and fr. May-Jun. Limestone mountains at 900-1000 m. Distributed in Hainan.

岩生远志 *Polygala saxicola*

坝王远志 *Polygala bawanglingensis*

新疆远志
Polygala hybrida DC.

多年生草本。茎通常多数丛生。单叶互生，无柄。总状花序顶生；萼片5，宿存，外面3枚小，椭圆状披针形，里面2枚大，花瓣状；花瓣3，紫红色，鸡冠状附属物条状微裂。蒴果长圆形，具翅。花期5-7月，果期6-9月。生海拔1200-1800米的林下、草地或河漫滩。产新疆。蒙古和俄罗斯亦有。

Herbs perennial. Stems caespitose. Leaves alternate, sessile. Racemes terminal; sepals 5, persistent, outer sepals 3, small, elliptic-lanceolate, inner sepals 2, large, petaloid; petals 3, purple-red, keel apex with slightly lobed appendages. Capsule oblong, winged. Fl. May-Jul. Fr. Jun-Sep. Forests on slopes of hills, grasslands or floodplain at 1200-1800 m. Distributed in Xinjiang. Also in Mongolia and Russia.

新疆远志 *Polygala hybrida*

小花远志 *Polygala arvensis*

小花远志
Polygala arvensis Willd.

一年生草本。茎多分枝。叶互生，叶厚纸质，倒卵形、长圆形或椭圆状长圆形。总状花序腋生或腋外生，长不及叶，花少，但密集；萼片5，宿存；花瓣3，白色或紫色；龙骨瓣盔状，较侧瓣长，顶端背部具2束多分枝的鸡冠状附属物。蒴果近圆形，几无翅。花果期7-10月。生中低海拔的山坡草地、海岸边。产华东、华南和西南和台湾。南亚、东南亚和澳大利亚亦有。

Herbs annual. Stems multi-branched. Leaves alternate, thickly papery, obovate, oblong or elliptic-oblong. Racemes axillary or extra-axillary, shorter than leaves, few flowered and clusterlike; sepals 5, persistent; petals 3, white or purple; keel cucullate, longer than lateral petals, apex with 2 indistinct groups of multifid appendages. Capsule suborbicular, minutely winged. Fl. and fr. Jul-Oct. Beaches, grassland slopes of hills near sea level to middle elevations. Distributed in E, S and SW China, and Taiwan. Also in S and SE Asia, and Australia.

远志
Polygala tenuifolia Willd.

多年生草本。茎多分枝，丛生。叶近无柄；叶线形至线状披针形，纸质。总状花序顶生，花稀少，偏向一侧；萼片5，宿存；花瓣3，紫色。蒴果球形，具狭翅。花果期5-9月。生海拔(200-)500-2300米的草地、林中、灌丛中或山坡。产中国西南、华中、西北、华北和东北。俄罗斯、蒙古和朝鲜半岛亦有。

Perennial herbs. Stems much-branched, caespitose. Leaves subsessile; leaves linear to linear-lanceolate, papery. Racemes terminal, sparsely flowered, secund; sepals 5, persistent; petals 3, purple. Capsules globose, narrowly winged. Fl. and fr. May-Sep. Grasslands, forests, thickets or slopes at (200-)500-2300 m. Distributed in SW, C, NW, N and NE China. Also in Russia, Mongolia and Korean Peninsula.

圆锥花远志
Polygala paniculata L.

一年生直立草本。茎上部多分枝，呈圆锥状。叶互生，或下部4-5假轮生，叶披针形至线状披针形，近无柄。总状花序顶生或与叶对生；萼片5，外面3

远志 *Polygala tenuifolia*

圆锥花远志 *Polygala paniculata*

香港远志 *Polygala hongkongensis*

枚小，内面2枚花瓣状，带紫色，椭圆状长圆形；花瓣3，白色或紫罗兰色，龙骨瓣具多裂(4-6)的鸡冠状附属物。蒴果长圆形，顶端具缺刻，几无翅，无毛。种子长圆形，黑色，密被白色短柔毛。生低海拔。热带广布；广东和台湾引种栽培。

Herbs annual, erect. Stems mostly much branched, paniculate. Leaves alternate, lowest 4-5 often in pseudowhorls; leaf blade lanceolate to linear-lanceolate, subsessile. Racemes terminal or opposite to leaves; sepals 5, outer sepals 3, small, inner sepals 2, petaloid, purple, elliptic-oblong; petals 3, white or violet, keel apex with multifid appendages. Capsule oblong, apex notched, not winged, glabrous. Seeds oblong, black, densely white pubescent. At low elevations. Widely distributed in the tropics; introduced and naturalized in Guangdong and Taiwan.

蓼叶远志

Polygala persicariifolia DC.

一年生草本。茎中部以上多分枝。叶薄纸质，披针形至线状披针形。总状花序枝叉生或顶生；萼片5，宿存，内萼片2枚，花瓣状，阔倒卵形或圆形；花瓣3，粉红色至紫色，龙骨瓣盔状，具缘毛，顶端具2束线状鸡冠状附属物。蒴果长圆形或圆形，具狭翅及缘毛，顶端微缺。花期7-9月，果期8-10月。生海拔1200-2200(-2800)米的山坡林下、草地或路旁。产广西、四川、贵州和云南。西南亚、南亚、东南亚和非洲亦有。

Herbs annual. Stems mostly branched. Leaf thinly papery, blade lanceolate to linear-lanceolate. Racemes in fork or terminal on lateral branches; sepals 5, persistent, inner sepals 2, petaloid, broadly obovate or orbicular; petals 3, pink to purple, keel cucullate, ciliate, apex with 2 bundles of linear appendages. Capsule oblong or orbicular, narrowly winged, ciliate, apex retuse. Fl. Jul-Sep. Fr. Aug-Oct. Forests, grasslands or roadsides on slopes of hills at 1200-2200(-2800) m. Distributed in Guangxi, Sichuan, Guizhou and Yunnan. Also in SW, S and SE Asia, and Africa.

香港远志

Polygala hongkongensis Hemsl.

直立草本或亚灌木。茎枝细。单叶互生，茎下部叶小，卵形，上部叶披针形。总状花序顶生；萼片5，宿存，内萼片2枚，花瓣状；花瓣3，白色或紫色，龙骨瓣顶端具流苏状鸡冠状附属物。蒴果近圆形，具阔翅，先端具缺刻。花期5-6月，果期6-7月，或花果期3-4月。生海拔300-1400米的沟谷林下或灌丛中。产华中、华南、华东、西南和新疆。

Herbs or subshrubs, erect. Stem and branches thin. Leaves alternate, lower leaves small, ovate, upper leaves lanceolate. Racemes terminal; sepals 5, persistent, inner sepals 2, petaloid; petals 3, white or purple, keel apex with broad fimbriate appendages. Capsule suborbicular, broadly winged, apex retuse. Fl. May-Jun. Fr. Jun-Jul, or fl. and fr. Mar-Apr. Forests, shrub forests, mountain valleys at 300-1400 m. Distributed in C, S, E and SW China, and Xinjiang.

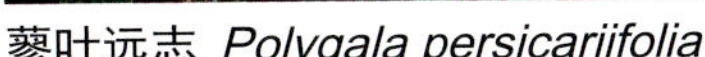
蓼叶远志 *Polygala persicariifolia*

西伯利亚远志 *Polygala sibirica*

华南远志 *Polygala glomerata*

西伯利亚远志
Polygala sibirica L.

多年生草本。茎丛生，通常直立。叶纸质或近革质；下部叶卵形，小；上部叶披针形或椭圆状披针形，大。总状花序生腋外或假顶生，少花，一般高出茎顶；萼片5，宿存；花瓣3，蓝紫色。蒴果近心形，具狭翅。花期4-7月，果期5-8月。生海拔1100-3300米的林缘、草地、草坡、石灰岩山地或路边。产中国大部分地区。西南亚、东北亚、澳大利亚和欧洲东部亦有。

Perennial herbs. Stems caespitose, often erect. Leaves papery or subleathery; lower leaves ovate, small; upper leaves lanceolate or elliptic-lanceolate, large. Racemes extra-axillary or pseudoterminal, few flowered, usually over apex of stem; sepals 5, persistent; petals 3, blue-purple. Capsules subcordate and narrowly winged. Fl. Apr-Jul. Fr. May-Aug. Forest edges, grasslands, grassy slopes, limestone mountains or roadsides at 1100-3300 m. Distributed in most parts of China. Also in SW and NE Asia, Australia and E Europe.

合叶草
Polygala subopposita S. K. Chen

一年生草本。茎和小枝圆柱状。叶近对生，近无柄；阔椭圆形或长圆状椭圆形，纸质，两面具柔毛。总状花序腋上生；花瓣3，白色或黄色。蒴果近球状，具翅。花果期8-11月。生海拔600-1400米的灌丛中、草坡或沟谷。产云南、四川和贵州。

Annual herbs. Stems and branches terete. Leaves subopposite, subsessile; leaves broadly elliptic or oblong-elliptic, papery, both surfaces villous. Racemes supra-axillary; petals 3, white or yellow. Capsules suborbicular, winged. Fl. and fr. Aug-Nov. Thickets, grassy slopes or valleys at 600-1400 m. Distributed in Yunnan, Sichuan and Guizhou.

华南远志
Polygala glomerata Lour.

一年生直立草本。叶纸质，互生。总状花序腋上生，稀腋生；苞片2枚；萼片5，绿色，宿存，具缘毛，外面3枚卵状披针形，长约2毫米；内面2枚花瓣状，镰刀形，长约4.5毫米；花瓣3，淡黄色或白粉色；雄蕊8。蒴果圆形，具窄翅和缘毛。花期4-10月，果期5-11月。生

合叶草 *Polygala subopposita*

齿果草 *Salomonia cantoniensis*

海拔500-1000米的山坡草地或灌丛中。产中国西南、华南和东南。南亚和东南亚亦有。

Annual herbs, erect. Leaves papery, alternate. Racemes supra-axillary, rarely axillary; bracts 2; sepals 5, green, persistent, ciliate, outer 3 ovate-lanceolate, ca. 2 mm long; inner 2 petaloid, falcate, ca. 4.5 mm long; petals 3, yellowish or pale red, stamens 8. Capsules orbicular, with narrow wings and ciliate hairs. Fl. Apr-Oct. Fr. May-Nov. Grassy slopes or thickets at 500-1000 m. Distributed in SW, S and SE China. Also in S and SE Asia.

齿果草
Salomonia cantoniensis Lour.

一年生直立草本。茎多分枝，具狭翅。叶卵状心形或心形，膜质，无毛，具三出脉。穗状花序顶生，花期后伸长；花瓣3，淡红色。蒴果肾形，具三角状齿。花期7-8月，果期8-10月。生海拔600-1500米的山坡林下、灌丛、山坡或草地。产中国西南和华南。南亚和东南亚亦有。

Annual herbs, erect. Stems multi-branched, narrowly winged. Leaves ovate-cordate or cordate, membranous, glabrous, 3-veined. Spikes terminal, elongated after anthesis; petals 3, pale red. Capsules reniform, with deltoid teeth. Fl. Jul-Aug. Fr. Aug-Oct. Forests on slopes, thickets, mountain slopes or grasslands at 600-1500 m. Distributed in SW and S China. Also in S and SE Asia.

寄生鳞叶草
Epirixanthes elongata Blume

寄生草本。茎纤细。叶鳞片状。穗状花序顶生，不分枝；苞片线形，宿存；花瓣3，淡黄色或淡红色。蒴果近倒心形，藏于宿存萼筒内。花果期7-10月。生海拔600-1100米的阴湿的林中或竹林中。产云南南部、海南和福建。印度、缅甸、泰国、越南、马来西亚和印度尼西亚亦有。

Herbs parasitic. Stems slender. Leaves scale-like. Spikes terminal, unbranched; bracts linear, persistent; petals 3, pale yellow or pale red. Capsules subobcordate, included in calyx tube. Fl. and fr. Jul-Oct. Wet and shady forests or bamboo forests at 600-1100 m. Distributed in S Yunnan, Hainan and Fujian. Also in India, Myanmar, Thailand, Vietnam, Malaysia and Indonesia.

寄生鳞叶草
Epirixanthes elongata

毒鼠子科 Dichapetalaceae

毒鼠子
Dichapetalum gelonioides (Roxb.) Engler

小乔木或灌木。叶椭圆形至长圆状椭圆形。花单性，组成聚伞花序或单生叶腋；花瓣5，宽匙形，先端微裂或近全缘。核果倒心形(2室)或斜椭圆体形(1室)，密被黄褐色至白色短柔毛。花期3-6月，果期7-11月。生海拔1500米以下的山地密林中。产广东、海南、广西和云南。南亚和东南亚亦有。

Small trees or shrubs. Leaves elliptic to oblong-elliptic, Flowers unisexual, in cymes or single, axillary; petals 5, broadly spatulate, apex slightly lobed or subentire. Drupes obcordate (2-loculed) or obliquely ellipsoid (1-loculed), densely yellow-brown to white pubescent. Fl. Mar-Jun. Fr. Jul-Nov. Dense montane forests below 1500 m. Distributed in Guangdong, Hainan, Guangxi and Yunnan. Also in S and SE Asia.

毒鼠子 *Dichapetalum gelonioides*

中文名索引

K

L

M

Y

拉丁学名索引

D

E

F

G

H

R

S